q-Rung Orthopair Fuzzy Sets

Harish Garg
Editor

q-Rung Orthopair Fuzzy Sets

Theory and Applications

Springer

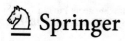

Editor
Harish Garg
Thapar Institute of Engineering
and Technology
Patiala, Punjab, India

ISBN 978-981-19-1451-5 ISBN 978-981-19-1449-2 (eBook)
https://doi.org/10.1007/978-981-19-1449-2

Mathematics Subject Classification: 03E72, 62A86, 90C27, 90C29, 90B50, 91A35, 91B06, 90C70

This Springer imprint is published by the registered company Springer Nature Singapore Pte Ltd.
The registered company address is: 152 Beach Road, #21-01/04 Gateway East, Singapore 189721, Singapore

Preface

Many decision-making problems in real-life scenarios depend on how to deal with uncertainty, which is typically a big challenge for decision-makers. Mathematical models are not common, but where the complexity is not usually probabilistic, various models emerged along with fuzzy logic approach. The classical approaches are typically based on crisp sets and crisp models that overlook vagueness, hesitancy, and uncertainties. Because of the uncertainty of today's decision-making issues, it is not possible for a decision-maker to interpret all of the information relevant to all decision-making strategies. For handling uncertain real-life problems, researchers have introduced various set-theoretic models. In 1965, Lotfi A. Zadeh introduced the fuzzy set theory as an extension of crisp sets to overcome the uncertainty of real-world problems.

Since the appearance of fuzzy set theory, many applications and extensions of it have been developed which are found in both theoretical and practical studies from engineering to arts and humanities, and from life sciences to physical sciences. In this book, a new extension of the fuzzy sets, entitled q-rung orthopair fuzzy set (q-ROFS), which is an extension of the crisp set, intuitionistic fuzzy set, and the Pythagorean fuzzy sets, is introduced by eminent researchers with several applications. In q-ROFS, each element is characterized by two degrees, namely membership degrees μ and non-membership degree υ, such that the restriction $\mu^q + \upsilon^q \leq 1$ for $\mu, \upsilon \in [0, 1]$, where $q > 0$. The objective of this book is to present some new advanced technologies and algorithms for solving real-life decision-making problems by using the features of the q-ROFSs.

This book consists of 19 chapters. The first two chapters present the mathematical aspects and topology of the q-rung orthopair fuzzy sets. The next five chapters relate to the implementation of the q-rung orthopair set to decision-making problems. These chapters include the application to medical diagnosis, inventory model, and multi-attribute decision-making. The next four chapters are related to generalized q-rung orthopair set and its application which include generalized orthopair fuzzy information measures, CODAS method, generalized orthopair fuzzy 2-tuple linguistic information, etc. Application of the q-rung orthopair fuzzy set to the real-life industrial problems involves five chapters related to green campus transportation, social responsibility evaluation pattern, insurance companies in the healthcare sectors, application to the cite selection of electric vehicles, and supplier selection problem in Industry 4.0 transition. Finally, the last three chapters are on the theory of the extension of q-rung orthopair fuzzy sets as well as their applications to the decision-making process. These extensions include pentagonal orthopair numbers, fuzzy soft sets, and interval-valued dual hesitant q-rung pairs.

Chapter 1 "q-Rung Orthopair Fuzzy Supra Topological Applications in Data Mining Process" introduces the q-rung orthopair fuzzy supra topological spaces and defines their certain mapping. To illustrate the stated topological structure, an algorithm related to multiple attribute decision-making problems is presented and applied to solve the problems related to medical diagnosis and data mining. Chapter 2 "q-Rung Orthopair Fuzzy Soft Topology with Multi-attribute Decision-Making" introduces the concept of q-rung orthopair fuzzy soft topology including sub-topology, interior, exterior, boundary, closure, etc. Moreover, grey relational analysis (GRA) approach, generalized choice-value method (GCVM), and aggregation operators-based techniques accompanied by three algorithms are stated to address q-rung orthopair fuzzy soft uncertain information. Numerical examples of these methods are also presented in real-life problems to discuss their impacts.

Chapter 3 "Decision-Making on Patients' Medical Status Based on a q-Rung Orthopair Fuzzy Max-Min-Max Composite Relation" presents a max–min–max composition relation under q-ROFSs with some theoretic characterizations. Based on these stated relations, authors have presented an algorithm for solving the decision-making problems in the field of medical diagnosis. Chapter 4 "Soergel Distance Measures for q-Rung Orthopair Fuzzy Sets and Their Applications" defines the Soergel-type distances and weighted distances to determine the metric between two q-ROFSs. Corresponding similarity measures are also derived from it. The validity and efficiency of the stated measures are demonstrated through a case study related to the multi-attribute decision-making process. Chapter 5 "TOPSIS Techniques on q-Rung Orthopair Fuzzy Sets and Its Extensions" integrates the concept of the TOPSIS method with q-ROF set for solving MCDM problems with hesitant fuzzy information. An example related to military aircraft overhaul effectiveness is taken to demonstrate it. Additionally, the authors have also introduced the TOPSIS technique for solving decision-making problems by using the q-rung orthopair fuzzy soft sets (q-ROFSfS) features. A case of a medical clinic under certain criteria is presented to illustrate the application of the proposed TOPSIS approach. In Chap. 6 "Knowledge Measure-Based q-Rung Orthopair Fuzzy Inventory Model", the authors have analysed the economic order quantity (EOQ) model with faulty products and screening errors under q-ROFS environment. In the study, the authors develop the EOQ model under two cases, such as replacement warranty claiming strategy with mending option and replacement warranty claiming strategy with emergency purchase option. For both cases, the q-ROF inventory model is framed by presuming the proportion of faulty products and the proportion of misclassification errors as q-rung orthopair fuzzy variables. The knowledge measure-based q-rung orthopair fuzzy inventory model is proposed by computing the knowledge measure of the variables. Finally, the sensitivity analysis with respect to the various parameters is provided for both the cases to strengthen the results. In Chap. 7 "Higher Type q-Rung Orthopair Fuzzy Sets: Interval Analysis", some cross-entropy and Hausdorff distance measures for q-rung interval-valued orthopair fuzzy set (q-RIVOFS) are formulated to scale the divergence and similarity measure between the pairs of q-RIVOFSs. In addition, an integrated q-RIVOFS-TODIM approach is constructed to display the valid applications of q-RIVOFSs in MADM areas. In the method, a linear programming model is established to derive the attribute weights, which can consider not only attributes' contribution to decision-making process but also the credibility of attribute evaluation information. A case of medical waste disposal method selection is presented to illustrate the advantages of q-RIVOFSs and the practicability of the proposed q-RIVOFS-TODIM approach.

In the context of the generalized q-ROFSs, Chap. 8 "Evidence-Based Cloud Vendor Assessment with Generalized Orthopair Fuzzy Information and Partial Weight Data" provides a new decision framework for cloud vendor (CV) selection based on the quality of service offered by the vendor. In this framework, the authors include methods for weight calculation and ranking based on decision-maker's (DM's) data. Also, preferences from each DM are given as input to the evidence-Bayesian approximation algorithm, which determines ordering of cloud vendors based on an individual's opinion. A case study related to CV adoption by an academic institution is provided to illustrate the model. In Chap. 9 "Supplier Selection Process Based on CODAS Method Using q-Rung Orthopair Fuzzy Information", the COmbinative Distance-based ASsessment (CODAS) method is extended to its q-rung orthopair CODAS version for handling the impreciseness and vagueness in decision-making process. A case study related to the supplier selection problem is taken to illustrate the approach. Chapter 10 "Group Decision-Making Framework with Generalized Orthopair Fuzzy 2-Tuple Linguistic Information" deals with the study of the Maclaurin symmetric mean (MSM) operator to the generalized orthopair fuzzy 2-tuple linguistic (GOFTL) set. In this chapter, GOFTL-MSM and the GOFTL weighted MSM, GOFTL dual MSM, and GOFTL weighted dual MSM operators are proposed along with desirable properties to aggregate the fuzzy information. Based on these proposed operators, a group decision-making algorithm is presented and illustrates them with a case study of the selection of the most preferable supplier(s) in enterprise framework group (EFG) of companies. In Chap. 11 "3PL Service Provider Selection with q-Rung Orthopair Fuzzy Based CODAS Method", authors have developed the q-ROF-CODAS method by adapting the CODAS method to q-ROFS. A case of third-party logistics (3PL) service provider selection for a retail company is presented to illustrate the application of the proposed q-ROF CODAS method.

Related to the application of q-ROFSs to real-life industrial problems, Chap. 12 "An Integrated Proximity Indexed Value and q-Rung Orthopair Fuzzy Decision-Making Model for Prioritization of Green Campus Transportation" considers the application of q-ROFSs to sustainable campus transportation. Global warming and air pollution are two of the most severe problems, requiring sustainable and environmentally friendly measures. To address this completely, the authors have proposed a hybrid multi-criteria framework based on the q-Rung orthopair proximity indexed value (q-ROF PIV) method and the logarithm methodology of additive weights (q-ROF PIV-LMAW). In this study, a q-ROF PIV method is combined with an algorithm for determining the weight coefficients of the criteria, based on the application of a logarithmic additive function to define the relationship between the criteria. Chapter 12 "An Integrated Proximity Indexed Value and Q-Rung Orthopair Fuzzy Decision-Making Model for Prioritization of Green Campus Transportation" deals with the application of q-ROFSs to the corporate social responsibility (CSR) evaluation mechanism which helps manage the platform-based enterprises. In Chap. 13 "Platform-Based Corporate Social Responsibility Evaluation with Three-Way Group Decisions under q-Rung Orthopair Fuzzy Environment", authors propose a multi-criteria decision-making method with three-way group decisions in q-rung orthopair fuzzy environment to assess and classify the CSR of platform-based enterprises. Chapter 14 "MARCOS Technique by Using q-Rung Orthopair Fuzzy Sets for Evaluating the Performance of Insurance Companies in Terms of Healthcare Services" extends the measurement of alternatives and ranking according to the compromise solution (MARCOS) technique under the consideration of q-rung orthopair fuzzy numbers and discuss their application to the performance of insurance companies in healthcare sectors. In Chap. 15 "Interval Complex q-Rung Orthopair Fuzzy Aggregation Operators and Their Applications in Cite Selection of Electric Vehicle", the authors have presented the study of the interval complex q-ROFS and investigated their properties. A decision-making algorithm is presented under the uncertain complex features and it validates the study through a case study of site selection of the electric vehicles. In Chap. 16 "A Novel Fermatean Fuzzy Analytic Hierarchy Process Proposition and Its Usage for Supplier Selection Problem in Industry 4.0 Transition", the authors have extended the analytic hierarchy process (AHP) to the Fermeatean fuzzy number, which is a special case of the q-ROFSs, and hence solves the decision-making process problems. A suggested process has been demonstrated through a case study of a real supplier selection problem for Industry 4.0 transition.

The book also deals with the theory of the extensions of the q-ROFS sets such as pentagonal q-ROFS, q-rung orthopair fuzzy soft set, dual hesitant, etc., and their applications to solve the decision-making problems in the last three chapters. In this context, Chap. 17 "Pentagonal q-Rung Orthopair Numbers and Their Applications" deals with the pentagonal q-Rung orthopair fuzzy numbers (Pq-ROFN) and normal Pq-ROFN by using the concept of the norm operations. Some operations on the stated concept are defined and based on it, a multi-attribute decision-making algorithm is presented to solve the problems. Chapter 18 "q-Rung Orthopair Fuzzy Soft Set-Based Multi-criteria Decision-Making" is on to introduce the hybrid structure named q-Rung orthopair fuzzy soft sets (q-ROFSSs) by combining the features of q-ROFSs and Molodtsov's soft sets. Various algebraic properties of the set are stated. Later, certain mathematical models for multi-criteria decision-making (MCDM) problems such as TOPSIS, VIKOR, and similarity measures are defined to solve the MCDM problems. A case study related to the selection of appropriate persons for key ministries of a country, selection of agricultural land, and diagnosis during COVID-19 are investigated. In Chap. 19 "Development of Heronian Mean-Based Aggregation Operators Under Interval-Valued Dual Hesitant q-Rung

Orthopair Fuzzy Environments for Multicriteria Decision-Making", some aggregation operators with Heronian mean concept are defined to aggregate the interval-valued dual hesitant q-rung orthopair fuzzy (IVDHq-ROF) environment. Based on the proposed operators, it presents an approach for multi-criteria decision-making problems to solve problems.

We hope that this book will provide a useful resource of ideas, techniques, and methods for the research on the theory and applications of q-rung orthopair fuzzy sets. We are grateful to the referees for their valuable and highly appreciated work contributed to select the high-quality chapters published in this book. We would like to also thank Springer Nature and the team for their supportive role throughout the process of editing this book.

Patiala, India Harish Garg

Contents

About the Editor

Harish Garg is Associate Professor at the School of Mathematics, Thapar Institute of Engineering and Technology, Patiala, India. He completed his Ph.D. in Mathematics from the Indian Institute of Technology Roorkee, India, in 2013. His research interests include soft computing, decision making, aggregation operators, evolutionary algorithm, expert systems, and decision support systems. He has authored around 330 papers, published in international journals of repute, and has supervised 7 Ph.D. dissertations. He is Recipient of the Top-Cited Paper by India-based Author (2015–2019) from Elsevier. He serves as Editor-in-Chief for *Journal of Computational and Cognitive Engineering; Annals of Optimization Theory & Practice* and Associate Editor for several renowned journals. His google citations are over 13000 with h-index 65. For more details, visit http://sites.google.com/site/harishg58iitr/.

Chapter 1
q-Rung Orthopair Fuzzy Supra Topological Applications in Data Mining Process

Mani Parimala, Cenap Ozel, M. A. Al Shumrani, and Aynur Keskin Kaymakci

Abstract The idea of q-rung orthopair fuzzy sets is an extension of intuitionistic and Pythagorean fuzzy sets. The main goal of this manuscript is to present the notion of q-rung orthopair fuzzy supra topological spaces (q-rofsts), a hybrid form of intuitionistic fuzzy supra topological spaces and Pythagorean fuzzy supra topological spaces. In addition, several contradictory examples and their assertions in fuzzy supra topological spaces of Abd El-Monsef and Ramadan (Indian J Pure Appl Math 18(4):322–329, 1987, [9]) are produced using q-rung orthopair fuzzy mappings. Finally, a new multiple attribute decision-making technique based on the q-rung orthopair fuzzy scoring function is suggested as an application to tackle medical diagnosis issues.

Keywords Fuzzy topology · Intuitionistic topology · Pythagorean fuzzy topology · q-rung orthopair fuzzy topology · q-rung orthopair fuzzy supra topology

---

M. Parimala (✉)
Department of Mathematics, Bannari Amman Institute of Technology, Sathyamangalam, Tamil Nadu, India
e-mail: rishwanthpari@gmail.com

C. Ozel · M. A. A. Shumrani
Department of Mathematics, King Abdulaziz University, 80203, Jeddah 21589, Saudi Arabia
e-mail: maalshmrani1@kau.edu.sa

A. K. Kaymakci
Department of Mathematics, Faculty of Sciences, Selcuk University, 42030 Konya, Turkey
e-mail: akeskin@selcuk.edu.tr

H. Garg (ed.), *q-Rung Orthopair Fuzzy Sets*,
https://doi.org/10.1007/978-981-19-1449-2_1

1.1 Introduction

In 1965, Zadeh [1] developed the notion of fuzzy set, an expansion of the crisp set for analyzing unreliable mathematical information. Fuzzy logic and fuzzy set theory concepts were used by Adlassnig [2] to establish medical connections in a computerized diagnosis system. The idea of [3–5] has been said in the disciplines of control theory, biology, artificial intelligence, economics and probability to name a few. Chang [6] proposed the concept of fuzzy topological spaces, and Lowen [7] investigated its characteristics further. In [8], Mashhour et al. developed supra topological space by weakening one topological postulate and analyzing its characteristics. Abd El-Monsef and Ramadan et al. [9] presented fuzzy supra topological spaces (fsts). Atanassov [10] and Yager [11] developed intuitionistic fuzzy sets and Pythagorean fuzzy sets, respectively, that addressed both membership degree and non-membership degree of an element. Intuitionistic fuzzy topology was presented by Coker [12]. Saadati and Park [13] went on to research the fundamental notion of intuitionistic fuzzy poiq-rofts. De et al. [14] were the first to explore intuitionistic fuzzy set applications in medical diagnosis. Several researchers [15–18] explored intuitionistic fuzzy sets in medical diagnostics in more depth. In 2013, Yager [11] proposed the concept of Pythagorean fuzzy sets. Later in 2019, the concept of Pythagorean fuzzy topological space was proposed by Olgun et al. [19].

The q-rung orthopair fuzzy set (q-ROFS) is a notion that may be used for real-world engineering and scientific applications. Yager [20] was the first to introduce and begin the q-rung orthopair fuzzy set in 2017. Since its appearance, researchers are engaged in the extensions and their applications to the decision-making process. For instance, Garg [21] presented a decision-making algorithm by defining the concept of sine-trigonometric operational laws. Wang et al. [22] discussed the green supplier selection problem using the concept of q-ROFS. Wei et al. [23] discussed the Heronian mean operators for the decision-making problems under the pairs of q-ROFSs. Garg [24] integrated the concept of the q-ROFS with the connection number (CN) of the set pair analysis and hence defined the idea of CN-qROFS. The applicability of this concept has been demonstrated through a decision-making process. Wang et al. [25] presented a group decision-making algorithm with q-ROFS and linguistic features. Related to the topological spaces, Turkarslan et al. [26] in 2021 introduced the concept of q-rung orthopair fuzzy topological spaces.

The rest of the chapter is summarized as follows: In Sect. 1.2, we presented some essential preliminaries of fuzzy, intuitionistic fuzzy, Pythagorean fuzzy and q-rung orthopair fuzzy sets and topological spaces. In Sect. 1.3, the idea of q-rung orthopair fuzzy supra topological spaces is defined. We introduce supra continuity and S^*-q-rung orthopair fuzzy continuity giving some contradicting examples in [9] in Sect. 1.4. In Sect. 1.5, an algorithm for data processing in a supra topological of q-rung orthopair fuzzy environment is provided as a real-world application. We solve a numerical example of the above-suggested technique in Sect. 1.6. The conclusion and future work of this article is stated in Sect. 1.7.

1.2 Preliminary

This part of the paper includes some of the preliminary definitions of fuzzy, intuitionistic, Pythagorean, q-rung orthopair fuzzy sets and respective topological spaces that will be utilized in this work.

Definition 1.2.1 ([1]) $A = \{(x, \mu_A(x)) : x \in X\}$ is said to be a fuzzy set on the non-void set X; for every $x \in X$, the membership function is $\mu_A(x) \in [0, 1]$ of A.

Definition 1.2.2 A set $A = \{(x, \mu_A(x), \nu_A(x)) : x \in X\}$, where X is a non-void set, is called

1. an intuitionistic fuzzy set (IFS) [10], where $0 \le \mu_A(x) + \nu_A(x) \le 1$. The set of all intuitionistic sets of X is denoted by $IFS(X)$.
2. a Pythagorean fuzzy set (PyFS) [11], where $0 \le \mu_A^2(x) + \nu_A^2(x) \le 1$. $PyFS(X)$ represents all the sets of Pythagorean fuzzy sets of X.
3. a q-rung orthopair fuzzy set (qROFS) [20], where $0 \le \mu_A^q(x) + \nu_A^q(x) \le 1$. The set of all q-rung orthopair fuzzy sets of X is denoted by $qROFS(X)$.

For all $x \in X$, membership function degree is $\mu_A(x)$ and non- membership function $\nu_A(x)$ of A.

We simply argue that qROFS may be divided into classes of orthopair fuzzy numbers with unique q values by definition.

When $q = 1$, for example, it becomes IFS and when $q = 2$, it turns into a PyFS. As a result, IFS and PyFS are subtypes of qROFS. The diagrammatic representation of these sets is given in Fig. 1.1.

Definition 1.2.3 ([20]) The below statements hold for qROFS sets A and B on X:

1. $A \cup B = \vee(\mu_A(x), \mu_B(x)), \wedge(\nu_A(x), \nu_B(x))$.
2. $A \cap B = \wedge(\mu_A(x), \mu_B(x)), \vee(\nu_A(x), \nu_B(x))$.
3. $A = B$ iff $A \subseteq B$ and $B \subseteq A$.
4. $A \oplus B = (\sqrt[q]{\mu_A^q(x) + \mu_B^q(x) - \mu_A^q(x)\mu_B^q(x)}, \nu_A(x)\nu_B(x))$.
5. $A \otimes B = (\mu_A(x)\mu_B(x), \sqrt[q]{\nu_A^q(x) + \nu_B^q(x) - \nu_A^q(x)\nu_B^q(x)})$.
6. $A^c = (\nu_A(x), \mu_A(x))$.
7. $\alpha A = \sqrt[q]{1 - (1 - \mu_A^q(x))^\alpha}, \nu_A^\alpha(x)$, for $\alpha \ge 1$.
8. $A^\alpha = \mu_A^\alpha(x), \sqrt[q]{1 - (1 - \nu_A^q(x))^\alpha}$, for $\alpha \ge 1$.

Notation 1.2.4 1. $1_q = (1, 0)$ is the qROFS whole set.
2. $0_q = (0, 1)$ is the qROFS empty set.

Definition 1.2.5 ([20]) Let $A = (\mu_A(x), \nu_A(x))$ be a q-ROF number, then the

1. score function is $S(A) = \frac{1+\mu_A^q(x)-\nu_A^q(x)}{2}$.
2. accuracy function is $H(A) = \mu_A^q(x) + \nu_A^q(x)$.

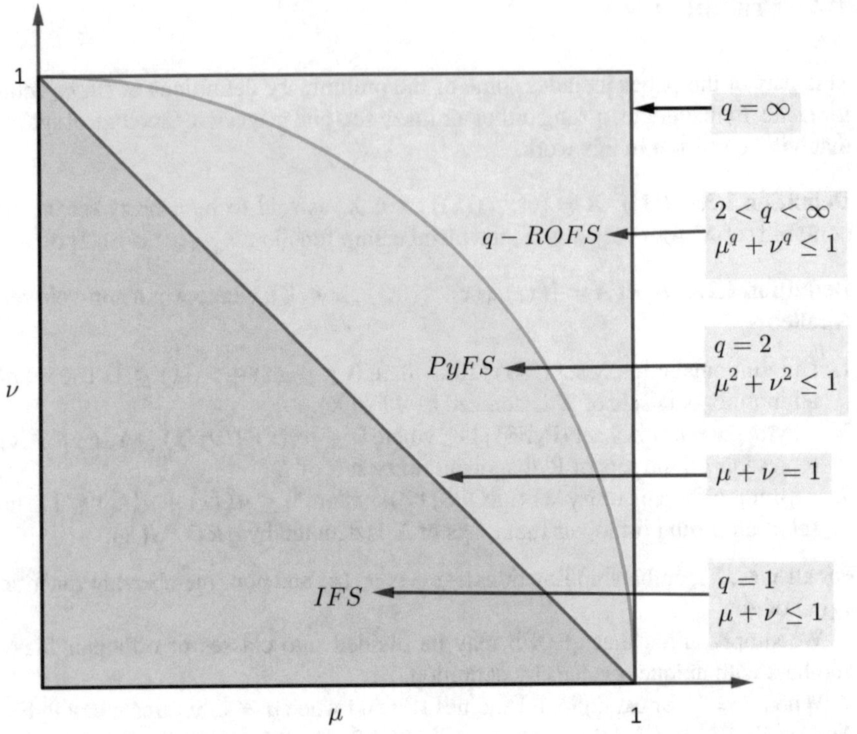

Fig. 1.1 Spaces of IFS, PyFS and qROFS

Definition 1.2.6 ([6]) A sub-collection τ_f of I^X is called a fuzzy topology on X, if

1. $X, \emptyset \in \tau_f$,
2. $\cup A_f \in \tau_f$ and
3. $\cap_{f=1}^{n} A_i \in \tau_f$.

Then (X, τ_f) is called fuzzy topological space (fts).

Definition 1.2.7 ([12]) A subfamily τ_i of $I^X$is called intuitionistic fuzzy topology on X if

1. $X, \emptyset \in \tau_i$,
2. $\cup A_i \in \tau_i$ and
3. $\cap_{i=1}^{n} A_i \in \tau_i$.

Then (X, τ_i) is called intuitionistic fuzzy topological space (ifts).

Definition 1.2.8 ([19]) A subfamily τ_p of I^X is said to be Pythagorean fuzzy topology on X if

1. $X, \emptyset \in \tau_p$,

2. $\cup A_p \in \tau_p$ and
3. $\cap_{p=1}^{n} A_p \in \tau_p$.

Then (X, τ_p) is called Pythagorean fuzzy topological space (Pyfts), elements of τ_p are called open Pythagorean fuzzy sets and their complements are closed Pythagorean fuzzy sets.

Definition 1.2.9 ([26]) A subfamily τ_q of $N(X)$ is said to be q-rung orthopair fuzzy topology (q-roft) on X if

1. $X, \emptyset \in \tau_q$,
2. $\cup A_q \in \tau_q$ and
3. $\cap_{q=1}^{n} A_q \in \tau_q$.

Then (X, τ_q) is known as q-rofts, each member of τ_q is called open q-rung orthopair fuzzy set and the complement q-roft of an open q-rung orthopair fuzzy set is called a closed q-rung orthopair fuzzy set of qROF.

If A is a qROFS, then A has the following closure and interior defined as follows:

1. $\text{int}_q(A) = \cup\{G : G \subseteq A, G \in \tau_q\}$.
2. $\text{cl}_q(A) = \cap\{F : A \subseteq F, F^c \in \tau_q\}$.

Definition 1.2.10 ([26]) Let $A = \{(x, \mu_A(x), \nu_A(x)) : x \in X\}$, $B = \{(y, \mu_B(y), \nu_B(y)) : y \in Y\}$ $f : X \to Y$ be a function between two qROFS.

1. B has a pre-image under the map f, then $f^{-1}(B) = \{(x, f^{-1}(\mu_B)(x), f^{-1}(\nu_B)(x)) : x \in X\}$ is a qROFS on X.
2. A has an image of the map f, then $f(A) = \{(y, f(\mu_A)(y), f(\nu_A)(y)) : y \in Y\}$ is a qROFS on Y, where

$$f(\mu_A)(y) = \begin{cases} \sup_{x \in f^{-1}(y)} \mu_A(x), & \text{if } f^{-1}(y) \neq \emptyset \\ 0, & \text{otherwise} \end{cases}$$

$$f(\nu_A)(y) = \begin{cases} \inf_{x \in f^{-1}(y)} \nu_A(x), & \text{if } f^{-1}(y) \neq \emptyset \\ 0, & \text{otherwise} \end{cases}$$

1.3 q-Rung Orthopair Fuzzy Supra Topological Spaces

We introduce q-rung orthopair fuzzy supra topological spaces (q-rofsts) and study their properties in this section.

Definition 1.3.1 A q-rofst on a non-void set X is a collection τ_q^* of qROFS of X having the following properties:

1. the sets $1_q, 0_q \in \tau_q^*$.
2. $\cup_{i=1}^{\infty} A_i \in \tau_q^*$ for $\{A_i\} \in \tau_q^*$.

Then (X, τ_q^*) is said to be q-rofsts on X. τ_q^*'s members are called q-rofs-open sets and q-rofs-closed set is a complement set. A q-rofsts τ_q^* on X is known to be an associated q-rofsts with qroft τ_q if $\tau_q \subseteq \tau_q^*$. Every qroft on X is q-rofst on X.

Definition 1.3.2 ([9]) A subfamily $\tau_f^*$of I^X is called fuzzy supra topology on X if

1. the sets $1_f, 0_f \in \tau_q^*$.
2. $\cup_{i=1}^{\infty} A_i \in \tau_f^*$ for $\{A_i\} \in \tau_f^*$.

Then (X, τ_f^*) is said to be fuzzy supra topological space (fsts) on X. The members of τ_f^* are called fuzzy supra open sets.

Definition 1.3.3 ([27]) A subfamily $\tau_i^*$of I^X is called intuitionistic supra fuzzy topology on X if

1. the sets $1_i, 0_i \in \tau_q^*$.
2. $\cup_{i=1}^{\infty} A_i \in \tau_i^*$ for $\{A_i\} \in \tau_i^*$.

Then (X, τ_i^*) is said to be intuitionistic supra fuzzy topological space (isfts) on X. The members of τ_i^* are called isf-open sets.

Proposition 1.3.4 *The $(\tau_q^*)^c$ of all q-rofs-closed sets in (X, τ_q^*) holds: $\emptyset, X \in (\tau_q^*)^c$ and $(\tau_q^*)^c$ is closed under the arbitrary intersection.*

Proof The result of the proposition is trivial. ■

Definition 1.3.5 The q-rofst interior $\text{int}_{\tau_q^*}(A)$ and closure $\text{cl}_{\tau_q^*}(A)$ operators of a q-rofs A are, respectively, defined as

1. $\text{cl}_{\tau_q^*}(A) = \cap\{F : A \subseteq F \text{ and } F^c \in \tau_q^*\}$ and
2. $\text{int}_{\tau_q^*}(A) = \cup\{G : G \subseteq A \text{ and } G \in \tau_q^*\}$.

Theorem 1.3.6 *Let (X, τ_q^*) be a q-rofsts. Let A and B be q-rofs in X. Then*

1. *$A = \text{cl}_{\tau_q^*}(A)$ iff A is q-rofs closed.*
2. *$A = \text{int}_{\tau_q^*}(A)$ iff A is q-rofs open.*
3. *$\text{cl}_{\tau_q^*}(A) \subseteq \text{cl}_{\tau_q^*}(B)$, if $A \subseteq B$.*
4. *$\text{int}_{\tau_q^*}(A) \subseteq \text{int}_{\tau_q^*}(B)$, if $A \subseteq B$.*
5. *$\text{cl}_{\tau_q^*}(A) \cup \text{cl}_{\tau_q^*}(B) \subseteq \text{cl}_{\tau_q^*}(A \cup B)$.*
6. *$\text{int}_{\tau_q^*}(A) \cup \text{int}_{\tau_q^*}(B) \subseteq \text{int}_{\tau_q^*}(A \cup B)$.*
7. *$\text{cl}_{\tau_q^*}(A) \cap \text{cl}_{\tau_q^*}(B) \supseteq (A \cap B)$.*
8. *$\text{int}_{\tau_q^*}(A) \cap \text{int}_{\tau_q^*}(B) \supseteq (A \cap B)$.*
9. *$\text{int}_{\tau_q^*}(A^c) = (\text{cl}_{\tau_q^*}(A))^c$.*

Proof Only we show (iii), (v) and (ix).

(iii): $\text{cl}_{\tau_q^*}(B) = \cap\{G : G^c \in \tau_q^*, B \subseteq G\} \supseteq \cap\{G : G^c \in \tau_q^*, A \subseteq G\} = \text{cl}_{\tau_q^*}(A)$. Thus $\text{cl}_{\tau_q^*}(A) \subseteq \text{cl}_{\tau_q^*}(B)$.

(v): Since $A \cup B \supseteq A, B$, then $\text{cl}_{\tau_q^*}(A) \cup \text{cl}_{\tau_q^*}(B) \subseteq \text{cl}_{\tau_q^*}(A \cup B)$.

(ix): $\text{cl}_{\tau_q^*}(A) = \cap\{G : G^c \in \tau_q^*, G \supseteq A\}$, $(\text{cl}_{\tau_q^*}(A))^c = \cup G^c : G^c$ is a q-rung orthopair fuzzy supra open in X and $G^c \subseteq A^c\} = \text{int}_{\tau_q^*}(A^c)$. Thus, $(\text{cl}_{\tau_q^*}(A))^c = \text{int}_{\tau_q^*}(A^c)$. ■

Remark 1.3.7 In q-rofts, we have $\mathrm{cl}_{\tau_q}(A \cup B) = \mathrm{cl}_{\tau_q}(A) \cup \mathrm{cl}_{\tau_q}(B)$ and $\mathrm{int}_{\tau_q}(A \cap B) = \mathrm{int}_{\tau_q}(A) \cap \mathrm{int}_{\tau_q}(B)$. The example below shows that these equalities are not true in q-rofsts.

Example 1.3.8 Let $X = \{a, b\}$ with q-rofs (q=3) $A = ((0.6, 0.85), (0.5, 0.9))$, $B = ((0.6, 0.9), (0.7, 0.85))$ and $C = ((0.7, 0.95), (0.8, 0.7))$, and let 3-rofst be $\tau_q^* = \{\emptyset, X, A, B, C\} = \{\emptyset, X, ((0.6, 0.85), (0.5, 0.9)), ((0.6, 0.9), (0.7, 0.85)), ((0.7, 0.95), (0.8, 0.7))\}$. Then $(\tau_q^*)^c = \{X, \emptyset, ((0.85, 0.6), (0.9, 0.5)), ((0.9, 0.6), (0.85, 0.7)), ((0.95, 0.7), (0.7, 0.8))\}$. Now $cl_{\tau_q^*}(A \cup B) = ((0.85, 0.6), (0.7, 0.85))$ and $cl_{\tau_q^*}(A) \cup cl_{\tau_q^*}(B) = ((0.9, 0.6), (0.9, 0.5))$. Therefore, $\mathrm{cl}_{\tau_q^*}(A \cup B) \neq \mathrm{cl}_{\tau_q^*}(A) \cup \mathrm{cl}_{\tau_q^*}(B)$. And now $int_{\tau_q^*}(A \cap B) = ((0.6, 0.85), (0.5, 0.85))$ and $int_{\tau_q^*}(A) \cap int_{\tau_q^*}(B) = ((0.6, 0.9), (0.5, 0.9))$. Therefore, $int_{\tau_q^*}(A \cap B) \neq int_{\tau_q^*}(A) \cap int_{\tau_q^*}(B)$.

1.4 Mappings of q-Rung Orthopair Fuzzy Spaces

The characteristics of various mappings in q-rung orthopair fuzzy supra topological spaces are defined and established in this section.

Definition 1.4.1 Let τ_q^* and σ_q^* be associated q-rofst to τ_q and σ_q, respectively. A mapping f from a q-rofts (X, τ_q) into q-rofts (Y, σ_q) is called S^*-q-rof-open if every image of q-rof-open set in (X, τ_q) is q-rofs-open in (Y, σ_q^*) and $f : X \to Y$ is known as S^*-q-rof continuous if each inverse image of q-rof-open set in (Y, σ_q^*) is q-rofs-open in (X, τ_q^*).

Definition 1.4.2 Let τ_q^* and σ_q^* be associated q-rofst to q-rofts's τ_q and σ_q, respectively. A mapping f from a q-rofts (X, τ_q) into a q-rofts (Y, σ_q) is called supra q-rof-open if every image of q-rofs-open set in (X, τ_q^*) is a q-rofs=open in (Y, σ_q^*) and $f : X \to Y$ is known as supra q-rof continuous if every inverse image of q-rofs-open set in (Y, σ_q^*) is q-rofs-open in (X, τ_q^*).

A mapping f of q-rofts (X, τ_q) into q-rofts (Y, σ_q) is known as a mapping of q-rof subspace $(A, (\tau_q)_A)$ into q-rof subspace $(B, (\sigma_q)_B)$ if $f(A) \subset B$.

Definition 1.4.3 f is a map from q-rof subspace $(A, (\tau_q)_A)$ of q-rofts (X, τ_q) into q-rof subspace $(B, (\sigma_q)_B)$ of q-rofts (Y, σ_q) is known as relatively q-rof continuous if $f^{-1}(O) \cap A \in (\tau_q)_A$ for each $O \in (\sigma_q)_B$. If $f(O') \in (\sigma_q)_B$ for each $O' \in (\tau_q)_A$, then f is called relatively q-rof-open.

Theorem 1.4.4 *If a mapping f is q-rung orthopair fuzzy continuous from q-rofts (X, τ_q) into q-rofts (Y, σ_q) and $f(A) \subset B$, then f is relatively q-rof continuous from q-rof subspace $(A, (\tau_q)_A)$ of q-rofts (X, τ_q) into q-rof subspace $(B, (\sigma_q)_B)$ of q-rofts (Y, σ_q).*

Proof Consider $O \in (\sigma_q)_B$, then $\exists\, G \in \sigma_q \ni O = B \cap G$ and $f^{-1}(G) \in \tau_q$. That is $f^{-1}(O) \cap A = f^{-1}(B) \cap f^{-1}(G) \cap A = f^{-1}(G) \cap A \in (\tau_q^*)_A$. ■

Remark 1.4.5 1. Each q-rof continuous (q-rof open) map is S^*-q-rof continuous (S^*-q-rof open). Generally, the converse of the statement is not true.
2. Each supra q-rof continuous (resp. supra q-rof open) mapping is S^*-q-rung orthopair fuzzy continuous (resp. S^*-q-rung orthopair fuzzy open). Generally, the converse of the statement is not true.
3. All mappings of supra q-rung orthopair fuzzy continuous and q-rof continuous are independent.
4. All mappings of supra q-rung orthopair fuzzy open and q-rof open are independent.

Proof This follows directly from the definitions. ■

Observation 1.4.6 For contradicting the statement q-rofts, we have the following few examples from [9]. In fsts, consider $Y = \{x, y, z\}$, $X = \{a, b, c\}$ with fuzzy topologies $\tau_f = \{\emptyset, X, (0.75, 0.85), (0.95, 0.65)\}$ and $\sigma_f = \{\emptyset, Y, (0.95, 0.65)\}$. Let $\tau_f^* = \{\emptyset, X, (0.75, 0.85), (0.95, 0.65), (0.85, 0.75)\}$ and $\sigma_f^* = \{\emptyset, Y, (0.95, 0.65), (0.3, 0.65), (0.75, 0.65)\}$ be associated fst of τ_f and σ_f, respectively. Let $h : X \to Y$ be a mapping defined by $h(c) = z$, $h(b) = y$, $h(a) = x$. Then $h^{-1}((0.75, 0.85)) = (0.75, 0.85) \notin \tau_f^*$. Hence, h is fuzzy continuous however it is not supra fuzzy continuous. If $g : Y \to X$ is a mapping defined by $g(z) = c$, $g(y) = b$, $g(x) = a$, then g is fuzzy open however it is not supra fuzzy open.

Theorem 1.4.7 *Let (X, τ_q) and (Y, σ_q) be q-rofts and let $f : X \to Y$. Then the following are equivalent.*

1. *$f : X \to Y$ is S^*-q-rung orthopair fuzzy continuous.*
2. *Every inverse image of q-rof closed set in (Y, σ_q) is q-rofs closed in (X, τ_q^*).*
3. *Every q-rof set A in Y, $\mathrm{cl}_{\tau_q^*}(f^{-1}(A)) \subseteq f^{-1}\mathrm{cl}_{\sigma_q^*}(A))$.*
4. *Every q-rof set B in X, $f(\mathrm{cl}_{\tau_q^*}(B)) \subseteq \mathrm{cl}_{\sigma_q^*}(f(B))$.*
5. *Every q-rof set A in Y, $\mathrm{int}_{\tau_q^*}(f \to (A)) \supseteq f^1(\mathrm{int}_{\sigma_q^*}(A))$.*

Proof (i) ⇒ (ii): Assume f is a S^*-q-rof continuous and A is a q-rof closed set in (Y, σ_q). $f^{-1}(Y - A) = X - f^{-1}(A)$ is q-rofs open in (X, τ_q^*) and so $f^{-1}(A)$ is q-rofs closed in (x, τ_q^*).
(ii) ⇒ (iii): $\mathrm{cl}_{\sigma_q}(A)$ is q-rung orthopair fuzzy closed in (Y, σ_q), for each q-rung orthopair fuzzy set $A \in Y$, then $f^{-1}(\mathrm{cl}_{\sigma_q}(A))$ is q-rung orthopair fuzzy supra closed in (X, τ_q^*). Thus $f^{-1}(\mathrm{cl}_{\sigma_q}(A)) = \mathrm{cl}_{\tau_q^*}(f^{-1}(\mathrm{cl}_{\sigma_q}(A))) \supseteq \mathrm{cl}_{\tau_q^*}(f^{-1}(A))$.
(iii) ⇒ (iv): $f^{-1}(\mathrm{cl}_{\sigma_q}(f(B))) \supseteq \mathrm{cl}_{\tau_q^*}(f^{-1}(f(B))) \supseteq \mathrm{cl}_{\tau_q^*}(B)$, for every q-rof set B in X and so $f(\mathrm{cl}_{\tau_q^*}(B)) \subseteq \mathrm{cl}_{\sigma_q}(f(B))$.
(iv) ⇒ (ii): Let $B = f^1(A)$, for every q-rof closed set A in Y, then $f(\mathrm{cl}_{\tau_q^*}(B)) \subseteq \mathrm{cl}_{\sigma_q}(f(B)) \subseteq \mathrm{cl}_{\sigma_q}(A) = A$ and $\mathrm{cl}_{\tau_q^*}(B) \subseteq f^{-1}(f(\mathrm{cl}_{\tau_q^* T}(B))) \subseteq f^{-1}(A) = B$. Hence, $B = f^{-1}(A)$ is q-rofs closed in X.
(ii) ⇒ (i): Let A be a q-rof open set in Y. Then $X - f^{-1}(A) = f^{-1}(Y - A)$ is q-rofs closed in X, since $Y - A$ is q-rof closed in Y. Therefore, $f^{-1}(A)$ is q-rofs open in X.

(i) ⇒ (v): $f^{-1}(\text{int}_{\sigma_q}(A))$ is q-rofs open in X, for every q-rof set A in Y and $\text{int}_{\tau_q^*}(f^{-1}(A)) \supseteq \text{int}_{\tau_q^*}(f^{-1}(\text{int}_{\sigma_q}(A))) = f^{-1}(\text{int}_{\sigma_q}(A))$.
(v) ⇒ (i): $f^{-1}(A) = f^{-1}(\text{int}_{\sigma_q}(A)) \subseteq \text{int}_{\tau_q^*}(f^{-1}(A))$, for every q-rof open set A in Y and so $f^{-1}(A)$ is q-rofs open in X. ■

Theorem 1.4.8 *Let (X, τ_q) and (Y, σ_q) be q-rofts and let $f : X \to Y$. Then the following are equivalent.*

1. *A mapping $f : (X, \tau_q^*) \to (Y, \sigma_q^*)$ is q-rofs continuous.*
2. *Every inverse image of q-rofs closed set in (Y, σ_q^*) is q-rofs closed in (X, τ_q^*).*
3. *Every q-rof set A in Y, $\text{cl}_{\tau_q^*}(f^{-1}(A)) \subseteq f^{-1}(\text{cl}_{\sigma_q^*}(A)) \subseteq f^{-1}(\text{cl}_{\tau_q}(A))$.*
4. *Every q-rof set B in X, $f(\text{cl}_{\tau_q^*}(B)) \subseteq \text{cl}_{\sigma_q^*}(f(B)) \subseteq \text{cl}_{\gamma_q^*}(f(B))$.*
5. *Every q-rof set A in Y, $\text{int}_{\tau_q^*}(f^{-1}(A)) \supseteq f^{-1}(\text{int}_{\sigma_q^*}(A)) \supseteq f^{-1}(\text{int}_{\sigma_q}(A))$.*

Proof The proof follows directly from Theorem 1.4.7 ■

Theorem 1.4.9 *If $f : X \to Y$ is S^*-q-rof continuous and $g : Y \to Z$ is q-rof continuous, then $g \circ f : X \to Z$ is S^*-q-rof continuous.*

Proof From the definitions, the proof follows. ■

Theorem 1.4.10 *If $f : X \to Y$ is supra q-rof continuous and $g : Y \to Z$ is S^*-q-rof continuous (or q-rof continuous), then $g \circ f : X \to Z$ is S^*-q-rof continuous.*

Proof From the definitions, the proof follows. ■

Theorem 1.4.11 *If the mappings $f : X \to Y$ and $g : Y \to Z$ are supra q-rof continuous (supra q-rof open), then $g \circ f : X \to Z$ is supra q-rof continuous (supra q-rof open).*

Proof From the definitions, the proof is achieved. ■

1.5 Algorithm for Data Mining Problem Via q-Rung Orthopair Fuzzy Supra Topology

We offer a systematic technique for a multi-attribute decision-making (MADM) problem using q-rung orthopair fuzzy information in this part. The methodical technique to select the appropriate qualities and alternatives in a decision-making scenario is proposed in the following phases.
Step 1: Problem selection:

Consider a MADM problem with m attributes $\alpha_1, \alpha_2, \ldots, \alpha_m$ and n alternatives $\beta_1, \beta_2, \ldots, \beta_n$ and p attributes $\gamma_1, \gamma_2, \ldots, \gamma_p$, $(n \leq p)$.

	β_1	β_2	.	.	.	β_n
α_1	(σ_{11})	(σ_{12})	.	.	.	(σ_{1n})
α_2	(σ_{21})	(σ_{22})	.	.	.	(σ_{2n})
.	.	.	.	.	.	.
.	.	.	.	.	.	.
.	.	.	.	.	.	.
α_m	(σ_{m1})	(σ_{m2})	.	.	.	(σ_{mn})

	α_1	α_2	.	.	.	α_m
γ_1	(ζ_{11})	(ζ_{12})	.	.	.	(ζ_{1m})
γ_2	(ζ_{21})	(ζ_{22})	.	.	.	(ζ_{2m})
.	.	.	.	.	.	.
.	.	.	.	.	.	.
.	.	.	.	.	.	.
γ_p	(ζ_{p1})	(ζ_{p2})	.	.	.	(ζ_{pm})

Here, all the attributes and ζ_{ki} $(i = 1, 2, \ldots, m$ and $k = 1, 2, \ldots, p)$ are q-rung orthopair fuzzy numbers.
Step 2: Form q-rung orthopair fuzzy supra topologies for (β_j) and (γ_k):

1. $\tau_j^* = A \cup B$, where $A = \{1_q, 0_q, \sigma_{1j}, \sigma_{2j}, \ldots, \sigma_{mj}\}$ and $B = \{\sigma_{1j} \cup \sigma_{2j}, \sigma_{1j} \cup \sigma_{3j}, \ldots, \sigma_{m-1j} \cup \sigma_{mj}\}$.
2. $\nu_k^* = C \cup D$, where $C = \{1_q, 0_q, \zeta_{kl}, \zeta_{k2}, \ldots, \zeta_{km}\}$ and $D = \{\zeta_{k1} \cup \zeta_{k2}, \zeta_{kl} \cup \zeta_{k3}, \ldots, \zeta_{km-i} \cup \zeta_{km}\}$.

Step 3: Find q-ROFSF: q-ROFSF of A, B, C, D, β_j and γ_k is stated in the following way.

1. $\text{q-ROFSF}(A) = \frac{1}{2(m+2)}\left[\sum_{i=1}^{m+2}[\frac{1+\mu_i^q-\nu_i^q}{2}]\right]$, and
   $\text{q-ROFSF}(B) = \frac{1}{2r}\left[\sum_{i=1}^{r}[\frac{1+\mu_i^q-\nu_i^q}{2}]\right]$. For $j = 1, 2, \ldots, n$,
   $$\text{q-ROFSF}(C_j) = \begin{cases} \text{q-ROFSF}(A), & \text{if q-ROFSF}(B) = 0 \\ \frac{1}{2}[\text{q-ROFSF}(A) + \text{q-ROFSF}(B)], & \text{otherwise} \end{cases}$$
2. $\text{q-ROFSF}(C) = \frac{1}{2(m+2)}\left[\sum_{i=1}^{m+2}[\frac{1+\mu_i^q-\nu_i^q}{2}]\right]$, and
   $\text{q-ROFSF}(D) = \frac{1}{2s}\left[\sum_{i=1}^{s}[\frac{1+\mu_i^q-\nu_i^q}{2}]\right]$. For $k = 1, 2, \ldots, p$,
   $$\text{q-ROFSF}(D_k) = \begin{cases} \text{q-ROFSF}(C), & \text{if q-ROFSF}(D) = 0 \\ \frac{1}{2}[\text{q-ROFSF}(C) + \text{q-ROFSF}(D)], & \text{otherwise} \end{cases}$$

Step 4: Final Decision
Arrange q-rung orthopair fuzzy score values for the alternatives $C_1 \le C_2 \le \cdots \le C_q$ and the attributes $D_1 \le D_2 \le \cdots \le D_p$. Choose the attributes D_p and D_{p-1} for the alternatives C_1 and C_2 and vice versa. If $n < p$ is true, D_k is ignored.

The proposed Algorithm for MADM via q-rung orthopair fuzzy supra topological spaces is represented as a flowchart in Fig. 1.2.

1.6 Numerical Example

New technologies in medical field have expanded the amount of information available to medical doctors, which includes uncertainty. The practice of categorizing diverse sets of indications under a specific pattern of a disease is a tough challenge in medical

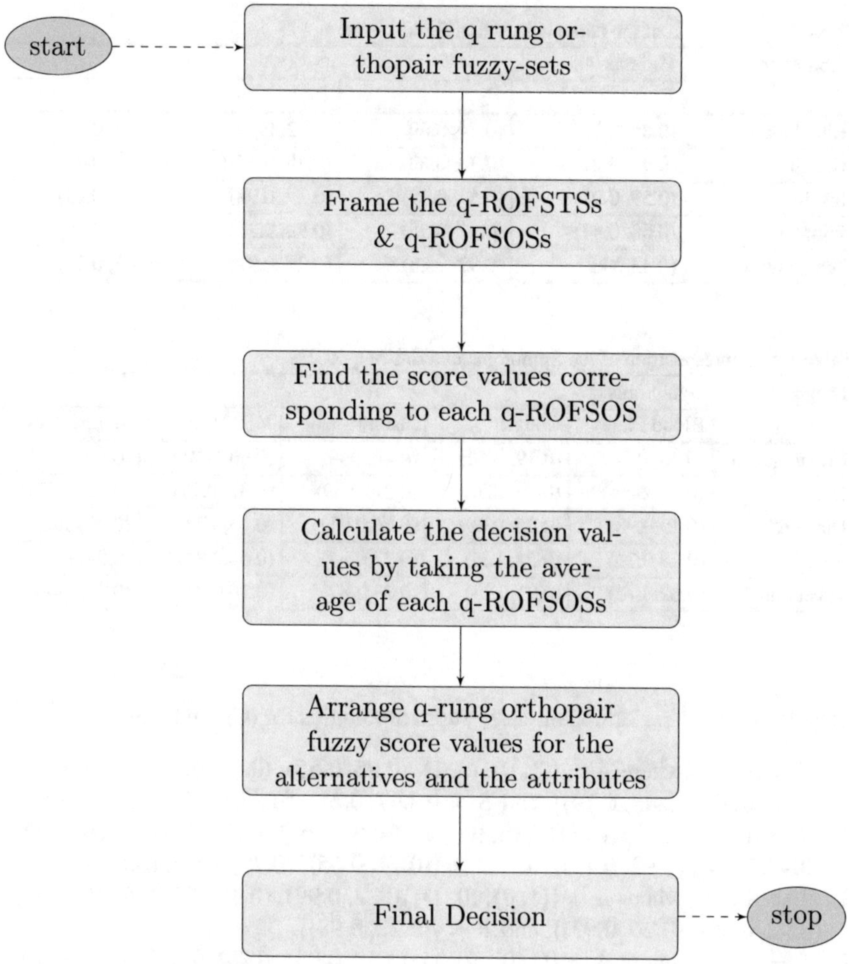

Fig. 1.2 Flowchart of the proposed Algorithm

diagnosis. In this part, we use a clinical diagnosis problem to demonstrate the efficacy and applicability of the above-mentioned technique.
Step 1: Problem field selection: Observe the tables below, which provide information on four individuals P_1, P_2, P_3, P_4 whose symptoms include blood plates, cough, insulin, joint pain and temperature. We need to locate the patient as well as the illness that the patient is suffering from, such as Chikungunya, Dengue, Diabetes, Swine Flu or Tuberculosis. Table 1.1 shows data by q-rung orthopair fuzzy numbers; here, for instance, we have q=3 with the functions of membership (μ) and non-membership (ν). Table 1.2 shows that cough is low for Chikungunya ($\mu = 0.19$, $\nu = 0.92$) but high for Tuberculosis ($\mu = 0.91$, $\nu = 0.23$).

Table 1.1 Input data of the patients in q-ROFS form

Symptoms	Patients			
	P_1	P_2	P_3	P_4
Blood Plates	(0.85, 0.53)	(0.74, 0.69)	(0.22, 0.99)	(0.32, 0.96)
Cough	(0.97, 0.27)	(0.78, 0.63)	(0.45, 0.65)	(0.81, 0.72)
Insulin	(0.59, 0.91)	(0.84, 0.59)	(0.19, 0.98)	(0.36, 0.89)
Joint Pain	(0.68, 0.47)	(0.47, 0.55)	(0.92, 0.36)	(0.29, 0.89)
Temperature	(0.43,0.79)	(0.92, 0.28)	(0.26, 0.97)	(0.43, 0.22)

Table 1.2 Representation of the Symptoms in terms of q-ROFS

Diagnosis	Symptoms				
	Blood Plates	Cough	Insulin	Joint Pain	Temperature
Chikungunya	(0.96, 0.54)	(0.19, 0.92)	(0.44, 0.86)	(0.96, 0.21)	(0.89, 0.25)
Dengue	(0.95, 0.54)	(0.26, 0.78)	(0.22, 0.89)	(0.37, 0.77)	(0.82, 0.37)
Diabetes	(0.19,0.98)	(0.25, 0.89)	(0.36,0.91)	(0.43,0.72)	(0.49,0.92)
Swine Flu	(0.28,0.75)	(0.54,0.84)	(0.27,0.94)	(0.64,0.83)	(0.94,0.23)
Tuberculosis	(0.62,0.73)	(0.91,0.23)	(0.23,0.87)	(0.16,0.99)	(0.22,0.91)

Step 2: Form q-rung orthopair fuzzy supra topologies for (Cj) and (D_k):

1. $\tau_1^* = A \cup B$, where $A = \{(1, 0), (0, 1), (0.85, 0.53), (0.97, 0.27), (0.59, 0.91), (0.68, 0.47), (0.43, 0.79)\}$, and $B = \{(0.97, 0.27), (0.85, 0.47), (0.59, 0.79)\}$.
2. $\tau_2^* = A \cup B$, where $A = \{(1, 0), (0, , 1), (0.74, 0.69), (0.78, 0.63), (0.84, 0.59), (0.47, 0.55), (0.92, 0.28)\}$, and $B = \{(0.74, 0.55), (0.78, 0.55), (0.84, 0.55)\}$.
3. $\tau_3^* = A \cup B$, where $A = \{(1, 0), (0, 1), (0.22, 0.99), (0.45, 0.65), (0.19, 0.98), (0.92, 0.36), (0.26, 0.97)\}$, and $B = \{(0.22, 0.98)\}$.
4. $\tau_4^* = A \cup B$, where $A = \{(1, 0), (0, 1), (0.32, 0.96), (0.32, 0.87), (0.92, 0.49), (0.23, 0.94), (0.22, 0.91)\}$, and $B = \{(0.32, 0.89), (0.81, 0.22)\}$.
5. $\nu_1^* = C \cup D$, here $C = \{(1, 0), (0, 1), (0.96, 0.54), (0.19, 0.92), (0.44, 0.86), (0.96, 0.21), (0.89, 0.25)\}$, and $D = \{(0.96, 0.25)\}$.
6. $\nu_2^* = C \cup D$, where $C = \{(1, 0), (0, 1), (0.95, 0.54), (0.26, 0.78), (0.22, 0.89), (0.37, 0.77), (0.82, 0.37)\}$, and $D = \{(0.95, 0.37)\}$.
7. $\nu_3^* = C \cup D$, where $C = \{(1, 0), (0, 1), (0.19, 0.98), (0.25, 0.89), (0.36, 0.91), (0.43, 0.72), (0.49, 0.92)\}$, and $D = \{(0.36, 0.89), (0.49, 0.89), (0.49, 0.91), (0.49, 0.72)\}$.
8. $\nu_4^* = C \cup D$, where $C = \{(1, 0), (0, 1), (0.28, 0.75), (0.54, 0.84), (0.27, 0.94), (0.64, 0.83), (0.94, 0.23)\}$, and $D = \{(0.64, 0.75), (0.54, 0.84)\}$.
9. $\nu_k^* = C \cup D$, here $C = \{(1, 0), (0, 1), (0.62, 0.73), (0.91, 0.23), (0.23, 0.87), (0.16, 0.99), (0.22, 0.91)\}$, and $D = \{\emptyset\}$.

Step 3: Find q-ROFSF:

1. q-ROFSF(A) = 0.2696 and q-ROFSF(B) = 0.265, where $r = 3$. q-ROFSF(C_1) = 0.2673.
2. q-ROFSF(A) = 0.2861 and q-ROFSF(B) = 0.3091, where $r = 3$. q-ROFSF(C_2) = 0.2976.
3. q-ROFSF(A) = 0.1818 and q-ROFSF(B) = 0.06, where $r = 1$. q-ROFSF(C_3) = 0.1209.
4. q-ROFSF(A) = 0.1975 and q-ROFSF(B) = 0.2525, where $r = 2$. q-ROFSF(C_4) = 0.225.
5. q-ROFSF(C) = 0.2736 and q-ROFSF(D) = 0.4275, where $s = 1$. q-ROFSF(D_1) = 0.3505.
6. q-ROFSF(C) = 0.2239 and q-ROFSF(D) = 0.395, where $s = 1$. q-ROFSF(D_2) = 0.3095.
7. q-ROFSF(C) = 0.1536 and q-ROFSF(D) = 0.1513, where s =4. q-ROFSF(D_3) = 0.1524.
8. q-ROFSF(C) = 0.2171 and q-ROFSF(D) = 0.1988, where $s = 2$. q-ROFSF(D_4) = 0.2079.
9. q-ROFSF(C) = 0.1932 and q-ROFSF(D) = 0, where $s = 0$. q-ROFSF(D_5) = 0.1932.

Step 4: Final Decision:

Arrange q-rung orthopair fuzzy score values for the alternatives C_1, C_2, C_3, C_4 and the attributes D_1, D_2, D_3, D_4, D_5 in ascending order. We get the following sequences $C_3 \leq C_4 \leq C_1 \leq C_2$ and $D_3 \leq D_5 \leq D_4 \leq D_2 \leq D_1$. Patients P_1, P_2, P_3 and P_4 have Diabetes, Swine Flu, Chikungunya and Dengue, respectively.

1.7 Conclusion and Future Work

The q-rofts with the idea of vagueness is one of the present studies in generic fuzzy topological spaces. We introduced the q-rorfsts as well as their real-world applications. Furthermore, certain mappings in q-rung orthopair fuzzy supra topological spaces were studied and some contradictory examples were presented. A multiple attribute decision-making algorithm via q-rung orthopair fuzzy supra topological spaces is presented and a numerical example in medical diagnosis data mining for the proposed algorithm is presented finally. In future, this work may be applied to various fields of topology research including soft, rough, digital topology and interval-valued sets [28, 29].

References

1. L.A. Zadeh, Probability measures of fuzzy events. J. Math. Anal. Appl. **23**(2), 421–427 (1968)
2. K.P. Adlassnig, Fuzzy set theory in medical diagnosis. IEEE Trans. Syst. Man Cybern. **16**(2), 260–265 (1986)
3. M. Sugeno, An Introductory survey of fuzzy control. Inf. Sci. **36**, 59–83 (1985)
4. P.R. Innocent, R.I. John, Computer aided fuzzy medical diagnosis. Inf. Sci. **162**, 81–104 (2004)
5. T.J. Roos, *Fuzzy Logic with Engineering Applications* (McGraw Hill P. C, New York, 1994)
6. C.L. Chang, Fuzzy topological spaces. J. Math. Anal. Appl. **24**, 182–190 (1968)
7. R. Lowen, Fuzzy topological spaces and fuzzy compactness. J. Math. Anal. Appl. **56**, 621–633 (1976)
8. A.S. Mashhour, A.A. Allam, F.S. Mohmoud, F.H. Khedr, On supra topological spaces. Indian J. Pure and Appl. Math. **14**(4), 502–510 (1983)
9. M.E. Abd El-Monsef, A.E. Ramadan, On fuzzy supra topological spaces. Indian J. Pure Appl. Math. 18(4), 322–329 (1987)
10. K. Atanassov, Intuitionistic fuzzy sets. Fuzzy Sets Syst. **20**, 87–96 (1986)
11. R.R. Yager, Pythagorean fuzzy subsets, in *Proceedings of the 2013 Joint IFSA World Congress and NAFIPS Annual Meeting (IFSA/NAFIPS), Edmonton, Canada, June 24–28, 2013* (IEEE, 2013), pp. 57–61
12. D. Coker, An introduction to intuitionistic Fuzzy topological spaces. Fuzzy Sets Syst. **88**(1), 81–89 (1997)
13. R. Saadati, J.H. Park, On the intuitionistic fuzzy topological space. Chaos, Solitons Fractals **27**(2), 331–344 (2006)
14. S.K. De, A. Biswas, R. Roy, An application of intuitionistic fuzzy sets in medical diagnosis. Fuzzy Sets Syst. **117**(2), 209–213 (2001)
15. E. Szmidt, J. Kacprzyk, Intuitionistic fuzzy sets in some medical applications, in *International Conference on Computational Intelligence* (Springer, Berlin, Heidelberg, 2001), pp. 148–151
16. P. Biswas, S. Pramanik, B.C. Giri, A study on information technology professionals' health problem based on intuitionistic fuzzy cosine similarity measure. Swiss J. Stat. Appl. Math. **2**(1), 44–50 (2014)
17. V. Khatibi, G.A. Montazer, Intuitionistic fuzzy set vs. fuzzy set application in medical pattern recognition. Artif. Intell. Med. **47**(1), 43–52 (2009)
18. K.C. Hung, H.W. Tuan, Medical diagnosis based on intuitionistic fuzzy sets revisited. J. Interdiscip. Math. **16**(6), 385–395 (2013)
19. M. Olgun, M. Unver, S. Yardimci, Pythagorean fuzzy topological spaces. Complex Intell. Syst. **5**, 177–183 (2019)
20. R.R. Yager, Generalized orthopair fuzzy sets. IEEE Trans. Fuzzy Syst. **25**, 1222–1230 (2017)
21. H. Garg, A novel trigonometric operation-based q-rung orthopair fuzzy aggregation operator and its fundamental properties. Neural Comput. Appl. **32**(18), 15077–15099 (2020)
22. R. Wang, Y. Li, A novel approach for green supplier selection under a q-rung orthopair fuzzy environment. Symmetry **10**(12), 687–712 (2018)
23. G. Wei, H. Gao, Y. Wei, Some q-rung orthopair fuzzy Heronian mean operators in multiple attribute decision making. Int. J. Intell. Syst. **33**, 1426–1458 (2018)
24. H. Garg, CN-q-ROFS: Connection number-based q-rung orthopair fuzzy set and their application to decision-making process. Int. J. Intell. Syst. **36**(7), 3106–3143 (2021)
25. H. Wang, Y. Ju, P. Liu, Multi-attribute group decision making methods based on q-rung orthopair fuzzy linguistic sets. Int. J. Intell. Syst. **34**(6), 1129–1157 (2019)
26. E. Turkarslan, M. Unver, M. Olgun, q-Rung Orthopair Fuzzy Topological Spaces. Lobachevskii J. Math. **42**, 470–478 (2021)
27. S.E. Abbas, Intuitionistic supra fuzzy topological spaces. Chaos, Solitons Fractals **21**, 1205–1214 (2004)
28. H. Garg, A new possibility degree measure for interval-valued q-rung orthopair fuzzy sets in decision-making. Int. J. Intell. Syst. **36**(1), 526–557 (2021)

29. H. Garg, New exponential operation laws and operators for interval-valued q-rung orthopair fuzzy sets in group decision making process. Neural Comput. Appl. **33**(20), 13927–13963 (2021)

Mani Parimala is an Assistant Professor in the Department of Mathematics at Bannari Amman Institute of Technology. Dr. Parimala received her Ph.D. in Mathematics in 2012 from Bharathiar University. She is one of the Editors in Chief of Asia Mathematika—An International Journal and Journal of Engineering Mathematics & Statistics. She is a reviewer in many reputable journals. She published more than 100 research articles in reputed international journals. Her major researches include general topology, algebraic structures and optimization techniques.

Cenap Ozel is a Professor at the King Abdulaziz University. He received his Ph.D. from the Glasgow University, Scotland, in 1998. His research interests include algebra, topology, differential geometry and mathematical physics. He has published several papers and books.

Mohammed A. Al Shumrani is a Professor of Mathematics at King Abdulaziz University, Jeddah, Saudi Arabia. He got his M.Sc. in Mathematics in 2002 from University of Missouri-Kansas City, USA. He got his Ph.D. in Mathematics in 2006 from University of Glasgow, UK. His research interests include General Topology, Algebraic Topology, Category Theory and Neutrosophic Theory. He has published several papers and one book.

Aynur Keskin Kaymakci is a Professor of Mathematics at Selcuk University, Konya, Turkey. Dr. Keskin Kaymakci got her Ph.D. from Selcuk University in 2003. She is one of the editors of Pakistan Academy of Sciences, Journal of Mathematics and Computational Intelligence, IJMTT, Adıyaman University Journal of Science. She is a reviewer in many reputable journals. Her research subjects are general topology, ideal topological spaces, soft topology and rough sets theory. She has published several papers.

Chapter 2
q-Rung Orthopair Fuzzy Soft Topology with Multi-attribute Decision-Making

Muhammad Tahir Hamid, Muhammad Riaz, and Khalid Naeem

Abstract In this chapter, the idea of q-rung orthopair fuzzy soft sets is extended to introduce the notion of q-rung orthopair fuzzy soft topology together with some interesting results. Certain properties of q-rung orthopair fuzzy soft topology are investigated for their practical applications in multi-attribute decision-making. For these objectives, grey relational analysis, generalized choice value method, and aggregation operators-based technique are proposed to address q-rung orthopair fuzzy soft uncertain information. Numerical examples of these methods are also presented from real-life situations. The validity and efficiency of proposed methods are analysed by their performing comparative analysis.

Keywords q-rung orthopair fuzzy soft topology · Multi-attribute decision-making · GRA · Generalized choice value method · Aggregation operators

2.1 Introduction

Information handling is an important aspect of our life. The classical approaches utilized for manipulating incomplete and inconsistent information are typically based on crisp sets and crisp models that overlook vagueness, hesitancy, and uncertainties. For handling uncertain real-life problems, researchers have introduced various set theoretic models. To handle inconsistent information, Fuzzy sets [1] by Zadeh, rough

M. T. Hamid
Department of Mathematics & Statistics, The University of Lahore, Lahore, Pakistan

M. Riaz (✉)
Department of Mathematics, University of the Punjab Lahore, Lahore, Pakistan
e-mail: mriaz.math@pu.edu.pk

K. Naeem
Department of Mathematics, FG Degree College, Lahore Cantt., Lahore, Pakistan

H. Garg (ed.), *q-Rung Orthopair Fuzzy Sets*,
https://doi.org/10.1007/978-981-19-1449-2_2

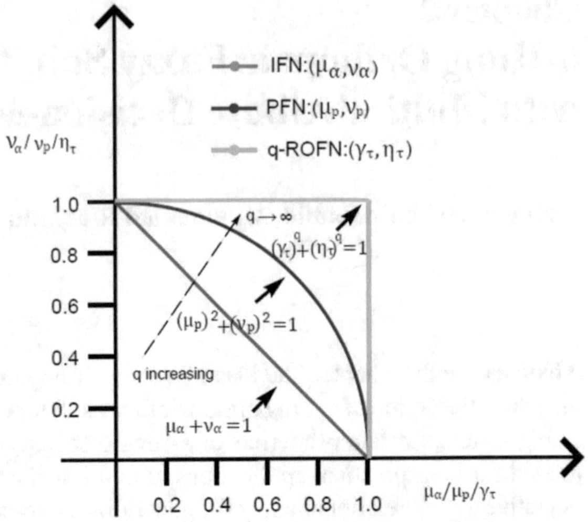

Fig. 2.1 Space for IFN, PFN, and q-ROFN. (The line in green colour may be observed as dotted)

sets [2] by Pawlak, and soft sets [3] by Molodstov are taken as fundamental tools in computational intelligence, artificial intelligence, social sciences, medical diagnosis, and engineering.

With the invention of new theories and models, researchers around the globe started working on different directions to explore new dimensions to cope with real-life problems. Atanassov [4–6], by expanding the notion of Zadeh's fuzzy set, proposed intuitionistic fuzzy sets (IFSs) and intuitionistic fuzzy numbers (IFNs) by supplementing non-membership function in addition to membership function. Atanassov imposed the condition on membership and non-membership functions that the sum of these two must lie in the unit closed interval. Xu [7] proposed IF-aggregation operators. Xu and Yager [8] rendered geometric aggregation operators for IFNs. Zhao et al. [9] presented generalized aggregation operators for IFNs. Yager [10, 11] and Yager and Abbasov [12] broadened the notion of IFSs to Pythagorean fuzzy sets (PFSs), by altering the condition on membership and non-membership function to that sum of their squares should fall in the unit closed interval. Later, in 2017, Yager [13] extended the idea to q-rung orthopair fuzzy sets (q-ROFSs) by amending the restriction on the two functions that sum of their qth powers should lie from 0 to 1. Thus, he broadened the space for the selection of values of the two functions and empowered the decision-makers to choose any values for the two functions from 0 to 1. Ali [14] developed certain features of q-ROFSs and their orbits. The spaces for intuitionistic fuzzy, Pythagorean fuzzy, and q-rung orthopair fuzzy numbers are depicted in Fig. 2.1.

Many researchers around the globe have studied the concepts of soft sets and soft topology. Aygunoglu et al. [15] studied fuzzy topological space; Cagman et al. [16] explored soft topology. Shabir and Naz [17] rendered more results on soft topological spaces. Bashir and Sabir [18] unveiled some structures on soft topological spaces. Roy and Samanta [19] established some important results on soft topological spaces.

Hazra et al. [20] presented definition of continuity of soft mappings in topology on soft subsets and soft topology. Varol et al. [21] studied new approach to soft topology. Riaz and Tehrim [22] initiated the notion of bipolar fuzzy soft topology in decision-making. Riaz et al. proposed N-soft topology [23], M-parameterized N-soft topology [24], m-polar neutrosophic topology [25], hesitant fuzzy topology [26], and Pythagorean fuzzy soft topology [27]. Li and Cui [28] and Tokat [29] introduced IFS-topology.

Fuzzy soft sets, Neutrosophic sets, rough sets, Pythagorean fuzzy sets, and q-ROFSs have been studied by many researchers; Eraslan and Karaaslan [30], Akram et al. [31–33], Feng et al. [34, 35], Kumar and Garg [36], Garg [37, 38], Peng and Selvachandran [39], Peng and Yang [40], Peng and Liu [41], Peng et al. [42], Zhang and Zhan [43], Zhang et al. [44], Zhan et al. [45, 46], and Zhang and Xu [47]. Garg [48] presented connection number-based q-rung orthopair fuzzy set and their application to decision-making process. Yang and Garg [49] studied interaction power partitioned Maclaurin symmetric mean operators under q-rung orthopair uncertain linguistic information. Garg [50] explored a new possibility degree measure for interval-valued q-rung orthopair fuzzy sets in decision-making. He et al. [51] unveiled a q-rung orthopair cloud-based multi-attribute decision-making algorithm considering the information error and multilayer heterogeneous relationship of attributes. Liu and Wang [52] presented some q-rung orthopair fuzzy aggregation operator and their application to multi-attribute decision-making. Liu et al. [53] studied multiple attribute group decision-making based on q-rung orthopair fuzzy Heronian mean operators.

Garg [54] presented a novel trigonometric operation-based q-rung orthopair fuzzy aggregation operator and its fundamental properties. Garg [55] further explored new exponential operation laws and operators for interval-valued q-rung orthopair fuzzy sets in group decision-making process. Liu and Liu [56] investigated some q-rung orthopair fuzzy Bonferroni mean operators and their application to multi-attribute group decision-making. Akram et al. [57] studied a hybrid decision-making framework under complex spherical fuzzy prioritized weighted aggregation operators. Akram et al. [58] presented extensions of Dombi aggregation operators for decision-making under m-polar fuzzy information. Farid and Riaz [59] unveiled some generalized q-rung orthopair fuzzy Einstein interactive geometric aggregation operators with improved operational laws. Riaz et al. [60] rendered novel q-rung orthopair fuzzy interaction aggregation operators and their application to low-carbon green supply chain management. Riaz et al. [61] discussed some q-rung orthopair fuzzy hybrid aggregation operators and TOPSIS method for multi-attribute decision-making. Riaz et al. [62] presented Pythagorean m-polar fuzzy weighted aggregation operators and algorithm for the investment strategic decision-making. Naeem et al. [63] presented Pythagorean fuzzy soft MCGDM methods based on TOPSIS, VIKOR, and aggregation operators. Naeem et al. [64] explored a mathematical approach to medical diagnosis via Pythagorean fuzzy soft TOPSIS, VIKOR, and generalized aggregation operators.

The ambition behind this work is to present some remarkable notions and results about q-rung orthopair fuzzy soft and q-rung orthopair fuzzy soft topology including sub-topology, interior, exterior, boundary, closure, basis, and separation axioms of a q-rung orthopair fuzzy soft topology. The usage of q-rung orthopair fuzzy soft sets in decision-making is also highlighted.

This chapter is designed and divided into five sections. In the second section, some fundamental concepts are presented. Next section deals with the basic theory of q-rung orthopair fuzzy soft sets (q-ROFSSs). The fourth section presents novel concepts of topological structure on q-ROFSSs. Certain properties of q-rung orthopair fuzzy soft topology (q-ROFST) are also investigated there. Section 2.5 is allocated for multi-attribute decision-making (MADM) based on grey relational analysis (GRA), general choice value method (GCVM), and aggregation operators-based technique accompanied by three algorithms. An application of proposed techniques is also developed there. Lastly, the conclusion is given in Sect. 2.6.

2.2 Some Elementary Models

This section is particularly established for various primary notions consisting of the soft set, fuzzy set, IFS, PFS, and q-ROFSs. These concepts are beneficial for the leftover part of this chapter.

Definition 2.2.1 (*Fuzzy Set* [1]) Let $\eta_A : \chi \to [0, 1]$ be the membership function and χ be the universe. A fuzzy set can be defined as

$$A = \{(\sigma, \eta_A(\sigma)) : \sigma \in \chi\},$$

where $\eta_A(\sigma)$ gives the membership degree (MD) for σ to A.

Definition 2.2.2 (*Soft Set* [3]) We consider a classical set χ and a set E of parameters or attributes, with $\mathbb{A} \subseteq E$. A soft set can be expressed as

$$\xi_{\mathbb{A}} = \Big\{(\epsilon, \eta(\epsilon)) : \epsilon \in \mathbb{A}, \eta(\epsilon) \in 2^{\chi}\Big\},$$

where $\eta : \mathbb{A} \to 2^{\chi}$ is a set-valued mapping.

Definition 2.2.3 (*Intuitionistic fuzzy set* [4, 6]) An intuitionistic fuzzy set (IFS) $\mathbb{A}$ on the universe χ may be written as

$$\mathbb{A} = \Big\{\Big(\sigma, \eta_{\mathbb{A}}(\sigma), \vartheta_{\mathbb{A}}(\sigma)\Big) : 0 \le \eta_{\mathbb{A}}(\sigma) + \vartheta_{\mathbb{A}}(\sigma) \le 1, \sigma \in \chi\Big\}.$$

The values $\eta_{\mathbb{A}}(\sigma)$ and $\vartheta_{\mathbb{A}}(\sigma)$ are called as MD and non-membership degrees (NMD), respectively.

For convenience, Xu [7] and Xu and Yager [8] named intuitionistic fuzzy number (IFN) to the ordered pair $\Big(\eta_{\mathbb{A}}(\sigma), \vartheta_{\mathbb{A}}(\sigma)\Big)$.

Definition 2.2.4 (*Pythagorean fuzzy set* [10–12]) An object in the form of Pythagorean fuzzy set (PFS) may be expressed as

$$\mathbb{P} = \big\{ < \sigma, \eta_P(\sigma), \vartheta_P(\sigma) >: 0 \leq \eta_P^2(\sigma) + \vartheta_P^2(\sigma) \leq 1, \sigma \in \chi \big\},$$

where η_P and ϑ_P are membership function (MF) and non-membership function, respectively.

For convenience, Zhang and Xu [47] suggested that a Pythagorean fuzzy number (PFN) can be written as $\Big(\eta_p(\sigma), \vartheta_p(\sigma)\Big)$.

Further properties of PFSs and PFNs are investigated by Peng and Selvachandran [39], Peng and Yang [40], and Peng et al. [42].

Definition 2.2.5 ([13]) A q-rung orthopair fuzzy set (q-ROFS) may be represented as

$$\mathbb{R} = \Big\{ \Big\langle \sigma, \eta_R(\sigma), \vartheta_R(\sigma) \Big\rangle : 0 \leq \eta_R(\sigma)^q + \vartheta_R(\sigma)^q \leq 1, \sigma \in \chi \Big\},$$

where $\eta_R(\sigma) \in [0, 1]$ and $\vartheta_R(\sigma) \in [0, 1]$ are the MD and NMD of $\rho \in \mathbb{R}$. The degree of hesitancy for σ to $\mathbb{R}$ may be given by

$$\Upsilon_R = \sqrt[q]{1 - \eta_R(\sigma)^q - \vartheta_R(\sigma)^q}.$$

The ordered pairs $\aleph = \Big(\eta_\aleph(\sigma), \vartheta_\aleph(\sigma)\Big)$, with $0 \leq \eta_\aleph(\sigma)^q + \vartheta_\aleph(\sigma)^q \leq 1, \rho \in \mathbb{R}, q > 1$, are known as q-rung orthopair fuzzy numbers (q-ROFNs). The spaces for IFN, PFN, and q-ROFN are already discussed in Fig. 2.1.

Further properties of q-ROFNs and q-ROFSs are investigated by numerous researchers mainly including Liu and Wang [52], Peng and Liu [41], Liu et al. [53], Garg [54], Garg [55], Liu and Liu [56], Akram et al. [57, 58], and Riaz et al. [59–61].

2.3 q-Rung Orthopair Fuzzy Soft Sets

We begin this section with the theory of q-rung orthopair fuzzy soft set (q-ROFSS) as a broadening of q-rung orthopair fuzzy set (q-ROFS) and soft set (SS). The proposed set is more powerful to cope with uncertain and vague statistics with parameterizations as well as ordered pairs of association and dissociation in a broader space. There is no denying the fact that a q-rung orthopair fuzzy soft set yields the broadened view of the Pythagorean fuzzy soft set [42, 63]. We consider χ as universe and $\mathbb{P}$ as set of attributes with $\mathbb{A} \subseteq \mathbb{P}$.

Definition 2.3.1 Let $qROF^\chi$ be the family of q-rung orthopair fuzzy sets on χ, and $\Gamma : \mathbb{A} \to qROF^\chi$ be a mapping. A *q-rung orthopair fuzzy soft set* (q-ROFSS) on χ, designated as (Γ, A) or Γ_A, is described as

Table 2.1 Tabular array of q-ROFSS Γ_A

Γ_A	ϵ_1	ϵ_2	$\cdots$	ϵ_n
σ_1	$(\eta 11, \vartheta_{11})$	$(\eta_{12}, \vartheta_{12})$	$\cdots$	$(\eta_{1n}, \vartheta_{1n})$
σ_2	$(\eta_{21}, \vartheta_{21})$	$(\eta_{22}, \vartheta_{22})$	$\cdots$	$(\eta_{2n}, \vartheta_{2n})$
$\vdots$	$\vdots$	$\vdots$	$\ddots$	$\vdots$
σ_m	$(\eta_{m1}, \vartheta_{m1})$	$(\eta_{m2}, \vartheta_{m2})$	$\cdots$	$(\eta_{mn}, \vartheta_{mn})$

$$
\begin{aligned}
(\Gamma, A) &= \left\{(\epsilon, \{\sigma, \eta_{\Gamma_A}(\sigma), \vartheta_{\Gamma_A}(\sigma)\}) : \epsilon \in \mathbb{A}, \sigma \in \chi\right\} \\
&= \left\{\left(\epsilon, \left\{\frac{\rho}{(\sigma_{\Gamma_A}(\sigma), \vartheta_{\Gamma_A}(\sigma))}\right\}\right) : \epsilon \in \mathbb{A}, \sigma \in \chi\right\} \\
&= \left\{\left(\epsilon, \left\{\frac{(\eta_{\Gamma_A}(\sigma), \vartheta_{\Gamma_A}(\sigma))}{\sigma}\right\}\right) : \epsilon \in \mathbb{A}, \sigma \in \chi\right\}.
\end{aligned}
$$

Here, we have two mappings, namely the membership mapping $\eta_{\Gamma_A} : \chi \to [0, 1]$ and the other is non-membership mapping $\vartheta_{\Gamma_A} : \chi \to [0, 1]$. Particularly, we can say that $\eta_{\Gamma_A}(\sigma)$ represent the membership grade and $\vartheta_{\Gamma_A}(\sigma)$ non-membership grade for the choice $\sigma \in \chi$ to the set (Γ, A) with the specification that the sum of their qth power will not go beyond unity. The value $\Gamma(\epsilon)$ for every parameter ϵ represents the approximate point.

The assemblage of all q-ROFSS on χ receiving from A is said to be a q-ROFS class and is represented by q-ROFS(χ, A).

Consider $\eta_{ij} = \eta_{\Gamma_A}(\epsilon_j)(\sigma_i)$ and $\vartheta_{ij} = \vartheta_{\Gamma_A}(\epsilon_j)(\sigma_i)$ where i and j vary, simultaneously, from 1 to m and from 1 to n. We can express q-ROFSS Γ_A in tabular array as exhibited in Table 2.1.

Its matrix form is

$$
\begin{aligned}
(\Gamma, A) &= [(\eta_{ij}, \vartheta_{ij})]_{m\times n} \\
&= \begin{pmatrix} (\eta_{11}, \vartheta_{11}) & (\eta_{12}, \vartheta_{12}) & \cdots & (\eta_{1n}, \vartheta_{1n}) \\ (\eta_{21}, \vartheta_{21}) & (\eta_{22}, \vartheta_{22}) & \cdots & (\eta_{2n}, \vartheta_{2n}) \\ \vdots & \vdots & \ddots & \vdots \\ (\eta_{m1}, \vartheta_{m1}) & (\eta_{m2}, \vartheta_{m2}) & \cdots & (\eta_{mn}, \vartheta_{mn}) \end{pmatrix}.
\end{aligned}
$$

The above shown matrix is called *q-rung orthopair fuzzy soft matrix* or in simple wording q-ROFS matrix.

Definition 2.3.2 Let $\Gamma^{(1)}_{A_1}$ and $\Gamma^{(2)}_{A_2}$ be two (q-ROFSSs) on χ. We express $\Gamma^{(1)}_{A_1}$ is *q-ROFS subset* of $\Gamma^{(2)}_{A_2}$, i.e. $\Gamma^{(1)}_{A_1} \widetilde{\subseteq} \Gamma^{(2)}_{A_2}$ if

(i) $A_1 \subseteq A_2$, and
(ii) $\Gamma^{(1)}(\epsilon)$ is q-ROFS subset of $\Gamma^{(2)}(\epsilon)$ for all $\epsilon \in A_1$.

The most noticeable point is that $\Gamma^{(1)}_{A_1} \tilde{\subseteq} \Gamma^{(2)}_{A_2}$ have no need that every member of $\Gamma^{(1)}_{A_1}$ must be a member of $\Gamma^{(2)}_{A_2}$, as we consider in ordinary set theory.

Definition 2.3.3 Let (Γ_1, A_1) and (Γ_2, A_2) be two q-ROFSSs on χ. Their *union* may be expressed as $(\Gamma, A) = (\Gamma_1, A_1) \tilde{\cup} (\Gamma_2, A_2)$, where $A = A_1 \cup A_2$ and for all $\epsilon \in A$

$$\Gamma(\epsilon) = \begin{cases} \Gamma_1(\epsilon), & \text{if } \epsilon \in A_1 \backslash A_2 \\ \Gamma_2(\epsilon), & \text{if } \epsilon \in A_2 \backslash A_1 \\ \Gamma_1(\epsilon) \cup \Gamma_2(\epsilon), & \text{if } \epsilon \in A_1 \cap A_2 \end{cases},$$

where $\Gamma_1(\epsilon) \cup \Gamma_2(\epsilon)$ is the union of two q-ROFSSs.

Definition 2.3.4 The *intersection* of two q-ROFSSs may be expressed as by taking (Γ_1, A_1) and (Γ_2, A_2) two q-ROFSS $(\Gamma, A) = (\Gamma_1, A_1) \tilde{\cap} (\Gamma_2, A_2)$, for $A = A_1 \cap A_2 \neq \phi$ and $\Gamma(\epsilon) = \Gamma_1(\epsilon) \cap \Gamma_2(\epsilon)$ for every $\epsilon \in A$.

Definition 2.3.5 For the *difference* (Γ, A) of two q-ROFSSs (Γ_1, A_1) and (Γ_2, A_2) on χ may be expressed as

$$(\Gamma, \epsilon) = (\Gamma_1, \epsilon) \setminus (\Gamma_2, \epsilon); \forall \epsilon \in A$$

and defined by

$$(\Gamma_1, A_1) \tilde{\setminus} (\Gamma_2, A_2) = \left\{ \left(\epsilon, \{\sigma, \min\{\eta_{\Gamma_1(\epsilon)}(\sigma), \vartheta_{\Gamma_2(\epsilon)}(\sigma)\}, \max\{\vartheta_{\Gamma_1(\epsilon)}(\sigma), \eta_{\Gamma_2(\epsilon)}(\sigma)\}\} \right) : \sigma \in \chi, \epsilon \in A \right\}.$$

Definition 2.3.6 For the *complement* of (Γ, A), we define $\Gamma^c : A \to q - ROFS^{\chi}$ expressed by $\Gamma^c(\epsilon) = [\Gamma(\epsilon)]^c$, for every $\epsilon \in A$. This may be represented as $(\Gamma, A)^c$ or (Γ^c, A). Thus, for any

$$\Gamma(\epsilon) = \{(\sigma, \eta_{\Gamma(\epsilon)}(\sigma), \vartheta_{\Gamma(\epsilon)}(\sigma)) : \sigma \in \chi\},$$

we have

$$\Gamma^c(\epsilon) = \{(\sigma, \vartheta_{\Gamma(\epsilon)}(\sigma), \eta_{\Gamma(\epsilon)}(\sigma)) : \sigma \in \chi\}$$

for every $\epsilon \in A$.

Example 2.3.7 Let $\chi = \{\varsigma_1, \varsigma_2, \ldots, \varsigma_6\}$ and $\mathbb{P} = \{\epsilon_1, \epsilon_2, \ldots, \epsilon_6\}$. Suppose that $A_1 = \{\epsilon_1, \epsilon_3\}$ and $A_2 = \{\epsilon_2, \epsilon_3, \epsilon_4\}$. Now we take the q-ROFS sets
$(\Gamma_1, A_1) = \{(\epsilon_1, \{\varsigma_1, 0.711, 0.359\}, \{\varsigma_3, 0.239, 0.959\}, \{\varsigma_6, 0.341, 0.490\}),$
$(\epsilon_3, \{\varsigma_2, 0.611, 0.819\}, \{\varsigma_6, 0.829, 0.409\})\}$, and
$(\Gamma_2, A_2) = \{(\epsilon_2, \{\varsigma_3, 0.460, 0.509\}, \{\varsigma_4, 0.611, 0.491\}, \{\varsigma_5, 0.940, 0.319\},$
$\{\varsigma_6, 0.870, 0.449\}), (\epsilon_3, \{\varsigma_1, 0.559, 0.661\}, \{\varsigma_6, 0.710, 0.549\}),$
$(\epsilon_4, \{\varsigma_2, 0.359, 0.870\}, \{\varsigma_3, 0.711, 0.389\}, \{\varsigma_6, 0.459, 0.630\})\}$
with $q = 3$.

(1) Their union is
$(\Gamma_1, A_1)\widetilde{\cup}(\Gamma_2, A_2) = \{(\epsilon_1, \{\varsigma_1, 0.711, 0.359\}, \{\varsigma_3, 0.239, 0.959\}, \{\varsigma_6, 0.341, 0.490\}), (\epsilon_2, \{\varsigma_3, 0.460, 0.509\}, \{\varsigma_4, 0.611, 0.491\}, \{\varsigma_5, 0.940, 0.319\}, \{\varsigma_6, 0.870, 0.449\}), (\epsilon_3, \{\varsigma_1, 0.559, 0.661\}, \{\varsigma_2, 0.611, 0.819\}, \{\varsigma_6, 0.829, 0.409\}), (\epsilon_4, \{\varsigma_2, 0.359, 0.870\}, \{\varsigma_3, 0.711, 0.389\}, \{\varsigma_6, 0.459, 0.630\})\}$.

(2) Their intersection is
$(\Gamma_1, A_1)\widetilde{\cap}(\Gamma_2, A_2) = \{(\epsilon_3, \{\varsigma_6, 0.710, 0.549\})\}$.

(3) The complements of (Γ_1, A_1) and (Γ_2, A_2) are
$(\Gamma_1, A_1)^c = \{(\epsilon_1, \{\varsigma_1, 0.359, 0.710\}, \{\varsigma_3, 0.959, 0.239\}, \{\varsigma_6, 0.490, 0.341\}), (\epsilon_3, \{\varsigma_2, 0.819, 0.611\}, \{\varsigma_6, 0.409, 0.829\})\}$
and
$(\Gamma_2, A_2)^c = \{(\epsilon_2, \{\varsigma_3, 0.509, 0.460\}, \{\varsigma_4, 0.491, 0.611\}, \{\varsigma_5, 0.319, 0.940\}, \{\varsigma_6, 0.449, 0.870\}), (e_3, \{\varsigma_1, 0.661, 0.559\}, \{\varsigma_6, 0.549, 0.710\}), (\epsilon_4, \{\varsigma_2, 0.870, 0.359\}, \{\varsigma_3, 0.389, 0.711\}, \{\varsigma_6, 0.630, 0.459\})\}$,
respectively.

(4) The difference of (Γ_1, A_1) and (Γ_2, A_2) is
$(\Gamma_1, A_1)\widetilde{\setminus}(\Gamma_2, A_2) = \{(\epsilon_1, \{\varsigma_1, 0.710, 0.359\}, \{\varsigma_3, 0.239, 0.959\}, \{\varsigma_6, 0.341, 0.490\}), (\epsilon_3, \{\varsigma_2, 0.611, 0.819\}, \{\varsigma_6, 0.710, 0.549\})\}$.

These notions may be remembered in a very simple way by using q-ROFS matrices as shown:

$$(\Gamma_1, A_1) = \begin{pmatrix} (0.711, 0.359) & (0, 1) & (0, 1) & (0, 1) & (0, 1) & (0, 1) \\ (0, 1) & (0, 1) & (0.611, 0.819) & (0, 1) & (0, 1) & (0, 1) \\ (0.239, 0.959) & (0, 1) & (0, 1) & (0, 1) & (0, 1) & (0, 1) \\ (0, 1) & (0, 1) & (0, 1) & (0, 1) & (0, 1) & (0, 1) \\ (0, 1) & (0, 1) & (0, 1) & (0, 1) & (0, 1) & (0., 1.) \\ (0.341, 0.490) & (0, 1) & (0.829, 0.409) & (0, 1) & (0, 1) & (0, 1) \end{pmatrix}$$

$$(\Gamma_2, A_2) = \begin{pmatrix} (0, 1) & (0, 1) & (0.559, 0.661) & (0, 1) & (0, 1) & (0, 1) \\ (0, 1) & (0, 1) & (0, 1) & (0.359, 0.870) & (0, 1) & (0, 1) \\ (0, 1) & (0.46, 0.51) & (0, 1) & (0.711, 0.389) & (0, 1) & (0, 1) \\ (0, 1) & (0.611, 0.491) & (0, 1) & (0, 1) & (0, 1) & (0, 1) \\ (0, 1) & (0.940, 0.319) & (0, 1) & (0, 1) & (0, 1) & (0, 1) \\ (0, 1) & (0.870, 0.449) & (0.710, 0.549) & (0.459, 0.630) & (0, 1) & (0, 1) \end{pmatrix}$$

$$(\Gamma_1, A_1)\widetilde{\cup}(\Gamma_2, A_2) = \begin{pmatrix} (0.711, 0.359) & (0, 1) & (0.559, 0.661) & (0, 1) & (0, 1) & (0, 1) \\ (0, 1) & (0, 1) & (0.61, 0.82) & (0.359, 0.870) & (0, 1) & (0, 1) \\ (0.239, 0.959) & (0.460, 0.509) & (0, 1) & (0.711, 0.389) & (0, 1) & (0, 1) \\ (0, 1) & (0.611, 0.491) & (0, 1) & (0, 1) & (0, 1) & (0, 1) \\ (0, 1) & (0.940, 0.319) & (0, 1) & (0, 1) & (0, 1) & (0, 1) \\ (0.341, 0.490) & (0.870, 0.449) & (0.829, 0.409) & (0.459, 0.630) & (0, 1) & (0, 1) \end{pmatrix}$$

$$(\Gamma_1, A_1)\widetilde{\cap}(\Gamma_2, A_2) = \begin{pmatrix} (0, 1) & (0, 1) & (0, 1) & (0, 1) & (0, 1) & (0, 1) \\ (0, 1) & (0, 1) & (0, 1) & (0, 1) & (0, 1) & (0, 1) \\ (0, 1) & (0, 1) & (0, 1) & (0, 1) & (0, 1) & (0, 1) \\ (0, 1) & (0, 1) & (0, 1) & (0, 1) & (0, 1) & (0, 1) \\ (0, 1) & (0, 1) & (0, 1) & (0, 1) & (0, 1) & (0, 1) \\ (0, 1) & (0, 1) & (0.710, 0.549) & (0, 1) & (0, 1) & (0, 1) \end{pmatrix}$$

$$(\Gamma_1, A_1)^c = \begin{pmatrix} (0.359, 0.711) & (0,1) & (0,1) & (0,1) & (0,1) & (0,1) \\ (0,1) & (0,1) & (0.819, 0.611) & (0,1) & (0,1) & (0,1) \\ (0.959, 0.239) & (0,1) & (0,1) & (0,1) & (0,1) & (0,1) \\ (0,1) & (0,1) & (0,1) & (0,1) & (0,1) & (0,1) \\ (0,1) & (0,1) & (0,1) & (0,1) & (0,1) & (0,1) \\ (0.490, 0.341) & (0,1) & (0.409, 0.829) & (0,1) & (0,1) & (0,1) \end{pmatrix}$$

$$(\Gamma_2, A_2)^c = \begin{pmatrix} (0,1) & (0,1) & (0.661, 0.559) & (0,1) & (0,1) & (0,1) \\ (0,1) & (0,1) & (0,1) & (0.870, 0.359) & (0,1) & (0,1) \\ (0,1) & (0.509, 0.460) & (0,1) & (0.389, 0.711) & (0,1) & (0,1) \\ (0,1) & (0.491, 0.611) & (0,1) & (0,1) & (0,1) & (0,1) \\ (0,1) & (0.319, 0.940) & (0,1) & (0,1) & (0,1) & (0,1) \\ (0,1) & (0.449, 0.870) & (0.549, 0.710) & (0.630, 0.459) & (0,1) & (0,1) \end{pmatrix}$$

$$(\Gamma_1, A_1)\widetilde{\setminus}(\Gamma_2, A_2) = \begin{pmatrix} (0.711, 0.359) & (0,1) & (0,1) & (0,1) & (0,1) & (0,1) \\ (0,1) & (0,1) & (0.611, 0.819) & (0,1) & (0,1) & (0,1) \\ (0.239, 0.959) & (0,1) & (0,1) & (0,1) & (0,1) & (0,1) \\ (0,1) & (0,1) & (0,1) & (0,1) & (0,1) & (0,1) \\ (0.341, 0.490) & (0,1) & (0,1) & (0,1) & (0,1) & (0,1) \\ (0,1) & (0,1) & (0.710, 0.549) & (0,1) & (0,1) & (0,1) \end{pmatrix}.$$

Definition 2.3.8 A q-ROFSS is termed as a *null q-ROFSS*, represented by Γ_ϕ or Φ, and it can be expressed as

$$\Gamma_\phi = \left\{\left(\epsilon, \left\{\frac{\sigma}{(0,1)}\right\}\right) : \epsilon \in A, \sigma \in \chi\right\}.$$

Definition 2.3.9 A q-ROFSS is said to be an *absolute q-ROFSS*, represented by $\breve{\chi}$, expressed as

$$\breve{\chi} = \left\{\left(\epsilon, \left\{\frac{\sigma}{(1,0)}\right\}\right) : \epsilon \in A, \rho \in \chi\right\}.$$

Definition 2.3.10 A q-ROFSS (Γ, A) is said to be a q-rung orthopair fuzzy soft point (q-ROFS-point), expressed as ϵ_Γ, if for $\epsilon \in A$ we have

(i) $\Gamma(\epsilon) \neq \Gamma_\phi$, and
(ii) $\Gamma(\epsilon') = \Gamma_\phi$, for all $\epsilon' \in A - \{\epsilon\}$.

Definition 2.3.11 A q-rung orthopair fuzzy soft point/number (q-ROFS point/number) ϵ_Γ is present in q-ROFSS (Γ_1, A), i.e. $\epsilon_\Gamma \widetilde{\in} (\Gamma_1, A)$ if $\epsilon \in A \Rightarrow \Gamma(\epsilon) \subseteq \Gamma_1(\epsilon)$.

Example 2.3.12 Let $\chi = \{\xi, \eta, \lambda\}$ and $A = \{\epsilon_1, \epsilon_2\}$, then

$$\epsilon_\Gamma^{(1)} = \{(\epsilon_1, \{(\xi, 0.49, 0.41), (\eta, 0.56, 0.21)\})\}$$

and

$$\epsilon_\Gamma^{(2)} = \{(\epsilon_2, \{(\xi, 0.68, 0.53)(\lambda, 0.77, 0.48)\})\}$$

are two distinct q-ROFS points that belong to the q-ROFSS

$$\Gamma_A = \{(\epsilon_1, \{(\xi, 0.49, 0.41), (\eta, 0.56, 0.21)\}), (\epsilon_2, \{(\xi, 0.68, 0.53), (\lambda, 0.77, 0.48)\})\}.$$

2.4 q-Rung Orthopair Fuzzy Soft Topology

In this part, we unveil some notions of q-rung orthopair fuzzy soft topology (q-ROFST) as an expansion of Pythagorean fuzzy soft topology [27], intuitionistic fuzzy soft topology [28, 29], and soft topology [15–21]. To define q-ROFST, we use null q-ROFSS denoted by Γ_ϕ and absolute q-ROFSS denoted by $\breve{\chi}_A$.

Definition 2.4.1 Let χ be the universe, E the set of parameters, and $A, B, A_i \subseteq E$. The aggregate $\widetilde{\mathcal{T}}$ of q-rung orthopair fuzzy sets on χ is called q-rung orthopair fuzzy soft topology (q-ROFST) if it satisfies the following:

(i) $\Gamma_\phi, \breve{\chi}_A \,\widetilde{\in}\, \widetilde{\mathcal{T}}$,
(ii) $\Gamma_A \widetilde{\cap} \Gamma_B \,\widetilde{\in}\, \widetilde{\mathcal{T}}$ whenever $\Gamma_A, \Gamma_B \widetilde{\in} \widetilde{\mathcal{T}}$, and
(iii) If $\Gamma_{A_i} \widetilde{\in} \widetilde{\mathcal{T}}, \forall\, i \in I$, then $\widetilde{\bigcup}_{i \in I} \Gamma_{A_i} \,\widetilde{\in}\, \widetilde{\mathcal{T}}$, I being some indexing set.

The ordered pair $(\breve{\chi}_E, \widetilde{\mathcal{T}})$ is said to be q-ROFS topological space. The elements of $\widetilde{\mathcal{T}}$ are called q-ROFS open sets and their complements are said to be a q-ROFS closed sets.

Definition 2.4.2 We consider $(\breve{\chi}_E, \widetilde{\mathcal{T}}_\chi)$ a q-ROFST and $\mathbb{Y} \subseteq \chi$. Then $\widetilde{\mathcal{T}}_{\mathbb{Y}}$ is said to be a q-ROFST on $\mathbb{Y}$ whose members are q-ROFS open sets determined by $\kappa_B = \kappa_A \widetilde{\cap} \breve{Y}_E$, where κ_A are q-ROFS open sets of $\widetilde{\mathcal{T}}_\chi$, Γ_B are q-ROFS open sets of $\widetilde{\mathcal{T}}_{\mathbb{Y}}$, and $\breve{\mathbb{Y}}_E$ is absolute q-ROFSS on $\mathbb{Y}$. Then,

$$\widetilde{\mathcal{T}}_{\mathbb{Y}} = \left\{ \Gamma_B : \Gamma_B = \Gamma_A \widetilde{\cap} \breve{\mathbb{Y}}_E, \Gamma_A \in \widetilde{\mathcal{T}}_\chi \right\}$$

is called q-ROFS subspace topology of q-ROFS topology.

Example 2.4.3 We consider $\chi = \{\rho_1, \rho_2, \rho_3\}$ as a universe and $E = \{\epsilon_i : i = 1, \ldots, 4\}$ be a family of attributes. Taking two sets $\mathbb{A} = \{\epsilon_1, \epsilon_2\}$ and $\mathbb{B} = \{\epsilon_1, \epsilon_2, \epsilon_3\}$ as sub collections of E, we suppose that

$$\Gamma_A^{(1)} = \left\{ \left(\epsilon_1, \left\{ \frac{\rho_1}{(0.762, 0.513)} \right\}, \left\{ \frac{\rho_2}{(0.794, 0.623)} \right\} \right), \left(\epsilon_2, \left\{ \frac{\rho_2}{(0.355, 0.830)} \right\}, \left\{ \frac{\rho_3}{(0.371, 0.832)} \right\} \right) \right\},$$

$$\Gamma_B^{(2)} = \left\{ \left(\epsilon_1, \left\{ \frac{\rho_1}{(0.850, 0.331)} \right\}, \left\{ \frac{\rho_2}{(0.812, 0.453)} \right\} \right), \left(\epsilon_2, \left\{ \frac{\rho_2}{(0.544, 0.633)} \right\}, \left\{ \frac{\rho_3}{(0.810, 0.311)} \right\} \right), \left(\epsilon_3, \left\{ \frac{\rho_1}{(0.644, 0.630)} \right\}, \left\{ \frac{\rho_2}{(0.412, 0.832)} \right\} \right) \right\}.$$

Then,

$$\widetilde{\mathcal{T}}_\chi = \{\Gamma_\phi, \breve{\chi}_E, \Gamma_A^{(1)}, \Gamma_B^{(2)}\}$$

becomes a q-ROFS topology on χ.
Now, absolute q-ROFSS on $\mathbb{Y} = \{\rho_2, \rho_3\} \subseteq \chi$ is

$$\breve{\mathbb{Y}}_E = \left\{ \left(\epsilon_i, \left\{ \frac{\rho_2}{(1, 0)} \right\}, \left\{ \frac{\rho_3}{(1, 0)} \right\} \right) : 1 \leq i \leq 4 \right\}.$$

Since

$$\begin{aligned}
\breve{\mathbb{Y}}_E \tilde{\cap} \Gamma_\phi &= \Gamma_\phi, \\
\breve{\mathbb{Y}}_E \tilde{\cap} \Gamma_A^{(1)} &= \left\{ \left(\epsilon_1, \left\{ \frac{\rho_2}{(0.794, 0.623)} \right\} \right), \left(\epsilon_2, \left\{ \frac{\rho_2}{(0.355, 0.830)} \right\}, \left\{ \frac{\rho_3}{(0.371, 0.832)} \right\} \right) \right\} \\
&= \Psi_A^{(1)}, \\
\breve{\mathbb{Y}}_E \tilde{\cap} \Gamma_B^{(2)} &= \left\{ \left(\epsilon_1, \left\{ \frac{\rho_2}{(0.812, 0.453)} \right\}, \left(\epsilon_2, \left\{ \frac{\rho_2}{(0.544, 0.630)} \right\}, \left\{ \frac{\rho_3}{(0.810, 0.311)} \right\} \right), \left(\epsilon_3, \left\{ \frac{\rho_2}{(0.412, 0.832)} \right\} \right) \right\} \\
&= \Psi_B^{(2)}, \\
\breve{\mathbb{Y}}_E \tilde{\cap} \breve{\mathbb{X}}_E &= \breve{\mathbb{Y}}_E
\end{aligned}$$

so

$$\tilde{\tau}_{\mathbb{Y}} = \{\Gamma_\phi, \Psi_A^{(1)}, \Psi_B^{(2)}, \breve{\mathbb{Y}}_E\}$$

is a q-ROFS sub-topology of $\tilde{\tau}_\chi$.

Definition 2.4.4 Let $(\tilde{\chi}_E, \tilde{\tau})$ be a q-ROFS topological space and $\Gamma_A \tilde{\subseteq} \tilde{\chi}_E$.

(1) The union of all q-ROFS open subsets of Γ_A is called *interior* Γ_A° of Γ_A. The so obtained Γ_A° will become the largest q-ROFS open subset of Γ_A.
(2) The *closure* $\overline{\Gamma_A}$ of Γ_A is the intersection of all q-ROFS closed supersets of Γ_A. The so obtained $\overline{\Gamma_A}$ will become the smallest q-ROFS closed superset of Γ_A.
(3) The *frontier* or *boundary* $Fr(\Gamma_A)$ of Γ_A may be represented as

$$Fr(\Gamma_A) = \overline{\Gamma_A} \tilde{\cap} \overline{\Gamma_A^c}.$$

(4) The *exterior* $Ext(\Gamma_A)$ of Γ_A may be written in the form

$$Ext(\Gamma_A) = (\Gamma_A^c)^\circ.$$

We elaborate these concepts in forthcoming example.

Example 2.4.5 We consider

$$\chi = \{\rho_1, \rho_2\}$$

as a universe. Taking $B = \{\epsilon_1, \epsilon_2\} \in E$ as a collection of characteristics. Also, we suppose that

$$\begin{aligned}
\Gamma_B^{(1)} &= \left\{ \left(\epsilon_1, \left\{ \frac{\rho_1}{(0.471, 0.531)} \right\}, \left\{ \frac{\rho_2}{(0.630, 0.271)} \right\} \right), \left(\epsilon_2, \left\{ \frac{\rho_1}{(0.333, 0.791)} \right\}, \left\{ \frac{\rho_2}{(0.460, 0.442)} \right\} \right) \right\} \\
\Gamma_B^{(2)} &= \left\{ \left(\epsilon_1, \left\{ \frac{\rho_1}{(0.677, 0.291)} \right\}, \left\{ \frac{\rho_2}{(0.710, 0.177)} \right\} \right), \left(\epsilon_2, \left\{ \frac{\rho_1}{(0.371, 0.630)} \right\}, \left\{ \frac{\rho_2}{(0.581, 0.390)} \right\} \right) \right\} \\
\Gamma_B^{(3)} &= \left\{ \left(\epsilon_1, \left\{ \frac{\rho_1}{(0.744, 0.211)} \right\}, \left\{ \frac{\rho_2}{(0.802, 0.072)} \right\} \right), \left(\epsilon_2, \left\{ \frac{\rho_1}{(0.640, 0.522)} \right\}, \left\{ \frac{\rho_2}{(0.606, 0.313)} \right\} \right) \right\}.
\end{aligned}$$

We here obtained an assemblage $\tilde{\mathcal{T}} = \{\Gamma_\phi, \breve{\chi}_B, \Gamma_B^{(1)}, \Gamma_B^{(2)}, \Gamma_B^{(3)}\}$ of q-ROFS subsets of $\breve{\chi}_B$ that becomes now a q-ROFS topology. Members of $\tilde{\mathcal{T}}$ act as q-ROFS open sets and the associated q-ROFS closed sets are expressed as

$$\left(\Gamma_\phi\right)^c = \breve{\chi}_B$$
$$\left(\breve{\chi}_B\right)^c = \Gamma_\phi$$
$$\left(\Gamma_B^{(1)}\right)^c = \left\{\left(\epsilon_1, \left\{\frac{\rho_1}{(0.531, 0.471)}\right\}, \left\{\frac{\rho_2}{(0.271, 0.630)}\right\}\right), \left(\epsilon_2, \left\{\frac{\rho_1}{(0.791, 0.333)}\right\}, \left\{\frac{\rho_2}{(0.442, 0.460)}\right\}\right)\right\}$$
$$\left(\Gamma_B^{(2)}\right)^c = \left\{\left(\epsilon_1, \left\{\frac{\rho_1}{(0.291, 0.677)}\right\}, \left\{\frac{\rho_2}{(0.177, 0.710)}\right\}\right), \left(\epsilon_2, \left\{\frac{\rho_1}{(0.630, 0.371)}\right\}, \left\{\frac{\rho_2}{(0.390, 0.581)}\right\}\right)\right\}$$
$$\left(\Gamma_B^{(3)}\right)^c = \left\{\left(\epsilon_1, \left\{\frac{\rho_1}{(0.211, 0.744)}\right\}, \left\{\frac{\rho_2}{(0.072, 0.802)}\right\}\right), \left(\epsilon_2, \left\{\frac{\rho_1}{(0.522, 0.640)}\right\}, \left\{\frac{\rho_2}{(0.313, 0.606)}\right\}\right)\right\}.$$

We consider q-ROFSS

$$\Gamma_B = \left\{\left(\epsilon_1, \left\{\frac{\rho_1}{(0.584, 0.326)}\right\}, \left\{\frac{\rho_2}{(0.735, 0.291)}\right\}\right), \left(\epsilon_2, \left\{\frac{\rho_1}{(0.717, 0.235)}\right\}, \left\{\frac{\rho_2}{(0.778, 0.095)}\right\}\right)\right\}$$

such that

$$\left(\Gamma_B\right)^c = \left\{\left(\epsilon_1, \left\{\frac{\rho_1}{(0.326, 0.584)}\right\}, \left\{\frac{\rho_2}{(0.291, 0.735)}\right\}\right), \left(\epsilon_2, \left\{\frac{\rho_1}{(0.235, 0.717)}\right\}, \left\{\frac{\rho_2}{(0.095, 0.778)}\right\}\right)\right\}.$$

The interior of q-ROFS Γ_B is

$$\begin{aligned}\Gamma_B^\circ &= \Gamma_\phi \tilde{\cup} \Gamma_B^{(1)} \tilde{\cup} \Gamma_B^{(3)} \\ &= \left\{\left(\epsilon_1, \left\{\frac{\rho_1}{(0.471, 0.531)}\right\}, \left\{\frac{\rho_2}{(0.802, 0.072)}\right\}\right), \left(\epsilon_2, \left\{\frac{\rho_1}{(0.735, 0.291)}\right\}, \left\{\frac{\rho_2}{(0.606, 0.313)}\right\}\right)\right\}.\end{aligned}$$

The closure of the q-ROFS Γ_B is

$$\overline{\Gamma_B} = \breve{\chi}_E.$$

Now,

$$\begin{aligned}\overline{\Gamma_B^c} &= \breve{\chi}_B \tilde{\cap} \left(\Gamma_B^{(1)}\right)^c \tilde{\cap} \left(\Gamma_B^{(3)}\right)^c \\ &= \left\{\left(\epsilon_1, \left\{\frac{\rho_1}{(0.531, 0.471)}\right\}, \left\{\frac{\rho_2}{(0.072, 0.802)}\right\}\right), \left(\epsilon_2, \left\{\frac{\rho_1}{(0.291, 0.735)}\right\}, \left\{\frac{\rho_2}{(0.313, 0.606)}\right\}\right)\right\}.\end{aligned}$$

The frontier of q-ROFS Γ_B is

$$Fr(\Gamma_B) = \overline{\Gamma_B} \tilde{\cap} \overline{\Gamma_B^c}$$
$$= \left\{ \left(\epsilon_1, \left\{ \frac{\rho_1}{(0.326, 0.584)} \right\}, \left\{ \frac{\rho_2}{(0.291, 0.735)} \right\} \right), \left(\epsilon_2, \left\{ \frac{\rho_1}{(0.235, 0.717)} \right\}, \left\{ \frac{\rho_2}{(0.095, 0.778)} \right\} \right) \right\}.$$

The exterior of q-ROFS Γ_B is

$$Ext(\Gamma_B) = (\Gamma_B^c)^\circ$$
$$= \Gamma_\phi.$$

Theorem 2.4.6 *We consider $(\tilde{\chi}_E, \tilde{\tau})$ a q-ROFS topological space and $\Gamma_B \tilde{\subseteq} \tilde{\chi}_E$, then*

(a) $(\Gamma_B^\circ)^c = \overline{(\Gamma_B^c)}$.
(b) $(\overline{\Gamma_B})^c = (\Gamma_B^c)^\circ$.

Theorem 2.4.7 *We consider $(\tilde{\chi}_E, \tilde{\tau})$ a q-ROFS topological space and $\Gamma_B \tilde{\subseteq} \tilde{\chi}_E$, then $Fr(\Gamma_B) = Fr(\Gamma_B^c)$*

Proof By definition,

$$\begin{aligned} Fr(\Gamma_B) &= \overline{(\Gamma_B)} \tilde{\cap} \overline{(\Gamma_B^c)} \\ &= \overline{(\Gamma_B^c)} \tilde{\cap} \overline{(\Gamma_B)} \\ &= \overline{(\Gamma_B^c)} \tilde{\cap} \overline{[(\Gamma_B^c)]^c} \\ &= Fr(\Gamma_B^c). \end{aligned}$$

□

Definition 2.4.8 Let χ be the universe and E the collection of attributes. Then,

(i) The discrete q-ROFS topology is defined by $\tilde{\ }_{\text{discrete}} =$ q-ROFS$(\chi, \mathbb{P})$, where q-ROFS(χ, E) is aggregate of all q-ROFS subsets of a q-ROFS (χ, E).
(ii) The indiscrete q-ROFS topology is defined by $\tilde{\ }_{\text{indiscrete}} = \{\Gamma_\phi, \breve{\chi}_E\}$.

We observe that $\tilde{\ }_{\text{discrete}}$ is the largest q-ROFT on χ and $\tilde{\ }_{\text{indiscrete}}$ is the smallest q-ROFT on χ, respectively.

The union of two q-ROFS topologies may or may not be a q-ROFS topology but their intersection is always a q-ROFS topology. The forthcoming example illustrates these concepts.

Example 2.4.9 Let $\chi = \{\rho_1, \rho_2\}$ be a classical set with $E = \{\epsilon_i : i = 1 \ldots 4\}$ as the set of characteristics. We also consider $\mathbb{A} = \{\epsilon_1, \epsilon_2\}$, $\mathbb{B} = \{\epsilon_3, \epsilon_4\} \subseteq E$.

$$\Gamma_A = \left\{ \left(\epsilon_1, \left\{ \frac{\rho_1}{(0.414, 0.710)} \right\}, \left\{ \frac{\rho_2}{(0.762, 0.291)} \right\} \right), \left(\epsilon_2, \left\{ \frac{\rho_1}{(0.762, 0.355)} \right\}, \left\{ \frac{\rho_2}{(0.582, 0.433)} \right\} \right) \right\}$$

$$\Gamma_B = \left\{ \left(\epsilon_3, \left\{ \frac{\rho_1}{(0.830, 0.373)} \right\}, \left\{ \frac{\rho_2}{(0.162, 0.630)} \right\} \right), \left(\epsilon_4, \left\{ \frac{\rho_1}{(0.591, 0.322)} \right\}, \left\{ \frac{\rho_2}{(0.794, 0.455)} \right\} \right) \right\}$$

then

$$\widetilde{\mathcal{T}}_1 = \{\Phi_E, \Gamma_A, \breve{\chi}_E\}$$

and

$$\widetilde{\mathcal{T}}_2 = \{\Phi_E, \Gamma_B, \breve{\chi}_E\}$$

are q-ROFS topologies on χ but

$$\widetilde{\mathcal{T}}_1 \tilde{\cup} \widetilde{\mathcal{T}}_2 = \{\Phi_E, \breve{\chi}_E, \Gamma_A, \Gamma_B\}$$

does not form a q-ROFS topology.

Definition 2.4.10 Let $\widetilde{\mathcal{T}}_1$ and $\widetilde{\mathcal{T}}_2$ be two q-ROFS topological spaces. If $\widetilde{\mathcal{T}}_1 \tilde{\subseteq} \widetilde{\mathcal{T}}_2$, then $\widetilde{\mathcal{T}}_1$ is said to be *weaker* or *coarser* than $\widetilde{\mathcal{T}}_2$ and $\widetilde{\mathcal{T}}_2$ is known as *stronger* or *finer* than $\widetilde{\mathcal{T}}_1$. Thus, $\widetilde{\mathcal{T}}_1$ and $\widetilde{\mathcal{T}}_2$ are called *comparable*.

Example 2.4.11 Let $\chi = \{\rho_1, \rho_2\}$ be a classical set and $E = \{\epsilon_1, \epsilon_2, \epsilon_3\}$ with $\mathbb{A} = \{e_1, e_2\} \subset E$. We now take q-ROFSSs as

$$\Gamma_A^{(1)} = \left\{\left(\epsilon_1, \left\{\frac{\rho_1}{(0.412, 0.674)}\right\}, \left\{\frac{\rho_2}{(0.610, 0.473)}\right\}\right), \left(\epsilon_2, \left\{\frac{\rho_1}{(0.515, 0.870)}\right\}, \left\{\frac{\rho_2}{(0.614, 0.771)}\right\}\right)\right\}$$

$$\Gamma_A^{(2)} = \left\{\left(\epsilon_1, \left\{\frac{\rho_1}{(0.716, 0.317)}\right\}, \left\{\frac{\rho_2}{(0.724, 0.372)}\right\}\right), \left(\epsilon_2, \left\{\frac{\rho_1}{(0.670, 0.661)}\right\}, \left\{\frac{\rho_2}{(0.725, 0.660)}\right\}\right)\right\}$$

$$\Gamma_A^{(3)} = \left\{\left(\epsilon_1, \left\{\frac{\rho_1}{(0.797, 0.276)}\right\}, \left\{\frac{\rho_2}{(0.842, 0.255)}\right\}\right), \left(\epsilon_2, \left\{\frac{\rho_1}{(0.866, 0.533)}\right\}, \left\{\frac{\rho_2}{(0.842, 0.550)}\right\}\right)\right\}.$$

Now, we have $\tilde{}_1 = \{\Phi_E, \Gamma_A^{(1)}, \breve{\chi}_E\}$ and $\tilde{}_2 = \{\Phi_E, \Gamma_A^{(1)}, \Gamma_A^{(2)}, \Gamma_A^{(3)}, \breve{\chi}_E\}$ as two q-ROFS topologies. As $\tilde{}_1 \tilde{\subset} \tilde{}_2$, so $\tilde{}_1$ is coarser than $\tilde{}_2$.

Remark It is worth noticing that "law of contradiction" and "law of excluded middle" that make sense in crisp sets are not felicitous in IFSSs and PFSSs. We substantiate them in q-ROFSS theory by considering a few results in q-ROFST that may vary from crisp topology. Now we discuss a few verifications in the commendation of our statement.

Example 2.4.12 We consider a classical set $\chi = \{\rho_1, \rho_2, \rho_3\}$ and a set of attributes $E = \{\epsilon_1, \epsilon_2, \epsilon_3, \epsilon_4\}$ by taking $\mathbb{A} = \{\epsilon_1, \epsilon_2\} \subseteq E$. We also take

$$\Gamma_A = \left\{\left(\epsilon_1, \left\{\frac{\rho_1}{(0.510, 0.271)}\right\}, \left\{\frac{\rho_2}{(0.615, 0.473)}\right\}\right), \left(\epsilon_2, \left\{\frac{\rho_1}{(0.510, 0.271)}\right\}, \left\{\frac{\rho_2}{(0.615, 0.473)}\right\}\right)\right\}.$$

Obviously,

$$\widetilde{\mathcal{T}} = \{\Phi_E, \breve{\chi}_E, \Gamma_A, \Gamma_A^c\}$$

is not a q-ROFST on X, for neither $\Gamma_A \tilde{\cap} \Gamma_A^c \tilde{\in} \widetilde{\mathcal{T}}$ nor $\Gamma_A \tilde{\cup} \Gamma_A^c \tilde{\in} \widetilde{\mathcal{T}}$.

Definition 2.4.13 Let $(\check{\chi}_E, \widetilde{\tau})$ be a q-ROFS topological space. Then, $\mathbb{B} \widetilde{\subseteq} \widetilde{\tau}$ is a *basis* for $\widetilde{\tau}$ if for each $\Gamma_A \widetilde{\in} \widetilde{\tau}$, there exists $B_i \in \mathbb{B}$ such that $\Gamma_A = \widetilde{\cup} B_i$, i.e. each q-ROFS open set in $\widetilde{\tau}$ is a union of members of $\mathbb{B}$.

Example 2.4.14 We consider $X = \{\varrho_1, \varrho_2\}$ the universal set and $A = \{\epsilon_1, \epsilon_2\} \subseteq E$ be characteristics' set. Picking q-ROFSSs

$$\Gamma_A^{(1)} = \left\{ \left(\epsilon_1, \left\{ \frac{\rho_1}{(0.264, 0.550)} \right\}, \left\{ \frac{\rho_2}{(0.171, 0.725)} \right\} \right), \left(\epsilon_2, \left\{ \frac{\rho_1}{(0.373, 0.860)} \right\}, \left\{ \frac{\rho_2}{(0.372, 0.701)} \right\} \right) \right\},$$

$$\Gamma_A^{(2)} = \left\{ \left(\epsilon_1, \left\{ \frac{\rho_1}{(0.481, 0.343)} \right\}, \left\{ \frac{\rho_2}{(0.174, 0.723)} \right\} \right), \left(\epsilon_2, \left\{ \frac{\rho_1}{(0.490, 0.675)} \right\}, \left\{ \frac{\rho_2}{(0.505, 0.642)} \right\} \right) \right\},$$

$$\Gamma_A^{(3)} = \left\{ \left(\epsilon_1, \left\{ \frac{\rho_1}{(0.630, 0.242)} \right\}, \left\{ \frac{\rho_2}{(0.544, 0.600)} \right\} \right), \left(\epsilon_2, \left\{ \frac{\rho_1}{(0.690, 0.606)} \right\}, \left\{ \frac{\rho_2}{(0.544, 0.600)} \right\} \right) \right\}.$$

Here, we got

$$\mathbb{B} = \left\{ \Gamma_A^{(1)}, \Gamma_A^{(2)}, \Gamma_A^{(3)} \right\}$$

that furnishes a q-ROFS basis for the q-ROFST $\widetilde{\tau} = \left\{ \Phi_E, \check{\chi}_E, \Gamma_A^{(1)}, \Gamma_A^{(2)}, \Gamma_A^{(3)} \right\}$.

Theorem 2.4.15 *Let $\mathbb{B}$ be a q-ROFS basis for a topology $\widetilde{\tau}$. Then, for every $\epsilon \in E$, $\mathbb{B}_\epsilon = \{\Gamma(\epsilon) : \Gamma_\mathbb{P} \in \mathbb{B}\}$ provides a q-ROFS basis for q-rung orthopair fuzzy soft topology $\widetilde{\tau}(\epsilon) = \{\Gamma(\epsilon) : \Gamma_E \in \widetilde{\tau}\}$.*

Proof We suppose that $\Gamma^{(1)}(\epsilon) \in (e)$, where $\Gamma_E^{(1)} \in$. As $\mathbb{B}$ is a q-ROFS basis for $\widetilde{\tau}$, then by definition, there will exist a $\mathbb{B}' \in \mathbb{B}$ such that $\Gamma_E^{(1)} = \cup \mathbb{B}'$ and thus $\Gamma^{(1)}(\epsilon) = \cup \mathbb{B}'_\epsilon$, where $\mathbb{B}'_\epsilon = \{\Gamma(e) : \Gamma_E \in \mathbb{B}'\} \subseteq \mathbb{B}_\epsilon$. $\qquad\square$

Theorem 2.4.16 *We consider $(\check{\chi}_E, \widetilde{\tau})$ a q-ROFS topological space. An assemblage*

$$\mathbb{B} = \{\Gamma_{B_\alpha} \mid \alpha \in \Omega\} \widetilde{\subseteq} \widetilde{\mathbb{B}\tau}$$

provides a q-ROFS basis for $\widetilde{\tau}$ in such a way that for each q-ROFSS open set Γ_U and a q-ROFS point $\mathbb{P}_\epsilon^x \in \Gamma_U$, there is a $\Gamma_{B_\alpha} \in \mathbb{B}$ such that $\mathbb{P}_\epsilon^x \in \Gamma_{B_\alpha} \widetilde{\subseteq} \Gamma_U$.

Proof Here, we take $\mathbb{B}$ as a q-ROFS basis for $\widetilde{\tau}$. Also for each q-ROFS open set Γ_U of $\widetilde{\tau}$, then there are q-ROFS sets Γ_{B_γ}, $\gamma \in \Omega' \widetilde{\subseteq} \Omega$ such that $\Gamma_U = \widetilde{\cup} \Gamma_{B_\gamma}$. Thus, for every $\mathbb{P}_e^x \in \Gamma_U$, we have a B_γ so that

$$\mathbb{P}_e^x \in \Gamma_{B_\gamma} \widetilde{\subseteq} \Gamma_U.$$

Conversely, we suppose that for each q-ROFS open set Γ_U and $\mathbb{P}_\epsilon^x \in \Gamma_U$, we have Γ_B in $\mathbb{B}$ so that $\mathbb{P}_\epsilon^x \in \Gamma_B \widetilde{\subseteq} \Gamma_U$. Thus,

$$\Gamma_U = \widetilde{\cup} \mathbb{P}_e^x \widetilde{\subseteq} \widetilde{\cup} \Gamma_{B_\gamma} \widetilde{\subseteq} \Gamma_U$$

such that

$$\Gamma_U = \tilde{\bigcup}\Gamma_{B_\gamma}.$$

Hence, each q-ROFS open set Γ_U is q-ROFS union of q-ROFSSs in $\mathbb{B}$. □

2.4.1 q-ROFS Separation Axioms

Definition 2.4.17 A q-rung orthopair fuzzy soft topological space $(\tilde{\chi}_E, \tilde{\mathcal{T}})$ is known as a q-ROFS $\mathbb{T}_0$-space if for two distinct q-ROFS points $\varpi_\Gamma^{(1)}$ and $\varpi_\Gamma^{(2)}$, there exists a q-ROFS open set Γ_A which contains one of the q-ROFS point but not other.

Example 2.4.18 Every discrete q-rung orthopair fuzzy soft topological space is a q-ROFS $\mathbb{T}_0$-space since there exists an open set $\{\varpi_\Gamma^{(1)}\}$ containing $\varpi_\Gamma^{(1)}$ only but not $\varpi_\Gamma^{(2)}$.

Theorem 2.4.19 *A q-ROFS topological space $(\tilde{\chi}_E, \tilde{\mathcal{T}})$ is a q-ROFS $\mathbb{T}_0$-space if and only if for every $\varpi_\Gamma^{(1)}, \varpi_\Gamma^{(2)} \tilde{\in} \check{\chi}_E$, $\varpi_\Gamma^{(1)} \neq \varpi_\Gamma^{(2)}$ gives us $\overline{\{\varpi_\Gamma^{(1)}\}} \neq \overline{\{\varpi_\Gamma^{(2)}\}}$.*

Proof Suppose that $(\tilde{\chi}_E, \tilde{\mathcal{T}})$ is a q-ROFS $\mathbb{T}_0$-space and $\varpi_\Gamma^{(1)}, \varpi_\Gamma^{(2)}$ are two different q-ROFS points. Thus, by definition of $\mathbb{T}_0$-space, there exists one q-ROFS open set Γ_C, say, that contains one of the q-ROFS points. Here, we consider that Γ_C containing $\varpi_\Gamma^{(1)}$ but not $\varpi_\Gamma^{(2)}$. Such that $\varpi_\Gamma^{(1)}$ may not be the accumulation point of $\{\varpi_\Gamma^{(2)}\}$. In this way, we get $\overline{\{\varpi_\Gamma^{(1)}\}} \neq \overline{\{\varpi_\Gamma^{(2)}\}}$.

Conversely, we take two distinct q-ROFS points $\varpi_\Gamma^{(1)}, \varpi_\Gamma^{(2)}$ having $\overline{\{\varpi_\Gamma^{(1)}\}} \neq \overline{\{\varpi_\Gamma^{(2)}\}}$. Also we take, on the contrary basis, that $(\tilde{\chi}_E, \tilde{\mathcal{T}})$ is not a q-ROFS $\mathbb{T}_0$-space, i.e. for each q-ROFS open set containing $\varpi_\Gamma^{(1)}$ as well as $\varpi_\Gamma^{(2)}$. Thus, by using property of accumulation, point $\varpi_\Gamma^{(1)}$ contains in $\overline{\{\varpi_\Gamma^{(2)}\}}$ such that $\overline{\{\varpi_\Gamma^{(1)}\}} \tilde{\subseteq} \overline{\{\varpi_\Gamma^{(2)}\}}$. In the same way, we know that every q-ROFS open set that contains $\varpi_\Gamma^{(2)}$ is also containing $\varpi_\Gamma^{(1)}$ (otherwise $(\tilde{\chi}_E, \tilde{\mathcal{T}})$ will become a q-ROFS $\mathbb{T}_0$-space), so by the property of limit, point $\varpi_\Gamma^{(2)}$ is in $\overline{\{\varpi_\Gamma^{(1)}\}}$, which implies that $\overline{\{\varpi_\Gamma^{(2)}\}} \tilde{\subseteq} \overline{\{\varpi_\Gamma^{(1)}\}}$. Then $\overline{\{\varpi_\Gamma^{(2)}\}} = \overline{\{\varpi_\Gamma^{(1)}\}}$, which is a contradiction against our hypothesis. □

Remark The characteristic of being a q-ROFS $\mathbb{T}_0$-space is a hereditary property.

Definition 2.4.20 A q-ROFSTS is reckoned as a q-ROFS $\mathbb{T}_1$-space if for any two distinct q-ROFS points $\varpi_\Gamma^{(1)}, \varpi_\Gamma^{(2)}$ of $\tilde{\chi}_E$, there exist two q-ROFS open sets Γ_C and Γ_D such that $\varpi_\Gamma^{(1)} \tilde{\in} \Gamma_C$, $\varpi_\Gamma^{(2)} \tilde{\notin} \Gamma_C$ and $\varpi_\Gamma^{(1)} \tilde{\notin} \Gamma_D$, $\varpi_\Gamma^{(2)} \tilde{\in} \Gamma_D$. (i.e. both q-ROFS open set contain exactly one of the q-ROFS points).

Example 2.4.21 Every discrete q-ROFSTS is a q-ROFS $\mathbb{T}_1$-space. Here for $\varpi_\Gamma^{(1)}$ and $\varpi_\Gamma^{(2)}$ are two distinct q-ROFS points in $\tilde{\chi}_E$, there are q-ROFS open sets $\{\varpi_\Gamma^{(1)}\}$ and $\{\varpi_\Gamma^{(2)}\}$ such that $\varpi_\Gamma^{(1)} \in \{\varpi_\Gamma^{(1)}\}$ whereas $\varpi_\Gamma^{(2)} \notin \{\varpi_\Gamma^{(1)}\}$.

Theorem 2.4.22 *The following statements are equivalent:*

(i) *Each q-ROFS (singleton) point of $\tilde{\chi}_E$ is q-ROFS closed.*
(ii) $(\tilde{\chi}_E, \tilde{\tau})$ *is a q-ROFS $\mathbb{T}_1$-space.*

Proof The proof of this theorem is obvious. □

Remark The characteristic of being a q-ROFS $\mathbb{T}_1$-space is transferred to all of its subspaces.

Definition 2.4.23 A q-ROFS space $(\tilde{\chi}_E, \tilde{\tau})$ is known as q-ROFS $\mathbb{T}_2$-space or q-ROFS Hausdorff space if for every two distinct q-ROFS points $\varpi_\Gamma^{(1)}$ and $\varpi_\Gamma^{(2)}$ of $\tilde{\chi}_E$, there are two q-ROFS open sets Γ_C and Γ_D such that

$$\varpi_\Gamma^{(1)} \tilde{\in} \Gamma_C, \ \ \varpi_\Gamma^{(2)} \tilde{\in} \Gamma_D \ \ \& \ \ \Gamma_C \tilde{\cap} \Gamma_D = \Gamma_\phi,$$

where Γ_ϕ is a null q-ROFSS.

Example 2.4.24 Consider a discrete q-ROFS topological space $(\tilde{\chi}_E, \mho)$. Take $\varpi_\Gamma^{(1)}$ and $\varpi_\Gamma^{(2)}$ any two distinct q-ROFS points in $\tilde{\chi}_E$. It is obvious now that $\{\varpi_\Gamma^{(1)}\}$ and $\{\varpi_\Gamma^{(2)}\}$ are disjoint q-ROFS open sets, such that

$$\varpi_\Gamma^{(1)} \tilde{\in} \{\varpi_\Gamma^{(1)}\}, \ \ \varpi_\Gamma^{(2)} \tilde{\in} \{\varpi_\Gamma^{(2)}\}, \ \text{ and } \ \{\varpi_\Gamma^{(1)}\} \tilde{\cap} \{\varpi_\Gamma^{(2)}\} = \Gamma_\phi.$$

Thus, $(\tilde{\chi}_E, \mho)$ is a q-ROFS $\mathbb{T}_2$-space.

Definition 2.4.25 A q-ROFSTS $(\tilde{\chi}_E, \tilde{\tau})$ is known as a q-ROFS regular space, for any q-ROFS closed set Γ_A and q-ROFS point $\varpi_\Gamma \notin \Gamma_A$, $\exists$ q-ROFS open sets Γ_C and Γ_D such that $\varpi_\Gamma \tilde{\in} \Gamma_C$, $\Gamma_A \tilde{\subseteq} \Gamma_D$, and $\Gamma_C \tilde{\cap} \Gamma_D = \Gamma_\phi$, where Γ_ϕ is the null q-ROFS.

A q-ROFS topological space $(\tilde{\chi}_E, \tilde{\tau})$ is known as a *q-ROFS $\mathbb{T}_3$-space* if it is both q-ROFS regular and q-ROFS $\mathbb{T}_1$-space.

Definition 2.4.26 A q-ROFS topological space $(\tilde{\chi}_E, \tilde{\tau})$ is said to be a q-ROFS normal space if there exist two q-ROFS closed disjoint subsets Γ_A and Γ_B of $(\tilde{\chi}_E, \tilde{\tau})$, $\exists$ two q-ROFS open sets Γ_C and Γ_D such that $\Gamma_A \tilde{\subseteq} \Gamma_C$, $\Gamma_B \tilde{\subseteq} \Gamma_D$, and $\Gamma_C \tilde{\cap} \Gamma_D = \phi_E$.

A q-ROFS normal $\mathbb{T}_1$-space is called a q-ROFS $\mathbb{T}_4$-space.

Theorem 2.4.27 *A q-rung orthopair fuzzy soft topological space $(\tilde{\chi}_E, \tilde{\tau})$ is a q-ROFS $\mathbb{T}_2$-space if for two distinct q-ROFS points $\varpi_\Gamma^{(1)}$ and $\varpi_\Gamma^{(2)}$, there exist q-ROFS closed sets Γ_{C_1} and Γ_{C_2} such that $\varpi_\Gamma^{(1)} \tilde{\in} \Gamma_{C_1}$, $\varpi_\Gamma^{(2)} \tilde{\notin} \Gamma_{C_1}$, $\varpi_\Gamma^{(1)} \tilde{\notin} \Gamma_{C_2}$, $\varpi_\Gamma^{(2)} \tilde{\in} \Gamma_{C_2}$, and $\Gamma_{C_1} \tilde{\cup} \Gamma_{C_2} = \check{\chi}_E$.*

Proof Let $(\tilde{\chi}_E, \tilde{\tau})$ be a q-ROFS $\mathbb{T}_2$-space, we also consider $\varpi_\Gamma^{(1)}$ and $\varpi_\Gamma^{(2)}$ be two distinct q-rung orthopair fuzzy soft points of $\tilde{\chi}_E$. By using definition, there exist two q-rung orthopair fuzzy soft open sets Γ_{U_1} and Γ_{U_2} in such a way that $\varpi_\Gamma^{(1)} \tilde{\in} \Gamma_{U_1}$,

$\varpi_\Gamma^{(2)}\widetilde{\notin}\Gamma_{U_1}$, $\varpi_\Gamma^{(1)}\widetilde{\notin}\Gamma_{U_2}$, $\varpi_\Gamma^{(2)}\widetilde{\in}\Gamma_{U_2}$, and $\Gamma_{U_1}\widetilde{\cap}\Gamma_{U_2} = \phi_E$. Thus, $\{\Gamma_{U_1}\}^c\widetilde{\cup}\{\Gamma_{U_2}\}^c = \breve{\chi}_E$ and also $\varpi_\Gamma^{(1)}\widetilde{\notin}\{\Gamma_{U_1}\}^c = \Gamma_{C_2}$, $\varpi_\Gamma^{(2)}\widetilde{\in}\{\Gamma_{U_1}\}^c = \Gamma_{C_2}$. $\varpi_\Gamma^{(1)}\widetilde{\in}\{\Gamma_{U_2}\}^c = \Gamma_{C_1}$, $\varpi_\Gamma^{(2)}\widetilde{\notin}\{\Gamma_{U_2}\}^c = \Gamma_{C_1}$.

Now, we take into consideration that for any two distinct q-rung orthopair fuzzy soft points $\varpi_\Gamma^{(1)}, \varpi_\Gamma^{(2)}\widetilde{\in}\tilde{\chi}_E$, we have q-rung orthopair fuzzy soft closed sets Γ_{C_1} and Γ_{C_2} such that $\varpi_\Gamma^{(1)}\widetilde{\in}\Gamma_{C_1}$, $\varpi_\Gamma^{(2)}\widetilde{\notin}\Gamma_{C_1}$, $\varpi_\Gamma^{(1)}\widetilde{\notin}\Gamma_{C_2}$, $\varpi_\Gamma^{(2)}\widetilde{\in}\Gamma_{C_2}$, and $\Gamma_{C_1}\widetilde{\cup}\Gamma_{C_2} = \breve{\chi}_E$. So $\{\Gamma_{C_1}\}^c$ and $\{\Gamma_{C_2}\}^c$ are q-ROFS open sets such that $\varpi_\Gamma^{(1)}\widetilde{\notin}\{\Gamma_{C_1}\}^c$, $\varpi_\Gamma^{(2)}\widetilde{\in}\{\Gamma_{C_1}\}^c$, $\varpi_\Gamma^{(1)}\widetilde{\in}\{\Gamma_{C_2}\}^c$, $\varpi_\Gamma^{(2)}\widetilde{\notin}\{\Gamma_{C_2}\}^c$, and $\{\Gamma_{C_1}\}^c\widetilde{\cap}\{\Gamma_{C_2}\}^c = \{\breve{\chi}_E\}^c = \phi_E$. So we can say that $\tilde{\chi}_E$ is a q-rung orthopair fuzzy soft (q-ROFS) $\mathbb{T}_2$-space. □

Remark A q-ROFS $\mathbb{T}_2$-space also fulfils the congenital property, that is every subspace of a q-ROFS $\mathbb{T}_2$-space is q-ROFS $\mathbb{T}_2$-space.

Theorem 2.4.28 *Every q-ROFS $\mathbb{T}_4$-space is q-ROFS regular, i.e. q-ROFS normal $\mathbb{T}_1$-space is a q-ROFS regular.*

Proof We suppose that $(\tilde{\chi}_E, \tilde{\mathcal{T}})$ is a q-ROFS $\mathbb{T}_4$-topological space. We also consider that ϖ_Γ be a q-ROFS point in $\tilde{\chi}_E$. By applying Theorem 2.4.22 $\{\varpi_\Gamma\}$, a closed q-ROFSS in $(\tilde{\chi}_E, \tilde{\mathcal{T}})$. Let Γ_A be a closed q-ROFSS that does not have ϖ_Γ. As $(\tilde{\chi}_E$ is q-ROFS normal, thus we may have an open q-ROFSS namely Γ_U, Γ_V such that

$$\{\varpi_\Gamma\}\widetilde{\subseteq}\Gamma_U, \Gamma_A\widetilde{\subseteq}\Gamma_V, \text{ and } \Gamma_U \cap \Gamma_V = \Gamma_\phi,$$

Thus,

$$\varpi_\Gamma\widetilde{\in}\Gamma_U, \Gamma_A\widetilde{\subseteq}\Gamma_V, \text{ and } \Gamma_U \cap \Gamma_V = \Gamma_\phi.$$

Hence, $(\tilde{\chi}_E, \tilde{\mathcal{T}})$ is q-ROFS regular. □

2.5 Multi-attribute Decision-Making

This section is particularly designed for the newly established approach for q-ROFST towards uncertainties. The judgement of every decision-maker (DM) is intimated by a q-ROFS open set. Moreover, to observe the concordant decision, the q-ROFST is established with the assistance of q-rung orthopair fuzzy soft (q-ROFS) open sets for expressing the judgement of DMs. A vigorous MADM approach is discussed here in addition to already existing MADM techniques. In the first step, we exercise an optimization technique known as "grey relational analysis" (GRA). To investigate the optimal choice, we utilize the concept of compromise solution (a solution that is nearest to the positive ideal solution while farthest from the negative ideal solution is known as compromise solution).

For decision alternatives, we allocate the linguistic terms as given in Table 2.2. We render a technique by discussing each and every step, as follows:

Algorithm 1 (Grey relational analysis)

Step 1: Recognize the situation and decide upon the course of action. Consider that $\mathbb{V} = \{\chi_j : j = 1, \ldots, n\}$ be a set (with some limitations) of alternatives under consideration and $\mathbb{D} = \{d_k : k = 1, \ldots, m\}$ be a team of DMs. By considering a family of attributes or benchmarks and the $(j, k)^{th}$ entry of the q-ROFS matrix demonstrate the weight suggested by k^{th} DM to j^{th} alternative.

Step 2: Evaluate matrix of weighted parameters as $\mathbb{A} = [w_{jk}]_{n\times m}$, where w_{jk} shows fuzzy weights assigned by the DM d_j to the attributes α_j by allocating lingual values given in Table 2.2.

Step 3: Evaluate the normalized matrix $\dot{\mathbb{A}} = [\dot{w}_{jk}]_{n\times m}$ by using $\dot{w}_{jk} = \frac{w_{jk}}{\sqrt{\Sigma_{j=1}^{n} w_{jk}^2}}$. Then calculate the weight vector of DMs $W = (\beth_1, \beth_2, \ldots, \beth_m)$, where $\beth_j = \frac{\sum_{j=1}^{n} \dot{w}_{jk}}{\sum_{k=1}^{m}\sum_{j=1}^{n} \dot{w}_{jk}}$.

Step 4: Construct q-ROFS Γ_D and evaluate the matrix $\mathbb{S}_D$ by replacing the q-ROFNs $(\eta_{\mathbb{R}}, \vartheta_{\mathbb{R}})$.

Step 5: Calculate the highest value h_j and the least value l_j of the matrix $\mathbb{S}_D$. Here, we have two options:

(i) Least-the-better (for non-beneficial objects) by using

$$^{-}(j) = \frac{h_j - d_j}{h_j - l_j} \in [0, 1]. \tag{2.1}$$

(ii) Highest-the-better (for beneficial objects) by making use of

$$^{+}(j) = \frac{d_j - l_j}{h_j - l_j} \in [0, 1]. \tag{2.2}$$

We maintain a q-ROFS decision matrix $\mathbb{B} = [\xi_{ik}]_{n\times m}$ with the help of either least-the-better or highest-the-better formulas.

Step 6: Now, calculate the grey relational coefficient (GRC) by using

$$GRC = \frac{1}{\Sigma_{k=1}^{m} w_k|\xi_{ik} - 1| + 1}, \; (i = 1, \ldots, n). \tag{2.3}$$

Step 7: Rank the choices in accordance with GRC values.

Table 2.2 Linguistic terms for deciding alternatives

Linguistic Terms	Fuzzy Weights
Least valuable (LtV)	[0.000, 0.250]
Less valuable (LV)	(0.251, 0.549]
Valuable(V)	(0.550, 0.700]
Much valuable (MV)	(0.701, 0.850]
Highly valuable (HV)	(0.851, 1.000]

2.5.1 *Numerical Application*

In this subsection, we render an application of the proposed technique, i.e. GRA based upon q-ROFS topology to determine the most appropriate investment strategy.

Example 2.5.1 Assume that an investment group is planning to invest its money in real estate by buying some residential plots from one of the four valuable housing societies.

Step 1: Suppose that $\chi = \{\alpha_i : i = 1, \ldots, 4\}$ is the set of four housing societies for financing, $E = \{\epsilon_i : i = 1, \ldots, 4\}$ is the family of attributes, and $\mathbb{D} = \{D_i : i = 1, \ldots, 4\}$ is the group of decision-makers. The attributes are characterized as follows:

$$\epsilon_1 = \text{Gated community, boundary wall, with good infrastructure.}$$
$$\epsilon_2 = \text{Legal approval by concerned authorities.}$$
$$\epsilon_3 = \text{Affordable prices, and facility of instalments.}$$
$$\epsilon_4 = \text{Amenities, facilities, and security.}$$

Step 2: Now keeping in view Table 2.2, form a weighted parameter matrix

$$\begin{aligned}\mathbb{P} &= [w_{ij}]_{4\times 4}\\ &= \begin{pmatrix} \text{LV} & \text{HV} & \text{MV} & \text{V} \\ \text{V} & \text{LtV} & \text{LV} & \text{MV} \\ \text{LtV} & \text{MV} & \text{LV} & \text{HV} \\ \text{LV} & \text{V} & \text{HV} & \text{MV} \end{pmatrix}\\ &= \begin{pmatrix} 0.301 & 0.899 & 0.749 & 0.550 \\ 0.550 & 0.149 & 0.301 & 0.749 \\ 0.149 & 0.749 & 0.301 & 0.899 \\ 0.301 & 0.550 & 0.899 & 0.749 \end{pmatrix}.\end{aligned}$$

Step 3: The normalized weighted matrix is

$$\hat{N} = [\hat{n}_{ij}]_{4\times 4} = \begin{pmatrix} 0.4220 & 0.6909 & 0.6021 & 0.3678 \\ 0.7735 & 0.1150 & 0.2410 & 0.5015 \\ 0.2110 & 0.5758 & 0.2410 & 0.6021 \\ 0.4220 & 0.4224 & 0.7225 & 0.5015 \end{pmatrix}.$$

The weight vector so obtained is $\mathbb{W} = (0.2464, 0.2436, 0.2437, 0.2663)$.

Step 4: The q-ROFS decision matrix D_i of each member is placed, in which alternatives are expressed row-wise and parameters are expressed column-wise such that the aggregate of all D_i's makes a q-ROFST.

$$D_1 = \begin{pmatrix} (0.899, 0.171) & (0.360, 0.772) & (0.951, 0.312) & (0.880, 0.462) \\ (0.103, 0.702) & (0.560, 0.661) & (0.582, 0.491) & (0.772, 0.622) \\ (0.633, 0.273) & (0.491, 0.462) & (0.440, 0.402) & (0.661, 0.464) \\ (0.522, 0.380) & (0.401, 0.661) & (0.695, 0.560) & (0.550, 0.560) \end{pmatrix}$$

$$D_2 = \begin{pmatrix} (1.000, 0.000) & (1.000, 0.000) & (1.000, 0.000) & (1.000, 0.000) \\ (1.000, 0.000) & (1.000, 0.000) & (1.000, 0.000) & (1.000, 0.000) \\ (1.000, 0.000) & (1.000, 0.000) & (1.000, 0.000) & (1.000, 0.000) \\ (1.000, 0.000) & (1.000, 0.000) & (1.000, 0.000) & (1.000, 0.000) \end{pmatrix}$$

$$D_3 = \begin{pmatrix} (0.655, 0.844) & (0.252, 0.884) & (0.810, 0.582) & (0.791, 0.682) \\ (0.075, 0.782) & (0.340, 0.791) & (0.522, 0.695) & (0.572, 0.711) \\ (0.550, 0.610) & (0.295, 0.550) & (0.333, 0.561) & (0.495, 0.531) \\ (0.381, 0.432) & (0.370, 0.731) & (0.604, 0.742) & (0.301, 0.652) \end{pmatrix}$$

$$D_4 = \begin{pmatrix} (0.000, 1.000) & (0.000, 1.000) & (0.000, 1.000) & (0.000, 1.000) \\ (0.000, 1.000) & (0.000, 1.000) & (0.000, 1.000) & (0.000, 1.000) \\ (0.000, 1.000) & (0.000, 1.000) & (0.000, 1.000) & (0.000, 1.000) \\ (0.000, 1.000) & (0.000, 1.000) & (0.000, 1.000) & (0.000, 1.000) \end{pmatrix}.$$

Now, the aggregated decision matrix becomes

$$\mathbb{A} = \frac{\sum_{i=1}^{4} D_i}{4} = \begin{pmatrix} (0.6385, 0.5038) & (0.4030, 0.6641) & (0.6902, 0.4735) & (0.6678, 0.5360) \\ (0.2945, 0.6210) & (0.4750, 0.6130) & (0.5260, 0.5465) & (0.5860, 0.5833) \\ (0.5457, 0.4707) & (0.4465, 0.5030) & (0.4432, 0.4908) & (0.5390, 0.4988) \\ (0.4758, 0.4530) & (0.4427, 0.5980) & (0.5747, 0.5755) & (0.4628, 0.5530) \end{pmatrix} = [\dot{v}_{jk}]_{4\times 4}.$$

Step 5: Here is the membership matrix to be selected for making decision

$$B = [\ddot{v}_{jk}]_{4\times 4} = \begin{pmatrix} 0.6385\ 0.4030\ 0.6902\ 0.6678 \\ 0.2945\ 0.4750\ 0.5260\ 0.5860 \\ 0.5457\ 0.4465\ 0.4432\ 0.5390 \\ 0.4758\ 0.4427\ 0.5747\ 0.4628 \end{pmatrix}.$$

Step 6: Calculate the highest value h_j of the matrix $\mathbb{S}_D$. For highest-the-better (for beneficial objects) by employing

$$^{+}(j) = \frac{d_j - l_j}{h_j - l_j}.$$

We maintain a q-ROFS decision matrix $\mathbb{B} = [\xi_{ik}]_{n\times m}$ with the help of highest-the-better formulas.

$$\mathbb{B} = [\ddot{v}_{jk}]_{4\times 4} = \begin{pmatrix} 0.8693\ 0.2742\ 1.0000\ 0.9434 \\ 0.0000\ 0.4562\ 0.5850\ 0.7367 \\ 0.6348\ 0.3841\ 0.3758\ 0.6179 \\ 0.4582\ 0.3745\ 0.7081\ 0.4253 \end{pmatrix}.$$

Step 6: Now Calculate GRC using

$$GRC = \frac{1}{\Sigma_{k=1}^{m} w_k |\xi_{ik} - 1| + 1}, (i = 1, \ldots, n).$$

Step 7: The preference order (Rank the alternatives according to (GRC)), with "highest-the-better", of the alternatives is

$$\alpha_1 \succ \alpha_3 \succ \alpha_4 \succ \alpha_2.$$

Since GRC of α_1 is highest, so the optimal choice is α_1. In view of this ranking, it may be inferred that the company should invest in α_1 (Table 2.3).

Table 2.3 Grey relational coefficients (GRC) of each alternative

Alternative (ξ_j)	GRC
α_1	0.2367
α_2	0.2198
α_3	0.2225
α_4	0.2217

2.5.2 *Generalized Choice Value Method*

Now we discuss generalized choice value method (GCVM) for MADM. For this objective, we develop an extension of existing choice value method towards qROFSSs. The first four steps are same as Algorithm 1, so we start with Step 5.

Algorithm 2 (Generalized choice value method)

Step 5: Find the matrix of choice values using $C = \frac{1}{\Sigma W}(\Gamma_D \times W^t)$.
Step 6: Determine the score function s by using $s = (\eta_{\mathbb{R}})^q - (\vartheta_{\mathbb{R}})^q$ $(q > 1)$. Alternatively, in case of tie, we can use the accuracy function $a(\rho_i)$ using $a = (\eta_{\mathbb{R}})^q + (\vartheta_{\mathbb{R}})^q$ $(q > 1)$.
Step 7: Determine the desired alternative which has greatest score value.

Example 2.5.2 We consider Example 2.5.1 to apply Algorithm 2 by using GCVM for MADM. We head forward with Step 5 as follows.
The choice values matrix $\mathcal{C}$ for q-ROFSs may be given by

$$\mathcal{C} = \frac{1}{\Sigma W_{ij}}(\Gamma_D \times W^t)$$

$$= \frac{1}{1.000}\begin{bmatrix} (0.6385, 0.5038) & (0.4030, 0.6641) & (0.6902, 0.4735) & (0.6678, 0.5360) \\ (0.5457, 0.4707) & (0.4465, 0.5030) & (0.4432, 0.4908) & (0.5390, 0.4988) \\ (0.2945, 0.6210) & (0.4750, 0.6130) & (0.5260, 0.5465) & (0.5860, 0.5833) \\ (0.4758, 0.4530) & (0.4427, 0.4980) & (0.5747, 0.4755) & (0.4628, 0.4530) \end{bmatrix} \begin{bmatrix} 0.2464 \\ 0.2436 \\ 0.2437 \\ 0.2663 \end{bmatrix}$$

$$= \begin{bmatrix} (0.6015, 0.5440) \\ (0.4948, 0.4909) \\ (0.4725, 0.5909) \\ (0.4884, 0.4694) \end{bmatrix}.$$

For $q = 6$, the score function for values is exhibited in Table 2.4.
Table 2.4 demonstrates that

$$\alpha_1 \succ \alpha_4 \succ \alpha_2 \succ \alpha_3.$$

Table 2.4 Score values for alternatives with $q = 6$

χ	$S = \eta^6 - \vartheta^6$	Ranking
α_1	0.0214	1
α_2	0.0007	3
α_3	−0.0314	4
α_4	0.0029	2

Hence, we obtain the same optimal choice as in Example 2.5.1.
It is worth mentioning that we obtain the same optimal choice for every admissible value of q.
We may set up a q-ROFS topology-based technique given in Algorithm 3, as under. First two steps are same as Algorithm 1, so we skip them:

Algorithm 3 (q-ROFS Topology Method)

Step 3: Compute the score values of the entries of each decision matrix using score function $S = \eta^q - \vartheta^q$.
Step 4: Construct q-ROFS topology τ by using decision matrices D_i such that each decision matrix is a q-ROFSS open set.
Step 5: Compute the aggregated sets of all q-ROFS open sets by using the formula

$$\mathfrak{C}^{\star}_{D_n} = \left\{ \frac{\mathcal{L}(\alpha_{\mathrm{i}})}{\alpha_{\mathrm{i}}} : u_{\mathrm{i}} \in \mathbb{V} \right\}, \text{ where } \mathcal{L}(\alpha_{\mathrm{i}}) = \sum_{j \in \mathcal{J} \tilde{\subseteq} \aleph} S(\alpha_{ij}). \tag{2.4}$$

Step 6: Rank the alternatives according to the aggregated values.

Example 2.5.3 We consider Example 2.5.1 to apply Algorithm 3 by using q-ROFST. Steps 1 and 2 of Algorithm 1 and Algorithm 3 are similar.
Step 3: Compute score values of each decision matrix by using score function.

$$(D_1)^* = \begin{pmatrix} 0.7125 & -0.1413 & 0.8297 & 0.5828 \\ -0.3448 & -0.1131 & 0.0787 & 0.2194 \\ 0.2332 & 0.0197 & 0.0202 & 0.1889 \\ 0.0873 & -0.2243 & 0.1600 & -0.0092 \end{pmatrix}$$

$$(D_2)^* = \begin{pmatrix} 1.000 & 1.000 & 1.000 & 1.000 \\ 1.000 & 1.000 & 1.000 & 1.000 \\ 1.000 & 1.000 & 1.000 & 1.000 \\ 1.000 & 1.000 & 1.000 & 1.000 \end{pmatrix}$$

$$(D_3)^* = \begin{pmatrix} -0.3202 & -0.6748 & 0.3343 & 0.1776 \\ -0.4777 & -0.4556 & -0.1934 & -0.1722 \\ -0.0606 & -0.1407 & -0.1396 & -0.0284 \\ -0.0253 & -0.3399 & -0.1881 & -0.2498 \end{pmatrix}$$

$$(D_4)^* = \begin{pmatrix} -1.000 & -1.000 & -1.000 & -1.000 \\ -1.000 & -1.000 & -1.000 & -1.000 \\ -1.000 & -1.000 & -1.000 & -1.000 \\ -1.000 & -1.000 & -1.000 & -1.000 \end{pmatrix}.$$

Step 3: Now construct q-ROFS topology

$$\tau = \{(D_i)^*, i = 1, 2, 3, 4\}.$$

Step 4:

$$(D_1)^* = \left\{ \frac{1.9927}{\alpha_1}, \frac{-0.1598}{\alpha_2}, \frac{0.2731}{\alpha_3}, \frac{0.0138}{\alpha_4} \right\}$$

$$(D_2)^* = \left\{ \frac{4.000}{\alpha_1}, \frac{4.000}{\alpha_2}, \frac{4.000}{\alpha_3}, \frac{4.000}{\alpha_4} \right\}$$

$$(D_3)^* = \left\{ \frac{-0.4831}{\alpha_1}, \frac{-1.2989}{\alpha_2}, \frac{-0.3693}{\alpha_3}, \frac{-0.8031}{\alpha_4} \right\}$$

$$(D_4)^* = \left\{ \frac{-4.000}{\alpha_1}, \frac{-4.000}{\alpha_2}, \frac{-4.000}{\alpha_3}, \frac{-4.000}{\alpha_4} \right\}.$$

Step 5: To find the final scores of each alternative, we compute the sum of aggregated values computed in Step 4.

$$\mathfrak{C}^\star_{\oplus(D_i)^*}(\alpha_i) = \mathfrak{C}^\star_{(D_1)^*}(\alpha_i) + \mathfrak{C}^\star_{(D_2)^*}(\alpha_i) + \mathfrak{C}^\star_{(D_3)^*}(\alpha_i) + \mathfrak{C}^\star_{(D_4)^*}(\alpha_i)$$

$$= \left\{ \frac{1.5096}{\alpha_1}, \frac{-1.4587}{\alpha_2}, \frac{-0.0962}{\alpha_3}, \frac{-0.7893}{\alpha_4} \right\}.$$

Step 6: The ranking of alternatives in a linear order becomes

$$\alpha_1 \succ \alpha_3 \succ \alpha_4 \succ \alpha_2.$$

Hence, once again, we obtain the same optimal choice as in Example 2.5.1.

The rankings obtained through the three suggested algorithms are depicted in Fig. 2.2.

The comparison of the rankings obtained through three algorithms and some existing methodologies is given in Table 2.5.

Table 2.5 advocates the reliability of the suggested methodologies.

2.6 Conclusion

The concepts of q-rung orthopair fuzzy sets and soft sets have a large number of applications in computational intelligence and multi-attribute decision-making (MADM) problems. Motivated by these efficient models, a new hybrid model of q-rung orthopair fuzzy soft sets is proposed. The idea of q-rung orthopair fuzzy soft topology is defined and practical applications in MCDM are presented. For these objectives, some properties of q-rung orthopair fuzzy soft topology are investigated and related results are derived. Some fundamentals of q-rung orthopair fuzzy soft

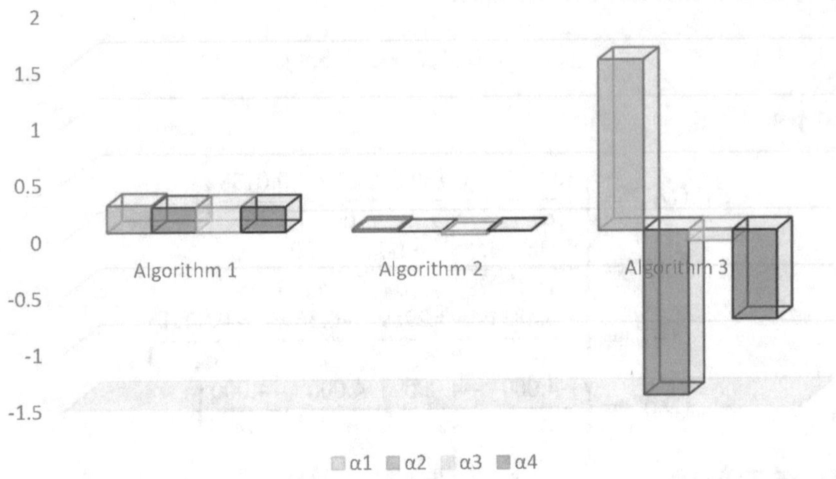

Fig. 2.2 Intra-comparison of three rankings

Table 2.5 Mutual comparison of rankings

Method	Optimal alternative
Garg [54]	α_1
Liu and Wang [52]	α_1
Farid and Riaz [59]	α_1
Proposed Algorithm 1 (GRA)	α_1
Proposed Algorithm 2 (GCVM)	α_1
Proposed Algorithm 3	α_1

separation axioms including q-ROFS $\mathbb{T}_i$-spaces ($i = 1, 2, 3, 4$), q-ROFS regular, and q-ROFS normal are proposed. The idea of score function and accuracy function is used for ranking of feasible alternatives. Moreover, grey relational analysis (GRA) approach, generalized choice value method (GCVM), and aggregation operators-based technique accompanied by three algorithms are proposed to address q-rung orthopair fuzzy soft uncertain information. Numerical examples of these methods are also presented in real-life problems. The validity and efficiency of proposed methods are analysed by performing a comparative analysis with some existing methods. It is demonstrated that the optimal choice remains unaltered.

In the future, we intend to explore more results from q-rung orthopair fuzzy soft topological spaces and some more methodologies to tackle more complex decision-making problems from daily life.

Authors' Contributions The authors contributed to each part of this paper equally.

References

1. L.A. Zadeh, Fuzzy sets. Inf. Control **8**, 338–356 (1965)
2. Z. Pawlak, Rough sets. Int. J. Comput. Inf. Sci. **11**, 341–356 (1982)
3. D. Molodtsov, Soft set theory-first results. Comput. Math. Appl. **37**(4–5), 19–31 (1999)
4. K.T. Atanassov, Intuitionistic fuzzy sets, in: *VII ITKRs Session*, Sofia, June 1983, ed. by V. Sgurev. (Central Sci. and Techn. Library, Bulg. Academy of Sciences, 1984)
5. K.T. Atanassov, Intuitionistic fuzzy sets. Fuzzy Sets Syst. **20**(1), 87–96 (1986)
6. K.T. Atanassov, Geometrical interpretation of the elements of the intuitionistic fuzzy objects. Int. J. Bioautom. **20**(S1), S27–S42 (2016)
7. Z.S. Xu, Intuitionistic fuzzy aggregation operators. IEEE Trans. Fuzzy Syst. **15**(6), 1179–1187 (2007)
8. Z.S. Xu, R.R. Yager, Some geometric aggregation operators based on intuitionistic fuzzy sets. Int. J. Gen Syst **35**(4), 417–433 (2006)
9. H. Zhao, Z.S. Xu, M.F. Ni, S.S. Lui, Generalized aggregation operators for intuitionistic fuzzy sets. Int. J. Intell. Syst. **25**(1), 1–30 (2010)
10. R. R. Yager, Pythagorean fuzzy subsets, in *IFSA World Congress and NAFIPS Annual Meeting 2013 Joint*, Edmonton, Canada I(EEE, 2013), pp. 57–61
11. R.R. Yager, Pythagorean membership grades in multi-criteria decision making. IEEE Trans. Fuzzy Syst. **22**(4), 958–965 (2014)
12. R.R. Yager, A.M. Abbasov, Pythagorean membership grades, complex numbers, and decision making. Int. J. Intell. Syst. **28**(5), 436–452 (2013)
13. R.R. Yager, Generalized orthopair fuzzy sets. IEEE Trans. Fuzzy Syst. **25**(5), 1220–1230 (2017)
14. M.I. Ali, Another view on q-rung orthopair fuzzy sets. Int. J. Intell. Syst. **33**(5), 2139–2153 (2018)
15. A. Aygunoglu, V. Cetkin, H. Aygun, An introduction to fuzzy soft topological spaces. Hacet. J. Math. Stat. **43**(2), 193–204 (2014)
16. N. Çağman, S. Karataş, S. Enginoglu, Soft topology. Comput. Math. Appl. **62**(1), 351–358 (2011)
17. M. Shabir, M. Naz, On soft topological spaces. Comput. Math. Appl. **61**(7), 1786–1799 (2011)
18. B. Ahmad, S. Hussain, On some structures of soft topology. Math. Sci. **6**(64), 1–7 (2012)
19. S. Roy, T.K. Samanta, A note on a soft topological space. Punjab Univ. J. Math. **46**(1), 19–24 (2014)
20. H. Hazra, P. Majumdar, S.K. Samanta, Soft topology. Fuzzy Inf. Eng. **4**(1), 105–115 (2012). https://doi.org/10.1007/s12543-012-0104-2
21. B.P. Varol, A. Shostak, H. Aygun, A new approach to soft topology. Hacet. J. Math. Stat. **41**(5), 731–741 (2012)
22. M. Riaz, S.T. Tehrim, On bipolar fuzzy soft topology with application. Soft. Comput. **24**(24), 18259–18272 (2020)
23. M. Riaz, N. Çağman, I. Zareef, M. Aslam, N-soft topology and its applications to multi-criteria group decision making. J. Intell. Fuzzy Syst. **36**(6), 6521–6536 (2019)
24. M. Riaz, M.A. Razzaq, M. Aslam, D. Pamucar, M-parameterized N-soft topology-based TOPSIS approach for multi-attribute decision making. Symmetry **13**(5), 1–31 (2021)
25. M.R. Hashmi, M. Riaz, F. Smarandache, *m*-polar neutrosophic topology with applications to multi-criteria decision-making in medical diagnosis and clustering analysis. Int. J. Fuzzy Syst. **22**(1), 273–292 (2020)
26. M. Riaz, B. Davvaz, A. Fakhar, A. Firdous, Hesitant fuzzy soft topology and its applications to multi-attribute group decision-making. Soft. Comput. **24**(21), 16269–16289 (2020)
27. M. Riaz, K. Naeem, M. Aslam, D. Afzal, F.A.A. Ahmed, S.S. Jamal, Multi-criteria group decision making with Pythagorean fuzzy soft topology. J. Intell. Fuzzy Syst. **39**(5), 6703–6720 (2020)
28. Z. Li, R. Cui, On the topological structure of intuitionistic fuzzy soft sets. Ann. Fuzzy Math. Inform. **5**(1), 229–239 (2013)

29. I. Osmanoglu, D. Tokat, On intuitionistic fuzzy soft topology. Gen. Math. Notes **19**(2), 59–70 (2013)
30. S. Eraslan, F. Karaaslan, A group decision making method based on TOPSIS under fuzzy soft environment. J. New Theory **3**, 30–40 (2015)
31. M. Akram, G. Ali, Hybrid models for decision-making based on rough Pythagorean fuzzy bipolar soft information. Granul. Comput. **5**(1), 1–15 (2020)
32. M. Akram, M. Arshad, A novel trapezoidal bipolar fuzzy TOPSIS method for group decision-making. Group Decis. Negot. **28**(3), 565–584 (2019)
33. M. Akram, A. Adeel, Novel TOPSIS method for group decision-making based on hesitant m-polar fuzzy model. J. Intell. Fuzzy Syst. **37**(6), 8077–8096 (2019)
34. F. Feng, Y.B. Jun, X. Liu, L. Li, An adjustable approach to fuzzy soft set based decision making. J. Comput. Appl. Math. **234**(1), 10–20 (2010)
35. F. Feng, C. Li, B. Davvaz, M.I. Ali, Soft sets combined with fuzzy sets and rough sets, a tentative approach. Soft. Comput. **14**(9), 899–911 (2010)
36. K. Kumar, H. Garg, TOPSIS method based on the connection number of set pair analysis under interval-valued intuitionistic fuzzy set environment. Comput. Appl. Math. **37**(2), 1319–1329 (2018)
37. H. Garg, A new generalized Pythagorean fuzzy information aggregation using Einstein operations and its application to decision making. Int. J. Intell. Syst. **31**(9), 886–920 (2018)
38. H. Garg, Generalized Pythagorean fuzzy geometric aggregation operators using Einstein t-norm and t-conorm for multicriteria decision-making process. Int. J. Intell. Syst. **32**(16), 597–630 (2017)
39. X.D. Peng, G. Selvachandran, Pythagorean fuzzy set: state of the art and future directions. Artif. Intell. Rev. **52**, 1873–1927 (2019)
40. X.D. Peng, Y. Yang, Some results for Pythagorean fuzzy sets. Int. J. Intell. Syst. **30**(11), 1133–1160 (2015)
41. X.D. Peng, L. Liu, Information measures for q-rung orthopair fuzzy sets. Int. J. Intell. Syst. **34**(8), 1795–1834 (2019)
42. X.D. Peng, Y.Y. Yang, J. Song, Y. Jiang, Pythagorean fuzzy soft set and its application. Comput. Eng. **41**(7), 224–229 (2015)
43. L. Zhang, J. Zhan, Fuzzy soft β-covering based fuzzy rough sets and corresponding decision-making applications. Int. J. Mach. Learn. Cybern. **10**, 1487–1502 (2019)
44. L. Zhang, J. Zhan, J.C.R. Alcantud, Novel classes of fuzzy soft β-coverings-based fuzzy rough sets with applications to multi-criteria fuzzy group decision making. Soft. Comput. **23**, 5327–5351 (2019)
45. J. Zhan, Q. Liu, T. Herawan, A novel soft rough set: soft rough hemirings and its multicriteria group decision making. Appl. Soft Comput. **54**, 393–402 (2017)
46. J. Zhan, M.I. Ali, N. Mehmood, On a novel uncertain soft set model: Z-soft fuzzy rough set model and corresponding decision making methods. Appl. Soft Comput. **56**, 446–457 (2017)
47. X.L. Zhang, Z.S. Xu, Extension of TOPSIS to multiple criteria decision making with Pythagorean fuzzy sets. Int. J. Intell. Syst. **29**(12), 1061–1078 (2014)
48. H. Garg, CN-q-ROFS: connection number-based q-rung orthopair fuzzy set and their application to decision-making process. Int. J. Intell. Syst. **36**(7), 3106–3143 (2021). https://doi.org/10.1002/int.22406
49. Z. Yang, H. Garg, Interaction power partitioned maclaurin symmetric mean operators under q-rung orthopair uncertain linguistic information. Int. J. Fuzzy Syst. 1–19 (2021). https://doi.org/10.1007/s40815-021-01062-5
50. H. Garg, A new possibility degree measure for interval-valued q-rung orthopair fuzzy sets in decision-making. Int. J. Intell. Syst. **36**(1), 526–557 (2021). https://doi.org/10.1002/int.22308
51. P. He, C. Li, H. Garg, J. Liu, Z. Yang, X. Guo, A q-rung orthopair cloud-based multi-attribute decision-making algorithm: considering the information error and multilayer heterogeneous relationship of attributes. IEEE Access **9**, 132541–132557 (2021). https://doi.org/10.1109/ACCESS.2021.3114330

52. P. Liu, P. Wang, Some q-rung orthopair fuzzy aggregation operator and their application to multi-attribute decision making. Int. J. Intell. Syst. **33**(2), 259–280 (2018)
53. Z. Liu, S. Wang, P. Liu, Multiple attribute group decision making based on q-rung orthopair fuzzy Heronian mean operators. J. Intell. Syst. **33**(12), 2341–2363 (2018)
54. H. Garg, A novel trigonometric operation-based q-rung orthopair fuzzy aggregation operator and its fundamental properties. Neural Comput. Appl. **32**, 15077–15099 (2020)
55. H. Garg, New exponential operation laws and operators for interval-valued q-rung orthopair fuzzy sets in group decision making process. Neural Comput. Appl. **33**(20), 13937–13963 (2021). https://doi.org/10.1007/s00521-021-06036-0
56. P. Liu, J. Liu, Some q-rung orthopair fuzzy Bonferroni mean operators and their application to multi-attribute group decision making. J. Intell. Syst. **33**(2), 315–347 (2018)
57. M. Akram, A. Khan, J.C.R. Alcantud, G. Santos-Garcia, A hybrid decision-making framework under complex spherical fuzzy prioritized weighted aggregation operators. Expert. Syst. (2021). https://doi.org/10.1111/exsy.12712
58. M. Akram, N. Yaqoob, G. Ali, W. Chammam, Extensions of Dombi aggregation operators for decision making under m-polar fuzzy information. J. Math. **6**, 1–20 (2020). https://doi.org/10.1155/2020/4739567
59. H.M.A. Farid, M. Riaz, Some generalized q-rung orthopair fuzzy Einstein interactive geometric aggregation operators with improved operational laws. Int. J. Intell. Syst. (2021). https://doi.org/10.1002/int.22587
60. M. Riaz, H. Garg, H.M.A. Farid, M. Aslam, Novel q-rung orthopair fuzzy interaction aggregation operators and their application to low-carbon green supply chain management. J. Intell. Fuzzy Syst. **41**(2), 4109–4126 (2021). https://doi.org/10.3233/JIFS-210506
61. M. Riaz, H.M.A. Farid, F. Karaaslan, M.R. Hashmi, Some q-rung orthopair fuzzy hybrid aggregation operators and TOPSIS method for multi-attribute decision-making. J. Intell. Fuzzy Syst. **39**(1), 1227–1241 (2020)
62. M. Riaz, K. Naeem, R. Chinram, A. Iampan, Pythagorean m-polar fuzzy weighted aggregation operators and algorithm for the investment strategic decision making. J. Math. (2021). https://doi.org/10.1155/2021/6644994
63. K. Naeem, M. Riaz, X.D. Peng, D. Afzal, Pythagorean fuzzy soft MCGDM methods based on TOPSIS, VIKOR and aggregation operators. J. Intell. Fuzzy Syst. **37**(5), 6937–6957 (2019)
64. K. Naeem, M. Riaz, F. Karaaslan, A mathematical approach to medical diagnosis via Pythagorean fuzzy soft TOPSIS, VIKOR and generalized aggregation operators. Complex Intell. Syst. (2021). https://doi.org/10.1007/s40747-021-00458-y

Dr. Muhammad Tahir Hamid received his M.Sc. degree in Mathematics from the Islamia University Bahawalpur in 1988. He received an M.Phil degree in Mathematics from the Lahore Garrison University, Lahore, in 2015. He received a Ph.D. degree in Mathematics from University of Lahore, 2021. He is working as an Associate Professor and Head of the Department of Mathematics, Garrison Post Graduate College, Sarfraz Rafiqui Road Lahore Cant. He is the author of 03+ SCI research papers. He is also a reviewer of international journals. Title of Ph.D. Thesis: Some Contributions to q-Rung Orthopair m-Polar Fuzzy Sets and Neutrosophic Sets with Applications. Supervised by Dr. Deeba Afzal and Dr. Muhammad Riaz. His research interests include q-Rung Orthopair Fuzzy sets, Neutrosophic sets, Linear Diophantine Fuzzy sets, Multi-Criteria Decision-Making Problems, Aggregation Operators, Information Measures, Information Fusion, fuzzy modeling, fuzzy set theory, and Computational Intelligence.

Dr. Muhammad Riaz has received M.Sc., M.Phil and Ph.D. degrees in Mathematics from the Department of Mathematics, University of the Punjab, Lahore. He has 23+ years of regular teaching and research experience. He has published 90+ research articles in international peer-reviewed SCIE & ESCI journals with 1500+ citations. He has supervised 05 Ph.D. students and 19 M.Phil students. Currently, he is supervising 05 M.Phil and 03 Ph.D. students. He is HEC Approved Supervisor.

His research interests include Pure Mathematics, Fuzzy Mathematics, Topology, Algebra, Artificial intelligence, Computational Intelligence, soft set theory, rough set theory, neutrosophic sets, linear Diophantine fuzzy sets, biomathematics, aggregation operators with applications in decision-making problems, medical diagnosis, information measures, image processing, network topology, and pattern recognition.
He has been a reviewer for 40+ SCI journals. He is a member of the editorial board of 07 journals. As the project leader, he has directed two research projects at the national or provincial level. He has delivered seminars/talks as invited speaker at many international conferences. He is the student's advisor in the department. He is a member of the board of studies, a member of the board of faculty, a member of the departmental doctoral program committee (DDPC), and a member of many university-level committees.

Dr. Khalid Naeem received his B. Sc. and M. Sc. degrees from the University of the Punjab, Lahore. He received his M. Phil and Ph.D. degrees in Mathematics from The University of Lahore, Lahore. He is working as an Associate Professor and Head of the Department of Mathematics, Federal Govt. Degree College, Lahore Cantt. He has more than 22 years of regular teaching and research experience. He is the author of 16 research papers in SCIE/ESCI International Journals. He is also a reviewer of various international journals. His research interests include Pythagorean fuzzy soft sets, q-Rung Orthopair Fuzzy sets, Neutrosophic sets, Multi-Criteria Decision-Making Problems, Aggregation Operators, Information Measures, Information Fusion, Fuzzy modeling, and Computational intelligence.

Chapter 3
Decision-Making on Patients' Medical Status Based on a q-Rung Orthopair Fuzzy Max-Min-Max Composite Relation

Paul Augustine Ejegwa

Abstract q-Rung orthopair fuzzy set (qROFS) is a family of generalized fuzzy sets including intuitionistic fuzzy set, Pythagorean fuzzy set, Fermatean fuzzy set among others. q-Rung orthopair fuzzy set has higher prospect of applications in decision science because it can conveniently tackle vague problems that are beyond the reach of the aforementioned generalized fuzzy sets. The concept of composite relation is a very important information measure use to determine multiple criteria decision-making problems. This chapter proposes max-min-max composite relation under q-Rung orthopair fuzzy sets. Some theorems are used to characterize certain salient properties of q-Rung orthopair fuzzy sets. An easy to follow algorithm and flowchart of the q-Rung orthopair fuzzy max-min-max composite relation are presented to illustrate the computational processes. A case of medical decision-making (MDM) is determined in q-Rung orthopair fuzzy environment to demonstrate the applicability of the proposed q-Rung orthopair fuzzy max-min-max composite relation where diseases and patients are presented as q-Rung orthopair fuzzy values in the feature space of certain symptoms. A comparative study of intuitionistic fuzzy set, Pythagorean fuzzy set, Fermatean fuzzy set and q-Rung orthopair fuzzy set based on max-min-max composite relation is carried out to ascertain the superiority of q-Rung orthopair fuzzy set in curbing uncertainties. It is gleaned from the findings of this chapter that (i) a q-Rung orthopair fuzzy set is an advanced soft computing construct with the ability to precisely curb uncertainty compare to intuitionistic fuzzy set, Pythagorean fuzzy set and Fermatean fuzzy set, (ii) a q-Rung orthopair fuzzy max-min-max composite relation is a reliable information measure for determining decision making problems with precision.

Keywords Decision-making · Intuitionistic fuzzy set · Pythagorean fuzzy set · q-Rung orthopair fuzzy sets · q-Rung orthopair fuzzy max-min-max composite relation

---

P. A. Ejegwa (✉)
Department of Mathematics, University of Agriculture, P.M.B. 2373, Makurdi, Nigeria
e-mail: ejegwa.augustine@uam.edu.ng

H. Garg (ed.), *q-Rung Orthopair Fuzzy Sets*,
https://doi.org/10.1007/978-981-19-1449-2_3

3.1 Introduction

Decision-making is perhaps the most important component of life. Decision-making which is the art of making choices by identifying a decision, gathering information, and assessing alternative resolutions plays a vital role in the planning process. Using a step-by-step decision-making process can be helpful in making more deliberate, thoughtful decisions by organizing relevant information and defining alternatives. This approach increases the chances that the most satisfying alternative would be chosen. Albeit, the approach could not curb uncertainties embedded in decision-making process. This setback seemed to be insurmountable until the introduction of fuzzy technology. Fuzzy set (FS) introduced by Zadeh [1] enhances solution of several practical problems including decision making, medical diagnosis, pattern recognition, etc. Fuzzy set theory though important is limited in the sense that it considers membership grade only. Sequel to this drawback, several generalized versions of FS were introduced such as intuitionistic fuzzy set(IFS) [2], Pythagorean fuzzy set (PFS) [3–5], Fermatean fuzzy set (FFS) and [6, 7]. To start with, the concept of IFSs has been applied in several applicative areas using some information measures. Similarity measures of IFSs have been used in pattern recognition [8, 9] and medical diagnostic reasoning [10]. Other information measures like relations, distance measures and correlation measures have been applied to solve some real life issues under IFSs [11–15]. Certain fuzzy structures have been discussed and applied in multiple criteria group decision-making (MCGDM) [16–18].

The limitation of the concept of IFSs is that it only models a case when the sum of membership degree (MD) μ and non-membership degree (NMD) ν is not more than one, otherwise it becomes handicap. Sequel to this shortcoming, the concept of intuitionistic fuzzy set of second type (IFSST) [3, 19] was proposed, popularly called Pythagorean fuzzy sets (PFSs) [4, 5]. PFS is used to characterize the uncertain information more sufficiently and accurately than IFS. Some distance and similarity measures between PFSs have been introduced and characterized. Li and Zeng [20] introduced a new distance measure between PFSs with real-life applications. Many studies have been carried out on Pythagorean fuzzy distance/similarity measures with applications to personnel appointment [21], pattern recognition [22–26] and multiple criteria group decision-making [27]. An approach for tackling multi-attribute decision-making with interval-valued Pythagorean fuzzy linguistic information was discussed in [28]. Certain aggregation operators using Einstein operator, Einstein t-conorm and Einstein t-norm defined in Pythagorean fuzzy environment for decision-making were discussed in [29, 30]. Linguistic PFSs was proposed by Garg [31] and applied to multi-attribute decision-making. Liang and Xu [32] discussed a new extension of TOPSIS method for multiple criteria decision-making with hesitant PFSs. Due to the resourcefulness of PFS, it has been used to solve some practical problems [33–39].

The idea of PFSs could not model some decision making problems, for instance if a decision maker wants to make a choice when $\mu = \frac{\sqrt[3]{6}}{2}$ and $\nu = \frac{1}{2}$. In a quest to solve such puzzle, Begum and Srinivasan [6] enlarged the scope of PFS to propose intuitionistic fuzzy set of third type (IFSTT) which Senapati and Yager [7] referred to as Fermatean fuzzy sets (FFSs). In a Fermatean fuzzy set, $\mu^2 + \nu^2 \geq 1$ and $\mu^3 + \nu^3 \leq 1$. FFS handles uncertain information more easily in the process of decision making. Silambarasan [40] discussed certain new operators on FFSs and discussed their several properties like necessity and possibility operators. Senapati and Yager [7] discussed the basic operations on FFSs, score/accuracy functions for ranking of FFSs, and introduced the Euclidean distance in Fermatean fuzzy setting with application in MCDM via TOPSIS method. Liu et al. [41] proposed Fermatean fuzzy linguistic term sets, discussed the basic operational laws, score/accuracy functions of Fermatean fuzzy linguistic numbers, and also introduced Fermatean fuzzy linguistic weighted aggregation operator, Fermatean fuzzy linguistic weighted geometric operator, and the Fermatean fuzzy linguistic distance measures with an extension of TOPSIS method on the introduced distance measures. Yang et al. [42] examined the properties of continuous Fermatean fuzzy information by proposing Fermatean fuzzy functions, subtraction and division operations on Fermatean fuzzy functions and discussed their properties. Senapati and Yager [43] introduced some new weighted aggregated operators on FFSs and discussed their corresponding desirable properties in details with application in solving multi-criteria decision-making problem. Senapati and Yager [44] introduced subtraction, division and Fermatean arithmetic mean operations on FFSs and developed a Fermatean fuzzy weighted product model to solve multi-criteria decision-making problem.

The concepts of IFSs, PFSs and FFSs are all referred to orthopair fuzzy sets. Yager [45] proposed the concept of generalized orthopair fuzzy sets also called q-Rung orthopair fuzzy set (qROFS) which is a generalization of both IFS, PFS and FFS. The condition for MD and NMD of qROFS is $0 \leq \mu^q + \nu^q \leq 1$ where $q \geq 1$. Silambarasan [46] discussed some new operations defined over qROFSs. Decision-making problem using orthopair fuzzy TOPSIS method has been discussed [47]. Khan et al. [48] proposed the knowledge measure for qROFSs with application. The concept of correlation coefficient for qROFSs has been discussed and applied to clustering analysis. The concept of aggregation operators under qROFSs has been studied with applications to multiple attribute decision-making and multiple attribute group decision-making. Pinar and Boran [49] applied qROFSs to supplier selection, and multiple criteria decision-making in robotic agricultural farming with q-Rung orthopair m-polar fuzzy sets has been presented [50]. The concept of mean or aggregation operators has been studied under q-Rung orthopair fuzzy environment with applications to decision-making [51–58].

Though the concept of composite relation is a very important information measure use to determine multiple criteria decision-making problems, it has not been investigated under q-Rung orthopair fuzzy sets. To better appreciate the applications of qROFSs, this work is motivated to introduce q-Rung orthopair fuzzy max-min-max composite relation as an information measure for decision-making with precision. The main contributions of this chapter include (i) characterizations of some properties of q-Rung orthopair fuzzy sets, (ii) an exploration of q-Rung orthopair fuzzy max-min-max composite relation as a reliable information measure with precision, (iii) a comparative study of intuitionistic fuzzy set, Pythagorean fuzzy set, Fermatean fuzzy set and q-Rung orthopair fuzzy set based on composite relation to ascertain their abilities in curbing uncertainties, and (iv) a demonstration of the application of q-Rung orthopair fuzzy sets in medical diagnosis where diseases and patients are presented as q-Rung orthopair fuzzy values. For the sake of clarity the chapter is outlined as follows: Section two presents some preliminaries, qROFS and its characterizations, Section three discusses q-Rung orthopair fuzzy max-min-max composite relation and its algorithm with flowchart, Section four presents the application of q-Rung orthopair fuzzy max-min-max composite relation in MDM where diseases and patients are presented as q-Rung orthopair fuzzy values in the feature space of some symptoms, and Section five summaries the chapter with recommendations for future endeavours.

3.2 Preliminaries

First and foremost, we recall the concepts of IFS and PFS as follow:

Definition 3.1 ([2]) An object in X denoted by A is called an intuitionistic fuzzy set if

$$A = \{\langle x, \mu_A(x), \nu_A(x)\rangle \mid x \in X\}, \tag{3.1}$$

where the functions $\mu_A,\ \nu_A : X \to [0, 1]$ define MD and NMD of $x \in X$ for

$$0 \leq \mu_A(x) + \nu_A(x) \leq 1.$$

For any IFS A in X, $\pi_A(x) \in [0, 1] = 1 - \mu_A(x) - \nu_A(x)$ is the IFS index or hesitation margin of A.

Definition 3.2 ([4]) A Pythagorean fuzzy set L of X is defined by

$$L = \{\langle x, \mu_L(x), \nu_L(x)\rangle \mid x \in X\}, \tag{3.2}$$

where the functions $\mu_L,\ \nu_L : X \to [0, 1]$ are the membership and non-membership degrees of $x \in X$, and $0 \leq \mu_L^2(x) + \nu_L^2(x) \leq 1$. For a PFS L in X, $\pi_L(x) \in [0, 1] = \sqrt{1 - \mu_L^2(x) - \nu_L^2(x)}$ is the PFS index or hesitation margin of L.

3.2.1 *q-Rung Orthopair Fuzzy Sets*

Some fundamentals of qROFSs have been presented as seen in [45, 46, 48, 49, 59]. Let X denote a non-empty set throughout the chapter.

Definition 3.3 ([45]) A qROFS M of X is a generalized orthopair fuzzy set characterized by

$$M = \{\langle x, \mu_M(x), \nu_M(x)\rangle \mid x \in X\}, \tag{3.3}$$

where the functions $\mu_M, \nu_M: X \to [0, 1]$ define MD and NMD of $x \in X$ for

$$0 \leq \mu_M^q(x) + \nu_M^q(x) \leq 1,$$

and $q \geq 1$. For a qROFS, M in X, $\pi_M(x) \in [0, 1] = \sqrt[q]{1 - \mu_M^q(x) - \nu_M^q(x)}$ represents the qROFS index or hesitation margin of M.

Definition 3.4 ([46]) Suppose M and N in X are qROFSs, then

(i) $M = N$ iff $\mu_M(x) = \mu_N(x), \nu_M(x) = \nu_N(x)\ \forall x \in X$.
(ii) $M \subseteq N$ iff $\mu_M(x) \leq \mu_N(x), \nu_M(x) \geq \nu_N(x)\ \forall x \in X$.
(iii) $\overline{M} = \{\langle x, \nu_M(x), \mu_M(x)\rangle | x \in X\}$.
(iv) $M \cup N = \{\langle x, \max(\mu_M(x), \mu_N(x)), \min(\nu_M(x), \nu_N(x))\rangle | x \in X\}$.
(v) $M \cap N = \{\langle x, \min(\mu_M(x), \mu_N(x)), \max(\nu_M(x), \nu_N(x))\rangle | x \in X\}$.

The following are some algebraic structures of qROFSs associated with operations.

Remark 3.1 Suppose M, N and O are qROFSs of X, then the following properties hold:

(a) Idempotent;

(i) $M \cap M = M$
(ii) $M \cup M = M$

(b) Commutativity;

(i) $M \cap N = N \cap M$
(ii) $M \cup N = N \cup M$

(c) Associativity;

(i) $M \cap (N \cap O) = (M \cap M) \cap O$
(ii) $M \cup (N \cup O) = (M \cup N) \cup O$

(d) Distributivity;

(i) $M \cap (N \cup O) = (M \cap N) \cup (M \cap O)$
(ii) $M \cup (N \cap O) = (M \cup N) \cap (M \cup O)$

(e) De Morgan's laws;

(i) $\overline{M \cap N} = \overline{M} \cup \overline{N}$
(ii) $\overline{M \cup N} = \overline{M} \cap \overline{N}$.

Now, we introduce q-Rung orthopair fuzzy values (qROFVs) as follows:

Definition 3.5 qROFVs are described by the form $\langle x, y\rangle$ such that $x^q + y^q \leq 1$ where $x, y \in [0, 1]$. A qROFV evaluates the qROFS for which the components (x and y) are interpreted as MD and NMD.

3.2.1.1 Differences Between qROFS and Other Generalized Fuzzy Sets

A qROFS is a generalized form of intuitionistic/Pythagorean/Fermatean fuzzy sets such that $IFS \subset PFS \subset FFS \subset qROFS$. The differences can be seen in Table 3.1.

In a nutshell, a qROFS is a generalized orthopair fuzzy set because IFS, PFS and FFS can be recovered from it. A qROFS, M of X is an IFS if $q = 1$, implies $0 \leq \mu_M(x) + \nu_M(x) \leq 1$ for $\pi_M(x) = 1 - \mu_M(x) - \nu_M(x)$. A qROFS, M of X is a PFS if $q = 2$, implies $0 \leq \mu_M^2(x) + \nu_M^2(x) \leq 1$ for $\pi_M(x) = \sqrt{1 - \mu_M^2(x) - \nu_M^2(x)}$. Similarly, A qROFS, M of X is a FFS if $q = 3$, implies $0 \leq \mu_M^3(x) + \nu_M^3(x) \leq 1$ for $\pi_M(x) = \sqrt[3]{1 - \mu_M^3(x) - \nu_M^3(x)}$.

3.2.1.2 Characterizations of qROFSs

We discuss some properties of qROFSs as follows:

Theorem 3.1 *Let M be qROFS of X and $\pi_M(x) = 0$, then the following hold:*

(i) $|\mu_M(x)| = \sqrt[q]{|\nu_M^q(x) - 1|}$

(ii) $|\nu_M(x)| = \sqrt[q]{|\mu_M^q(x) - 1|}$.

Proof Assume $\pi_M(x) = 0$ for $x \in X$, we have

$$\begin{aligned} \mu_M^q(x) + \nu_M^q(x) = 1 &\Rightarrow -\mu_M^q(x) = \nu_M^q(x) - 1 \\ &\Rightarrow |\mu_M^q(x)| = |\nu_M^q(x) - 1| \\ &\Rightarrow |\mu_M(x)|^q = |\nu_M^q(x) - 1| \\ &\Rightarrow |\mu_M(x)| = \sqrt[q]{|\nu_M^q(x) - 1|}, \end{aligned}$$

which proves (i). The proof of (ii) is similar.

Definition 3.6 The score function α of a qROFS M is defined by $\alpha(M) = \mu_M^q(x) - \nu_M^q(x)$, where $\alpha(M) \in [-1, 1]$. The accuracy function β of M is defined by $\beta(M) = \mu_M^q(x) + \nu_A^q(x)$ for $\beta(M) \in [0, 1]$. Consequently, the qROFS index of M is $\pi_M(x) = \sqrt[q]{1 - \beta(M)}$.

Table 3.1 IFS, PFS, FFS and qROFS

IFS	PFS	FFS	qROFS
$0 \leq \mu + \nu \leq 1$	$0 \leq \mu^2 + \nu^2 \leq 1$	$0 \leq \mu^3 + \nu^3 \leq 1$	$0 \leq \mu^q + \nu^q \leq 1$, $q \geq 1$
$\pi = 1 - \mu - \nu$	$\pi = \sqrt{1 - \mu^2 - \nu^2}$	$\pi = \sqrt[3]{1 - \mu^3 - \nu^3}$	$\pi = \sqrt[q]{1 - \mu^q - \nu^q}$, $q \geq 1$
$\mu + \nu + \pi = 1$	$\mu^2 + \nu^2 + \pi^2 = 1$	$\mu^3 + \nu^3 + \pi^3 = 1$	$\mu^q + \nu^q + \pi^q = 1$, $q \geq 1$

Theorem 3.2 *Let M be qROFS of X, then the following hold* $\forall x \in X$.

(i) $\alpha(M) = 0 \Leftrightarrow \mu_M(x) = \nu_M(x)$.
(ii) $\alpha(M) = 1 \Leftrightarrow |\nu_M(x)| = \sqrt[q]{|1 - \mu_M^q(x)|}$.
(iii) $\alpha(M) = -1 \Leftrightarrow \mu_M(x) = \sqrt[q]{|\nu_M^q(x) - 1|}$.

Proof (i) Suppose $\alpha(M) = 0$, then $\mu_M^q(x) = \nu_M^q(x) \Rightarrow \mu_M(x) = \nu_M(x)$.

Conversely, assume $\mu_M(x) = \nu_M(x)$, then $\mu_M^q(x) = \nu_M^q(x)$ implies that $\mu_M^q(x) - \nu_M^q(x) = 0$. Hence $\alpha(M) = 0$.

(ii) Suppose $\alpha(M) = 1$, then

$$\begin{aligned} 1 - \mu_M^q(x) = -\nu_M^q(x) &\Rightarrow |1 - \mu_M^q(x)| = |\nu_M(x)|^q \\ &\Rightarrow |\nu_M(x)| = \sqrt[q]{|1 - \mu_M^q(x)|}. \end{aligned}$$

Conversely, if $|\nu_M(x)| = \sqrt[q]{|1 - \mu_M^q(x)|}$, then we get

$$|\nu_M(x)|^q = |1 - \mu_M^q(x)| \Rightarrow \nu_M^q(x) = -1 + \mu_M^q(x)$$

or

$$|\nu_M(x)|^q = |1 - \mu_M^q(x)| \Rightarrow -\nu_M^q(x) = 1 - \mu_M^q(x).$$

Take $-\nu_M^q(x) = 1 - \mu_M^q(x)$ then $\mu_M^q(x) - \nu_M^q(x) = 1$, hence $\alpha(M) = 1$.

(iii) Suppose $\alpha(M) = -1$. Then

$$\nu_M^q(x) - 1 = \mu_M^q(x) \Rightarrow \mu_M(x) = \sqrt[q]{\nu_M^q(x) - 1}.$$

Conversely, suppose $\mu_M(x) = \sqrt[q]{\nu_M^q(x) - 1}$. Then

$$\begin{aligned} \mu_M^q(x) = \nu_M^q(x) - 1 &\Rightarrow \mu_M^q(x) - \nu_M^q(x) = -1 \\ &\Rightarrow \alpha(M) = -1. \end{aligned}$$

Theorem 3.3 *Suppose M is a qROFS of X, then the following statements hold* $\forall x \in X$.

(i) $\beta(M) = 1 \Leftrightarrow \pi_M(x) = 0$.
(ii) $\beta(M) = 0 \Leftrightarrow |\mu_M(x)| = |\nu_M(x)|$.

Proof (i) Suppose $\beta(M) = 1$, then $\mu_M^q(x) + \nu_M^q(x) = 1$. Hence $\pi_M(x) = 0$ since $\pi_M(x) = \sqrt{1 - \beta(M)}$.

Conversely, suppose that $\pi_M(x) = 0$, then it follows that

$$\mu_M^q(x) + \nu_M^q(x) = 1 \Rightarrow \beta(M) = 1.$$

(ii) Suppose $\beta(M) = 0$, then $\mu_M^q(x) = -\nu_M^q(x)$ or $\nu_M^q(x) = -\mu_M^q(x) \Leftrightarrow |\mu_M^q(x)| = |\nu_M^q(x)| \Leftrightarrow |\mu_M(x)| = |\nu_M(x)|$.

Proposition 3.1 *Suppose M and N are qROFSs of X, then*

(i) $\alpha(M) = \alpha(N) \Leftrightarrow M = N$.
(ii) $\alpha(M) \leq \alpha(N) \Leftrightarrow M \subseteq N$.
(iii) $\alpha(M) < \alpha(N) \Leftrightarrow M \subseteq N$ *and* $M \neq N$.

Proof The proofs are obvious from Definition 3.6.

Proposition 3.2 *Suppose M and N are qROFSs of X, then*

(i) $\beta(M) = \beta(N) \Leftrightarrow M = N$.
(ii) $\beta(M) \leq \beta(N) \Leftrightarrow M \subseteq N$.
(iii) $\beta(M) < \beta(N) \Leftrightarrow M \subseteq N$ *and* $M \neq N$.

Proof The proofs are obvious from Definition 3.6.

3.3 q-Rung Orthopair Fuzzy Max-Min-Max Composite Relation

To enhance the applications of qROFSs, a novel information measure is proposed called q-Rung orthopair fuzzy max-min-max composite relation (qROFMMMCR) as follows.

Suppose X and Y are non-empty sets. A q-Rung orthopair fuzzy relation (qROFR) R from X to Y is a qROFS of $X \times Y$ consisting of MD μ_R and NMD ν_R. A qROFR from X to Y is denoted by $R(X \to Y)$.

Definition 3.7 Suppose M is a qROFS of X, then a qROFMMMCR of $R(X \to Y)$ with M is a qROFS N of Y denoted by $N = R \circ M$, such that its MD and NMD are defined by

$$\left.\begin{aligned} \mu_N(y) &= \bigvee \Big(\min(\mu_M(x), \mu_R(x, y))\Big) \\ \nu_N(y) &= \bigwedge \Big(\max(\nu_M(x), \nu_R(x, y))\Big) \end{aligned}\right\}, \tag{3.4}$$

$\forall x \in X$ and $y \in Y$, where $\bigvee$ = maximum, $\bigwedge$ = minimum.

Definition 3.8 Suppose $Q(X \to Y)$ and $R(Y \to Z)$ are qROFRs, then the qROFMMMCR, $R \circ Q$ is a qROFR from X to Z such that its MD and NMD are defined by

$$\left.\begin{aligned} \mu_{R\circ Q}(x, z) &= \bigvee \Big(\min(\mu_Q(x, y), \mu_R(y, z))\Big) \\ \nu_{R\circ Q}(x, z) &= \bigwedge \Big(\max(\nu_Q(x, y), \nu_R(y, z))\Big) \end{aligned}\right\}. \tag{3.5}$$

Synthesizing Definitions 3.7 and 3.8, the qROFMMMCR N or $R \circ Q$ can be computed by

$$N = \mu_N(y) - \nu_N(y)\pi_N(y), \ \forall y \in Y \text{ or} \tag{3.6}$$

$$R \circ Q = \mu_{R \circ Q}(x, z) - \nu_{R \circ Q}(x, z)\pi_{R \circ Q}(x, z), \ \forall (x, z) \in X \times Z. \tag{3.7}$$

Proposition 3.3 *If R_a and R_b are two qROFRs on $X \times Y$ and $Y \times Z$, respectively. Then*

(i) $(R_a^{-1})^{-1} = R_a$, $(R_b^{-1})^{-1} = R_b$,
(ii) $(R_a \circ R_b)^{-1} = R_b^{-1} \circ R_a^{-1}$,
(iii) $(R_a \cap R_b)^{-1} = R_a^{-1} \cap R_b^{-1}$,
(iv) $(R_a \cup R_b)^{-1} = R_a^{-1} \cup R_b^{-1}$.

Proof The proofs are obvious from Definition 3.8.

Proposition 3.4 *Suppose R, R_a and R_b are qROFRs, then we have*

(i) $R_a \cap R_b \subseteq R_a \Rightarrow R_a \cap R_b \subseteq R_b$.
(ii) $R_a \subseteq R_b \Rightarrow R_a^{-1} \subseteq R_b^{-1}$.
(iii) $R_a \subseteq R$, $R_b \subseteq R \Rightarrow R_a \cup R_b \subseteq R$.
(iv) $R \subseteq R_a$, $R \subseteq R_b \Rightarrow R \subseteq R_a \cup R_b$.

Proof The proofs are obvious from Definition 3.8.

3.3.1 Numerical Application

We apply the proposed qROFMMMCR to numerical example as follows. Let L and M be qROFSs in $X = \{x_1, x_2, x_3\}$ defined by

$$L = \{\langle x_1, 0.6, 0.2\rangle, \langle x_2, 0.4, 0.6\rangle, \langle x_3, 0.5, 0.3\rangle\}$$

and

$$M = \{\langle x_1, 0.8, 0.1\rangle, \langle x_2, 0.7, 0.3\rangle, \langle x_3, 0.6, 0.1\rangle\}.$$

We find the composite relation N using Definitions 3.7 and 3.8, respectively:

$$\min(\mu_R(l_i, x_j), \mu_S(x_j, m_k)) = 0.6, 0.4, 0.5$$

implying that

$$\mu_N(l_i, m_k) = \bigvee(0.6, 0.4, 0.5) = 0.6.$$

Clearly, $\min[\mu_R(l_i, x_j), \mu_S(x_j, m_k)]$ is gotten by using Definitions 3.7 and 3.8. Applying this to L and M, the minimum value of the membership values of the elements of X in L and M are 0.6, 0.4, 0.5.

Table 3.2 Values of qROFMMMCR

$q=1$	$q=2$	$q=3$	$q=4$
0.56	0.4451	0.4162	0.4069

Again,

$$\max[\nu_R(l_i, x_j), \nu_S(x_j, m_k)] = 0.2, 0.6, 0.3$$

implying that

$$\nu_N(l_i, m_k) = \bigwedge(0.2, 0.6, 0.3) = 0.2.$$

By explanation, $\max[\nu_R(l_i, x_j), \nu_S(x_j, m_k)]$ is gotten by using Definitions 3.7 and 3.8. Applying this to L and M, the maximum value of the non-membership values of the elements of X in L and M are 0.2, 0.6, 0.3. Now, the composite relation between L and M using qROFMMMCR for $q=1$ is $N = 0.6 - (0.2 \times 0.2) = 0.56$. The composite relation between L and M using qROFMMMCR for $q=2$ is $N = 0.6 - (0.2 \times 0.7746) = 0.4451$. The composite relation between L and M using qROFMMMCR for $q=3$ is $N = 0.6 - (0.2 \times 0.9189) = 0.4162$. The composite relation between L and M using qROFMMMCR for $q=4$ is $N = 0.6 - (0.2 \times 0.9655) = 0.4069$. Table 3.2 contains the values of the composite relation, N for $q = 1, 2, 3, 4$ for easy comparison.

Although the results show that the qROFMMMCR for $q=1$ gives the best measure of the relationship between L and M, it is not true because of the presence of some uncertainties (i.e., setback of IFS). It is observed that as q increases, the relationship decreases. Nonetheless, since it has been established that vagueness or uncertainties could be best tackled as q increases, hence the proposed qROFMMMCR is more reasonable and reliable for $q=4$. In other hand, as q increases, the qROFMMMCR is more equipped to curb uncertainties.

One of the prominent advantage of qROFMMMCR is that it forestalls information loss because all the conventional parameters of qROFSs are included in the computation unlike some distance and similarity measures under qROFSs.

3.4 Application of qROFMMMCR in Disease Diagnosis

In this section, we discuss an application of qROFSs to medical diagnosis using the proposed qROFMMMCR via simulation of a disease diagnosis. Assume S is a set of symptoms, D is a set of diseases, and P is a set of patients. We define a q-Rung orthopair fuzzy medical knowledge based on a qRFOR R from S to D denoted by $S \times D$ to reveal the degree of association and the degree of non-association between S and D. In the q-Rung orthopair fuzzy medical diagnosis process, we determine the symptoms of the disease, formulate the medical knowledge based on qROFR, and determine the diagnosis on the basis of the composition via qROFMMMCR.

3.4.1 qROFMMMCR in Terms of Patients and Diseases with Respect to Symptoms

Suppose L is a qROFS of the set S and R is a qROFR from S to D. Then the qROFMMMCR N for L with $R(S \rightarrow D)$ denoted by $N = L \circ R$ signifies the medical state of the patient as a qROFS N of D with the membership function given by

$$\mu_N(d) = \bigvee \Big(\min(\mu_L(s), \mu_R(s, d)) \Big)$$

and the non-membership function given by

$$\nu_N(d) = \bigwedge \Big(\max(\nu_L(s), \nu_R(s, d)) \Big)$$

for all the diseases. If the medical state of a given patient, P is described in terms of a qROFS L of S, then P is assumed to be assigned diagnosis with regards to qROFS N, through a qROFR R of q-Rung orthopair fuzzy medical knowledge from S to D as simulated by medical knowledge in terms of degrees of association and non-association between symptoms and diseases.

For a qROFR R (i.e., $S \rightarrow D$), we construct a qROFR Q from P to S denoted by $P \rightarrow S$. Then the qROFMMMCR N of qROFRs R and Q (i.e., $N = R \circ Q$) depicts the medical state of P with regards to the diagnosis as a qROFR from P to D given by the membership and non-membership functions in Eq. (3.8)

$$\left.\begin{aligned} \mu_N(p, d) &= \bigvee \Big(\min(\mu_Q(p, s), \mu_R(s, d)) \Big) \\ \nu_N(p, d) &= \bigwedge \Big(\max(\nu_Q(p, s), \nu_R(s, d)) \Big) \end{aligned}\right\}, \tag{3.8}$$

for all the patients and diseases. For qROFRs R and Q, the qROFMMMCR $N = R \circ Q$ can be determined. The result for which $R \circ Q = \mu_{R \circ Q}(p, d) - \nu_{R \circ Q}(p, d)\pi_{R \circ Q}(p, d)$ is the greatest decides the disease D the patient P is suffering from.

Before the experiment of disease diagnosis, we present an algorithm and a flow chart based on the proposed qROFMMMCR discussed in Sect. 3.3 to be used in Sect. 3.4.2.

3.4.1.1 Algorithm for qROFMMMCR in Disease Diagnosis

The disease diagnosis algorithm based on the proposed qROFMMMCR is made up of the following steps.

1. Establish a relation between symptoms S and diseases D under q-Rung orthopair fuzzy environment.

2. Establish a relation between patients P and symptoms S under q-Rung orthopair fuzzy environment.
3. Find the membership and non-membership functions of $R \circ Q$ between each of the patients and diseases with respect to the symptoms.
4. Compute $N = R \circ Q$ between each of the patients and diseases with respect to the symptoms.
5. Determine the diagnosis based on the result for which $R \circ Q$ is the greatest.

The flowchart of the algorithm follows.

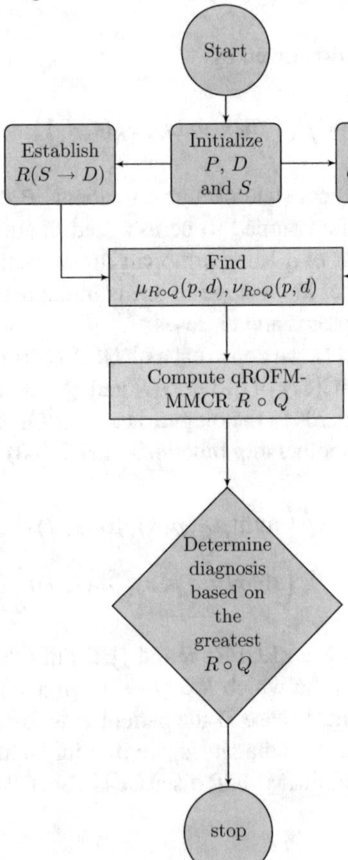

3.4.2 Experiment of Disease Diagnosis

Suppose four patients $P = \{P_1, P_2, P_3, P_4\}$ visit a certain medical laboratory for disease diagnosis. From clinical examination, the following symptoms, S are observed

Table 3.3 $Q(P \to S)$

Q	Temperature	Headache	Stomach pain	Cough	Chest pain
P_1	$\langle 0.8, 0.1\rangle$	$\langle 0.6, 0.1\rangle$	$\langle 0.2, 0.8\rangle$	$\langle 0.6, 0.1\rangle$	$\langle 0.1, 0.6\rangle$
P_2	$\langle 0.0, 0.8\rangle$	$\langle 0.4, 0.4\rangle$	$\langle 0.6, 0.1\rangle$	$\langle 0.1, 0.7\rangle$	$\langle 0.1, 0.8\rangle$
P_3	$\langle 0.8, 0.1\rangle$	$\langle 0.8, 0.1\rangle$	$\langle 0.0, 0.6\rangle$	$\langle 0.2, 0.7\rangle$	$\langle 0.0, 0.5\rangle$
P_4	$\langle 0.6, 0.1\rangle$	$\langle 0.5, 0.4\rangle$	$\langle 0.3, 0.4\rangle$	$\langle 0.7, 0.2\rangle$	$\langle 0.3, 0.4\rangle$

Table 3.4 $R(S \to D)$

R	Viral fever	Malaria fever	Typhoid fever	Stomach problem	Chest problem
Temperature	$\langle 0.4, 0.0\rangle$	$\langle 0.7, 0.0\rangle$	$\langle 0.3, 0.3\rangle$	$\langle 0.1, 0.7\rangle$	$\langle 0.1, 0.8\rangle$
Headache	$\langle 0.3, 0.5\rangle$	$\langle 0.2, 0.6\rangle$	$\langle 0.6, 0.1\rangle$	$\langle 0.2, 0.4\rangle$	$\langle 0.0, 0.8\rangle$
Stomach pain	$\langle 0.1, 0.7\rangle$	$\langle 0.0, 0.9\rangle$	$\langle 0.2, 0.7\rangle$	$\langle 0.8, 0.0\rangle$	$\langle 0.2, 0.8\rangle$
Cough	$\langle 0.4, 0.3\rangle$	$\langle 0.7, 0.0\rangle$	$\langle 0.2, 0.6\rangle$	$\langle 0.2, 0.7\rangle$	$\langle 0.2, 0.8\rangle$
Chest pain	$\langle 0.1, 0.7\rangle$	$\langle 0.1, 0.8\rangle$	$\langle 0.1, 0.9\rangle$	$\langle 0.2, 0.7\rangle$	$\langle 0.8, 0.1\rangle$

Table 3.5 $\mu_{R\circ Q}(p, d)$ and $\nu_{R\circ Q}(p, d)$

$\mu_{R\circ Q}, \nu_{R\circ Q}$	Viral fever	Malaria fever	Typhoid fever	Stomach problem	Chest problem
P_1	$\langle 0.4, 0.1\rangle$	$\langle 0.7, 0.1\rangle$	$\langle 0.6, 0.1\rangle$	$\langle 0.2, 0.4\rangle$	$\langle 0.2, 0.6\rangle$
P_2	$\langle 0.3, 0.5\rangle$	$\langle 0.2, 0.6\rangle$	$\langle 0.4, 0.4\rangle$	$\langle 0.6, 0.1\rangle$	$\langle 0.2, 0.8\rangle$
P_3	$\langle 0.4, 0.1\rangle$	$\langle 0.7, 0.1\rangle$	$\langle 0.6, 0.1\rangle$	$\langle 0.2, 0.4\rangle$	$\langle 0.2, 0.5\rangle$
P_4	$\langle 0.4, 0.1\rangle$	$\langle 0.7, 0.1\rangle$	$\langle 0.5, 0.3\rangle$	$\langle 0.3, 0.4\rangle$	$\langle 0.3, 0.4\rangle$

viz; high temperature, headache, stomach pain, cough, and chest pain. The qROFR $Q(P \to S)$ is given hypothetically, in Table 3.3.

The patients P are suspected to be suffering from the following diseases $D =$ {viral fever, malaria, typhoid, stomach problem, and heart problem} due to the symptoms S. The qROFR $R(S \to D)$ is given hypothetically, in Table 3.4. The membership and non-membership functions of the qROFMMMCR $R \circ Q$ are in Table 3.5. The data in Tables 3.3 and 3.4 are extracted from [11], which we present under q-Rung orthopair fuzzy environment.

3.4.2.1 Medical Decision-Making on the Patients Medical Conditions

After computing the qROFS indexes, the qROFMMMCR $R \circ Q$ is given in Tables 3.6, 3.7, 3.8 and 3.9.

The medical diagnosis from Table 3.6 for $q = 1$ shows that

Table 3.6 $R \circ Q, q = 1$

$R \circ Q$	Viral fever	Malaria fever	Typhoid fever	Stomach problem	Chest problem
P_1	0.35	0.68	0.57	0.04	0.08
P_2	0.20	0.08	0.32	0.57	−0.60
P_3	0.35	0.68	0.57	0.04	0.05
P_4	0.35	0.68	0.44	0.18	0.18

Table 3.7 $R \circ Q, q = 2$

$R \circ Q$	Viral fever	Malaria fever	Typhoid fever	Stomach problem	Chest problem
P_1	0.3089	0.6293	0.5206	−0.1578	−0.2648
P_2	−0.1062	−0.2648	0.0702	0.5206	−0.2526
P_3	0.3089	0.6293	0.5206	−0.1578	−0.2213
P_4	0.3089	0.6293	0.2563	−0.0464	−0.0464

(i) patient P_1 is diagnosed with malaria fever (0.68) with a reasonable proportion of typhoid fever (0.57),
(ii) patient P_2 is diagnosed with stomach problem (0.57),
(iii) patient P_3 is diagnosed with malaria fever (0.68) with a reasonable proportion of typhoid fever (0.57), and
(iv) patient P_4 is diagnosed with malaria fever (0.68).

None of the patient is diagnosed with viral fever and chest problem. In fact, the relationship between patient P_2 and chest problem is negative (−0.60). From the diagnosis, it is advisable the medical practitioner place Patients P_1 and P_3 on the same treatment because they have the same infection load of malaria fever and typhoid fever.

The medical diagnosis from Table 3.7 for $q = 2$ shows that

(i) patient P_1 is diagnosed with malaria fever (0.6293) with a reasonable proportion of typhoid fever (0.5206),
(ii) patient P_2 is diagnosed with stomach problem (0.5206),
(iii) patient P_3 is diagnosed with malaria fever (0.6293) with a reasonable proportion of typhoid fever (0.5206), and
(iv) patient P_4 is diagnosed with malaria fever (0.6293).

None of the patient is diagnosed with viral fever and chest problem. In fact, the relationship between patient P_1 with stomach problem and chest problem are negative (−0.1578, −0.2648), the relationship between patient P_2 with viral fever, malaria fever and chest problem are negative (−0.1062, −0.2648, −0.2526), the relationship between patient P_3 with stomach problem and chest problem are negative (−0.1578, −0.2213), and the relationship between patient P_4 with stomach problem and chest

Table 3.8 $R \circ Q, q = 3$

$R \circ Q$	Viral fever	Malaria fever	Typhoid fever	Stomach problem	Chest problem
P_1	0.3022	0.6131	0.5078	−0.1902	−0.3513
P_2	−0.1733	−0.3513	0.0178	0.5078	−0.4264
P_3	0.3022	0.6131	0.5078	−0.1902	−0.2768
P_4	0.3022	0.6131	0.2161	−0.0875	−0.0875

Table 3.9 $R \circ Q, q = 4$

$R \circ Q$	Viral fever	Malaria fever	Typhoid fever	Stomach problem	Chest problem
P_1	0.3007	0.6066	0.5034	−0.1972	−0.3793
P_2	−0.1910	−0.3793	0.0052	0.5034	−0.5008
P_3	0.3007	0.6066	0.5034	−0.1972	−0.2918
P_4	0.3007	0.6066	0.2054	−0.0966	−0.0966

problem are negative(−0.0464, −0.0464). From the diagnosis, it is advisable the medical practitioner place Patients P_1 and P_3 on the same treatment because they have the same infection load of malaria fever and typhoid fever.

The medical diagnosis from Table 3.8 for $q = 3$ shows that

(i) patient P_1 is diagnosed with malaria fever (0.6131) with a reasonable proportion of typhoid fever (0.5078),
(ii) patient P_2 is diagnosed with stomach problem (0.5078),
(iii) patient P_3 is diagnosed with malaria fever (0.6131) with a reasonable proportion of typhoid fever(0.5078), and
(iv) patient P_4 is diagnosed with malaria fever (0.6131).

None of the patient is diagnosed with viral fever and chest problem. In fact, the relationship between patient P_1 with stomach problem and chest problem are negative (−0.1902, −0.3513), the relationship between patient P_2 with viral fever, malaria fever and chest problem are negative (−0.1733, −0.3513, −0.4264), the relationship between patient P_3 with stomach problem and chest problem are negative (−0.1902, −0.2768), and the relationship between patient P_4 with stomach problem and chest problem are negative (−0.0875, −0.0875). From the diagnosis, it is advisable the medical practitioner place Patients P_1 and P_3 on the same treatment because they have the same infection load of malaria fever and typhoid fever.

The medical diagnosis from Table 3.9 for $q = 4$ shows that

(i) patient P_1 is diagnosed with malaria fever (0.6066) with a reasonable proportion of typhoid fever (0.5034),
(ii) patient P_2 is diagnosed with stomach problem (0.5034),

(iii) patient P_3 is diagnosed with malaria fever (0.6066) with a reasonable proportion of typhoid fever (0.5034), and
(iv) patient P_4 is diagnosed with malaria fever (0.6066).

None of the patient is diagnosed with viral fever and chest problem. In fact, the relationship between patient P_1 with stomach problem and chest problem are negative (−0.1972, −0.3793), the relationship between patient P_2 with viral fever, malaria fever and chest problem are negative (−0.1910, −0.3793, −0.5008), the relationship between patient P_3 with stomach problem and chest problem are negative (−0.1972, −0.2918), and the relationship between patient P_4 with stomach problem and chest problem are negative (−0.0966, −0.0966). From the diagnosis, it is advisable the medical practitioner place Patients P_1 and P_3 on the same treatment because they have the same infection load of malaria fever and typhoid fever.

3.4.2.2 Some Observations

The discussions so far show concord of diagnoses for $q = 1$, $q = 2$, $q = 3$ and $q = 4$. However, as q increases the severity of the infections decreases because the qROFSs become more equipped to curb the embedded fuzziness hence give reliable diagnosis. Whereas the diagnoses when $q = 1$ could lead to maladministration of dosage and drugs side effects, the diagnoses for $q > 1$ will enhance appropriate drugs administration with no or minimal side effects.

3.5 Conclusion

A qROFS is a generalized orthopair fuzzy set equipped with the facilities to control uncertainties that are beyond the capacity of IFS, PFS, FFS among others. In a way to better appreciate the applicability of qROFSs, this chapter introduced a qROFMMMCR as a reliable information measure. Specifically, we have differentiated qROFSs from other generalized fuzzy sets with some characterizations. An easy to follow algorithm with its flowchart were presented. To demonstrate the application of qROFS via qROFMMMCR, a case of disease diagnosis was carried out where the diseases and the patients presented are viewed as qROFVs. The diagnoses gotten through this process will enhance appropriate drugs administration and treatment. The findings of this chapter suggest that (i) a q-Rung orthopair fuzzy set is an advanced soft computing construct with the ability to precisely curb uncertainty compare to intuitionistic fuzzy set, Pythagorean fuzzy set and Fermatean fuzzy set, (ii) a q-Rung orthopair fuzzy max-min-max composite relation is a reliable information measure for determining decision making problems with precision. The limitation of qROFMMMCR is that it cannot be extended to any construct that has more than three parameters such as picture fuzzy sets without modification. In future research, (i) the application of the proposed qROFMMMCR could be considered in multiple attributes group decision-making and (ii) the proposed information measure could be enhanced for better result.

References

1. L.A. Zadeh, Fuzzy sets. Inf. Control **8**, 338–353 (1965)
2. K.T. Atanassov, Intuitionistic fuzzy sets. Fuzzy Set Syst. **20**, 87–96 (1986)
3. K.T. Atanassov, *Intuitionistic Fuzzy Sets: Theory and Applications* (Physica-Verlag, Heidelberg, 1999)
4. R.R. Yager, Pythagorean membership grades in multicriteria decision making. Technical Report MII-3301 (Machine Intelligence Institute, Iona College, New Rochelle, NY, 2013)
5. R.R. Yager, A.M. Abbasov, Pythagorean membership grades, complex numbers and decision making. Int. J. Intell. Syst. **28**(5), 436–452 (2013)
6. S.S. Begum, R. Srinivasan, Some properties on intuitionistic fuzzy sets of third type. Ann. Fuzzy Math. Inform. **10**(5), 799–804 (2015)
7. T. Senapati, R.R. Yager, Fermatean fuzzy sets. J. Amb. Intell. Human Comput. **11**, 663–674 (2020)
8. F.E. Boran, D. Akay, A biparametric similarity measure on intuitionistic fuzzy sets with applications to pattern recognition. Inf. Sci. **255**(10), 45–57 (2014)
9. S.M. Chen, C.H. Chang, A novel similarity measure between Atanassov's intuitionistic fuzzy sets based on transformation techniques with applications to pattern recognition. Inf. Sci. **291**, 96–114 (2015)
10. E. Szmidt, J. Kacprzyk, Medical diagnostic reasoning using a similarity measure for intuitionistic fuzzy sets. Note IFS **10**(4), 61–69 (2004)
11. S.K. De, R. Biswas, A.R. Roy, An application of intuitionistic fuzzy sets in medical diagnosis. Fuzzy Set Syst. **117**(2), 209–213 (2001)
12. P.A. Ejegwa, Novel correlation coefficient for intuitionistic fuzzy sets and its application to multi-criteria decision-making problems. Int. J. Fuzzy Syst. Appl. **10**(2), 39–58 (2021)
13. P.A. Ejegwa, I.C. Onyeke, Intuitionistic fuzzy statistical correlation algorithm with applications to multi-criteria based decision-making processes. Int. J. Intell. Syst. **36**(3), 1386–1407 (2021)
14. A.G. Hatzimichailidis, A.G. Papakostas, V.G. Kaburlasos, A novel distance measure of intuitionistic fuzzy sets and its application to pattern recognition problems. Int. J. Intell. Syst. **27**, 396–409 (2012)
15. E. Szmidt, J. Kacprzyk, Intuitionistic fuzzy sets in some medical applications. Note IFS **7**(4), 58–64 (2001)
16. H. Kamaci, Linear Diophantine fuzzy algebraic structures. J. Amb. Intell. Human Comput. (2021). https://doi.org/10.1007/s12652-020-02826-x
17. H. Kamaci, H. Garg, S. Petchimuthu, Bipolar trapezoidal neutrosophic sets and their Dombi operators with applications in multicriteria decision making. Soft Comput. **25**, 8417–8440 (2021)
18. S. Petchimuthu, H. Garg, H. Kamaci, A.O. Atagun, The mean operators and generalized products of fuzzy soft matrices and their applications in MCGDM. Comput. Appl. Math **39**(2020). https://doi.org/10.1007/s40314-020-1083-2
19. R. Parvathi, N. Palaniappan, Some operations on IFSs of second type. Note IFS **10**(2), 1–19 (2004)
20. D.Q. Li, W.Y. Zeng, Distance measure of Pythagorean fuzzy sets. Int. J. Intell. Syst. **33**, 348–361 (2018)

21. P.A. Ejegwa, Personnel appointments: a Pythagorean fuzzy sets approach using similarity measure. J. Inf. Comput. Sci. **14**(2), 94–102 (2019)
22. P.A. Ejegwa, Modified Zhang and Xu's distance measure of Pythagorean fuzzy sets and its application to pattern recognition problems. Neural Comput. Appl. **32**(14), 10199–10208 (2020)
23. P.A. Ejegwa, New similarity measures for Pythagorean fuzzy sets with applications. Int. J. Fuzzy Comput. Model **3**(1), 75–94 (2020)
24. P.A. Ejegwa, J.A. Awolola, Novel distance measures for Pythagorean fuzzy sets with applications to pattern recognition problems. Granul. Comput. **6**, 181–189 (2021)
25. P.A. Ejegwa, S. Wen, Y. Feng, W. Zhang, N. Tang, Novel Pythagorean fuzzy correlation measures via Pythagorean fuzzy deviation, variance and covariance with applications to pattern recognition and career placement. IEEE Trans. Fuzzy Syst. (2021). https://doi.org/10.1109/TFUZZ.2021.3063794
26. P.A. Ejegwa, S. Wen, Y. Feng, W. Zhang, J. Chen, Some new Pythagorean fuzzy correlation techniques via statistical viewpoint with applications to decision-making problems. J. Intell. Fuzzy Syst. **40**(5), 9873–9886 (2021)
27. W. Zeng, D. Li, Q. Yin, Distance and similarity measures of Pythagorean fuzzy sets and their applications to multiple criteria group decision making. Int. J. Intell. Syst. **33**(11), 2236–2254 (2018)
28. Y.Q. Du, F. Hou, W. Zafar, Q. Yu, Y. Zhai, A novel method for multiattribute decision making with interval-valued Pythagorean fuzzy linguistic information. Int. J. Intell. Syst. **32**(10), 1085–1112 (2017)
29. H. Garg, A new generalized Pythagorean fuzzy information aggregation using Einstein operations and its application to decision making. Int. J. Intell. Syst. **31**(9), 886–920 (2016)
30. H. Garg, Generalized Pythagorean fuzzy geometric aggregation operators using Einstein t-norm and t-conorm for multicriteria decision making process. Int. J. Intell. Syst. **32**(6), 597–630 (2017)
31. H. Garg, Linguistic Pythagorean fuzzy sets and its applications in multiattribute decision making process. Int. J. Intell. Syst. **33**(6), 1234–1263 (2018)
32. D. Liang, Z. Xu, The new extension of TOPSIS method for multiple criteria decision making with hesitant Pythagorean fuzzy sets. Appl. Soft Comput. **60**, 167–179 (2017)
33. P.A. Ejegwa, Generalized triparametric correlation coefficient for Pythagorean fuzzy sets with application to MCDM problems. Granul. Comput. **6**(3), 557–566 (2021)
34. P.A. Ejegwa, V. Adah, I.C. Onyeke, Some modified Pythagorean fuzzy correlation measures with application in determining some selected decision-making problems. Granul. Comput. (2021). https://doi.org/10.1007/s41066-021-00272-4
35. P.A. Ejegwa, J.A. Awolola, Real-life decision making based on a new correlation coefficient in Pythagorean fuzzy environment. Ann. Fuzzy Math. Inform. **21**(1), 51–67 (2021)
36. P.A. Ejegwa, Y. Feng, W. Zhang, Pattern recognition based on an improved Szmidt and Kacprzyk's correlation coefficient in Pythagorean fuzzy environment, in *Advances in Neural Networks-ISNN 2020*, ed. by H. Min, Q. Sitian, Z. Nian. Lecture Notes in Computer Science (LNCS), vol. 12557 (Springer, 2020), pp. 190–206
37. P.A. Ejegwa, C. Jana, Some new weighted correlation coefficients between Pythagorean fuzzy sets and their applications, in *Pythagorean Fuzzy Sets*, ed. by H. Garg. (Springer, 2021), pp. 39–64
38. P.A. Ejegwa, I.C. Onyeke, V. Adah, A Pythagorean fuzzy algorithm embedded with a new correlation measure and its application in diagnostic processes. Granul. Comput. (2020). https://doi.org/10.1007/s41066-020-00246-y
39. P.A. Ejegwa, S. Wen, Y. Feng, W. Zhang, Determination of pattern recognition problems based on a Pythagorean fuzzy correlation measure from statistical viewpoint, in *Proceedings of the 13th International Conference of Advanced Computational Intelligence* (Wanzhou, China, 2021), pp. 132–139
40. I. Silambarasan, New operators for Fermatean fuzzy sets. Ann. Commun. Math **3**(2), 116–131 (2020)

41. D. Liu, Y. Liu, X. Chen, Fermatean fuzzy linguistic set and its application in multicriteria decision making. Int. J. Intell. Syst. **34**(5), 878–894 (2019)
42. Z. Yang, H. Garg, X. Li, Differential calculus of Fermatean fuzzy functions: continuities, derivatives, and differentials. Int. J. Comput. Intell. Syst. **14**(1), 282–294 (2021)
43. T. Senapati, R.R. Yager, Fermatean fuzzy weighted averaging/geometric operators and its application in multi-criteria decision-making methods. Eng. Appl. Artif. Intell. **85**, 112–121 (2019)
44. T. Senapati, R.R. Yager, Some new operations over Fermatean fuzzy numbers and application of Fermatean fuzzy WPM in multiple criteria decision making. Informatica **30**(2), 391–412 (2019)
45. R.R. Yager, Generalized orthopair fuzzy sets. IEEE Trans. Fuzzy Syst. **25**(5), 1222–1230 (2017)
46. I. Silambarasan, New operations defined over the q-Rung orthopair fuzzy sets. J. Int. Math Virtual Inst. **10**(2), 341–359 (2020)
47. E. Dogu, A decision-making approach with q-Rung orthopair fuzzy sets: orthopair fuzzy TOPSIS method. Acad. Platf. J. Eng. Sci. **9**(1), 214–222 (2021)
48. M.J. Khan, P. Kumam, M. Shutaywi, Knowledge measure for the q-Rung orthopair fuzzy sets. Int. J. Intell. Syst. **36**(2), 628–655 (2021)
49. A. Pinar, F.E. Boran, A q-Rung orthopair fuzzy multi-criteria group decision making method for supplier selection based on a novel distance measure. Int. J. Mach. Learn Cybernet **11**, 1749–1780 (2020)
50. M. Riaz, M.T. Hamid, D. Afzal, D. Pamucar, Y.M. Chu, Multi-criteria decision making in robotic agri-farming with q-Rung orthopair m-polar fuzzy sets. PLoS One **16**(2), e0246485 (2021)
51. H. Garg, CN-q-ROFS: connection number-based q-Rung orthopair fuzzy set and their application to decision-making process. Int. J. Intell. Syst. **36**(7), 3106–3143 (2021)
52. H. Garg, A new possibility degree measure for interval-valued q-Rung orthopair fuzzy sets in decision-making. Int. J. Intell. Syst. **36**(1), 526–557 (2021)
53. H. Garg, A novel trigonometric operation-based q-Rung orthopair fuzzy aggregation operator and its fundamental properties. Neural Comput. Appl. **32**, 15077–15099 (2020)
54. H. Garg, New exponential operation laws and operators for interval-valued q-Rung orthopair fuzzy sets in group decision making process. Neural Comput. Appl. **33**(20), 13937–13963 (2021). https://doi.org/10.1007/s00521-021-06036-0
55. H. Garg, S.M. Chen, Multiattribute group decision making based on neutrality aggregation operators of q-Rung orthopair fuzzy sets. Inf. Sci. **517**, 427–447 (2020)
56. X. Peng, J. Dai, H. Garg, Exponential operation and aggregation operator for q-Rung orthopair fuzzy set and their decision-making method with a new score function. Int. J. Intell. Syst. **33**(11), 2255–2282 (2018)
57. M. Riaz, H. Garg, H.M.A. Farid, M. Aslam, Novel q-Rung orthopair fuzzy interaction aggregation operators and their application to low-carbon green supply chain management. J. Intell. Fuzzy Syst. **41**(2), 4109–4126 (2021). https://doi.org/10.3233/JIFS-210506
58. Z. Yang, H. Garg, Interaction power partitioned Maclaurin symmetric mean operators under q-Rung orthopair uncertain linguistic information. Int. J. Fuzzy Syst. **1–19**(2021). https://doi.org/10.1007/s40815-021-01062-5
59. P. Liu, P. Wang, Some q-Rung orthopair fuzzy aggregation operators and their applications to multiple-attribute decision making. Int. J. Intell. Syst. **33**(2), 259–280 (2018)

Paul Augustine Ejegwa received a B.Sc. degree in mathematics from Benue State University, Makurdi, Nigeria in 2008 and a M.Sc. degree in mathematics from Ahmadu Bello University, Zaria, Nigeria in 2013. He also received a Ph.D. degree in mathematics from Ahmadu Bello University, Zaria, Nigeria in 2018. Dr. Ejegwa was a post doctorate research fellow with the Key Laboratory of Intelligent Information Processing and Control, Chongqing Three Gorges University, Wanzhou, Chongqing 404100, China from 11th September, 2020 to 10th September, 2021. His research interest includes nonclassical algebra, fuzzy algebra, computational intelligence, and decision making techniques. He has published many papers in the listed research areas.

Chapter 4
Soergel Distance Measures for q-Rung Orthopair Fuzzy Sets and Their Applications

Hüseyin Kamacı and Subramanian Petchimuthu

Abstract The virtue of the q-rung orthopair fuzzy set inherits those of the intuitionistic fuzzy set and the Pythagorean fuzzy set in loosening the constraint on support and counter-support. The very lax requirement gives the evaluators great freedom in expressing their beliefs about membership degrees and non-membership degrees, which makes q-rung orthopair fuzzy sets having a wide scope of application in practice. A distance measure is an important mathematical tool for distinguishing the difference between q-rung orthopair fuzzy sets and allows to deal with problems such as multi-criteria decision-making, medical diagnosis, and pattern recognition under a q-rung orthopair fuzzy environment. Unfortunately, many of the existing q-rung orthopair fuzzy distance measures have their limitations. To eliminate such restrictions, in this chapter, the Soergel-type distances of q-rung orthopair fuzzy sets are introduced and the basis on which the orthopairs can be ranked is established. The weighted types of the proposed Soergel distances and their corresponding similarity coefficients are derived. In addition, the validity of the emerging distance measures is shown by comparing them with the distance measures described in some recent research studies through numerical examples. Some charts are provided to visually display the various characteristics and to analyze the properties of the proposed distance measures. The outputs verify that these Soergel distance measures of q-rung orthopair fuzzy sets outperform other existing metrics in measuring uncertainty and avoiding counterintuitive cases. Some illustrative examples of decision-making in real life are presented, demonstrating the strong discrimination capability and effectiveness of the proposed Soergel distance measures.

Keywords Fuzzy set · q-Rung orthopair fuzzy set · Soergel distance measure · Medical diagnosis · Pattern recognition · Benchmarking

---

H. Kamacı (✉)
Faculty of Science and Arts, Department of Mathematics, Yozgat Bozok University, 66100 Yozgat, Turkey
e-mail: huseyin.kamaci@hotmail.com

S. Petchimuthu
Department of Science and Humanities (Mathematics), University College of Engineering, Nagercoil 629004, Tamil Nadu, India

H. Garg (ed.), *q-Rung Orthopair Fuzzy Sets*,
https://doi.org/10.1007/978-981-19-1449-2_4

4.1 Introduction

Fuzzy set (FS) theory originated by Zadeh [81] has achieved great success in handling ambiguity and uncertainty in various fields. In FS theory, only the grade of membership is given and the grade of non-membership is equal to one minus the grade of membership. For instance, if the grade to which an object satisfies the concept is 0.6, the grade of non-membership is merely set to 0.4. The concepts of union, intersection, inclusion, complement, convexity, relation, etc., were extended to such sets, and various properties of these concepts in the context of FSs were established. These sets were advanced in different directions such as graph [65, 68], matrix [13, 38, 56–58], algebra and statistics [1, 34, 60]. Rough set (RS) theory is another mathematical approach to imperfect knowledge [49–52]. Generally, an FS may be viewed as a class with unsharp boundaries, whereas an RS is a coarsely described crisp set [79]. Among the many mathematical theories designed to model various uncertain concepts, FS and RS have received much attention and are actively studied by many researchers worldwide. To tackle more complex problems involving vague information in the real world, the theory of intuitionistic fuzzy sets (IFSs) has been initiated by Atanassov [2]. In the structure of this set, an object can have a degree of membership and a degree of non-membership such that its sum is less than or equal to 1. In the following years, by improving IFSs, many researchers studied their extended types such as interval-valued intuitionistic fuzzy sets (IVIFSs) [3, 9, 28, 30, 87], intuitionistic fuzzy multisets (IFMSs) [15, 67], refined intuitionistic fuzzy set (RIFS) [59], and triangular and trapezoidal forms [66, 69, 71]. The Pythagorean fuzzy set (PyFS) [76], initiated by Yager, is a generalization of the IFS. The PyFS is of great significance for the decision-making problem because it promotes the domain of IFS. Some explorations of the theory of PyFSs can be found in [19, 21, 70, 85, 86]. The orthopair $\langle 0.7, 0.3\rangle$ is a fuzzy membership degree. If the grade of support against is 0.2, then $\langle 0.7, 0.2\rangle$ is an intuitionistic membership degree since $0.7 + 0.2 \leq 1$. If the grade of support against is 0.4, then due to $0.7^2 + 0.4^2 \leq 1$, it is a Pythagorean membership degree (but not an intuitionistic membership degree). However, if the grade of support against is 0.8, this situation cannot be described by employing either IFSs or PyFSs.

Recently, Yager [77] proposed a generalization of IFSs and PyFSs, which are referred to as q-rung orthopair fuzzy sets (q-ROFSs, also named the IFSs of qth type) in which the sum of the qth power of the support and the qth power of the support against is less than or equal to 1. This provides the systems modelers more freedom in expressing their beliefs about membership and non-membership degrees. For example, $\langle 0.7, 0.8\rangle$ is a q-rung orthopair membership degree ($q \geq 3$) since $0.7^3 + 0.8^3 \leq 1$, and so, the q-ROFS is a suitable tool to treat such a situation. To understand the theoretical aspects of q-ROFSs deeply, one can refer [18, 22, 23, 46, 53]. Furthermore, Liu et al. [45, 46] discussed some new q-rung orthopair fuzzy weighted averaging and geometric operators to deal with the decision-making. Xu et al. [75] improved the existing q-rung orthopair fuzzy aggregation operators and applied them to multi-attribute group decision-making. Wei et al. [73] defined some

Maclaurin symmetric mean operators for q-ROFSs and investigated their applications to potential assessment of emerging technology commercialization. In the last two years, a large number of aggregation operators on q-ROFSs have been developed. Especially, Riaz et al. [61, 63, 64] described some aggregation operators such as geometric, hybrid, Einstein to fuse the q-rung orthopair fuzzy information. Yang and Garg [78] derived interaction power partitioned Maclaurin symmetric mean operators to aggregate q-rung orthopair uncertain linguistic information. In [62], the authors developed novel q-rung orthopair fuzzy interaction aggregation operators and applied them to low-carbon green supply chain management. Garg [20, 24] introduced some q-rung orthopair fuzzy aggregation operators based on trigonometric operation and neutrality.

A distance measure is an objective score that summarizes the relative difference between two objects in a problem domain. Similarity measures may then be derived from the distance. Distance/similarity measures between FSs are used to indicate the difference (dissimilarity)/similarity degree between the information carried by fuzzy sets. Measuring the distance and similarity of two objects/points is an important topic in FS theory as it allows to tackle problems in various fields such as decision-making, pattern recognition, and medical diagnosis. In the last decades, many researchers studied the fuzzy information measures [12, 39, 47, 48], intuitionistic fuzzy similarity measures [11, 27, 40, 44, 80] (vague similarity measures [10, 26, 42]), intuitionistic fuzzy possibility measures [25], intuitionistic fuzzy correlation coefficients [16], intuitionistic fuzzy distance measures [7], Pythagorean fuzzy similarity/distance measures [41, 43, 55, 74, 83, 84], and special parametric cases [4, 5, 8, 31–33]. To follow these developments, the scientific studies of q-ROFSs in recent years have mainly focused on distance and similarity measures. Du [14] proposed some Minkowski-type approaches to determine the distance between two q-ROFSs. Zeng et al. [82] derived weighted induced logarithmic distance coefficients to measure q-rung orthopair fuzzy information and investigated their application in multiple attribute decision-making. Wang et al. [72] discussed similarity measures of q-ROFSs based on cosine function and their applicability to real-life issues. Jan et al. [29] studied the generalized dice similarity measures for q-ROFSs with their applications. Khan et al. [35], and Peng and Liu [54] investigated the relationships between distance measures and similarity measures of q-ROFSs, improved the systematic transformation of information measures (distance measure and similarity measure) for q-ROFSs, and presented some new formulas of information measures of q-ROFSs. In the last few years, various information measures of q-ROFSs have been derived (see [6, 17, 35–37]), and are still being derived. The information measures studied in the literature hitherto may not always be sufficient to measure the relationship between any two q-ROFSs. (In the Comparison Analysis section of this chapter, the inadequacies of many of the existing q-ROFS-based information measures are explained in detail.) Therefore, new types of information measures for q-ROFSs are being derived to eliminate such drawbacks. The main objective of this chapter is to contribute to this research domain with new concepts based on Soergel distance for q-ROFSs. To achieve this aim, this chapter introduces the Soergel-type distance measures of q-ROFSs, improving the existing methodologies. It also presents the validation and merits of these developed distance measures (and their corresponding similarity coefficients) by comparing them with existing q-ROFS-based distance (and similarity) approaches. These new distance measurement types are used for the best selections of items on the problems of investment, medical diagnosis, pattern recognition, and benchmarking, and their acquired consequences are analyzed.

The layout of this chapter is systematized as follows: In Sect. 4.2, some essential notions of q-ROFSs and some existing q-ROF information measures are briefly reviewed. Section 4.3 introduces (weighted) Soergel-type distance/similarity measures for q-ROFSs and discusses their efficiency. In Sect. 4.4, a decision-making algorithm based on Soergel-type distance/similarity measures of q-ROFSs is elaborated and illustrative examples are presented to state this developed algorithm. In Sect. 4.5, a comparative analysis with existing distance/similarity measures of q-ROFSs is given to demonstrate the validity of the proposed Soergel-type distance/similarity measures. Section 4.6 is devoted to sensitivity analysis and the advantages of the developed q-ROF information measures. Section 4.7 gives some conclusions.

4.2 Background

4.2.1 q-Rung Orthopair Fuzzy Sets

In the year 1965, Zadeh [81] introduced the FS as an effective approach to model uncertain information. By the budding of FSs, he described a membership function from a reference set to the range [0, 1], and thereby determined a membership degree for each element in the reference set. In the year 1986, Atanassov [2] extended the concept of FS to IFSs, in which each IFN is allocated a membership degree and a non-membership degree such that the sum of them is less than or equal to 1. Nevertheless, the IFSs are incapable of expressing the membership degree and non-membership degree (especially when their sum is greater than 1) of each element in the universe of discourse belonging to an IFS. To further expand the IFSs, Yager [76] proposed the PyFSs, which somehow extended the space of membership and non-membership degrees by introducing a new constraint that the sum of squares is less than or equal to 1. Thus, Yager slightly relaxed the constraint in the structure of Atanassov's IFS. However, this relaxation in PyFSs was sometimes insufficient to represent the information collected under IFS environment. Fortifying the idea of PyFS, Yager [77] introduced the q-RPFSs, which somehow extended the space of membership and non-membership degrees by proposing a new lax constraint that the sum of qth power ($q > 1$) of the membership degree and non-membership degree is equal to or less than 1. The definition of q-ROFS is given as follows.

Definition 4.2.1 ([77]) Let Λ be a universe of discourse, a q-ROFS Φ on Λ is defined as

$$\Phi = \{(\lambda_i, \langle \wp_\Phi(\lambda_i), \eta_\Phi(\lambda_i)\rangle) : \lambda_i \in \Lambda\} \quad (4.2.1)$$

where $\wp_\Phi, \eta_\Phi : \Lambda \to [0, 1]$ are said to be the membership function and non-membership function, respectively. $\wp_\Phi(\lambda_i)$ and $\eta_\Phi(\lambda_i)$ represent the degrees of membership and non-membership of $\lambda_i \in \Lambda$ to the set Φ and satisfy the following condition

$$0 \le \wp_\Phi^q(\lambda_i) + \eta_\Phi^q(\lambda_i) \le 1 \tag{4.2.2}$$

for all $\lambda_i \in \Lambda$, where $q \ge 1$.
Furthermore,

$$\tau_\Phi(\lambda_i) = \sqrt[q]{1 - \left(\wp_\Phi^q(\lambda_i) + \eta_\Phi^q(\lambda_i)\right)} \tag{4.2.3}$$

is called a q-rung orthopair fuzzy index or degree of hesitation membership of $\lambda_i \in \Lambda$ to the set Φ.

The pair $\langle \wp_\Phi(\lambda_i), \eta_\Phi(\lambda_i) \rangle$ is called a q-rung orthopair fuzzy number (q-ROFN), which can be simply denoted by

$$\phi = \langle \wp_\Phi, \eta_\Phi \rangle \tag{4.2.4}$$

Note that the q-ROFS Φ is reduced to

- the IFS when $q = 1$,
- the PyFS when $q = 2$.

Therefore, it is obvious that the space of q-ROFS is larger than IFS and PyFS. The following Fig. 4.1 illustrates the relationship among the q-ROFN, IFN, and PyFN.

Definition 4.2.2 Let $\Phi = \{(\lambda_i, \langle \wp_\Phi(\lambda_i), \eta_\Phi(\lambda_i)\rangle) : \lambda_i \in \Lambda\}$ and $\Psi = \{(\lambda_i, \langle \wp_\Psi(\lambda_i), \eta_\Psi(\lambda_i)\rangle) : \lambda_i \in \Lambda\}$ be two q-ROFSs on Λ. Then

(a) Φ is described as being a subset of Ψ, or contained in Ψ, written $\Phi \subseteq \Psi$, if $\wp_\Phi(\lambda_i) \le \wp_\Psi(\lambda_i)$ and $\eta_\Phi(\lambda_i) \ge \eta_\Psi(\lambda_i)$ for all $\lambda_i \in \Lambda$.
(b) Φ and Ψ are equal, written $\Phi = \Psi$, if $\wp_\Phi(\lambda_i) = \wp_\Psi(\lambda_i)$ and $\eta_\Phi(\lambda_i) = \eta_\Psi(\lambda_i)$ for all $\lambda_i \in \Lambda$.

Some fundamental operations such as complement, intersection, and the union that construct new sets from given crisp sets are also adapted for the q-ROFSs. Such operations of q-ROFSs are described as follows:

Definition 4.2.3 ([77]) Let $\Phi = \{(\lambda_i, \langle \wp_\Phi(\lambda_i), \eta_\Phi(\lambda_i)\rangle) : \lambda_i \in \Lambda\}$ and $\Psi = \{(\lambda_i, \langle \wp_\Psi(\lambda_i), \eta_\Psi(\lambda_i)\rangle) : \lambda_i \in \Lambda\}$ be two q-ROFSs on Λ. Then

(a) the complement of Φ, symbolized by Φ^c, is defined as

$$\Phi^c = \{(\lambda_i, \langle \eta_\Phi(\lambda_i), \wp_\Phi(\lambda_i)\rangle) : \lambda_i \in \Lambda\}.$$

(b) the intersection of Φ and Ψ, symbolized by $\Phi \cap \Psi$,

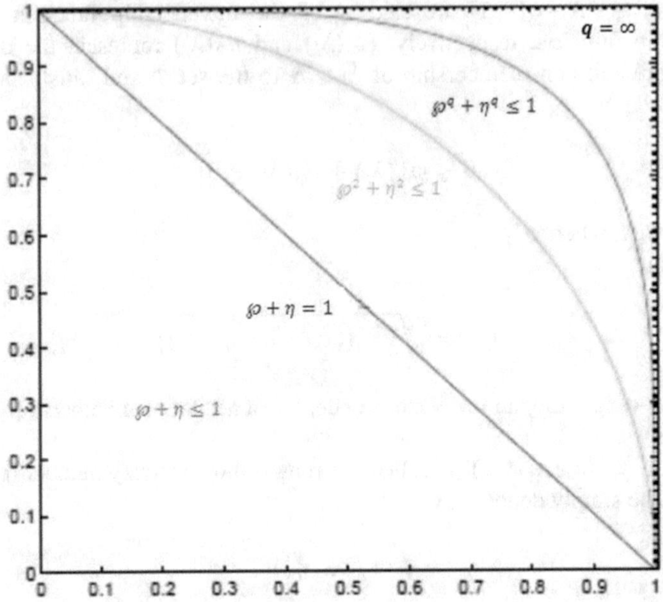

Fig. 4.1 The relationship among the q-ROFN, IFN, and PyFN

$$\Phi \cap \Psi = \{(\lambda_i, \langle \min\{\wp_\Phi(\lambda_i)\wp_\Psi(\lambda_i)\}, \max\{\eta_\Phi(\lambda_i), \eta_\Psi(\lambda_i)\}\rangle) : \lambda_i \in \Lambda\}.$$

(c) the union of Φ and Ψ, symbolized by $\Phi \cup \Psi$,

$$\Phi \cup \Psi = \{(\lambda_i, \langle \max\{\wp_\Phi(\lambda_i)\wp_\Psi(\lambda_i)\}, \min\{\eta_\Phi(\lambda_i), \eta_\Psi(\lambda_i)\}\rangle) : \lambda_i \in \Lambda\}.$$

Several operational rules for algebraic sum and algebraic product on the q-ROFNs are presented below.

Definition 4.2.4 ([77]) Let $\phi = \langle \wp_\Phi, \eta_\Phi \rangle$, and $\psi = \langle \wp_\Psi, \eta_\Psi \rangle$ be two q-ROFNs on Λ and $\xi > 0$ be any real number. Then

(a) $\phi \oplus \psi = \langle \sqrt[q]{\wp_\Phi^q + \wp_\Psi^q - \wp_\Phi^q \wp_\Psi^q}, \eta_\Phi \eta_\Psi \rangle$,
(b) $\phi \otimes \psi = \langle \wp_\Phi \wp_\Psi, \sqrt[q]{\eta_\Phi^q + \eta_\Psi^q - \eta_\Phi^q \eta_\Psi^q} \rangle$,
(c) $\xi \phi = \langle \sqrt[q]{1 - (1 - \wp_\Phi^q)^\xi}, \eta_\Phi^\xi \rangle$,
(d) $\phi^\xi = \langle \wp_\Phi^\xi, \sqrt[q]{1 - (1 - \eta_\Phi^q)^\xi} \rangle$.

4.2.2 *Some Existing Information Measures for q-ROFSs*

Many distances and similarity measures for q-ROFSs were proposed in the literature to determine the distance and similarity among q-ROFSs. The most famous of those are offered in this section. Peng et al. [54] defined distance measures and distance-based similarity measures ($\mathfrak{S}(\Phi, \Psi) = 1 - \mathfrak{D}(\Phi, \Psi)$) among q-ROFSs Φ and Ψ as in Table 4.1. Wang et al. [72] defined correlation coefficient, cosine, and cotangent-based similarity measures and their corresponding weighted similarity measures among q-ROFSs Φ and Ψ as in Table 4.2.

4.3 Soergel Distance Measures for q-ROFSs and Their Validation/Efficiency

The predominant goal of this component is to develop new twelve types of Soergel distance and weighted Soergel distance for q-ROFSs and to derive their corresponding similarity coefficients. Also, to present their special cases and desirable properties. Eventually, the validation of the proposed (weighted) Soergel distances and their corresponding similarities for q-ROFSs are checked.

From now on, Φ and Ψ are two q-ROFSs on Λ such that for each $\lambda_i \in \Lambda$, at least one of the grades $\wp_\Phi(\lambda_i)$, $\eta_\Phi(\lambda_i)$, $\wp_\Psi(\lambda_i)$, and $\eta_\Psi(\lambda_i)$ is nonzero. Moreover, the hesitancy degrees of $\lambda_i \in \Lambda$ for the q-ROFSs Φ and Ψ are $\tau_\Phi(\lambda_i) = \sqrt[q]{1 - \left((\wp_\Phi(\lambda_i))^q + (\eta_\Phi(\lambda_i))^q\right)}$ and $\tau_\Psi(\lambda_i) = \sqrt[q]{1 - \left((\wp_\Psi(\lambda_i))^q + (\eta_\Psi(\lambda_i))^q\right)}$ for all $\lambda_i \in \Lambda$, respectively.

4.3.1 *Twelve Types of Soergel Distance Measures for q-ROFSs*

The twelve types of Soergel distance measures (SoDMs) determine the distance between two q-ROFSs, are proposed as follows:

(1) The type-1 SoDM for two q-ROFSs Φ and Ψ is denoted and defined by

$$ {}^{1}\mathfrak{D}_{q-ROFS}^{Soergel}(\Phi, \Psi) = \frac{1}{n}\sum_{i=1}^{n} \frac{|\wp_\Phi^q(\lambda_i) - \wp_\Psi^q(\lambda_i)| + |\eta_\Phi^q(\lambda_i) - \eta_\Psi^q(\lambda_i)|}{\max\{\wp_\Phi^q(\lambda_i), \wp_\Psi^q(\lambda_i)\} + \max\{\eta_\Phi^q(\lambda_i), \eta_\Psi^q(\lambda_i)\}} \tag{4.3.1} $$

(2) The type-2 SoDM for two q-ROFSs Φ and Ψ is denoted and defined by

Table 4.1 Peng et al. [54]'s distance and similarity measures for q-ROFSs

Distance measure	Similarity measure
$\mathfrak{D}_1(\Phi,\Psi)=\frac{1}{2n}\sum_{i=1}^{n}(\lvert\wp_\Phi^q(\lambda_i)-\wp_\Psi^q(\lambda_i)\rvert+\lvert\eta_\Phi^q(\lambda_i)-\eta_\Psi^q(\lambda_i)\rvert+\lvert\tau_\Phi^q(\lambda_i)-\tau_\Psi^q(\lambda_i)\rvert)$	$\mathfrak{S}_1(\Phi,\Psi)=1-\frac{1}{2n}\sum_{i=1}^{n}(\lvert\wp_\Phi^q(\lambda_i)-\wp_\Psi^q(\lambda_i)\rvert+\lvert\eta_\Phi^q(\lambda_i)-\eta_\Psi^q(\lambda_i)\rvert+\lvert\tau_\Phi^q(\lambda_i)-\tau_\Psi^q(\lambda_i)\rvert)$
$\mathfrak{D}_2(\Phi,\Psi)=\frac{1}{2n}\sum_{i=1}^{n}\lvert\wp_\Phi^q(\lambda_i)-\wp_\Psi^q(\lambda_i)-(\eta_\Phi^q(\lambda_i)-\eta_\Psi^q(\lambda_i))\rvert$	$\mathfrak{S}_2(\Phi,\Psi)=1-\frac{1}{2n}\sum_{i=1}^{n}\lvert\wp_\Phi^q(\lambda_i)-\wp_\Psi^q(\lambda_i)-(\eta_\Phi^q(\lambda_i)-\eta_\Psi^q(\lambda_i))\rvert$
$\mathfrak{D}_3(\Phi,\Psi)=\frac{1}{4n}(\sum_{i=1}^{n}(\lvert\wp_\Phi^q(\lambda_i)-\wp_\Psi^q(\lambda_i)\rvert+\lvert\eta_\Phi^q(\lambda_i)-\eta_\Psi^q(\lambda_i)\rvert+\lvert\tau_\Phi^q(\lambda_i)-\tau_\Psi^q(\lambda_i)\rvert)+$ $\sum_{i=1}^{n}\lvert\wp_\Phi^q(\lambda_i)-\wp_\Psi^q(\lambda_i)-(\eta_\Phi^q(\lambda_i)-\eta_\Psi^q(\lambda_i))\rvert)$	$\mathfrak{S}_3(\Phi,\Psi)=1-\frac{1}{4n}(\sum_{i=1}^{n}(\lvert\wp_\Phi^q(\lambda_i)-\wp_\Psi^q(\lambda_i)\rvert+\lvert\eta_\Phi^q(\lambda_i)-\eta_\Psi^q(\lambda_i)\rvert+\lvert\tau_\Phi^q(\lambda_i)-\tau_\Psi^q(\lambda_i)\rvert)+$ $\sum_{i=1}^{n}\lvert\wp_\Phi^q(\lambda_i)-\wp_\Psi^q(\lambda_i)-(\eta_\Phi^q(\lambda_i)-\eta_\Psi^q(\lambda_i))\rvert)$
$\mathfrak{D}_4(\Phi,\Psi)=\frac{1}{n}\sum_{i=1}^{n}\max(\lvert\wp_\Phi^q(\lambda_i)-\wp_\Psi^q(\lambda_i)\rvert,\lvert\eta_\Phi^q(\lambda_i)-\eta_\Psi^q(\lambda_i)\rvert)$	$\mathfrak{S}_4(\Phi,\Psi)=1-\frac{1}{n}\sum_{i=1}^{n}\max(\lvert\wp_\Phi^q(\lambda_i)-\wp_\Psi^q(\lambda_i)\rvert,\lvert\eta_\Phi^q(\lambda_i)-\eta_\Psi^q(\lambda_i)\rvert)$
$\mathfrak{D}_5(\Phi,\Psi)=\frac{2}{n}\sum_{i=1}^{n}\frac{max(\lvert\wp_\Phi^q(\lambda_i)-\wp_\Psi^q(\lambda_i)\rvert,\lvert\eta_\Phi^q(\lambda_i)-\eta_\Psi^q(\lambda_i)\rvert)}{1+\max(\lvert\wp_\Phi^q(\lambda_i)-\wp_\Psi^q(\lambda_i)\rvert,\lvert\eta_\Phi^q(\lambda_i)-\eta_\Psi^q(\lambda_i)\rvert)}$	$\mathfrak{S}_5(\Phi,\Psi)=\frac{1}{n}\sum_{i=1}^{n}\frac{1-\max(\lvert\wp_\Phi^q(\lambda_i)-\wp_\Psi^q(\lambda_i)\rvert,\lvert\eta_\Phi^q(\lambda_i)-\eta_\Psi^q(\lambda_i)\rvert)}{1+\max(\lvert\wp_\Phi^q(\lambda_i)-\wp_\Psi^q(\lambda_i)\rvert,\lvert\eta_\Phi^q(\lambda_i)-\eta_\Psi^q(\lambda_i)\rvert)}$
$\mathfrak{D}_6(\Phi,\Psi)=\frac{2\sum_{i=1}^{n}\max(\lvert\wp_\Phi^q(\lambda_i)-\wp_\Psi^q(\lambda_i)\rvert,\lvert\eta_\Phi^q(\lambda_i)-\eta_\Psi^q(\lambda_i)\rvert)}{\sum_{i=1}^{n}(1+\max(\lvert\wp_\Phi^q(\lambda_i)-\wp_\Psi^q(\lambda_i)\rvert,\lvert\eta_\Phi^q(\lambda_i)-\eta_\Psi^q(\lambda_i)\rvert))}$	$\mathfrak{S}_6(\Phi,\Psi)=\frac{\sum_{i=1}^{n}(1-\max(\lvert\wp_\Phi^q(\lambda_i)-\wp_\Psi^q(\lambda_i)\rvert,\lvert\eta_\Phi^q(\lambda_i)-\eta_\Psi^q(\lambda_i)\rvert))}{\sum_{i=1}^{n}(1+\max(\lvert\wp_\Phi^q(\lambda_i)-\wp_\Psi^q(\lambda_i)\rvert,\lvert\eta_\Phi^q(\lambda_i)-\eta_\Psi^q(\lambda_i)\rvert))}$
$\mathfrak{D}_7(\Phi,\Psi)=1-\alpha\frac{\sum_{i=1}^{n}\min(\wp_\Phi^q(\lambda_i),\wp_\Psi^q(\lambda_i))}{\sum_{i=1}^{n}\max(\wp_\Phi^q(\lambda_i),\wp_\Psi^q(\lambda_i))}-\beta\frac{\sum_{i=1}^{n}\min(\eta_\Phi^q(\lambda_i),\eta_\Psi^q(\lambda_i))}{\sum_{i=1}^{n}\max(\eta_\Phi^q(\lambda_i),\eta_\Psi^q(\lambda_i))},\alpha+\beta=1,\alpha,\beta\in[0,1]$	$\mathfrak{S}_7(\Phi,\Psi)=\alpha\frac{\sum_{i=1}^{n}\min(\wp_\Phi^q(\lambda_i),\wp_\Psi^q(\lambda_i))}{\sum_{i=1}^{n}\max(\wp_\Phi^q(\lambda_i),\wp_\Psi^q(\lambda_i))}+\beta\frac{\sum_{i=1}^{n}\min(\eta_\Phi^q(\lambda_i),\eta_\Psi^q(\lambda_i))}{\sum_{i=1}^{n}\max(\eta_\Phi^q(\lambda_i),\eta_\Psi^q(\lambda_i))},\alpha+\beta=1,\alpha,\beta\in[0,1]$
$\mathfrak{D}_8(\Phi,\Psi)=1-\frac{\alpha}{n}\frac{\sum_{i=1}^{n}\min(\wp_\Phi^q(\lambda_i),\wp_\Psi^q(\lambda_i))}{\sum_{i=1}^{n}\max(\wp_\Phi^q(\lambda_i),\wp_\Psi^q(\lambda_i))}-\frac{\beta}{n}\frac{\sum_{i=1}^{n}\min(\eta_\Phi^q(\lambda_i),\eta_\Psi^q(\lambda_i))}{\sum_{i=1}^{n}\max(\eta_\Phi^q(\lambda_i),\eta_\Psi^q(\lambda_i))},\alpha+\beta=1,$ $\alpha,\beta\in[0,1]$	$\mathfrak{S}_8(\Phi,\Psi)=\frac{\alpha}{n}\frac{\sum_{i=1}^{n}\min(\wp_\Phi^q(\lambda_i),\wp_\Psi^q(\lambda_i))}{\sum_{i=1}^{n}\max(\wp_\Phi^q(\lambda_i),\wp_\Psi^q(\lambda_i))}+\frac{\beta}{n}\frac{\sum_{i=1}^{n}\min(\eta_\Phi^q(\lambda_i),\eta_\Psi^q(\lambda_i))}{\sum_{i=1}^{n}\max(\eta_\Phi^q(\lambda_i),\eta_\Psi^q(\lambda_i))},\alpha+\beta=1,\alpha,\beta\in[0,1]$
$\mathfrak{D}_9(\Phi,\Psi)=1-\frac{1}{n}\sum_{i=1}^{n}\frac{(\min(\wp_\Phi^q(\lambda_i),\wp_\Psi^q(\lambda_i))+\min(\eta_\Phi^q(\lambda_i),\eta_\Psi^q(\lambda_i)))}{(\max(\wp_\Phi^q(\lambda_i),\wp_\Psi^q(\lambda_i))+\max(\eta_\Phi^q(\lambda_i),\eta_\Psi^q(\lambda_i)))}$	$\mathfrak{S}_9(\Phi,\Psi)=\frac{1}{n}\sum_{i=1}^{n}\frac{(\min(\wp_\Phi^q(\lambda_i),\wp_\Psi^q(\lambda_i))+\min(\eta_\Phi^q(\lambda_i),\eta_\Psi^q(\lambda_i)))}{(\max(\wp_\Phi^q(\lambda_i),\wp_\Psi^q(\lambda_i))+\max(\eta_\Phi^q(\lambda_i),\eta_\Psi^q(\lambda_i)))}$
$\mathfrak{S}_{10}(\Phi,\Psi)=1-\frac{\sum_{i=1}^{n}(\min(\wp_\Phi^q(\lambda_i),\wp_\Psi^q(\lambda_i))+\min(\eta_\Phi^q(\lambda_i),\eta_\Psi^q(\lambda_i)))}{\sum_{i=1}^{n}(\max(\wp_\Phi^q(\lambda_i),\wp_\Psi^q(\lambda_i))+\max(\eta_\Phi^q(\lambda_i),\eta_\Psi^q(\lambda_i)))}$	$\mathfrak{S}_{10}(\Phi,\Psi)=\frac{\sum_{i=1}^{n}(\min(\wp_\Phi^q(\lambda_i),\wp_\Psi^q(\lambda_i))+\min(\eta_\Phi^q(\lambda_i),\eta_\Psi^q(\lambda_i)))}{\sum_{i=1}^{n}(\max(\wp_\Phi^q(\lambda_i),\wp_\Psi^q(\lambda_i))+\max(\eta_\Phi^q(\lambda_i),\eta_\Psi^q(\lambda_i)))}$
$\mathfrak{S}_{11}(\Phi,\Psi)=1-\frac{1}{n}\sum_{i=1}^{n}\frac{\min(\wp_\Phi^q(\lambda_i),\wp_\Psi^q(\lambda_i))+\min(1-\eta_\Phi^q(\lambda_i),1-\eta_\Psi^q(\lambda_i))}{\max(\wp_\Phi^q(\lambda_i),\wp_\Psi^q(\lambda_i))+\max(1-\eta_\Phi^q(\lambda_i),1-\eta_\Psi^q(\lambda_i))}$	$\mathfrak{S}_{11}(\Phi,\Psi)=\frac{1}{n}\sum_{i=1}^{n}\frac{\min(\wp_\Phi^q(\lambda_i),\wp_\Psi^q(\lambda_i))+\min(1-\eta_\Phi^q(\lambda_i),1-\eta_\Psi^q(\lambda_i))}{\max(\wp_\Phi^q(\lambda_i),\wp_\Psi^q(\lambda_i))+\max(1-\eta_\Phi^q(\lambda_i),1-\eta_\Psi^q(\lambda_i))}$
$\mathfrak{D}_{12}(\Phi,\Psi)=1-\frac{\sum_{i=1}^{n}(\min(\wp_\Phi^q(\lambda_i),\wp_\Psi^q(\lambda_i))+\min(1-\eta_\Phi^q(\lambda_i),1-\eta_\Psi^q(\lambda_i)))}{\sum_{i=1}^{n}(\max(\wp_\Phi^q(\lambda_i),\wp_\Psi^q(\lambda_i))+\max(1-\eta_\Phi^q(\lambda_i),1-\eta_\Psi^q(\lambda_i)))}$	$\mathfrak{S}_{12}(\Phi,\Psi)=\frac{\sum_{i=1}^{n}(\min(\wp_\Phi^q(\lambda_i),\wp_\Psi^q(\lambda_i))+\min(1-\eta_\Phi^q(\lambda_i),1-\eta_\Psi^q(\lambda_i)))}{\sum_{i=1}^{n}(\max(\wp_\Phi^q(\lambda_i),\wp_\Psi^q(\lambda_i))+\max(1-\eta_\Phi^q(\lambda_i),1-\eta_\Psi^q(\lambda_i)))}$

Table 4.2 Wang et al. [72]'s similarity and weighted similarity measures for q-ROFSs

Similarity measure	Weighted similarity measure
$\mathfrak{S}_{13}(\Phi,\Psi)=\frac{1}{n}\sum_{i=1}^{n}\frac{\wp^q_\Phi(\lambda_i)\wp^q_\Psi(\lambda_i)+\eta^q_\Phi(\lambda_i)\eta^q_\Psi(\lambda_i)}{\sqrt{\wp^{2q}_\Phi(\lambda_i)+\eta^{2q}_\Phi(\lambda_i)}\sqrt{\wp^{2q}_\Psi(\lambda_i)+\eta^{2q}_\Psi(\lambda_i)}}$	$\mathfrak{W}\mathfrak{S}_{13}(\Phi,\Psi)=\sum_{i=1}^{n}\omega_i\frac{\wp^q_\Phi(\lambda_i)\wp^q_\Psi(\lambda_i)+\eta^q_\Phi(\lambda_i)\eta^q_\Psi(\lambda_i)}{\sqrt{\wp^{2q}_\Phi(\lambda_i)+\eta^{2q}_\Phi(\lambda_i)}\sqrt{\wp^{2q}_\Psi(\lambda_i)+\eta^{2q}_\Psi(\lambda_i)}}$
$\mathfrak{S}_{14}(\Phi,\Psi)=\frac{1}{n}\sum_{i=1}^{n}\frac{\wp^q_\Phi(\lambda_i)\wp^q_\Psi(\lambda_i)+\eta^q_\Phi(\lambda_i)\eta^q_\Psi(\lambda_i)+\tau^q_\Phi(\lambda_i)\tau^q_\Psi(\lambda_i)}{\sqrt{\wp^{2q}_\Phi(\lambda_i)+\eta^{2q}_\Phi(\lambda_i)+\tau^{2q}_\Phi(\lambda_i)}\sqrt{\wp^{2q}_\Psi(\lambda_i)+\eta^{2q}_\Psi(\lambda_i)+\tau^{2q}_\Psi(\lambda_i)}}$	$\mathfrak{W}\mathfrak{S}_{14}(\Phi,\Psi)=\sum_{i=1}^{n}\omega_i\frac{\wp^q_\Phi(\lambda_i)\wp^q_\Psi(\lambda_i)+\eta^q_\Phi(\lambda_i)\eta^q_\Psi(\lambda_i)+\tau^q_\Phi(\lambda_i)\tau^q_\Psi(\lambda_i)}{\sqrt{\wp^{2q}_\Phi(\lambda_i)+\eta^{2q}_\Phi(\lambda_i)+\tau^{2q}_\Phi(\lambda_i)}\sqrt{\wp^{2q}_\Psi(\lambda_i)+\eta^{2q}_\Psi(\lambda_i)+\tau^{2q}_\Psi(\lambda_i)}}$
$\mathfrak{S}_{15}(\Phi,\Psi)=\frac{1}{n}\sum_{i=1}^{n}\cos[(\frac{\pi}{2})\max(\lvert\wp^q_\Phi(\lambda_i)-\wp^q_\Psi(\lambda_i)\rvert,\lvert\eta^q_\Phi(\lambda_i)-\eta^q_\Psi(\lambda_i)\rvert)]$	$\mathfrak{W}\mathfrak{S}_{15}(\Phi,\Psi)=\sum_{i=1}^{n}\omega_i\cos[(\frac{\pi}{2})\max(\lvert\wp^q_\Phi(\lambda_i)-\wp^q_\Psi(\lambda_i)\rvert,\lvert\eta^q_\Phi(\lambda_i)-\eta^q_\Psi(\lambda_i)\rvert)]$
$\mathfrak{S}_{16}(\Phi,\Psi)=\frac{1}{n}\sum_{i=1}^{n}cos[(\frac{\pi}{4})(\lvert\wp^q_\Phi(\lambda_i)-\wp^q_\Psi(\lambda_i)\rvert+\lvert\eta^q_\Phi(\lambda_i)-\eta^q_\Psi(\lambda_i)\rvert)]$	$\mathfrak{W}\mathfrak{S}_{16}(\Phi,\Psi)=\sum_{i=1}^{n}\omega_i cos[(\frac{\pi}{4})(\lvert\wp^q_\Phi(\lambda_i)-\wp^q_\Psi(\lambda_i)\rvert+\lvert\eta^q_\Phi(\lambda_i)-\eta^q_\Psi(\lambda_i)\rvert)]$
$\mathfrak{S}_{17}(\Phi,\Psi)=$ $\frac{1}{n}\sum_{i=1}^{n}\cos[(\frac{\pi}{2})\max(\lvert\wp^q_\Phi(\lambda_i)-\wp^q_\Psi(\lambda_i)\rvert,\lvert\eta^q_\Phi(\lambda_i)-\eta^q_\Psi(\lambda_i)\rvert,\lvert\tau^q_\Phi(\lambda_i)-\tau^q_\Psi(\lambda_i)\rvert)]$	$\mathfrak{W}\mathfrak{S}_{17}(\Phi,\Psi)=$ $\sum_{i=1}^{n}\omega_i\cos[(\frac{\pi}{2})\max(\lvert\wp^q_\Phi(\lambda_i)-\wp^q_\Psi(\lambda_i)\rvert,\lvert\eta^q_\Phi(\lambda_i)-\eta^q_\Psi(\lambda_i)\rvert,\lvert\tau^q_\Phi(\lambda_i)-\tau^q_\Psi(\lambda_i)\rvert)]$;
$\mathfrak{S}_{18}(\Phi,\Psi)=$ $\frac{1}{n}\sum_{i=1}^{n}cos[(\frac{\pi}{4})(\lvert\wp^q_\Phi(\lambda_i)-\wp^q_\Psi(\lambda_i)\rvert+\lvert\eta^q_\Phi(\lambda_i)-\eta^q_\Psi(\lambda_i)\rvert+\lvert\tau^q_\Phi(\lambda_i)-\tau^q_\Psi(\lambda_i)\rvert)]$	$\mathfrak{W}\mathfrak{S}_{18}(\Phi,\Psi)=$ $\sum_{i=1}^{n}\omega_i cos[(\frac{\pi}{4})(\lvert\wp^q_\Phi(\lambda_i)-\wp^q_\Psi(\lambda_i)\rvert+\lvert\eta^q_\Phi(\lambda_i)-\eta^q_\Psi(\lambda_i)\rvert+\lvert\tau^q_\Phi(\lambda_i)-\tau^q_\Psi(\lambda_i)\rvert)]$
$\mathfrak{S}_{19}(\Phi,\Psi)=\frac{1}{n}\sum_{i=1}^{n}\cot[\frac{\pi}{4}+\frac{\pi}{4}\max(\lvert\wp^q_\Phi(\lambda_i)-\wp^q_\Psi(\lambda_i)\rvert,\lvert\eta^q_\Phi(\lambda_i)-\eta^q_\Psi(\lambda_i)\rvert)]$	$\mathfrak{W}\mathfrak{S}_{19}(\Phi,\Psi)=\sum_{i=1}^{n}\omega_i\cot[\frac{\pi}{4}+\frac{\pi}{4}\max(\lvert\wp^q_\Phi(\lambda_i)-\wp^q_\Psi(\lambda_i)\rvert,\lvert\eta^q_\Phi(\lambda_i)-\eta^q_\Psi(\lambda_i)\rvert)]$
$\mathfrak{S}_{20}(\Phi,\Psi)=\frac{1}{n}\sum_{i=1}^{n}cot[\frac{\pi}{4}+\frac{\pi}{8}(\lvert\wp^q_\Phi(\lambda_i)-\wp^q_\Psi(\lambda_i)\rvert+\lvert\eta^q_\Phi(\lambda_i)-\eta^q_\Psi(\lambda_i)\rvert)]$	$\mathfrak{W}\mathfrak{S}_{20}(\Phi,\Psi)=\sum_{i=1}^{n}\omega_i cot[\frac{\pi}{4}+\frac{\pi}{8}(\lvert\wp^q_\Phi(\lambda_i)-\wp^q_\Psi(\lambda_i)\rvert+\lvert\eta^q_\Phi(\lambda_i)-\eta^q_\Psi(\lambda_i)\rvert)]$
$\mathfrak{S}_{21}(\Phi,\Psi)=$ $\frac{1}{n}\sum_{i=1}^{n}\cot[\frac{\pi}{4}+\frac{\pi}{4}\max(\lvert\wp^q_\Phi(\lambda_i)-\wp^q_\Psi(\lambda_i)\rvert,\lvert\eta^q_\Phi(\lambda_i)-\eta^q_\Psi(\lambda_i)\rvert,\lvert\tau^q_\Phi(\lambda_i)-\tau^q_\Psi(\lambda_i)\rvert)]$	$\mathfrak{W}\mathfrak{S}_{21}(\Phi,\Psi)=$ $\sum_{i=1}^{n}\omega_i\cot[\frac{\pi}{4}+\frac{\pi}{4}\max(\lvert\wp^q_\Phi(\lambda_i)-\wp^q_\Psi(\lambda_i)\rvert,\lvert\eta^q_\Phi(\lambda_i)-\eta^q_\Psi(\lambda_i)\rvert,\lvert\tau^q_\Phi(\lambda_i)-\tau^q_\Psi(\lambda_i)\rvert)]$
$\mathfrak{S}_{22}(\Phi,\Psi)=$ $\frac{1}{n}\sum_{i=1}^{n}cot[\frac{\pi}{4}+\frac{\pi}{8}(\lvert\wp^q_\Phi(\lambda_i)-\wp^q_\Psi(\lambda_i)\rvert+\lvert\eta^q_\Phi(\lambda_i)-\eta^q_\Psi(\lambda_i)\rvert+\lvert\tau^q_\Phi(\lambda_i)-\tau^q_\Psi(\lambda_i)\rvert)]$	$\mathfrak{W}\mathfrak{S}_{22}(\Phi,\Psi)=$ $\sum_{i=1}^{n}\omega_i cot[\frac{\pi}{4}+\frac{\pi}{8}(\lvert\wp^q_\Phi(\lambda_i)-\wp^q_\Psi(\lambda_i)\rvert+\lvert\eta^q_\Phi(\lambda_i)-\eta^q_\Psi(\lambda_i)\rvert+\lvert\tau^q_\Phi(\lambda_i)-\tau^q_\Psi(\lambda_i)\rvert)]$

$$^{2}\mathfrak{D}^{Soergel}_{q-ROFS}(\Phi,\Psi)=\frac{\sum\limits_{i=1}^{n}(|\wp^{q}_{\Phi}(\lambda_i)-\wp^{q}_{\Psi}(\lambda_i)|+|\eta^{q}_{\Phi}(\lambda_i)-\eta^{q}_{\Psi}(\lambda_i)|)}{\sum\limits_{i=1}^{n}(\max\{\wp^{q}_{\Phi}(\lambda_i),\wp^{q}_{\Psi}(\lambda_i)\}+\max\{\eta^{q}_{\Phi}(\lambda_i),\eta^{q}_{\Psi}(\lambda_i)\})} \tag{4.3.2}$$

From the Eqs. (4.3.1) and (4.3.2), it is obvious that $^{1}\mathfrak{D}^{Soergel}_{q-ROFS}(\Phi,\Psi)={}^{1}\mathfrak{D}^{Soergel}_{q-ROFS}(\Phi^*,\Psi^*)$ and $^{2}\mathfrak{D}^{Soergel}_{q-ROFS}(\Phi,\Psi)={}^{2}\mathfrak{D}^{Soergel}_{q-ROFS}(\Phi^*,\Psi^*)$ whenever $\Phi=\Phi^*$ and $\Psi=\Psi^*$ (or $\Phi=\Psi^*$ and $\Psi=\Phi^*$). Consider the q-ROFSs $\Phi=\{(\lambda_1,\langle 0.6,0.4\rangle),(\lambda_2,\langle 0.3,0.1\rangle)\}$, $\Psi=\{(\lambda_1,\langle 0.3,0.1\rangle),(\lambda_2,\langle 0.6,0.4\rangle)\}$, $\Phi^*=\{(\lambda_1,\langle 0.2,0.2\rangle),(\lambda_2,\langle 0.7,0.3\rangle)\}$ and $\Psi^*=\{(\lambda_1,\langle 0.7,0.3\rangle),(\lambda_2,\langle 0.2,0.2\rangle)\}$ (where $q=1$). Although $\Phi\neq\Phi^*$, $\Psi\neq\Psi^*$, $\Phi\neq\Psi^*$ and $\Psi\neq\Phi^*$, it is obtained as $^{1}\mathfrak{D}^{Soergel}_{q-ROFS}(\Phi,\Psi)={}^{1}\mathfrak{D}^{Soergel}_{q-ROFS}(\Phi^*,\Psi^*)=0.6$ and $^{2}\mathfrak{D}^{Soergel}_{q-ROFS}(\Phi,\Psi)={}^{2}\mathfrak{D}^{Soergel}_{q-ROFS}(\Phi^*,\Psi^*)=0.6$. To deal with such situations or constraints, the equations of $^{1}\mathfrak{D}^{Soergel}_{q-ROFS}$ and $^{2}\mathfrak{D}^{Soergel}_{q-ROFS}$ can be improved by multiplying or adding some terms. By integrating some terms into the numerator parts in the formula of the above SoDMs for q-ROFSs can be improved as follows:

(3) The type-3 SoDM for two q-ROFSs Φ and Ψ is denoted and defined by

$$^{3}\mathfrak{D}^{Soergel}_{q-ROFS}(\Phi,\Psi)=\frac{1}{n}\sum_{i=1}^{n}\frac{\sqrt{\begin{pmatrix}(|\wp^{q}_{\Phi}(\lambda_i)-\wp^{q}_{\Psi}(\lambda_i)|+|\eta^{q}_{\Phi}(\lambda_i)-\eta^{q}_{\Psi}(\lambda_i)|)\times\\ 2\min\{|\wp^{q}_{\Phi}(\lambda_i)-\wp^{q}_{\Psi}(\lambda_i)|,|\eta^{q}_{\Phi}(\lambda_i)-\eta^{q}_{\Psi}(\lambda_i)|\}\end{pmatrix}}}{\max\{\wp^{q}_{\Phi}(\lambda_i),\wp^{q}_{\Psi}(\lambda_i)\}+\max\{\eta^{q}_{\Phi}(\lambda_i),\eta^{q}_{\Psi}(\lambda_i)\}} \tag{4.3.3}$$

(4) The type-4 SoDM for two q-ROFSs Φ and Ψ is denoted and defined by

$$^{4}\mathfrak{D}^{Soergel}_{q-ROFS}(\Phi,\Psi)=\frac{\sum\limits_{i=1}^{n}\sqrt{\begin{pmatrix}(|\wp^{q}_{\Phi}(\lambda_i)-\wp^{q}_{\Psi}(\lambda_i)|+|\eta^{q}_{\Phi}(\lambda_i)-\eta^{q}_{\Psi}(\lambda_i)|)\times\\ 2\min\{|\wp^{q}_{\Phi}(\lambda_i)-\wp^{q}_{\Psi}(\lambda_i)|,|\eta^{q}_{\Phi}(\lambda_i)-\eta^{q}_{\Psi}(\lambda_i)|\}\end{pmatrix}}}{\sum\limits_{i=1}^{n}(\max\{\wp^{q}_{\Phi}(\lambda_i),\wp^{q}_{\Psi}(\lambda_i)\}+\max\{\eta^{q}_{\Phi}(\lambda_i),\eta^{q}_{\Psi}(\lambda_i)\})} \tag{4.3.4}$$

Note that by type-3 SoDM and type-4 SoDM can overcome the difficulties mentioned above. However, these SoDMs do not produce consistent outputs when either $\wp_{\Phi}(\lambda_i)=\wp_{\Psi}(\lambda_i)$ or $\eta_{\Phi}(\lambda_i)=\eta_{\Psi}(\lambda_i)$ for $\lambda_i\in\Lambda$. For example, consider the above q-ROFSs Φ and Ψ, and also $\Phi^*=\{(\lambda_1,\langle 0,0.4\rangle),(\lambda_2,\langle 0.4,0.6\rangle)\}$ and $\Psi^*=\{(\lambda_1,\langle 0.6,0.4\rangle),(\lambda_2,\langle 0.4,0\rangle)\}$ (where $q=1$). Although $\Phi\neq\Phi^*$, $\Psi\neq\Psi^*$, $\Phi\neq\Psi^*$ and $\Psi\neq\Phi^*$, it is obtained as $^{1}\mathfrak{D}^{Soergel}_{q-ROFS}(\Phi,\Psi)={}^{1}\mathfrak{D}^{Soergel}_{q-ROFS}(\Phi^*,\Psi^*)=0.6$ and $^{2}\mathfrak{D}^{Soergel}_{q-ROFS}(\Phi,\Psi)={}^{2}\mathfrak{D}^{Soergel}_{q-ROFS}(\Phi^*,\Psi^*)=0.6$. Also, $^{3}\mathfrak{D}^{Soergel}_{q-ROFS}(\Phi^*,\Psi^*)={}^{4}\mathfrak{D}^{Soergel}_{q-ROFS}(\Phi^*,\Psi^*)=0$, and so these results are inconsistent. To eliminate this limitation, the following types of SoDM can be proposed.

(5) The type-5 SoDM for two q-ROFSs Φ and Ψ is denoted and defined by

$$ {}^{5}\mathfrak{D}^{Soergel}_{q-ROFS}(\Phi,\Psi)=\frac{1}{n}\sum_{i=1}^{n}\frac{\frac{1}{2}\begin{pmatrix}(|\wp^{q}_{\Phi}(\lambda_i)-\wp^{q}_{\Psi}(\lambda_i)|+|\eta^{q}_{\Phi}(\lambda_i)-\eta^{q}_{\Psi}(\lambda_i)|)+\\(\ln(1+|\wp^{q}_{\Phi}(\lambda_i)-\wp^{q}_{\Psi}(\lambda_i)|)+\ln(1+|\eta^{q}_{\Phi}(\lambda_i)-\eta^{q}_{\Psi}(\lambda_i)|))\end{pmatrix}}{\max\{\wp^{q}_{\Phi}(\lambda_i),\wp^{q}_{\Psi}(\lambda_i)\}+\max\{\eta^{q}_{\Phi}(\lambda_i),\eta^{q}_{\Psi}(\lambda_i)\}} \tag{4.3.5} $$

(6) The type-6 SoDM for two q-ROFSs Φ and Ψ is denoted and defined by

$$ {}^{6}\mathfrak{D}^{Soergel}_{q-ROFS}(\Phi,\Psi)=\frac{\frac{1}{2}\sum_{i=1}^{n}\begin{pmatrix}(|\wp^{q}_{\Phi}(\lambda_i)-\wp^{q}_{\Psi}(\lambda_i)|+|\eta^{q}_{\Phi}(\lambda_i)-\eta^{q}_{\Psi}(\lambda_i)|)+\\(\ln(1+|\wp^{q}_{\Phi}(\lambda_i)-\wp^{q}_{\Psi}(\lambda_i)|)+\ln(1+|\eta^{q}_{\Phi}(\lambda_i)-\eta^{q}_{\Psi}(\lambda_i)|))\end{pmatrix}}{\sum_{i=1}^{n}(\max\{\wp^{q}_{\Phi}(\lambda_i),\wp^{q}_{\Psi}(\lambda_i)\}+\max\{\eta^{q}_{\Phi}(\lambda_i),\eta^{q}_{\Psi}(\lambda_i)\})} \tag{4.3.6} $$

Consider the q-ROFSs $\Phi=\{(\lambda_1,\langle 0.4,0.5\rangle)\}$, $\Psi=\{(\lambda_1,\langle 0.3,0.2\rangle)\}$, $\Phi^*=\{(\lambda_1,\langle 0.3,0.3\rangle)\}$ and $\Psi^*=\{(\lambda_1,\langle 0.2,0.6\rangle)\}$ (where $q=1$). Although $\Phi\neq\Phi^*$, $\Psi\neq\Psi^*$, $\Phi\neq\Psi^*$ and $\Psi\neq\Phi^*$, it is deduced as

$$ \begin{aligned} {}^{1}\mathfrak{D}^{Soergel}_{q-ROFS}(\Phi,\Psi)&={}^{1}\mathfrak{D}^{Soergel}_{q-ROFS}(\Phi^*,\Psi^*)=0.444,\\ {}^{2}\mathfrak{D}^{Soergel}_{q-ROFS}(\Phi,\Psi)&={}^{2}\mathfrak{D}^{Soergel}_{q-ROFS}(\Phi^*,\Psi^*)=0.444,\\ {}^{3}\mathfrak{D}^{Soergel}_{q-ROFS}(\Phi,\Psi)&={}^{3}\mathfrak{D}^{Soergel}_{q-ROFS}(\Phi^*,\Psi^*)=0.314,\\ {}^{4}\mathfrak{D}^{Soergel}_{q-ROFS}(\Phi,\Psi)&={}^{4}\mathfrak{D}^{Soergel}_{q-ROFS}(\Phi^*,\Psi^*)=0.314,\\ {}^{5}\mathfrak{D}^{Soergel}_{q-ROFS}(\Phi,\Psi)&={}^{5}\mathfrak{D}^{Soergel}_{q-ROFS}(\Phi^*,\Psi^*)=0.421,\\ {}^{6}\mathfrak{D}^{Soergel}_{q-ROFS}(\Phi,\Psi)&={}^{6}\mathfrak{D}^{Soergel}_{q-ROFS}(\Phi^*,\Psi^*)=0.421. \end{aligned} $$

These equalities are illogical, and they can be overcome by integrating the degree of hesitancy. Therefore, it can be recreated the above SoDMs of q-ROFSs by considering the "hesitancy degree" as follows:

(7) The type-7 SoDM for two q-ROFSs Φ and Ψ is denoted and defined by

$$ {}^{7}\mathfrak{D}^{Soergel}_{q-ROFS}(\Phi,\Psi)=\frac{1}{n}\sum_{i=1}^{n}\frac{|\wp^{q}_{\Phi}(\lambda_i)-\wp^{q}_{\Psi}(\lambda_i)|+|\eta^{q}_{\Phi}(\lambda_i)-\eta^{q}_{\Psi}(\lambda_i)|+|\tau^{q}_{\Phi}(\lambda_i)-\tau^{q}_{\Psi}(\lambda_i)|}{\max\{\wp^{q}_{\Phi}(\lambda_i),\wp^{q}_{\Psi}(\lambda_i)\}+\max\{\eta^{q}_{\Phi}(\lambda_i),\eta^{q}_{\Psi}(\lambda_i)\}+\max\{\tau^{q}_{\Phi}(\lambda_i),\tau^{q}_{\Psi}(\lambda_i)\}} \tag{4.3.7} $$

(8) The type-8 SoDM for two q-ROFSs Φ and Ψ is denoted and defined by

$$ {}^{8}\mathfrak{D}^{Soergel}_{q-ROFS}(\Phi,\Psi)=\frac{\sum_{i=1}^{n}(|\wp^{q}_{\Phi}(\lambda_i)-\wp^{q}_{\Psi}(\lambda_i)|+|\eta^{q}_{\Phi}(\lambda_i)-\eta^{q}_{\Psi}(\lambda_i)|+|\tau^{q}_{\Phi}(\lambda_i)-\tau^{q}_{\Psi}(\lambda_i)|)}{\sum_{i=1}^{n}(\max\{\wp^{q}_{\Phi}(\lambda_i),\wp^{q}_{\Psi}(\lambda_i)\}+\max\{\eta^{q}_{\Phi}(\lambda_i),\eta^{q}_{\Psi}(\lambda_i)\}+\max\{\tau^{q}_{\Phi}(\lambda_i),\tau^{q}_{\Psi}(\lambda_i)\})} \tag{4.3.8} $$

(9) The type-9 SoDM for two q-ROFSs Φ and Ψ is denoted and defined by

$$ {}^{9}\mathfrak{D}^{Soergel}_{q-ROFS}(\Phi,\Psi)=\frac{1}{n}\sum_{i=1}^{n}\frac{\sqrt{\begin{pmatrix}(|\wp^q_\Phi(\lambda_i)-\wp^q_\Psi(\lambda_i)|+|\eta^q_\Phi(\lambda_i)-\eta^q_\Psi(\lambda_i)|+|\tau^q_\Phi(\lambda_i)-\tau^q_\Psi(\lambda_i)|)\times\\ 3\min\{|\wp^q_\Phi(\lambda_i)-\wp^q_\Psi(\lambda_i)|,|\eta^q_\Phi(\lambda_i)-\eta^q_\Psi(\lambda_i)|,|\tau^q_\Phi(\lambda_i)-\tau^q_\Psi(\lambda_i)|\}\end{pmatrix}}}{\max\{\wp^q_\Phi(\lambda_i),\wp^q_\Psi(\lambda_i)\}+\max\{\eta^q_\Phi(\lambda_i),\eta^q_\Psi(\lambda_i)\}+\max\{\tau^q_\Phi(\lambda_i),\tau^q_\Psi(\lambda_i)\}} \tag{4.3.9}$$

(10) The type-10 SoDM for two q-ROFSs Φ and Ψ is denoted and defined by

$$ {}^{10}\mathfrak{D}^{Soergel}_{q-ROFS}(\Phi,\Psi)=\frac{\sum_{i=1}^{n}\sqrt{\begin{pmatrix}(|\wp^q_\Phi(\lambda_i)-\wp^q_\Psi(\lambda_i)|+|\eta^q_\Phi(\lambda_i)-\eta^q_\Psi(\lambda_i)|+|\tau^q_\Phi(\lambda_i)-\tau^q_\Psi(\lambda_i)|)\times\\ 3\min\{|\wp^q_\Phi(\lambda_i)-\wp^q_\Psi(\lambda_i)|,|\eta^q_\Phi(\lambda_i)-\eta^q_\Psi(\lambda_i)|,|\tau^q_\Phi(\lambda_i)-\tau^q_\Psi(\lambda_i)|\}\end{pmatrix}}}{\sum_{i=1}^{n}(\max\{\wp^q_\Phi(\lambda_i),\wp^q_\Psi(\lambda_i)\}+\max\{\eta^q_\Phi(\lambda_i),\eta^q_\Psi(\lambda_i)\}+\max\{\tau^q_\Phi(\lambda_i),\tau^q_\Psi(\lambda_i)\})} \tag{4.3.10}$$

(11) The type-11 SoDM for two q-ROFSs Φ and Ψ is denoted and defined by

$$ {}^{11}\mathfrak{D}^{Soergel}_{q-ROFS}(\Phi,\Psi) = \frac{1}{n}\sum_{i=1}^{n}\frac{\frac{1}{2}\begin{pmatrix}(|\wp^q_\Phi(\lambda_i)-\wp^q_\Psi(\lambda_i)|+|\eta^q_\Phi(\lambda_i)-\eta^q_\Psi(\lambda_i)|+|\tau^q_\Phi(\lambda_i)-\tau^q_\Psi(\lambda_i)|)+\\ (\ln(1+|\wp^q_\Phi(\lambda_i)-\wp^q_\Psi(\lambda_i)|)+\ln(1+|\eta^q_\Phi(\lambda_i)-\eta^q_\Psi(\lambda_i)|)+\ln(1+|\tau^q_\Phi(\lambda_i)-\tau^q_\Psi(\lambda_i)|))\end{pmatrix}}{\max\{\wp^q_\Phi(\lambda_i),\wp^q_\Psi(\lambda_i)\}+\max\{\eta^q_\Phi(\lambda_i),\eta^q_\Psi(\lambda_i)\},\max\{\tau^q_\Phi(\lambda_i),\tau^q_\Psi(\lambda_i)\}} \tag{4.3.11}$$

(12) The type-12 SoDM for two q-ROFSs Φ and Ψ is denoted and defined by

$$ {}^{12}\mathfrak{D}^{Soergel}_{q-ROFS}(\Phi,\Psi) = \frac{\frac{1}{2}\sum_{i=1}^{n}\begin{pmatrix}(|\wp^q_\Phi(\lambda_i)-\wp^q_\Psi(\lambda_i)|+|\eta^q_\Phi(\lambda_i)-\eta^q_\Psi(\lambda_i)|+|\tau^q_\Phi(\lambda_i)-\tau^q_\Psi(\lambda_i)|)+\\ (\ln(1+|\wp^q_\Phi(\lambda_i)-\wp^q_\Psi(\lambda_i)|)+\ln(1+|\eta^q_\Phi(\lambda_i)-\eta^q_\Psi(\lambda_i)|)+\ln(1+|\tau^q_\Phi(\lambda_i)-\tau^q_\Psi(\lambda_i)|))\end{pmatrix}}{\sum_{i=1}^{n}(\max\{\wp^q_\Phi(\lambda_i),\wp^q_\Psi(\lambda_i)\}+\max\{\eta^q_\Phi(\lambda_i),\eta^q_\Psi(\lambda_i)\},\max\{\tau^q_\Phi(\lambda_i),\tau^q_\Psi(\lambda_i)\})} \tag{4.3.12}$$

Theorem 4.3.1 *The type-κ SoDM ${}^{\kappa}\mathfrak{D}^{Soergel}_{q-ROFS}$ $(\kappa = 1, 2, \ldots, 12)$ satisfies the following conditions:*

1. $0 \leq {}^{\kappa}\mathfrak{D}^{Soergel}_{q-ROFS}(\Phi,\Psi) \leq 1$,
2. ${}^{\kappa}\mathfrak{D}^{Soergel}_{q-ROFS}(\Phi,\Psi) = {}^{\kappa}\mathfrak{D}^{Soergel}_{q-ROFS}(\Psi,\Phi)$,
3. ${}^{\kappa}\mathfrak{D}^{Soergel}_{q-ROFS}(\Phi,\Psi) = 0$ *if* $\Phi = \Psi$.

Proof For $\kappa = 1$, that is, for type-1 SoDM, the properties are verified. For other SoDMs, they can be demonstrated by similar techniques and so are omitted.

Let Φ and Ψ be two q-ROFSs on Λ.

1. Obviously, $0 \leq |\wp^q_\Phi(\lambda_i)-\wp^q_\Psi(\lambda_i)| \leq \max\{\wp^q_\Phi(\lambda_i),\wp^q_\Psi(\lambda_i)\}$ and $0 \leq |\eta^q_\Phi(\lambda_i) - \eta^q_\Psi(\lambda_i)| \leq \max\{\eta^q_\Phi(\lambda_i),\eta^q_\Psi(\lambda_i)\}$ for all $\lambda_i \in \Lambda$. By adding two inequalities side-by-side, it can be written as

$$0 \leq |\wp_{\Phi}^{q}(\lambda_i) - \wp_{\Psi}^{q}(\lambda_i)| + |\eta_{\Phi}^{q}(\lambda_i) - \eta_{\Psi}^{q}(\lambda_i)| \leq \max\{\wp_{\Phi}^{q}(\lambda_i), \wp_{\Psi}^{q}(\lambda_i)\} + \max\{\eta_{\Phi}^{q}(\lambda_i), \eta_{\Psi}^{q}(\lambda_i)\} \tag{4.3.13}$$

for all $\lambda_i \in \Lambda$. This implies that

$$0 \leq \frac{|\wp_{\Phi}^{q}(\lambda_i) - \wp_{\Psi}^{q}(\lambda_i)| + |\eta_{\Phi}^{q}(\lambda_i) - \eta_{\Psi}^{q}(\lambda_i)|}{\max\{\wp_{\Phi}^{q}(\lambda_i), \wp_{\Psi}^{q}(\lambda_i)\} + \max\{\eta_{\Phi}^{q}(\lambda_i), \eta_{\Psi}^{q}(\lambda_i)\}} \leq 1 \tag{4.3.14}$$

for all $\lambda_i \in \Lambda$. Thus, it is deduced that $0 \leq {}^{1}\mathfrak{D}_{q-ROFS}^{Soergel}(\Phi, \Psi) \leq 1$.

2. To demonstrate the property on commutativity, it is written

$$\begin{aligned}
{}^{1}\mathfrak{D}_{q-ROFS}^{Soergel}(\Phi, \Psi) &= \frac{1}{n}\sum_{i=1}^{n} \frac{|\wp_{\Phi}^{q}(\lambda_i) - \wp_{\Psi}^{q}(\lambda_i)| + |\eta_{\Phi}^{q}(\lambda_i) - \eta_{\Psi}^{q}(\lambda_i)|}{\max\{\wp_{\Phi}^{q}(\lambda_i), \wp_{\Psi}^{q}(\lambda_i)\} + \max\{\eta_{\Phi}^{q}(\lambda_i), \eta_{\Psi}^{q}(\lambda_i)\}} \\
&= \frac{1}{n}\sum_{i=1}^{n} \frac{|\wp_{\Psi}^{q}(\lambda_i) - \wp_{\Phi}^{q}(\lambda_i)| + |\eta_{\Psi}^{q}(\lambda_i) - \eta_{\Phi}^{q}(\lambda_i)|}{\max\{\wp_{\Psi}^{q}(\lambda_i), \wp_{\Phi}^{q}(\lambda_i)\} + \max\{\eta_{\Psi}^{q}(\lambda_i), \eta_{\Phi}^{q}(\lambda_i)\}} \\
&= {}^{1}\mathfrak{D}_{q-ROFS}^{Soergel}(\Psi, \Phi)
\end{aligned}$$

3. Suppose that $\Phi = \Psi$. Then, it is known from the definition of the equality of q-ROFSs that $\wp_{\Phi}(\lambda_i) = \wp_{\Psi}(\lambda_i)$ and $\eta_{\Phi}(\lambda_i) = \eta_{\Psi}(\lambda_i)$ for all $\lambda_i \in \Lambda$. By Eq. (4.3.1),

$$\begin{aligned}
{}^{1}\mathfrak{D}_{q-ROFS}^{Soergel}(\Phi, \Psi) &= \frac{1}{n}\sum_{i=1}^{n} \frac{|\wp_{\Phi}^{q}(\lambda_i) - \wp_{\Psi}^{q}(\lambda_i)| + |\eta_{\Phi}^{q}(\lambda_i) - \eta_{\Psi}^{q}(\lambda_i)|}{\max\{\wp_{\Phi}^{q}(\lambda_i), \wp_{\Psi}^{q}(\lambda_i)\} + \max\{\eta_{\Phi}^{q}(\lambda_i), \eta_{\Psi}^{q}(\lambda_i)\}} \\
&= \frac{1}{n}\sum_{i=1}^{n} \frac{|\wp_{\Phi}^{q}(\lambda_i) - \wp_{\Phi}^{q}(\lambda_i)| + |\eta_{\Phi}^{q}(\lambda_i) - \eta_{\Phi}^{q}(\lambda_i)|}{\max\{\wp_{\Phi}^{q}(\lambda_i), \wp_{\Phi}^{q}(\lambda_i)\} + \max\{\eta_{\Phi}^{q}(\lambda_i), \eta_{\Phi}^{q}(\lambda_i)\}} \\
&= 0.
\end{aligned}$$

Note 1. Since type-κ SoDM ($\kappa = 1, 2, \ldots, 12$) are bounded in the range [0, 1], the corresponding similarities (i.e., type-κ SoSM ${}^{\kappa}\mathfrak{S}_{q-ROFS}^{Soergel}$) are obtained simply as

$${}^{\kappa}\mathfrak{S}_{q-ROFS}^{Soergel}(\Phi, \Psi) = 1 - {}^{\kappa}\mathfrak{D}_{q-ROFS}^{Soergel}(\Phi, \Psi) \tag{4.3.15}$$

For the Eq. (4.3.15), by considering Theorem 4.3.1, it is obvious the following properties:

1. $0 \leq {}^{\kappa}\mathfrak{S}_{q-ROFS}^{Soergel}(\Phi, \Psi) \leq 1$,
2. ${}^{\kappa}\mathfrak{S}_{q-ROFS}^{Soergel}(\Phi, \Psi) = {}^{\kappa}\mathfrak{S}_{q-ROFS}^{Soergel}(\Psi, \Phi)$,
3. ${}^{\kappa}\mathfrak{S}_{q-ROFS}^{Soergel}(\Phi, \Psi) = 1$ if $\Phi = \Psi$.

Example 4.3.2 Let $\Phi = \{(\lambda_1, \langle 0.7, 0.5\rangle), (\lambda_2, \langle 0.4, 0.5\rangle)\}$ and $\Psi = \{(\lambda_1, \langle 0.5, 0.9\rangle), (\lambda_2, \langle 0.3, 0.3\rangle)\}$ be two q-ROFSs. Since $0.5^q + 0.9^q \leq 1$ for all

Table 4.3 Type-κ SoDM and SoSM between q-ROFSs Φ and Ψ

κ	1	2	3	4	5	6	7	8	9	10	11	12
${}^{\kappa}\mathfrak{D}^{Soergel}_{q-ROFS}(\Phi, \Psi)$	0.741	0.759	0.544	0.554	0.698	0.696	0.496	0.54	0.353	0.388	0.459	0.498
${}^{\kappa}\mathfrak{S}^{Soergel}_{q-ROFS}(\Phi, \Psi)$	0.259	0.241	0.456	0.446	0.302	0.304	0.504	0.46	0.647	0.612	0.541	0.502

$q \geq 3$. Assume that $q = 3$. The type-κ SoDM ($\kappa = 1, 2, \ldots, 12$) and type-$\kappa$ SoSM ($\kappa = 1, 2, \ldots, 12$) are calculated as in Table 4.3.

Consequently, the twelve different types of SoDMs of q-ROFSs Φ and Ψ are values between 0.241 and 0.759. The causes for these different distances are that some of the proposed distance and similarity measures pay attention to the minimum values of the absolute difference of membership degrees and absolute difference of non-membership degrees, some of them measure the distance between two q-ROFSs, taking into account the degree of hesitation in addition to membership and non-membership degrees. Some also include both. Or other reasons. The range of SoDM coefficients allows the evaluation of real-world problems from different perspectives. The different types of SoSMs can be interpreted similarly.

The above SoDMs and SoSMs are remarkable techniques for determining the distance and similarity between two q-ROFSs without weights. However, given the complexity of issues in the real world, it is inevitable to include weights in the structure of q-ROFSs. When the q-ROFSs Φ and Ψ have the evaluation weights, the above SoDMs and SoSMs are insufficient to measure the distance and similarity between these q-ROFSs. By eliminating such limitations, the weights need to be included in Eqs. (4.3.1)–(4.3.12).

4.3.2 *Twelve Types of Weighted Soergel Distance Measures for q-ROFSs*

The twelve types of weighted Soergel distance measures (WSoDMs) determine the weighted distance between two q-ROFSs, are proposed as follows:

Assume that $\omega = (\omega_1, \omega_2, \ldots, \omega_n)^T$ is the weight vector such that $\omega_i \in (0, 1]$ for $i = 1, 2, \ldots, n$ and $\sum_{i=1}^{n} \omega_i = 1$.

(1) The type-1 WSoDM for two q-ROFSs Φ and Ψ is denoted and defined by

$$
{}^{\omega 1}\mathfrak{D}^{Soergel}_{q-ROFS}(\Phi, \Psi) = \sum_{i=1}^{n} \omega_i \frac{|\wp^q_\Phi(\lambda_i) - \wp^q_\Psi(\lambda_i)| + |\eta^q_\Phi(\lambda_i) - \eta^q_\Psi(\lambda_i)|}{\max\{\wp^q_\Phi(\lambda_i), \wp^q_\Psi(\lambda_i)\} + \max\{\eta^q_\Phi(\lambda_i), \eta^q_\Psi(\lambda_i)\}} \tag{4.3.16}
$$

(2) The type-2 WSoDM for two q-ROFSs Φ and Ψ is denoted and defined by

$$^{\omega 2}\mathfrak{D}^{Soergel}_{q-ROFS}(\Phi, \Psi) = \frac{\sum_{i=1}^{n} \omega_i (|\wp^q_\Phi(\lambda_i) - \wp^q_\Psi(\lambda_i)| + |\eta^q_\Phi(\lambda_i) - \eta^q_\Psi(\lambda_i)|)}{\sum_{i=1}^{n} \omega_i (\max\{\wp^q_\Phi(\lambda_i), \wp^q_\Psi(\lambda_i)\} + \max\{\eta^q_\Phi(\lambda_i), \eta^q_\Psi(\lambda_i)\})} \tag{4.3.17}$$

(3) The type-3 WSoDM for two q-ROFSs Φ and Ψ is denoted and defined by

$$^{\omega 3}\mathfrak{D}^{Soergel}_{q-ROFS}(\Phi, \Psi) = \sum_{i=1}^{n} \omega_i \frac{\sqrt{\begin{pmatrix} (|\wp^q_\Phi(\lambda_i) - \wp^q_\Psi(\lambda_i)| + |\eta^q_\Phi(\lambda_i) - \eta^q_\Psi(\lambda_i)|)\times \\ 2\min\{|\wp^q_\Phi(\lambda_i) - \wp^q_\Psi(\lambda_i)|, |\eta^q_\Phi(\lambda_i) - \eta^q_\Psi(\lambda_i)|\} \end{pmatrix}}}{\max\{\wp^q_\Phi(\lambda_i), \wp^q_\Psi(\lambda_i)\} + \max\{\eta^q_\Phi(\lambda_i), \eta^q_\Psi(\lambda_i)\}} \tag{4.3.18}$$

(4) The type-4 WSoDM for two q-ROFSs Φ and Ψ is denoted and defined by

$$^{\omega 4}\mathfrak{D}^{Soergel}_{q-ROFS}(\Phi, \Psi) = \frac{\sum_{i=1}^{n} \omega_i \sqrt{\begin{pmatrix} (|\wp^q_\Phi(\lambda_i) - \wp^q_\Psi(\lambda_i)| + |\eta^q_\Phi(\lambda_i) - \eta^q_\Psi(\lambda_i)|)\times \\ 2\min\{|\wp^q_\Phi(\lambda_i) - \wp^q_\Psi(\lambda_i)|, |\eta^q_\Phi(\lambda_i) - \eta^q_\Psi(\lambda_i)|\} \end{pmatrix}}}{\sum_{i=1}^{n} \omega_i (\max\{\wp^q_\Phi(\lambda_i), \wp^q_\Psi(\lambda_i)\} + \max\{\eta^q_\Phi(\lambda_i), \eta^q_\Psi(\lambda_i)\})} \tag{4.3.19}$$

(5) The type-5 WSoDM for two q-ROFSs Φ and Ψ is denoted and defined by

$$^{\omega 5}\mathfrak{D}^{Soergel}_{q-ROFS}(\Phi, \Psi) = \sum_{i=1}^{n} \omega_i \frac{\frac{1}{2}\begin{pmatrix} (|\wp^q_\Phi(\lambda_i) - \wp^q_\Psi(\lambda_i)| + |\eta^q_\Phi(\lambda_i) - \eta^q_\Psi(\lambda_i)|)+ \\ (\ln(1 + |\wp^q_\Phi(\lambda_i) - \wp^q_\Psi(\lambda_i)|) + \ln(1 + |\eta^q_\Phi(\lambda_i) - \eta^q_\Psi(\lambda_i)|)) \end{pmatrix}}{\max\{\wp^q_\Phi(\lambda_i), \wp^q_\Psi(\lambda_i)\} + \max\{\eta^q_\Phi(\lambda_i), \eta^q_\Psi(\lambda_i)\}} \tag{4.3.20}$$

(6) The type-6 WSoDM for two q-ROFSs Φ and Ψ is denoted and defined by

$$^{\omega 6}\mathfrak{D}^{Soergel}_{q-ROFS}(\Phi, \Psi) = \frac{\frac{1}{2}\sum_{i=1}^{n} \omega_i \begin{pmatrix} (|\wp^q_\Phi(\lambda_i) - \wp^q_\Psi(\lambda_i)| + |\eta^q_\Phi(\lambda_i) - \eta^q_\Psi(\lambda_i)|)+ \\ (\ln(1 + |\wp^q_\Phi(\lambda_i) - \wp^q_\Psi(\lambda_i)|) + \ln(1 + |\eta^q_\Phi(\lambda_i) - \eta^q_\Psi(\lambda_i)|)) \end{pmatrix}}{\sum_{i=1}^{n} \omega_i (\max\{\wp^q_\Phi(\lambda_i), \wp^q_\Psi(\lambda_i)\} + \max\{\eta^q_\Phi(\lambda_i), \eta^q_\Psi(\lambda_i)\})} \tag{4.3.21}$$

(7) The type-7 WSoDM for two q-ROFSs Φ and Ψ is denoted and defined by

$$^{\omega 7}\mathfrak{D}^{Soergel}_{q-ROFS}(\Phi, \Psi) = \sum_{i=1}^{n} \omega_i \frac{|\wp^q_\Phi(\lambda_i) - \wp^q_\Psi(\lambda_i)| + |\eta^q_\Phi(\lambda_i) - \eta^q_\Psi(\lambda_i)| + |\tau^q_\Phi(\lambda_i) - \tau^q_\Psi(\lambda_i)|}{\max\{\wp^q_\Phi(\lambda_i), \wp^q_\Psi(\lambda_i)\} + \max\{\eta^q_\Phi(\lambda_i), \eta^q_\Psi(\lambda_i)\} + \max\{\tau^q_\Phi(\lambda_i), \tau^q_\Psi(\lambda_i)\}} \tag{4.3.22}$$

(8) The type-8 WSoDM for two q-ROFSs Φ and Ψ is denoted and defined by

$$ {}^{\omega 8}\mathfrak{D}_{q-ROFS}^{Soergel}(\Phi, \Psi) = \frac{\sum_{i=1}^{n} \omega_i (|\wp_{\Phi}^{q}(\lambda_i) - \wp_{\Psi}^{q}(\lambda_i)| + |\eta_{\Phi}^{q}(\lambda_i) - \eta_{\Psi}^{q}(\lambda_i)| + |\tau_{\Phi}^{q}(\lambda_i) - \tau_{\Psi}^{q}(\lambda_i)|)}{\sum_{i=1}^{n} \omega_i (\max\{\wp_{\Phi}^{q}(\lambda_i), \wp_{\Psi}^{q}(\lambda_i)\} + \max\{\eta_{\Phi}^{q}(\lambda_i), \eta_{\Psi}^{q}(\lambda_i)\} + \max\{\tau_{\Phi}^{q}(\lambda_i), \tau_{\Psi}^{q}(\lambda_i)\})} \tag{4.3.23} $$

(9) The type-9 WSoDM for two q-ROFSs Φ and Ψ is denoted and defined by

$$ {}^{\omega 9}\mathfrak{D}_{q-ROFS}^{Soergel}(\Phi, \Psi) = \sum_{i=1}^{n} \omega_i \frac{\sqrt{\begin{pmatrix} (|\wp_{\Phi}^{q}(\lambda_i) - \wp_{\Psi}^{q}(\lambda_i)| + |\eta_{\Phi}^{q}(\lambda_i) - \eta_{\Psi}^{q}(\lambda_i)| + |\tau_{\Phi}^{q}(\lambda_i) - \tau_{\Psi}^{q}(\lambda_i)|) \times \\ 3\min\{|\wp_{\Phi}^{q}(\lambda_i) - \wp_{\Psi}^{q}(\lambda_i)|, |\eta_{\Phi}^{q}(\lambda_i) - \eta_{\Psi}^{q}(\lambda_i)|, |\tau_{\Phi}^{q}(\lambda_i) - \tau_{\Psi}^{q}(\lambda_i)|\} \end{pmatrix}}}{\max\{\wp_{\Phi}^{q}(\lambda_i), \wp_{\Psi}^{q}(\lambda_i)\} + \max\{\eta_{\Phi}^{q}(\lambda_i), \eta_{\Psi}^{q}(\lambda_i)\} + \max\{\tau_{\Phi}^{q}(\lambda_i), \tau_{\Psi}^{q}(\lambda_i)\}} \tag{4.3.24} $$

(10) The type-10 WSoDM for two q-ROFSs Φ and Ψ is denoted and defined by

$$ {}^{\omega 10}\mathfrak{D}_{q-ROFS}^{Soergel}(\Phi, \Psi) = \frac{\sum_{i=1}^{n} \omega_i \sqrt{\begin{pmatrix} (|\wp_{\Phi}^{q}(\lambda_i) - \wp_{\Psi}^{q}(\lambda_i)| + |\eta_{\Phi}^{q}(\lambda_i) - \eta_{\Psi}^{q}(\lambda_i)| + |\tau_{\Phi}^{q}(\lambda_i) - \tau_{\Psi}^{q}(\lambda_i)|) \times \\ 3\min\{|\wp_{\Phi}^{q}(\lambda_i) - \wp_{\Psi}^{q}(\lambda_i)|, |\eta_{\Phi}^{q}(\lambda_i) - \eta_{\Psi}^{q}(\lambda_i)|, |\tau_{\Phi}^{q}(\lambda_i) - \tau_{\Psi}^{q}(\lambda_i)|\} \end{pmatrix}}}{\sum_{i=1}^{n} \omega_i (\max\{\wp_{\Phi}^{q}(\lambda_i), \wp_{\Psi}^{q}(\lambda_i)\} + \max\{\eta_{\Phi}^{q}(\lambda_i), \eta_{\Psi}^{q}(\lambda_i)\} + \max\{\tau_{\Phi}^{q}(\lambda_i), \tau_{\Psi}^{q}(\lambda_i)\})} \tag{4.3.25} $$

(11) The type-11 WSoDM for two q-ROFSs Φ and Ψ is denoted and defined by

$$ {}^{\omega 11}\mathfrak{D}_{q-ROFS}^{Soergel}(\Phi, \Psi) = \sum_{i=1}^{n} \omega_i \frac{\frac{1}{2}\begin{pmatrix} (|\wp_{\Phi}^{q}(\lambda_i) - \wp_{\Psi}^{q}(\lambda_i)| + |\eta_{\Phi}^{q}(\lambda_i) - \eta_{\Psi}^{q}(\lambda_i)| + |\tau_{\Phi}^{q}(\lambda_i) - \tau_{\Psi}^{q}(\lambda_i)|) + \\ (\ln(1 + |\wp_{\Phi}^{q}(\lambda_i) - \wp_{\Psi}^{q}(\lambda_i)|) + \ln(1 + |\eta_{\Phi}^{q}(\lambda_i) - \eta_{\Psi}^{q}(\lambda_i)|) + \ln(1 + |\tau_{\Phi}^{q}(\lambda_i) - \tau_{\Psi}^{q}(\lambda_i)|)) \end{pmatrix}}{\max\{\wp_{\Phi}^{q}(\lambda_i), \wp_{\Psi}^{q}(\lambda_i)\} + \max\{\eta_{\Phi}^{q}(\lambda_i), \eta_{\Psi}^{q}(\lambda_i)\}, \max\{\tau_{\Phi}^{q}(\lambda_i), \tau_{\Psi}^{q}(\lambda_i)\}} \tag{4.3.26} $$

(12) The type-12 WSoDM for two q-ROFSs Φ and Ψ is denoted and defined by

$$ {}^{\omega 12}\mathfrak{D}_{q-ROFS}^{Soergel}(\Phi, \Psi) = \frac{\frac{1}{2}\sum_{i=1}^{n} \omega_i \begin{pmatrix} (|\wp_{\Phi}^{q}(\lambda_i) - \wp_{\Psi}^{q}(\lambda_i)| + |\eta_{\Phi}^{q}(\lambda_i) - \eta_{\Psi}^{q}(\lambda_i)| + |\tau_{\Phi}^{q}(\lambda_i) - \tau_{\Psi}^{q}(\lambda_i)|) + \\ (\ln(1 + |\wp_{\Phi}^{q}(\lambda_i) - \wp_{\Psi}^{q}(\lambda_i)|) + \ln(1 + |\eta_{\Phi}^{q}(\lambda_i) - \eta_{\Psi}^{q}(\lambda_i)|) + \ln(1 + |\tau_{\Phi}^{q}(\lambda_i) - \tau_{\Psi}^{q}(\lambda_i)|)) \end{pmatrix}}{\sum_{i=1}^{n} \omega_i (\max\{\wp_{\Phi}^{q}(\lambda_i), \wp_{\Psi}^{q}(\lambda_i)\} + \max\{\eta_{\Phi}^{q}(\lambda_i), \eta_{\Psi}^{q}(\lambda_i)\}, \max\{\tau_{\Phi}^{q}(\lambda_i), \tau_{\Psi}^{q}(\lambda_i)\})} \tag{4.3.27} $$

Remark 4.3.3 If it is taken as $\omega = (\frac{1}{n}, \frac{1}{n}, \ldots, \frac{1}{n})^T$ then the type-κ WSoDM of q-ROFSs is reduced to the type-κ SoDM of q-ROFS for each $\kappa \in \{1, 2, \ldots, 12\}$, i.e., ${}^{\omega\kappa}\mathfrak{D}_{q-ROFS}^{Soergel}(\Phi, \Psi) = {}^{\kappa}\mathfrak{D}_{q-ROFS}^{Soergel}(\Phi, \Psi)$ for each $\kappa \in \{1, 2, \ldots, 12\}$ if $\omega = (\frac{1}{n}, \frac{1}{n}, \ldots, \frac{1}{n})^T$.

Theorem 4.3.4 *For the type-κ WSoDM, ${}^{\omega\kappa}\mathfrak{D}_{q-ROFS}^{Soergel}$, the following properties are satisfied:*

1. $0 \leq {}^{\omega\kappa}\mathfrak{D}_{q-ROFS}^{Soergel}(\Phi, \Psi) \leq 1$,

2. ${}^{\omega\kappa}\mathfrak{D}^{Soergel}_{q-ROFS}(\Phi, \Psi) = {}^{\omega\kappa}\mathfrak{D}^{Soergel}_{q-ROFS}(\Psi, \Phi)$,
3. ${}^{\omega\kappa}\mathfrak{D}^{Soergel}_{q-ROFS}(\Phi, \Psi) = 0$ *if* $\Phi = \Psi$.

Proof For $\kappa = 10$, that is, for type-2 WSoDM, the properties are verified. For other WSoDMs, they can be proved by proceeding with similar techniques and so are omitted.

Let Φ and Ψ be two q-ROFSs on Λ and $\omega = (\omega_1, \omega_2, \ldots, \omega_n)^T$ be the weight vector such that $\omega_i \in (0, 1]$ for $i = 1, 2, \ldots, n$ and $\sum_{i=1}^{n} \omega_i = 1$.

1. Since $0 \le |\wp^q_\Phi(\lambda_i) - \wp^q_\Psi(\lambda_i)| \le \max\{\wp^q_\Phi(\lambda_i), \wp^q_\Psi(\lambda_i)\}$, $0 \le |\eta^q_\Phi(\lambda_i) - \eta^q_\Psi(\lambda_i)| \le \max\{\eta^q_\Phi(\lambda_i), \eta^q_\Psi(\lambda_i)\}$ and $0 \le |\tau^q_\Phi(\lambda_i) - \tau^q_\Psi(\lambda_i)| \le \max\{\tau^q_\Phi(\lambda_i), \tau^q_\Psi(\lambda_i)\}$ for all $\lambda_i \in \Lambda$, it is deduced that

$$0 \le \frac{\sum\limits_{i=1}^{n} \omega_i (|\wp^q_\Phi(\lambda_i) - \wp^q_\Psi(\lambda_i)| + |\eta^q_\Phi(\lambda_i) - \eta^q_\Psi(\lambda_i)| + |\tau^q_\Phi(\lambda_i) - \tau^q_\Psi(\lambda_i)|)}{\sum\limits_{i=1}^{n} \omega_i (\max\{\wp^q_\Phi(\lambda_i), \wp^q_\Psi(\lambda_i)\} + \max\{\eta^q_\Phi(\lambda_i), \eta^q_\Psi(\lambda_i)\} + \max\{\tau^q_\Phi(\lambda_i), \tau^q_\Psi(\lambda_i)\})} \le 1 \tag{4.3.28}$$

On the other hand, it is clear that

$$0 \le 3\min\{|\wp^q_\Phi(\lambda_i) - \wp^q_\Psi(\lambda_i)|, |\eta^q_\Phi(\lambda_i) - \eta^q_\Psi(\lambda_i)|, |\tau^q_\Phi(\lambda_i) - \tau^q_\Psi(\lambda_i)|\} \le$$
$$|\wp^q_\Phi(\lambda_i) - \wp^q_\Psi(\lambda_i)| + |\eta^q_\Phi(\lambda_i) - \eta^q_\Psi(\lambda_i)| + |\tau^q_\Phi(\lambda_i) - \tau^q_\Psi(\lambda_i)|) \le 1$$

for all $\lambda_i \in \Lambda$ and so

$$0 \le \sqrt{\begin{pmatrix} (|\wp^q_\Phi(\lambda_i) - \wp^q_\Psi(\lambda_i)| + |\eta^q_\Phi(\lambda_i) - \eta^q_\Psi(\lambda_i)| + |\tau^q_\Phi(\lambda_i) - \tau^q_\Psi(\lambda_i)|)\times \\ 3\min\{|\wp^q_\Phi(\lambda_i) - \wp^q_\Psi(\lambda_i)|, |\eta^q_\Phi(\lambda_i) - \eta^q_\Psi(\lambda_i)|, |\tau^q_\Phi(\lambda_i) - \tau^q_\Psi(\lambda_i)|\} \end{pmatrix}} \le$$
$$\sqrt{\left((|\wp^q_\Phi(\lambda_i) - \wp^q_\Psi(\lambda_i)| + |\eta^q_\Phi(\lambda_i) - \eta^q_\Psi(\lambda_i)| + |\tau^q_\Phi(\lambda_i) - \tau^q_\Psi(\lambda_i)|)^2 \right)}$$

for all $\lambda_i \in \Lambda$. This implies that

$$0 \le \sum_{i=1}^{n} \omega_i \sqrt{\begin{pmatrix} (|\wp^q_\Phi(\lambda_i) - \wp^q_\Psi(\lambda_i)| + |\eta^q_\Phi(\lambda_i) - \eta^q_\Psi(\lambda_i)| + |\tau^q_\Phi(\lambda_i) - \tau^q_\Psi(\lambda_i)|)\times \\ 3\min\{|\wp^q_\Phi(\lambda_i) - \wp^q_\Psi(\lambda_i)|, |\eta^q_\Phi(\lambda_i) - \eta^q_\Psi(\lambda_i)|, |\tau^q_\Phi(\lambda_i) - \tau^q_\Psi(\lambda_i)|\} \end{pmatrix}} \le$$
$$\sum_{i=1}^{n} \omega_i \left(|\wp^q_\Phi(\lambda_i) - \wp^q_\Psi(\lambda_i)| + |\eta^q_\Phi(\lambda_i) - \eta^q_\Psi(\lambda_i)| + |\tau^q_\Phi(\lambda_i) - \tau^q_\Psi(\lambda_i)| \right) \tag{4.3.29}$$

From Eqs. (4.3.28) and (4.3.29), it is obtained as

$$0 \le \frac{\sum\limits_{i=1}^{n} \omega_i \sqrt{\begin{pmatrix} (|\wp^q_\Phi(\lambda_i) - \wp^q_\Psi(\lambda_i)| + |\eta^q_\Phi(\lambda_i) - \eta^q_\Psi(\lambda_i)| + |\tau^q_\Phi(\lambda_i) - \tau^q_\Psi(\lambda_i)|)\times \\ 3\min\{|\wp^q_\Phi(\lambda_i) - \wp^q_\Psi(\lambda_i)|, |\eta^q_\Phi(\lambda_i) - \eta^q_\Psi(\lambda_i)|, |\tau^q_\Phi(\lambda_i) - \tau^q_\Psi(\lambda_i)|\} \end{pmatrix}}}{\sum\limits_{i=1}^{n} \omega_i (\max\{\wp^q_\Phi(\lambda_i), \wp^q_\Psi(\lambda_i)\} + \max\{\eta^q_\Phi(\lambda_i), \eta^q_\Psi(\lambda_i)\} + \max\{\tau^q_\Phi(\lambda_i), \tau^q_\Psi(\lambda_i)\})} \le 1$$

i.e., $0 \leq {}^{\omega 10}\mathfrak{D}^{Soergel}_{q-ROFS}(\Phi, \Psi) \leq 1$.

2. This property is demonstrated as

$$
\begin{aligned}
{}^{\omega 10}\mathfrak{D}^{Soergel}_{q-ROFS}(\Phi, \Psi) &= \frac{\sum_{i=1}^{n} \omega_i \sqrt{\begin{pmatrix} (|\wp^q_\Phi(\lambda_i) - \wp^q_\Psi(\lambda_i)| + |\eta^q_\Phi(\lambda_i) - \eta^q_\Psi(\lambda_i)| + |\tau^q_\Phi(\lambda_i) - \tau^q_\Psi(\lambda_i)|) \times \\ 3\min\{|\wp^q_\Phi(\lambda_i) - \wp^q_\Psi(\lambda_i)|, |\eta^q_\Phi(\lambda_i) - \eta^q_\Psi(\lambda_i)|, |\tau^q_\Phi(\lambda_i) - \tau^q_\Psi(\lambda_i)|\} \end{pmatrix}}}{\sum_{i=1}^{n} \omega_i (\max\{\wp^q_\Phi(\lambda_i), \wp^q_\Psi(\lambda_i)\} + \max\{\eta^q_\Phi(\lambda_i), \eta^q_\Psi(\lambda_i)\} + \max\{\tau^q_\Phi(\lambda_i), \tau^q_\Psi(\lambda_i)\})} \\
&= \frac{\sum_{i=1}^{n} \omega_i \sqrt{\begin{pmatrix} (|\wp^q_\Psi(\lambda_i) - \wp^q_\Phi(\lambda_i)| + |\eta^q_\Psi(\lambda_i) - \eta^q_\Phi(\lambda_i)| + |\tau^q_\Psi(\lambda_i) - \tau^q_\Phi(\lambda_i)|) \times \\ 3\min\{|\wp^q_\Psi(\lambda_i) - \wp^q_\Phi(\lambda_i)|, |\eta^q_\Psi(\lambda_i) - \eta^q_\Phi(\lambda_i)|, |\tau^q_\Psi(\lambda_i) - \tau^q_\Phi(\lambda_i)|\} \end{pmatrix}}}{\sum_{i=1}^{n} \omega_i (\max\{\wp^q_\Psi(\lambda_i), \wp^q_\Phi(\lambda_i)\} + \max\{\eta^q_\Psi(\lambda_i), \eta^q_\Phi(\lambda_i)\} + \max\{\tau^q_\Psi(\lambda_i), \tau^q_\Phi(\lambda_i)\})} \\
&= {}^{\omega 10}\mathfrak{D}^{Soergel}_{q-ROFS}(\Psi, \Phi).
\end{aligned}
$$

3. Suppose that $\Phi = \Psi$, i.e., $\wp_\Phi(\lambda_i) = \wp_\Psi(\lambda_i)$ and $\eta_\Phi(\lambda_i) = \eta_\Psi(\lambda_i)$ for all $\lambda_i \in \Lambda$. Then, it is obvious that $\tau_\Phi(\lambda_i) = \tau_\Psi(\lambda_i)$ for all $\lambda_i \in \Lambda$. By Eq. (4.3.25), it is obtained that

$$
\begin{aligned}
{}^{\omega 10}\mathfrak{D}^{Soergel}_{q-ROFS}(\Phi, \Psi) &= \frac{\sum_{i=1}^{n} \omega_i \sqrt{\begin{pmatrix} (|\wp^q_\Phi(\lambda_i) - \wp^q_\Psi(\lambda_i)| + |\eta^q_\Phi(\lambda_i) - \eta^q_\Psi(\lambda_i)| + |\tau^q_\Phi(\lambda_i) - \tau^q_\Psi(\lambda_i)|) \times \\ 3\min\{|\wp^q_\Phi(\lambda_i) - \wp^q_\Psi(\lambda_i)|, |\eta^q_\Phi(\lambda_i) - \eta^q_\Psi(\lambda_i)|, |\tau^q_\Phi(\lambda_i) - \tau^q_\Psi(\lambda_i)|\} \end{pmatrix}}}{\sum_{i=1}^{n} \omega_i (\max\{\wp^q_\Phi(\lambda_i), \wp^q_\Psi(\lambda_i)\} + \max\{\eta^q_\Phi(\lambda_i), \eta^q_\Psi(\lambda_i)\} + \max\{\tau^q_\Phi(\lambda_i), \tau^q_\Psi(\lambda_i)\})} \\
&= \frac{\sum_{i=1}^{n} \omega_i \sqrt{\begin{pmatrix} (|\wp^q_\Phi(\lambda_i) - \wp^q_\Phi(\lambda_i)| + |\eta^q_\Phi(\lambda_i) - \eta^q_\Phi(\lambda_i)| + |\tau^q_\Phi(\lambda_i) - \tau^q_\Phi(\lambda_i)|) \times \\ 3\min\{|\wp^q_\Phi(\lambda_i) - \wp^q_\Phi(\lambda_i)|, |\eta^q_\Phi(\lambda_i) - \eta^q_\Phi(\lambda_i)|, |\tau^q_\Phi(\lambda_i) - \tau^q_\Phi(\lambda_i)|\} \end{pmatrix}}}{\sum_{i=1}^{n} \omega_i (\max\{\wp^q_\Phi(\lambda_i), \wp^q_\Phi(\lambda_i)\} + \max\{\eta^q_\Phi(\lambda_i), \eta^q_\Phi(\lambda_i)\} + \max\{\tau^q_\Phi(\lambda_i), \tau^q_\Phi(\lambda_i)\})} \\
&= 0.
\end{aligned}
$$

Note 2. Since type-κ WSoDM ($\kappa = 1, 2, \ldots, 12$) are bounded in the range $[0, 1]$, the corresponding similarities (i.e., type-κ WSoSM ${}^{\omega\kappa}\mathfrak{S}^{Soergel}_{q-ROFS}$) are achieved simply as

$$
{}^{\omega\kappa}\mathfrak{S}^{Soergel}_{q-ROFS}(\Phi, \Psi) = 1 - {}^{\omega\kappa}\mathfrak{D}^{Soergel}_{q-ROFS}(\Phi, \Psi) \tag{4.3.30}
$$

For Eq. (4.3.30), by considering Theorem 4.3.4, the following properties can be derived:

1. $0 \leq {}^{\omega\kappa}\mathfrak{S}^{Soergel}_{q-ROFS}(\Phi, \Psi) \leq 1$,
2. ${}^{\omega\kappa}\mathfrak{S}^{Soergel}_{q-ROFS}(\Phi, \Psi) = {}^{\omega\kappa}\mathfrak{S}^{Soergel}_{q-ROFS}(\Psi, \Phi)$,
3. ${}^{\omega\kappa}\mathfrak{S}^{Soergel}_{q-ROFS}(\Phi, \Psi) = 1$ if $\Phi = \Psi$.

Example 4.3.5 Consider the q-ROFSs Φ and Ψ in Example 4.3.2 (where $q = 3$). Assume that $\omega_1 = 0.3$ and $\omega_2 = 0.7$ are the weights of λ_1 and λ_2, respectively. By considering these weights, the type-κ WSoDM ($\kappa = 1, 2, \ldots, 12$) and the type-$\kappa$ WSoSM ($\kappa = 1, 2, \ldots, 12$) are obtained as in Table 4.4.

Table 4.4 Type-κ WSoDM and WSoSM between q-ROFSs Φ and Ψ

κ	1	2	3	4	5	6	7	8	9	10	11	12
${}^{\omega\kappa}\mathfrak{D}^{Soergel}_{q-ROFS}(\Phi, \Psi)$	0.73	0.751	0.538	0.55	0.699	0.697	0.392	0.432	0.273	0.304	0.368	0.404
${}^{\omega\kappa}\mathfrak{S}^{Soergel}_{q-ROFS}(\Phi, \Psi)$	0.27	0.249	0.462	0.45	0.301	0.303	0.608	0.568	0.727	0.696	0.632	0.596

4.3.3 The Validation/Efficiency of SoDMs and SoSMs for q-ROFSs

To display the validity/efficiency of proposed SoDMs and SoSMs, consider 6 sorts of instances Φ and Ψ as in Table 4.5. Assume that $q = 7$. Utilize the existing Peng et al. [54]'s distance distance measures $\mathfrak{D}_\kappa$ $(\kappa = 3, 4, \ldots, 12)$ (Table 4.1) towards the proposed SoDMs ${}^{\kappa}\mathfrak{D}^{Soergel}_{q-ROFS}$ $(\kappa = 7, 8, \ldots, 12)$ between Φ and Ψ. On the other hand, make use of existing Wang et al. [72]'s similarity measures (Table 4.2) $\mathfrak{S}_\kappa$ $(\kappa = 1, 2, 3, 4, 7, 8)$ towards the proposed SoSMs ${}^{\kappa}\mathfrak{S}^{Soergel}_{q-ROFS}$ $(\kappa = 7, 8, \ldots, 12)$ between Φ and Ψ. The measure values are supplied in Table 4.5. From Table 4.5, it's far found that the measure values for Instance 2 and Instance 4 are identical with admire to the existing Peng et al. [54]'s distance measures $\mathfrak{D}_\kappa$ $(\kappa = 3, 4, \ldots, 12)$ (Table 4.1), the measure values for Instance 5 and Instance 6 are identical with admire to the existing Peng et al. [54]'s distance measures $\mathfrak{D}_\kappa$ $(\kappa = 4, 5, \ldots, 10)$ (Table 4.1), and also, the measure values for all the Instances are identical as 1 with admire to the existing Peng et al. [54]'s distance measures $\mathfrak{D}_\kappa$ $(\kappa = 7, 8, \ldots, 10)$ (Table 4.1). It leads to produce an inconsistent result for all the Peng et al. [54]'s distance measures $\mathfrak{D}_\kappa$ $(\kappa = 3, 4, \ldots, 12)$ (Table 4.1). However, the proposed SoDMs ${}^{\kappa}\mathfrak{D}^{Soergel}_{q-ROFS}$ $(\kappa = 7, 8, \ldots, 12)$ produce the consistent results through the different measure values for all the Instances. On the same way, from Table 4.5, it is understand that the measure values for existing Wang et al. [72]'s similarity measures (Table 4.2) $\mathfrak{S}_\kappa$ $(\kappa = 1, 2, 3, 4, 7, 8)$ are inconsistent. However, the proposed SoSMs ${}^{\kappa}\mathfrak{D}^{Soergel}_{q-ROFS}$ $(\kappa = 7, 8, \ldots, 12)$ produce the consistent results for all Instances. It suggests that the proposed SoDMs and SoSMs are advanced and effective than the existing Peng et al. [54]'s distance measures $\mathfrak{D}_\kappa$ $(\kappa = 3, 4, \ldots, 12)$ (Table 4.1) and Wang et al. [72]'s similarity measures (Table 4.2) $\mathfrak{S}_\kappa$ $(\kappa = 1, 2, 3, 4, 7, 8)$.

4.4 Applications of SoDMs

This section focuses on the decision-making method using the proposed SoDMs and WSoDMs and their applications under the q-rung orthopair fuzzy environment.

Table 4.5 Comparison of proposed SoDMs and SoSMs with existing measures ($q = 7$)

		Instance 1	Instance 2	Instance 3	Instance 4	Instance 5	Instance 6
	Φ	$\{(\lambda, \langle 0.8, 0\rangle)\}$	$\{(\lambda, \langle 0.5, 0\rangle)\}$	$\{(\lambda, \langle 0, 0\rangle)\}$	$\{(\lambda, \langle 0, 0\rangle)\}$	$\{(\lambda, \langle 0.9, 0.9\rangle)\}$	$\{(\lambda, \langle 0.7, 0.9\rangle)\}$
	Ψ	$\{(\lambda, \langle 0, 0.6\rangle)\}$	$\{(\lambda, \langle 0, 0.9\rangle)\}$	$\{(\lambda, \langle 0.7, 0.6\rangle)\}$	$\{(\lambda, \langle 0.5, 0.9\rangle)\}$	$\{(\lambda, \langle 0, 0\rangle)\}$	$\{(\lambda, \langle 0, 0\rangle)\}$
Peng et al. [54]'s distance measure (Table 4.1)	$\mathfrak{D}_3(\Phi, \Psi)$	0.1642848	0.3606758	0.0687641	0.3606758	0.4782969	0.3793113
	$\mathfrak{D}_4(\Phi, \Psi)$	0.2097152	0.4782969	0.0823543	0.4782969	0.4782969	0.4782969
	$\mathfrak{D}_5(\Phi, \Psi)$	0.3467183	0.6470918	0.1521762	0.6470918	0.6470918	0.6470918
	$\mathfrak{D}_6(\Phi, \Psi)$	0.3467183	0.6470918	0.1521762	0.6470918	0.6470918	0.6470918
	$\mathfrak{D}_7(\Phi, \Psi)$	1	1	1	1	1	1
	$\mathfrak{D}_8(\Phi, \Psi)$	1	1	1	1	1	1
	$\mathfrak{D}_9(\Phi, \Psi)$	1	1	1	1	1	1
	$\mathfrak{D}_{10}(\Phi, \Psi)$	1	1	1	1	1	1
	$\mathfrak{D}_{11}(\Phi, \Psi)$	0.1964998	0.4823411	0.1019517	0.4823411	0.6470918	0.5179923
	$\mathfrak{D}_{12}(\Phi, \Psi)$	0.1964998	0.4823411	0.1019517	0.4823411	0.6470918	0.5179923
Proposed SoDM	${}^{7}\mathfrak{D}^{Soergel}_{q-ROFS}(\Phi, \Psi)$	0.3467183	0.6470918	0.1987627	0.6542041	0.9778154	0.7184837
	${}^{8}\mathfrak{D}^{Soergel}_{q-ROFS}(\Phi, \Psi)$	0.3467183	0.6470918	0.1987627	0.6542041	0.9778154	0.7184837
	${}^{9}\mathfrak{D}^{Soergel}_{q-ROFS}(\Phi, \Psi)$	0.1551447	0.1012879	0.1226106	0.101575	0.846813	0.3372557
	${}^{10}\mathfrak{D}^{Soergel}_{q-ROFS}(\Phi, \Psi)$	0.1551447	0.1012879	0.1226106	0.101575	0.846813	0.3372557
	${}^{11}\mathfrak{D}^{Soergel}_{q-ROFS}(\Phi, \Psi)$	0.3324736	0.5888054	0.1945862	0.5945233	0.8602128	0.6524311
	${}^{12}\mathfrak{D}^{Soergel}_{q-ROFS}(\Phi, \Psi)$	0.3324736	0.5888054	0.1945862	0.5945233	0.8602128	0.6524311
Wang et al. [72]'s similarity measure (Table 4.2)	$\mathfrak{S}_{13}(\Phi, \Psi)$	0	0	NaN	NaN	NaN	NaN
	$\mathfrak{S}_{14}(\Phi, \Psi)$	0.5	0.5	NaN	NaN	NaN	NaN
	$\mathfrak{S}_{15}(\Phi, \Psi)$	0.9462302	0.7307973	0.9916444	0.7307973	0.7307973	0.7307973
	$\mathfrak{S}_{16}(\Phi, \Psi)$	0.9826229	0.9279994	0.9962468	0.9279994	0.7307973	0.9046092
	$\mathfrak{S}_{17}(\Phi, \Psi)$	0.7149485	0.4343277	0.8783377	0.4343277	0.4343277	0.4343277
	$\mathfrak{S}_{18}(\Phi, \Psi)$	0.8287886	0.6760978	0.9168827	0.6760978	0.4343277	0.6342606

(continued)

Table 4.5 (continued)

		Instance 1	Instance 2	Instance 3	Instance 4	Instance 5	Instance 6
Proposed SoSMs	${}^{7}\mathfrak{S}_{q-ROFS}^{Soergel}(\Phi, \Psi)$	0.6532817	0.3529082	0.8012373	0.3457959	0.0221846	0.2815163
	${}^{8}\mathfrak{S}_{q-ROFS}^{Soergel}(\Phi, \Psi)$	0.6532817	0.3529082	0.8012373	0.3457959	0.0221846	0.2815163
	${}^{9}\mathfrak{S}_{q-ROFS}^{Soergel}(\Phi, \Psi)$	0.8448553	0.8987121	0.8773894	0.898425	0.153187	0.6627443
	${}^{10}\mathfrak{S}_{q-ROFS}^{Soergel}(\Phi, \Psi)$	0.8448553	0.8987121	0.8773894	0.898425	0.153187	0.6627443
	${}^{11}\mathfrak{S}_{q-ROFS}^{Soergel}(\Phi, \Psi)$	0.6675264	0.4111946	0.8054138	0.4054767	0.1397872	0.3475689
	${}^{12}\mathfrak{S}_{q-ROFS}^{Soergel}(\Phi, \Psi)$	0.6675264	0.4111946	0.8054138	0.4054767	0.1397872	0.3475689

In the table, NaN stands for Not a Number

4.4.1 Proposed Decision-Making Method

The following notations or assumptions are used to present the multi-attribute decision-making problems for their evaluation with a q-rung orthopair fuzzy environment. Let $\mathfrak{A} = \{\mathcal{A}_1, \mathcal{A}_2, \ldots, \mathcal{A}_n\}$ be the set of n different alternatives that should be evaluated under the set of m different attributes (parameters, criteria) $\mathfrak{P} = \{\mathcal{P}_1, \mathcal{P}_2, \ldots, \mathcal{P}_m\}$. The weight vector information corresponding to each attribute (parameter, criterion) is indicated by $\omega = (\omega_1, \omega_2, \ldots, \omega_m)^T$ such that $\omega_i \in (0, 1]$ and $\sum_{j=1}^{m} \omega_j = 1$. Suppose that these alternatives $\mathcal{A}_i$ $(i = 1, 2, \ldots, n)$ are evaluated by an expert who will take full responsibility for the whole process and present their preferences under the q-rung orthopair fuzzy environment. These preference values can be represented in the form of the decision matrix $\mathcal{M} = (\mu_{ij})_{n \times m}$ where $\mu_{ij} = \langle \wp_{ij}, \eta_{ij} \rangle$ denotes the priority values (membership degree and non-membership degree) of alternative $\mathcal{A}_i$ concerning the parameter $\mathcal{P}_j$ given by expert such that $\wp_{ij}, \eta_{ij} \in [0, 1]$ and $\wp_{ij}^q + \eta_{ij}^q \leq 1$ for $q \geq 1$. Then, the proposed method is outlined in the various steps described below.

Step 1. Create the q-rung orthopair fuzzy decision matrix $\mathcal{M} = (\mu_{ij})_{n \times m}$ as follows:

$$\mathcal{M} = (\mu_{ij})_{n \times m} = \begin{array}{c} \\ \mathcal{A}_1 \\ \mathcal{A}_2 \\ . \\ . \\ . \\ \mathcal{A}_n \end{array} \begin{array}{c} \begin{array}{cccc} \mathcal{P}_1(\omega_1) & \mathcal{P}_2(\omega_2) & \ldots & \mathcal{P}_m(\omega_m) \end{array} \\ \begin{bmatrix} \langle \wp_{11}, \eta_{11} \rangle & \langle \wp_{12}, \eta_{12} \rangle & \ldots & \langle \wp_{1m}, \eta_{1m} \rangle \\ \langle \wp_{21}, \eta_{21} \rangle & \langle \wp_{22}, \eta_{22} \rangle & \ldots & \langle \wp_{2m}, \eta_{2m} \rangle \\ . & . & . & . \\ . & . & . & . \\ . & . & . & . \\ \langle \wp_{n1}, \eta_{n1} \rangle & \langle \wp_{n2}, \eta_{n2} \rangle & \ldots & \langle \wp_{nm}, \eta_{nm} \rangle \end{bmatrix} \end{array}$$

Step 2. Determine the measure values between the alternatives $\mathcal{A}_i$ $(i = 1, 2, \ldots, n)$ and the ideal alternative $\mathcal{B}$ by using the proposed type-κ SoDMs or type-κ SoSMs (type-κ WSoDMs or type-κ WSoSMs if the parameters have weights).

Step 3. Rank all the alternatives based on an index derived from α=arg max$\{{}^{\kappa}\mathfrak{D}_{q-ROFS}^{Soergel}\}$ or α=arg max$\{{}^{\omega\kappa}\mathfrak{D}_{q-ROFS}^{Soergel}\}$ (if the parameters have weights). According to the phenomena of multi-attribute decision-making problem

- The higher the measure index, the better the alternative $\mathcal{A}_i$ $(i = 1, 2, \ldots, n)$.
- The lower the measure index, the better the alternative $\mathcal{A}_i$ $(i = 1, 2, \ldots, n)$. (Put another way, if rank all the alternatives based on an index as obtained from $\beta = \arg\max\{{}^{\kappa}\mathfrak{S}_{q-ROFS}^{Soergel}\}$ or $\beta = \arg\max\{{}^{\omega\kappa}\mathfrak{S}_{q-ROFS}^{Soergel}\}$ then the higher the measure index, the better the alternative $\mathcal{A}_i$.)

Moreover, a flowchart is drawn in Fig. 4.2 for a visual understanding of the proposed method.

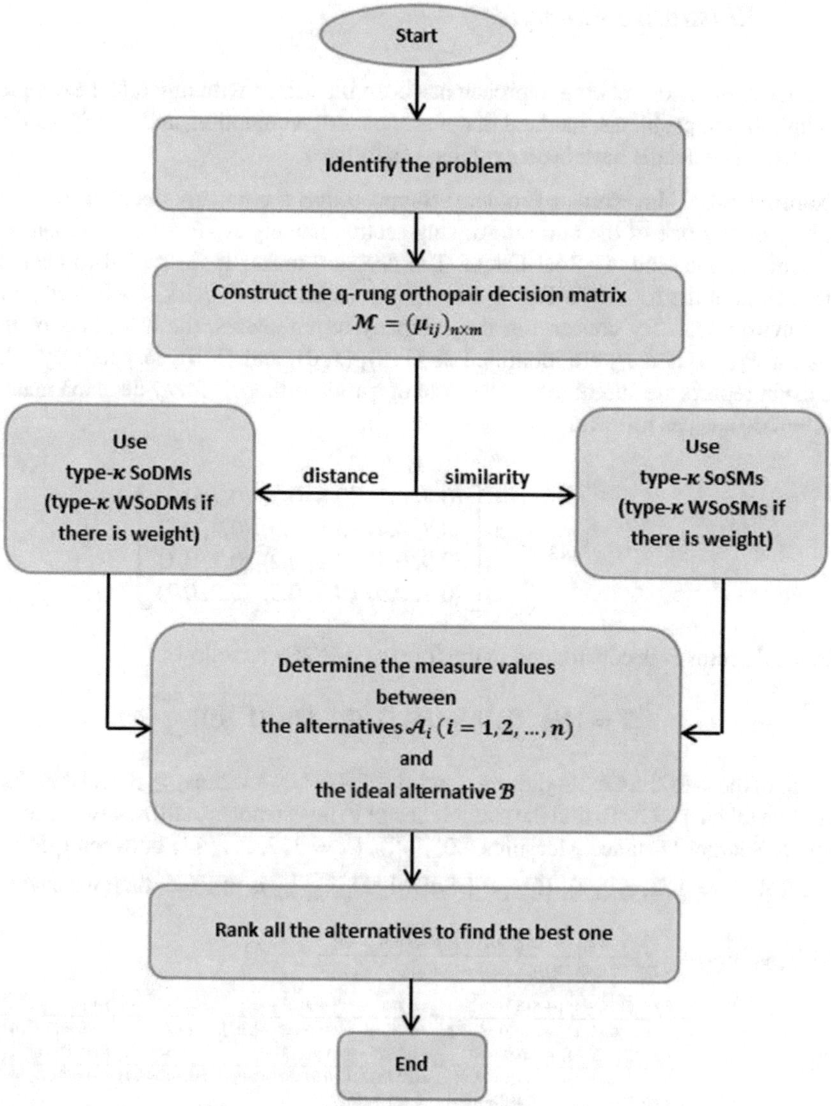

Fig. 4.2 Flowchart of the proposed decision-making method

4.4.2 Illustrative Examples

The above decision-making approach has been illustrated with numerical examples on investment problems, medical diagnosis, pattern recognition, and benchmarking problem. The details have been explained as follows.

Example 4.4.1 (Investment Problem) Suppose that a company decided to invest money in any one of the current growing sectors namely A_1-Power, A_2-Financial, A_3-Automobile, and A_4-Real Estate. The decision maker is instructed to analyze the sectors in the form of q-ROFS concerning the criteria P_1-Risk, P_2-Growth, and P_3-Environment. By concerning the company requirements, the q-ROFNs of the criteria P_1, P_2, and P_3 are identified as $\langle 1, 0\rangle$, $\langle 1, 0\rangle$, and $\langle 1, 0\rangle$, respectively. The decision reports are submitted in the form of q-rung orthopair fuzzy decision matrix $\mathcal{M} = (\mu_{ij})_{n\times m}$ as follows:

$$\mathcal{M} = (\mu_{ij})_{4\times 3} = \begin{array}{c} \\ A_1 \\ A_2 \\ A_3 \\ A_4 \end{array}\begin{array}{c} \begin{array}{ccc} P_1 & P_2 & P_3 \end{array} \\ \begin{bmatrix} \langle 0.7, 0.3\rangle & \langle 0.8, 0.2\rangle & \langle 0.7, 0.3\rangle \\ \langle 0.8, 0.4\rangle & \langle 0.5, 0.9\rangle & \langle 0.4, 0.4\rangle \\ \langle 0.9, 0.1\rangle & \langle 0.8, 0.2\rangle & \langle 0.7, 0.1\rangle \\ \langle 0.1, 0.9\rangle & \langle 0.5, 0.5\rangle & \langle 0.2, 0.7\rangle \end{bmatrix} \end{array}$$

An ideal sector $\mathcal{B}$ is constructed in the form of q-ROFS as follows.

$$\mathcal{B} = \{(P_1, \langle 1, 0\rangle), (P_2, \langle 1, 0\rangle), (P_3, \langle 1, 0\rangle)\}$$

For all of the q-ROFSs A_1, A_2, A_3, A_4, and $\mathcal{B}$, $\wp^q + \eta^q \leq 1$ when $q \geq 3$, and therefore it can be taken $q = 3$. To find the suitable sector to invest money, utilize any one of the type-κ Soergel Distance Measures ${}^{\kappa}\mathfrak{D}^{Soergel}_{q-ROFS}(\kappa = 1, 2, \ldots, 12)$ between q-ROFS $\mathcal{B}$ and A_i $(i = 1, 2, \ldots, 4)$. If type-1 SoDM ${}^{1}\mathfrak{D}^{Soergel}_{q-ROFS}$ is utilized, then we have

$$\begin{aligned}
{}^{1}\mathfrak{D}^{Soergel}_{q-ROFS}(A_1, \mathcal{B}) &= \frac{1}{n}\sum_{i=1}^{n} \frac{|\wp^q_{\Phi}(\lambda_i) - \wp^q_{\Psi}(\lambda_i)| + |\eta^q_{\Phi}(\lambda_i) - \eta^q_{\Psi}(\lambda_i)|}{\max\{\wp^q_{\Phi}(\lambda_i), \wp^q_{\Psi}(\lambda_i)\} + \max\{\eta^q_{\Phi}(\lambda_i), \eta^q_{\Psi}(\lambda_i)\}} \\
&= \frac{1}{3}\Big(\frac{|0.7^3 - 1^3| + |0.3^3 - 0^3|}{\max\{0.7^3, 1^3\} + \max\{0.3^3, 0^3\}} + \frac{|0.8^3 - 1^3| + |0.2^3 - 0^3|}{\max\{0.8^3, 1^3\} + \max\{0.2^3, 0^3\}} + \frac{|0.7^3 - 1^3| + |0.3^3 - 0^3|}{\max\{0.7^3, 1^3\} + \max\{0.3^3, 0^3\}}\Big) \\
&= \frac{1}{3}\Big(\frac{|0.343 - 1| + |0.027 - 0|}{\max\{0.343, 1\} + \max\{0.027, 0\}} + \frac{|0.512 - 1| + |0.008 - 0|}{\max\{0.512, 1\} + \max\{0.008, 0\}} + \frac{|0.343 - 1| + |0.027 - 0|}{\max\{0.343, 1\} + \max\{0.027, 0\}}\Big) \\
&= \frac{1}{3}\Big(\frac{0.657 + 0.027}{1 + 0.027} + \frac{0.488 + 0.008}{1 + 0.008} + \frac{0.657 + 0.027}{1 + 0.027}\Big) \\
&= \frac{1}{3}\Big(\frac{0.684}{1.027} + \frac{0.496}{1.008} + \frac{0.684}{1.027}\Big) \\
&= \frac{1}{3}\Big(0.6660175 + 0.4920635 + 0.6660175\Big) \\
&= 0.6080328
\end{aligned}$$

Similarly, the remaining type-1 Soergel distance indices are calculated as follows. ${}^{1}\mathfrak{D}^{Soergel}_{q-ROFS}(A_2, \mathcal{B}) = 0.7954502$, ${}^{1}\mathfrak{D}^{Soergel}_{q-ROFS}(A_3, \mathcal{B}) = 0.4737115$ and ${}^{1}\mathfrak{D}^{Soergel}_{q-ROFS}(A_4, \mathcal{B}) = 0.9607846$.

Then, according to the above type-1 Soergel distance index, the ranking order of sectors is $A_4 \succ_{({}^1\mathfrak{D}^{Soergel})} A_2 \succ_{({}^1\mathfrak{D}^{Soergel})} A_1 \succ_{({}^1\mathfrak{D}^{Soergel})} A_3$. Here we consider the higher the distance measure coefficient, the better the alternative. It means that the Real Estate (A_4) sector is closely satisfying the company requirements among the other three sectors. Hence, the company may invest money in the Real Estate (A_4) sector by concerning the type-1 SoDM for the q-ROF information. It is clear that the ranking order of sectors is the same as $A_4 \succ A_2 \succ A_1 \succ A_3$ for all remaining proposed type-κ SoDMs ${}^{\kappa}\mathfrak{D}^{Soergel}_{q-ROFS}(\kappa = 2, \ldots, 12)$ and existing Peng et al. [54]'s distance approaches $\mathfrak{D}_\kappa(\kappa = 1, 2, 3, 4, 5, 6, 9, 10, 11, 12)$ (Table 4.1). However, Peng et al. [54]'s distance approaches $\mathfrak{D}_\kappa(\kappa = 7, 8)$ (Table 4.1) are inconsistent. It proves that the proposed type-κ SoDMs ${}^{\kappa}\mathfrak{D}^{Soergel}_{q-ROFS}(\kappa = 2, \ldots, 12)$ are effective in compare with existing Peng et al. [54]'s distance approaches $\mathfrak{D}_\kappa(\kappa = 2, \ldots, 12)$ (Table 4.1). It is shown in Table 4.6.

Example 4.4.2 (Medical Diagnosis Problem) Consider the following set of diagnoses

$$D = \{D_1\,(COVID - 19),\; D_2\,(Stomach\ problem),\; D_3\,(Viral\ fever),\; D_4\,(Chest\ problem)\}$$

and a set of symptoms

$$S = \{S_1\,(Cough),\; S_2\,(Headache),\; S_3\,(Temperature),\; S_4\,(Stomach\ pain),\; S_5\,(Chest\ pain)\}$$

which are presented in the decision matrix form of q-ROFSs as follows.

$$\mathcal{M} = (\mu_{ij})_{4\times5} = \begin{array}{c} \\ D_1 \\ D_2 \\ D_3 \\ D_4 \end{array} \begin{array}{c} \begin{array}{ccccc} S_1 & S_2 & S_3 & S_4 & S_5 \end{array} \\ \begin{bmatrix} \langle 0.8, 0.2\rangle & \langle 0.9, 0.3\rangle & \langle 0.9, 0.1\rangle & \langle 0.8, 0.2\rangle & \langle 0.7, 0.1\rangle \\ \langle 0.6, 0.9\rangle & \langle 0.7, 0.9\rangle & \langle 0.6, 0.9\rangle & \langle 0.1, 0.2\rangle & \langle 0.5, 0.2\rangle \\ \langle 0.5, 0.2\rangle & \langle 0.7, 0.5\rangle & \langle 0.9, 0.5\rangle & \langle 0.4, 0.7\rangle & \langle 0.6, 0.9\rangle \\ \langle 0.5, 0.8\rangle & \langle 0.6, 0.7\rangle & \langle 0.6, 0.4\rangle & \langle 0.3, 0.5\rangle & \langle 1, 0\rangle \end{bmatrix} \end{array}$$

Assume that a patient P was evaluated by an expert (doctor) to determine which diseases are the most affected by the person. For this purpose, they treated this patient as a reference set and achieved their preferences for all the symptoms in terms of q-ROFNs, and presented with the following set:

$$P = \{(S_1, \langle 0.7, 0.1\rangle), (S_2, \langle 0.8, 0.4\rangle), (S_3, \langle 0.8, 0\rangle), (S_4, \langle 0.2, 0.9\rangle), (S_5, \langle 0.8, 0.8\rangle)\}.$$

For all of the q-ROFSs D_1, D_2, D_3, D_4, and P, $\wp^q + \eta^q \leq 1$ when $q \geq 4$, and therefore it can be taken $q = 4$.

Now, the target is to classify the patient P in one of the D_1, D_2, D_3 and D_4. For this, the type-1 WSoDM as given in Eq. (4.3.16) is used for the weight vector $\omega = (0.3, 0.1, 0.6, 0, 0)^T$, then we have

Table 4.6 Comparison of proposed SoDMs with Peng et al. [54]'s distance approaches (Table 4.1) for Example 4.4.1

Distance measures	κ	A_1	A_2	A_3	A_4	Ranking order
Proposed SoDM ${}^{\kappa}\mathfrak{D}_{q-ROFS}^{Soergel}$	1	0.6080328	0.7954502	0.4737115	0.9607846	$A_4 \succ A_2 \succ A_1 \succ A_3$
	2	0.6087524	0.8182525	0.4737542	0.9680724	$A_4 \succ A_2 \succ A_1 \succ A_3$
	3	0.1542156	0.4901833	0.0493061	0.6916801	$A_4 \succ A_2 \succ A_1 \succ A_3$
	4	0.1546241	0.5581648	0.0493969	0.7253393	$A_4 \succ A_2 \succ A_1 \succ A_3$
	5	0.5436049	0.6962704	0.4282306	0.832637	$A_4 \succ A_2 \succ A_1 \succ A_3$
	6	0.5442035	0.7148405	0.4282746	0.8379351	$A_4 \succ A_2 \succ A_1 \succ A_3$
	7	0.7473043	0.8520632	0.6251164	0.9762723	$A_4 \succ A_2 \succ A_1 \succ A_3$
	8	0.7505206	0.8677109	0.6413043	0.9771565	$A_4 \succ A_2 \succ A_1 \succ A_3$
	9	0.1655434	0.3558406	0.0574908	0.5952421	$A_4 \succ A_2 \succ A_1 \succ A_3$
	10	0.1677496	0.3600505	0.0580068	0.5985461	$A_4 \succ A_2 \succ A_1 \succ A_3$
	11	0.668208	0.7499372	0.5664274	0.8517359	$A_4 \succ A_2 \succ A_1 \succ A_3$
	12	0.6707522	0.7617507	0.5796736	0.8524914	$A_4 \succ A_2 \succ A_1 \succ A_3$
Peng et al. [54]'s distance approach $\mathfrak{D}_{\kappa}$ (Table 4.1)	1	0.6006667	0.7663333	0.472	0.9553333	$A_4 \succ A_2 \succ A_1 \succ A_3$
	2	0.3106667	0.526	0.2376667	0.6771667	$A_4 \succ A_2 \succ A_1 \succ A_3$
	3	0.4556667	0.6461667	0.3548333	0.81625	$A_4 \succ A_2 \succ A_1 \succ A_3$
	4	0.6006667	0.7663333	0.472	0.9553333	$A_4 \succ A_2 \succ A_1 \succ A_3$
	5	0.7473043	0.8520632	0.6251164	0.9762723	$A_4 \succ A_2 \succ A_1 \succ A_3$
	6	0.7505206	0.8677109	0.6413043	0.9771565	$A_4 \succ A_2 \succ A_1 \succ A_3$
	7	1	1	1	1	$A_4 = A_2 = A_1 = A_3$
	8	1	1	1	1	$A_4 = A_2 = A_1 = A_3$
	9	0.6080328	0.7954502	0.4737115	0.9607846	$A_4 \succ A_2 \succ A_1 \succ A_3$
	10	0.6087524	0.8182525	0.4737542	0.9680724	$A_4 \succ A_2 \succ A_1 \succ A_3$
	11	0.3106667	0.526	0.2376667	0.6771667	$A_4 \succ A_2 \succ A_1 \succ A_3$
	12	0.3106667	0.526	0.2376667	0.6771667	$A_4 \succ A_2 \succ A_1 \succ A_3$

$$
\begin{aligned}
{}^{\omega 1}\mathfrak{D}^{Soergel}_{q-ROFS}(D_1, P) &= \sum_{i=1}^{n} \omega_i \frac{|\wp^q_\Phi(\lambda_i) - \wp^q_\Psi(\lambda_i)| + |\eta^q_\Phi(\lambda_i) - \eta^q_\Psi(\lambda_i)|}{\max\{\wp^q_\Phi(\lambda_i), \wp^q_\Psi(\lambda_i)\} + \max\{\eta^q_\Phi(\lambda_i), \eta^q_\Psi(\lambda_i)\}} \\
&= (0.3)\frac{|0.8^4 - 0.7^4| + |0.2^4 - 0.1^4|}{\max\{0.8^4, 0.7^4\} + \max\{0.2^4, 0.1^4\}} + (0.1)\frac{|0.9^4 - 0.8^4| + |0.3^4 - 0.4^4|}{\max\{0.9^4, 0.8^4\} + \max\{0.3^4, 0.4^4\}} \\
&+ (0.6)\frac{|0.9^4 - 0.8^4| + |0.1^4 - 0^4|}{\max\{0.9^4, 0.8^4\} + \max\{0.1^4, 0^4\}} \\
&= (0.3)\frac{|0.4096 - 0.2401| + |0.0016 - 0.0001|}{\max\{0.4096, 0.2401\} + \max\{0.0016, 0.0001\}} + (0.1)\frac{|0.6561 - 0.4096| + |0.0081 - 0.0256|}{\max\{0.6561, 0.4096\} + \max\{0.0081, 0.0256\}} \\
&+ (0.6)\frac{|0.6561 - 0.4096| + |0.0001 - 0|}{\max\{0.6561, 0.4096\} + \max\{0.0001, 0\}} \\
&= (0.3)\Big(\frac{0.1695 + 0.0015}{0.4096 + 0.0016}\Big) + (0.1)\Big(\frac{0.2465 + 0.0175}{0.6561 + 0.0256}\Big) + (0.6)\Big(\frac{0.2465 + 0.0001}{0.6561 + 0.0001}\Big) \\
&= (0.3)\Big(\frac{0.171}{0.4112}\Big) + (0.1)\Big(\frac{0.264}{0.6817}\Big) + (0.6)\Big(\frac{0.2466}{0.6562}\Big) \\
&= 0.1247568 + 0.0387267 + 0.2254800 \\
&= 0.3889635
\end{aligned}
$$

Similarly, the remaining all type-1 weighted Soergel distance indices are calculated as follows.

${}^{\omega 1}\mathfrak{D}^{Soergel}_{q-ROFS}(D_2, P) = 0.8586853$, ${}^{\omega 1}\mathfrak{D}^{Soergel}_{q-ROFS}(D_3, P) = 0.5240216$ and ${}^{\omega 1}\mathfrak{D}^{Soergel}_{q-ROFS}(D_4, P) = 0.7685299$.

Then, according to the above type-1 weighted Soergel distance index, the ranking order of diseases is $D_2 \succ_{({}^{\omega 1}\mathfrak{D}^{Soergel})} D_4 \succ_{({}^{\omega 1}\mathfrak{D}^{Soergel})} D_3 \succ_{({}^{\omega 1}\mathfrak{D}^{Soergel})} D_1$. Here we consider the lower the distance measure value, the better the alternative. That is, according to the corresponding weighted similarity index, the ranking order is $D_1 \succ_{({}^{\omega 1}\mathfrak{S}^{Soergel})} D_3 \succ_{({}^{\omega 1}\mathfrak{S}^{Soergel})} D_4 \succ_{({}^{\omega 1}\mathfrak{S}^{Soergel})} D_2$. So the patient P suffers from D_1 (COVID-19) disease. On the other hand, if the weighted similarity indices are computed by using the determined weights of symptoms $\omega = (0.3, 0.1, 0.6, 0, 0)^T$ and $\omega = (0.02, 0.4, 0.2, 0.3, 0.08)^T$ then their corresponding results are in Table 4.7. From Table 4.7, it is observed that the ranking order of proposed WSoSMs is changed due to the change of weight vector. Moreover, a radar chart (see Fig. 4.3) is drawn for a visual understanding of weight vectors reaction on WSoSMs.

Example 4.4.3 (Pattern Recognition Problem) Given classes of building material, these known patterns are represented in the decision matrix form of q-ROFSs K_1, K_2, and K_3 in the feature space $\mathfrak{X} = \{X_1, X_2, X_3\}$ as below.

$$
\mathcal{M} = (\mu_{ij})_{3\times 3} = \begin{array}{c} \\ K_1 \\ K_2 \\ K_3 \end{array} \begin{array}{c} \begin{array}{ccc} X_1 & X_2 & X_3 \end{array} \\ \begin{bmatrix} \langle 0.6, 0.2\rangle & \langle 0.5, 0.3\rangle & \langle 0.7, 0.9\rangle \\ \langle 0.7, 0.2\rangle & \langle 0.8, 0.1\rangle & \langle 0.9, 0.2\rangle \\ \langle 0.6, 0.3\rangle & \langle 0.8, 0.1\rangle & \langle 0.9, 0.3\rangle \end{bmatrix} \end{array}
$$

Given another kind of building material (i.e., unknown q-ROFS pattern) as

$$
U = \{(X_1, \langle 0.8, 0.7\rangle), (X_2, \langle 0.9, 0.9\rangle), (X_3, \langle 0.4, 0.5\rangle)\},
$$

to which field does this kind of building material U most probably belong?

For all of the q-ROFSs K_1, K_2, K_3, and U, $\wp^q + \eta^q \leq 1$ when $q \geq 7$, and therefore

Table 4.7 Results of WSoSMs according to different weight vectors for Example 4.4.2

Weight vectors	k	Outputs of proposed WSoSMs				Ranking order
		D_1	D_2	D_3	D_4	
$\omega = (0.3, 0.1, 0.6, 0, 0)^T$	1	0.6110364	0.1413147	0.4759784	0.2314701	$D_1 \succ D_3 \succ D_4 \succ D_2$
	2	0.6144212	0.141144	0.5284454	0.2150864	$D_1 \succ D_3 \succ D_4 \succ D_2$
	3	0.9629534	0.4057243	0.7809904	0.5457902	$D_1 \succ D_3 \succ D_4 \succ D_2$
	4	0.9647654	0.3989586	0.7509161	0.5045899	$D_1 \succ D_3 \succ D_4 \succ D_2$
	5	0.629641	0.2269054	0.4978864	0.2771985	$D_1 \succ D_3 \succ D_4 \succ D_2$
	6	0.6333916	0.2266265	0.548792	0.2626142	$D_1 \succ D_3 \succ D_4 \succ D_2$
	7	0.6354502	0.2095745	0.5972545	0.5194329	$D_1 \succ D_3 \succ D_4 \succ D_2$
	8	0.6341071	0.2095482	0.5928259	0.5164727	$D_1 \succ D_3 \succ D_4 \succ D_2$
	9	0.9711884	0.4510473	0.8172212	0.7362747	$D_1 \succ D_3 \succ D_4 \succ D_2$
	10	0.9712762	0.450955	0.8123584	0.7302378	$D_1 \succ D_3 \succ D_4 \succ D_2$
	11	0.6533551	0.284818	0.6172822	0.5469793	$D_1 \succ D_3 \succ D_4 \succ D_2$
	12	0.6521376	0.2847958	0.6132285	0.5442915	$D_1 \succ D_3 \succ D_4 \succ D_2$
$\omega = (0.02, 0.4, 0.2, 0.3, 0.08)^T$	1	0.4059739	0.1339796	0.4911443	0.2092361	$D_3 \succ D_1 \succ D_4 \succ D_2$
	2	0.343596	0.152612	0.4864755	0.2102116	$D_3 \succ D_1 \succ D_4 \succ D_2$
	3	0.6349722	0.5685321	0.7366125	0.5536272	$D_3 \succ D_1 \succ D_2 \succ D_4$
	4	0.554686	0.529908	0.7233657	0.5278006	$D_3 \succ D_1 \succ D_2 \succ D_4$
	5	0.4523804	0.2221276	0.5186734	0.268863	$D_3 \succ D_1 \succ D_4 \succ D_2$
	6	0.3983995	0.2376021	0.5156572	0.2722165	$D_3 \succ D_1 \succ D_4 \succ D_2$
	7	0.4608059	0.2098147	0.572326	0.4429885	$D_3 \succ D_1 \succ D_4 \succ D_2$
	8	0.4347099	0.2093741	0.5620704	0.4270404	$D_3 \succ D_1 \succ D_4 \succ D_2$
	9	0.7284744	0.5977202	0.8143724	0.7856604	$D_3 \succ D_4 \succ D_1 \succ D_2$
	10	0.6957552	0.5967685	0.81351	0.7880513	$D_3 \succ D_4 \succ D_1 \succ D_2$
	11	0.5002389	0.289437	0.5970579	0.4861675	$D_3 \succ D_1 \succ D_4 \succ D_2$
	12	0.4776232	0.2890786	0.5880334	0.4730237	$D_3 \succ D_1 \succ D_4 \succ D_2$

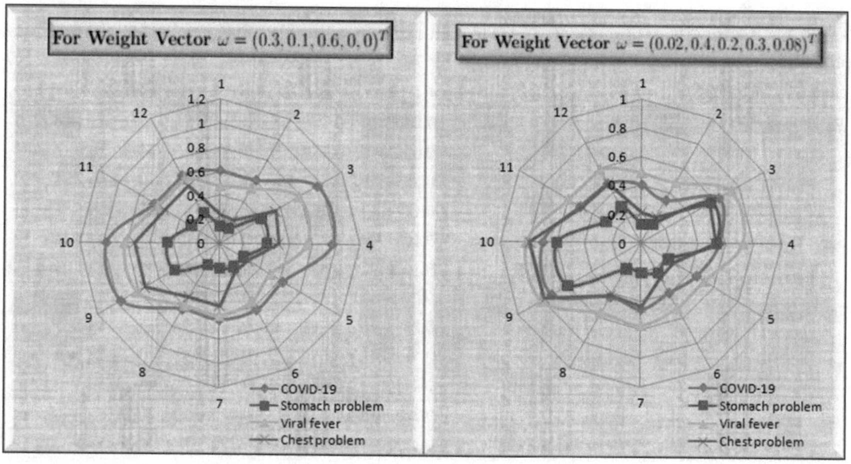

Fig. 4.3 Impact of weight vectors $\omega = (0.3, 0.1, 0.6, 0, 0)^T$ and $\omega = (0.02, 0.4, 0.2, 0.3, 0.08)^T$ for Example 4.4.2

it can be taken $q = 7$. The proposed type-1 SoDM is utilized to classify the pattern U in one of the classes K_1, K_2, and K_3 similar to Example 4.4.1. The obtained results are as follows.

$$^1\mathfrak{D}^{Soergel}_{q-ROFS}(K_1, U) = 0.9596193, {}^1\mathfrak{D}^{Soergel}_{q-ROFS}(K_2, U) = 0.8317866 \text{ and}$$
$$^1\mathfrak{D}^{Soergel}_{q-ROFS}(K_3, U) = 0.8934513$$

Then, according to the type-1 Soergel distance index, the ranking order based on the recognition principle is $K_1 \succ_{(^1\mathfrak{D}^{Soergel})} K_3 \succ_{(^1\mathfrak{D}^{Soergel})} K_2$. Here we consider the lower the distance measure value, the better the alternative. That is, according to the corresponding similarity index, the ranking order is $K_2 \succ_{(^1\mathfrak{S}^{Soergel})} K_3 \succ_{(^1\mathfrak{S}^{Soergel})} K_1$. So the pattern U should be classified with K_2. On the other hand, the type-κ SoSMs ($\kappa = 1, 2, \ldots, 12$) are calculated. The results obtained by type-κ SoSMs ($\kappa = 1, 2, \ldots, 12$) are compared with the existing Peng et al. [54]'s similarity approaches $\mathfrak{S}_\kappa(\kappa = 1, \ldots, 12)$ (Table 4.1) and Wang et al. [72]'s similarity approaches $\mathfrak{S}_\kappa(\kappa = 13, \ldots, 22)$ (Table 4.2) as in Table 4.8. The observation is visually presented in Fig. 4.4.

Example 4.4.4 (Benchmarking Problem) The COVID-19 pandemic has affected the world since 2019 and caused the death of many people. To deal with this epidemic, the governments (or health departments) have followed different protection measures, political strategies, or medical methods. While some governments (or health departments) could achieve success, albeit partially, some of them failed and could not save the country from the effect of the pandemic.

Suppose that, under the set $\mathcal{F} = \{F_1 = Cleaning, F_2 = Social\ Distancing, F_3 = Movement\ Constraint\}$ of the most likely factors for COVID-19 outbreak

Table 4.8 Comparison of proposed SoSMs with existing similarity measures in [54, 72] for Example 4.4.3

Similarity measures	k	K_1	K_2	K_3	Ranking order
Proposed SoSM ${}^{k}\mathfrak{S}^{Soergel}_{q-ROFS}$	1	0.0403807	0.1682134	0.1065487	$K_2 \succ K_3 \succ K_1$
	2	0.0251413	0.1693211	0.1382225	$K_2 \succ K_3 \succ K_1$
	3	0.2554427	0.507586	0.4828736	$K_2 \succ K_3 \succ K_1$
	4	0.1977127	0.4776451	0.4654332	$K_2 \succ K_3 \succ K_1$
	5	0.1067944	0.2246883	0.1671615	$K_2 \succ K_3 \succ K_1$
	6	0.1020148	0.2314882	0.2024794	$K_2 \succ K_3 \succ K_1$
	7	0.2993071	0.3842026	0.3605886	$K_2 \succ K_3 \succ K_1$
	8	0.2594918	0.3534121	0.3370789	$K_2 \succ K_3 \succ K_1$
	9	0.513801	0.6682517	0.6622865	$K_2 \succ K_3 \succ K_1$
	10	0.4722251	0.6462383	0.64232	$K_2 \succ K_3 \succ K_1$
	11	0.3659075	0.4369745	0.4155131	$K_2 \succ K_3 \succ K_1$
	12	0.3331249	0.4107563	0.3957006	$K_2 \succ K_3 \succ K_1$
Peng et al. [54]'s similarity approach $\mathfrak{S}_k$	1	0.412058	0.5222535	0.5042019	$K_2 \succ K_3 \succ K_1$
	2	0.9172096	0.8768012	0.8677411	$K_1 \succ K_2 \succ K_3$
	3	0.6646338	0.6995274	0.6859715	$K_2 \succ K_3 \succ K_1$
	4	0.6232386	0.6392279	0.6211077	$K_2 \succ K_1 \succ K_3$
	5	0.4685501	0.4937908	0.4665879	$K_2 \succ K_1 \succ K_3$
	6	0.4526845	0.4697539	0.4504396	$K_2 \succ K_1 \succ K_3$
	7	0.0933295	0.4181889	0.3030106	$K_2 \succ K_3 \succ K_1$
	8	0.116311	0.362292	0.3254007	$K_2 \succ K_3 \succ K_1$
	9	0.0403807	0.1682134	0.1065487	$K_2 \succ K_3 \succ K_1$
	10	0.0251413	0.1693211	0.1382225	$K_2 \succ K_3 \succ K_1$
	11	0.5423317	0.664568	0.6496645	$K_2 \succ K_3 \succ K_1$
	12	0.5311869	0.6541188	0.6411344	$K_2 \succ K_3 \succ K_1$
Wang et al. [72]'s similarity approach $\mathfrak{S}_k$	13	0.8856436	0.6144143	0.6154782	$K_1 \succ K_3 \succ K_2$
	14	0.9442389	0.818709	0.8193594	$K_1 \succ K_3 \succ K_2$
	15	0.8098954	0.814468	0.8076283	$K_2 \succ K_1 \succ K_3$
	16	0.8738024	0.9159262	0.913325	$K_2 \succ K_3 \succ K_1$
	17	0.6819059	0.7913072	0.7817246	$K_2 \succ K_3 \succ K_1$
	18	0.6164575	0.7429429	0.7330173	$K_2 \succ K_3 \succ K_1$
	19	0.5416272	0.5625887	0.5396316	$K_2 \succ K_1 \succ K_3$
	20	0.6296791	0.6868813	0.6749539	$K_2 \succ K_3 \succ K_1$
	21	0.4329663	0.5230555	0.5018535	$K_2 \succ K_3 \succ K_1$
	22	0.3928376	0.4860158	0.4650301	$K_2 \succ K_3 \succ K_1$

control, the decision matrix based on q-ROF information for the countries C_1, C_2, C_3, and C_4 are created as follows.

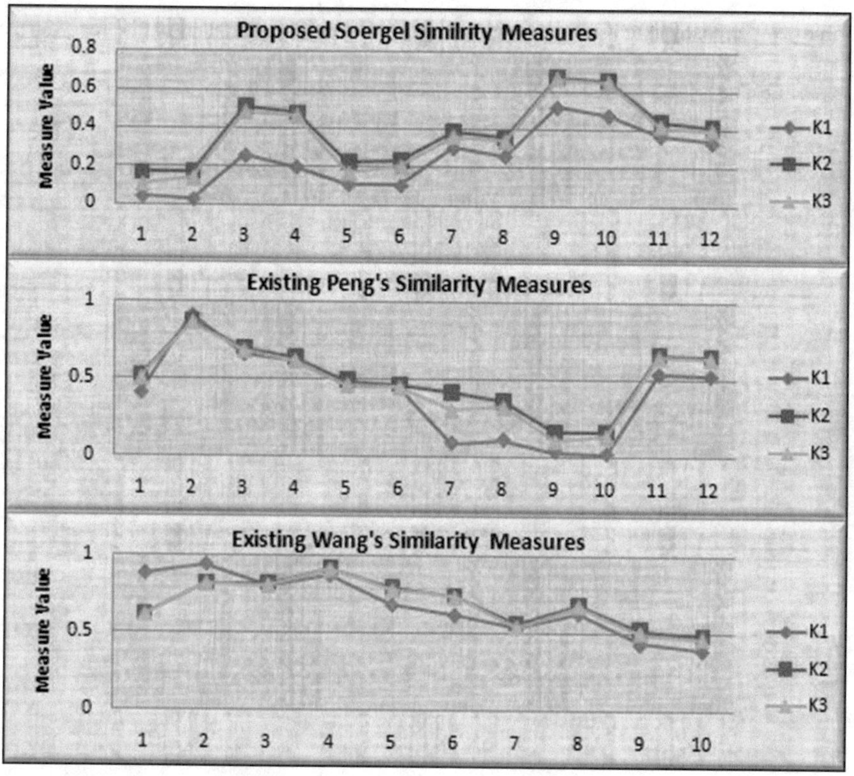

Fig. 4.4 Comparison of proposed SoSMs with similarity approaches in [54, 72] for Example 4.4.3

$$\mathcal{M} = (\mu_{ij})_{4\times 3} = \begin{array}{c} \\ C_1 \\ C_2 \\ C_3 \\ C_4 \end{array} \begin{array}{c} \begin{array}{ccc} F_1 & F_2 & F_3 \end{array} \\ \begin{bmatrix} \langle 0.9, 0.1\rangle & \langle 0.2, 0.9\rangle & \langle 0.7, 0.9\rangle \\ \langle 0.4, 0.8\rangle & \langle 0.1, 0.4\rangle & \langle 0.1, 0.7\rangle \\ \langle 0.4, 0.5\rangle & \langle 0.8, 0.2\rangle & \langle 0.4, 0.4\rangle \\ \langle 0.7, 0.8\rangle & \langle 0.1, 0.4\rangle & \langle 0.5, 0.8\rangle \end{bmatrix} \end{array}$$

For another country B, which has not been able to control the COVID-19 outbreak, the q-ROFS is presented as below.

$$B = \{(F_1, \langle 0.5, 0.6\rangle), (F_2, \langle 0.9, 0.1\rangle), (F_3, \langle 0.3, 0.3\rangle)\}.$$

For all of the q-ROFSs C_1, C_2, C_3, C_4, and B, $\wp^q + \eta^q \leq 1$ when $q \geq 4$, and therefore it can be taken $q = 4$. The government of country B has made a new decision: They will benchmark with countries C_1, C_2, C_3, and C_4 which are also successful in combating the pandemic and thus will follow the COVID-19 outbreak control strategy of the country in which they are most dissimilar. The proposed type-1

Table 4.9 Comparison of proposed SoDMs with existing distance measures in [54] for Example 4.4.4

Proposed SoDM ${}^{k}\mathfrak{D}^{Soergel}_{q-ROFS}$	1	0.9669847	0.8793083	0.5340167	0.8899056	$C_1 \succ C_4 \succ C_2 \succ C_3$
	2	0.9731138	0.8833096	0.4295228	0.8843915	$C_1 \succ C_4 \succ C_2 \succ C_3$
	3	0.7539801	0.2823604	0.3937071	0.4552562	$C_1 \succ C_4 \succ C_3 \succ C_2$
	4	0.7954013	0.2862536	0.1663536	0.4504063	$C_1 \succ C_4 \succ C_2 \succ C_3$
	5	0.8678948	0.8145247	0.5239552	0.8175576	$C_1 \succ C_4 \succ C_2 \succ C_3$
	6	0.8712929	0.8087332	0.4132719	0.8110449	$C_1 \succ C_4 \succ C_2 \succ C_3$
	7	0.8244736	0.5354646	0.217182	0.6821438	$C_1 \succ C_4 \succ C_2 \succ C_3$
	8	0.8303182	0.5604607	0.2277359	0.6869461	$C_1 \succ C_4 \succ C_2 \succ C_3$
	9	0.353759	0.1571607	0.0779352	0.3117806	$C_1 \succ C_4 \succ C_2 \succ C_3$
	10	0.3631382	0.1610648	0.07664	0.3065673	$C_1 \succ C_4 \succ C_2 \succ C_3$
	11	0.7339189	0.4920779	0.208891	0.6212832	$C_1 \succ C_4 \succ C_2 \succ C_3$
	12	0.7385651	0.5127048	0.2187554	0.6248759	$C_1 \succ C_4 \succ C_2 \succ C_3$
Peng et al. [54]'s distance measure $\mathfrak{D}_k$	1	0.7098667	0.3893333	0.1285	0.5231667	$C_1 \succ C_4 \succ C_2 \succ C_3$
	2	0.4082667	0.2064	0.0463667	0.1885	$C_1 \succ C_2 \succ C_4 \succ C_3$
	3	0.5590667	0.2978667	0.0874333	0.3558333	$C_1 \succ C_4 \succ C_2 \succ C_3$
	4	0.6325333	0.3893333	0.1103667	0.4458333	$C_1 \succ C_4 \succ C_2 \succ C_3$
	5	0.7745527	0.5354646	0.1852223	0.6009094	$C_1 \succ C_4 \succ C_2 \succ C_3$
	6	0.7749102	0.5604607	0.1987932	0.6167147	$C_1 \succ C_4 \succ C_2 \succ C_3$
	7	0.9867303	0.6459516	0.1388407	0.6599121	$C_1 \succ C_4 \succ C_2 \succ C_3$
	8	0.9955768	0.8819839	0.7129469	0.8866374	$C_1 \succ C_4 \succ C_2 \succ C_3$
	9	0.9669847	0.8793083	0.5340167	0.8899056	$C_1 \succ C_4 \succ C_2 \succ C_3$
	10	0.9731138	0.8833096	0.4295228	0.8843915	$C_1 \succ C_4 \succ C_2 \succ C_3$
	11	0.647435	0.3304091	0.0960522	0.4186597	$C_1 \succ C_4 \succ C_2 \succ C_3$
	12	0.6411972	0.3450639	0.1053491	0.4174409	$C_1 \succ C_4 \succ C_2 \succ C_3$

WSoDM is utilized for the weight vector $\omega = (0.5, 0.2, 0.3)^T$ to benchmark country B with the countries C_1, C_2, C_3, and C_4 under the COVID-19 outbreak control factors similar to Example 4.4.2. The calculated results are as follows.

$$
{}^{\omega 1}\mathfrak{D}^{Soergel}_{q-ROFS}(C_1, B) = 0.9544806,\ {}^{\omega 1}\mathfrak{D}^{Soergel}_{q-ROFS}(C_2, B) = 0.825658,\\
{}^{\omega 1}\mathfrak{D}^{Soergel}_{q-ROFS}(C_3, B) = 0.5511848 \text{ and } {}^{\omega 1}\mathfrak{D}^{Soergel}_{q-ROFS}(C_4, B) = 0.8418094.
$$

Then, according to the type-1 weighted Soergel distance index, the ranking order is obtained as $C_1 \succ_{\omega 1 \mathfrak{D}^{Soergel}} C_4 \succ_{\omega 1 \mathfrak{D}^{Soergel}} C_2 \succ_{\omega 1 \mathfrak{D}^{Soergel}} C_3$. Here we consider the higher the distance measure coefficient, the better the alternative. Hence, the government of country B should follow the strategies implemented by the government of country C_1 to control the COVID-19 outbreak. On the other hand, the type-κ Soergel distance index ($\kappa = 1, 2, \ldots, 12$) are computed by using the equal weights $\omega = (1/3, 1/3, 1/3)^T$ (that is, ignoring the weights as per Remark 1) for COVID-19 outbreak control factors and the corresponding results are compared with the existing Peng et al. [54]'s distance measures as in Table 4.9. From Table 4.9, it is observed that the ranking order $C_1 \succ_{\omega 1 \mathfrak{D}^{Soergel}} C_4 \succ_{\omega 1 \mathfrak{D}^{Soergel}} C_2 \succ_{\omega 1 \mathfrak{D}^{Soergel}} C_3$ is the most common for the proposed type-κ Soergel distance measures ${}^{k}\mathfrak{D}^{Soergel}_{q-ROFS}$ ($\kappa = 1, 2, \ldots, 12$) and existing Peng et al. [54]'s distance measures $\mathfrak{D}_\kappa$ ($\kappa = 1, 2, \ldots, 12$).

4.5 Comparison Analysis

The intuitionistic fuzzy set (IFS) and Pythagorean fuzzy set (PyFS) may be regarded as a q-ROFS through assuming $q = 1$ and $q = 2$, respectively. It allows evaluating the existing similarity measures in IFS and PyFS with the proposed Soergel similarity measures in q-ROFS with $q = 1$ and $q = 2$, respectively. From Table 4.10, the ranking of the existing similarity measures in IFS, S_L [42], S_{CC} [11], $S_{HY1}, S_{HY2}, S_{HY3}$ [27], $S_{LS1}, S_{LS2}, S_{LS3}$ [44] coincide with the ranking of the proposed similarity measures ${}^{\kappa}\mathfrak{S}^{Soergel}_{q-ROFS}$ ($\kappa = 1, 2, \ldots, 12$) for Example 2 [54]. Moreover, the ranking of existing similarity measure S_{BA} [8] can not be classified for the same Example 2 [54]. The ranking of existing weighted similarity measure WGD^4_{TSFS} ($\rho = 0.5$) [29] in IFS coincide with the proposed WSoSMs ${}^{\omega\kappa}\mathfrak{S}^{Soergel}_{q-ROFS}$ $\forall\, \kappa = 1$ to 12 except $\kappa = 9$ and 10 for Table 7 [29]. The motive for the exception of $\kappa = 9$ and 10 may be the influence of parameter ρ in WGD^4_{TSFS} [29]. In PyFS, the ranking of existing similarity measures S_{P2}, S_{P3} [55], and S_Z [83] are precisely matching with the ranking of proposed SoSMs ${}^{\kappa}\mathfrak{S}^{Soergel}_{q-ROFS}$ ($\kappa = 1, 2, \ldots, 12$) for Example 3 [54]. Also, the ranking of proposed WSoSMs ${}^{\omega\kappa}\mathfrak{S}^{Soergel}_{q-ROFS}$ $\forall\, \kappa = 1$ to 12 except $\kappa = 3, 9$, and 10 are same with the ranking of existing weighted similarity measure S [43] ($\tau = 1, 2, 3, 4$) for Table 5.1 [43]. The non-matching for $\kappa = 3, 9$, and 10 is happened because of the influence of τ in S [43]. In q-ROFS (q=3), the ranking of proposed SoSMs ${}^{\kappa}\mathfrak{S}^{Soergel}_{q-ROFS}$ $\forall\, \kappa = 1$ to 12 except $\kappa = 4, 9$, and 10 are matching with the ranking of existing similarity measure S_{13} ($p = 1, t_1 = 2, t_2 = 2$) [54] for Example 3 [54]. The reason for exception $\kappa = 4, 9$, and 10 is the impact of parameters $p = 1, t_1 = 2$, and $t_2 = 2$ in the existing similarity measure S_{13} [54]. In q-ROFS (q=5), the ranking of proposed WSoSMs ${}^{\omega\kappa}\mathfrak{S}^{Soergel}_{q-ROFS}$ $\forall\, \kappa = 1, 2, 5, 6$ and $\kappa = 10$ are matching with the ranking of existing weighted similarity measures WGD^1_{TSFS} ($\rho = 0, 0.3, 0.5, 0.7$) [29] and WGD^4_{TSFS} ($\rho = 0.5$) [29], respectively, for Table 1 [29]. But, the ranking of proposed WSoSMs ${}^{\omega\kappa}\mathfrak{S}^{Soergel}_{q-ROFS}$ $\forall\, \kappa = 3, 4, 7, 8, 9, 11$, and 12 are little deviate from the ranking of existing weighted similarity measure WGD^1_{TSFS} ($\rho = 0, 0.3, 0.5, 0.7$) [29] due to the influence of the parameter ρ in WGD^1_{TSFS} [29]. From the overall comparison, the proposed SoSMs ${}^{\kappa}\mathfrak{S}^{Soergel}_{q-ROFS}$ $\forall\, \kappa = 1, 2, \ldots 12$ are greater efficient than the prevailing one.

4.6 Sensitivity Analysis and Advantages of SoDMs

4.6.1 Sensitivity Analysis of SoDMs for the Value of q

This section presents a sensitivity evaluation to understand the behavior of measure values and the ranking order of alternatives when the values of the parameter q vary from small to large. For the experiment, the proposed SoDMs ${}^{\kappa}\mathfrak{D}^{Soergel}_{q-ROFS}$($\kappa = 1, 2, \ldots, 12$) are utilized for the Example 4.4.1 by taking $q = 5, 10, 70$. The test

Table 4.10 Comparison of proposed (W)SoSMs ${}^{\kappa}\mathfrak{S}^{Soergel}_{q-ROFS}$ ($\kappa = 1, 2, \ldots, 12$) with existing similarity measures

Notion	Problem	Existing similarity measures	Ranking order	Proposed (W)SoSMs ${}^{\kappa}\mathfrak{S}^{Soergel}_{q-ROFS}$ ($\kappa = 1, 2, \ldots, 12$)	Ranking order
IFS ($q-ROFS, q = 1$)	Example 2 [54]	S_L [42] S_{CC} [11] $S_{HY1}, S_{HY2}, S_{HY3}$ [27] $S_{LS1}, S_{LS2}, S_{LS3}$ [44] S_{BA} [8]	$C_3 \succ C_1 \succ C_2$ $C_3 \succ C_1 \succ C_2$ $C_3 \succ C_1 \succ C_2$ $C_3 \succ C_1 \succ C_2$ $C_3 = C_1 \succ C_2$	$\kappa = 1, 2, \ldots, 12$	$C_3 \succ C_1 \succ C_2$
IFS ($q-ROFS, q = 1$)	Table 7 [29]	WGD^4_{TSFS} ($\rho = 0.5$) [29]	$S_4 \succ S_2 \succ S_1 \succ S_3$	$\kappa = 1, 2, 3, 4, 5, 6, 7, 8, 11, 12$ (W) $\kappa = 9, 10$ (W)	$S_4 \succ S_2 \succ S_3 \succ S_1$ $S_4 \succ S_2 \succ S_1 \succ S_3$
$PyFS$ ($q-ROFS, q = 2$)	Example 3 [54]	S_{P1}, S_{P2}, S_{P3} [55] S_Z [83]	$C_2 \succ C_1 \succ C_3$ $C_2 \succ C_1 \succ C_3$	$\kappa = 1, 2, \ldots, 12$	$C_2 \succ C_1 \succ C_3$
$PyFS$ ($q-ROFS, q = 2$)	Table 5.1 [43]	S [43] ($\tau = 1, 2, 3, 4$)	$S_2 \succ S_4 \succ S_3 \succ S_1$	$\kappa = 1, 2, 4, 5, 6, 7, 8, 11, 12$ (W) $\kappa = 3$ (W) $\kappa = 9, 10$ (W)	$S_2 \succ S_4 \succ S_3 \succ S_1$ $S_2 \succ S_3 \succ S_4 \succ S_1$ $S_3 \succ S_2 \succ S_4 \succ S_1$
$q-ROFS$ ($q = 3$)	Example 3 [54]	S_{13} ($p = 1, t_1 = 2, t_2 = 2$) [54]	$C_2 \succ C_1 \succ C_3$	$\kappa = 1, 2, 3, 5, 6, 7, 8, 11, 12$ $\kappa = 4, 9, 10$	$C_2 \succ C_1 \succ C_3$ $C_1 \succ C_2 \succ C_3$
$q-ROFS$ ($q = 5$)	Table 1 [29]	WGD^1_{TSFS} ($\rho = 0, 0.3, 0.5, 0.7$) [29]	$S_2 \succ S_4 \succ S_1 \succ S_3$	$\kappa = 1, 2, 5, 6$ (W) $\kappa = 3, 4, 9$ (W) $\kappa = 7, 8, 11, 12$ (W)	$S_2 \succ S_4 \succ S_1 \succ S_3$ $S_2 \succ S_4 \succ S_3 \succ S_1$ $S_2 \succ S_1 \succ S_4 \succ S_3$
$q-ROFS$ ($q = 5$)	Table 1 [29]	WGD^4_{TSFS} ($\rho = 0.5$) [29]	$S_4 \succ S_2 \succ S_1 \succ S_3$	$\kappa = 10$ (W)	$S_4 \succ S_2 \succ S_1 \succ S_3$

In the table, (W) stands for WSoSMs

result is offered in Table 4.11. From Table 4.11, it's far visible that the measure values of alternatives approaching 1 for the SoDMs ${}^{\kappa}\mathfrak{D}^{Soergel}_{q-ROFS}(\kappa = 1, 2, 7, 8)$, approaching 0 for the SoDMs ${}^{\kappa}\mathfrak{D}^{Soergel}_{q-ROFS}(\kappa = 3, 4, 9, 10)$, neither approaching 0 nor approaching 1 for the SoDMs ${}^{\kappa}\mathfrak{D}^{Soergel}_{q-ROFS}(\kappa = 5, 6, 11, 12)$. Also, the proposed SoDMs ${}^{\kappa}\mathfrak{D}^{Soergel}_{q-ROFS}(\kappa = 1, 2, \ldots, 12)$ produce consistent ranking order for a small value of q ($q = 5, 10$), moderate consistent and inconsistent ranking order for a large value of q ($q = 70$). Consequently, the use of a large value of q is not in practice for Example 4.4.1 (considering precision of 10^{-7}), so forget about it.

Table 4.11 Sensitivity analysis of SoDMs for Example 4.4.1 concerning varies q

q	Proposed SoDM	Measure value from $\mathcal{B}$ to				Ranking order
		A_1	A_2	A_3	A_4	
5	1	0.7790332	0.8819524	0.6379575	0.9898056	$A_4 \succ A_2 \succ A_1 \succ A_3$
	2	0.7791081	0.8977643	0.637961	0.9916671	$A_4 \succ A_2 \succ A_1 \succ A_3$
	3	0.0492635	0.3706278	0.0092275	0.546831	$A_4 \succ A_2 \succ A_1 \succ A_3$
	4	0.0492835	0.4481715	0.0092287	0.5929163	$A_4 \succ A_2 \succ A_1 \succ A_3$
	5	0.6773524	0.7622888	0.5628104	0.8516612	$A_4 \succ A_2 \succ A_1 \succ A_3$
	6	0.6774113	0.7762239	0.5628136	0.8538587	$A_4 \succ A_2 \succ A_1 \succ A_3$
	7	0.8735224	0.9276791	0.7644597	0.994654	$A_4 \succ A_2 \succ A_1 \succ A_3$
	8	0.8756001	0.9344377	0.7789391	0.9947088	$A_4 \succ A_2 \succ A_1 \succ A_3$
	9	0.0472409	0.332875	0.0096192	0.5007736	$A_4 \succ A_2 \succ A_1 \succ A_3$
	10	0.0480113	0.3432213	0.0097584	0.5022616	$A_4 \succ A_2 \succ A_1 \succ A_3$
	11	0.759772	0.8062072	0.6760626	0.8594522	$A_4 \succ A_2 \succ A_1 \succ A_3$
	12	0.7612742	0.8113767	0.6871829	0.8595544	$A_4 \succ A_2 \succ A_1 \succ A_3$
10	1	0.945377	0.963936	0.8385666	0.9996748	$A_4 \succ A_2 \succ A_1 \succ A_3$
	2	0.9453771	0.9676144	0.8385666	0.9997109	$A_4 \succ A_2 \succ A_1 \succ A_3$
	3	0.002401	0.2489915	0.000151	0.3325413	$A_4 \succ A_2 \succ A_1 \succ A_3$
	4	0.002401	0.2978936	0.000151	0.3715364	$A_4 \succ A_2 \succ A_1 \succ A_3$
	5	0.8053243	0.8263855	0.7223606	0.8548192	$A_4 \succ A_2 \succ A_1 \succ A_3$
	6	0.8053243	0.8306468	0.7223606	0.8561283	$A_4 \succ A_2 \succ A_1 \succ A_3$
	7	0.9715383	0.9809087	0.9059299	0.9998371	$A_4 \succ A_2 \succ A_1 \succ A_3$
	8	0.9719216	0.9815913	0.9121961	0.9998372	$A_4 \succ A_2 \succ A_1 \succ A_3$
	9	0.0021143	0.2494207	0.0001385	0.3224386	$A_4 \succ A_2 \succ A_1 \succ A_3$
	10	0.0021377	0.2536639	0.0001422	0.3224848	$A_4 \succ A_2 \succ A_1 \succ A_3$
	11	0.8276791	0.8428029	0.781347	0.8566066	$A_4 \succ A_2 \succ A_1 \succ A_3$
	12	0.8279363	0.8434188	0.7857867	0.8566082	$A_4 \succ A_2 \succ A_1 \succ A_3$
70	1	0.9999999	0.9999999	0.9997911	1	$A_4 \succ A_2 = A_1 \succ A_3$
	2	0.9999999	0.9999999	0.9997911	1	$A_4 \succ A_2 = A_1 \succ A_3$
	3	0	0.0117963	0	0.0117981	$A_4 \succ A_2 \succ A_1 = A_3$
	4	0	0.0118012	0	0.011803	$A_4 \succ A_2 \succ A_1 = A_3$
	5	0.8465735	0.8466055	0.8464169	0.8466056	$A_4 \succ A_2 \succ A_1 \succ A_3$
	6	0.8465735	0.8466056	0.8464169	0.8466056	$A_4 \succ A_2 \succ A_1 \succ A_3$
	7	1	1	0.9998955	1	$A_4 = A_2 = A_1 \succ A_3$
	8	1	1	0.9998955	1	$A_4 = A_2 = A_1 \succ A_3$
	9	0	0.0102191	0	0.0102206	$A_4 \succ A_2 \succ A_1 = A_3$
	10	0	0.0102191	0	0.0102206	$A_4 \succ A_2 \succ A_1 = A_3$
	11	0.8465736	0.8465997	0.8465053	0.8465997	$A_4 = A_2 \succ A_1 \succ A_3$
	12	0.8465736	0.8465997	0.8465053	0.8465997	$A_4 = A_2 \succ A_1 \succ A_3$

4.6.2 Advantages of Proposed Approaches

(1) The q-ROFS expresses the decision information more widely than that of the IFS and the PFS, it gives the decision maker more freedom in expressing their belief about membership degree.
(2) Each of the proposed types of Soergel distances based on q-ROFSs (and their corresponding similarity coefficients) has its advantages. By these approaches, the distance (and similarity) between any two q-ROFSs can be easily determined. We can assert this for almost all q-ROFSs.
(3) Some distance measures, distance-based similarity measures, correlation coefficient, cosine, and cotangent-based similarity measures for q-ROFSs are available in the literature. As already mentioned, however, some special cases cannot be dealt with the existing measures under q-rung orthopair fuzzy information. Therefore, the proposed Soergel measures of distance and similarity are presented as a different alternative way to treat real-life problems that cannot be dealt with by other q-ROFS-based information measures.

4.6.3 Limitations of Proposed Approaches

(1) The proposed SoDMs/SoSMs of q-ROFSs are not suitable to deal with the problems under the environments of interval-valued, picture, and hesitant types of q-ROFSs. However, the proposed SoDMs/SoSMs can be adapted for these extended q-ROFSs.
(2) The range of q in the proposed SoDMs/SoSMs can sometimes drastically change the outputs. Therefore, the choice of q is important for the persuasiveness of the result. That is, in some cases, the arbitrariness of q can cause a disadvantage.

4.7 Conclusion

One of the most important research topics is how to represent the information based on uncertainty and vague in human evaluation. The IFSs and PyFSs are all excellent mathematical models in handling uncertain-based information when human beings regarding responses not only yes but also no. The q-ROFS theory is a more powerful tool than the intuitionistic fuzzy models and Pythagorean fuzzy models, as membership and membership degrees are more inclusive. Soergel distance measures are complementary concepts quantifying the difference and closeness between two objects or points. The Soergel distance measures have been studied and improved in many scientific studies. Regarding the distance measures of IFSs and PyFSs, it was noted that these tools could not analyze the data available in the form of q-ROFNs. So far, many distance measures for q-ROFSs have been established, which generalize each of the distance measures based on IFS and PyFS. However, these tools also have several limitations. This chapter proposed twelve kinds of Soergel distance to measure q-rung orthopair fuzzy information and then analyzed their properties. Each of these measures has its advantages in handling real-life problems. The efficiency of the proposed SoDMs and WSoDMs was demonstrated by solving the multi-attribute decision-making problems. Thereby, it is pointed out that these new distance measure approaches can be useful in dealing with many real-life issues such as network problems, shortest path problems, signal systems, and image segmentation enhancement.

An interesting aspect for future research is to improve new types of Soergel distance coefficients measuring the q-rung orthopair fuzzy information, and further to adopt some more complicated applications from the fields of clustering analysis, forecasting, mathematical programming, uncertain programming, etc. The other is to try to utilize these distance measures in the environments of complex q-ROFSs, bipolar q-ROFSs, hesitant q-ROFSs, and cubic q-ROFSs. These are important directions for future research.

References

1. P. Albert, The algebra of fuzzy logic. Fuzzy Sets Syst. **1**(3), 203–230 (1978)
2. K.T. Atanassov, Intuitionistic fuzzy sets. Fuzzy Sets Syst. **20**(1), 87–96 (1986)
3. K.T. Atanassov, Interval valued intuitionistic fuzzy sets. Fuzzy Sets Syst. **31**(3), 343–349 (1989)
4. E. Aygün, H. Kamacı, Some generalized operations in soft set theory and their role in similarity and decision making. J. Intell. Fuzzy Syst. **36**(6), 6537–6547 (2019)
5. E. Aygün, H. Kamacı, Some new algebraic structures of soft sets. Soft Comput. **25**(13), 8609–8626 (2021)
6. H. Bashir, S. Inayatullah, A. Alsanad, R. Anjum, M. Mosleh, P. Ashraf, Some improved correlation coefficients for q-rung orthopair fuzzy sets and their applications in cluster analysis. Math. Probl. Eng. **2021**, 11 (2021)
7. K. Bhattacharya, S.K. De, Decision making under intuitionistic fuzzy metric distances. Ann. Optim. Theory Pract. **3**(2), 49–64 (2020)
8. F.E. Boran, D.A. Akay, A biparametric similarity measure on intuitionistic fuzzy sets with applications to pattern recognition. Inf. Sci. **255**, 45–57 (2014)
9. S. Broumi, F. Smarandache, New operations over interval valued intuitionistic hesitant fuzzy set. Math. Stat. **2**(2), 62–71 (2014)
10. S.-M. Chen, Similarity measures between vague sets and between elements. IEEE Trans. Syst. Man Cybern. **27**, 153–158 (1997)
11. S.-M. Chen, C.-H. Chang, A novel similarity measure between Atanssov's intuitionistic fuzzy sets based on transformation techniques with applications to pattern recognition. Inf. Sci. **291**, 96–114 (2015)
12. I. Couso, L. Garrido, L. Sánchez, Similarity and dissimilarity measures between fuzzy sets: a formal relational study. Inf. Sci. **229**, 122–141 (2013)
13. M. Dehghan, M. Ghatee, B. Hashemi, Inverse of a fuzzy matrix of fuzzy numbers. Int. J. Comput. Math. **86**(8), 1433–1452 (2009)
14. W.S. Du, Minkowski-type distance measures for generalized orthopair fuzzy sets. Int. J. Intell. Syst. **33**, 802–817 (2018)
15. P.A. Ejegwa, New operations on intuitionistic fuzzy multisets. J. Math. Inf. **3**, 17–23 (2015)

16. P.A. Ejegwa, I.C. Onyeke, V. Adah, An algorithm for an improved intuitionistic fuzzy correlation measure with medical diagnostic application. Ann. Optim. Theory Pract. **3**(3), 51–68 (2020)
17. B. Farhadinia, S. Effati, F. Chiclana, A family of similarity measures for q-rung orthopair fuzzy sets and their applications to multiple criteria decision making. Int. J. Intell. Syst. **36**(4), 1535–1559 (2021)
18. H. Garg, A new possibility degree measure for interval-valued q-rung orthopair fuzzy sets in decision-making. Int. J. Intell. Syst. **36**(1), 526–557 (2021)
19. H. Garg, A novel accuracy function under interval-valued Pythagorean fuzzy environment for solving multicriteria decision making problem. J. Intell. Fuzzy Syst. **31**(1), 529–540 (2016)
20. H. Garg, A novel trigonometric operation-based q-rung orthopair fuzzy aggregation operator and its fundamental properties. Neural Comput. Appl. **32**(18), 15077–15099 (2020)
21. H. Garg, Confidence levels based Pythagorean fuzzy aggregation operators and its application to decision-making process. Comput. Math. Organ. Theory **23**(4), 546–571 (2017)
22. H. Garg, New exponential operation laws and operators for interval-valued q-rung orthopair fuzzy sets in group decision making process. Neural Comput. Appl. **33**(20), 13937–13963 (2021). https://doi.org/10.1007/s00521-021-06036-0
23. H. Garg, CN-q-ROFS: connection number-based q-rung orthopair fuzzy set and their application to decision-making process. Int. J. Intell. Syst. **36**(7), 3106–3143 (2021)
24. H. Garg, S.-M. Chen, Multiattribute group decision making based on neutrality aggregation operators of q-rung orthopair fuzzy sets. Inf. Sci. **517**, 427–447 (2020)
25. H. Garg, K. Kumar, A novel possibility measure to interval-valued intuitionistic fuzzy set using connection number of set pair analysis and its applications. Neural Comput. Appl. **32**, 3337–3348 (2020)
26. D.H. Hong, C. Kim, A note on similarity measures between vague sets and between elements. Inf. Sci. **115**, 83–96 (1999)
27. W.-L. Hung, M.-S. Yang, Similarity measures of intuitionistic fuzzy sets based on Hausdorff distance. Pattern Recogn. Lett. **25**, 1603–1611 (2004)
28. M. Imtiaz, M. Saqlain, M. Saeed, TOPSIS for multi criteria decision making in octagonal intuitionistic fuzzy environment by using accuracy function. J. New Theory **31**, 32–40 (2020)
29. N. Jan, L. Zedam, T. Mahmood, E. Rak, Z. Ali, Generalized dice similarity measures for q-rung orthopair fuzzy sets with applications. Complex Intell. Syst. **6**, 545–558 (2020)
30. H. Kamacı, Interval-valued fuzzy parameterized intuitionistic fuzzy soft sets and their applications. Cumhur. Sci. J. **40**(2), 317–331 (2019)
31. H. Kamacı, A novel approach to similarity of soft sets. Adıyaman Univ. J. Sci. **9**(1), 23–35 (2019)
32. H. Kamacı, Similarity measure for soft matrices and its applications. J. Intell. Fuzzy Syst. **36**(4), 3061–3072 (2019)
33. H. Kamacı, Simplified neutrosophic multiplicative refined sets and their correlation coefficients with application in medical pattern recognition. Neutrosophic Sets Syst. **41**, 270–285 (2021)
34. A. Kandel, W.J. Byatt, Fuzzy sets, fuzzy algebra, and fuzzy statistics. Proc. IEEE **66**(12), 1619–1639 (1978)
35. M.J. Khan, P. Kumam, M. Shutaywi, Knowledge measure for the q-rung orthopair fuzzy sets. Int. J. Intell. Syst. **36**, 628–655 (2021)
36. M.J. Khan, P. Kumam, M. Shutaywi, W. Kumam, Improved knowledge measures for q-rung orthopair fuzzy sets. Int. J. Comput. Intell. Syst. **14**(1), 1700–1713 (2021)
37. M.J. Khan, P. Kumam, N.A. Alreshidi, W. Kumam, Improved cosine and cotangent function-based similarity measures for q-rung orthopair fuzzy sets and TOPSIS method. Complex Intell. Syst. (2021), in press. https://doi.org/10.1007/s40747-021-00425-7
38. K.H. Kim, F.W. Roush, Generalized fuzzy matrices. Fuzzy Sets Syst. **4**(3), 293–315 (1980)
39. M. Kon, Fuzzy distance and fuzzy norm. J. Oper. Res. Soc. Jpn. **60**(2), 66–77 (2017)
40. D. Li, C. Cheng, New similarity measures of intuitionistic fuzzy sets and application to pattern recognitions. Pattern Recogn. Lett. **23**, 221–225 (2002)

41. D. Li, W. Zeng, Distance measure of Pythagorean fuzzy sets. Int. J. Intell. Syst. **33**, 348–361 (2018)
42. Y. Li, D.L. Olson, Z. Qin, Similarity measures between intuitionistic fuzzy (vague) sets a comparative analysis. Pattern Recogn. Lett. **28**, 278–285 (2007)
43. Z. Li, M. Lu, Some novel similarity and distance measures of Pythagorean fuzzy sets and their applications. J. Intell. Fuzzy Syst. **37**(2), 1781–1799 (2019)
44. Z. Liang, P. Shi, Similarity measures on intuitionistic fuzzy sets. Pattern Recogn. Lett. **24**, 2687–2693 (2003)
45. P. Liu, P. Wang, Some q-rung orthopair fuzzy aggregation operators and their applications to multiple-attribute decision making. Int. J. Intell. Syst. **33**, 259–280 (2018)
46. P. Liu, T. Mahmood, Z. Ali, Complex q-rung orthopair fuzzy aggregation operators and their applications in multi-attribute group decision making. Information **11**(1), 1–27 (2020)
47. H.B. Mitchell, On the Dengfeng-Chuntian similarity measure and its application to pattern recognition. Pattern Recogn. Lett. **24**, 3101–3104 (2003)
48. A. Pal, B. Mondal, N. Bhattacharyya, S. Raha, Similarity in fuzzy systems. J. Uncertain. Anal. Appl. **2**, 18 (2014)
49. Z. Pawlak, Rough sets. Int. J. Comput. Inf. Sci. **11**, 341–356 (1982)
50. Z. Pawlak, Rough sets and intelligent data analysis. Inf. Sci. **147**, 1–12 (2002)
51. Z. Pawlak, A. Skowron, Rough sets and Boolean reasoning. Inf. Sci. **177**, 41–73 (2007)
52. Z. Pawlak, A. Skowron, Rudiments of rough sets. Inf. Sci. **177**, 3–27 (2007)
53. X. Peng, J. Dai, H. Garg, Exponential operation and aggregation operator for q-rung orthopair fuzzy set and their decision-making method with a new score function. Int. J. Intell. Syst. **33**(11), 2255–2282 (2018)
54. X. Peng, L. Liu, Information measures for q-rung orthopair fuzzy sets. Int. J. Intell. Syst. **34**, 1795–1834 (2019)
55. X. Peng, H. Yuan, Y. Yang, Pythagorean fuzzy information measures and their applications. Int. J. Intell. Syst. **32**, 991–1029 (2017)
56. S. Petchimuthu, H. Garg, H. Kamacı, A.O. Atagün, The mean operators and generalized products of fuzzy soft matrices and their applications in MCGDM. Comput. Appl. Math. **39**(2), 1–32 (2020). Article number 68
57. S. Petchimuthu, H. Kamacı, The adjustable approaches to multi-criteria group decision making based on inverse fuzzy soft matrices. Sci. Iran. (in press). https://doi.org/10.24200/sci.2020.54294.3686
58. S. Petchimuthu, H. Kamacı, The row-products of inverse soft matrices in multicriteria decision making. J. Intell. Fuzzy Syst. **36**(6), 6425–6441 (2019)
59. A.U. Rahman, M.R. Ahmad, M. Saeed, M. Ahsan, M. Arshad, M. Ihsan, A study on fundamentals of refined intuitionistic fuzzy set with some properties. J. Fuzzy Ext. Appl. **1**(4), 300–314 (2020)
60. A.U. Rahman, M. Saeed, M. Arshad, M. Ihsan, M.R. Ahmad, (m, n)-convexity-cum-concavity on fuzzy soft set with applications in first and second sense. Punjab Univ. J. Math. **53**(1), 19–33 (2021)
61. M. Riaz, A. Razzaq, H. Kalsoom, D. Pamučar, H.M. A. Farid, Y.-M. Chu, q-rung orthopair fuzzy geometric aggregation operators based on generalized and group-generalized parameters with application to water loss management. Symmetry **12**(8), 1236 (2020). https://doi.org/10.3390/sym12081236
62. M. Riaz, H. Garg, H.M.A. Farid, M. Aslam, Novel q-rung orthopair fuzzy interaction aggregation operators and their application to low-carbon green supply chain management. J. Intell. Fuzzy Syst. **41**(2), 4109–4126 (2021). https://doi.org/10.3233/JIFS-210506
63. M. Riaz, H.M.A. Farid, F. Karaaslan, M.R. Hashmi, Some q-rung orthopair fuzzy hybrid aggregation operators and TOPSIS method for multi-attribute decision-making. J. Intell. Fuzzy Syst. **39**(1), 1227–1241 (2020)
64. M. Riaz, W. Salabun, H.M.A. Farid, N. Ali, J. Watróbski, A robust q-rung orthopair fuzzy information aggregation using Einstein operations with application to sustainable energy planning decision management. Energies **13**(9), 2155 (2020). https://doi.org/10.3390/en13092155

65. S. Samanta, B. Sarkar, Representation of competitions by generalized fuzzy graphs. Int. J. Comput. Intell. Syst. **11**(1), 1005–1015 (2018)
66. A.K. Shaw, T.K. Roy, Trapezoidal intuitionistic fuzzy number with some arithmetic operations and its application on reliability evaluation. Int. J. Math. Oper. Res. **5**(1), 55–73 (2013)
67. T.K. Shinoj, S.J. John, Intuitionistic fuzzy multisets and its application in medical diagnosis. World Acad. Sci. Eng. Technol. **6**, 1–28 (2012)
68. M. Sitara, M. Akram, M.Y. Bhatti, Fuzzy graph structures with application. Mathematics **7**(1), 63 (2019)
69. R. Taghaodi, A novel solution approach for solving intuitionistic fuzzy transportation problem of type-2. Ann. Optim. Theory Pract. **2**(2), 11–24 (2019)
70. P. Talukdar, S. Goala, P. Dutta, B. Limboo, Fuzzy multicriteria decision making in medical diagnosis using an advanced distance measure on linguistic Pythagorean fuzzy sets. Ann. Optim. Theory Pract. **3**(4), 113–131 (2020)
71. V. Uluçay, I. Deli, M. Şahin, Intuitionistic trapezoidal fuzzy multi-numbers and its application to multi-criteria decision-making problems. Complex Intell. Syst. **5**, 65–78 (2019)
72. P. Wang, J. Wang, G. Wei, C. Wei, Similarity measures of q-rung orthopair fuzzy sets based on cosine function and their applications. Mathematics **7**(4), 340 (2019)
73. G. Wei, C. Wei, J. Wang, H. Gao, Y. Wei, Some q-rung orthopair fuzzy Maclaurin symmetric mean operators and their applications to potential evaluation of emerging technology commercialization. Int. J. Intell. Syst. **34**, 50–81 (2019)
74. G. Wei, Y. Wei, Similarity measures of Pythagorean fuzzy sets based on the cosine function and their applications. Int. J. Intell. Syst. **33**, 634–652 (2018)
75. L. Xu, Y. Liu, H. Liu, Some improved q-rung orthopair fuzzy aggregation operators and their applications to multiattribute group decision-making. Math. Probl. Eng. **2019**, 2019, 18 pages
76. R.P. Yager, Pythagorean fuzzy subsets, in *Proceedings of the IFSA World Congress and NAFIPS Annual Meeting, Edmonton, Canada* (2013), pp. 57–61
77. R.P. Yager, Generalized orthopair fuzzy sets. IEEE Trans. Fuzzy Syst. **25**, 1222–1230 (2017)
78. Z. Yang, H. Garg, Interaction power partitioned Maclaurin symmetric mean operators under q-rung orthopair uncertain linguistic information. Int. J. Fuzzy Syst. 1–19 (2021). https://doi.org/10.1007/s40815-021-01062-5
79. Y.Y. Yao, A comparative study of fuzzy sets and rough sets. Inf. Sci. **109**, 227–242 (1998)
80. J. Ye, Cosine similarity measures for intuitionistic fuzzy sets and their applications. Math. Comput. Modell. **53**, 91–97 (2011)
81. L.A. Zadeh, Fuzzy sets. Inf. Control **8**(3), 338–353 (1965)
82. S. Zeng, Y. Hu, X. Xie, Q-rung orthopair fuzzy weighted induced logarithmic distance measures and their application in multiple attribute decision making. Eng. Appl. Artif. Intell. **100**, 104–167 (2021)
83. X. Zhang, A novel approach based on similarity measure for Pythagorean fuzzy multiple criteria group decision making. Int. J. Intell. Syst. **31**, 593–611 (2016)
84. F. Zhou, T.-Y. Chen, A novel distance measure for Pythagorean fuzzy sets and its applications to the technique for order preference by similarity to ideal solutions. Int. J. Comput. Intell. Syst. **12**(2), 955–969 (2019)
85. R.M. Zulqarnain, X.L. Xin, H. Garg, W.A. Khan, Aggregation operators of Pythagorean fuzzy soft sets with their application for green supplier chain management. J. Intell. Fuzzy Syst. **40**(3), 5545–5563 (2021)
86. R.M. Zulqarnain, X.L. Xin, H. Garg, R. Ali, Interaction aggregation operators to solve multi criteria decision making problem under Pythagorean fuzzy soft environment. J. Intelli. Fuzzy Syst. **41**(1), 1151–1171 (2021)
87. R.M. Zulqarnain, X.L. Xin, M. Saqlain, W.A. Khan, TOPSIS method based on the correlation coefficient of interval-valued intuitionistic fuzzy soft sets and aggregation operators with their application in decision-making. J. Math. 12 pages (2021)

Hüseyin Kamacı is an Associate Professor at the Mathematics Department in the Science and Arts Faculty of Yozgat Bozok University, Turkey. He received his M.Sc. and Ph.D. degrees in Mathematics from Bozok University, Yozgat, Turkey in 2014 and 2018, respectively. His research interests include mathematical logic, fuzzy logic, neutrosophic logic, intuitionistic fuzzy set, q-rung orthopair fuzzy set, rough set, soft computing, operational research, computational intelligence, multi-criteria decision-making, game theory. He has published many valuable articles on these scientific topics in international academic journals including Neural Computing and Applications, Soft Computing, Computational and Applied Mathematics, Applied Soft Computing, Journal of Intelligent and Fuzzy Systems, Journal of Ambient Intelligence and Humanized Computing, Turkish Journal of Mathematics, Scientia Iranica, Neutrosophic Sets and Systems, Filomat, International Journal of Machine Learning and Cybernetics, etc.

Subramanian Petchimuthu is an Assistant Professor in the Department of Mathematics at University College of Engineering Nagercoil (A Constituent College of Anna University, Chennai), Kanyakumari district, Tamilnadu, India. He has received M.Sc. degree from Manonmaniam Sundaranar University, Tamilnadu in 2005 and Ph.D. from Periyar University, Tamilnadu in 2009. His main areas of research interests are fuzzy sets, q-rung fuzzy sets, soft sets, soft matrices, multi-criteria group decision-making, group theory, ring theory, near-ring, gamma ring theory. He has many valuable publications on these issues in different scientific journals.

Chapter 5
TOPSIS Techniques on q-Rung Orthopair Fuzzy Sets and Its Extensions

V. Salsabeela, Aparna Sivadas, Bibin Mathew, and Sunil Jacob John

Abstract An effective technique for dealing with multiple-criteria decision-making (MCDM) problems of real world is the Technique for Order of Preference by Similarity to Ideal Solution (TOPSIS). q-Rung Orthopair Fuzzy Sets (q-ROFs) were introduced by Yager as a generalization of intuitionistic fuzzy sets, in which the sum of the qth powers of the (membership and non-membership) degrees is restricted to one. As the value of q increases, the feasible region for orthopair also increases. This results in more orthopairs satisfying the limitations and hence broadening the scope of representation of fuzzy information. This chapter makes an attempt to integrate the TOPSIS technique with q-ROF sets with some of its generalizations and extensions. Initially, we explore the concept of TOPSIS technique to solve MCDM problems under q-Rung Orthopair Fuzzy (q-ROF) environment and illustrate it with its application in solving a transport policy problem. Then we consider the TOPSIS technique to solve MCDM problems under q-Rung Orthopair Hesitant Fuzzy (q-ROHF) settings. To explain it we have mentioned an illustration of military aircraft overhaul effectiveness evaluation. In addition to the above-mentioned methods, we present the TOPSIS technique for solving decision-making problems under the newly introduced q-rung orthopair fuzzy soft set (q-ROFS$_f$S). Here, we tackle a problem of selection of a medical clinic utilizing the q-ROFS$_f$ TOPSIS method.

Keywords TOPSIS · q-ROFS · q-ROFNs · q-ROHFS · MCDM

5.1 Introduction

A fuzzy set $\mathbf{A}$, introduced by Zadeh on the universal set $\eth$ can be represented as $\mathbf{A} = \{\langle \varepsilon, \varkappa_{\mathbf{A}}(\varepsilon) \rangle \mid \varepsilon \in \eth\}$, where $\varkappa_{\mathbf{A}} : \eth \to [0, 1]$ is called the the membership

V. Salsabeela · A. Sivadas · B. Mathew · S. J. John (✉)
Department of Mathematics, National Institute of Technology Calicut, Calicut 673601, India
e-mail: sunil@nitc.ac.in

B. Mathew
Department of Basic Science, Amal Jyothi College of Engineering, Kanjirappally, Kottayam, Kerala 686518, India

H. Garg (ed.), *q-Rung Orthopair Fuzzy Sets*,
https://doi.org/10.1007/978-981-19-1449-2_5

function. Atanassov [1] modified this concept by proposing intuitionistic fuzzy set (IFS), in which A in $\eth$ can be described as $A = \{\langle \varepsilon, \varkappa_{\mathbf{A}}(\varepsilon), \varphi_{\mathbf{A}}(\varepsilon)\rangle \mid \varepsilon \in \eth\}$ where the functions; $\varkappa_{\mathbf{A}} : \eth \to [0, 1]$ and $\varphi_{\mathbf{A}} : \eth \to [0, 1]$ are called the membership and the non-membership functions, respectively, satisfying the condition $0 \leq \varkappa_{\mathbf{A}}(\varepsilon) + \varphi_{\mathbf{A}}(\varepsilon) \leq 1$. Further, the hesitation degree of ε can be defined as $\pi_{\mathbf{A}}(\varepsilon) = 1 - \varkappa_{\mathbf{A}}(\varepsilon) - \varphi_{\mathbf{A}}(\varepsilon)$ and $0 \leq \pi_{\mathbf{A}}(\varepsilon) \leq 1$. Later, Ronald [25] generalized IFS to get Pythagorean fuzzy sets (PFS) in which the sum of the squares of the membership and the non-membership values is bounded by 1. With this additional convenience, the set of feasible values for membership and non-membership degrees have been significantly widened. As a continuation of these works, Yager [32] introduced the q-rung orthopair fuzzy sets (q-ROFS) where the sum of the qth powers of the membership degree and the non-membership degree is restricted to one. A q-rung orthopair fuzzy set $\mathbf{A}$ on the universal set $\eth$ can be represented as $\mathbf{A} = \{\langle \varepsilon, \varkappa_{\mathbf{A}}(\varepsilon), \varphi_{\mathbf{A}}(\varepsilon)\rangle \mid \varepsilon \in \eth\}$ where $\varkappa_{\mathbf{A}} : \eth \to [0, 1]$ is membership function and $\varphi_{\mathbf{A}} : \eth \to [0, 1]$ is non-membership function of $\mathbf{A}$ with $(\varkappa_{\mathbf{A}}(\varepsilon))^q + (\varphi_{\mathbf{A}}(\varepsilon))^q \leq 1$. The hesitation degree of each $\varepsilon \in \eth$ is defined to be $\pi_{\mathbf{A}}(\varepsilon) = \big(1 - (\varkappa_{\mathbf{A}}(\varepsilon))^q - (\varphi_{\mathbf{A}}(\varepsilon))^q\big)^{1/q}$. Obviously, the catchment area for membership degrees is more and hence the flexibility in processing comprehensive information is larger than the previous types of fuzzy sets.

TOPSIS (Technique for Order Preference by Similarity to Ideal Solution) introduced by Hwang and Yoon [11] is an efficient technique in handling MCDM problems. This method is helpful in organizing, solving, analysing, comparing and ranking of alternatives in MCDM problems.

This chapter is an attempt to review the existing TOPSIS techniques in the framework of q-ROFS and two of its extensions, namely, the q-rung orthopair hesitant fuzzy set and the q-rung orthopair fuzzy soft set. Suitable examples fitting in these contexts are also adapted from [23, 30].

TOPSIS is a key component of the vast array of techniques available in the literature for decision-making. As an application, we review TOPSIS method based on q-ROFNs for MADM in transport policy problem [23]. For understanding the outstanding characteristics to deal with the uncertain information and decision-making processes in various contexts related with q-rung orthopair fuzzy sets, we refer [4, 6, 8]. For further information and more details of the q-ROFS, we refer to the studies [5, 7, 13, 22, 34].

The researchers have provided several strategies for resolving complicated MADM problems using q-ROF data. Aggregation of q-ROF data is crucial in solving such problems. Uncertainty measures are inseparable from the study of theories dealing with uncertainty [16, 30]. Entropy is one such measure that effectively quantifies the amount of uncertainty depicted by that particular structure. While similarity measures allow a comparison between two such structures. Many axiomatic definitions for entropy and similarity measures have been developed. Based on these axioms, various valid measures are proposed in the literature. Entropy measures are extensively used in MCDM problems in finding the weights of the attributes involved. In [30], several entropy measures of q-rung orthopair hesitant fuzzy sets are provided.

Based on them, a multi-attribute decision-making model using TOPSIS in q-rung orthopair hesitant fuzzy environment is presented. Further, a numerical example for decision-making, precisely a military aircraft overhaul efficacy assessment using the suggested TOPSIS framework is furnished.

Soft set is an emerging mathematical tool to deal with uncertainty. Molodtsov [18] defines a soft set as a parameterized family of subsets of the universal set where each element is considered as a set of approximate elements of the soft set. There exist various approaches to depict and manipulate uncertainty but each of them has their own difficulties as identified by Molodtsov [18]. The advantage of soft set theory in data analysis is that it does not require any grade of membership as in the case of fuzzy set theory. Thus, the theory of soft sets is highly applied in various disciplines. Molodtsov [18], Molodtsov et al. [19] applied the soft sets to fields such as game theory, operations research, Riemann integration, Perron integration, probability and so on. For the better understanding of soft sets, we refer [12, 17–19].

Being a hybrid structure of q-rung orthopair fuzzy set and soft set, q-ROFS$_f$S is more efficient, influential, and expressive in resolving the difficulty, uncertainty of membership, and non-membership in MCDM problems. In this paper, we present the TOPSIS technique in solving MCGDM problems under q-ROFS$_f$ environment [21]. Finally, examples of selecting the right medical clinic for disease diagnosis in an emergency using the TOPSIS method and evaluating psycho linguistic schools based on multiple criteria group decision-making method are provided in q-rung orthopair fuzzy soft settings.

The organization of the chapter is as follows: In Sect. 5.2, we briefly review some basic concepts of q-ROFS, q-rung orthopair hesitant fuzzy set and a hybrid structure involving soft set and q-ROFS namely q-rung orthopair fuzzy soft set. In Sect. 5.3, we have compiled TOPSIS algorithms proposed for the structures mentioned in Sect. 5.2. In Sect. 5.4, we have compiled the illustrative examples for these algorithms. TOPSIS Techniques on q-rung orthopair fuzzy soft sets are discussed in Sect. 5.5. Then illustrative examples corresponding to each cases are enlisted in Sect. 5.6. Finally, we end this chapter with a concrete conclusion and some future directions in Sect. 5.7.

5.2 Preliminaries

Definition 5.1 [32] Consider the universe of discourse $\eth$, a q-ROFS $\mathbf{A}$ on $\eth$ is given as

$$\mathbf{A} = \{\langle \varepsilon, \varkappa_{\mathbf{A}}(\varepsilon), \varphi_{\mathbf{A}}(\varepsilon)\rangle \mid \varepsilon \in \eth\}$$

where $\varkappa_{\mathbf{A}} : \eth \to [0, 1]$ represents the degree of membership of the element ε to $\mathbf{A}$ and $\varphi_{\mathbf{A}} : \eth \to [0, 1]$ represents the degree of non-membership of ε to $\mathbf{A}$ satisfying the condition: $0 \leq \varkappa_{\mathbf{A}}(\varepsilon)^q + \varphi_{\mathbf{A}}(\varepsilon)^q \leq 1, (q \geq 1)$.

The degree of indeterminacy of an element $\varepsilon \in \eth$ is defined as $\pi_A(\varepsilon) = \left(1 - (\varkappa_A(\varepsilon))^q - (\varphi_A(\varepsilon))^q\right)^{1/q}$.

Definition 5.2 [21] Consider a q-ROFS $A = \{\langle \varepsilon, \varkappa_A(\varepsilon), \varphi_{\mathbf{A}}(\varepsilon)\rangle \mid \varepsilon \in \eth\}$ be a q-ROFS, Then each pair $\langle \varkappa_{\mathbf{A}}(\varepsilon), \varphi_{\mathbf{A}}(\varepsilon)\rangle$ is termed as a q-rung orthopair fuzzy number (q-ROFN) and is signified as $\langle \varkappa, \varphi\rangle$.

Basic Operations of q-ROFNs

Definition 5.3 [15] Consider three q-ROFNs, $\alpha = \langle \varkappa, \varphi\rangle$, $\alpha_1 = \langle \varkappa_1, \varphi_1\rangle$, and $\alpha_2 = \langle \varkappa_2, \varphi_2\rangle$, then the basic operations are listed below:

(i) $\alpha^c = \langle \varphi, \varkappa\rangle$ where α^c denotes the complement of α
(ii) $\alpha_1 \vee \alpha_2 = \langle \max\{\varkappa_1, \varkappa_2\}, \min\{\varphi_1, \varphi_2\}\rangle$
(iii) $\alpha_1 \wedge \alpha_2 = \langle \min\{\varkappa_1, \varkappa_2\}, \max\{\varphi_1, \varphi_2\}\rangle$
(iv) $\alpha_1 \preceq \alpha_2$ if $\varkappa_1 \leq \varkappa_2, \varphi_1 \geq \varphi_2$
(v) $\alpha_1 \oplus \alpha_2 = \left\langle \left(\varkappa_1^q + \varkappa_2^q - \varkappa_1^q \varkappa_2^q\right)^{1/q}, \varphi_1\varphi_2 \right\rangle$
(vi) $\alpha_1 \otimes \alpha_2 = \left\langle \varkappa_1\varkappa_2, \left(\varphi_1^q + \varphi_2^q - \varphi_1^q\varphi_2^q\right)^{1/q} \right\rangle$
(vii) $\lambda\alpha_1 = \left\langle \left(1 - (1 - \varkappa_1^q)^\lambda\right)^{1/q}, \varphi_1^\lambda \right\rangle$
(viii) $\alpha_1^\lambda = \left\langle \varkappa_1^\lambda, \left(1 - (1 - \varphi_1^q)^\lambda\right)^{1/q} \right\rangle$
(ix) $\alpha_1 \ominus \alpha_2 = \left\langle \sqrt[q]{\frac{\varkappa_1^q - \varkappa_2^q}{1 - \varkappa_2^q}}, \frac{\varphi_1}{\varphi_2} \right\rangle$, if $\varkappa_1 \geq \varkappa_2, \varphi_1 \leq \min\left\{\varphi_2, \frac{\varphi_2\pi_1}{\pi_2}\right\}$
(x) $\alpha_1 \oslash \alpha_2 = \left\langle \frac{\varkappa_1}{\varkappa_2}, \sqrt[q]{\frac{\varphi_1^q - \varphi_2^q}{1 - \varphi_2^q}} \right\rangle$ if $\varphi_1 \geq \varphi_2, \varkappa_1 \leq \min\left\{\varkappa_2, \frac{\varkappa_2\pi_1}{\pi_2}\right\}$.

Definition 5.4 [32] Suppose $\alpha = \langle \varkappa, \varphi\rangle$ is a q-ROFN, score function of α, $score(\alpha) = \varkappa^q - \varphi^q$.

Definition 5.5 [32] Consider the q-ROFN $\alpha = \langle \varkappa, \varphi\rangle$, the accuracy function is defined as $acc(\alpha) = \varkappa^q + \varphi^q$.

Evidently $s(\alpha) \in [-1, 1]$ and $acc(\alpha) \in [0, 1]$. Using the values of score function and accuracy function q-ROFNs are ordered.

Definition 5.6 [15] Suppose $\alpha_1 = \langle \varkappa_1, \varphi_1\rangle$ and $\alpha_2 = \langle \varkappa_2, \varphi_2\rangle$ be two q-ROFNs, if

(1) $score(\alpha_1) > score(\alpha_2)$, then $\alpha_1 > \alpha_2$
(2) $score(\alpha_1) = score(\alpha_2)$, then consider $acc(\alpha_1)$ and $acc(\alpha_2)$, if $acc(\alpha_1) > acc(\alpha_2)$, $\alpha_1 > \alpha_2$ and if $acc(\alpha_1) = acc(\alpha_2)$, then $\alpha_1 = \alpha_2$.

In Sect. 5.5, we are discussing q-rung orthopair fuzzy soft sets (q-ROFS$_f$S), which is a combination of q-rung orthopair fuzzy sets and soft sets. A soft set is a parameterized family of subsets of the universal set and gives a collection of approximate descriptions of an object. The basic definitions of soft sets and fuzzy soft sets are given below.

Definition 5.7 [2] Consider $\eth$ as the universe of discourse, $\mathcal{P}$ denotes a set of parameters and $P \subseteq \mathcal{P}$. The pair (F, P) is known to be a soft set over $\eth$, where F is a function defined by $F: P \to \mathbb{P}(\eth)$ where $\mathbb{P}(\eth)$ represents the power set of $\eth$.

Definition 5.8 [3] Consider $\eth$ as the universe of discourse, a pair $(\hat{F}, P)$ is known to be a fuzzy soft set $(FS_f S)$ over $\eth$, where $\hat{F}$ is a function defined by $\hat{F}: P \to F^{(\eth)}$; $F^{(\eth)}$ stand for the collections of all fuzzy subsets of $\eth$.

Definition 5.9 [10] Consider $\eth$ as the universe of discourse, $P \subseteq \mathcal{P}$. A pair $(\mathcal{H}, P)$ is called a q-rung orthopair fuzzy soft set (q-ROFS$_f$S) over $\eth$, where $\mathcal{H}$ is a function given by $\mathcal{H}: P \to q - ROFS^{(\eth)}$; $q - ROFS^{(\eth)}$ represents the collections of all q-rung orthopair fuzzy sets (q-ROFS) of $\eth$, and is given as

$$\mathcal{H}_{p_j}(\varepsilon_i) = \left\{\left\langle \varepsilon_i, u_j(\varepsilon_i), v_j(\varepsilon_i)\right\rangle \mid \varepsilon_i \in \eth, q \geq 1\right\}$$

Here, $u_j(\varepsilon_i), v_j(\varepsilon_i)$ denotes the membership and non-membership degrees of $\varepsilon_i \in \eth$ to $\mathcal{H}_{p_j}$, respectively, and satisfying the condition that $0 \leq (u_j(\varepsilon_i))^q + (v_j(\varepsilon_i))^q \leq 1$ and $q \geq 1$.

Definition 5.10 [29] Consider the universe of discourse $\eth$, a hesitant fuzzy set on $\eth$ is defined using a function h on $\eth$ whose image is a subset of [0,1].

$$h: \eth \to \mathbb{P}([0, 1]).$$

Employing the HFSs and q-ROFs, Liu et al. [14] introduced q-rung orthopair hesitant fuzzy set (q-ROHFS).

Definition 5.11 [14] Consider the universe of discourse $\eth$, a q-ROHFS H on $\eth$ is given as:

$$\mathrm{H} = \left\{< \varepsilon, M_{\mathrm{H}}(\varepsilon), N_{\mathrm{H}}(\varepsilon) >_q \mid \varepsilon \in \eth\right\}$$

where $M_{\mathrm{H}}(\varepsilon)$ and $N_{\mathrm{H}}(\varepsilon)$ are non-empty finite subsets of [0, 1], and correspondingly denotes the possible membership and non-membership degrees of ε and for any $\varepsilon \in \eth$, for every $u_{\mathrm{H}}(\varepsilon) \in M_{\mathrm{H}}(\varepsilon)$ and $v_{\mathrm{H}}(\varepsilon) \in N_{\mathrm{H}}(\varepsilon)$ satisfies:

$$0 \leq u_{\mathrm{H}}(\varepsilon), v_{\mathrm{H}}(\varepsilon) \leq 1, \quad 0 \leq u_{\mathrm{H}}^q(\varepsilon) + v_{\mathrm{H}}^q(\varepsilon) \leq 1.$$

Definition 5.12 If $\mathrm{H} = \left\{\langle \varepsilon, M_{\mathrm{H}}(\varepsilon), N_{\mathrm{H}}(\varepsilon)\rangle_q \mid \varepsilon \in \eth\right\}$ is q-ROHFS then each $\langle M_{\mathrm{H}}(\varepsilon), N_{\mathrm{H}}(\varepsilon)\rangle_q$ is termed as the q-rung orthopair hesitant fuzzy number (q-ROHFN) and is denoted as
$\hbar = \langle M_{\hbar}, N_{\hbar}\rangle_q$.

The degree of indeterminacy of q-ROHFN is expressed as: $\pi_{\hbar}(\varepsilon) = [1 - (u_{\hbar}(\varepsilon)^q + v_{\hbar}(\varepsilon)^q)]^{\frac{1}{q}}$.

Definition 5.13 [30] Let $\hbar_1 = \langle M_{\hbar_1}, N_{\hbar_1}\rangle_q$, $\hbar_2 = \langle M_{\hbar_2}, N_{\hbar_2}\rangle_q$, $\hbar_3 = \langle M_{\hbar_3}, N_{\hbar_3}\rangle_q$ be three q-ROHFNs in the collection of q-ROHFNs (q-ROHFN($\eth$)). Consider a function $\diamondsuit: q - ROHFN(\eth) \times q - ROHFN(\eth) \to [0, 1]$ satisfying the following :

(i) $0 \leq \diamondsuit(\hbar_1, \hbar_2) \leq 1$
(ii) $\diamondsuit(\hbar_1, \hbar_2) = \diamondsuit(\hbar_2, \hbar_1)$
(iii) $\diamondsuit(\hbar_1, \hbar_2) = 0 \Leftrightarrow \hbar_1 = \hbar_2$
(iv) $\diamondsuit(\hbar_1, \hbar_3) + \diamondsuit(\hbar_2, \hbar_3) \geq \diamondsuit(\hbar_1, \hbar_2)$.
then $\diamondsuit(\hbar_1, \hbar_2)$ is the distance measure between $\hbar_1$ and $\hbar_2$.

For the calculation of distance measures and entropy measures of q-ROHFNs, the following assumptions are made:

(1) The elements in $M_{\hbar_1}$ and $N_{\hbar_1}$ of q-ROHFN $\hbar_1(\varepsilon)$ are arranged in ascending order. Also $u_{\hbar_1}^{(i)}$ and $v_{\hbar_1}^{(i)}$, respectively, represent the ith element values from smaller to greater in $M_{\hbar_1}$ and $N_{\hbar_1}$.
(2) Let $\hbar_1 = \langle M_{\hbar_1}, N_{\hbar_1}\rangle_q$, $\hbar_2 = \langle M_{\hbar_2}, N_{\hbar_2}\rangle_q$ be two q-ROHFNs and $|M_{\hbar_1}|$ and $|M_{\hbar_2}|$ denote the number of elements in $M_{\hbar_1}$ and $M_{\hbar_2}$, respectively. If $|M_{\hbar_1}| \neq |M_{\hbar_2}|$ and $|M_{\hbar_1}| > |M_{\hbar_2}|$ then, $M_{\hbar_2}$ are added with elements until $|M_{\hbar_1}| = |M_{\hbar_2}|$. If $|M_{\hbar_1}| < |M_{\hbar_2}|$, then $M_{\hbar_1}$ are added with elements until $|M_{\hbar_1}| = |M_{\hbar_2}|$. The same process is repeated for $N_{\hbar_1}$ and $N_{\hbar_2}$.

Definition 5.14 [30] Consider a q-ROHFN $\hbar_1 = \langle M_{\hbar_1}, N_{\hbar_1}\rangle_q$ and let $\hbar = \{\lambda_i \mid i = 1, 2, \ldots, l_n\}$ denote either $M_{\hbar_1}$ or $N_{\hbar_1}$ then elements are added to $\hbar$ using the equation: $\theta\hbar^+ + (1-\theta)\hbar^-, 0 \leq \theta \leq 1$ where $\hbar^+ = \max\{\lambda_i \mid i = 1, 2, \ldots, l_n\}$ and $\hbar^- = \min\{\lambda_i \mid i = 1, 2, \ldots, l_n\}$.

Definition 5.15 [30] Let $\hbar_1 = \langle M_{\hbar_1}, N_{\hbar_1}\rangle_q$, $\hbar_2 = \langle M_{\hbar_2}, N_{\hbar_2}\rangle_q$ denote q-ROHFNs, following are some distance measures for them:

(i) q-ROHF hamming distance:

$$\diamondsuit_1(\hbar_1, \hbar_2) = \frac{1}{2l}\sum_{i=1}^{l}\left\{\left|u_1^{(i)}(\varepsilon) - u_2^{(i)}(\varepsilon)\right| + \left|v_1^{(i)}(\varepsilon) - v_2^{(i)}(\varepsilon)\right| \left|\pi_1^{(i)}(\varepsilon) - \pi_2^{(i)}(\varepsilon)\right|\right\}$$

(ii) q-ROHF Euclidean distance:

$$\diamondsuit_2(\hbar_1, \hbar_2) = \left\{\frac{1}{2l}\sum_{i=1}^{l}\left\{\left| u_1^{(i)}(\varepsilon) - u_2^{(i)}(\varepsilon)\right|^2 + \left|v_1^{(i)}(\varepsilon) - v_2^{(i)}(\varepsilon)\right|^2 + \left|\pi_1^{(i)}(\varepsilon) - \pi_2^{(i)}(\varepsilon)\right|^2\right\}\right\}^{\frac{1}{2}}$$

(iii) q-ROHF generalized Euclidean distance:

$$\diamondsuit_3 (\hbar_1, \hbar_2) = \left\{ \frac{1}{2l} \sum_{i=1}^{l} \left\{ \left| u_1^{(i)}(\varepsilon) - u_2^{(i)}(\varepsilon) \right|^k + \left| v_1^{(i)}(\varepsilon) - v_2^{(i)}(\varepsilon) \right|^k + \left| \pi_1^{(i)}(\varepsilon) - \pi_2^{(i)}(\varepsilon) \right|^k \right\} \right\}^{\frac{1}{k}}, \quad k > 0.$$

Definition 5.16 [30] Consider q-ROHFSs H_1 and H_2 with q-ROHFNs $\hbar_1 = \langle M_{\hbar_1}, N_{\hbar_1} \rangle_q$ and $\hbar_2 = \langle M_{\hbar_2}, N_{\hbar_2} \rangle_q$, following are some distance measures for H_1 and H_2:

(i) The Hamming distance

$$\diamondsuit_1' (\mathrm{H}_1, \mathrm{H}_2) = \frac{1}{n} \sum_{j=1}^{n} \left[\frac{1}{2l_j} \sum_{i=1}^{l_j} \left\{ \left| u_1^{(i)} (\varepsilon_j) - u_2^{(i)} (\varepsilon_j) \right| + \left| v_1^{(i)} (\varepsilon_j) - v_2^{(i)} (\varepsilon_j) \right| + \left| \pi_1^{(i)} (\varepsilon_j) - \pi_2^{(i)} (\varepsilon_j) \right| \right\} \right].$$

(ii) The Euclidean distance

$$\diamondsuit_2' (\mathrm{H}_1, \mathrm{H}_2) = \left\{ \frac{1}{n} \sum_{j=1}^{n} \left[\frac{1}{2l_j} \sum_{i=1}^{l_j} \left\{ \left| u_1^{(i)} (\varepsilon_j) - u_2^{(i)} (\varepsilon_j) \right|^2 + \left| v_1^{(i)} (\varepsilon_j) - v_2^{(i)} (\varepsilon_j) \right|^2 + \left| \pi_1^{(i)} (\varepsilon_j) - \pi_2^{(i)} (\varepsilon_j) \right|^2 \right\} \right] \right\}^{\frac{1}{2}}.$$

(iii) The generalized Euclidean distance

$$\diamondsuit_3' (\mathrm{H}_1, \mathrm{H}_2) = \left\{ \frac{1}{n} \sum_{j=1}^{n} \left[\frac{1}{2l_j} \sum_{i=1}^{l_j} \left\{ u_1^{(i)} (\varepsilon_j) - u_2^{(i)} (\varepsilon_j) \right|^k + \left| v_1^{(i)} (\varepsilon_j) - v_2^{(i)} (\varepsilon_j) \right|^k + \left| \pi_1^{(i)} (\varepsilon_j) - \pi_2^{(i)} (\varepsilon_j) \right|^k \right\} \right] \right\}^{\frac{1}{k}}, \quad k > 0.$$

Definition 5.17 [30] For q-ROHFSs, H_1 and H_2 with q-ROHFNs $\hbar_1 = \langle M_{\hbar_1}, N_{\hbar_1} \rangle_q$ and $\hbar_2 = \langle M_{\hbar_2}, N_{\hbar_2} \rangle_q$, respectively, the weighted generalized distance between H_1 and H_2 is given as:

$$\Diamond_w^1(\mathrm{H}_1,\mathrm{H}_2)=\left\{\sum_{j=1}^{n}w_j\left[\frac{1}{2l_j}\sum_{i=1}^{l_j}\left\{\left|u_1^{(i)}(\varepsilon_j)-u_2^{(i)}(\varepsilon_j)\right|^k\right.\right.\right.$$
$$\left.\left.\left.+\left|v_1^{(i)}(\varepsilon_j)-v_2^{(i)}(\varepsilon_j)\right|^k+\left|\pi_1^{(i)}(\varepsilon_j)-\pi_2^{(i)}(\varepsilon_j)\right|^k\right\}\right]\right\}^{\frac{1}{k}}$$

where $k > 0$ is a constant.

Definition 5.18 [30] Let q-ROHFS(Ə) denote the set of all q-ROHFSs over Ə and let $\mathrm{H}_1, \mathrm{H}_2 \in$ q-ROHFS(Ə), a function $e : q - ROHFS(Ə) \to [0, 1]$ satisfying the following properties is an entropy measure of q-ROHFS.

1. $e(\mathrm{H}) = 0 \Leftrightarrow \mathrm{H}$ is a distinct set
2. $e(\mathrm{H}) = 1 \Leftrightarrow M_{\mathrm{H}}(\varepsilon) = N_{\mathrm{H}}(\varepsilon) = \{0\}, \forall \varepsilon \in Ə$
3. $e(\mathrm{H}) = e(\mathrm{H}^c)$
4. $e(\mathrm{H}_1) \geq e(\mathrm{H}_2)$, if $\Diamond(\mathrm{H}_1, \mathrm{H}^0) \leq \Diamond(\mathrm{H}_2, \mathrm{H}^0)$ where $\mathrm{H}^0 = \langle \varepsilon, \{0\}, \{0\} \mid \varepsilon \in Ə \rangle$.

Some entropy measures for q-ROHFS are enlisted below [30]:

(1) $e_1(\mathrm{H}) = 1 - \frac{1}{n}\sum_{j=1}^{n}\left[\frac{1}{2l_j}\sum_{i=1}^{l_j}\left\{\left|u_{\mathrm{H}}^{(i)}(\varepsilon_j)\right| + \left|v_{\mathrm{H}}^{(i)}(\varepsilon_j)\right| + \left|\pi_{\mathrm{H}}^{(i)}(\varepsilon_j) - 1\right|\right\}\right]$

(2) $e_2(\mathrm{H}) = 1 - \sqrt{\frac{1}{n}\sum_{j=1}^{n}\left[\frac{1}{2l_j}\sum_{i=1}^{l_j}\left\{\left|u_{\mathrm{H}}^{(i)}(\varepsilon_j)\right|^2 + \left|v_{\mathrm{H}}^{(i)}(\varepsilon_j)\right|^2 + \left|\pi_{\mathrm{H}}^{(i)}(\varepsilon_j) - 1\right|^2\right\}\right]}$

(3) $e_3(\mathrm{H}) = 1 - \left(\frac{1}{n}\sum_{j=1}^{n}\left[\frac{1}{2l_j}\sum_{i=1}^{l_j}\left\{\left|u_{\mathrm{H}}^{(i)}(\varepsilon_j)\right|^k + \left|v_{\mathrm{H}}^{(i)}(\varepsilon_j)\right|^k + \left|\pi_{\mathrm{H}}^{(i)}(\varepsilon_j) - 1\right|^k\right\}\right]\right)^{\frac{1}{k}}, k > 0$

(4) $e_4(\mathrm{H}) = 1 - \frac{1}{n}\sum_{j=1}\left[\frac{1}{2l_j}\sum_{i=1}\left\{\left|\left(u_{\mathrm{H}}^{(i)}(\varepsilon_j)\right)^2\right| + \left|\left(v_{\mathrm{H}}^{(i)}(\varepsilon_j)\right)^2\right| + \left|\left(\pi_{\mathrm{H}}^{(i)}(\varepsilon_j)\right)^2 - 1\right|\right\}\right]$.

5.2.1 TOPSIS

Eraslan [2] suggested that the best alternative be determined by the shortest distance from the positive ideal solution (PIS) and the greatest distance from the negative ideal solution (NIS). This method regards the distances to both PIS and NIS at the same time, and a preferential order is ranked based on their proximity. The PIS aims to maximize benefit criteria while minimizing cost criteria, whereas the NIS aims to maximize cost criteria while minimizing benefit criteria. TOPSIS primarily uses Euclidean distances to measure the alternatives with their PIS and NIS [28].

The TOPSIS process includes the following processes: Decision matrix normalization, distance measures, and aggregation operators [28]. Ahead of the start of the procedure, a decision matrix is usually needed. The decision matrix lists competitive alternatives row by row, with attribute ratings or scores column by column. Assume that the available data, including quantitative and qualitative information, are filled out in the given decision matrix. Normalization is the process of conforming or reducing these scores to a norm or standard. Normalization of quantifiable or linguistic data could begin with a transformation to a linear scale. The normalized process is usually performed column-wise to compare the alternatives on each attribute, and the normalized value will be a positive value between 0 and 1. Computational issues caused by different measurements in the decision matrix are thus eliminated. The procedural steps of TOPSIS are listed below [2]:

Step i. Defining the problem (purpose of determining and identifying assessment criteria).
Step ii. Creating a scorecard and criterion weights, as well as a decision matrix for each decision-maker.
Step iii. Creating the weighted standard (normalized) decision matrix.
Step iv. Creating the weighted normalized decision matrix.
Step v. Identifying the positive and negative ideal solutions (PIS) and (NIS).
Step vi. Calculating the separation measures from positive ideal solution and negative ideal solution.
Step vii. Calculating the alternative's proximity to the ideal solution.
Step viii. Ranking the preference order.

5.3 TOPSIS Techniques on q-ROFS

Being a major technique in solving MADM problems, TOPSIS has also been extended to q-ROFS. Following is the TOPSIS algorithm presented by Riaz et al. [23] based on q-ROFNs.

Algorithm I

Step i. Determine the problem
Let $\vartheta = \{\vartheta_i : \quad i = 1, 2, 3, \ldots n\}$ be a set of decision-makers, $\digamma = \{\digamma_i : i = 1, 2, 3, \ldots n\}$ is the finite set of alternatives and $\mho = \{\mho_i : i = 1, 2, 3, \ldots n\}$, the set of parameters.
Step ii. Using the provided linguistic terms, create a weighted parameter matrix $\mathcal{Q}$ as

$$\mathcal{Q} = \left[y_{ij}\right]_{n\times m} = \begin{pmatrix} y_{11} & y_{12} & \cdots & y_{1m} \\ y_{21} & y_{22} & \cdots & y_{2m} \\ \vdots & \vdots & \ddots & \vdots \\ y_{i1} & y_{i2} & \cdots & y_{im} \\ \vdots & \vdots & \ddots & \vdots \\ y_{n1} & y_{n2} & \cdots & y_{nm} \end{pmatrix}$$

where y_{ij} denote the weight provided by the expert ϑ_i for the alternative $\digamma i$ in view of the linguistic variables.

Step iii. Formulate the normalized weighted matrix

$$\hat{N} = \left[\hat{n}_{ij}\right]_{n\times m} = \begin{pmatrix} \hat{n}_{11} & \hat{n}_{12} & \cdots & \hat{n}_{1m} \\ \hat{n}_{21} & \hat{n}_{22} & \cdots & \hat{n}_{2m} \\ \vdots & \vdots & \ddots & \vdots \\ \hat{n}_{i1} & \hat{n}_{i2} & \cdots & \hat{n}_{im} \\ \vdots & \vdots & \ddots & \vdots \\ \hat{n}_{n1} & \hat{n}_{n2} & \cdots & \hat{n}_{nm} \end{pmatrix}$$

where $\hat{n}_{ij} = \frac{y_{ij}}{\sqrt{\sum_{i=1}^{n} y_{ij}^2}}$ and determine the weighted vector $\mathfrak{y} = (\mathfrak{y}_1, \mathfrak{y}_2, \ldots, \mathfrak{y}_m)$, where $\mathfrak{y}_i = \frac{y_i}{\sum_{\alpha=1}^{n} y_{\alpha i}}$ and $y_j = \frac{\sum_{i=1}^{n} \hat{n}_{ij}}{n}$. Notice that $\Sigma_{\alpha=1}^{n} y_{\alpha i}$ gives the aggregate of weights of *ith* column in $\mathcal{Q}$.

Step iv. Create q-ROFS decision matrix

$$Z_i = \left[z_{jk}^i\right]_{l\times m} = \begin{pmatrix} z_{11}^i & z_{12}^i & \cdots & z_{1m}^i \\ z_{21}^i & z_{22}^i & \cdots & z_{2m}^i \\ \vdots & \vdots & \ddots & \vdots \\ z_{j1}^i & z_{j2}^i & \cdots & z_{jm}^i \\ \vdots & \vdots & \ddots & \vdots \\ z_{l1}^i & z_{l2}^i & \cdots & z_{lm}^i \end{pmatrix}$$

where z_{ik}^i is a q-ROFS element, for *ith* decision-maker. After that obtain the aggregating matrix

$$A = \frac{Z_1 + Z_2 + \cdots + Z_n}{n} = \left[\dot{z}_{jk}\right]_{l\times m}$$

Step v. Obtain the weighted q-ROFS decision matrix.

$$\daleth = \left[\daleth_{jk}\right]_{l\times m} = \begin{pmatrix} \daleth_{11} & \daleth_{12} & \cdots & \daleth_{1m} \\ \daleth_{21} & \daleth_{22} & \cdots & \daleth_{2m} \\ \vdots & \vdots & \ddots & \vdots \\ \daleth_{j1} & \daleth_{j2} & \cdots & \daleth_{jm} \\ \vdots & \vdots & \ddots & \vdots \\ \daleth_{l1} & \daleth_{l2} & \cdots & \daleth_{lm} \end{pmatrix}$$

where $\daleth_{jk} = \mathfrak{y}_k \times \dot{z}_{jk}$

Step vi. Create q-ROFS-valued positive ideal solution (q-ROFSV-PIS) and q-ROFS-valued negative ideal solution (q-ROFSV-NIS) by using in order

$$\begin{aligned} \text{q-ROFSV-PIS} &= \left\{\daleth_1^+, \daleth_2^+, \ldots, \daleth_l^+\right\} \\ &= \left\{\left(\vee_k \daleth_{jk}, \wedge_k \daleth_{jk}\right) : k = 1, 2, \ldots, m\right\} \end{aligned}$$

and

$$\begin{aligned}\text{q-ROFSV-NIS} &= \left\{\gimel_1^-, \gimel_2^-, \ldots, \gimel_l^-\right\} \\ &= \left\{\left(\wedge_k \gimel_{jk}, \vee_k \gimel_{jk}\right) : k = 1, 2, \ldots, m\right\}\end{aligned}$$

where $\vee$ and $\wedge$, respectively, indicate the union and intersection of q-ROFS .
Step vii. Compute q-ROFS-Euclidean distances of each $\gimel_{ij}$ from ideal solutions, using

$$\begin{aligned}\left(\varrho_j^+\right)^2 = \sum_{k=1}^{m} \Bigg\{ & \frac{1}{2}\left(\gimel_{jk}^+ - {}^+\gimel_k^+\right)^2 + \frac{1}{2}\left(\daleth_{jk}^+ - {}^+\right. \\ & \left.\daleth_k^+\right)^2 + \frac{1}{2}\left(\gimel_{jk}^- - {}^+\gimel_k^-\right)^2 + \frac{1}{2}\left(\daleth_{jk}^- - {}^+\daleth_k^-\right)^2 \Bigg\}\end{aligned}$$

and

$$\begin{aligned}\left(\varrho_j^-\right)^2 = \sum_{k=1}^{m} \Bigg\{ & \frac{1}{2}\left(\gimel_{jk}^+ - {}^-\gimel_k^+\right)^2 + \frac{1}{2}\left(\daleth_{jk}^+ - {}^-\daleth_k^+\right)^2 + \frac{1}{2}\left(\gimel_{jk}^- - \gimel_k^-\right)^2 \\ & + \frac{1}{2}\left(\daleth_{jk}^- - {}^-\daleth_k^-\right)^2 \Bigg\}.\end{aligned}$$

Step viii. Compute the closeness coefficient of each alternative with ideal solution by using

$$\partial^*\left(\gimel_j\right) = \frac{\varrho_j^-}{\varrho_j^+ + \varrho_j^-} \quad \text{which lies in} \quad [0, 1].$$

Step ix. Order the alternatives.
Figure 5.1 depicts the steps in the algorithm [23].

5.4 Combined Weighting TOPSIS MADM Using *q*-ROHFS

Being a structure involving both q-rung orthopair fuzzy sets (q-ROFSs) and hesitant fuzzy sets (HFSs), q-rung orthopair hesitant fuzzy sets (q-ROHFSs) are more operative in managing the ambiguity and expert hesitancy of membership and non-membership in MADM problems. A multiple attribute decision-making method using the q-rung orthopair hesitant fuzzy sets was developed in [35]. Viewing the advantages of q-ROHFSs, Wang et al. [30] suggested the following improved TOPSIS model for MADM problems under q-rung orthopair hesitant fuzzy setting.

Algorithm II

Phase I: Finding Attribute Weights Based on Combined Weighting

Following are the steps involved in the attribute weight solution model using combined weighting.

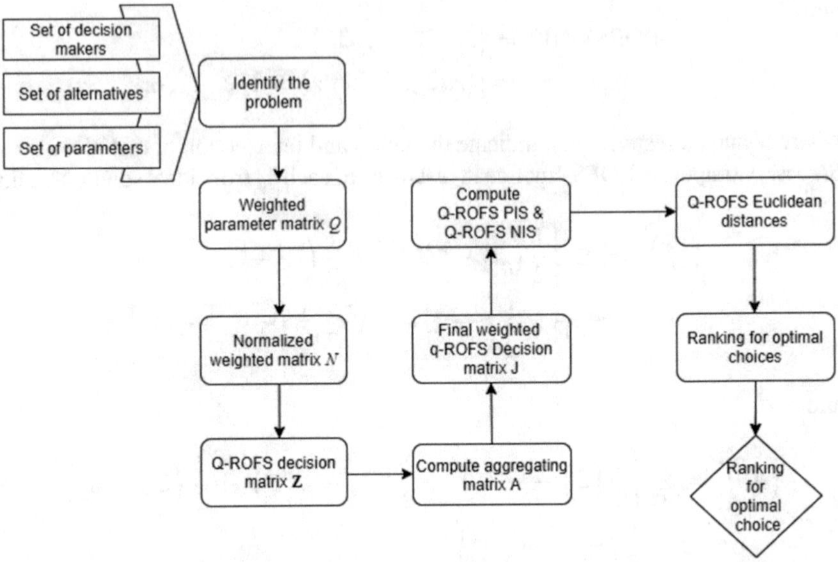

Fig. 5.1 Flowchart of Algorithm I

Step i. Formulating a *q*-rung orthopair hesitant fuzzy decision matrix.
Consider the set of alternatives, $Z = \{z_1, z_2, \ldots, z_i\}$ and the set of attributes $B = \{b_1, b_2, \ldots, b_j\}$, the assessment values of each alternative under the mentioned criteria are represented as a q-ROHFN $\hbar_{ij} = \langle M_\hbar, N_\hbar \rangle$. Hence, the decision matrix is obtained as $Q = \left(\hbar_{ij}\right)_{m \times n}$ taking $i = 1, 2, \ldots, m$ and $j = 1, 2, \ldots, n$. Using the assumptions listed for calculating distance measures, the decision matrix $Q = \left(\hbar_{ij}\right)_{m \times n}$ is developed into $Q' = \left(\hbar'_{ij}\right)_{m \times n}$.
Step ii. Constituting the initial judgment matrix.
Experts assess the attribute set B involved, and compare their inputs with each other using values in between 1 and 9, and provide the initial judgment matrix B' as:

$$B' = \begin{vmatrix} b_{11} & b_{12} & \cdots & b_{1n} \\ b_{21} & b_{22} & \cdots & b_{2n} \\ \vdots & \ddots & \vdots & \vdots \\ b_{n1} & b_{n2} & \cdots & b_{nn} \end{vmatrix}$$

Step iii. Optimizing the initial judgment matrix.
Since the judgment matrix B' is excessively subjective, an improved method from [31] is utilized to compute the attribute weights to lessen the subjective impact. In this method, the original judgment matrix is optimized using Eq. (5.1) to obtain the matrix F.

$$F_{ij} = \begin{Bmatrix} \frac{r_i - r_j}{r_{\max} - r_{\min}}(\lambda - 1) + 1, r_i \geq r_j \\ \left[\frac{r_j - r_i}{r_{\max} - r_{\min}}(\lambda - 1) + 1\right]^{-1}, r_i < r_j \end{Bmatrix} \tag{5.1}$$

$$r_i = \sum_{j=1}^{n} b_{ij} \tag{5.2}$$

$$\lambda = \frac{r_{\max}}{r_{\min}} \tag{5.3}$$

where b_{ij} is an entry in B', $r_{\max}$ and $r_{\min}$, respectively, denote the maximum and the minimum values of r_i; n is the order of B'.

To further limit the subjective effect of experts, the matrix F is optimized using the Eq. (5.5). Here, matrix G is obtained as the best transfer matrix of F and matrix P as the optimized matrix.

$$G_{ij} = \frac{1}{n}\sum_{k=1}^{n}\left(\log_{10} f_{ik} - \log_{10} f_{jk}\right) \tag{5.4}$$

$$p_{ij} = 10^{g_{ij}}. \tag{5.5}$$

Step iv. Obtaining the weights of attributes using Firefly algorithm.

The determination of attribute weights is not objective and exact while using the conventional AHP. Here, the weights of attributes, denoted as ρ_i are obtained by solving a non-linear multi-objective optimization problem constructed by employing the Firefly algorithm [33]. Figure 5.2 depicts the flowchart of the algorithm [30]. Considering that the judgment matrix meets the conditions for consistency, the required objective function in Firefly algorithm could be generated as in Eq. (5.6)

$$\begin{cases} \min \text{CIF}(n) = \frac{1}{n}\sum_{i=1}^{n}\left|\sum_{l=1}^{n}(p_{il}\rho_l) - n\rho_i\right| \\ \text{s.t. } \rho_l > 0, l = 1, 2, \ldots, n; \sum_{l=1}^{n}\rho_l = 1 \end{cases} \tag{5.6}$$

in which CIF(n) is the consistency index function and ρ_l is the optimization variable. The judgment matrix is made more accurate by examining the consistency of this matrix, which is analogous to solving the Eq. (5.6).

Step v. Obtaining the objective weights (ρ_j) of attributes.

The objective weights are calculated using the Formula (5.7), which uses the entropy of q-ROHFSs in the decision matrix Q'.

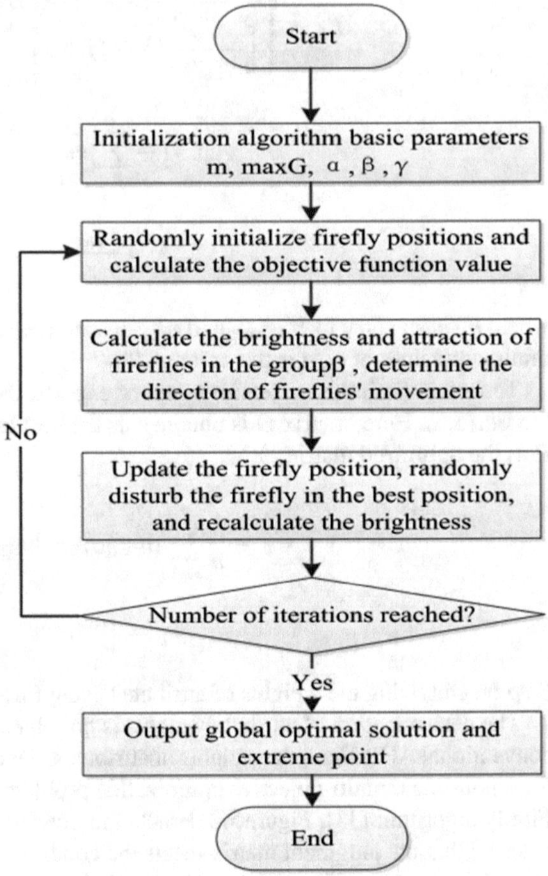

Fig. 5.2 Flowchart of firefly algorithm

$$\rho_j = \frac{1 - e\left(b_j\right)}{n - \sum_{j=1}^{n} e\left(b_j\right)} \tag{5.7}$$

where

$$e(b) = 1 - \left(\frac{1}{n}\sum_{j=1}^{n}\left[\frac{1}{2l_j}\sum_{i=1}^{l_j}\left\{\left|u_b^{(i)}\left(\varepsilon_j\right)\right|^k + \left|v_b^{(i)}\left(\varepsilon_j\right)\right|^k + \left|\pi_b^{(i)}\left(\varepsilon_j\right) - 1\right|^k\right\}\right]\right)^{\frac{1}{k}}.$$

Step vi. Combining the weights to get attribute weights

The subjective and objective weights obtained in ***Step iv*** and ***Step v***, respectively, are combined using the formula depicted in Eq. (5.8).

$$\rho_t = c\rho_j + (1-c)\rho_i \tag{5.8}$$

where: $0 \le c \le 1$.

Phase II: Extending TOPSIS to MADM with q-ROHFS

Step vii. Obtaining the positive and negative ideal solutions

The extended decision matrix Q' is employed to construct positive ideal solution $Q'^{+} = (\hbar_1^+, \hbar_2^+, \ldots, \hbar_m^+)$ and the negative ideal solution $Q'^{-} = (\hbar_1^-, \hbar_2^-, \ldots, \hbar_m^-)$. The attributes involved are classified as benefit type and cost type and the ideal solutions are attained with regard to the Formulas (5.9) and (5.10).

$$\hbar_j^+ = \begin{cases} \max_i \hbar'_{ij}, j \in U_1 \\ \min_i \hbar'_{ij}, j \in U_2 \end{cases} \tag{5.9}$$

$$\hbar_j^- = \begin{cases} \min_i \hbar'_{ij}, j \in U_1 \\ \max_i \hbar'_{ij}, j \in U_2 \end{cases} \tag{5.10}$$

where U_1 depicts the set of benefit-type attributes and U_2 represents the set of cost-type attributes.
Step viii. Computing the combined weighted distance between each alternative and the positive and negative ideal solutions.

The weighted distance formula along with the combined weights w_t is utilized to calculate the positive combined weighted distance measure $\diamond_i^+$ between the evaluation $\hbar'_{ij}$ and the positive ideal solution $\hbar_j^+$. Analogously the negative combined weighted distance measure $\diamond_i^-$ between the evaluation value $\hbar'_{ij}$ and the negative ideal solution $\hbar_j^-$ is calculated.

$$\diamond_{\mathrm{w}}^{1}(H_1, H_2) = \left\{ \sum_{j=1}^{n} w_j \left[\frac{1}{2l_j} \sum_{i=1}^{l_j} \left\{ \left| u_1^{(i)}(\varepsilon_j) - u_2^{(i)}(\varepsilon_j) \right|^k \right.\right.\right.$$

$$\left.\left.\left. + \left| v_1^{(i)}(\varepsilon_j) - v_2^{(i)}(\varepsilon_j) \right|^k + \left| \pi_1^{(i)}(\varepsilon_j) - \pi_2^{(i)}(\varepsilon_j) \right|^k \right\} \right] \right\}^{\frac{1}{k}}$$

$$\diamond^+ = \sum_{t=1}^{m} \rho_t \diamond_{ij}^+ \tag{5.11}$$

$$\diamond^- = \sum_{t=1}^{m} \rho_t \diamond_{ij}^- . \tag{5.12}$$

Step xi. Finding the relative closeness rc_i for each alternative.
rc_i is obtained by considering the risk preference coefficient α.

$$rc_i = \frac{(1-\alpha)\diamondsuit_i^-}{(1-\alpha)\diamondsuit_i^- + \alpha\diamondsuit_i^+} \tag{5.13}$$

where $0 \le \alpha \le 1$. $\alpha > 0.5$ suggests the decision-maker to be a risk acceptor; $\alpha < 0.5$, the decision-maker to be risk-relluctant; $\alpha = 0.5$ mark that is the decision-maker to be risk-balanced.
Step x. Ordering the alternatives in accordance with the rc_i values. The greater the value of rc_i, better the alternative z_i.

5.5 TOPSIS Techniques on q-ROFS$_f$S

TOPSIS is being utilized to choose on a supreme elective to compromise solutions. The solution closest to the ideal solution and the furthest from the negative ideal solution is recognized as a compromise solution [26]. In the article [26], we can develop q-ROFS$_f$S from the Fermatean Fuzzy Soft Sets and can study, how q-ROFS$_f$S can be used in MCGDM by making use of TOPSIS. Firstly, we will expand TOPSIS to the q-ROFS$_f$S and afterwards study a problem related to the selection of medical clinic. This example is taken from a new paper, which aids decision-makers in reaching a definitive decision and evaluating it in a systemic way without any partiality. Here, we can see that q-rung orthopair fuzzy soft sets are more efficient, strong, and substantive in solving the complexity, uncertainty, and expert hesitancy of membership and non- membership in multiple criteria group decision-making (MCGDM) problems as a combination of q-rung orthopair fuzzy sets and soft sets. The procedural steps for the proposed method are given in the following.

Algorithm III

Step i. Defining the problem.
Formulate a MCGDM problem with a set of decision-makers $G = \{G_i, i \in I_m.\}$, let $A = \{A_i, i \in I_m\}$ corresponds the set of alternatives and $P = \{P_i, i \in I_m\}$ constitute the family of parameters or criterion.
Step ii. Constructing the weighted parameter matrix.
Construct the weighted parameter matrix $\hat{B} = \hat{b}_{ij}$ using the linguistic terms described in that context. Here $\hat{b}_{ij}$ denotes the weight given to the alternative A_i by the decision-maker G_i.
Step iii. Constructing the normalized weighted matrix.
Construct the normalized weighted matrix $\hat{Q} = [\hat{q}_{ij}]_{n\times m}$, where $\hat{q}_{ij} = \frac{\hat{b}_{ij}}{\sqrt{\sum_{i=1}^{n} \hat{b}_{ij}^2}}$ and acquiring the weighted matrix $W = (w_1, w_2 \ldots, w_m)$ where $w_i = \frac{p_i}{\sum_{i=1}^{n} p_i}$ and $p_j = \frac{\sum_{i=1}^{n} \hat{q}_{ij}}{n}$.

Step iv. Constructing the q-ROFS decision matrix.
q-rung orthopair fuzzy set (q-ROFS) decision matrix can be constructed as $\hat{F} = [\hat{f}_{rs}]$, where $\hat{f}_{rs}$ is a q-ROFS element, for rth decision-maker so that G_r assign rth alternative with respect to the sth criteria. The aggregation matrix is then obtained using

$$\tilde{\mathrm{F}} = \frac{\hat{F}_1 + \hat{F}_2 + \cdots \hat{F}_n}{n} = [\ddot{F}_{rs}]_{l\times m}.$$

Step v. Constructing the weighted q-ROFS decision matrix.
The following formula can be used to calculate the weighted q-ROFS decision matrix.
$\hat{G} = [\hat{g}_{rs}]_{l\times m}$, where $\hat{g}_{rs} = w_s \times \ \ddot{F}_{rs}$.
Step vi. Constructing the q-ROFS-valued positive ideal solution (q-ROFSV-PIS) and q-ROFS-valued negative ideal solution (q-ROFSV-NIS).

$$FFSV - PIS = \{\hat{g}_1^+, \hat{g}_2^+, \ldots, \hat{g}_l^+\} = \{(\vee_s \hat{g}_{rs}, \wedge_s \hat{g}_{rs}) : s \in I_m\}$$

and

$$FFSV - NIS = \{\hat{g}_1^-, \hat{g}_2^-, \ldots, \hat{g}_l^-\} = \{(\wedge_s \hat{g}_{rs}, \vee_s \hat{g}_{rs}) : s \in I_m\}$$

where $\vee$ and $\wedge$, respectively, indicate the union and intersection of q-ROFS
Step vii. Computing the Euclidean distances.
Determine the Euclidean distances of q-ROFS for alternatives from q-Rung Orthopair Fuzzy Soft Valued ideal solutions using

$$\hat{O}_r^+ = \frac{1}{2 \mid \mathscr{S} \mid} \sum_{s=1}^{m} \Big(\mid \mu_{rs}^q - \mu_r^{+q} \mid + \mid \gamma_{rs}^q - \gamma_r^{+q} \mid + \mid \pi_{rs}^q - \pi_r^{+q} \mid \Big)$$

$$\hat{O}_r^- = \frac{1}{2 \mid \mathscr{S} \mid} \sum_{s=1}^{m} \Big(\mid \mu_{rs}^q - \mu_r^{-q} \mid + \mid \gamma_{rs}^q - \gamma_r^{-q} \mid + \mid \pi_{rs}^q - \pi_r^{-q} \mid \Big)$$

where $\pi_{rs} = \big(1 - (\mu_{rs}^q + \gamma_{rs}^q)\big)^{\frac{1}{q}}$.
Step viii. Computing Relative closeness
The degree to which each alternative is similar to the best solution is determined by,

$$K^*(\hat{g}_j) = \frac{\hat{O}_j^-}{\hat{O}_j^+ + \hat{O}_j^-} \in [0, 1].$$

Step xi. Ranking the order of preferences.
Here, a collection of alternatives can be ranked corresponding to the value of $K^*(\hat{g}_j)$. As the integrated model has outlined, We'll use an example to demonstrate the procedure.

Flowchart III

A flowchart is also included to graphically depict the decision-making process (Fig. 5.3).

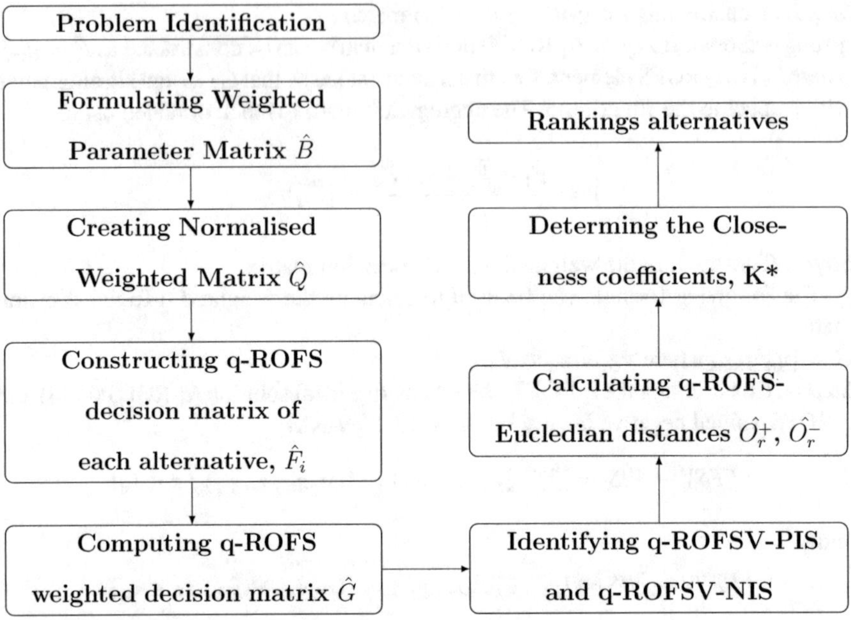

Fig. 5.3 Flowchart representation of Algorithm III

5.6 Applications

In this section, we enlist the illustrative examples to validate the applicability of the various TOPSIS models discussed in above sections. First we discuss Example 5.1 related with TOPSIS Techniques on q-ROFS taken from [23]. In Example 5.2, we can observe the computational efficiency of the TOPSIS model in the q-rung orthopair hesitant fuzzy environment [30]. Finally an illustrating example related with TOPSIS Techniques on q-ROFS$_f$S [26] settings is discussed as Example 5.3.

Example 5.1 Here, TOPSIS method based on q-Rung Orthopair Fuzzy numbers(q-ROFNs) for Multi Attribute Decision-Making (MADM) in transport policy problem is presented.

A city transportation committee intends to implement a transportation system that is best appropriate for the city. There are three different types of systems to choose from: Car-share, Ride-share, and Park and ride.

Step i: Constituting the problem: Suppose $D = \{M_i : i = 1, 2, 3, 4\}$ be the set decision-makers, $A = \{z_i : i = 1, 2, 3\}$ is the collection of available options, $C = \{C_i : i = 1, 2, \ldots 13\}$ is the set of criteria. In Table 5.2, the criteria pertaining to transport system problem are listed.
Step ii: Making the weighted parameter matrix, wherein every decision-maker des-

Table 5.1 Linguistic terms to determine alternatives

Linguistic terms	Fuzzy weights
Very High Importance (*VHI*)	0.95
High Importance (*HI*)	0.80
Average Importance (*AI*)	0.50
Low Importance (*LI*)	0.35

Table 5.2 Criterions for transport system problem

	Criteria	Description	Type
C_1	Rates	The cost of running the transportation system	Cost
C_2	Protection	The transportation ensures safety	Benefit
C_3	Reliability	Capability to carry out the service	Benefit
C_4	Air pollution	Transport-related air contaminants	Cost
C_5	Noise	Noise produced by transportation	Cost
C_6	Fuels	Fuels used include gasoline, CNG, and diesel	Cost
C_7	Travel costs	Travel expenses between stations	Cost
C_8	Energy	Transportation's energy consumption	Cost
C_9	Economy	Such as labour employment, etc.	Benefit
C_{10}	Competency	Technology has advanced significantly	Benefit
C_{11}	Mobility	Capability to provide transportation area	Benefit
C_{12}	Productivity	Capability to meet goals	Benefit
C_{13}	Excellence	Service of the highest quality	Benefit

ignates linguistic terms from Table 5.1 to every parameter. The significance assigned by decision-makers to every criterion is shown in Table 5.2. We create a weighted parameter matrix $\hat{Q}$ using the data from Table 5.3.

Table 5.3 Linguistic terms for criteria C_i

Criteria	Decision-maker			
	D_1	D_2	D_3	D_4
C_1	*VHI*	*HI*	*HI*	*VHI*
C_2	*HI*	*HI*	*VHI*	*AI*
C_3	*AI*	*VHI*	*LI*	*HI*
C_4	*VHI*	*VHI*	*HI*	*VHI*
C_5	*VHI*	*HI*	*HI*	*HI*
C_6	*HI*	*LI*	*AI*	*AI*
C_7	*HI*	*AI*	*AI*	*LI*
C_8	*LI*	*LI*	*AI*	*AI*
C_9	*HI*	*HI*	*AI*	*HI*
C_{10}	*VHI*	*AI*	*VHI*	*LI*
C_{11}	*AI*	*VHI*	*LI*	*VHI*
C_{12}	*LI*	*HI*	*AI*	*LI*
C_{13}	*VHI*	*LI*	*AI*	*VHI*

$$
\begin{aligned}
\hat{Q} &= [y_{ij}]_{4\times 13} \\
&= \begin{pmatrix}
VHI & HI & AI & VHI & VHI & HI & HI & LI & HI & VHI & AI & LI & VHI \\
HI & HI & VHI & VHI & HI & LI & AI & LI & HI & AI & VHI & HI & LI \\
HI & VHI & LI & HI & HI & AI & AI & AI & AI & VHI & LI & AI & AI \\
VHI & AI & HI & VHI & HI & AI & LI & AI & HI & LI & VHI & LI & VHI
\end{pmatrix} \\
&= \begin{pmatrix}
0.95 & 0.80 & 0.50 & 0.95 & 0.95 & 0.80 & 0.80 & 0.80 & 0.80 & 0.80 & 0.80 & 0.80 & 0.80 \\
0.50 & 0.35 & 0.95 & 0.50 & 0.35 & 0.35 & 0.80 & 0.80 & 0.80 & 0.80 & 0.80 & 0.80 & 0.80 \\
0.35 & 0.35 & 0.50 & 0.35 & 0.95 & 0.80 & 0.80 & 0.80 & 0.80 & 0.80 & 0.80 & 0.80 & 0.80 \\
0.50 & 0.80 & 0.95 & 0.35 & 0.35 & 0.50 & 0.80 & 0.80 & 0.80 & 0.80 & 0.80 & 0.80 & 0.80
\end{pmatrix}
\end{aligned}
$$

where y_{ij} is the weight provided by the decision-maker D_i (row-wise) to each parameter C_j.

Step iii: The normalized weighted matrix is

$$
\begin{aligned}
\hat{N} &= [\hat{n}_{ij}]_{4\times 13} \\
&= \begin{pmatrix}
0.5409 & 0.5129 & 0.3613 & 0.5129 & 0.5655 & 0.7119 & 0.7119 & 0.4055 & 0.5774 & 0.6438 & 0.3388 & 0.3285 & 0.6438 \\
0.4555 & 0.5129 & 0.6865 & 0.5129 & 0.4762 & 0.3115 & 0.4449 & 0.4055 & 0.5774 & 0.3388 & 0.6438 & 0.7509 & 0.2372 \\
0.4555 & 0.5771 & 0.2529 & 0.4372 & 0.4762 & 0.4449 & 0.4449 & 0.5793 & 0.3609 & 0.6438 & 0.2372 & 0.4693 & 0.3388 \\
0.5409 & 0.3206 & 0.5781 & 0.5129 & 0.4762 & 0.4449 & 0.3115 & 0.5793 & 0.5774 & 0.2372 & 0.6438 & 0.3285 & 0.6438
\end{pmatrix}
\end{aligned}
$$

Therefore, the weighted vector can obtained as $W = (0.08, 0.08, 0.07, 0.08, 0.08, 0.08, 0.08, 0.08, 0.08, 0.07, 0.07, 0.08, 0.07)$.

Step iv: With the track record of the companies under consideration in mind, the q-ROFS decision matrix $\daleth_i$ of each expert is offered, with alternatives depicted row-wise and parameters expressed column-wise. As a result, the aggregated decision

Table 5.4 Distance measures and closeness coefficient

$\gimel_i$	$\left(\varrho_j^{+}\right)$	$\left(\varrho_j^{-}\right)$	$\eth^{*}\left(\gimel_j\right)$
$\gimel_1$	0.16171	0.12728	0.44043
$\gimel_2$	0.13248	0.10247	0.43614
$\gimel_3$	0.14491	0.11467	0.44175

matrix is shown in Appendix 1.
Step v: The weighted q-ROFS decision matrix is shown in Appendix 2.
Step vi: Now, we find the q-ROFS valued positive ideal solution (q-ROFSV-PIS) and the q-ROFS valued negative ideal solution (q-ROFSV-NIS) and list them as
q-ROFSV-PIS $= \{\gimel_1^{+}, \gimel_2^{+}, \gimel_3^{+}\} =$(0.08, 0.01), (0.07, 0.02), (0.07, 0.01)
and q-ROFSV-NIS $= \{\gimel_1^{-}, \gimel_2^{-}, \gimel_3^{-}\} =$ (0.01, 0.07), (0.02, 0.06), (0.01, 0.06).
Step vii, viii: Table 5.4 shows the q-ROFS-Euclidean distances of each alternative from q-ROFSV-PIS and q-ROFSV-NIS, as well as the closeness coefficients.
Step ix: The order of the alternatives is $\gimel_3 \succ \gimel_1 \succ \gimel_2$. So $\gimel_3$ is the best and it is to be selected.

Example 5.2 The following numerical example verifies the applicability of TOPSIS method to MADM under q-rung orthopair hesitant fuzzy environment.
During the renewal of the hardware of aviation machinery, the military aircraft overhaul exhibits many features like "different kinds, long cycles, sophisticated process", etc. Scientific evaluation of the efficacy of military aircraft overhaul is necessary for increasing the maintenance potential, boosting overhaul capacities, judiciously allotting resources, reducing the overhaul cycle, alleviating the backlog problems, and also for assuring the smooth conduct of warfare training. Here, the primary objective is to select an efficient aircraft overhaul plant to reduce the expenses of aircraft missions. For an effective assessment of the efficiency of aircraft overhaul, the numerical values of the involved parameters were considered from overhaul plants Z_1, Z_2, Z_3 and Z_4. For assessment, values from fixing the same type of aircraft are used. A review of overhaul-related literature and field investigations of the net workflow of overhaul plants, it is identified that four major factors that influence the overhaul efficiency are: equipment assistance b_1, maintenance quality b_2, maintenance tasks b_3, and maintenance resources b_4, which are elaborated in Table 5.5.

Throughout this problem, it is taken that $q = 3$.
Step i. Formulating a q-rung orthopair hesitant fuzzy decision matrix.
The evaluation of an alternative corresponding to each attribute is a q-ROHFS and the decision matrix $Q = \left(\hbar_{ij}\right)_{m\times n}$ is obtained as in Table 5.6.

Table 5.5 Attribute characterization table

Attribute	Attribute description
Equipment assistance b_1	The equipment assistance is primarily reflective of the overhaul's efficiency in terms of the indicators such as service parts promptness as well as the turnover rate of inventory of the machinery
Maintenance quality b_2	The rate of rework occurrences, the count of quality problems that arise amidst the entire equipment repair, and the pass rate of flight tests are all indicators of maintenance quality
Maintenance tasks b_3	Repair volume, capacity utilization rate, repair cycle, and in-plant cycle are all indicators of maintenance tasks
Maintenance resources b_4	The influence of repair resources may be measured by the rate of matching for spare parts variety, production equipment utilization, working time utilization, staff technical skill level, machine location, circulation rate, and so on

Table 5.6 q-ROHF decision matrix Q

	b_1	b_2	b_3	b_4
Z_1	$\langle\{0.5, 0.6\}, \{0.4\}\rangle$	$\langle\{0.7, 0.9\}, \{0.1, 0.3\}\rangle$	$\langle\{0.7, 0.8\}, \{0.3\}\rangle$	$\langle\{0.3, 0.5, 0.6\}, \{0.4, 0.5\}\rangle$
Z_2	$\langle\{0.4, 0.5\}, \{0.5, 0.6\}\rangle$	$\langle\{0.7, 0.9\}, \{0.1, 0.2\}\rangle$	$\langle\{0.6, 0.8\}, \{0.2, 0.4\}\rangle$	$\langle\{0.2, 0.5\}, \{0.5, 0.7\}\rangle$
Z_3	$\langle\{0.3, 0.5, 0.7\}, \{0.6, 0.8\}\rangle$	$\langle\{0.7, 0.8\}, \{0.2\}\rangle$	$\langle\{0.7\}, \{0.3, 0.4\}\rangle$	$\langle\{0.3, 0.5\}, \{0.3, 0.4, 0.6\}\rangle$
Z_4	$\langle\{0.4, 0.6\}, \{0.3, 0.5\}\rangle$	$\langle\{0.7, 0.8\}, \{0.2, 0.3\}\rangle$	$\langle\{0.8\}, \{0.1, 0.4\}\rangle$	$\langle\{0.4, 0.6\}, \{0.5\}\rangle$

Table 5.7 Expanded decision matrix Q'

	b_1	b_2	b_3	b_4
Z_1	$\langle\{0.5, 0.6\}, \{0.4, 0.4\}\rangle$	$\langle\{0.7, 0.9\}, \{0.1, 0.3\}\rangle$	$\langle\{0.7, 0.8\}, \{0.3, 0.3\}\rangle$	$\langle\{0.3, 0.5, 0.6\}, \{0.4, 0.5, 0.5\}\rangle$
Z_2	$\langle\{0.4, 0.5\}, \{0.5, 0.6\}\rangle$	$\langle\{0.7, 0.9\}, \{0.1, 0.2\}\rangle$	$\langle\{0.6, 0.8\}, \{0.2, 0.4\}\rangle$	$\langle\{0.2, 0.5\}, \{0.5, 0.7\}\rangle$
Z_3	$\langle\{0.3, 0.5, 0.7\}, \{0.6, 0.8, 0.8\}\rangle$	$\langle\{0.7, 0.8\}, \{0.2, 0.2\}\rangle$	$\langle\{0.7, 0.7\}, \{0.3, 0.4\}\rangle$	$\langle\{0.3, 0.5, 0.5\}, \{0.3, 0.4, 0.6\}\rangle$
Z_4	$\langle\{0.4, 0.6\}, \{0.3, 0.5\}\rangle$	$\langle\{0.7, 0.8\}, \{0.2, 0.3\}\rangle$	$\langle\{0.8, 0.8\}, \{0.1, 0.4\}\rangle$	$\langle\{0.4, 0.6\}, \{0.5, 0.5\}\rangle$

As shown in Table 5.7, the decision matrix is expanded as $Q' = \left(\hbar'_{ij}\right)_{m\times n}$ to accommodate the calculations involved in the distance formula. It is assumed that $\theta = 1$.

Step ii. Constituting the initial judgment matrix.
Here, the initial judgment matrix is denoted as B'.

$$B' = \begin{array}{|c|cccc|} & b_1 & b_2 & b_3 & b_4 \\ b_1 & 1 & 1/4 & 1/3 & 2 \\ b_2 & 4 & 1 & 2 & 4 \\ b_3 & 3 & 1/2 & 1 & 3 \\ b_4 & 1/2 & 1/4 & 1/3 & 1 \end{array}$$

Step iii. Optimizing the initial judgment matrix B'.
B' is optimized to obtain the matrix **P** by using Formulas from (5.1) to (5.5).

$$\mathbf{P} = \begin{bmatrix} 1 & 0.3926 & 0.6148 & 1.1839 \\ 2.5474 & 1 & 1.5660 & 3.0158 \\ 1.6267 & 0.6386 & 1 & 1.9258 \\ 0.8447 & 0.3316 & 0.5193 & 1 \end{bmatrix}$$

Step iv. Obtaining the weights of indicators using Firefly Algorithm.

The basic parameters of the algorithm are enlisted as: the number of fireflies under consideration, $m = 20$, maximum number of iterations $maxG = 200$, disturbance factor $\alpha = 0.2$, $\beta = 0.97$, and the light attraction factor $\lambda = 1$. Considering the flow of the Firefly algorithm and applying the matrix **P** in (5.6), the subjective weights ρ_i are computed as $\rho_i = (0.1664, 0.4232, 0.2703, 0.1401)$.

Step v. Obtaining the objective weights (ρ_j) of attributes.

Initially, for each attribute b_i, the entropy e^i is calculated. The computed entropy values are $e^1 = 0.4352$, $e^2 = 0.3335$, $e^3 = 0.3648$, $e^4 = 0.4902$. Then using (5.7) the objective weights are obtained as $\rho_j = (0.2377, 0.2805, 0.2673, 0.2155)$ (Table 5.8).

Step vi. Combining the objective and subjective weights to obtain the attribute weights.

Putting c = 0.5 in (5.8), the combined weight ρ_t of the attributes are obtained as $\rho_t = (0.1950, 0.3603, 0.2785, 0.1662)$.

Step vii. Finding the positive ideal solution and the negative ideal solutions.

Referring the expanded decision matrix Q' and Formulas (5.9) and (5.10), the positive and negative ideal solutions are obtained as:

$$Q'^{+} = (\langle\{0.7\}, \{0.3\}\rangle, \langle\{0.9\}, \{0.1\}\rangle, \langle\{0.8\}, \{0.1\}\rangle, \langle\{0.6\}, \{0.3\}\rangle)$$
$$Q'^{-} = (\langle\{0.3\}, \{0.8\}\rangle, \langle\{0.7\}, \{0.3\}\rangle, \langle\{0.6\}, \{0.4\}\rangle, \langle\{0.2\}, \{0.7\}\rangle).$$

Step viii. Computing the combined weighted distances $\diamondsuit^+$ and $\diamondsuit^-$.

Table 5.8 Subjective and objective weights of each attribute

	b_1	b_2	b_3	b_4
ρ_j	0.2377	0.2805	0.2673	0.2155
ρ_i	0.1309	0.4814	0.2954	0.0923

Table 5.9 The distances of each alternative from the ideal solutions

	Z_1	Z_2	Z_3	Z_4
$\diamondsuit^+$	0.1843	0.2549	0.2911	0.2056
$\diamondsuit^-$	0.2654	0.2044	0.2238	0.2786

Table 5.10 Linguistic variables for evaluating alternatives

Linguistic variables	Fuzzy weights
Outstanding (O)	0.29
Extremely Good (EG)	0.18
Medium Good (MG)	0.31
Medium (M)	0.22

Employing the attribute weights ρ_t into Formulas (5.11) and (5.12), the combined weighted distances $\diamondsuit^+$ and $\diamondsuit^-$ are obtained as listed in Table 5.9.

Step ix. Computing the relative closeness rc_i for each alternative.

Assuming $\alpha = 0.5$ and using it in Formula (5.13) to the rc_i values for each alternative Z_i is determined:

$$rc_1 = 0.5902, \quad rc_2 = 0.4451, \quad rc_3 = 0.4347, \quad rc_4 = 0.5754$$

Step x. Ordering the alternatives in accordance with the rc_i values to obtain $Z_1 \succ Z_4 \succ Z_2 \succ Z_3$ which shows that Z_1 is the optimal solution that is the overhaul plant Z_1 is the most efficient one.

Example 5.3 To demonstrate the calculation process of the approaches q-ROFS$_f$S, one example is taken from a new paper [27] provided below.

The choice of medical clinics is an important part of our lives when it comes to disease diagnosis. In the following study, we apply the TOPSIS approach to choose the best medical clinic for disease diagnosis in any emergency situation. Now, using the algorithm of this new proposed approach, we will solve the following problem in the following steps. For problem solving, we use the following linguistic terminology.

For renowned celebrities, the government of a nation needs to choose the best hospital (medical clinic). The government recruits a committee of four professionals (decision-makers) to choose the right medical facility. To begin, the experts choose four hospitals from the country's capital, as follows: $H = \{H_1, H_2, H_3, H_4\}$, and choose one of the best hospitals out of four using the four assessment $\{T_1, T_2, T_3, T_4\}$,

- T_1 : Security
- T_2 : Environment
- T_3 : Qualified staff
- T_4 : Expenses

Step i. Consider the set $\{G = G_i, i \in I_4\}$ consisting of members of a committee of professionals or decision-makers $E = \{E_i, i \in I_5\}$ is the collection of hospitals/alternatives and $T = \{T_i, i \in I_4\}$ as a family of parameters.
Step ii. Create the weighted parameter matrix using the information in Table 5.10 consisting of linguistic terms used for the relative importance rating of experts. Experts examine the four hospitals based on the parameters under consideration and present their findings in the following matrix.

$$\hat{B} = [\hat{b}_{ij}]_{4\times 4} = \begin{pmatrix} G & MG & O & O \\ M & EG & MG & M \\ O & MG & EG & EG \\ EG & O & EG & O \end{pmatrix} = \begin{pmatrix} 0.31 & 0.31 & 0.29 & 0.29 \\ 0.22 & 0.18 & 0.31 & 0.22 \\ 0.29 & 0.31 & 0.18 & 0.18 \\ 0.18 & 0.29 & 0.18 & 0.29 \end{pmatrix}$$

where $\hat{b}_{ij}$ is the weight assigned to each parameter T_j by the decision-maker G_i.
Step iii. Compute the normalized weighted matrix by using above weighted parameter matrix.

$$\hat{Q} = [\hat{q}_{ij}]_{4\times 4} = \begin{pmatrix} 0.61 & 0.55 & 0.59 & 0.59 \\ 0.43 & 0.32 & 0.63 & 0.45 \\ 0.57 & 0.55 & 0.37 & 0.37 \\ 0.35 & 0.52 & 0.37 & 0.59 \end{pmatrix}$$

As a result of which the weighted vector is $W = (\,0.08, 0.30, 0.31, 0.31)$.
Step iv. Assume that four decision-makers generate the q-ROFS matrix below, in which the alternatives are expressed row-wise and the parameters are represented column-wise. Without loss of generality, we take $q = 4$

$$F_1 = \begin{pmatrix} (0.95, 0.58) & (0.89, 0.78) & (0.98, 0.49) & (0.88, 0.69) \\ (0.59, 0.94) & (0.75, 0.85) & (0.92, 0.65) & (0.76, 0.89) \\ (0.68, 0.89) & (0.84, 0.79) & (0, 1) & (0.78, 0.89) \\ (0.83, 0.76) & (0.61, 0.93) & (0.57, 0.96) & (0.62, 0.92) \end{pmatrix}$$
$$F_2 = \begin{pmatrix} (0.84, 0.76) & (0.37, 0.99) & (0.72, 0.86) & (0.81, 0.78) \\ (0, 1) & (0.94, 0.59) & (0.97, 0.57) & (0.82, 0.77) \\ (0.63, 0.92) & (0.75, 0.90) & (0.80, 0.80) & (0.69, 0.89) \\ (0.45, 0.97) & (0.63, 0.91) & (0.73, 0.86) & (0.44, 0.98) \end{pmatrix}$$
$$F_3 = \begin{pmatrix} (0.86, 0.74) & (0.58, 0.94) & (0.34, 0.99) & (0.55, 0.95) \\ (0.42, 0.98) & (0.86, 0.72) & (0.67, 0.89) & (0, 1) \\ (0, 1) & (0.96, 0.52) & (0.65, 0.95) & (0.80, 0.90) \\ (0.77, 0.88) & (0.38, 0.96) & (0.61, 0.95) & (0.48, 0.98) \end{pmatrix}$$
$$F_4 = \begin{pmatrix} (0.71, 0.87) & (0.67, 0.89) & (0.45, 0.97) & (0.82, 0.78) \\ (0.55, 0.95) & (0.92, 0.72) & (0.64, 0.91) & (0.52, 0.96) \\ (0.36, 0.99) & (0.60, 0.93) & (0.73, 0.87) & (0.81, 0.81) \\ (0.49, 0.98) & (0.59, 0.93) & (0.68, 0.89) & (0, 1) \end{pmatrix}$$

$$\text{Average decision matrix is } \tilde{\mathrm{F}} = \begin{pmatrix} (0.84, 0.74) & (0.63, 0.9) & (0.62, 0.83) & (0.77, 0.8) \\ (0.39, 0.97) & (0.87, 0.72) & (0.8, 0.76) & (0.53, 0.91) \\ (0.42, 0.95) & (0.79, 0.79) & (0.55, 0.91) & (0.77, 0.87) \\ (0.64, 0.89) & (0.55, 0.93) & (0.65, 0.92) & (0.39, 0.97) \end{pmatrix}$$

Step v. Using the weighted vector W, construct the weighted q-ROFS decision matrix as follows

$$\hat{G} = [\hat{G}_{rs}]_{5\times 4} = \begin{pmatrix} (0.07, 0.06) & (0.19, 0.27) & (0.19, 0.26) & (0.24, 0.25) \\ (0.03, 0.08) & (0.26, 0.22) & (0.25, 0.24) & (0.16, 0.28) \\ (0.03, 0.08) & (0.24, 0.24) & (0.17, 0.28) & (0.24, 0.27) \\ (0.05, 0.07) & (0.17, 0.28) & (0.21, 0.29) & (0.12, 0.30) \end{pmatrix}$$

Step vi. q-ROFS-valued positive ideal solution (q-ROFSV-PIS) and q-ROFS-valued negative ideal solution (q-ROFSV-NIS) are located as follows:

$$\begin{aligned} \text{q-ROFSV-PIS} &= \{\hat{g}_1^+, \hat{g}_2^+, \hat{g}_3^+, \hat{g}_4^+\} \\ &= \{(0.24, 0.06), (0.26, 0.08), (0.24, 0.08), (0.21, 0.07)\} \\ \text{q-ROFSV-NIS} &= \{\hat{g}_1^-, \hat{g}_2^-, \hat{g}_3^-, \hat{g}_4^-\} \\ &= \{(0.07, 0.27), (0.03, 0.28), (0.03, 0.28), (0.05, 0.30)\}. \end{aligned}$$

Step vii. The distances between each alternative and q-ROFSV-PIS as well as the q-ROFSV-NIS by using q-ROFS Euclidean distances and relative closeness index are computed in Table 5.11.

Step viii: According to relative closeness measure of each alternative $\{H_1, H_2, H_3, H_4\}$, we get the ranking of alternatives as,

$$H_2 > H_3 > H_1 > H_4$$

As a result, with the above assessment requirements, H_2 is the safest diagnostic clinic for disease diagnosis.

Table 5.11 Distance measures and closeness coefficient

Alternative (E_i)	O_i^+	O_i^-	K_i^*
H_1	0.0043	0.0028	0.3944
H_2	0.0035	0.0038	0.5205
H_3	0.0045	0.0034	0.4304
H_4	0.0057	0.0030	0.3448

5.7 Conclusions

This chapter is an overview of the existing TOPSIS techniques applied to q-rung orthopair fuzzy set and some of its extensions. We presented the approach of q-rung orthopair fuzzy sets (q-ROFSs). The proposed models are practical technique to handle uncertainties by means of parametrization together with membership and non-membership. We compiled three algorithms for decision-making in various contexts. One of them is the use of the TOPSIS method based on q-ROFNs for MADM in transport policy problem. The algorithm developed for this MADM method shows the efficiency of TOPSIS method in dealing with q-rung orthopair decision information. Another one is a TOPSIS model under the q-ROHF setting, which is illustrated with a numerical example showing the assessment of military aircraft overhaul plants. This model could provide greater precision when expressing fuzzy and ambiguous data. Numerical examples that describe real situations depict the practical applications of this method. In addition, TOPSIS approach based on q-ROFS$_f$S shows the effectiveness of group decision-making method in the real world. This review confirms that the existing TOPSIS approaches are efficient, and it also provides an insight to extend TOPSIS to solve MADM problems using other hybrid structures of fuzzy sets.

Appendix 1

$$
\begin{aligned}
A &= Z_1 + Z_2 + Z_3 + Z_4 \\
&= \begin{pmatrix}
(0.83, 0.53) & (0.84, 0.21) & (0.32, 0.16) & (0.93, 0.14) & (0.81, 0.23) & (0.73, 0.21) \\
 & (0.36, 0.79) & (0.26, 0.16) & (0.43, 0.71) & (0.16, 0.63) & (0.36, 0.71) \\
 & (0.96, 0.12) & (0.86, 0.32) & & & \\
(0.72, 0.63) & (0.56, 0.72) & (0.32, 0.63) & (0.51, 0.71) & (0.26, 0.71) & (0.36, 0.36) \\
 & (0.24, 0.84) & (0.69, 0.34) & (0.49, 0.63) & (0.26, 0.61) & (0.61, 0.36) \\
 & (0.71, 0.36) & (0.26, 0.37) & & & \\
(0.39, 0.79) & (0.61, 0.72) & (0.23, 0.26) & (0.56, 0.36) & (0.26, 0.71) & (0.31, 0.76) \\
 & (0.63, 0.49) & (0.39, 0.71) & (0.82, 0.26) & (0.71, 0.30) & (0.71, 0.21) \\
 & (0.38, 0.81) & (0.18, 0.47) & & &
\end{pmatrix} \\
&= \left[\dot{z}_{jk}\right]_{3\times 13}.
\end{aligned}
$$

Appendix 2

$$J = \ddot{z}_{jk_{3\times 13}}$$
$$= \begin{pmatrix} (0.04, 0.07) & (0.07, 0.02) & (0.01, 0.02) & (0.01, 0.07) & (0.02, 0.06) & (0.02, 0.06) \\ & (0.06, 0.03) & (0.01, 0.02) & (0.03, 0.06) & (0.01, 0.04) & (0.03, 0.05) \\ & (0.08, 0.01) & (0.07, 0.03) & & & \\ (0.05, 0.02) & (0.04, 0.06) & (0.04, 0.02) & (0.06, 0.04) & (0.06, 0.02) & (0.03, 0.03) \\ & (0.07, 0.02) & (0.03, 0.06) & (0.04, 0.05) & (0.01, 0.04) & (0.03, 0.05) \\ & (0.08, 0.01) & (0.07, 0.03) & & & \\ (0.06, 0.03) & (0.05, 0.06) & (0.02, 0.02) & (0.03, 0.04) & (0.06, 0.02) & (0.06, 0.02) \\ & (0.05, 0.04) & (0.06, 0.03) & (0.07, 0.02) & (0.05, 0.02) & (0.04, 0.01) \\ & (0.03, 0.06) & (0.01, 0.04) & & & \end{pmatrix}$$

where $\ddot{z}_{jk} = \mathfrak{y}_k \times \dot{z}_{jk}$.

References

1. K.T. Atanassov, Intuitionistic fuzzy sets, in *Intuitionistic Fuzzy Sets*. (Springer, 1999), pp. 1–137
2. S. Eraslan, A decision making method via TOPSIS on soft sets. J. New Results Sci. **4**(8), 57–70 (2015)
3. S. Eraslan, F. Karaaslan, A group decision making method based on topsis under fuzzy soft environment. J. New Theory **3**, 30–40 (2015)
4. H. Garg, A new possibility degree measure for interval-valued q-rung orthopair fuzzy sets in decision-making. Int. J. Intell. Syst. **36**(1), 526–557 (2021)
5. H. Garg, A novel trigonometric operation-based q-rung orthopair fuzzy aggregation operator and its fundamental properties. Neural Comput. Appl. **32**(18), 15077–15099 (2020)
6. H. Garg, CN-q-ROFS: connection number-based q-rung orthopair fuzzy set and their application to decision-making process. Int. J. Intell. Syst. **36**(7), 3106–3143 (2021)
7. H. Garg, et al., Multi-criteria decision-making algorithm based on aggregation operators under the complex interval-valued q-rung orthopair uncertain linguistic information. J. Intell. Fuzzy Syst. pp. 1–30 (2021)
8. H. Garg, New exponential operation laws and operators for interval-valued q-rung orthopair fuzzy sets in group decision making process. Neural Comput. Appl. 1–27 (2021)
9. H. Garg, Z. Ali, T. Mahmood, Algorithms for complex interval-valued q-rung orthopair fuzzy sets in decision making based on aggregation operators, AHP, and TOPSIS. Exp. Syst. **38**(1), e12609 (2021)
10. A. Hussain, et al., q-Rung orthopair fuzzy soft average aggregation operators and their application in multicriteria decision-making. Int. J. Intell. Syst. **35**(4), 571–599 (2020)
11. C.-L. Hwang, K. Yoon, Methods for multiple attribute decision making, in *Multiple Attribute Decision Making* (Springer, 1981), pp. 58–191
12. S.J. John, *Soft Sets: Theory and Applications*, vol. 400 (Springer Nature, 2020)
13. D. Liu, Y. Liu, L. Wang, The reference ideal TOPSIS method for linguistic q-rung orthopair fuzzy decision making based on linguistic scale function. J. Intell. Fuzzy Syst. Preprint, 1–21 (2020)

14. D. Liu, D. Peng, Z. Liu, The distance measures between q-rung orthopair hesitant fuzzy sets and their application in multiple criteria decision making. Int. J. Intell. Syst. **34**(9), 2104–2121 (2019)
15. P. Liu, P. Wang, Some q-rung orthopair fuzzy aggregation operators and their applications to multiple-attribute decision making. Int. J. Intell. Syst. **33**(2), 259–280 (2018)
16. T. Mahmood, Z. Ali, Entropy measure and TOPSIS method based on correlation coefficient using complex q-rung orthopair fuzzy information and its application to multi-attribute decision making. Soft Comput. **25**(2), 1249–1275 (2021)
17. P.K. Maji, R. Biswas, A. Ranjan Roy, Soft set theory. Comput. Math. Appl. **45**(4–5), 555–562 (2003)
18. D. Molodtsov, Soft set theory-first results. Comput. Math. Appl. **37**(4–5), 19–31 (1999)
19. D. Molodtsov, V. Yu Leonov, D.V. Kovkov, Soft sets technique and its application, in (2006), pp. 8–39
20. K. Naeem, et al., Pythagorean fuzzy soft MCGDM methods based on TOPSIS, VIKOR and aggregation operators. J. Intell. Fuzzy Syst. **37**(5), 6937–6957 (2019)
21. X. Peng, L. Liu, Information measures for q-rung orthopair fuzzy sets. Int. J. Intell. Syst. **34**(8), 1795–1834 (2019)
22. M. Riaz, et al., Novel q-rung orthopair fuzzy interaction aggregation operators and their application to low-carbon green supply chain management. J. Intell. Fuzzy Syst. 1–18 (2021)
23. M. Riaz, et al., Some q-rung orthopair fuzzy hybrid aggregation operators and TOPSIS method for multi-attribute decision-making. J. Intell. Fuzzy Syst. Preprint, 1–15 (2020)
24. M. Riaz, K. Naeem, D. Afzal, Pythagorean m-polar fuzzy soft sets with TOPSIS method for MCGDM. Punjab Univ. J. Math. **52**(3), 21–46 (2020)
25. R. Ronald, Yager, Pythagorean fuzzy subsets, in Joint IFSA World Congress and NAFIPS Annual Meeting (IFSA/NAFIPS), vol. 2013 (IEEE, 2013), pp. 57–61
26. V. Salsabeela, S.J. John, TOPSIS techniques on Fermatean fuzzy soft sets, in *AIP Conference Proceedings*, vol. 2336, no. 1, p. 040022 (2021)
27. V. Salsabeela, S.J. John, TOPSIS techniques on q-Rung Orthopair fuzzy soft sets, in Communicated (2021)
28. H.-S. Shih, H.-J. Shyur, E. Stanley Lee, An extension of TOPSIS for group decision making. Math. Comput. Model. **45**(7–8), 801–813 (2007)
29. V. Torra, Hesitant fuzzy sets. Int. J. Intell. Syst. **25**(6), 529–539 (2010)
30. Y. Wang, Z. Shan, L. Huang, The extension of TOPSIS method for multiattribute decision-making with q-Rung orthopair hesitant fuzzy sets. IEEE Access **8**, 165151–165167 (2020)
31. Q. Xie, J.-Q. Ni, S. Zhongbin, Fuzzy comprehensive evaluation of multiple environmental factors for swine building assessment and control. J. Hazard. Mater. **340**, 463–471 (2017)
32. R.R. Yager, Generalized orthopair fuzzy sets. IEEE Trans. Fuzzy Syst. **25**(5), 1222–1230 (2016)
33. X.-S. Yang, Fire y algorithm, stochastic test functions and design optimisation. Int. J. Bio-inspired Comput. **2**(2), 7884 (2010)
34. Z. Yang, H. Garg, Interaction power partitioned maclaurin symmetric mean operators under q-rung orthopair uncertain linguistic information. Int. J. Fuzzy Syst., 1–19 (2021)
35. W. Yang, Y. Pang, New q-Rung orthopair hesitant fuzzy decision making based on linear programming and TOPSIS. IEEE Access **8**, 221299–221311 (2020)
36. E.K. Zavadskas, et al., Development of TOPSIS method to solve complicated decision-making problems—an overview on developments from 2000 to 2015. Int. J. Inform. Technol. Decis. Making **15**(3), 645–682 (2016)

V. Salsabeela is a research scholar in the department of Mathematics at National Institute of Technology Calicut. Her area of research is TOPSIS Techniques on Soft Sets.

Aparna Sivadas is a research scholar in the department of mathematics at National Institute of Technology Calicut. Her area of research includes fuzzy sets and its extensions.

Bibin Mathew is a lecturer in the Department of Mathematics of the D B College Thalayolaparambu. He received his Ph.D from National Institute of Technology Calicut. His area of interest includes Rough sets and different hybrid structures.

Sunil Jacob John is a Faculty in the department of Mathematics at National Institute of Technology Calicut. His areas of interest include Topology, Fuzzy Mathematics, Rough Sets, Multisets and Soft sets.

Chapter 6
Knowledge Measure-Based q-Rung Orthopair Fuzzy Inventory Model

C. Sugapriya, S. Rajeswari, D. Nagarajan, and K. Jeganathan

Abstract As a generalization of intuitionistic fuzzy set and Pythagorean fuzzy set, the q-Rung Orthopair Fuzzy Set (qROFS) and its application are implemented in some decision-making problems. So far the notion of qROFS is not yet applied in any inventory management problems. This chapter analyzed an inventory model under the q-rung orthopair fuzzy environment and utilized the knowledge measure of qROFS to the proposed model. This study explores an economic order quantity model with faulty products and screening errors. To satisfy the customers' demand with perfect items, the proposed research utilized the product warranty claim strategy from the supplier. Since some of the faulty products are returned from the supplier without being replaced as good ones, this study examines under two cases say, the warranty unclaimed products are restored by mending option and the warranty unclaimed products are recovered by the emergency purchase option. For both cases, the q-Rung Orthopair Fuzzy (qROF) inventory model is framed by presuming the proportion of faulty products and the proportion of misclassification errors as q-Rung Orthopair Fuzzy Variables (qROFVs). The Knowledge Measure-based q-Rung Orthopair Fuzzy (KM-qROF) inventory model is proposed by computing the knowledge measure of qROFVs. The proposed model is illustrated with a numerical example along with the sensitivity analysis.

Keywords qROFS · Knowledge measure for qROFS · Inventory model · Warranty · Repair · Emergency purchase

C. Sugapriya · S. Rajeswari
Department of Mathematics, Queen Mary's College, University of Madras, Chennai, Tamil Nadu, India

D. Nagarajan (✉)
Department of Mathematics, Rajalakshmi Institute of Technology, Chennai, Tamil Nadu, India
e-mail: dnrmsu2002@yahoo.com

K. Jeganathan
Amanujan Institute for Advanced Study in Mathematics, University of Madras, Chennai, Tamil Nadu, India

H. Garg (ed.), *q-Rung Orthopair Fuzzy Sets*,
https://doi.org/10.1007/978-981-19-1449-2_6

6.1 Introduction

The production administrator implements many effective manufacturing procedures to avoid faulty products being produced. But still, substandard products are manufactured and supplied to the vendors. The main issue of an Economic Order Quantity (EOQ) model is to receive the substandard products in every lot, and the quantities of the received substandard products are unpredictable. The development of business throughout the globe leads the vendors to utilize some ethics to satisfy their customers. This study explores an EOQ model with faulty products, and in this model, the inspection errors that arise during the screening procedure are considered. According to Khan et al. [12], the screening procedure has been done to separate the substandard products from a replenished lot in both the warehouses. Since the screening process is improper, two types of misclassification errors arise during this process, these errors misclassified the perfect products as substandard and misclassified the substandard product as perfect ones. Due to the second type of misclassification error, the substandard products that are sold as perfect items are returned back from the customer throughout the cycle. Except for the return of products due to type II error, a great impact on customer satisfaction with the product occurs. This error leads to losing a customer and so the customer may switch to another vendor. Also, it affects the vendor's goodwill concern to the customer.

To satisfy the customer demand with perfect items, the proposed research analyzed the notion of mending and emergency purchase options along with the product warranty claim from the supplier. It is considered that the vendor utilizes the product replacement warranty option for the faulty products claim from the supplier. The proposed model examines two cases; the first case considers the warranty unclaimed products are restored by mending option using nearby repair shop while the second case presumes the warranty unclaimed products are recovered by emergency purchase option. Because the fraction of faulty products on every lot and the fraction of the two types of misclassification errors are unpredictable, this chapter utilized the notion of q-Rung Orthopair Fuzzy Set (qROFS) for an effective outcome. As a generalization of intuitionistic fuzzy sets and Pythagorean fuzzy sets, the qROFSs are more flexible and better suitable for the unpredictable situation. In this study, the proportion of faulty products and the proportion of two misclassification errors are uncertain. Thus, the application of qROFS on these variables leads the model to attain more approximate results. The q-Rung Orthopair Fuzzy (qROF) inventory model is framed by presuming the proportion of faulty products and the proportion of misclassification errors as q-Rung Orthopair Fuzzy Variables (qROFVs). The knowledge measure for qROFS introduced by Khan et al. [14] is adopted in this study to compute the KM-qROF inventory model.

6.1.1 Literature Review

Some of the research works under the various topics related to the proposed model are discussed as follows. In the view of the retailer to attain maximum profit in an EOQ model as well as to reduce the issue of substandard items that sold to the customer because of defective products received from a supplier, Salameh and Jaber [33] suggests the screening procedure in the EOQ model with flawed products. They designed an inventory model by considering the defective items separated and sold at scrap price. Khan et al. [12] extended the work of [33] with screening errors. Also, the research works [10, 15, 22, and 47] elongated the study [33] under various situations. As the restoring strategy of flawed products, Jaber et al. [11] enlarged the work [33] with the mending and emergency purchase option, and the study [11] has extended by Pimsap and Srisodaphol [25] with misclassification errors. Sarkar and Saren [34] formulated an EPQ model with deterioration due to the production of defective products. They assumed that the model has been performed by product inspection policy and because of that inspection errors may occur. The examined imperfect products are recovered at a constant price which includes the warranty price for unscreened imperfect products. Yeo and Yuan [44] studied the optimal guarantee scheme for a model with poor mending and investigated the fundamental guaranty policy is the lowest repair up to a particular warranty period.

The uncertainty situation leads the researchers of various fields to use the approach of the fuzzy set and its applications. Initially, the fuzzy set and its approach were initiated by Zadeh [45]. Then Atanassov [2] elongated the fuzzy set as Intuitionistic Fuzzy Set (IFS) and some authors classified the intuitionistic fuzzy set under various types and applied it to various situations. As a generalization of IFS, the Pythagorean Fuzzy Set (PyFS) is initiated by [37, 38] which helps the decision-makers to reach better outcomes than IFS. Further, in 2017, a generalization of IFS and PyFS, Yager [39] and Yager and Alajlan [40] initiated the notion of qROFS, and also Ali [1] provided a different approach to qROFS. The research works [4–9, 16–19, 24, 32, 35, 36, 41–43, 46] analyzed various operators such as neutrality aggregation operators, power aggregation operators and vikor methods, power Maclaurin symmetric mean operators, exponential operator, power Muirhead mean operators, Maclaurin symmetric mean operators, Heronian mean operators, Hamy mean operators, Dombi power partitioned Heronian mean operators for qROFS and implemented to decision-making problems under various situations. Recently, in 2020, Farhadinia and Liao [3] utilized the qROFS to decision-making problem with a new score value. Khan et al. [13] provided the latest ranking methodology for qROFS. Khan et al. [14] established the knowledge measure and entropy measure of qROFS and solved the multi-attribute decision-making problem by utilizing knowledge measure based qROFS. Riaz et al. [30] have analyzed the qROF prioritized aggregation operator to a supply chain management problem, and Riaz et al. [31] studied a robust qROF einstein prioritized aggregation operators and implemented to a decision-making problem.

Most Relatively to this study, because of the unpredictability of defective proportion, the works [20, 23, 26] designed the fuzzy inventory model by pursuing the proportion of defectives as fuzzy variables. On the use of cloud-type Intuitionistic Fuzzy parameters, Maity et al. [21] examined the EOQ model with shortage backordering under the intuitionistic dense fuzzy situation. Rajeswari et al. [27] presented a container management model with customer charge-sensitive demand utilizing the ECR and the renting option instead of buying new containers under the fuzzy environment. Recently, the studies [28, 29] examined the notion of prepayment option to the EOQ model with substandard items under different situations. Because the proportion of substandard products and the proportion of two misclassification errors are unpredictability, the current model was analyzed under qROF environment to attain the most approximate outcomes.

6.1.2 Research Gap and the Contribution

Many researchers analyzed the EOQ model with the substandard product under various strategies say model with screening errors, the model with scarcity, replaced the flawed product by repair option and emergency purchase option, and so on. But this study utilizes the replacement warranty claiming option to restore the flawed products. Also, so far no one has implemented the notion of qROFS in the inventory management problems. The proposed model examined an inventory model under the q-rung orthopair fuzzy environment and utilized the knowledge measure of qROFS to the proposed model.

The main objective of the research is to fulfill the customers' expectations with standard items so that the vendor adopts various strategies to make the defectives into good quality items. The motivation of this study is to help the vendor attain better profit by utilizing mending strategy and emergency purchase option along with the product replacement warranty option instead of selling the substandard items at a lower price. By computing the KM-qROF inventory model, the vendor obtains approximate profit for unpredictable parameters.

The elementary definitions of fuzzy set, IFS, PyFS, qROFS, and the KM-qROFS are provided in the below section. Followed by the sales revenue and the relevant costs, the total profit of the EOQ model is formulated in Sect. 6.3. The substandard proportion and misclassification errors are presumed as qROFVs and the KM-qROF inventory model is performed by providing the knowledge measure for qROFVs. In Sect. 6.4, the numerical computation and sensitivity analysis is performed with the comparative studies. Finally, the results of this research are deliberated in Sect. 6.5.

6.2 Preliminaries

Some of the fundamental definitions are provided in this section.

Definition 6.2.1 Zadeh [45] A fuzzy set $\tilde{\xi}$ in the universal set $\mathscr{U}$ is defined as a set of ordered pairs and it is expressed as.

$$\tilde{\xi} = \left(z', \mu_{\tilde{\xi}}\left(z'\right)\right) : z' \in \mathscr{U}$$

in which $\mu_{\tilde{\xi}}\left(z'\right)$ is a membership function of z' which assumes values in the range from 0 to 1 (ie.,) $\mu_{\tilde{\xi}}\left(z'\right) \in [0, 1]$.

Definition 6.2.2 Atanassov [2]: An intuitionistic fuzzy set $\tilde{\xi}$ in $\mathscr{U}$ (finite universe of discourse) is given as $\tilde{\xi} = \left\{\left(\varphi_{\tilde{\xi}}(z_x), \psi_{\tilde{\xi}}(z_x)\right) | z_x \in \mathscr{U}\right\}$, where $\varphi_{\tilde{\xi}} : \mathscr{U} \to [0, 1]$ and $\psi_{\tilde{\xi}} : \mathscr{U} \to [0, 1]$ satisfying the condition $0 \leq \varphi_{\tilde{\xi}}(z_x) + \psi_{\tilde{\xi}}(z_x) \leq 1$ and $\varphi_{\tilde{\xi}}(z_x)$ denotes the membership degree of $z_x \in \mathscr{U}$ in $\tilde{\xi}$ and $\psi_{\tilde{\xi}}(z_x)$ represents the non-membership function of $z_x \in \mathscr{U}$ in $\tilde{\xi}$. Along with this, the hesitance degree of $z_x \in \mathscr{U}$ in $\tilde{\xi}$ is denoted as $\Delta_{\tilde{\xi}}(z_x)$ and is given as $\Delta_{\tilde{\xi}}(z_x) = 1 - \varphi_{\tilde{\xi}}(z_x) - \psi_{\tilde{\xi}}(z_x)$.

Definition 6.2.3 Yager [37] and Yager and Abbasov [38]: A pythagorean fuzzy set $\tilde{\xi}$ in $\mathscr{U}$ (finite universe of discourse) is given as $\tilde{\xi} = \left\{\left(\varphi_{\tilde{\xi}}(z_x), \psi_{\tilde{\xi}}(z_x)\right) | z_x \in \mathscr{U}\right\}$, where $\varphi_{\tilde{\xi}}$ denotes the membership degree of $z_x \in \mathscr{U}$ in $\tilde{\xi}$ and $\psi_{\tilde{\xi}}$ represents the non-membership function of $z_x \in \mathscr{U}$ in $\tilde{\xi}$ such that $\varphi_{\tilde{\xi}} : \mathscr{U} \to [0, 1]$ and $\psi_{\tilde{\xi}} : \mathscr{U} \to [0, 1]$ and satisfying the condition $\varphi^2_{\tilde{\xi}}(z_x) + \psi^2_{\tilde{\xi}}(z_x) \leq 1$. Along with this, the hesitance degree of $z_x \in \mathscr{U}$ in $\tilde{\xi}$ is denoted as $\Delta_{\tilde{\xi}}(z_x)$ and is given as $\Delta_{\tilde{\xi}}(z_x) = \sqrt{1 - \left(\varphi^2_{\tilde{\xi}}(z_x) + \psi^2_{\tilde{\xi}}(z_x)\right)}$.

Definition 6.2.4 Yager [39] and Yager and Alajlan [40]: A q-Rung Orthopair Fuzzy Set (qROFS) $\tilde{\xi}$ in $\mathscr{U}$ (finite universe of discourse) is given as $\tilde{\xi} = \left\{\left(\varphi_{\tilde{\xi}}(z_x), \psi_{\tilde{\xi}}(z_x)\right) | z_x \in \mathscr{U}\right\}$, where $\varphi_{\tilde{\xi}}$ and $\psi_{\tilde{\xi}}$ represents the membership and non-membership function of $z_x \in \mathscr{U}$ in $\tilde{\xi}$ such that $\varphi_{\tilde{\xi}} : \mathscr{U} \to [0, 1]$ and $\psi_{\tilde{\xi}} : \mathscr{U} \to [0, 1]$ and satisfying the condition $\varphi^q_{\tilde{\xi}}(z_x) + \psi^q_{\tilde{\xi}}(z_x) \leq 1$. Along with this, the hesitance degree of $z_x \in \mathscr{U}$ in $\tilde{\xi}$ is denoted as $\Delta_{\tilde{\xi}}(z_x)$ and is given as $\Delta_{\tilde{\xi}}(z_x) = 1 - \left(\varphi^q_{\tilde{\xi}}(z_x) + \psi^q_{\tilde{\xi}}(z_x)\right)^{\frac{1}{q}}$. For convenience, we can write $\varphi_{\tilde{\xi}}(z_x) = \varphi_{\tilde{\xi}_x}$, $\psi_{\tilde{\xi}}(z_x) = \psi_{\tilde{\xi}_x}$ and $\Delta_{\tilde{\xi}}(z_x) = \Delta_{\tilde{\xi}_x}$.

Definition 6.2.5 Liu and Wang [18]: The score function $\mathbb{S}_q\left(\tilde{\xi}\right)$ and the accuracy function $\mathbb{A}_q\left(\tilde{\xi}\right)$ of a qROFS $\tilde{\xi} = \left(\varphi_{\tilde{\xi}_x}, \psi_{\tilde{\xi}_x}\right)$ are defined as $\mathbb{S}_q\left(\tilde{\xi}\right) = \varphi^q_{\tilde{\xi}} - \psi^q_{\tilde{\xi}}$ and $\mathbb{A}_q\left(\tilde{\xi}\right) = \varphi^q_{\tilde{\xi}} + \psi^q_{\tilde{\xi}}$, respectively, where $-1 \leq \mathbb{S}_q\left(\tilde{\xi}\right) \leq 1$ and $0 \leq \mathbb{A}_q\left(\tilde{\xi}\right) \leq 1$.

Definition 6.2.6 Khan et al. [14]: Let $\widetilde{\xi_1} = \left(\varphi_{\widetilde{\xi_1}}, \psi_{\widetilde{\xi_1}}\right)$ and $\widetilde{\xi_2} = \left(\varphi_{\widetilde{\xi_2}}, \psi_{\widetilde{\xi_2}}\right)$ be any two qROFNs then the basic operators are given as follows:

(i) $\widetilde{\xi_1}^c = \left(\psi_{\widetilde{\xi_1}}, \varphi_{\widetilde{\xi_1}}\right)$
(ii) $\widetilde{\xi_1} \vee \widetilde{\xi_2} = \left(\max\left\{\varphi_{\widetilde{\xi_1}}, \varphi_{\widetilde{\xi_2}}\right\}, \min\left\{\psi_{\widetilde{\xi_1}}, \psi_{\widetilde{\xi_2}}\right\}\right)$
(iii) $\widetilde{\xi_1} \wedge \widetilde{\xi_2} = \left(\min\left\{\varphi_{\widetilde{\xi_1}}, \varphi_{\widetilde{\xi_2}}\right\}, \max\left\{\psi_{\widetilde{\xi_1}}, \psi_{\widetilde{\xi_2}}\right\}\right)$
(iv) $\widetilde{\xi_1} \oplus \widetilde{\xi_2} = \left(\left(\varphi^q_{\widetilde{\xi_1}} + \varphi^q_{\widetilde{\xi_2}} - \varphi^q_{\widetilde{\xi_1}}.\varphi^q_{\widetilde{\xi_2}}\right)^{\frac{1}{q}}, \psi_{\widetilde{\xi_1}}.\psi_{\widetilde{\xi_2}}\right)$
(v) $\widetilde{\xi_1} \otimes \widetilde{\xi_2} = \left(\varphi_{\widetilde{\xi_1}}.\varphi_{\widetilde{\xi_2}}, \left(\psi^q_{\widetilde{\xi_1}} + \psi^q_{\widetilde{\xi_2}} - \psi^q_{\widetilde{\xi_1}}.\psi^q_{\widetilde{\xi_2}}\right)^{\frac{1}{q}}\right)$.

Definition 6.2.7 Khan et al. [14]: Let $\tilde{\xi}$ be a qROFS in $\mathscr{U}$. Then, the knowledge measure of $\tilde{\xi}$ is given as below

$$\mathbb{I}^q_{\mathsf{z}}\left(\tilde{\xi}\right) = \left(\frac{1}{N}\sum_{x=1}^{N}\left\{1 - 0.5\left(1 - \left(\varphi^{2q}_{\tilde{\xi}_x} + \psi^{2q}_{\tilde{\xi}_x}\right) + \frac{4}{\pi}\tan^{-1}\left[\Delta^{2q}_{\tilde{\xi}_x}\right]\right)\right\}\right)^{\frac{1}{q}}. \quad (6.1)$$

where $\Delta^{2q}_{\tilde{\xi}_x}$ is the hesitance degree of z_x, i.e., $\Delta_{\tilde{\xi}}(\mathsf{z}_x) = 1 - \left(\varphi^q_{\tilde{\xi}}(\mathsf{z}_x) + \psi^q_{\tilde{\xi}}(\mathsf{z}_x)\right)^{\frac{1}{q}}$. Thus, $\mathbb{I}^q_{\mathsf{z}}\left(\tilde{\xi}\right)$ can be written as

$$\mathbb{I}^q_{\mathsf{z}}\left(\tilde{\xi}\right) = \left(\frac{1}{N}\sum_{x=1}^{N}\left\{1 - 0.5\left(1 - \left(\varphi^{2q}_{\tilde{\xi}_x} + \psi^{2q}_{\tilde{\xi}_x}\right) + \frac{4}{\pi}\tan^{-1}\left[\left(1 - \varphi^q_{\tilde{\xi}_x} - \psi^q_{\tilde{\xi}_x}\right)^2\right]\right)\right\}\right)^{\frac{1}{q}}. \quad (6.2)$$

where $\mathbb{I}^q_{\mathsf{z}}\left(\tilde{\xi}\right) \in [0, 1]$.

6.3 Model Formulation

Figure 6.1 and Table 6.1. In this EOQ model, it is considered that when every cycle $\mathscr{T}'$ begins, a retailer purchased a lot l units at the cost of h per item which diminishes at a constant rate of demand d. Due to imperfect manufacturing procedures, lack of maintenance, careless handling, etc., the replenished products have some proportion of faulty products say σ. Hence, the screening procedure proceeds during $[0, \mathsf{t}_1]$ at the rate of β and separates the faulty products ($\mathscr{Q}_1$) at the end of t_1. Since the product sale also runs at the time of inspection, $\beta > \mathsf{d}$ is considered. Due to the imperfect screening process, two kinds of misclassification errors say, perfect items are misclassified as defectives and the defective products are misclassified as perfect items. According to these misclassification errors, four different categories may arise

Table 6.1 Notations used in the proposed model

$\mathscr{I}$	Order size
d	Demand rate (unit/year)
β	Screening rate (unit/year)
C	Fixed ordering price
k	Purchasing price/unit
i	Inspection price/unit
s_1	Price for selling perfect items/unit
s_2	Selling price for scrap items per unit that are unable to claim the warranty
k_a	Price of accepting a substandard product
k_r	Price of rejecting a perfect product
k_e	Purchasing price of an emergency order/unit
k_t	Transportation price per unit
k_l	Required stuff and Labor charge for repair per product
σ	Proportion of substandard products
η_1	Proportion of type I misclassification error [misclassify a perfect item as imperfect]
η_2	Proportion of type II misclassification error [misclassify an imperfect item as perfect]
λ	Proportion of warranty unclaimed product
$\mathscr{W}$	Rate of claim product warranty (unit/year)
$\mathscr{T}'$	Cycle length
t_1	Time to screen a lot size I
t_2	Time duration of warranty claiming process
t_3	Time duration of mending process for case (i) or Time duration of selling scrap items and emergency purchase for case (ii)
t_w	Time duration of transportation of warranty claiming process
$\mathscr{S}$	Repair setup cost
$\mathscr{A}$	Fixed transportation cost
h_R'	Holding charge at the repair shop
L	Handling cost/unit
K	Fixed charge for processing the warranty
h	Holding price/unit
h_W	Holding price of a warranty claimed product/unit
h_R	Holding price of a repaired product/unit
h_E	Holding price of an emergency purchased product
$\mathscr{Q}_1$	Number of products categorized as substandard per cycle
$\mathscr{Q}_2$	Number of substandard products rejected by the customer per cycle
$\tilde{\sigma}$	q-rung orthopair fuzzy proportion of faulty products
$\tilde{\eta}_1$	q-rung orthopair fuzzy proportion of type I misclassification error
$\tilde{\eta}_2$	q-rung orthopair fuzzy proportion of type II misclassification error

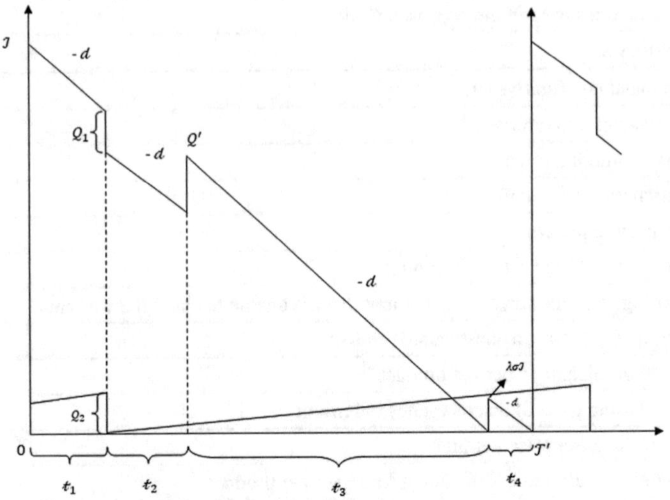

Fig. 6.1 Variation of stock level throughout the cycle

say, misclassify a good product as faulty [$\mathsf{I}(1-\sigma)\eta_1$], classify faulty product as faulty [$\mathsf{I}\sigma(1-\eta_2)$], classify a perfect item as perfect [$\mathsf{I}(1-\sigma)(1-\eta_1)$], and misclassify a faulty product as good [$\mathsf{I}\sigma\eta_2$], where η_1 and η_2 are the proportion of type I error and type II error, respectively, also due to misclassification errors, the charge of accepting a flawed product (k_a) and the charge of dismissing a good product (k_r) arises. Because of type II error, the faulty products ($\mathcal{Q}_2$) that are sold to customers are sent back to the retailer throughout the cycle $\mathcal{T}'$ and gathered at the end of t_1.

The products $\mathcal{Q}_1$ and $\mathcal{Q}_2$ are sent to the supplier to claim product replacement warranty during the time [t_1, t_2] and stocked the warranty claimed products with the retailer's inventory. The warranty claiming rate units per year is presumed as $\mathcal{W}$ and the total transportation time for claiming the replacement warranty is represented as t_w. Due to transportation damage, the proportion λ of flawed products say, $\lambda\sigma\mathcal{I}$ is unable to claim warranty. The restoring strategy of these products leads the study to analyze two cases such as restoring using mending strategy and restoring using emergency purchase option. The first case considers the products $\lambda\sigma\mathsf{I}$ are repaired by a local repair shop and restored as good items during the time t_3 and stored when the inventory level reaches zero and then started to satisfy the customer at the same rate. The second case presumes that the products $\lambda\sigma\mathsf{I}$ are replaced using emergency purchase option, that is, $\lambda\sigma\mathsf{I}$ products are sold at scrap price s_2 and replaced by purchasing the same quantity of good items from the nearby shop at the cost k_e at the end of t_3 and then started to satisfy the customer at the same rate. Since defective items are substituted as good items by the mending option or the emergency purchase option, shortages are not occurring in both cases (Fig. 6.2).

In this EOQ model, it is considered that all the ordered products I are satisfied at the demand rate d. Thus, the total cycle duration is given as $\mathcal{T}' = \frac{\mathsf{I}}{\mathsf{d}}$.

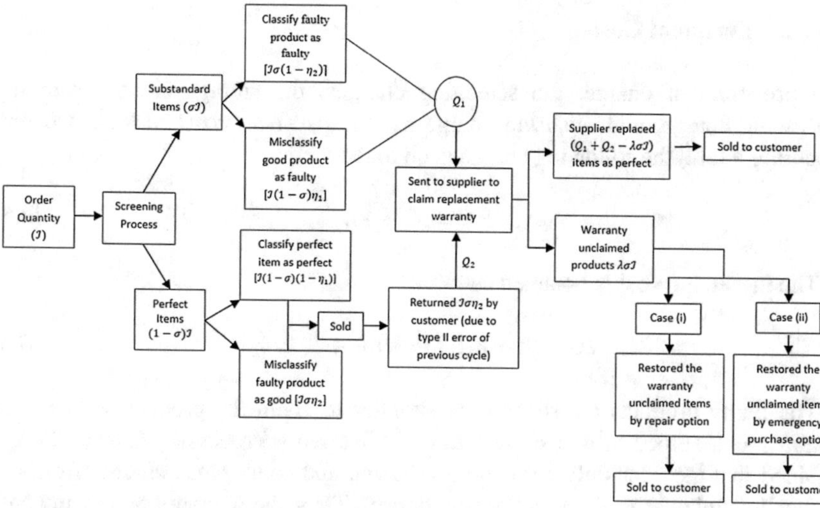

Fig. 6.2 Flow chart of inventory flows throughout the cycle

The sales revenue and pertinent costs are as follows. For simplification, it is assumed as below

$$\varepsilon_1 = (1-\sigma)(1-\eta_1) + \sigma\eta_2, \varepsilon_2 = (1-\sigma)\eta_1 + \sigma(1-\eta_2) \text{ and } \varepsilon_3 = \sigma\eta_2. \quad (6.3)$$

6.3.1 *Case (I): Replacement of the Faulty Option by Warranty Claiming and Repair Option*

According to the hypothesis of the current study, all the products $\mathcal{Q}_1 + \mathcal{Q}_2$ are sent back to the supplier to claim a replacement warranty. At the end of t_2, the replaced products are received from the supplier with some unclaimed products. In this case, it is presumed that the unclaimed products are repaired by a local repair shop and restored as good items and satisfy the customer demand.

6.3.1.1 Total Sales Revenue

The sales revenue is obtained by selling the perfect items at the unit selling price s_1. Since all the substandard products are restored as perfect ones, the total inventory I is sold. Thus, the sales income is given as

$$TR = \mathsf{s}_1\mathsf{I} \quad (6.4)$$

6.3.1.2 Pertinent Costs

The procurement charge, the screening charges, the storage charge, warranty processing charge, and repairing charge are the pertinent costs in this case. For procuring a lot l, the following charge is obtained.

$$PC = C + \mathsf{kl}. \tag{6.5}$$

The inspecting cost is obtained as

$$IC = \mathsf{il} + \mathsf{k}_r\mathsf{l}(1-\sigma)\eta_1 + \mathsf{k}_a\mathsf{l}\sigma\eta_2. \tag{6.6}$$

The faulty products are sent to the supplier to claim the product replacement warranty is assumed in this study. Thus, the Warranty Processing Cost (WPC) is obtained that includes only the handling charge and conveyance charge (from the retailer's warehouse to the supplier warehouse). Thus, the warranty processing cost is given as

$$WPC = K + (L + \mathsf{k}_t)(\mathscr{Q}_1 + \mathscr{Q}_2) = K + \mathsf{l}(L + \mathsf{k}_t)(\varepsilon_2 + \varepsilon_3). \tag{6.7}$$

The mending charge is the cost incurred by the mending store that incorporates the mending setup charge ($\mathscr{S}$), fixed conveyance price ($\mathscr{A}$), transportation charge per unit (k_t) (from warehouses to mending store and mending store to warehouses), storage price at the repairing store (h'_R), and the labor cost along with the spare parts (h_l) for the warranty unclaimable items $\lambda\sigma\mathsf{l}$.

Thus, the mending charge is obtained as

$$\begin{aligned} RC &= \mathscr{S} + 2\mathscr{A} + \lambda\sigma\mathsf{l}\big(2\mathsf{k}_t + \mathsf{k}_l + \mathsf{h}'_R\mathsf{t}_3\big) \\ &= \mathscr{S} + 2\mathscr{A} + \lambda\sigma\mathsf{l} \\ &\left(2\mathsf{k}_t + \mathsf{k}_l + \mathsf{h}'_R\left[\frac{\mathsf{l}(1+\varepsilon_3-\lambda\sigma)}{\mathsf{d}} - \left(\frac{\mathsf{l}}{\beta} + \frac{(\varepsilon_2+\varepsilon_3-\lambda\sigma)\mathsf{l}}{\mathscr{W}} + \mathsf{t}_{\mathrm{w}}\right)\right]\right) \end{aligned} \tag{6.8}$$

The storage charge includes the storage price of perfect products, the storage price of warranty claimed product, the storage price of faulty products sent back from the customer, and the storage charge of mended products. Thus, the storage price, in this case, is provided as follows: (For derivation see Appendix)

$$\begin{aligned} HC &= \frac{\mathsf{h}}{2}\mathsf{l}^2\left(\frac{2}{\beta} - \frac{\mathsf{d}}{\beta^2} + \frac{(1-\varepsilon_2-\frac{\mathsf{d}}{\beta})^2}{\mathsf{d}}\right) + \frac{\mathsf{hl}^2\varepsilon_3}{2\mathsf{d}} \\ &+ \mathsf{h}_W\mathsf{l}^2(\varepsilon_2+\varepsilon_3-\lambda\sigma)\left[\frac{1}{\mathsf{d}} - \frac{1}{\beta} - \frac{(\varepsilon_2+\varepsilon_3-\lambda\sigma)}{\mathscr{W}} - \frac{(\varepsilon_2-\varepsilon_3+\lambda\sigma)}{2\mathsf{d}}\right] \\ &- \mathsf{h}_W\mathsf{l}(\varepsilon_2+\varepsilon_3-\lambda\sigma)\mathsf{t}_{\mathrm{w}} + \frac{\mathsf{h}_R(\lambda\sigma)^2\mathsf{l}^2}{2\mathsf{d}}. \end{aligned} \tag{6.9}$$

6.3.1.3 Total Cost

The total cost for each cycle $\mathscr{T}'$ is the sum of all costs which is given as

$$TC(\mathsf{l}) = PC + IC + HC + WPC + RC. \tag{6.10}$$

$$\begin{aligned}TC(\mathsf{l}) &= C + \mathsf{kl} + \mathsf{il} + \mathsf{k}_r\mathsf{l}(1-\sigma)\eta_1 + \mathsf{k}_a\mathsf{l}\sigma\eta_2 + K + \mathsf{l}(L+\mathsf{k}_t)(\varepsilon_2+\varepsilon_3) + \mathscr{S} + 2\mathscr{A}\\ &+ \lambda\sigma\mathsf{l}\left(2\mathsf{k}_t + \mathsf{k}_l + \mathsf{h}'_R\left[\frac{\mathsf{l}(1+\varepsilon_3-\lambda\sigma)}{\mathsf{d}} - \left(\frac{\mathsf{l}}{\beta} + \frac{(\varepsilon_2+\varepsilon_3-\lambda\sigma)\mathsf{l}}{\mathscr{W}} + \mathsf{t_w}\right)\right]\right)\\ &+ \frac{\mathsf{h}}{2}\mathsf{l}^2\left(\frac{2}{\beta} - \frac{\mathsf{d}}{\beta^2} + \frac{(1-\varepsilon_2-\frac{\mathsf{d}}{\beta})^2}{\mathsf{d}}\right) + \frac{\mathsf{hl}^2\varepsilon_3}{2\mathsf{d}}\\ &+ \mathsf{h}_W\mathsf{l}^2(\varepsilon_2+\varepsilon_3-\lambda\sigma)\left[\frac{1}{\mathsf{d}} - \frac{1}{\beta} - \frac{(\varepsilon_2+\varepsilon_3-\lambda\sigma)}{\mathscr{W}} - \frac{(\varepsilon_2-\varepsilon_3+\lambda\sigma)}{2\mathsf{d}}\right]\\ &- \mathsf{h}_W\mathsf{l}(\varepsilon_2+\varepsilon_3-\lambda\sigma)\mathsf{t_w} + \frac{\mathsf{h}_R(\lambda\sigma)^2\mathsf{l}^2}{2\mathsf{d}}\end{aligned} \tag{6.11}$$

6.3.1.4 Total Profit Per Unit Time

The total profit per cycle is derived by reducing the pertinent costs from the total sales revenue, which is given below

$$TP(\mathsf{l}) = TR(\mathsf{l}) - TC(\mathsf{l}). \tag{6.12}$$

The total profit per unit time is $TPU(\mathsf{l}) = \frac{TP(\mathsf{l})}{\mathscr{T}}$.

$$\begin{aligned}TPU(\mathsf{l}) = \mathsf{l}\Bigg[&-\mathsf{d}\left(\frac{\mathsf{h}}{2}\left(\frac{2}{\beta} - \frac{\mathsf{d}}{\beta^2} + \frac{\left(1-\varepsilon_2-\frac{\mathsf{d}}{\beta}\right)^2}{\mathsf{d}}\right) + \frac{\mathsf{h}\varepsilon_3}{2\mathsf{d}}\right.\\ &+ \mathsf{h}_W(\varepsilon_2+\varepsilon_3-\lambda\sigma)\left[\frac{1}{\mathsf{d}} - \frac{1}{\beta} - \frac{(\varepsilon_2+\varepsilon_3-\lambda\sigma)}{\mathscr{W}} - \frac{(\varepsilon_2-\varepsilon_3+\lambda\sigma)}{2\mathsf{d}}\right]\\ &\left.+ \frac{\mathsf{h}_R(\lambda\sigma)^2}{2\mathsf{d}} + \mathsf{h}'_R\lambda\sigma\left[\frac{(1+\varepsilon_3-\lambda\sigma)}{\mathsf{d}} - \frac{(\varepsilon_2+\varepsilon_3-\lambda\sigma)}{\mathscr{W}} - \frac{1}{\beta}\right]\right)\Bigg]\\ &+ \frac{1}{\mathsf{l}}(-\mathsf{d}[C + K + \mathscr{S} + 2\mathscr{A}]) + \mathsf{d}[\mathsf{s_l} - \mathsf{k} - \mathsf{i} - \mathsf{k}_r(1-\sigma)\eta_1 - \mathsf{k}_a\sigma\eta_2\\ &+ \mathsf{h}_W(\varepsilon_2+\varepsilon_3-\lambda\sigma)\mathsf{t_w} - (L+\mathsf{k}_t)(\varepsilon_2+\varepsilon_3) - \lambda\sigma\left(2\mathsf{k}_t + \mathsf{k}_l - \mathsf{h}'_R\mathsf{t_w}\right).\end{aligned} \tag{6.13}$$

6.3.2 Case (ii): Replacement of the Faulty Option by Warranty Claiming and the Emergency Purchase Option

As a strategy of replacing the faulty products that are unable to claim replacement warranty, this case examines the notion of an emergency purchase option to replace the unclaimed products by a nearby dealer. The unclaimable faulty products are sold at scrap price s_2 once it received from the supplier.

6.3.2.1 Total Sales Revenue

The total sales revenue in this case is the sum of the sales revenue from perfect products, the sales income from faulty products, and the sales income from the emergency purchases products. Thus, the sales incomes are obtained as

$$R_1 = \mathsf{s}_1(1-\lambda\sigma)\mathsf{l}$$

$$R_2 = \mathsf{s}_2\lambda\sigma\mathsf{l}$$

$$R_3 = \mathsf{s}_1\lambda\sigma\mathsf{l}$$

$$TR = R_1 + R_2 + R_3 = \mathsf{s}_1\mathsf{l} + \mathsf{s}_2\lambda\sigma\mathsf{l} \tag{6.14}$$

6.3.2.2 Pertinent Costs

It is clear that PC, IC, and WPC are similar to case (i). Thus, the emergency purchase charge and the storage charge are derived as follows.

$$EPC = \mathsf{k}_e\lambda\sigma\mathsf{l}. \tag{6.15}$$

$$\begin{aligned} HC = {} & \frac{\mathsf{h}}{2}\mathsf{l}^2\left(\frac{2}{\beta} - \frac{\mathsf{d}}{\beta^2} + \frac{\left(1-\varepsilon_2-\frac{\mathsf{d}}{\beta}\right)^2}{\mathsf{d}}\right) + \frac{\mathsf{h}\mathsf{l}^2\varepsilon_3}{2\mathsf{d}} \\ & + \mathsf{h}_W\mathsf{l}^2(\varepsilon_2+\varepsilon_3-\lambda\sigma)\left[\frac{1}{\mathsf{d}} - \frac{1}{\beta} - \frac{(\varepsilon_2+\varepsilon_3-\lambda\sigma)}{\mathscr{W}} - \frac{(\varepsilon_2-\varepsilon_3+\lambda\sigma)}{2\mathsf{d}}\right] \\ & - \mathsf{h}_W\mathsf{l}(\varepsilon_2+\varepsilon_3-\lambda\sigma)\mathsf{t}_\mathsf{w} + \frac{\mathsf{h}_E(\lambda\sigma)^2\mathsf{l}^2}{2\mathsf{d}}. \end{aligned} \tag{6.16}$$

6.3.2.3 Total Cost

The total cost for this case is obtained as

$$
\begin{aligned}
TC(\mathsf{l}) &= C + \mathsf{kl} + \mathsf{il} + \mathsf{k}_r\mathsf{l}(1-\sigma)\eta_1 + \mathsf{k}_a\mathsf{l}\sigma\eta_2 \\
&\quad + K + \mathsf{l}(L + \mathsf{k}_t)(\varepsilon_2 + \varepsilon_3) + \mathsf{k}_e\lambda\sigma\mathsf{l} \\
&\quad + \frac{\mathsf{h}}{2}\mathsf{l}^2\left(\frac{2}{\beta} - \frac{\mathsf{d}}{\beta^2} + \frac{\left(1-\varepsilon_2-\frac{\mathsf{d}}{\beta}\right)^2}{\mathsf{d}}\right) + \frac{\mathsf{hl}^2\varepsilon_3}{2\mathsf{d}} \\
&\quad + \mathsf{h}_W\mathsf{l}^2(\varepsilon_2+\varepsilon_3-\lambda\sigma)\left[\frac{1}{\mathsf{d}} - \frac{1}{\beta} - \frac{(\varepsilon_2+\varepsilon_3-\lambda\sigma)}{\mathscr{W}} - \frac{(\varepsilon_2-\varepsilon_3+\lambda\sigma)}{2\mathsf{d}}\right] \\
&\quad - \mathsf{h}_W\mathsf{l}(\varepsilon_2+\varepsilon_3-\lambda\sigma)\mathsf{t_w} + \frac{\mathsf{h}_E(\lambda\sigma)^2\mathsf{l}^2}{2\mathsf{d}}.
\end{aligned} \tag{6.17}
$$

6.3.2.4 Total Profit Per Unit Time

The total profit per unit time for case (ii) is given below

$$
\begin{aligned}
TPU(\mathsf{l}) &= \mathsf{l}\left[-\mathsf{d}\left(\frac{\mathsf{h}}{2}\left(\frac{2}{\beta} - \frac{\mathsf{d}}{\beta^2} + \frac{\left(1-\varepsilon_2-\frac{\mathsf{d}}{\beta}\right)^2}{\mathsf{d}}\right)\right.\right. \\
&\quad + \frac{\mathsf{h}\varepsilon_3}{2\mathsf{d}} + \mathsf{h}_W(\varepsilon_2+\varepsilon_3-\lambda\sigma) \\
&\quad \left.\left.\left[\frac{1}{\mathsf{d}} - \frac{1}{\beta} - \frac{(\varepsilon_2+\varepsilon_3-\lambda\sigma)}{\mathscr{W}} - \frac{(\varepsilon_2-\varepsilon_3+\lambda\sigma)}{2\mathsf{d}}\right] + \frac{\mathsf{h}_E(\lambda\sigma)^2}{2\mathsf{d}}\right)\right] \\
&\quad + \frac{1}{\mathsf{l}}(-\mathsf{d}[C+K]) + \mathsf{d}[\mathsf{s}_1 + \mathsf{s}_2\lambda\sigma - \mathsf{k} - \mathsf{i} - \mathsf{k}_r(1-\sigma)\eta_1 - \mathsf{k}_a\sigma\eta_2 \\
&\quad + \mathsf{h}_W(\varepsilon_2+\varepsilon_3-\lambda\sigma)\mathsf{t_w} - (L+\mathsf{k}_t)(\varepsilon_2+\varepsilon_3) - \mathsf{k}_e\lambda\sigma\mathsf{l}
\end{aligned} \tag{6.18}
$$

6.3.3 *Inventory Model with q-Rung Orthopair Fuzzy Variables*

The proposed EOQ model is framed under the qROF environment by considering the proportion of faulty products, probability of misclassification errors into qROFVs. That is, the parameters $\tilde{\sigma}$, $\tilde{\eta_1}$ and $\tilde{\eta_2}$ are written as $\tilde{\sigma} = \left(\varphi_{\tilde{\sigma}_x}, \psi_{\tilde{\sigma}_x}\right)$, $\tilde{\eta_1} = \left(\varphi_{\widetilde{(\eta_1)}_x}, \psi_{\widetilde{(\eta_1)}_x}\right)$ and $\tilde{\eta_2} = \left(\varphi_{\widetilde{(\eta_2)}_x}, \psi_{\widetilde{(\eta_2)}_x}\right)$. Using these parameters, Eq. (6.3) implies

$$\widetilde{\varepsilon_1} = (1-\tilde{\sigma})(1-\widetilde{\eta_1}) + \tilde{\sigma}\widetilde{\eta_2};\ \widetilde{\varepsilon_2} = (1-\tilde{\sigma})\widetilde{\eta_1} + \tilde{\sigma}(1-\widetilde{\eta_2});\ \text{and } \widetilde{\varepsilon_3} = \tilde{\sigma}\widetilde{\eta_2}. \tag{6.19}$$

Thus, the total profit in qROF environment for both case (i) and case (ii) are obtained as

Case (i):

$$\begin{aligned} TPU_{qROF}(\mathsf{l}) = \mathsf{l}\Bigg[& -\mathsf{d}\left(\frac{\mathsf{h}}{2}\left(\frac{2}{\beta} - \frac{\mathsf{d}}{\beta^2} + \frac{\left(1-\widetilde{\varepsilon_2}-\frac{\mathsf{d}}{\beta}\right)^2}{\mathsf{d}}\right)\right. \\ & + \frac{\mathsf{h}\widetilde{\varepsilon_3}}{2\mathsf{d}} \quad + \mathsf{h}_W(\widetilde{\varepsilon_2}+\widetilde{\varepsilon_3}-\lambda\tilde{\sigma}) \\ & \left[\frac{1}{\mathsf{d}} - \frac{1}{\beta} - \frac{(\widetilde{\varepsilon_2}+\widetilde{\varepsilon_3}-\lambda\tilde{\sigma})}{\mathscr{W}} - \frac{(\widetilde{\varepsilon_2}-\widetilde{\varepsilon_3}+\lambda\tilde{\sigma})}{2\mathsf{d}}\right] + \frac{\mathsf{h}_R(\lambda\tilde{\sigma})^2}{2\mathsf{d}} \\ & \left. + \mathsf{h}'_R\lambda\tilde{\sigma}\left[\frac{(1+\widetilde{\varepsilon_3}-\lambda\tilde{\sigma})}{\mathsf{d}} - \frac{(\widetilde{\varepsilon_2}+\widetilde{\varepsilon_3}-\lambda\tilde{\sigma})}{\mathscr{W}} - \frac{1}{\beta}\right]\right)\Bigg] \\ & + \frac{1}{\mathsf{l}}(-\mathsf{d}[C+K+\mathscr{S}+2\mathscr{A}]) \quad + \mathsf{d}[\mathsf{s}_1 - \mathsf{k} - \mathsf{i} - \mathsf{k}_r(1-\tilde{\sigma})\widetilde{\eta_1} \\ & - \mathsf{k}_a\tilde{\sigma}\widetilde{\eta_2} + \mathsf{h}_W(\widetilde{\varepsilon_2}+\widetilde{\varepsilon_3}-\lambda\tilde{\sigma})\mathsf{t_w} \quad - (L+\mathsf{k}_t)(\widetilde{\varepsilon_2}+\widetilde{\varepsilon_3}) - \lambda\tilde{\sigma}(2\mathsf{k}_t+\mathsf{k}_l-\mathsf{h}'_R\mathsf{t_w}). \end{aligned} \tag{6.20}$$

Using the knowledge measure for qROFS, the measures $\mathbb{I}^q_{\mathsf{z}}(\tilde{\sigma})$, $\mathbb{I}^q_{\mathsf{z}}(\widetilde{\eta_1})$ and $\mathbb{I}^q_{\mathsf{z}}(\widetilde{\eta_2})$ are obtained as follows.

$$\mathbb{I}^q_{\mathsf{z}}(\tilde{\sigma}) = \left(\frac{1}{N}\sum_{x=1}^{N}\left\{1 - 0.5\left(1 - \left(\varphi^{2q}_{\tilde{\sigma}_x} + \psi^{2q}_{\tilde{\sigma}_x}\right) + \frac{4}{\pi}\tan^{-1}\left[\left(1-\varphi^{q}_{\tilde{\sigma}_x} - \psi^{q}_{\tilde{\sigma}_x}\right)^2\right]\right)\right\}\right)^{\frac{1}{q}} \tag{6.21}$$

$$\begin{aligned} & \mathbb{I}^q_{\mathsf{z}}(\widetilde{\eta_1}) \\ & = \left(\frac{1}{N}\sum_{x=1}^{N}\left\{1 - 0.5\left(1 - \left(\varphi^{2q}_{\widetilde{(\eta_1)_x}} + \psi^{2q}_{\widetilde{(\eta_1)_x}}\right) + \frac{4}{\pi}\tan^{-1}\left[\left(1-\varphi^{q}_{\widetilde{(\eta_1)_x}} - \psi^{q}_{\widetilde{(\eta_1)_x}}\right)^2\right]\right)\right\}\right)^{\frac{1}{q}} \end{aligned} \tag{6.22}$$

$$\begin{aligned} & \mathbb{I}^q_{\mathsf{z}}(\widetilde{\eta_2}) \\ & = \left(\frac{1}{N}\sum_{x=1}^{N}\left\{1 - 0.5\left(1 - \left(\varphi^{2q}_{\widetilde{(\eta_2)_x}} + \psi^{2q}_{\widetilde{(\eta_2)_x}}\right) + \frac{4}{\pi}\tan^{-1}\left[\left(1-\varphi^{q}_{\widetilde{(\eta_2)_x}} - \psi^{q}_{\widetilde{(\eta_2)_x}}\right)^2\right]\right)\right\}\right)^{\frac{1}{q}} \end{aligned} \tag{6.23}$$

Thus, Eq. (6.19) becomes

$\mathbb{I}(\widetilde{\varepsilon_1}) = \left(1-\mathbb{I}^q_{\mathsf{z}}(\tilde{\sigma})\right)\left(1-\mathbb{I}^q_{\mathsf{z}}(\widetilde{\eta_1})\right) + \mathbb{I}^q_{\mathsf{z}}(\tilde{\sigma})\mathbb{I}^q_{\mathsf{z}}(\widetilde{\eta_2})$,

$\mathbb{I}(\widetilde{\varepsilon_2}) = \left(1-\mathbb{I}^q_{\mathsf{z}}(\tilde{\sigma})\right)\mathbb{I}^q_{\mathsf{z}}(\widetilde{\eta_1}) + \mathbb{I}^q_{\mathsf{z}}(\tilde{\sigma})\left(1-\mathbb{I}^q_{\mathsf{z}}(\widetilde{\eta_2})\right)$ and

$$\mathbb{I}(\widetilde{\varepsilon_3}) = \mathbb{I}^q_{\mathsf{z}}(\tilde{\sigma})\mathbb{I}^q_{\mathsf{z}}(\widetilde{\eta_2})$$

Therefore, the KM-qROF inventory model is written as

$$TPU_{qROF}(\mathsf{l}) = \Delta_1\mathsf{l} + \Delta_2\frac{1}{\mathsf{l}} + \Delta_3 \tag{6.24}$$

where

$$\Delta_1 = -\mathsf{d}\left(\frac{\mathsf{h}}{2}\left(\frac{2}{\beta} - \frac{\mathsf{d}}{\beta^2} + \frac{\left(1 - \mathbb{I}(\widetilde{\varepsilon_2}) - \frac{\mathsf{d}}{\beta}\right)^2}{\mathsf{d}}\right) + \frac{\mathsf{h}\mathbb{I}(\widetilde{\varepsilon_3})}{2\mathsf{d}}\right.$$
$$+ \mathsf{h}_W\left(\mathbb{I}(\widetilde{\varepsilon_2}) + \mathbb{I}(\widetilde{\varepsilon_3}) - \lambda\mathbb{I}_{\mathsf{z}}^q(\tilde{\sigma})\right)\left[\frac{1}{\mathsf{d}} - \frac{1}{\beta} - \frac{\left(\mathbb{I}(\widetilde{\varepsilon_2}) + \mathbb{I}(\widetilde{\varepsilon_3}) - \lambda\mathbb{I}_{\mathsf{z}}^q(\tilde{\sigma})\right)}{\mathscr{W}}\right.$$
$$\left.-\frac{\left(\mathbb{I}(\widetilde{\varepsilon_2}) - \mathbb{I}(\widetilde{\varepsilon_3}) + \lambda\mathbb{I}_{\mathsf{z}}^q(\tilde{\sigma})\right)}{2\mathsf{d}}\right] + \frac{\mathsf{h}_R\left(\lambda\mathbb{I}_{\mathsf{z}}^q(\tilde{\sigma})\right)^2}{2\mathsf{d}}$$
$$\left.+\mathsf{h}'_R\lambda\tilde{\sigma}\left[\frac{\left(1 + \mathbb{I}(\widetilde{\varepsilon_3}) - \lambda\mathbb{I}_{\mathsf{z}}^q(\tilde{\sigma})\right)}{\mathsf{d}} - \frac{\left(\mathbb{I}(\widetilde{\varepsilon_2}) + \mathbb{I}(\widetilde{\varepsilon_3}) - \lambda\mathbb{I}_{\mathsf{z}}^q(\tilde{\sigma})\right)}{\mathscr{W}} - \frac{1}{\beta}\right]\right);$$

$$\Delta_2 = -\mathsf{d}[C + K + \mathscr{S} + 2\mathscr{A}];$$

$$\Delta_3 = \mathsf{d}[\mathsf{s}_1 - \mathsf{k} - \mathsf{i} - \mathsf{k}_r\left(1 - \mathbb{I}_{\mathsf{z}}^q(\tilde{\sigma})\right)\mathbb{I}_{\mathsf{z}}^q(\widetilde{\eta_1}) - \mathsf{k}_a\mathbb{I}_{\mathsf{z}}^q(\tilde{\sigma})\mathbb{I}_{\mathsf{z}}^q(\widetilde{\eta_2})$$
$$+ \mathsf{h}_W\left(\mathbb{I}(\widetilde{\varepsilon_2}) + \mathbb{I}(\widetilde{\varepsilon_3}) - \lambda\mathbb{I}_{\mathsf{z}}^q(\tilde{\sigma})\right)\mathsf{t}_\mathrm{w} - (L + \mathsf{k}_t)(\mathbb{I}(\widetilde{\varepsilon_2}) + \mathbb{I}(\widetilde{\varepsilon_3}))$$
$$- \lambda\mathbb{I}_{\mathsf{z}}^q(\tilde{\sigma})\left(2\mathsf{k}_t + \mathsf{k}_l - \mathsf{h}'_R\mathsf{t}_\mathrm{w}\right).$$

Case (ii):

$$TPU_{qROF}(\mathsf{l}) = \Delta'_1\mathsf{l} + \Delta'_2\frac{1}{\mathsf{l}} + \Delta'_3 \tag{6.25}$$

where

$$\Delta'_1 = -\mathsf{d}\left(\frac{\mathsf{h}}{2}\left(\frac{2}{\beta} - \frac{\mathsf{d}}{\beta^2} + \frac{\left(1 - \mathbb{I}(\widetilde{\varepsilon_2}) - \frac{\mathsf{d}}{\beta}\right)^2}{\mathsf{d}}\right) + \frac{\mathsf{h}\mathbb{I}(\widetilde{\varepsilon_3})}{2\mathsf{d}}\right.$$
$$+ \mathsf{h}_W\left(\mathbb{I}(\widetilde{\varepsilon_2}) + \mathbb{I}(\widetilde{\varepsilon_3}) - \lambda\mathbb{I}_{\mathsf{z}}^q(\tilde{\sigma})\right)\left[\frac{1}{\mathsf{d}} - \frac{1}{\beta} - \frac{\left(\mathbb{I}(\widetilde{\varepsilon_2}) + \mathbb{I}(\widetilde{\varepsilon_3}) - \lambda\mathbb{I}_{\mathsf{z}}^q(\tilde{\sigma})\right)}{\mathscr{W}}\right.$$
$$\left.\left.-\frac{\left(\mathbb{I}(\widetilde{\varepsilon_2}) - \mathbb{I}(\widetilde{\varepsilon_3}) + \lambda\mathbb{I}_{\mathsf{z}}^q(\tilde{\sigma})\right)}{2\mathsf{d}}\right] + \frac{\mathsf{h}_E\left(\lambda\mathbb{I}_{\mathsf{z}}^q(\tilde{\sigma})\right)^2}{2\mathsf{d}}\right);$$

$$\Delta'_2 = -\mathsf{d}[C + K];$$

$$\Delta_3' = \mathsf{d}[\mathsf{s}_1 + \mathsf{s}_2\lambda\mathbb{I}_{\mathsf{z}}^q(\tilde{\sigma}) - \mathsf{k} - \mathsf{i} - \mathsf{k}_r\left(1 - \mathbb{I}_{\mathsf{z}}^q(\tilde{\sigma})\right)\mathbb{I}_{\mathsf{z}}^q(\widetilde{\eta_1}) - \mathsf{k}_a\mathbb{I}_{\mathsf{z}}^q(\tilde{\sigma})\mathbb{I}_{\mathsf{z}}^q(\widetilde{\eta_2}) + \mathsf{h}_W\left(\mathbb{I}(\widetilde{\varepsilon_2}) + \mathbb{I}(\widetilde{\varepsilon_3}) - \lambda\mathbb{I}_{\mathsf{z}}^q(\tilde{\sigma})\right)\mathsf{t}_{\mathsf{w}} - (L + \mathsf{k}_t)(\mathbb{I}(\widetilde{\varepsilon_2}) + \mathbb{I}(\widetilde{\varepsilon_3})) - \mathsf{k}_e\lambda\mathbb{I}_{\mathsf{z}}^q(\tilde{\sigma}).$$

6.3.4 Vendor's Optimal Policy

The seller's objective is to identify the optimal order policy which increases the total profit per unit time. Hence, the concavity of the objective function is proved and the optimal order quantity (l^*) is attained by the following lemmas.

Lemma 6.1: The function of the total profit per unit time is strictly concave.

Proof: Differentiating Eq. (6.24) twice partially with respect to l can be obtained as

$$\frac{\partial TPU_{qROF}(\mathsf{l})}{\partial \mathsf{l}} = \Delta_1 - \Delta_2\frac{1}{\mathsf{l}^2}. \tag{6.26}$$

$$\frac{\partial^2 TPU_{qROF}(\mathsf{l})}{\partial \mathsf{l}^2} = \Delta_2\frac{1}{\mathsf{l}^3}. \tag{6.27}$$

Since, $\Delta_2 < 0$, then $\frac{\partial^2 TPU_{qROF}(\mathsf{l})}{\partial \mathsf{l}^2} < 0$ for every $\mathsf{l} > 0$.
Similarly, for case (ii)

$$\frac{\partial TPU_{qROF}(\mathsf{l})}{\partial \mathsf{l}} = \Delta_1' - \Delta_2'\frac{1}{\mathsf{l}^2}. \tag{6.28}$$

$$\frac{\partial^2 TPU_{qROF}(\mathsf{l})}{\partial \mathsf{l}^2} = \Delta_2'\frac{1}{\mathsf{l}^3}. \tag{6.29}$$

Since, $\Delta_2' < 0$, then $\frac{\partial^2 TPU_{qROF}(\mathsf{l})}{\partial \mathsf{l}^2} < 0$ for every $\mathsf{l} > 0$.

Thus, for case (i) and case (ii) the total profit per unit time is concave. Figure 6.3 shows the concavity of the total profit with respect to the order quantity. In this model, the concavity of the total profit function is proved by taking the order quantity level as input variables. It is clear that when the order quantity increases, the total profit also increases, and the total profit decreases once reached the optimal order level.

Lemma 6.2: The optimal order quantity which increases the expected total profit for a single unit time is

$$\text{For case(i) } \mathsf{l}^* = \sqrt{\frac{\Delta_2}{\Delta_1}};\ \text{case (ii) } \mathsf{l}^* = \sqrt{\frac{\Delta_2'}{\Delta_1'}}. \tag{6.30}$$

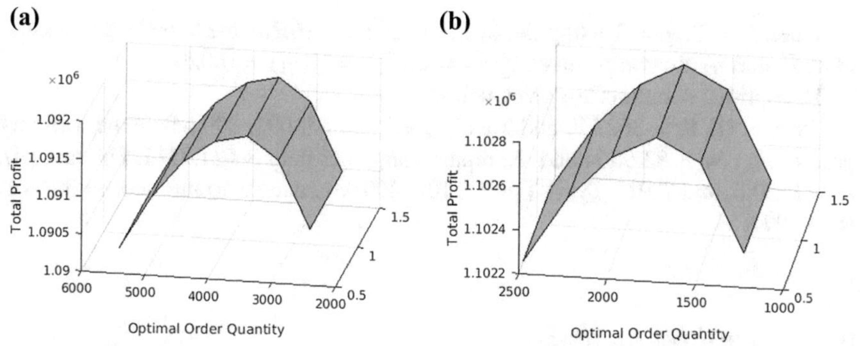

Fig. 6.3 Concavity of profit function

Proof: By equating the first-order derivative of the objective function to zero for both the cases, it is obtained as

$$\text{Case (i)}: \frac{\partial TPU_{qROF}(\mathsf{l})}{\partial \mathsf{l}} = \Delta_1 - \Delta_2 \frac{1}{\mathsf{l}^2} = 0 \text{ implies that } \mathsf{l}^* = \sqrt{\frac{\Delta_2}{\Delta_1}}.$$

$$\text{Case (ii)}: \frac{\partial TPU_{qROF}(\mathsf{l})}{\partial \mathsf{l}} = \Delta_1' - \Delta_2' \frac{1}{\mathsf{l}^2} = 0 \text{ implies that } \mathsf{l}^* = \sqrt{\frac{\Delta_2'}{\Delta_1'}}.$$

This completes the proof.

6.4 Numerical Computation

This part adopts the values of the parameter from [11] and [12].

$\mathsf{d} = 50,000$ (unit/year), $\beta = 1,75,200$ (unit/year), $\mathscr{W} = 50,000$ (unit/year), $\mathsf{s}_1 = \$50$ /unit, $\mathsf{s}_2 = \$30$/unit, $\mathsf{k} = \$25$ /unit, $\mathsf{h} = \$5$ /unit, $\mathsf{h}_R = \$6$ /unit, $\mathsf{h}_R' = \$4$ /unit, $\mathsf{h}_W = \$4$ /unit, $\mathsf{h}_E = \$1$ /unit, $\mathsf{i} = \$0.5$ /unit, $\lambda = 0.4$, $\mathsf{k}_a = \$500$ /unit, $\mathsf{k}_r = \$100$ /unit, $\mathsf{k}_l = \$5$ /unit, $\mathsf{k}_t = \$2$ /unit, $\mathsf{k}_e = \$30$ /unit $\mathscr{S} = \$100$ /unit, $\mathscr{A} = \$200$, $C = \$100$, $K = \$50$, $L = \$1$ /unit and $\mathsf{t_w} = 2/220$ year.

The qROF defective proportion ($\tilde{\sigma}$), the qROF proportion of misclassification errors ($\widetilde{\eta_1}$), and ($\widetilde{\eta_2}$) are given as

$$\tilde{\sigma} = \{(\mathsf{z}_1, 0.01, 0.015), (\mathsf{z}_2, 0.025, 0.02), (\mathsf{z}_3, 0.015, 0.02)\}$$

$$\widetilde{\eta_1} = \{(\mathsf{z}_1, 0.01, 0.015), (\mathsf{z}_2, 0.025, 0.02), (\mathsf{z}_3, 0.015, 0.02)\}$$

$$\widetilde{\eta_2} = \{(\mathsf{z}_1, 0.01, 0.015), (\mathsf{z}_2, 0.025, 0.02), (\mathsf{z}_3, 0.015, 0.02)\}$$

When $N = 3, q = 3$, using Eq. (6.21), (6.22), and (6.23), the knowledge measure of $\tilde{\sigma}$, $\widetilde{\eta_1}$ and $\widetilde{\eta_2}$ are computed as $\mathbb{I}_{\mathsf{z}}^{3}(\tilde{\sigma}) = \mathbb{I}_{\mathsf{z}}^{3}(\widetilde{\eta_2}) = \mathbb{I}_{\mathsf{z}}^{3}(\widetilde{\eta_1}) = 0.0203$.

The optimal solutions are given below

For case (i), $\mathsf{l}^* = 3627.7$, and $TPU_{qROF}(\mathsf{l}^*) = \$1,091,800$ where the warranty processing cost is \$2.9364 and the repair charge per item is \$21.2311. For case (ii), $\mathsf{l}^* = 1750.6$, and $TPU_{qROF}(\mathsf{l}^*) = \$1,102,800$ where the warranty processing cost is \$3.3974.

6.4.1 Sensitive Analysis

The sensitivity analysis explores the effect of the optimal solutions by changing some of the input variables. In this section, the impact of qROFVs $\tilde{\sigma}$, $\widetilde{\eta_1}$, and $\widetilde{\eta_2}$ on the optimal solutions is given. The effect of the variation of the proportion of the flawed products that are unable to claim warranty on the optimal solutions for both the cases and the variation of emergency purchase charge and its impact on the optimal solution of the second case is conferred.

Tables 6.2 and 6.3 show that, when qROFVs $\tilde{\sigma}$ and $\widetilde{\eta_1}$ raise, the optimal order quantity (l^*) increases for both cases whereas when $\widetilde{\eta_2}$ increases, the quantity l^* reduces for both the cases. When the parameters $\tilde{\sigma}$, $\widetilde{\eta_1}$, and $\widetilde{\eta_2}$ raise, the total profit reduces for both cases. Also, for case (i), when $\tilde{\sigma}$ and $\widetilde{\eta_2}$ raise, the warranty processing charge upticks slightly, but when $\widetilde{\eta_1}$ raises, the warranty processing charge reduces whereas when the parameters $\tilde{\sigma}$, $\widetilde{\eta_1}$, and $\widetilde{\eta_2}$ increase, the WPC for case (ii) decreases. The mending charge for case (i) decreases when $\tilde{\sigma}$ and $\widetilde{\eta_1}$ raise which upticks slightly when $\widetilde{\eta_2}$ increases. These analyses are clearly shown in Fig. 6.4 (Table 6.4).

It is observed that from the above table, for case (i): when the fraction of warranty unclaimed product increases, the WPC also increases and the quantity l^* upticks slightly but both the total profit and the mending charge per product reduce. For case (ii): when the fraction of warranty unclaimed product increases, both the WPC and the quantity l^* increase whereas the total profit remains unchanged. Also, it is noted that from Fig. 6.5, when the emergency purchase charge per product increases, the total profit reduces.

6.4.2 Comparison Study

The comparative study is given to show the better outcomes obtained among the two cases in the proposed model. Also, the comparison between the best option of the proposed model and the related studies is developed. In this study, when comparing both cases, along with the replacement warranty option the emergency purchase option is better than the repairing option. Since the emergency purchase option is a better one, the comparison study with the related works is performed by considering the second case (Table 6.5).

Table 6.2 Effect of qROF proportion of substandard products and misclassification errors on optimal solutions for case (i)

Parameter	qROFV	Knowledge measure	l*	TPU_{qROF}(l*)(USD)	$WPC/unit$(USD)	$RC/unit$(USD)
$\tilde{\sigma}$	$\{(z_1, 0.006, 0.011), (z_2, 0.021, 0.016), (z_3, 0.011, 0.016)\}$	0.0163	3625.6	1,094,100	2.9020	25.4290
	$\{(z_1, 0.008, 0.013), (z_2, 0.023, 0.018), (z_3, 0.013, 0.018)\}$	0.0183	3626.6	1,093,000	2.9197	23.1163
	$\{(z_1, 0.01, 0.015), (z_2, 0.025, 0.02), (z_3, 0.015, 0.02)\}$	0.0203	3627.7	1,091,800	2.9364	21.2311
	$\{(z_1, 0.012, 0.017), (z_2, 0.027, 0.022), (z_3, 0.017, 0.022)\}$	0.0223	3628.9	1,090,600	2.9521	19.6705
	$\{(z_1, 0.014, 0.019), (z_2, 0.029, 0.024), (z_3, 0.019, 0.024)\}$	0.0244	3630	1,089,400	2.9668	18.3608
$\tilde{\eta}_1$	$\{(z_1, 0.006, 0.011), (z_2, 0.021, 0.016), (z_3, 0.011, 0.016)\}$	0.0163	3625	1,111,800	3.0676	21.2426
	$\{(z_1, 0.008, 0.013), (z_2, 0.023, 0.018), (z_3, 0.013, 0.018)\}$	0.0183	3626.4	1,101,900	2.9982	21.2370
	$\{(z_1, 0.01, 0.015), (z_2, 0.025, 0.02), (z_3, 0.015, 0.02)\}$	0.0203	3627.7	1,091,800	2.9364	21.2311
	$\{(z_1, 0.012, 0.017), (z_2, 0.027, 0.022), (z_3, 0.017, 0.022)\}$	0.0223	3629.2	1,081,600	2.8811	21.2251
	$\{(z_1, 0.014, 0.019), (z_2, 0.029, 0.024), (z_3, 0.019, 0.024)\}$	0.0244	3630.6	1,071,300	2.8315	21.2189
$\tilde{\eta}_2$	$\{(z_1, 0.006, 0.011), (z_2, 0.021, 0.016), (z_3, 0.011, 0.016)\}$	0.0163	3628.1	1,093,900	2.9363	21.2294
	$\{(z_1, 0.008, 0.013), (z_2, 0.023, 0.018), (z_3, 0.013, 0.018)\}$	0.0183	3627.9	1,092,800	2.9364	21.2302
	$\{(z_1, 0.01, 0.015), (z_2, 0.025, 0.02), (z_3, 0.015, 0.02)\}$	0.0203	3627.7	1,091,800	2.9364	21.2311
	$\{(z_1, 0.012, 0.017), (z_2, 0.027, 0.022), (z_3, 0.017, 0.022)\}$	0.0223	3627.6	1,090,800	2.9364	21.2320
	$\{(z_1, 0.014, 0.019), (z_2, 0.029, 0.024), (z_3, 0.019, 0.024)\}$	0.0244	3627.4	1,089,700	2.9364	21.2328

Table 6.3 Effect of qROF proportion of substandard products and misclassification errors on optimal solutions for case (ii)

Parameter	qROFV	Knowledge measure of qROFV	l*	TPU_{qROF}(l*)(USD)	$WPC/unit$(USD)
$\tilde{\sigma}$	$\{(z_1, 0.006, 0.011), (z_2, 0.021, 0.016), (z_3, 0.011, 0.016)\}$	0.0163	1748	1,104,800	3.4002
	$\{(z_1, 0.008, 0.013), (z_2, 0.023, 0.018), (z_3, 0.013, 0.018)\}$	0.0183	1749.2	1,103,800	3.3988
	$\{(z_1, 0.01, 0.015), (z_2, 0.025, 0.02), (z_3, 0.015, 0.02)\}$	0.0203	1750.6	1,102,800	3.3974
	$\{(z_1, 0.012, 0.017), (z_2, 0.027, 0.022), (z_3, 0.017, 0.022)\}$	0.0223	1751.9	1,101,800	3.3962
	$\{(z_1, 0.014, 0.019), (z_2, 0.029, 0.024), (z_3, 0.019, 0.024)\}$	0.0244	1753.2	1,100,700	3.3950
$\widetilde{\eta_1}$	$\{(z_1, 0.006, 0.011), (z_2, 0.021, 0.016), (z_3, 0.011, 0.016)\}$	0.0163	1749.3	1,122,800	3.5935
	$\{(z_1, 0.008, 0.013), (z_2, 0.023, 0.018), (z_3, 0.013, 0.018)\}$	0.0183	1749.9	1,112,900	3.4899
	$\{(z_1, 0.01, 0.015), (z_2, 0.025, 0.02), (z_3, 0.015, 0.02)\}$	0.0203	1750.6	1,102,800	3.3974
	$\{(z_1, 0.012, 0.017), (z_2, 0.027, 0.022), (z_3, 0.017, 0.022)\}$	0.0223	1751.2	1,092,600	3.3148
	$\{(z_1, 0.014, 0.019), (z_2, 0.029, 0.024), (z_3, 0.019, 0.024)\}$	0.0244	1751.9	1,082,200	3.2408
$\widetilde{\eta_2}$	$\{(z_1, 0.006, 0.011), (z_2, 0.021, 0.016), (z_3, 0.011, 0.016)\}$	0.0163	1750.7	1,104,800	3.3974
	$\{(z_1, 0.008, 0.013), (z_2, 0.023, 0.018), (z_3, 0.013, 0.018)\}$	0.0183	1750.6	1,103,800	3.3974
	$\{(z_1, 0.01, 0.015), (z_2, 0.025, 0.02), (z_3, 0.015, 0.02)\}$	0.0203	1750.6	1,102,800	3.3974
	$\{(z_1, 0.012, 0.017), (z_2, 0.027, 0.022), (z_3, 0.017, 0.022)\}$	0.0223	1750.5	1,101,700	3.3975
	$\{(z_1, 0.014, 0.019), (z_2, 0.029, 0.024), (z_3, 0.019, 0.024)\}$	0.0244	1750.4	1,100,700	3.3975

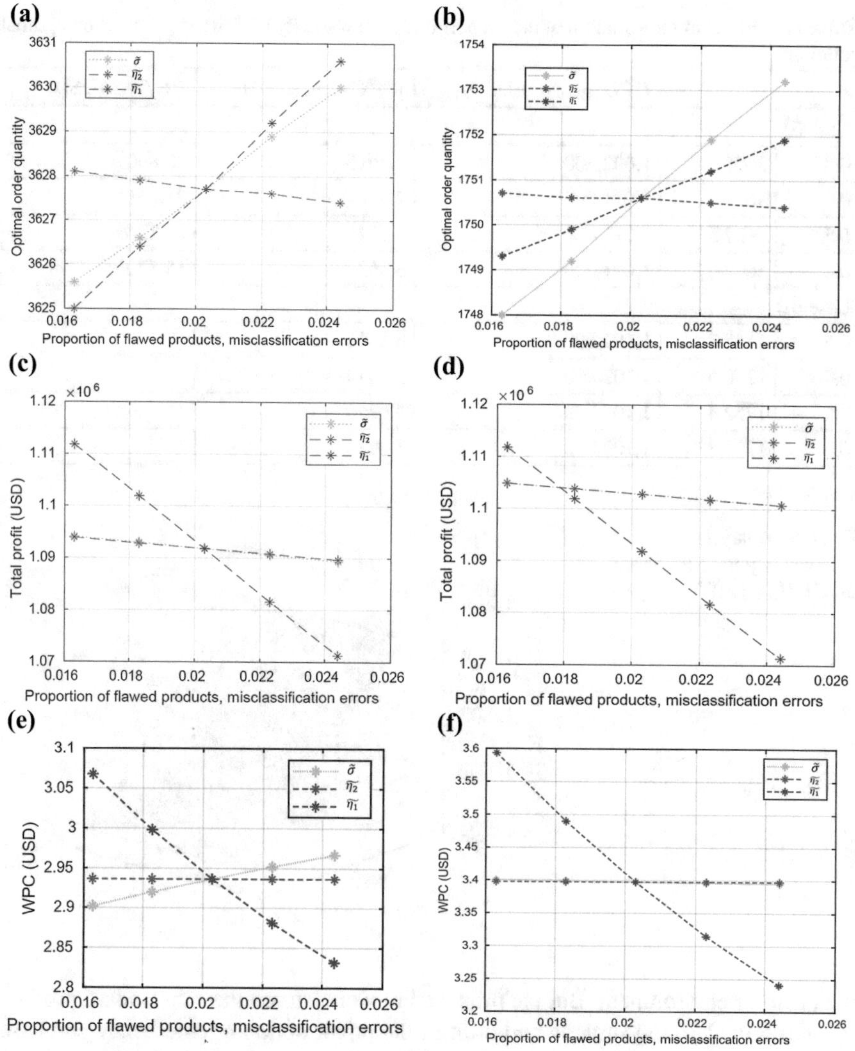

Fig. 6.4 Analysis of qROFVs $\tilde{\sigma}$, $\widetilde{\eta_1}$ and $\widetilde{\eta_2}$ on $TPU_{qROF}(l^*)$

Thus, by considering the proposed model without misclassification errors, that is by equating $\boldsymbol{\eta_1}$ as well as $\boldsymbol{\eta_2}$ to zero, the total profit in the proposed study is better than the total profit of studies Salameh and Jaber [33], Papachristos and Konstantaras [22], Jaber et al. [11], and Rajeswari and Sugapriya [26]. The studies Salameh and Jaber [33] and Papachristos and Konstantaras [22] developed the models by selling the substandard product at a lower price and the work Jaber et al. [11] assumes to replace the substandard products by repair and emergency purchase option and also Rajeswari and Sugapriya [26] restored the substandard products by repair option

Table 6.4 Effect of the variation of the proportion of the warranty unclaimed products on optimal solutions

λ	l*	TPU_{qROF}(l*)(USD)	$WPC/unit$(USD)	$RC/unit$(USD)
Case (i)				
0.3	3628	1,092,200	2.7615	26.8907
0.4	3627.7	1,091,800	2. 9364	21.2311
0.5	3627.5	1,091,400	3.1749	17.8349
0.6	3627.2	1,091,000	3.3621	15.5703
Case (ii)				
0.3	1748.7	1,102,800	3.1961	
0.4	1750.6	1,102,800	3.3974	
0.5	1752.4	1,102,800	3.6260	
0.6	1754.3	1,102,800	3.8878	

Fig. 6.5 Analysis emergency purchase charge on TPU_{qROF}(l*)

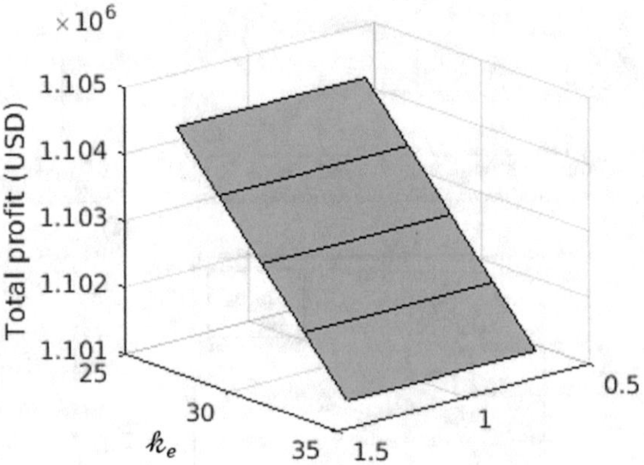

under fuzzy environment. But the proposed model restored the substandard product by warranty claiming strategy and utilized the repair option or emergency purchase option for warranty unclaimed products. Since the proportion of faulty items are replaced by the supplier using the warranty option, the least amount is only spent, and hence the current model is better when compared to these studies. And so, by presuming a model with misclassification errors, the total profit in this study is better than the total profit of studies Khan et al. [12] and Pimsap and Srisodaphol [25].

Table 6.5 The comparison study of the proposed model with some related works

References	Total profit (USD)	Total profit of the proposed model (USD)
Model without screening errors		
Salameh and Jaber [33]	1,212,235	1,214,400
Papachristos and Konstantaras [22]	1,121,235.036	1,214,400
Jaber et al. [11]	1,198,026	1,214,400
Rajeswari and Sugapriya [26]	1,197,300	1,214,400
Model with screening errors		
Khan et al. [12]	1,095,090	1,102,800
Pimsap and Srisodaphol [25]	1,038,250.83	1,102,800

6.5 Conclusion

The proposed study analyzed an EOQ model with substandard items and misclassification errors under two cases such as replacement warranty claiming strategy with mending option and replacement warranty claiming strategy with emergency purchase option. The strategy of restoring the faulty products in an EOQ model helps the vendor to increase the profit despite the sale of the flawed products at a lower cost and this option prevents the vendor from scarcity of products. The unpredictability situation leads the model to the qROF environment by presuming the uncertain parameters say, the substandard proportion and the fraction of misclassification errors as qROFVs. The KM-qROF inventory model is formulated by computing the knowledge measure of qROFVs $\tilde{\sigma}$, $\widetilde{\eta_1}$, and $\widetilde{\eta_2}$. Case (ii) that is, the warranty claiming and the emergency purchase option resulting in the better profit when compared to the first case. Practically, this model is applicable for spare parts of electronic products. The replacement warranty option proposed in this study is more pertinent for these products. The numerical analysis is provided and both cases of the proposed model show that the total profit is concave with respect to the order quantity. The sensitivity analysis of various parameters on the optimal solution is demonstrated and the comparison study shows the present study is a better one when compared to some relative research works Jaber et al. [11], Khan et al. [12], Papachristos and Konstantaras [22], Pimsap and Srisodaphol [25], Rajeswari and Sugapriya [26], and Salameh and Jaber [33]. In the future, the notion of this study can be enlarged to a two-warehouse EOQ model and it can also extend with the advance payment strategy, price-sensitive demand. The proposed inventory model can extend by considering the ordered product as deteriorating items, non-instantaneous deteriorating items also can extend by utilizing the trade credit option.

Appendix

Computation of storage charge.

The holding charge is obtained by adding the five parts (a), (b), (c), (d), and (e) in Fig. 6.6.

Using the substitutions, $\mathsf{I}_1 = \mathsf{I} - d\mathsf{t}_1$, $\mathsf{I}_2 = \mathsf{I}_1 - \mathscr{Q}_1$, $\mathsf{I}_3 = \mathsf{I} - d(\mathsf{t}_1 + \mathsf{t}_2) - \mathscr{Q}_1$, $\mathscr{Q}' = \mathsf{I}_3 + (\mathscr{Q}_1 + \mathscr{Q}_2 - \lambda\sigma\mathsf{I})$; where $\mathsf{t}_1 = \frac{\mathsf{I}}{\beta}$, $\mathsf{t}_2 = \frac{\mathscr{Q}_1+\mathscr{Q}_2}{\mathscr{W}} + \mathsf{t}_\mathrm{w}$, $\mathsf{t}_3 = \frac{\mathscr{Q}'}{d}$, $\mathsf{t}_4 = \frac{\lambda\sigma\mathsf{I}}{d}$ $\mathscr{Q}_1 = \varepsilon_2\mathsf{I}$ and $\mathscr{Q}_2 = \varepsilon_3\mathsf{I}$

$$(a) = \frac{\mathsf{h}}{2}\mathsf{I}^2\left(\frac{2}{\beta} - \frac{\mathsf{d}}{\beta^2}\right)$$

$$(b) = \frac{\mathsf{h}}{2}\mathsf{I}^2\left(\frac{\left(1-\varepsilon_2-\frac{\mathsf{d}}{\beta}\right)^2}{\mathsf{d}}\right)$$

$$(c) = \frac{\mathsf{h}_W}{2}\left(\frac{\mathscr{Q}'^{\,2} - \mathsf{I}_3^2}{\mathsf{d}}\right)$$

$$= \mathsf{h}_W\mathsf{I}^2(\varepsilon_2 + \varepsilon_3 - \lambda\sigma)\left[\frac{1}{\mathsf{d}} - \frac{1}{\beta} - \frac{(\varepsilon_2 + \varepsilon_3 - \lambda\sigma)}{\mathscr{W}} - \frac{(\varepsilon_2 - \varepsilon_3 + \lambda\sigma)}{2\mathsf{d}}\right]$$

$$- \mathsf{h}_W\mathsf{I}(\varepsilon_2 + \varepsilon_3 - \lambda\sigma)\mathsf{t}_\mathrm{w}$$

$$(d) = \frac{\mathsf{h}\mathscr{Q}_2\mathscr{T}'}{2} = \frac{\mathsf{h}\mathsf{I}^2\varepsilon_3}{2\mathsf{d}}$$

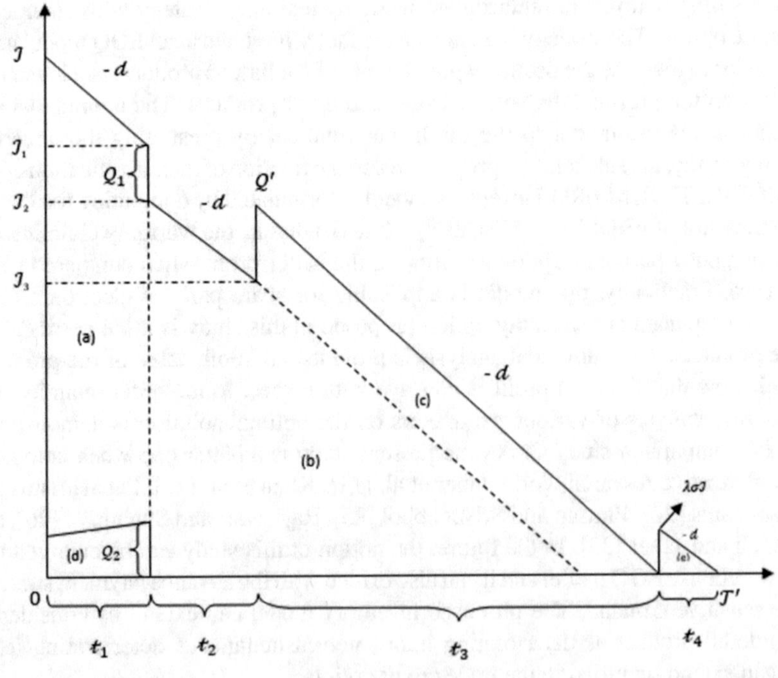

Fig. 6.6 Holding Charge

(e) $= \frac{h_R \lambda \sigma l t_4}{2} = \frac{h_R(\lambda\sigma)^2 l^2}{2d}$ [For case(i)]
(e) $= \frac{h_E \lambda \sigma l t_4}{2} = \frac{h_E(\lambda\sigma)^2 l^2}{2d}$ [For case (ii)]

On adding (a), (b), (c), (d), and (e), the holding charge for case (i) and case (ii) are given as follows:

$$HC = \frac{h}{2} l^2 \left(\frac{2}{\beta} - \frac{d}{\beta^2} + \frac{\left(1 - \varepsilon_2 - \frac{d}{\beta}\right)^2}{d} \right) + \frac{h l^2 \varepsilon_3}{2d}$$
$$+ h_W l^2 (\varepsilon_2 + \varepsilon_3 - \lambda\sigma) \left[\frac{1}{d} - \frac{1}{\beta} - \frac{(\varepsilon_2 + \varepsilon_3 - \lambda\sigma)}{\mathscr{W}} - \frac{(\varepsilon_2 - \varepsilon_3 + \lambda\sigma)}{2d} \right]$$
$$- h_W l (\varepsilon_2 + \varepsilon_3 - \lambda\sigma) t_w + \frac{h_R(\lambda\sigma)^2 l^2}{2d}.$$

and

$$HC = \frac{h}{2} l^2 \left(\frac{2}{\beta} - \frac{d}{\beta^2} + \frac{\left(1 - \varepsilon_2 - \frac{d}{\beta}\right)^2}{d} \right) + \frac{h l^2 \varepsilon_3}{2d}$$
$$+ h_W l^2 (\varepsilon_2 + \varepsilon_3 - \lambda\sigma) \left[\frac{1}{d} - \frac{1}{\beta} - \frac{(\varepsilon_2 + \varepsilon_3 - \lambda\sigma)}{\mathscr{W}} - \frac{(\varepsilon_2 - \varepsilon_3 + \lambda\sigma)}{2d} \right]$$
$$- h_W l (\varepsilon_2 + \varepsilon_3 - \lambda\sigma) t_w + \frac{h_E(\lambda\sigma)^2 l^2}{2d}.$$

References

1. M.I. Ali, Another view on q-rung orthopair fuzzy sets. Int. J. Intell. Syst. **33**, 2139–2153 (2018). https://doi.org/10. 1002/int.22007
2. K.T. Atanassov, Intuitionistic fuzzy sets. Fuzzy Sets Syst. **20**, 87–96 (1986)
3. B. Farhadinia, H. Liao, Score-based multiple criteria decision making process by using p-rung orthopair fuzzy sets. INFORMATICA (2020). doi:https://doi.org/10.15388/20-INFOR412
4. H. Garg, A novel trigonometric operation-based *q*-rung orthopair fuzzy aggregation operator and its fundamental properties. Neural Comput. Appl. **32**, 15077–15099 (2020). https://doi.org/10.1007/s00521-020-04859-x
5. H. Garg, S.M. Chen, Multiattribute group decision making based on neutrality aggregation operators of q- rung orthopair fuzzy sets. Inf Sci. **517**, 427–447 (2020)
6. H. Garg, J. Gwak, T. Mahmood, Z. Ali, Power aggregation operators and vikor methods for complex q-rung orthopair fuzzy sets and their applications. Mathematics **8**, 538 (2020)
7. H. Garg, A new possibility degree measure for interval-valued q-rung orthopair fuzzy sets in decision-making. Int J Intell Syst. **36**, 526–557 (2021). https://doi.org/10.1002/int.22308
8. H. Garg, CN- *q*q-ROFS: Connection number-based *q*-rung orthopair fuzzy set and their application to decision-making process. Int. J. Intell. Syst. **36**(7), 3106–3143 (2021). doi:https://doi.org/10.1002/int.22406

9. H. Garg, New exponential operation laws and operators for interval-valued *q*-rung orthopair fuzzy sets in group decision making process. Neural Comput. Appl. **33**(20), 13937–13963 (2021). https://doi.org/10.1007/s00521-021-06036-0
10. J.T. Hsu, L.F. Hsu, An EOQ model with imperfect quality items, inspection errors, shortage backordering, and sales returns. Int. J. Prod. Econ. **143**, 162–170 (2013)
11. M.Y. Jaber, S. Zanoni, L.E. Zanavella, Economic order quantity models for imperfect items with buy and repair options. Int. J. Prod. Econ. **155**, 126–131 (2014)
12. M. Khan, M.Y. Jaber, M. Bonney, An economic order quantity (EOQ) for items with imperfect quality and inspection errors. Int. J. Prod. Econ. **133**, 113–118 (2011)
13. M.J. Khan, M.I. Ali, M. Shutaywi, A new ranking technique for q-rung orthopair fuzzy values. Int. J. Intell. Syst. **36**, 558–592 (2020). https://doi.org/10.1002/int.22311
14. M.J. Khan, P. Kumam, M. Shutaywi, Knowledge measure for the q-rung orthopair fuzzy sets. Int. J. Intell. Syst. **36**(2), 628–655 (2020). https://doi.org/10.1002/int.22313
15. T.Y. Lin, M.T. Chen, An economic order quantity model with screening errors, returned cost, and shortages under quantity discounts. Afr. J. Bus. Manage. **5**(4), 1129–1135 (2011)
16. P. Liu, S. Chen, P. Wang, Multiple-attribute group decision-making based on q-rung orthopair fuzzy power maclaurin symmetric mean operators. IEEE Trans. Syst. Man Cybern Syst. **50**, 3741–3756 (2018). https://doi.org/10.1109/TSMC.2018.2852948
17. P. Liu, W. Liu, Multiple-attribute group decision-making method of linguistic q-rung orthopair fuzzy power Muirhead mean operators based on entropy weight. Int. J. Intell. Syst. **34**, 1755–1794 (2019). https://doi.org/10.1002/int.22114
18. P. Liu, P. Wang, Some q-rung orthopair fuzzy aggregation operators and their applications to multiple- attribute decision making. Int. J. Intell. Syst. **33**(2), 259–280 (2018)
19. P. Liu, Y. Wang, Multiple attribute decision making based on q-Rung Orthopair Fuzzy generalized Maclaurin symmetric mean operators. Inform. Sci. **518**, 181–210 (2020)
20. J. Liu, H. Zheng, Fuzzy economic order quantity model with imperfect items, shortage and inspection errors. Syst. Eng. Proc. **4**, 282–289 (2012)
21. S. Maity, S.K. De, S.P. Mondal, A study of a backorder EOQ model for cloud-type intuitionistic dense fuzzy demand rate. Int. J. Fuzzy Syst. **22**, 201–211 (2019). https://doi.org/10.1007/s40815-019-00756-1
22. S. Papachristos, I. Konstantaras, Economic ordering quantity models for items with imperfect quality. Int. J. Prod. Econ. **100**(1), 148–154 (2006)
23. R. Patro, M.M. Nayak, M. Acharya, An EOQ model for fuzzy defective rate with allowable proportionate discount. Opsearch **56**, 191–215 (2019)
24. X. Peng, J. Dai, H. Garg, Exponential operation and aggregation operator for q-rung orthopair fuzzy set and their decision-making method with a new score function. Int. J. Intell. Syst. **33**(11), 2255–2282 (2018). https://doi.org/10.1002/int.22028
25. P. Pimsap, W. Srisodaphol, Economic order quantity model for imperfect items under repair option and inspection errors. Brupa J Sci. **23**(1), 92–104 (2018)
26. S. Rajeswari, C. Sugapriya, Fuzzy economic order quantity model with imperfect quality items under repair option. J. Res. Lepidoptera **51**(1), 627–643 (2020)
27. S. Rajeswari, C. Sugapriya, D. Nagarajan, Fuzzy inventory model for NVOCC's returnable containers under empty container repositioning with leasing option. Compl. Intell. Syst. **7**, 753–764 (2021). https://doi.org/10.1007/s40747-020-00229-1
28. S. Rajeswari, C. Sugapriya, D. Nagarajan, An analysis of uncertain situation and advance payment system on a double-storage fuzzy inventory model. Opsearch (2021). https://doi.org/10.1007/s12597-021-00530-8
29. S. Rajeswari, C. Sugapriya, D. Nagarajan, J. Kavikumar, Optimization in fuzzy economic order quantity model involving pentagonal fuzzy parameter. Int. J. Fuzzy Syst. (2021). https://doi.org/10.1007/s40815-021-01111-z
30. M. Riaz, D. Pamucar, H.M.A. Farid, M.R. Hashmi, q-rung orthopair fuzzy prioritized aggregation operators and their application towards green supplier chain management. Symmetry **12**, 976 (2020)

31. M. Riaz, H.M.A. Farid, H. Kalsoom, D. Pamucar, Y.-M. Chu, A robust q-rung orthopair fuzzy einstein prioritized aggregation operators with application towards MCGDM. Symmetry **12**, 1058 (2020)
32. M. Riaz, H. Garg, H.M.A. Farid, M. Aslam, Novel Q-rung Orthopair fuzzy interaction aggregation operators and their application to low-carbon green supply Chain management. J. Intell. Fuzzy Syst. **41**(2), 4109–4126 (2021). https://doi.org/10.3233/JIFS-210506
33. M.K. Salameh, M.Y. Jaber, Economic production quantity model for items with imperfect quality. Int. J. Prod. Econ. **64**(1), 59–64 (2000)
34. B. Sarkar, S. Saren, Product inspection policy for an imperfect production system with inspection errors and warranty cost. European J. Operat. Res. **248**, 263–271 (2016)
35. G. Wei, H. Gao, Y. Wei, Some q-rung orthopair fuzzy Heronian mean operators in multiple attribute decision making. Int. J. Intell. Syst. **33**(7), 1426–1458 (2018). https://doi.org/10.1002/int.21985
36. Y. Xing, R. Zhang, J. Wang, K. Bai, J. Xue, A new multi-criteria group decision-making approach based on q- rung orthopair fuzzy interaction Hamy mean operators. Neural Comput. Appl. **32**, 7465–7488 (2019). https://doi.org/10.1007/s00521-019-04269-8
37. R.R. Yager, Pythagorean fuzzy subsets. In Proceedings of Joint IFSA World Congress and NAFIPS, Annual Meeting, Edmonton, Canada, June 24–28, pp. 57–61 (2013). doi:https://doi.org/10.1109/IFSA-NAFIPS.2013.6608375
38. R.R. Yager, A.M. Abbasov, Pythagorean membership grades, complex numbers, and decision making: pythagorean membership grades and fuzzy subsets. Int. J. Intell. Syst. **28**, 436–452 (2013)
39. R.R. Yager, Generalized orthopair fuzzy sets. IEEE Trans. Fuzzy Syst. **25**(5), 1222–1230 (2017)
40. R.R. Yager, N. Alajlan, Approximate reasoning with generalized orthopair fuzzy sets. Inf. Fusion. **38**, 65–73 (2017). https://doi.org/10.1016/j.inffus.2017.02.005
41. Z. Yang, H. Garg, Interaction power partitioned Maclaurin symmetric mean operators under q-Rung orthopair uncertain linguistic information. Int. J. Fuzzy Syst. (2021). https://doi.org/10.1007/s40815-021-01062-5
42. Z. Yang, H. Garg, J. Li, Investigation of multiple heterogeneous relationships using a q-rung orthopair fuzzy multi-criteria decision algorithm. Neural Comput. Appl. **33**(17), 10771–10786 (2021). https://doi.org/10.1007/s00521-020-05003-5
43. Z. Yang, X. Li, Z. Cao, J. Li, Q-rung orthopair normal fuzzy aggregation operators and their application in multi-attribute decision-making. Mathematics **7**(12), 1142 (2019). https://doi.org/10.3390/math7121142
44. W.M. Yeo, X.M. Yuan, Optimal warranty policies for systems with imperfect repair. European J. Operat. Res. **199**, 187–197 (2009)
45. L. Zadeh, Fuzzy sets. Inform. Control **8**, 338–353 (1965)
46. Y. Zhongs, H. Gao, X. Guo, Y. Qin, M. Huang, X. Luo, Dombi power partitioned Heronian mean operators of q- rung orthopair fuzzy numbers for multiple attribute group decision making. PLoS ONE **14**(10), e0222007 (2019). https://doi.org/10.1371/journal.pone.0222007
47. Y. Zhou, C. Chen, C. Li, Y. Zhong, A synergic economic order quantity model with trade credit, shortages, imperfect quality and inspection errors. Appl. Math. Model. **40**, 1012–1028 (2016)

Dr. C. Sugapriya is working as an Assistant Professor in the Department of Mathematics, Queen Mary's College, Chennai, Tamilnadu, India. Her academic experience includes teaching and research for more than 16 years of service. Her research interest includes image processing, fuzzy logic, optimization techniques, and inventory models. She has published her research articles more than 20 for various international journals, national journals indexed by SCI, SCOPUS, UGC, and google scholar.

S. Rajeswari is a Ph. D Research Scholar [Full-time] under the guidance of Dr. C Sugapriya at Queen Mary's college, Chennai, Tamilnadu, India. She has done her M.Sc. and M.Phil Degree from D. G. Vaishnav College, Chennai. She has worked as an assistant professor for five years. Her research interest includes fuzzy logic, optimization techniques, and inventory models. She has published her research articles in reputed journals indexed by SCI, SCOPUS, UGC, and google scholar.etc.

Dr. D. Nagarajan is currently working as a professor in Department of Mathematics, Rajalakshmi Institute of Technology, Chennai, India. He received his Ph.D. in Mathematics and Statistics at Manonmaniam Sundaranar univeristy,Tirunelveli, India. His research interests include stochastic modeling, Neutrosophic sets and system, Inventory systems. He had a granted a patent on Oxygen injection IoT device and method. He has published more than 100 research papers in reputed high impact journals and serving as an editorial member of some international journals and also Guest editor of two international journals.

K. Jeganathan is currently working as an Assistant Professor of Ramanujan Institute for Advanced Study in Mathematics at the University of Madras, Chennai, India. He received his Ph.D. in Mathematics for Stochastic Inventory Modeling at Alagappa University, Karaikudi, India. He holds his M.Phil in Mathematics from the Madurai Kamaraj University, Madurai, India and M.Sc. in Mathematics from the Cardamom Planters Association College, Bodinayakanur, India. His research interests include stochastic modeling, inventory and queuing systems. He has published many papers in reputed journals like Mathematics and Computers in Simulation, Electronics, AIMS Mathematics, OPSEARCH, etc

Chapter 7
Higher Type q-Rung Orthopair Fuzzy Sets: Interval Analysis

Yu-Dou Yang and Xue-Feng Ding

Abstract As a crucial extension of q-rung orthopair fuzzy sets, in recent years, the exploration of q-rung interval-valued orthopair fuzzy set (q-RIVOFS) is necessary and meaningful for portraying complex decision-making systems. The objective of this chapter is to review several basic concepts of q-RIVOFS, propose some novel measures for q-RIVOFS and present its linkages with the applied areas by establishing a novel q-RIVOFS-TODIM decision-making model. In this chapter, at the beginning, several theoretical knowledge of q-RIVOFSs including basic definition, score and accuracy function, comparison laws, and fundamental operation rules are reviewed. Subsequently, some q-RIVOF cross-entropy and Hausdorff distance are formulated to measure specific divergence degree between two q-RIVOFSs. In addition, an integrated q-RIVOFS-TODIM approach is constructed to display the valid applications of q-RIVOFSs in MADM areas. After that, a medical waste disposal method selection illustrative example is conducted to demonstrate the advantages of q-RIVOFSs and the practicability of the proposed q-RIVOFS-TODIM approach. Finally, sensitivity analysis and comparative analysis are further performed to show the superiority and robustness of the proposed approach. The experimental results reveal that q-RIVOFSs as a fuzzy language possess far-ranging application space; the proposed q-RIVOFS-TODIM approach has an ability to reflect on the risk aversion psychological behaviors of decision-makers; and compared to methods based on the closeness of each alternative to ideal solutions, the proposed approach is able to circumvent the occurrence of inverse sequence.

Keywords q-rung interval-valued orthopair fuzzy sets · Cross-entropy · Distance measure · TODIM · Multi-attribute decision-making

Y.-D. Yang · X.-F. Ding (✉)
Shanghai University, 99 Shangda Road, Baoshan District, Shanghai 201900, China
e-mail: athena_tju@sina.com

Y.-D. Yang
e-mail: yyd1365770727@126.com

H. Garg (ed.), *q-Rung Orthopair Fuzzy Sets*,
https://doi.org/10.1007/978-981-19-1449-2_7

7.1 Introduction

q-rung orthopair fuzzy sets (q-ROFSs), developed by Yager [1], efficaciously remedied the non-negligible limitations of some classical sets such as fuzzy sets (FSs) [2], intuitionistic fuzzy sets (IFSs) [3, 4], interval IFSs [5, 6], Pythagorean fuzzy sets (PFSs) [7] and interval PFSs [8, 9]. Thereafter, as a valid mathematical tool, q-ROFSs have been extensively applied to many research fields. For example, Yang et al. [10] proposed q-rung orthopair fuzzy (q-ROF) interaction weighted Heronian mean operators for aggregating cross-decision information. Tang et al. [11] primitively extended q-ROF numbers to decision-theoretic rough sets and established a three-way decision-making model. Garg [12] introduced a novel concept, connection number-based q-ROFS, to devise a group decision-making method. Garg and Chen [13] innovatively constructed some weighted averaging neutral aggregation operators to fuse q-ROF information. Riaz et al. [14] developed a sequence of q-ROF interaction operators. Peng et al. [15] coined a new exponential operational law about q-ROF numbers. Garg [16] presented some sine trigonometric operations laws for q-ROFSs and defined some sine trigonometry weighted averaging and geometric operators.

However, as the application progress of q-ROFSs has been developing, q-ROFSs have shown some limitations gradually. While handling with much real decision-making practice, it is hard and not appropriate for decision-makers to precisely quantify their judgments with an exact value due to the complexity and scarcity of fuzzy provided information. In such cases, it is more suitable and expedient for decision-makers to express their evaluations via the subset of the closed interval [0, 1]. Hence, Joshi, Singh, Bhatt, Vaisla [17] and Ju et al. [18] defined interval-valued membership and non-membership functions to develop q-rung interval-valued orthopair fuzzy sets (q-RIVOFSs). Therewith, q-RIVOFSs, generalizations of FSs, IFSs, PFSs, interval-valued IFSs, interval-valued PFSs and q-ROFSs, have gained intensive attention from academicians and practitioners, and there have been many studies focusing on q-RIVOFSs. These studies can be categorized into the following three directions:

(a) Basic properties research: Joshi, Singh, Bhatt, Vaisla [17] developed some important operations of q-RIVOFSs, such as negation, union, and intersection operational laws. Gao and Xu [19] introduced the elementary arithmetic of q-rung interval-valued orthopair fuzzy number (q-RIVOFN), including addition, multiplication, and their inverse. In addition, they also discussed and proved the operation properties of q-RIVOFN. Gao et al. [20] proposed some q-RIVOF operational rules on the basis of Archimedean t-conorm and t-norm. Garg [21] introduced a novel possibility degree measure for q-RIVOFSs and applied it to tackle multi-attribute decision-making (MADM) problems. Garg [22] firstly defined an original concept of q-connection number for q-RIVOFSs and established some q-exponential operation laws over q-connection number in order to solve fuzzy MADM problems.

(b) Fusion method research: Yang and Chang [23] integrated power average and Muirhead mean operators to introduce a new power Muirhead mean operator in interval q-RIVOF environments. Li et al. [24] characterized and investigated the q-RIVOF weighted geometric operator to fuse complex preference information. Wang et al. [25] presented some extended aggregation operators such as the q-RIVOF Hamy mean operator, q-RIVOF weighted Hamy mean operator, q-RIVOF dual Hamy mean operator and q-RIVOF weighted dual Hamy mean operator. Gao et al. [20] defined the q-RIVOF Archimedean Muirhead mean operator as well as weighted Archimedean Muirhead mean operator. They synchronously discussed some crucial properties and special cases of the two aforementioned operators. Ju et al. [18] gave some q-RIVOF weighted averaging operators. Wang et al. [26] developed some new q-RIVOF Maclaurin symmetric mean operators which take into account the interrelationship of arguments.

(c) Application research: Gao et al. [27] extended the classical VIKOR to q-RIVOF circumstances for dealing with supplier selection of medical consumer products problems. Jin et al. [28] formulated a scientific risk evaluation approach combining q-RIVOF-deviation maximization method with q-RIVOF-additive ratio assessment method. Liu et al. [29] developed a new green supplier selection method in a q-RIVOF environment to investigate the green supplier selection issue with a large-scale group of decision-makers.

Based on the foregoing literature analyses, evidently, most of the existing studies regarding q-RIVOFSs have mainly concentrated on fusion methods, and relatively ignored the research direction of combining q-RIVOFSs with the classical traditional decision models such as the TODIM model [30], the ELECTRE III model [31], the TOPSIS model [32]. Among these prominent decision models, in light of the merit of the TODIM method in portraying the psychological behaviors of DMs under risk ground on prospect theory [30], it has been viewed as an efficient and vital tool that can be applied into many decision domains. Lin et al. [33] established a risk evaluation model of excavation system in view of an extended TODIM method. To tackle green supplier selection problems, Celik et al. [34] merged best worst method and TODIM method under an improved interval type-2 fuzzy contexts. Arya and Kumar [35] proposed a novel picture fuzzy multi-criteria decision-making method by integrating VIKOR method into TODIM method. Liao et al. [36] introduced an extended TODIM method on the strength of cumulative prospect theory in probabilistic hesitant fuzzy environments. Liu et al. [37] developed an occupational health and safety risk assessment framework based on the TODIM and PROMETHEE methods. Su et al. [38] improved the classical TODIM method and reconstructed it by incorporating the prospect theory. Hong et al. [39] came up with an extension of TODIM method based on fuzzy evaluation and Shapley index for decision-makers to imitate the psychological behavior characteristics of failure mode and effect analysis team members. Liu et al. [40] designed a novel multi-attribute group decision-making approach based on Dempster-Shafer evidence theory and TODIM under the double hierarchy hesitant fuzzy linguistic term sets.

Nevertheless, up to now, there is no literature on the TODIM model with q-RIVOF information. Thus, it is blank but essential to extend the traditional TODIM into q-RIVOF circumstances. This is also a golden opportunity to manifest the applicability and validity of q-RIVOFSs. Therefore, this chapter aims to propose an integrated q-RIVOFS-TODIM decision-making model for validating the availability and accuracy of q-RIVOFSs and expanding the application range of q-RIVOFSs.

The contributions and novelty of this chapter mainly include:

(i) Some novel cross-entropy and Hausdorff distance measures for q-RIVOFSs are firstly formulated to scale the divergence and similarity degrees between any two q-RIVOFSs.
(ii) A q-RIVOF linear programing model is established to determine the weights of attributes, which can consider not only attributes' contribution to decision-making process but also the credibility of attribute evaluation information.
(iii) An integrated q-RIVOF-TODIM MADM method is proposed for highlighting the superiorities and key application directions of q-RIVOFSs, which is able to fill the current research gaps in relevant fields.
(iv) Additionally, an emergency response scheme selection problem is exemplified to reveal the efficacies of q-RIVOFSs in expressing complex information and verify the potencies of the proposed q-RIVOFS-TODIM approach.

This chapter is organized as follows. Section 7.2 retrospects several definitions and properties of q-RIVOFSs. In Sect. 7.3, some novel cross-entropy and Hausdorff distance measure for q-RIVOFSs are constructed. Section 7.4 proposes an integrated q-RIVOFS-TODIM approach for MADM problems. An illustrative example concerning COVID-19 epidemic, a sensitivity analysis and three comparative analyses are conducted in Sect. 7.5. In the end, some conclusions and future research are added.

7.2 Basic Concepts of *q*-RIVOFSs

The q-RIVOFS [17] was proposed to remedy the deficiencies of interval-valued IFSs and interval-valued PFSs in describing complicated information. The general definitions and properties of q-RIVOFSs are reviewed in this section.

Definition 7.1 [17] Suppose a q-RIVOFS A on a non-empty fixed set $X = \{x_1, x_2, \ldots, x_n\}$ can be expressed as

$$A = \{x, \mu_A(x), \upsilon_A(x) | x \in X\}, \tag{7.1}$$

where $\mu_A(x) = \left[\mu_A^L(x), \mu_A^U(x)\right] \in [0, 1]$ and $\upsilon_A(x) = \left[\upsilon_A^L(x), \upsilon_A^U(x)\right] \in [0, 1]$ are interval numbers, indicating the "support for membership" and "support against membership" of the element x to A, severally, subject to $(\mu_A^U(x_i))^q + (\upsilon_A^U(x_i))^q \leq 1$.

The hesitancy degree of a q-RIVOFS A is defined by $\pi_A(x)=\left[\pi_A^L(x), \pi_A^U(x)\right] = \left[(1-(\mu_A^U(x))^q-(\upsilon_A^U(x))^q)^{1/q}, (1-(\mu_A^L(x))^q-(\upsilon_A^L(x))^q)^{1/q}\right]$. For convenience, $\left(\left[\mu_A^L(x_i), \mu_A^U(x_i)\right], \left[\upsilon_A^L(x_i), \upsilon_A^U(x_i)\right]\right)$, named a q-RIVOFN [25], is briefed as $a = \left(\left[\mu_a^L, \mu_a^U\right], \left[\upsilon_a^L, \upsilon_a^U\right]\right)$.

Obviously, if $\mu_A^L(x) = \mu_A^U(x)$ and $\upsilon_A^L(x) = \upsilon_A^U(x)$ for any $x \in X$, then q-RIVOFSs will reduce to q-ROFSs proposed by Yager [1].

Definition 7.2 [25] Suppose $a = \left(\left[\mu_a^L, \mu_a^U\right], \left[\upsilon_a^L, \upsilon_a^U\right]\right)$ is a q-RIVOFN, the score function $S(a)$ and the accuracy function $H(a)$ of a are defined as follows:

$$S(a) = \frac{1+(\mu_a^U)^q-(\upsilon_a^U)^q+1+(\mu_a^L)^q-(\upsilon_a^L)^q}{4}, \tag{7.2}$$

$$H(a) = \frac{(\mu_a^U)^q+(\upsilon_a^U)^q+(\mu_a^L)^q+(\upsilon_a^L)^q}{2}. \tag{7.3}$$

Definition 7.3 [25] Let $a_1 = \left(\left[\mu_1^L, \mu_1^U\right], \left[\upsilon_1^L, \upsilon_1^U\right]\right)$ and $a_2 = \left(\left[\mu_2^L, \mu_2^U\right], \left[\upsilon_2^L, \upsilon_2^U\right]\right)$ be two q-RIVOFNs, then:

If $S(a_1) > S(a_2)$, $a_1 > a_2$;

If $S(a_1) = S(a_2)$, then:

(1) if $H(a_1) > H(a_2)$, $a_1 > a_2$;
(2) if $H(a_1) = H(a_2)$, $a_1 = a_2$;
(3) if $H(a_1) < H(a_2)$, $a_1 < a_2$.

Definition 7.4 [25] Suppose $a_1 = \left(\left[\mu_1^L, \mu_1^U\right], \left[\upsilon_1^L, \upsilon_1^U\right]\right)$, $a_2 = \left(\left[\mu_2^L, \mu_2^U\right], \left[\upsilon_2^L, \upsilon_2^U\right]\right)$ and $a = \left(\left[\mu_a^L, \mu_a^U\right], \left[\upsilon_a^L, \upsilon_a^U\right]\right)$ are three q-RIVOFNs, some fundamental operation rules for them are as shown below:

(i)
$$a_1 \oplus a_2 = \left(\begin{bmatrix} ((\mu_1^L)^q+(\mu_2^L)^q-(\mu_1^L)^q\times(\mu_2^L)^q)^{1/q}, \\ ((\mu_1^U)^q+(\mu_2^U)^q-(\mu_1^U)^q\times(\mu_2^U)^q)^{1/q} \end{bmatrix}, \left[(\upsilon_1^L)\times(\upsilon_2^L), (\upsilon_1^U)\times(\upsilon_2^U)\right]\right);$$

(ii)
$$a_1 \otimes a_2 = \left(\left[(\mu_1^L)\times(\mu_2^L), (\mu_1^U)\times(\mu_2^U)\right], \begin{bmatrix} ((\upsilon_1^L)^q+(\upsilon_2^L)^q-(\upsilon_1^L)^q\times(\upsilon_2^L)^q)^{1/q}, \\ ((\upsilon_1^U)^q+(\upsilon_2^U)^q-(\upsilon_1^U)^q\times(\upsilon_2^U)^q)^{1/q} \end{bmatrix}\right);$$

(iii) $\lambda a = \left(\left[(1-(1-(\mu_a^L)^q)^\lambda)^{1/q}, (1-(1-(\mu_a^U)^q)^\lambda)^{1/q}\right], \left[(\upsilon_a^L)^\lambda, (\upsilon_a^U)^\lambda\right]\right)$, $\lambda > 0$;

(iv) $a^\lambda = \left(\left[(\mu_a^L)^\lambda, (\mu_a^U)^\lambda\right], \left[(1-(1-(\upsilon_a^L)^q)^\lambda)^{1/q}, (1-(1-(\upsilon_a^U)^q)^\lambda)^{1/q}\right]\right)$, $\lambda > 0$;

(v) $a^c = \left(\left[\upsilon_a^L, \upsilon_a^U\right], \left[\mu_a^L, \mu_a^U\right]\right)$.

7.3 Some Novel Measures for *q*-RIVOFSs

Cross-entropy and distance measures for fuzzy sets play important roles in fuzzy system fields [41–43]. However, to our best knowledge, there is still no research developing *q*-RIVOF cross-entropy, and distance measures for *q*-RIVOFSs are lack of systematic studies. Therefore, some novel cross-entropy and Hausdoff distance measures are constructed in this section.

7.3.1 *Cross-Entropy Measure for q-RIVOFSs*

In this subsection, the axiomatic definitions of cross-entropy measure for *q*-RIVOFSs are proposed and exemplified. It's worth noting that the proposed cross-entropy integrates the membership, non-membership and hesitate degree of *q*-RIVOFSs synchronously.

Definition 7.5 Suppose A and B are two *q*-RIVOFSs in universe $X = \{x_1, x_2, \ldots, x_n\}$, and then the cross-entropy $CE(A, B)$ between A and B should meet the following conditions:

(i) $CE(A, B) \geq 0$;
(ii) $CE(A, B) = 0$ if and only if $A = B$;
(iii) $CE(A^c, B^c) = CE(A, B)$, where A^c and B^c are defined as the complementary sets of A and B.

Definition 7.6 For two *q*-RIVOFSs $\widetilde{A}$ and $\widetilde{B}$ in universe $X = \{x_1, x_2, \ldots, x_n\}$, $\widetilde{A}(x_i) = \left(\left[\mu_{\widetilde{A}}^L(x_i), \mu_{\widetilde{A}}^U(x_i)\right], \left[\upsilon_{\widetilde{A}}^L(x_i), \upsilon_{\widetilde{A}}^U(x_i)\right], \left[\pi_{\widetilde{A}}^L(x_i), \pi_{\widetilde{A}}^U(x_i)\right]\right)$ and $\widetilde{B}(x_i) = \left(\left[\mu_{\widetilde{B}}^L(x_i), \mu_{\widetilde{B}}^U(x_i)\right], \left[\upsilon_{\widetilde{B}}^L(x_i), \upsilon_{\widetilde{B}}^U(x_i)\right], \left[\pi_{\widetilde{B}}^L(x_i), \pi_{\widetilde{B}}^U(x_i)\right]\right)$ the cross-entropy of $\widetilde{A}$ and $\widetilde{B}$ can be formulized as below:

$$
\begin{aligned}
CE(\widetilde{A}, \widetilde{B}) = & \sum_{i=1}^{n} \frac{(\mu_A^L)^q + (\mu_A^U)^q}{2} \log_2 \frac{\frac{(\mu_A^L)^q + (\mu_A^U)^q}{2}}{\frac{1}{2}\left(\frac{(\mu_A^L)^q + (\mu_A^U)^q}{2} + \frac{(\mu_B^L)^q + (\mu_B^U)^q}{2}\right)} \\
& + \sum_{i=1}^{n} \frac{(\upsilon_A^L)^q + (\upsilon_A^U)^q}{2} \log_2 \frac{\frac{(\upsilon_A^L)^q + (\upsilon_A^U)^q}{2}}{\frac{1}{2}\left(\frac{(\upsilon_A^L)^q + (\upsilon_A^U)^q}{2} + \frac{(\upsilon_B^L)^q + (\upsilon_B^U)^q}{2}\right)} \\
& + \sum_{i=1}^{n} \frac{(\pi_A^L)^q + (\pi_A^U)^q}{2} \log_2 \frac{\frac{(\pi_A^L)^q + (\pi_A^U)^q}{2}}{\frac{1}{2}\left(\frac{(\pi_A^L)^q + (\pi_A^U)^q}{2} + \frac{(\pi_B^L)^q + (\pi_B^U)^q}{2}\right)}.
\end{aligned} \tag{7.4}
$$

Proposition 7.1 The measure defined in Eq. (7.4) is a cross-entropy of *q*-RIVOFSs, and satisfies the conditions (i)–(iii) given in Definition 7.5.

Proof Condition (iii) is obvious. The proof of conditions (i) and (ii) is shown below.

Equation (7.4) can be transformed into the following form:

$$CE(\tilde{A},\tilde{B}) = \sum_{i=1}^{n}\left(\frac{(\mu_A^L)^q+(\mu_A^U)^q}{2}\log_2\frac{(\mu_A^L)^q+(\mu_A^U)^q}{2} - \frac{(\mu_A^L)^q+(\mu_A^U)^q}{2}\log_2\left(\frac{(\mu_A^L)^q+(\mu_A^U)^q}{4}+\frac{(\mu_B^L)^q+(\mu_B^U)^q}{4}\right)\right)$$
$$+\sum_{i=1}^{n}\left(\frac{(\upsilon_A^L)^q+(\upsilon_A^U)^q}{2}\log_2\frac{(\upsilon_A^L)^q+(\upsilon_A^U)^q}{2} - \frac{(\upsilon_A^L)^q+(\upsilon_A^U)^q}{2}\log_2\left(\frac{(\upsilon_A^L)^q+(\upsilon_A^U)^q}{4}+\frac{(\upsilon_B^L)^q+(\upsilon_B^U)^q}{4}\right)\right)$$
$$+\sum_{i=1}^{n}\left(\frac{(\pi_A^L)^q+(\pi_A^U)^q}{2}\log_2\frac{(\pi_A^L)^q+(\pi_A^U)^q}{2} - \frac{(\pi_A^L)^q+(\pi_A^U)^q}{2}\log_2\left(\frac{(\pi_A^L)^q+(\pi_A^U)^q}{4}+\frac{(\pi_B^L)^q+(\pi_B^U)^q}{4}\right)\right)$$

In line with Shannon's inequality [44],

$$-\sum_{k=1}^{n} p_k \log p_k \le -\sum_{k=1}^{n} p_k \log q_k,$$

where $p_k \in \Gamma_n (k = 1, 2, \ldots, n)$, $q_k \in \Delta_n^\circ (k = 1, 2, \ldots, n)$, Γ_n and Δ_n° are two probability distributions.

Because $\frac{(\mu_A^L)^q+(\mu_A^U)^q}{2}+\frac{(\upsilon_A^L)^q+(\upsilon_A^U)^q}{2}+\frac{(\pi_A^L)^q+(\pi_A^U)^q}{2} = 1$, it is evidently $CE(\widetilde{A}, \widetilde{B}) \ge 0$, and $CE(\widetilde{A}, \widetilde{B}) = 0$ when $\widetilde{A} = \widetilde{B}$.

Therefore, the conditions (i) and (ii) are proved.

It can be seen that $CE(\widetilde{A}, \widetilde{B})$ is not symmetrical with respect to its arguments. Hence, a modified symmetric discrimination information measure based on $CE(\widetilde{A}, \widetilde{B})$ can be defined as

$$CE^*(\widetilde{A}, \widetilde{B}) = CE(\widetilde{A}, \widetilde{B}) + CE(\widetilde{B}, \widetilde{A}). \tag{7.5}$$

The divergence degree between $\widetilde{A}$ and $\widetilde{B}$ increases in $CE^*(\widetilde{A}, \widetilde{B})$.

Proposition 7.2 The measure defined in Eq. (7.5) is also a cross-entropy of q-RIVOFSs, and satisfies the conditions (i)–(iii) given in Definition 7.5.

The proof of Proposition 2 is omitted due to its similarity to Proposition 1.

Example 7.1 Assume two q-RIVOFSs $A = ([0.4, 0.6], [0.5, 0.7], [0.77, 0.39])$ and $B = ([0.3, 0.7], [0.5, 0.7], [0.81, 0.14])$, then the following results can be obtained by applying Eqs. (7.4) and (7.5):

$$CE(\widetilde{A}, \widetilde{B}) = 0.001, CE(\widetilde{B}, \widetilde{A}) = 0.001, CE^*(\widetilde{A}, \widetilde{B}) = 0.002.$$

7.3.2 Hausdorff Distance for q-RIVOFSs

The Hausdorff distance measures how far two non-empty compact subsets A and B resemble mutually with their positions in a Banach space S, which is initially putted forward by Nadler [45]. Let $d(a, b)$ be a metric for S. $d(z, A) = \min\{d(z, a)|a \in A\}$. The Hausdorff measure $H^*(A, B) = \max_{a\in A} d(a, B)$ is one-way. $H(A, B)$ is defined by $H(A, B) = \max\{H^*(A, B), H^*(B, A)\}$. If $S = \Re$, for any two intervals $A = [a_1, a_2]$ and $B = [b_1, b_2]$, the Hausdorff distance $H(A, B)$ is given by $H(A, B) = \max\{|a_1 - b_1|, |a_2 - b_2|\}$.

A definition of the Hausdorff distance for q-RIVOFSs is proposed as below.

Let $\widetilde{A}$ and $\widetilde{B}$ be two q-RIVOFNs in a finite set $X = \{x_1, x_2, \ldots, x_n\}$. $I_{\widetilde{A}}(x_i)$ and $I_{\widetilde{B}}(x_i)$ are two subintervals in [0, 1] with $I_{\widetilde{A}}(x_i) = \left(\left[(\mu^L_{\widetilde{A}}(x_i))^q, (\mu^U_{\widetilde{A}}(x_i))^q\right], \left[(\upsilon^L_{\widetilde{A}}(x_i))^q, (\upsilon^U_{\widetilde{A}}(x_i))^q\right]\right)$ and $I_{\widetilde{B}}(x_i) = \left(\left[(\mu^L_{\widetilde{B}}(x_i))^q, (\mu^U_{\widetilde{B}}(x_i))^q\right], \left[(\upsilon^L_{\widetilde{B}}(x_i))^q, (\upsilon^U_{\widetilde{B}}(x_i))^q\right]\right)$, $i = 1, 2, \ldots, n$. $H(I_{\widetilde{A}}(x_i), I_{\widetilde{B}}(x_i))$ is defined as the Hausdorff distance between $I_{\widetilde{A}}(x_i)$ and $I_{\widetilde{B}}(x_i)$.The Hausdorff distance between two q-RIVOFSs $\widetilde{A}$ and $\widetilde{B}$ is

$$\begin{aligned} d_H(\widetilde{A}, \widetilde{B}) &= H(I_{\widetilde{A}}(x_i), I_{\widetilde{B}}(x_i)) \\ &= \max\left\{\begin{array}{l} \left|(\mu^L_{\widetilde{A}}(x_i))^q - (\mu^L_{\widetilde{B}}(x_i))^q\right|, \left|1 - (\upsilon^L_{\widetilde{A}}(x_i))^q - (1 - (\upsilon^L_{\widetilde{B}}(x_i))^q)\right|, \\ \left|(\mu^U_{\widetilde{A}}(x_i))^q - (\mu^U_{\widetilde{B}}(x_i))^q\right|, \left|1 - (\upsilon^U_{\widetilde{A}}(x_i))^q - (1 - (\upsilon^U_{\widetilde{B}}(x_i))^q)\right| \end{array}\right\} \\ &= \max\left\{\begin{array}{l} \left|(\mu^L_{\widetilde{A}}(x_i))^q - (\mu^L_{\widetilde{B}}(x_i))^q\right|, \left|(\mu^U_{\widetilde{A}}(x_i))^q - (\mu^U_{\widetilde{B}}(x_i))^q\right|, \\ \left|(\upsilon^L_{\widetilde{A}}(x_i))^q - (\upsilon^L_{\widetilde{B}}(x_i))^q\right|, \left|(\upsilon^U_{\widetilde{A}}(x_i))^q - (\upsilon^U_{\widetilde{B}}(x_i))^q\right| \end{array}\right\} \end{aligned} \tag{7.6}$$

The normalized Hausdorff distance $d_{NH}(\widetilde{A}, \widetilde{B})$ from $\widetilde{A}$ to $\widetilde{B}$ is as the following formula:

$$\begin{aligned} d_{NH}(\widetilde{A}, \widetilde{B}) &= \frac{1}{n}\sum_{i=1}^{n} H(I_{\widetilde{A}}(x_i), I_{\widetilde{B}}(x_i)) \\ &= \frac{1}{n}\sum_{i=1}^{n} \max\left\{\begin{array}{l} \left|(\mu^L_{\widetilde{A}}(x_i))^q - (\mu^L_{\widetilde{B}}(x_i))^q\right|, \left|(\mu^U_{\widetilde{A}}(x_i))^q - (\mu^U_{\widetilde{B}}(x_i))^q\right|, \\ \left|(\upsilon^L_{\widetilde{A}}(x_i))^q - (\upsilon^L_{\widetilde{B}}(x_i))^q\right|, \left|(\upsilon^U_{\widetilde{A}}(x_i))^q - (\upsilon^U_{\widetilde{B}}(x_i))^q\right| \end{array}\right\}. \end{aligned} \tag{7.7}$$

Example 7.2 Suppose two q-RIVOFSs $A = ([0.4, 0.6], [0.5, 0.7], [0.77, 0.39])$ and $B = ([0.3, 0.7], [0.5, 0.7], [0.81, 0.14])$, then according to Eq. (7.7), the normalized Hausdorff distance $d_{NH}(A, B)$ between A and B is

$$d_{NH}(A, B) = \frac{1}{1}\max\{\left|(0.4)^2 - (0.3)^2\right|, \left|(0.6)^2 - (0.7)^2\right|,$$

$$\left|(0.5)^2-(0.5)^2\right|,\left|(0.7)^2-(0.7)^2\right|\}=0.13.$$

The following proposition is given to guarantee the reasonability of Eq. (7.7).

Proposition 7.3 The proposed normalized distance $d_{NH}(\widetilde{A},\widetilde{B})$ between two q-RIVOFNs $\widetilde{A}$ and $\widetilde{B}$ has the following properties (P_1)–(P_4):

(P_1) $0\le d_{NH}(\widetilde{A},\widetilde{B})\le 1$;
(P_2) $d_{NH}(\widetilde{A},\widetilde{B})=0$ if and only if $\widetilde{A}=\widetilde{B}$;
(P_3) $d_{NH}(\widetilde{A},\widetilde{B})=d_{NH}(\widetilde{B},\widetilde{A})$;
(P_4) If $\widetilde{A}\subseteq\widetilde{B}\subseteq\widetilde{C}$, then $d_{NH}(\widetilde{A},\widetilde{B})\le d_{NH}(\widetilde{A},\widetilde{C})$ and $d_{NH}(\widetilde{B},\widetilde{C})\le d_{NH}(\widetilde{A},\widetilde{C})$.

Proof Obviously, $d_{NH}(\widetilde{A},\widetilde{B})$ confirms to the properties P_1-P_3. (P_4)

$$H(I_{\widetilde{A}}(x_i),I_{\widetilde{B}}(x_i))=\max\left\{\begin{array}{l}\left|(\mu^L_{\widetilde{A}}(x_i))^q-(\mu^L_{\widetilde{B}}(x_i))^q\right|,\left|(\mu^U_{\widetilde{A}}(x_i))^q-(\mu^U_{\widetilde{B}}(x_i))^q\right|,\\ \left|(\upsilon^L_{\widetilde{A}}(x_i))^q-(\upsilon^L_{\widetilde{B}}(x_i))^q\right|,\left|(\upsilon^U_{\widetilde{A}}(x_i))^q-(\upsilon^U_{\widetilde{B}}(x_i))^q\right|\end{array}\right\},$$

$$H(I_{\widetilde{A}}(x_i),I_{\widetilde{C}}(x_i))=\max\left\{\begin{array}{l}\left|(\mu^L_{\widetilde{A}}(x_i))^q-(\mu^L_{\widetilde{C}}(x_i))^q\right|,\left|(\mu^U_{\widetilde{A}}(x_i))^q-(\mu^U_{\widetilde{C}}(x_i))^q\right|,\\ \left|(\upsilon^L_{\widetilde{A}}(x_i))^q-(\upsilon^L_{\widetilde{C}}(x_i))^q\right|,\left|(\upsilon^U_{\widetilde{A}}(x_i))^q-(\upsilon^U_{\widetilde{C}}(x_i))^q\right|\end{array}\right\},$$

$$H(I_{\widetilde{B}}(x_i),I_{\widetilde{C}}(x_i))=\max\left\{\begin{array}{l}\left|(\mu^L_{\widetilde{B}}(x_i))^q-(\mu^L_{\widetilde{C}}(x_i))^q\right|,\left|(\mu^U_{\widetilde{B}}(x_i))^q-(\mu^U_{\widetilde{C}}(x_i))^q\right|,\\ \left|(\upsilon^L_{\widetilde{B}}(x_i))^q-(\upsilon^L_{\widetilde{C}}(x_i))^q\right|,\left|(\upsilon^U_{\widetilde{B}}(x_i))^q-(\upsilon^U_{\widetilde{C}}(x_i))^q\right|\end{array}\right\}.$$

Due to $\widetilde{A}\subseteq\widetilde{B}\subseteq\widetilde{C}$, for $\forall x_i\in X$, the following inequalities can be deduced:

$$\begin{array}{l}(\mu^L_{\widetilde{A}}(x_i))^q\le(\mu^L_{\widetilde{B}}(x_i))^q\le(\mu^L_{\widetilde{C}}(x_i))^q,\ (\mu^U_{\widetilde{A}}(x_i))^q\le(\mu^U_{\widetilde{B}}(x_i))^q\le(\mu^U_{\widetilde{C}}(x_i))^q,\\ (\upsilon^L_{\widetilde{A}}(x_i))^q\ge(\upsilon^L_{\widetilde{B}}(x_i))^q\ge(\upsilon^L_{\widetilde{C}}(x_i))^q,\ (\upsilon^U_{\widetilde{A}}(x_i))^q\ge(\upsilon^U_{\widetilde{B}}(x_i))^q\ge(\upsilon^U_{\widetilde{C}}(x_i))^q,\\ \left|(\mu^L_{\widetilde{A}}(x_i))^q-(\mu^L_{\widetilde{C}}(x_i))^q\right|\ge\left|(\mu^L_{\widetilde{A}}(x_i))^q-(\mu^L_{\widetilde{B}}(x_i))^q\right|,\\ \left|(\mu^L_{\widetilde{A}}(x_i))^q-(\mu^L_{\widetilde{C}}(x_i))^q\right|\ge\left|(\mu^L_{\widetilde{B}}(x_i))^q-(\mu^L_{\widetilde{C}}(x_i))^q\right|,\\ \left|(\mu^U_{\widetilde{A}}(x_i))^q-(\mu^U_{\widetilde{C}}(x_i))^q\right|\ge\left|(\mu^U_{\widetilde{A}}(x_i))^q-(\mu^U_{\widetilde{B}}(x_i))^q\right|,\\ \left|(\mu^U_{\widetilde{A}}(x_i))^q-(\mu^U_{\widetilde{C}}(x_i))^q\right|\ge\left|(\mu^U_{\widetilde{B}}(x_i))^q-(\mu^U_{\widetilde{C}}(x_i))^q\right|,\\ \left|(\upsilon^L_{\widetilde{A}}(x_i))^q-(\upsilon^L_{\widetilde{C}}(x_i))^q\right|\ge\left|(\upsilon^L_{\widetilde{A}}(x_i))^q-(\upsilon^L_{\widetilde{B}}(x_i))^q\right|,\\ \left|(\upsilon^L_{\widetilde{A}}(x_i))^q-(\upsilon^L_{\widetilde{C}}(x_i))^q\right|\ge\left|(\upsilon^L_{\widetilde{B}}(x_i))^q-(\upsilon^L_{\widetilde{C}}(x_i))^q\right|,\\ \left|(\upsilon^U_{\widetilde{A}}(x_i))^q-(\upsilon^U_{\widetilde{C}}(x_i))^q\right|\ge\left|(\upsilon^U_{\widetilde{A}}(x_i))^q-(\upsilon^U_{\widetilde{B}}(x_i))^q\right|,\\ \left|(\upsilon^U_{\widetilde{A}}(x_i))^q-(\upsilon^U_{\widetilde{C}}(x_i))^q\right|\ge\left|(\upsilon^U_{\widetilde{B}}(x_i))^q-(\upsilon^U_{\widetilde{C}}(x_i))^q\right|.\end{array}$$

Case 1: $H(I_{\widetilde{A}}(x_i),I_{\widetilde{C}}(x_i))=\left|(\mu^L_{\widetilde{A}}(x_i))^q-(\mu^L_{\widetilde{C}}(x_i))^q\right|$.

In this case, although the magnitude relations among $\left|(\mu^L_{\widetilde{A}}(x_i))^q - (\mu^L_{\widetilde{C}}(x_i))^q\right|$, $\left|(\mu^U_{\widetilde{A}}(x_i))^q - (\mu^U_{\widetilde{C}}(x_i))^q\right|$, $\left|(\upsilon^L_{\widetilde{A}}(x_i))^q - (\upsilon^L_{\widetilde{C}}(x_i))^q\right|$ and $\left|(\upsilon^U_{\widetilde{A}}(x_i))^q - (\upsilon^U_{\widetilde{C}}(x_i))^q\right|$ have six possible scenarios, the value of $\left|(\mu^L_{\widetilde{A}}(x_i))^q - (\mu^L_{\widetilde{C}}(x_i))^q\right|$ is bigger than the other three. Combined the above inequalities, $H(I_{\widetilde{A}}(x_i), I_{\widetilde{B}}(x_i)) \le H(I_{\widetilde{A}}(x_i), I_{\widetilde{C}}(x_i))$ and $H(I_{\widetilde{B}}(x_i), I_{\widetilde{C}}(x_i)) \le H(I_{\widetilde{A}}(x_i), I_{\widetilde{C}}(x_i))$ can be obtained without a hitch. Hence, $d_{NH}(\widetilde{A}, \widetilde{B}) \le d_{NH}(\widetilde{A}, \widetilde{C})$ and $d_{NH}(\widetilde{B}, \widetilde{C}) \le d_{NH}(\widetilde{A}, \widetilde{C})$.

Case 2: $H(I_{\widetilde{A}}(x_i), I_{\widetilde{C}}(x_i)) = \left|(\mu^U_{\widetilde{A}}(x_i))^q - (\mu^U_{\widetilde{C}}(x_i))^q\right|$.

In this case, although the magnitude relations among $\left|(\mu^L_{\widetilde{A}}(x_i))^q - (\mu^L_{\widetilde{C}}(x_i))^q\right|$, $\left|(\mu^U_{\widetilde{A}}(x_i))^q - (\mu^U_{\widetilde{C}}(x_i))^q\right|$, $\left|(\upsilon^L_{\widetilde{A}}(x_i))^q - (\upsilon^L_{\widetilde{C}}(x_i))^q\right|$ and $\left|(\upsilon^U_{\widetilde{A}}(x_i))^q - (\upsilon^U_{\widetilde{C}}(x_i))^q\right|$ have six possible scenarios, the value of $\left|(\mu^U_{\widetilde{A}}(x_i))^q - (\mu^U_{\widetilde{C}}(x_i))^q\right|$ is bigger than the other three. Combined the above inequalities, $H(I_{\widetilde{A}}(x_i), I_{\widetilde{B}}(x_i)) \le H(I_{\widetilde{A}}(x_i), I_{\widetilde{C}}(x_i))$ and $H(I_{\widetilde{B}}(x_i), I_{\widetilde{C}}(x_i)) \le H(I_{\widetilde{A}}(x_i), I_{\widetilde{C}}(x_i))$ can be obtained without a hitch. Hence, $d_{NH}(\widetilde{A}, \widetilde{B}) \le d_{NH}(\widetilde{A}, \widetilde{C})$ and $d_{NH}(\widetilde{B}, \widetilde{C}) \le d_{NH}(\widetilde{A}, \widetilde{C})$.

Case 3: $H(I_{\widetilde{A}}(x_i), I_{\widetilde{C}}(x_i)) = \left|(\upsilon^L_{\widetilde{A}}(x_i))^q - (\upsilon^L_{\widetilde{C}}(x_i))^q\right|$.

In this case, although the magnitude relations among $\left|(\mu^L_{\widetilde{A}}(x_i))^q - (\mu^L_{\widetilde{C}}(x_i))^q\right|$, $\left|(\mu^U_{\widetilde{A}}(x_i))^q - (\mu^U_{\widetilde{C}}(x_i))^q\right|$, $\left|(\upsilon^L_{\widetilde{A}}(x_i))^q - (\upsilon^L_{\widetilde{s}}(x_i))^q\right|$ and $\left|(\upsilon^U_{\widetilde{A}}(x_i))^q - (\upsilon^U_{\widetilde{C}}(x_i))^q\right|$ have six possible scenarios, the value of $\left|(\upsilon^L_{\widetilde{A}}(x_i))^q - (\upsilon^L_{\widetilde{C}}(x_i))^q\right|$ is bigger than the other three. Combined the above inequalities, $H(I_{\widetilde{A}}(x_i), I_{\widetilde{B}}(x_i)) \le H(I_{\widetilde{A}}(x_i), I_{\widetilde{C}}(x_i))$ and $H(I_{\widetilde{B}}(x_i), I_{\widetilde{C}}(x_i)) \le H(I_{\widetilde{A}}(x_i), I_{\widetilde{C}}(x_i))$ can be obtained without a hitch. Hence, $d_{NH}(\widetilde{A}, \widetilde{B}) \le d_{NH}(\widetilde{A}, \widetilde{C})$ and $d_{NH}(\widetilde{B}, \widetilde{C}) \le d_{NH}(\widetilde{A}, \widetilde{C})$.

Case 4: $H(I_{\widetilde{A}}(x_i), I_{\widetilde{C}}(x_i)) = \left|(\upsilon^U_{\widetilde{A}}(x_i))^q - (\upsilon^U_{\widetilde{C}}(x_i))^q\right|$.

In this case, although the magnitude relations among $\left|(\mu^L_{\widetilde{A}}(x_i))^q - (\mu^L_{\widetilde{C}}(x_i))^q\right|$, $\left|(\mu^U_{\widetilde{A}}(x_i))^q - (\mu^U_{\widetilde{C}}(x_i))^q\right|$, $\left|(\upsilon^L_{\widetilde{A}}(x_i))^q - (\upsilon^L_{\widetilde{C}}(x_i))^q\right|$ and $\left|(\upsilon^U_{\widetilde{A}}(x_i))^q - (\upsilon^U_{\widetilde{C}}(x_i))^q\right|$ have six possible scenarios, the value of $\left|(\upsilon^U_{\widetilde{A}}(x_i))^q - (\upsilon^U_{\widetilde{C}}(x_i))^q\right|$ is bigger than the other three. Combined the above inequalities, $H(I_{\widetilde{A}}(x_i), I_{\widetilde{B}}(x_i)) \le H(I_{\widetilde{A}}(x_i), I_{\widetilde{s}}(x_i))$ and $H(I_{\widetilde{B}}(x_i), I_{\widetilde{C}}(x_i)) \le H(I_{\widetilde{A}}(x_i), I_{\widetilde{C}}(x_i))$ can be obtained without a hitch. Hence, $d_{NH}(\widetilde{A}, \widetilde{B}) \le d_{NH}(\widetilde{A}, \widetilde{C})$ and $d_{NH}(\widetilde{B}, \widetilde{C}) \le d_{NH}(\widetilde{A}, \widetilde{C})$.

Thus, the containment property (P_4) is proved.

7.4 Multi-Attribute Decision-Making Method Under q-RIVOF Circumstances

Multi-attribute decision-making is a significant application field of q-RIVOFSs. There have been a few crucial investigations regarding the applications of q-RIVOFSs in MADM domain [18, 24, 29]. In this section, a novel q-RIVOFS-TODIM approach is conducted to display how q-RIVOFSs contribute to handle MADM problems efficiently.

7.4.1 *TODIM Method with q-RIVOFSs*

Consider a MADM problem where m alternatives $A = \{A_1, A_2, \ldots, A_m\}$ are evaluated with respect to n attributes $C = \{C_1, C_2, \ldots, C_n\}$. Assume $w = (w_1, w_2, \ldots, w_n)^T$ is the corresponding attribute weight vector, meeting the conditions that $w_j \in [0, 1]$ and $\sum_{j=1}^{n} w_j = 1$. A q-RIVOFN $\tilde{r}_{ij} = (\tilde{\mu}_{ij}, \tilde{\upsilon}_{ij}, \tilde{\pi}_{ij}) = \left(\left[\mu_{ij}^L, \mu_{ij}^U\right], \left[\upsilon_{ij}^L, \upsilon_{ij}^U\right], \left[\pi_{ij}^L, \pi_{ij}^U\right]\right)$ is utilized to stand for the evaluation value of A_i in regard to C_j, where $\tilde{\mu}_{ij} = \left[\mu_{ij}^L, \mu_{ij}^U\right]$ indicates the range of degree to which alternative A_i satisfies attribute C_j, while $\tilde{\upsilon}_{ij} = \left[\upsilon_{ij}^L, \upsilon_{ij}^U\right]$ indicates the range of degree to which alternative A_i dissatisfies attribute C_j and $\tilde{\pi}_{ij} = \left[(1 - (\mu_{ij}^U)^q - (\upsilon_{ij}^U)^q)^{1/q}, (1 - (\mu_{ij}^L)^q - (\upsilon_{ij}^L)^q)^{1/q}\right]$ indicates the hesitation degree range. The MADM problem can be solved by the following three stages: (1) Stage 1. The overall decision matrix $\widehat{R} = (\hat{r}_{ij})_{m\times n}$, evaluated by a committee of experts based on their knowledge and problem-related information, is compactly depicted as

$$\widehat{R} = (\hat{r}_{ij})_{m\times n} = \begin{pmatrix} (\hat{\mu}_{11}, \upsilon_{11}, \hat{\pi}_{11}) & (\hat{\mu}_{12}, \hat{\upsilon}_{12}, \hat{\pi}_{12}) & \cdots & (\hat{\mu}_{1n}, \hat{\upsilon}_{1n}, \hat{\pi}_{1n}) \\ (\hat{\mu}_{21}, \hat{\upsilon}_{21}, \hat{\pi}_{21}) & (\hat{\mu}_{22}, \hat{\upsilon}_{22}, \hat{\pi}_{22}) & \cdots & (\hat{\mu}_{2n}, \hat{\upsilon}_{2n}, \hat{\pi}_{2n}) \\ \vdots & \vdots & \vdots & \vdots \\ (\hat{\mu}_{m1}, \hat{\upsilon}_{m1}, \hat{\pi}_{m1}) & (\hat{\mu}_{m2}, \hat{\upsilon}_{m2}, \hat{\pi}_{m2}) & \cdots & (\hat{\mu}_{mn}, \hat{\upsilon}_{mn}, \hat{\pi}_{mn}) \end{pmatrix};$$

(2) Stage 2. The attribute weights are calculated via a novel q-RIVOF cross-entropy programming model established resort to the deviation value of A_i and the fuzziness degree of C_j simultaneously. (3) Stage 3. On the basis of the normalized decision matrix and the attribute weights computed in Stage 2, the ranking results of all alternatives are obtained with the help of TODIM method. Figure 7.1 presents the basic conceptual graph of the above three stages.

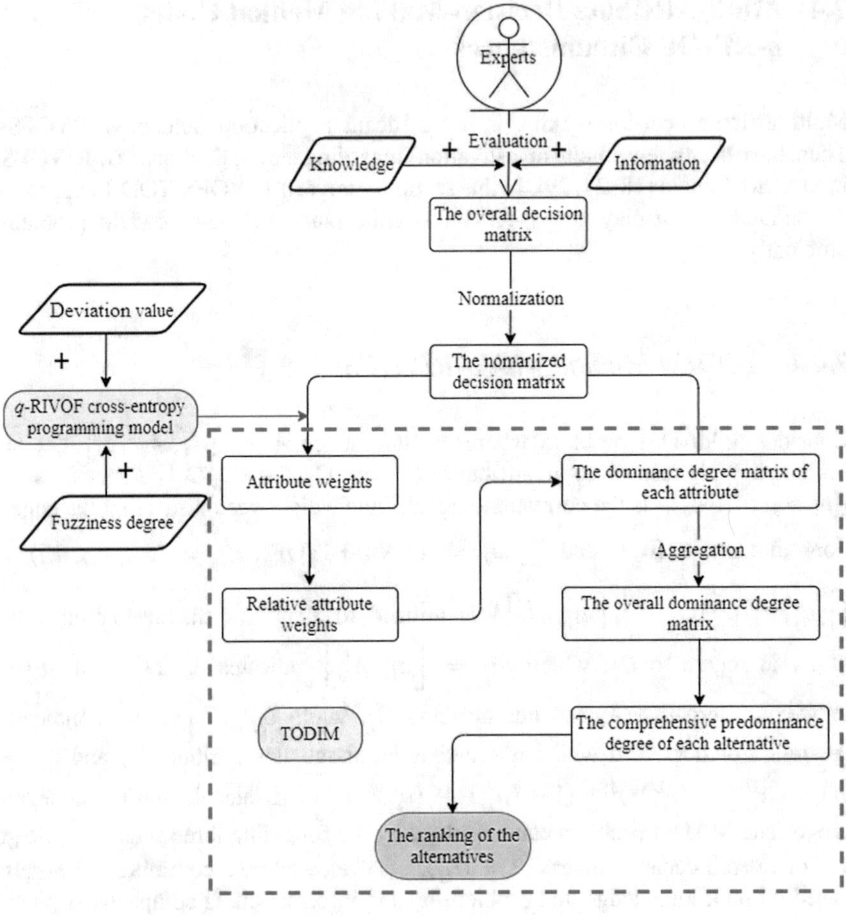

Fig. 7.1 Conceptual graph of the proposed q-RIVOFS-TODIM approach

The procedures of the q-RIVOF TODIM decision-making approach are as follows.

Step 1. Normalize the decision matrix $\widehat{R} = (\hat{r}_{ij})_{m\times n}$

Attributes are divided into two types: benefit and cost attributes. To keep attributes types consistent, cost attributes are supposed to be transformed into benefit ones. The initial decision matrix $\widehat{R} = (\hat{r}_{ij})_{m\times n}$ can then be converted into $\widetilde{R} = (\tilde{r}_{ij})_{m\times n}$ by the following formula:

$$\tilde{r}_{ij} = \begin{cases} \hat{r}_{ij} = \left(\left[\mu_{ij}^L, \mu_{ij}^U\right], \left[\upsilon_{ij}^L, \upsilon_{ij}^U\right], \left[\pi_{ij}^L, \pi_{ij}^U\right]\right), & for\ benefit \text{ attribute } C_j \\ (\hat{r}_{ij})^c = \left(\left[\upsilon_{ij}^L, \upsilon_{ij}^U\right], \left[\mu_{ij}^L, \mu_{ij}^U\right], \left[\pi_{ij}^L, \pi_{ij}^U\right]\right), & for\ cost\ attribute\ C_j \end{cases} \tag{7.8}$$

where $(\hat{r}_{ij})^c$ is the complement of $\hat{r}_{ij}$.

Step 2. Calculate the attribute weights resort to q-RIVOF cross-entropy programming model

For attribute $C_j (j = 1, 2, \ldots, n)$, the deviation value (DV) of $A_i (i = 1, 2, \ldots, m)$ over all the other alternatives can be determined by

$$DV_j = \frac{1}{n-1} \sum_{k=1, k \neq i}^{m} CE(\tilde{r}_{ij}, \tilde{r}_{kj}),\ j = 1, 2, \ldots, n, \tag{7.9}$$

where

$$CE(\tilde{r}_{ij}, \tilde{r}_{kj}) = \sum_{i=1}^{n} \frac{(\mu_{ij}^L)^q + (\mu_{ij}^U)^q}{2} \log_2 \frac{\frac{(\mu_{ij}^L)^q + (\mu_{ij}^U)^q}{2}}{\frac{1}{2}\left(\frac{(\mu_{ij}^L)^q + (\mu_{ij}^U)^q}{2} + \frac{(\mu_{kj}^L)^q + (\mu_{kj}^U)^q}{2}\right)}$$
$$+ \sum_{i=1}^{n} \frac{(\upsilon_{ij}^L)^q + (\upsilon_{ij}^U)^q}{2} \log_2 \frac{\frac{(\upsilon_{ij}^L)^q + (\upsilon_{ij}^U)^q}{2}}{\frac{1}{2}\left(\frac{(\upsilon_{ij}^L)^q + (\upsilon_{ij}^U)^q}{2} + \frac{(\upsilon_{kj}^L)^q + (\upsilon_{kj}^U)^q}{2}\right)}$$
$$+ \sum_{i=1}^{n} \frac{(\pi_{ij}^L)^q + (\pi_{ij}^U)^q}{2} \log_2 \frac{\frac{(\pi_{ij}^L)^q + (\pi_{ij}^U)^q}{2}}{\frac{1}{2}\left(\frac{(\pi_{ij}^L)^q + (\pi_{ij}^U)^q}{2} + \frac{(\pi_{kj}^L)^q + (\pi_{kj}^U)^q}{2}\right)}.$$

Because a q-RIVOFS A is considered as the fuzziest q-RIVOFS when $\tilde{\mu}_A(x_i) = \tilde{\upsilon}_A(x_i)$ for all $x_i \in X$, $\hat{r}_f = ([0.5, 0.5], [0.5, 0.5])$ can be referred to as the fuzziest q-RIVOFN. In the light of Definition 7.6, the fuzziness degree (FD) of C_j can be computed by

$$FD_j = \sum_{i=1}^{m} CE(\tilde{r}_{ij}, \tilde{r}_f),\ j = 1, 2, \ldots, n \tag{7.10}$$

in which $CE(\tilde{r}_{ij}, \tilde{r}_f)$ is the q-RIVOF cross-entropy measure between all assessments under attribute C_j and the fuzziest q-RIVOFN $\tilde{r}_f$. The credibility of evaluations under C_j increases in the value of FD_j.

According to the general rationale that if the correlative assessments of an attribute are more effective at distinguishing the characteristics of the problem, then it should be assigned a bigger weight, model (M-1) can be formulated to determine the optimal attribute weights:

$$Max\ F(w) = \sum_{j=1}^{n} F_j(w) = \sum_{j=1}^{n}\sum_{i=1}^{m}\left(\frac{1}{n-1}\sum_{k=1, k\neq i}^{m} CE(\tilde{r}_{ij}, \tilde{r}_{kj}) + CE(\tilde{r}_{ij}, \tilde{r}_f)\right) w_j$$

$$s.t. \sum_{j=1}^{n} w_j^2 = 1, w_j \geq 0, j = 1, 2, \ldots, n \tag{M-1}$$

The following Lagrange function is established by solving (M-1):

$$L(w_j, \lambda) = \sum_{j=1}^{n} \sum_{i=1}^{m} \left(\frac{1}{n-1} \sum_{k=1,k\neq i}^{m} CE(\tilde{r}_{ij}, \tilde{r}_{kj}) + CE(\tilde{r}_{ij}, \tilde{r}_f) \right) w_j + \frac{\lambda}{2} \left(\sum_{j=1}^{n} w_j^2 - 1 \right) \tag{7.11}$$

where λ is the Lagrangian multiplier.

Take the derivatives of Eq. (7.11) with respect to $w_j (j = 1, 2, \ldots, n)$ and λ, severally, and set the two partial derivatives equal to zero, then the following equations are obtained:

$$\begin{cases} \frac{\partial L}{\partial w_j} = \sum\limits_{i=1}^{m} (\frac{1}{n-1} \sum\limits_{k=1,k\neq i}^{m} CE(\tilde{r}_{ij}, \tilde{r}_{kj}) + CE(\tilde{r}_{ij}, \tilde{r}_f)) + \lambda w_j = 0 \\ \frac{\partial L}{\partial w_j} = \sum\limits_{j=1}^{n} w_j^2 - 1 = 0 \end{cases} \tag{7.12}$$

In virtue of Eq. (7.12), a formula for determining attribute weights is formulated as

$$w_j^* = \frac{\sum\limits_{i=1}^{m} \left(\frac{1}{n-1} \sum\limits_{k=1,k\neq i}^{m} CE(\tilde{r}_{ij}, \tilde{r}_{kj}) + CE(\tilde{r}_{ij}, \tilde{r}_f) \right)}{\sqrt{\sum\limits_{j=1}^{n} \left(\sum\limits_{i=1}^{m} \left(\frac{1}{n-1} \sum\limits_{k=1,k\neq i}^{m} CE(\tilde{r}_{ij}, \tilde{r}_{kj}) + CE(\tilde{r}_{ij}, \tilde{r}_f) \right) \right)^2}} \tag{7.13}$$

The normalized attribute weights can be derived by

$$w_j = \frac{\sum\limits_{i=1}^{m} \left(\frac{1}{n-1} \sum\limits_{k=1,k\neq i}^{m} CE(\tilde{r}_{ij}, \tilde{r}_{kj}) + CE(\tilde{r}_{ij}, \tilde{r}_f) \right)}{\sum\limits_{j=1}^{n} \left(\sum\limits_{i=1}^{m} \left(\frac{1}{n-1} \sum\limits_{k=1,k\neq i}^{m} CE(\tilde{r}_{ij}, \tilde{r}_{kj}) + CE(\tilde{r}_{ij}, \tilde{r}_f) \right) \right)^2} \tag{7.14}$$

Step 3. Compute the relative weights of attributes

The relative weight of each attribute C_j can be computed by

$$w_{jr} = \frac{w_j}{w_r}, \; j, r = 1, 2, \ldots, n, \tag{7.15}$$

where w_j is the weight of the attribute C_j, $w_r = \max\{w_j | j = 1, 2, \ldots, n\}$, and $0 \leq w_{jr} \leq 1$.

Step 4. Compute the dominance degree of A_i over A_t under C_j

The following formula can be employed to calculate the dominance degree of A_i over each alternative A_t under C_j:

$$\Phi_j(A_i, A_t) = \begin{cases} \sqrt{\frac{w_{jr} d(\overline{r}_{ij}, \overline{r}_{tj})}{\sum_{j=1}^{n} w_{jr}}} & if \; \overline{r}_{ij} > \overline{r} \\ 0 & if \; \overline{r}_{ij} = \overline{r}_{tj} \\ -\frac{1}{\theta}\sqrt{\frac{\sum_{j=1}^{n} w_{jr} d(\overline{r}_{ij}, \overline{r}_{tj})}{w_{jr}}} & if \; \overline{r}_{ij} < \overline{r}_{tj} \end{cases}, \tag{7.16}$$

where θ represents the attenuation factor of the losses, and $d(\overline{r}_{ij}, \overline{r}_{tj})$ is the q-RIVOF Hausdorrf distance measure between $\overline{r}_{ij}$ and $\overline{r}_{tj}$. If $\overline{r}_{ij} < \overline{r}_{tj}$, then $\Phi_j(A_i, A_t)$ signifies a loss; if $\overline{r}_{ij} > \overline{r}_{tj}$, then $\Phi_j(A_i, A_t)$ represents a gain.

The dominance degree matrix of C_j is

$$\Phi_j = [\Phi_j(A_i, A_t)]_{m \times m} = \begin{bmatrix} 0 & \Phi_j(A_1, A_2) & \cdots & \Phi_j(A_1, A_m) \\ \Phi_j(A_2, A_1) & 0 & \cdots & \Phi_j(A_2, A_m) \\ \vdots & \vdots & \cdots & \vdots \\ \Phi_j(A_m, A_1) & \Phi_j(A_m, A_2) & \cdots & 0 \end{bmatrix}, j = 1, 2, \ldots, n$$

Step 5. Obtain the overall dominance degree of A_i over each alternative A_t

The following equation can be applied to derive the overall dominance degree of A_i over each alternative A_t

$$\delta(A_i, A_t) = \sum_{j=1}^{n} \Phi_j(A_i, A_t), \; (i, t = 1, 2, \ldots, m). \tag{7.17}$$

Then, the overall dominance degree matrix δ is

$$\delta = [\delta(A_i, A_t)]_{m \times m} = \begin{bmatrix} 0 & \delta(A_1, A_2) & \cdots & \delta(A_1, A_m) \\ \delta(A_2, A_1) & 0 & \cdots & \delta(A_2, A_m) \\ \vdots & \vdots & \cdots & \vdots \\ \delta(A_m, A_1) & \delta(A_m, A_2) & \cdots & 0 \end{bmatrix}.$$

Step 6. Rank the alternatives according to the comprehensive predominance degree of A_i

The comprehensive predominance degree of A_i can be obtained by the following equation

$$\nu_i = \frac{\sum_{l=1}^{m} \delta(A_i, A_l) - \min_{1 \le i \le m} \left\{ \sum_{l=1}^{m} \delta(A_i, A_l) \right\}}{\max_{1 \le i \le m} \left\{ \sum_{l=1}^{m} \delta(A_i, A_l) \right\} - \min_{1 \le i \le m} \left\{ \sum_{l=1}^{m} \delta(A_i, A_l) \right\}}. \quad (7.18)$$

Rank the alternatives by ν_i in descending orders, the optimal choice is the alternative with the maximum ν_i.

7.5 Illustrative Example

Generally, emergency events will bring massive losses to human. For example, up to August 14, 2021, a major communicable disease, called COVID-19 epidemic, has caused 206,157,208 cases and 4,344,673 deaths around the world [46]. In order to reduce the adverse impacts of emergency events, selecting the most suitable response scheme in the shortest time is vital and requisite. Under this background, the proposed q-RIVOFS-TODIM MADM approach is utilized to choose the best response scheme for an emergency event in this section.

7.5.1 *Case Description*

To dispose the dramatically increased medical waste, a city named Y decided to entrust a company H as a temporary emergency disposal point to cope with the daily output of medical waste during COVID-19 epidemic. In accordance with the practical situation, the company H needs to choose the best medical waste disposal method from the following six alternatives: A_1 (pyrolysis, gasification, and incineration), A_2 (rotary kiln incineration), A_3 (high-temperature steam sterilization), A_4 (microwave disinfection), A_5 (chemical disinfection) and A_6 (safe centralized landfill). To assess the above six alternatives, five experts have a certain level of knowledge about various medical waste disposal methods are invited, including a professional from a medical waste disposal equipment company, an expert in environmental engineering, an administrator of a medical waste disposal enterprise, a staff member of the emergency response office of the Ministry of Ecology and Environment and an expert in emergency management. A decision-making committee consisting of these five experts assesses the above six alternatives according to four aspects: (1) economy (transformation cost C_1 and net cost per ton C_2); (2) environment (Waste residue C_3, noise C_4, and health effects C_5); (3) technology (reliability C_6, disposal efficiency C_7, and automation level C_8); (4) society (public acceptance C_9). In the decision-making process, to ensure the committee members master sufficient information, they will be briefed in detail about the actual conditions of company H and the advantages and disadvantages of different medical waste disposal methods. Then, the q-RIVOF overall decision matrix, as shown in Table 7.1, is obtained by the evaluation of the committee. In virtue of the proposed q-RIVOF-TODIM decision-making approach, the company H is supposed to sort out the most desirable medical waste disposal method in the end.

7.5.2 *Illustration of the Proposed Q-RIVOFS-TODIM Approach*

Step 1: Normalize the decision matrix $\widetilde{R} = (\tilde{r}_{ij})_{m \times n}$

Cost attributes C_1, C_2, C_3, C_4, and C_5 should be transformed into benefit attributes by Eq. (7.8). Table 7.2 shows the normalized decision matrix.

Step 2: Calculate the attribute weights resort to q-RIVOF cross-entropy programming model

Apply Eqs. (7.9), (7.10), and (7.14), the attribute weights are $w_1 = 0.0589$, $w_2 = 0.0819$, $w_3 = 0.0579$, $w_4 = 0.0798$, $w_5 = 0.1549$, $w_6 = 0.2206$, $w_7 = 0.1120$, $w_8 = 0.1038$, and $w_9 = 0.1302$.

Step 3. Compute the relative weights of attributes

Based on the results of Step 2, $w_6 = \max\{w_1, w_2, w_3, w_4, w_5, w_6, w_7, w_8, w_9\}$. Hence, C_6 is the reference attribute. In virtue of Eq. (7.15), the relative weights of each attribute are $w_{1r} = 0.2668$, $w_{2r} = 0.3712$, $w_{3r} = 0.2625$, $w_{4r} = 0.3616$, $w_{5r} = 0.7022$, $w_{6r} = 1.0000$, $w_{7r} = 0.5075$, $w_{8r} = 0.4706$, and $w_{9r} = 0.5901$.

Step 4: Compute the dominance degree of A_i over A_t under C_j

Set $\theta = 2.5$, by Eq. (7.16), the dominance degree matrices under $C_j (j = 1, 2, \ldots, 9)$ are

$$\Phi_1 = \begin{bmatrix} 0 & -0.7139 & -0.9471 & -0.8077 & 0.0653 & -1.0460 \\ 0.1051 & 0 & -0.7139 & -0.4663 & 0.1237 & -0.8407 \\ 0.1394 & 0.1051 & 0 & 0.0875 & 0.1471 & -0.4663 \\ 0.1189 & 0.0686 & -0.5945 & 0 & 0.1356 & -0.7419 \\ -0.4439 & -0.8407 & -0.9995 & -0.9217 & 0 & -1.0936 \\ 0.1539 & 0.1237 & 0.0686 & 0.1092 & 0.1609 & 0 \end{bmatrix},$$

Table 7.1 The q-RIVOF overall decision matrix

	A_1	A_2	A_3	A_4	A_5	A_6
C_1	([0.55, 0.70], [0.35, 0.40])	([0.45, 0.55], [0.45, 0.55])	([0.30, 0.45], [0.55, 0.70])	([0.35, 0.50], [0.50, 0.60])	([0.60, 0.75], [0.30, 0.35])	([0.25, 0.35], [0.60, 0.75])
C_2	([0.25, 0.40], [0.70, 0.80])	([0.30, 0.45], [0.65, 0.75])	([0.50, 0.60], [0.55, 0.70])	([0.50, 0.65], [0.50, 0.65])	([0.65, 0.75], [0.35, 0.50])	([0.20, 0.35], [0.70, 0.85])
C_3	([0.20, 0.35], [0.65, 0.80])	([0.30, 0.45], [0.65, 0.75])	([0.40, 0.55], [0.50, 0.65])	([0.50, 0.65], [0.40, 0.55])	([0.40, 0.50], [0.55, 0.70])	([0.15, 0.30], [0.70, 0.80])
C_4	([0.25, 0.40], [0.70, 0.85])	([0.40, 0.55], [0.60, 0.75])	([0.50, 0.60], [0.45, 0.55])	([0.50, 0.60], [0.45, 0.55])	([0.30, 0.45], [0.70, 0.80])	([0.15, 0.30], [0.75, 0.85])
C_5	([0.35, 0.50], [0.60, 0.75])	([0.60, 0.70], [0.35, 0.50])	([0.30, 0.40], [0.65, 0.75])	([0.20, 0.35], [0.70, 0.80])	([0.70, 0.80], [0.30, 0.45])	([0.75, 0.85], [0.10, 0.15])
C_6	([0.65, 0.80], [0.15, 0.25])	([0.60, 0.75], [0.25, 0.40])	([0.70, 0.85], [0.10, 0.20])	([0.80, 0.95], [0.05, 0.15])	([0.55, 0.70], [0.30, 0.45])	([0.05, 0.10], [0.70, 0.85])
C_7	([0.65, 0.75], [0.35, 0.45])	([0.75, 0.90], [0.20, 0.35])	([0.50, 0.55], [0.45, 0.50])	([0.50, 0.65], [0.40, 0.50])	([0.40, 0.50], [0.55, 0.65])	([0.15, 0.25], [0.65, 0.80])
C_8	([0.60, 0.70], [0.30, 0.45])	([0.60, 0.70], [0.30, 0.45])	([0.65, 0.80], [0.25, 0.40])	([0.65, 0.80], [0.25, 0.40])	([0.40, 0.55], [0.60, 0.75])	([0.20, 0.30], [0.65, 0.80])
C_9	([0.50, 0.55], [0.60, 0.65])	([0.55, 0.65], [0.40, 0.45])	([0.65, 0.80], [0.20, 0.35])	([0.75, 0.85], [0.15, 0.25])	([0.50, 0.55], [0.60, 0.70])	([0.15, 0.30], [0.65, 0.80])

Table 7.2 The normalized q-RIVOF overall decision matrix

	A_1	A_2	A_3	A_4	A_5	A_6
C_1	([0.35, 0.40], [0.55, 0.70])	([0.45, 0.55], [0.45, 0.55])	([0.55, 0.70], [0.30, 0.45])	([0.50, 0.60], [0.35, 0.50])	([0.30, 0.35], [0.60, 0.75])	([0.60, 0.75], [0.25, 0.35])
C_2	([0.70, 0.80], [0.25, 0.40])	([0.65, 0.75], [0.30, 0.45])	([0.55, 0.70], [0.50, 0.60])	([0.50, 0.65], [0.50, 0.65])	([0.35, 0.50], [0.65, 0.75])	([0.70, 0.85], [0.20, 0.35])
C_3	([0.65, 0.80], [0.20, 0.35])	([0.65, 0.75], [0.30, 0.45])	([0.50, 0.65], [0.40, 0.55])	([0.40, 0.55], [0.50, 0.65])	([0.55, 0.70], [0.40, 0.50])	([0.70, 0.80], [0.15, 0.30])
C_4	([0.70, 0.85], [0.25, 0.40])	([0.60, 0.75], [0.40, 0.55])	([0.45, 0.55], [0.50, 0.60])	([0.45, 0.55], [0.50, 0.60])	([0.70, 0.80], [0.30, 0.45])	([0.75, 0.85], [0.15, 0.30])
C_5	([0.60, 0.75], [0.35, 0.50])	([0.35, 0.50], [0.60, 0.70])	([0.65, 0.75], [0.30, 0.40])	([0.70, 0.80], [0.20, 0.35])	([0.30, 0.45], [0.70, 0.80])	([0.10, 0.15], [0.75, 0.85])
C_6	([0.65, 0.80], [0.15, 0.25])	([0.60, 0.75], [0.25, 0.40])	([0.70, 0.85], [0.10, 0.20])	([0.80, 0.95], [0.05, 0.15])	([0.55, 0.70], [0.30, 0.45])	([0.05, 0.10], [0.70, 0.85])
C_7	([0.65, 0.75], [0.35, 0.45])	([0.75, 0.90], [0.20, 0.35])	([0.50, 0.55], [0.45, 0.50])	([0.50, 0.65], [0.40, 0.50])	([0.40, 0.50], [0.55, 0.65])	([0.15, 0.25], [0.65, 0.80])
C_8	([0.60, 0.70], [0.30, 0.45])	([0.60, 0.70], [0.30, 0.45])	([0.65, 0.80], [0.25, 0.40])	([0.65, 0.80], [0.25, 0.40])	([0.40, 0.55], [0.60, 0.75])	([0.20, 0.30], [0.65, 0.80])
C_9	([0.50, 0.55], [0.60, 0.65])	([0.55, 0.65], [0.40, 0.45])	([0.65, 0.80], [0.20, 0.35])	([0.75, 0.85], [0.15, 0.25])	([0.50, 0.55], [0.60, 0.70])	([0.15, 0.30], [0.65, 0.80])

$$\Phi_2 = \begin{bmatrix} 0 & 0.0797 & 0.1280 & 0.1466 & 0.1816 & -0.4014 \\ -0.3891 & 0 & 0.1145 & 0.1342 & 0.1717 & -0.5591 \\ -0.6250 & -0.5591 & 0 & 0.0744 & 0.1402 & -0.6811 \\ -0.7161 & -0.6556 & -0.3631 & 0 & 0.1189 & -0.7655 \\ -0.8867 & -0.8386 & -0.6847 & -0.5805 & 0 & -0.9607 \\ 0.0822 & 0.1145 & 0.1395 & 0.1568 & 0.1967 & 0 \end{bmatrix},$$

$$\Phi_3 = \begin{bmatrix} 0 & 0.0681 & 0.1122 & 0.1398 & 0.0932 & -0.4318 \\ -0.4701 & 0 & 0.1000 & 0.1233 & 0.0834 & -0.5575 \\ -0.7752 & -0.6903 & 0 & 0.0834 & -0.4318 & -0.8143 \\ -0.9656 & -0.8516 & -0.5758 & 0 & -0.7197 & -0.9656 \\ -0.6437 & -0.5758 & 0.0625 & 0.1042 & 0 & -0.7197 \\ 0.0625 & 0.0807 & 0.1179 & 0.1398 & 0.1042 & 0 \end{bmatrix},$$

$$\Phi_4 = \begin{bmatrix} 0 & 0.1130 & 0.1830 & 0.1830 & 0.0811 & -0.3813 \\ -0.5665 & 0 & 0.1440 & 0.1440 & -0.5106 & -0.6529 \\ -0.9178 & -0.7222 & 0 & 0 & -0.8228 & -0.9178 \\ -0.9178 & -0.7222 & 0 & 0 & -0.82228 & -0.9178 \\ -0.4068 & 0.1018 & 0.1641 & 0.1641 & 0 & -0.4750 \\ 0.0760 & 0.1302 & 0.1830 & 0.1830 & 0.0947 & 0 \end{bmatrix},$$

$$\Phi_5 = \begin{bmatrix} 0 & 0.2200 & -0.3049 & -0.3664 & 0.2458 & 0.2892 \\ -0.56814 & 0 & -0.5838 & -0.6346 & 0.1524 & 0.1898 \\ 0.1181 & 0.2261 & 0 & -0.2829 & 0.2727 & 0.2952 \\ 0.1419 & 0.2458 & 0.1096 & 0 & 0.2832 & 0.3049 \\ -0.6346 & -0.3936 & -0.7041 & -0.7311 & 0 & 0.1670 \\ -0.7468 & -0.4900 & -0.7622 & -0.7986 & -0.4312 & 0 \end{bmatrix},$$

$$\Phi_6 = \begin{bmatrix} 0 & 0.1819 & -0.2446 & -0.4363 & 0.1819 & 0.3816 \\ -0.3298 & 0 & -0.4106 & -0.5469 & 0.1126 & 0.3523 \\ 0.1349 & 0.2265 & 0 & -0.3613 & 0.2265 & 0.3880 \\ 0.2407 & 0.3017 & 0.1993 & 0 & 0.3017 & 0.4438 \\ -0.3298 & -0.2042 & -0.4106 & -0.5469 & 0 & 0.3387 \\ -0.6918 & -0.6387 & -0.7035 & -0.8045 & -0.6141 & 0 \end{bmatrix},$$

$$\Phi_7 = \begin{bmatrix} 0 & -0.5947 & 0.1714 & 0.1390 & 0.1871 & 0.2366 \\ 0.1665 & 0 & 0.2383 & 0.2083 & 0.2504 & 0.2893 \\ -0.6095 & -0.8516 & 0 & -0.4141 & 0.1390 & 0.2090 \\ -0.4965 & -0.7441 & 0.1159 & 0 & 0.1390 & 0.2090 \\ -0.6682 & -0.8945 & -0.4965 & -0.4965 & 0 & 0.1561 \\ -0.8453 & -1.0335 & -0.7465 & -0.7465 & -0.5575 & 0 \end{bmatrix},$$

$$\Phi_8 = \begin{bmatrix} 0 & 0 & -0.4808 & -0.4808 & 0.1933 & 0.2131 \\ 0 & 0 & -0.4808 & -0.4808 & 0.1933 & 0.2131 \\ 0.1248 & 0.1248 & 0 & 0 & 0.2044 & 0.2390 \\ 0.1248 & 0.1248 & 0 & 0 & 0.2044 & 0.2390 \\ -0.7449 & -0.7449 & -0.7876 & -0.7876 & 0 & 0.1485 \\ -0.8211 & -0.8211 & -0.9207 & -0.9207 & -0.5723 & 0 \end{bmatrix},$$

$$\Phi_9 = \begin{bmatrix} 0 & -0.5200 & -0.6440 & -0.7184 & -0.0937 & 0.1721 \\ 0.1692 & 0 & -0.5170 & -0.6072 & 0.1935 & 0.2387 \\ 0.2096 & 0.1683 & 0 & -0.4148 & 0.2187 & 0.2676 \\ 0.2338 & 0.1976 & 0.1350 & 0 & 0.2359 & 0.2870 \\ -0.2880 & -0.5944 & -0.6720 & -0.7248 & 0 & 0.1721 \\ -0.5288 & -0.7333 & -0.8222 & -0.8817 & -0.5288 & 0 \end{bmatrix}.$$

Step 5: Obtain the overall dominance degree of A_i over each alternative A_t

According to Eq. (7.17), the overall dominance degree of A_i over each alternative A_t is

$$\delta = \begin{bmatrix} 0 & -1.1659 & -2.0267 & -2.2012 & 1.3231 & -0.9679 \\ -1.8829 & 0 & -2.1093 & -2.1261 & 0.7704 & -1.3269 \\ -2.2008 & -1.9724 & 0 & -1.2279 & 0.0940 & -1.4808 \\ -2.2359 & -2.0349 & -0.9736 & 0 & -0.1238 & -1.9074 \\ -5.0467 & -4.9848 & -4.5284 & -4.5208 & 0 & -2.2667 \\ -3.2591 & -3.2675 & -3.4460 & -3.5631 & -2.1472 & 0 \end{bmatrix}.$$

Step 6: Rank the alternatives according to the comprehensive predominance degree of A_i

By Eq. (7.18), the comprehensive predominance degree of alternatives are

$$v_1 = 1,\ v_2 = 0.8997,\ v_3 = 0.8927,\ v_4 = 0.8628,\ v_5 = 0,\ v_6 = 0.3473.$$

Hence, the ranking result obtained by the q-RIVOF TODIM is $A_1 > A_2 > A_3 > A_4 > A_6 > A_5$, which means the alternative A_1 is the best medical waste disposal scheme. From Table 7.2, it can be easily seen that A_1 has comparatively higher reliability, higher disposal efficiency, and lower net cost per ton, which are important factors in the disposing process of hazardous medical waste. Hence, the most desirable disposal method A_1 obtained by the proposed q-RIVOFS-TODIM approach is reasonable. This means that by synthesizing the actual decision-making environment and the merits and demerits of each alternative, the company H should select the alternative A_1 as the best medical waste disposal method and A_2 is the second-best solution.

7.5.3 Sensitivity Analysis

It is evident that the attenuation factor of the losses θ plays a vital role in the calculation process of the proposed q-RIVOF TODIM method. Therefore, to investigate how θ influences the comprehensive predominance degree of each alternative and decision-making results, let the value of θ vary from 0.25 to 10 based on the careful analysis of the existing literature, and the corresponding results are listed in Table 7.3 and shown in Fig. 7.2.

As observed from Table 7.3, the final sorting results may change with the variation of parameter value of θ. The different ranking orders of A_1, A_2, A_3, and A_4 are obtained under altering θ. Particularly, A_4 replaces A_1 and becomes the new best alternative when $9.7 \leq \theta \leq 10$. The most probable reason for this change is described as follows. When $\theta < 1$, the losses are amplified and the degree of amplification decreases with the increase of θ while the losses are weakened and the degree of weakening increases in the increase of θ when $1 \leq \theta$.

Figure 7.2 illustrates the comprehensive predominance degree of each alternative under different values of θ. As θ increases gradually, the comprehensive predominance degrees of A_3 and A_4 are always on the rise when $0.25 \leq \theta \leq 10$, and A_4 even have the maximum value compared to other five alternatives when $9.7 \leq \theta \leq 10$.

Table 7.3 The ranking results under different θ

θ	Order
$0.25 \leq \theta < 3.3$	$A_1 > A_2 > A_3 > A_4 > A_6 > A_5$
$3.3 \leq \theta < 4.3$	$A_1 > A_4 > A_2 > A_3 > A_6 > A_5$
$4.3 \leq \theta < 9.7$	$A_1 > A_4 > A_3 > A_2 > A_6 > A_5$
$9.7 \leq \theta \leq 10$	$A_4 > A_1 > A_3 > A_2 > A_6 > A_5$

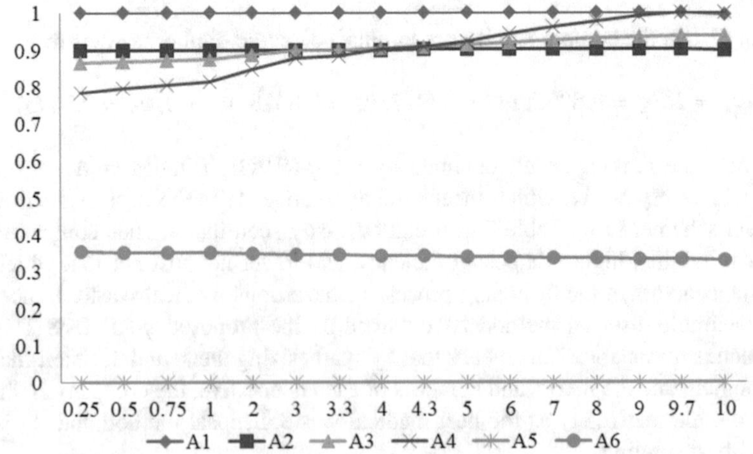

Fig. 7.2 The comprehensive predominance degrees with different θ

The figure for A_1 is constant when $0.25 \leq \theta < 9.7$ and then slightly decreases. Besides, the comprehensive predominance degrees of A_5 are fixed under varying θ. Meanwhile, the attenuation factor of the losses θ poses little impacts on the results of A_2 and A_6.

In this case, there are obvious effects of changing θ on the comprehensive predominance degree of each alternative and sorting results, which reflects the risk preference of decision-makers is able to exert influence on the final decision-making results. Hence, the solving process of proposed q-RIVOFS-TODIM approach, taking into account the psychological behaviors of decision-makers, is close to the real decision-making practice.

7.5.4 Comparative Analysis

In this subsection, the rationality and superiority of the proposed q-RIVOFS-TODIM approach are demonstrated by comparing it with other existing methods.

7.5.4.1 Comparison with the *Q*-RIVOF-VIKOR Method

The q-RIVOF-VIKOR method constructed by Gao et al. [27] is a typical and meaningful tool to tackle decision-making problems under q-RIVOF circumstances. The above illustrative example is solved by the q-RIVOF-VIKOR method, and the detailed procedures are listed as follows:

Step 1. Similar to step 1 in Sect. 7.4

Step 2. Obtain the results of (PIS) A^+ and (NIS) A^-

$$\begin{aligned} A^+ = \{&([0.6, 0.75], [0.6, 0.75]), ([0.7, 0.85], [0.65, 0.75]),\\ &([0.7, 0.8], [0.5, 0.65]), ([0.75, 0.85], [0.5, 0.6]),\\ &([0.7, 0.8], [0.75, 0.85]), ([0.8, 0.95], [0.7, 0.85]),\\ &([0.75, 0.9], [0.65, 0.8]), ([0.65, 0.8], [0.65, 0.8]),\\ &([0.75, 0.85], [0.65, 0.8])\}, \end{aligned}$$

$$\begin{aligned} A^- = \{&([0.3, 0.35], [0.25, 0.35]), ([0.35, 0.5], [0.2, 0.35]),\\ &([0.4, 0.55], [0.15, 0.3]), ([0.45, 0.55], [0.15, 0.3]),\\ &([0.1, 0.15], [0.2, 0.35]), ([0.05, 0.1], [0.05, 0.15]),\\ &([0.15, 0.25], [0.2, 0.35]), ([0.2, 0.3], [0.25, 0.4]),\\ &([0.15, 0.3], [0.15, 0.25])\}. \end{aligned}$$

Step 3. Calculate the mean group score Ψ_i and worst group score Υ_i of the alternative A_i

$$\Psi_1 = 0.6322,\ \Psi_2 = 0.6170,\ \Psi_3 = 0.6897,$$
$$\Psi_4 = 0.7130,\ \Psi_5 = 0.6191,\ \Psi_6 = 0.8058,$$
$$\Upsilon_1 = 0.1527,\ \Upsilon_2 = 0.1409,\ \Upsilon_3 = 0.1534,$$
$$\Upsilon_4 = 0.1482,\ \Upsilon_5 = 0.1336,\ \Upsilon_6 = 0.1910.$$

Step 4. Calculate the results of Θ_i

$$\Theta_1 = 0.2324,\ \Theta_2 = 0.0763,\ \Theta_3 = 0.3608,$$
$$\Theta_4 = 0.3566,\ \Theta_5 = 0.0045,\ \Theta_6 = 1.0000.$$

Step 5. Select the best alternative according to Θ_i

The smaller value the Θ_i is, the best alternative A_i is. Apparently, the ranking result obtained by the q-RIVOF-VIKOR method is $A_5 > A_2 > A_1 > A_4 > A_3 > A_6$, which means that A_5 is the most desirable alternative. This is not in accord with the real practice due to the fact that A_5 is with low reliability, low disposal efficiency and huge health effects.

Hence, the decision-making results gained by the q-RIVOF-VIKOR method is unreliable. The main reason for this error is that the q-RIVOF-VIKOR method is established resort to the closeness of each alternative to ideal solutions which may lead to inverted sequence engendered in VIKOR method. Furthermore, the q-RIVOF-VIKOR method neglects the decision-makers' psychological behaviors, and this is an another source of the decision-making fault.

7.5.4.2 Comparison with the *Q*-RIVOF-Hamy Mean Operator Decision-Making Method

Wang et al. [25] proposed the q-RIVOF-Hamy mean (q-RIVOF-HM) operator to cope with multi-attribute group decision-making problems with high uncertain information. As a representative of fusion method, q-RIVOF-HM operator is applied to handle the q-RIVOF decision matrix shown in Table 7.2.

Step 1. Obtain the fused value of all alternatives

Let $x = 2$, $q = 2$, the fused values of all alternatives are

$$\widetilde{A}_1 = ([0.6036, 0.7216], [0.3436, 0.4756]),$$

$$\widetilde{A}_2 = ([0.5813, 0.7012], [0.3705, 0.4997]),$$
$$\widetilde{A}_3 = ([0.5797, 0.7113], [0.3500, 0.4682]),$$
$$\widetilde{A}_4 = ([0.5892, 0.7233], [0.3423, 0.4744]),$$
$$\widetilde{A}_5 = ([0.4520, 0.5718], [0.5409, 0.6638]),$$
$$\widetilde{A}_6 = ([0.3888, 0.5130], [0.4839, 0.6365]).$$

Step 2. Compute the score value of each fused q-RIVOFNs

$$S_1 = 0.6352,\ S_2 = 0.6106,\ S_3 = 0.6251,$$
$$S_4 = 0.6320,\ S_5 = 0.4495,\ S_6 = 0.4438.$$

Step 3. Rank all alternatives according to the score values

The bigger value the S_i is, the best alternative A_i is. Obviously, the sorting result of alternatives is $A_1 > A_4 > A_3 > A_2 > A_5 > A_6$.

Same as the q-RIVOF-VIKOR method, the q-RIVOF-HM method also neglects decision-makers' attitudes to losses. Although the ranking order acquired by the q-RIVOF-HM method is highly similar with the results, $A_1 > A_4 > A_3 > A_2 > A_6 > A_5$, obtained by the proposed q-RIVOF-TODIM approach when $4.3 \leq \eta < 9.7$, the q-RIVOF-HM method may be inferior to the proposed approach in some particular situations where the psychological behaviors of decision-makers cannot be ignored.

7.5.4.3 Comparison with the TODIM Method Under Other Fuzzy Circumstances

To highlight the preponderance of q-RIVOFSs in describing the sophisticated decision-making problems, the proposed approach is compared with some extended TODIM in other fuzzy environments including IFS [47], interval IFS [48], PFS [49], interval PFS [50], q-ROFS [51], and T-spherical fuzzy sets (T-SFS) [52]. The corresponding comparative results are summarized as below:

- As special cases of q-RIVOFSs, IFS, PFS and q-ROFS are incapable of describing situations where $\mu_A^L(x_i)$, $\mu_A^U(x_i)$, $\upsilon_A^L(x_i)$, $\upsilon_A^U(x_i)$ are interval numbers.
- Interval IFS and interval PFS are also invalid when $(\mu_A^U(x_i)) + (\upsilon_A^U(x_i)) > 1$ and $(\mu_A^U(x_i))^2 + (\upsilon_A^U(x_i))^2 > 1$, respectively.
- T-SFS, a reformative extension of q-ROFS, are also out of work when $\mu_A^L(x_i)$, $\mu_A^U(x_i)$, $\upsilon_A^L(x_i)$, $\upsilon_A^U(x_i)$ are interval numbers.

With the help of the aforesaid comparisons among fuzzy languages, it is effortless to conclude that q-RIVOFSs has remarkable strengths in portraying complex systems.

7.6 Conclusion

This chapter proposes a novel q-RIVOF-TODIM approach for handling intricate decision-making problems. A case about choosing the best medical waste disposal method during COVID-19 epidemic is conducted to demonstrate the effectiveness and rationality of the proposed approach. The sensitivity and comparative analysis results show that this decision-making model has the following major merits:

(i) q-RIVOFSs, generations of many high-efficiency fuzzy languages including FS, IFS, interval IFS, PFS, interval PFS, q-ROFS and T-SFS, are able to deliver more complicated information and have a wider range of applications. It can also acclimatize itself to the varying environments by setting an appropriate value to the parameter q.

(ii) The proposed q-RIVOF-TODIM decision-making model is capable of reflecting decision-makers' risk aversion psychological behaviors by adjusting the parameter value of θ. To be specific, increasing the parameter θ can decrease the degree of amplification or increase the degree of weakening.

(iii) Besides, compared to decision-making methods based on the closeness of each alternative to ideal solutions, the proposed approach is competent in circumventing the occurrence of inverse sequence.

On the other hand, the proposed q-RIVOF-TODIM approach still has some restrictions: (1) This method capitalizes on linear scale transformation to weight all alternatives; however, rescaling via linear transformation may be invalidated occasionally. (2) To better focus on the main research points, the proposed approach pays no attention to time factors, whereas the real decision-making process is likely to be dynamic.

In the future, time factors are suggested to be blended in the proposed q-RIVOF-TODIM approach. Exploring other fusion methods under q-RIVOF circumstances, such as Dombi aggregation [53–55], is another interesting further research direction.

References

1. R.R. Yager, Generalized orthopair fuzzy sets. IEEE Trans Fuzzy Syst. **25**(5), 1222–1230 (2016)
2. L.A. Zadeh, Fuzzy sets. Inf Control. **8**(3), 338–353 (1965)
3. K.T. Atanassov, Intuitionistic fuzzy sets. Fuzzy Sets Syst. **20**(1), 87–96 (1986)
4. M. Suresh, K.A. Prakash, S. Vengataasalam, A new approach for ranking of intuitionistic fuzzy numbers. J. Fuzzy. Ext. Appl. **1**(1), 15–26 (2020)
5. K. Atanassov, G. Gargov, Interval-valued intuitionistic fuzzy sets. Fuzzy. Sets. Syst. **31**, 343–349 (1989)
6. C. Jana, M. Pal, A dynamical hybrid method to design decision making process based on GRA approach for multiple attributes problem. Eng. Appl. Artif. Intell. **100**, 104203–104212 (2021)
7. R.R. Yager, Pythagorean fuzzy subsets, in Proceedings of the Joint IFSA World Congress and NAFIPS Annual Meeting, (IEEE, Edmonton, Canada, 2013), pp. 57–61
8. H. Garg, New exponential operational laws and their aggregation operators for interval-valued Pythagorean fuzzy multicriteria decision-making. Int. J. Intell. Syst. **33**(3), 653–683 (2018)

9. V. Chinnadurai, A. Selvam, Interval valued Pythagorean fuzzy ideals in semigroups. J. Fuzzy. Ext. Appl. **4**(1), 313–322 (2020)
10. Z.L. Yang, T.X. Ouyang, X.L. Fu, X.D. Peng, A decision-making algorithm for online shopping using deep-learning-based opinion pairs mining and q-rung orthopair fuzzy interaction Heronian mean operators. Int. J. Intell. Syst. **35**(5), 783–825 (2020)
11. G.L. Tang, F. Chiclana, P.D. Liu, A decision-theoretic rough set model with q-rung orthopair fuzzy information and its application in stock investment evaluation. Appl. Soft. Comput. **91**, 106212–106226 (2020)
12. H. Garg, CN-q-ROFS: Connection number-based q-rung orthopair fuzzy set and their application to decision-making process. Int. J. Intell. Syst. **36**(7), 3106–3143 (2021)
13. H. Garg, S.M. Chen, Multiattribute group decision making based on neutrality aggregation operators of q-rung orthopair fuzzy sets. Inf. Sci. **517**, 427–447 (2020)
14. M. Riaz, H. Garg, H. Farid, M. Aslam, Novel q-rung orthopair fuzzy interaction aggregation operators and their application to low-carbon green supply chain management. J. Intell. Fuzzy. Syst. **41**(2), 4109–4126 (2021). https://doi.org/10.3233/JIFS-210506
15. X.D. Peng, J.G. Dai, H. Garg, Exponential operation and aggregation operator for q-rung orthopair fuzzy set and their decision-making method with a new score function. Int. J. Intell. Syst. **33**(11), 2255–2282 (2018)
16. H. Garg, A novel trigonometric operation-based q-rung orthopair fuzzy aggregation operator and its fundamental properties. Neural Comput. Appl. **32**(2), 15077–15099 (2020)
17. B.P. Joshi, A. Singh, P.K. Bhatt, K.S. Vaisla, Interval valued q-rung orthopair fuzzy sets and their properties. J. Intell. Fuzzy. Syst. **35**, 5225–5230 (2018)
18. Y. Ju, C. Luo, J. Ma, H. Gao, E.D. Santibanez Gonzalez, A. Wang, Some interval-valued q-rung orthopair weighted averaging operators and their applications to multiple-attribute decision making. Int. J. Intell. Syst. **34**(10), 2584–2606 (2019)
19. J. Gao, Z. Xu, Differential calculus of interval-valued q-rung orthopair fuzzy functions and their applications. Int. J. Intell. Syst. **34**(12), 3190–3219 (2019)
20. H. Gao, Y. Ju, W. Zhang, D. Ju, Multi-attribute decision-making method based on interval-valued q-rung orthopair fuzzy archimedean muirhead mean operators. IEEE Access. **7**, 74300–74315 (2019)
21. H. Garg, A new possibility degree measure for interval-valued q-rung orthopair fuzzy sets in decision-making. Int. J. Intell. Syst. **36**, 526–557 (2021)
22. H. Garg, New exponential operation laws and operators for interval-valued q-rung orthopair fuzzy sets in group decision making process. Neural Comput. Appl. **33**(20), 13937–13963 (2021)
23. Z.L. Yang, J.P. Chang, A multi-attribute decision-making-based site selection assessment algorithm for garbage disposal plant using interval q-rung orthopair fuzzy power Muirhead mean operator. Environ. Res. **193**(2), 110385–110394 (2020)
24. H.X. Li, Y. Yang, Y.C. Zhang, Interval-valued q-rung orthopair fuzzy weighted geometric aggregation operator and its application to multiple criteria decision-making, in 2020 IEEE 15th International Conference of System of Systems Engineering (SoSE), (IEEE, Budapest, HUNGARY, 2020), pp. 429–432
25. J. Wang, H. Gao, G. Wei, Y. Wei, Methods for multiple-attribute group decision making with q-rung interval-valued orthopair fuzzy information and their applications to the selection of green suppliers. Symmetry **11**(1), 56–82 (2019)
26. J. Wang, G.W. Wei, R. Wang, F.E. Alsaadi, T. Hayat, C. Wei, Y. Zhang, J. Wu, Some q-rung interval-valued orthopair fuzzy Maclaurin symmetric mean operators and their applications to multiple attribute group decision making. Int. J. Intell. Syst. **34**(6), 2769–2806 (2019)
27. H. Gao, L.G. Ran, G.W. Wei, C. Wei, J. Wu, VIKOR method for MAGDM based on q-rung interval-valued orthopair fuzzy information and its application to supplier selection of medical consumption products. Int. J. Environ. Res. Public Health **17**(2), 525–538 (2020)
28. C.X. Jin, Y. Ran, G.B. Zhang, Interval-valued q-rung orthopair fuzzy FMEA application to improve risk evaluation process of tool changing manipulator. Appl. Soft. Comput. **104**, 107192–107212 (2021)

29. L.M. Liu, W.Z. Cao, B. Shi, M. Tang, Large-scale green supplier selection approach under a q-rung interval-valued orthopair fuzzy environment. Processes **7**(9), 573–596 (2019)
30. Q. Ding, Y.M. Wang, M. Goh, TODIM dynamic emergency decision-making method based on hybrid weighted distance under probabilistic hesitant fuzzy information. Int. J. Intell. Syst. **23**, 474–491 (2021)
31. A. Mohamadghasemi, A. Hadi-Vencheh, F.H. Lotfi, M. Khalilzadeh, An integrated group FWA-ELECTRE III approach based on interval type-2 fuzzy sets for solving the MCDM problems using limit distance mean. Complex Intell. Syst. **6**, 355–389 (2020)
32. M. Khodadadi-Karimvand, H. Shirouyehzad, Well drilling fuzzy risk assessment using fuzzy FMEA and fuzzy TOPSIS. J. Fuzzy. Ext. Appl. **2**(2), 144–155 (2021)
33. S.S. Lin, S.L. Shen, N. Zhang, A.N. Zhou, An extended TODIM-based model for evaluating risks of excavation system. Acta Geotech. (2021), Published online
34. E. Celik, M. Yucesan, M. Gul, Green supplier selection for textile industry: a case study using BWM-TODIM integration under interval type-2 fuzzy sets. Environ Sci Pollut Res. (2021), Published online
35. V. Arya, S. Kumar, A picture fuzzy multiple criteria decision-making approach based on the combined TODIM-VIKOR and entropy weighted method. Cogn Comput. (2021), Published online
36. N. Liao, G. Wei, X. Chen, TODIM method based on cumulative prospect theory for multiple attributes group decision making under probabilistic hesitant fuzzy setting. Int. J. Fuzzy Syst. (2021), Published online
37. R. Liu, Y.J. Zhu, Y Chen, H.C. Liu, Occupational health and safety risk assessment using an integrated TODIM-PROMETHEE model under linguistic spherical fuzzy environment. Int. J. Intell. Syst. 1–23 (2021)
38. Y. Su, M. Zhao, C. Wei, X.D. Chen, PT-TODIM method for probabilistic linguistic MAGDM and application to industrial control system security supplier selection. Int. J. Fuzzy Syst. (2021), Published online
39. X.P. Hong, X.Y. Bai, Y. Song, Selection of product recycling channels based on extended TODIM method. Expert Syst. Appl. **168**, 114295–114305 (2021)
40. P. Liu, M. Shen, F. Teng, B.Y. Zhu, L.L. Rong, Y.S. Geng, Double hierarchy hesitant fuzzy linguistic entropy-based TODIM approach using evidential theory. Inf. Sci. **547**(1), 223–243 (2020)
41. Z.P. Tian, H.Y. Zhang, J. Wang, J.Q. Wang, X.H. Chen, Multi-criteria decision-making method based on a cross-entropy with interval neutrosophic sets. Int. J. Syst. Sci. **47**(15), 3598–3608 (2015)
42. L. Wang, H. Xue, Integrated decision-making method for heterogeneous attributes based on probabilistic linguistic cross-entropy and priority relations. Entropy **22**(9), 1009–1026 (2020)
43. Z. Hussian, M.S. Yang, Distance and similarity measures of Pythagorean fuzzy sets based on the Hausdorff metric with application to fuzzy TOPSIS. Int. J. Intell. Syst. **34**(10), 2633–2654 (2019)
44. C.E. Shannon, A mathematical theory of communication. Bell Syst. Tech. J. **27**(3), 379–423 (1948)
45. S.B. Nadler, *Hyperspaces of sets* (Marcel Dekker, New York, NY, 1978)
46. Coronavirus Resource Center at Johns Hopkins University of Medicine. COVID-19 Dashboard by the Center for Systems Science and Engineering (CSSE) at Johns Hopkins University (JHU) (2021). https://coronavirus.jhu.edu/map.html
47. R.A. Krohling, A.G.C. Pacheco, A.L.T. Siviero, IF-TODIM: an intuitionistic fuzzy TODIM to multi-criteria decision making. Knowl. Based Syst. **53**, 142–146 (2013)
48. R.A. Krohling, A.G.C. Pacheco, Interval-valued Intuitionistic Fuzzy TODIM. Proced. Comput. Sci. **31**, 236–244 (2014)
49. P.J. Ren, Z.S. Xu, X.J. Gou, Pythagorean fuzzy TODIM approach to multi-criteria decision making. Appl. Soft. Comput. **42**, 246–259 (2016)
50. A. Biswas, B. Sarkar, Interval-valued Pythagorean fuzzy TODIM approach through point operator-based similarity measures for multicriteria group decision making. Kybernetes **48**(3), 496–519 (2019)

51. X. Tian, M. Niu, W. Zhang, L.H. Li, E. Herrera-Viedma, A novel TODIM based on prospect theory to select green supplier with q-rung orthopair fuzzy set. Technol. Econ. Dev. Econ. **27**(2), 1–27 (2020)
52. Y. Ju, Y. Liang, C. Luo, P.W. Dong, E.D.R. Santibanez Gonzalez, A.H. Wang, T-spherical fuzzy TODIM method for multi-criteria group decision-making problem with incomplete weight information. Soft Comput. **25**, 2981–3001 (2021)
53. C. Jana, G. Muhiuddin, M. Pal, Some Dombi aggregation of q-rung orthopair fuzzy numbers in multiple-attribute decision making. Int. J. Intell. Syst. **34**(12), 3220–3240 (2019)
54. C. Jana, G. Muhiuddin, M. Pal, Multi-criteria decision making approach based on SVTrN Dombi aggregation functions. Artif. Intell. Rev. **54**, 3685–3723 (2021)
55. C. Jana, M. Pal, Multi-criteria decision making process based on some single-valued neutrosophic Dombi power aggregation operators. Soft. Comput. **25**, 5055–5072 (2021)

Yu-Dou Yang received the B.S. degree in Financial Management from the School of Economics and Management, Anhui Normal University, Wuhu, China, in 2019. She is currently pursuing the M.S. degree in Management Science and Engineering with Department of Management, Shanghai University, Shanghai, China. Her research interests include fuzzy game theory, uncertainty theory, emergency management, multi-attribute decision making and electric vehicle eco-chain management. She is proficient in the skills of utilizing a variety of computer software (STATA, Maple, Python) for data analysis. She has published 4 academic papers in important academic journals such as *Annals of Operations Research* and *Complex & Intelligent Systems.*

Xue-Feng Ding is an associate professor of Department of Management at Shanghai University. She received the B.S. degree and the M.S. degree from School of Computing at Changchun University of Science and Technology, Changchun, China, and the Ph.D degree from School of Management at University of Shanghai for Science and Technology, Shanghai, China. She was a Postdoctoral Fellow at Tongji University, Shanghai, China. Her research mainly focuses on emergency decision making and crisis management, uncertain reasoning and risk decision making, smart business eco-chain management. She, as the first or corresponding author, has published more than 40 academic papers in several crucial academic journals including *Applied Soft Computing, International Journal of Intelligent Systems, Soft Computing, Computer & Industrial Engineering* and so forth. She has leaded and participated in many national and provincial research projects and obtained high-quality achievements.

Chapter 8
Evidence-Based Cloud Vendor Assessment with Generalized Orthopair Fuzzy Information and Partial Weight Data

R. Krishankumar, Dragan Pamucar, and K. S. Ravichandran

Abstract As the information technology (IT) market booms globally, the urge for technological advancement grows. Cloud computing is a sophisticated technology that offers resources on demand. Due to the increase in computation, firms rely on cloud technology for resource management. Attracted by the abundant need, many cloud vendors evolve in the market, and selecting an apt vendor (CV) becomes complex due to the multiple service factors. Previous studies on CV selection incur lacunae viz., (i) uncertainty was not handled flexibly and (ii) personalized ranking was unavailable based on agent-driven data. Motivated by these lacunae and to glue the same, a scientific model is developed in this paper. A generalized orthopair fuzzy set is adopted for the flexible management of uncertainty and ease of preference sharing. Furthermore, a new mathematical model is formulated for factors' significance assessment, and an evidence-based approximation approach is proposed for ranking CVs based on agent-driven data. Finally, a real case study of CV adoption by an academic institution is provided with a discussion on the merits and limitations of the model from theoretical and statistical perspectives.

Keywords Cloud assessment · Evidence theory · Generalized fuzzy information · Mathematical model

R. Krishankumar
Department of Computer Science and Engineering, Amrita School of Engineering, Amrita Vishwa Vidyapeetham, Coimbatore, TN, India

D. Pamucar
Department of Logistics, Military Academy, University of Defence, 11000 Belgrade, Serbia

K. S. Ravichandran (✉)
Rajiv Gandhi National Institute of Youth Development, Sriperumbudur, TN, India
e-mail: ravichandran20962@gmail.com

H. Garg (ed.), *q-Rung Orthopair Fuzzy Sets*,
https://doi.org/10.1007/978-981-19-1449-2_8

8.1 Introduction

With abundant growth in the IT sector, there is an appetite for resources to handle high-end computation. Cloud computing is an attractive technology that offers resources on-demand through an internet-enabled platform. The pay-as-you-go and pay-on-usage paradigms have turned IT focus into cloud computing. Formally, *cloud computing* is internet-enabled technology that satisfies users' requirements by providing hardware/software as services via a self-reliable independent mode [1]. Generally, the traditional IT ideas cannot adequately satisfy user needs pertaining to resource acquisition, flexible implementation, and security. Modern IT ideas overcome these issues by gaining the advantage of cloud computing [1, 2]. Cloud researchers developed three categories for sharing resources viz., software, infrastructure, and platform as a service, which is generally termed as '*X*' as a service (*X*-aaS) [3]. *Infrastructure* denotes the hardware resources the user needs; *the platform* sits on top of infrastructure and aids in application building; finally, the *software* is a complete application given to the user based on their need. A recent report from the cloud computing market predicted that the compound annual growth rate of cloud by 2025 would be 17.50%, and the idea of work from home (due to the recent pandemic) contributes a significant share in the growth rate. The pandemic boosted the SaaS by 300%, which eventually increased the profit of the cloud market. A recent report claimed that the pandemic increased the revenue for the cloud market from USD 233 billion to USD 295 billion, which indicates the urge for cloud technologies around the globe (https://www.marketsandmarkets.com/Article no. 86,614,844; Dated 31.07.2021). A survey made by Garrison et al. [4] turns out to be apt as close to 50% of the government organization have focused on cloud adoption. Further, internet sources such as IDC and ENISA have claimed substantial revenue growth and cost-cutting IT sectors based on cloud adoption.

These positive notions trigger IT firms to actively adopt cloud technologies, but at the same time, there are some negative notions pertaining to cloud adoption, such as Martens and Teuteberg [5] identified that selection of apt CV is a complex decision problem that is driven by human mindset and other critical factors viz., technological adaptability, environment, and cost. Misra and Mondal [6] stated that nearly 45% of the global stakeholders are perplexed regarding cloud adoption, as they fear integrity, job loss, and trust-related issues. Based on these competing arguments, researchers have strived hard to develop frameworks for CV selection that can maintain a balance between positive and negative notions. Two popular reviews on CV selection from Whaiduzzaman et al. [7] and Sun et al. [8] inferred that (i) uncertainty is an integral part of CV selection owing to the conflicting behavior of factors; (ii) the analytical hierarchy process is commonly used approach for ranking; (iii) factors pose varying significance values that are either directly obtained or determined under unknown information conditions.

Based on these inferences, the following *research lacunae* can be identified viz., (i) flexibility in preference elicitation and uncertainty management are open problems in CV selection; (ii) utilization of partial grading information from agents is another issue; and (iii) finally, personalized ranking of CVs based on agent-driven data lacks in the earlier CV selection frameworks.

Motivated by these research gaps and to glue these lacunae, the following contributions are put forward:

- A generalized orthopair fuzzy set (GOFS) [9] is a generic preference structure that provides a wide window for preference sharing that promotes flexibility to agents' scope and viewpoints. As a specific case, GOFS at $q = 1$ is an intuitionistic fuzzy set, and $q = 2$ is a Pythagorean fuzzy set. In general, the factor q acts as a window that controls the preference flow and the grading from agents. This could facilitate ease of preference sharing on each CV based on a specific factor.
- Sometimes agents have a certain opinion on each factor individually or cumulatively, which are potential information that must be modeled in the framework for rational determination of the significance of each factor. Besides, the agents' reliability (importance) is also a vital parameter to be modeled in the formulation for rational calculation of significance values. For achieving this moto, a mathematical model is formulated that acquires such partial information as inequality constraints.
- Based on the agents' view, a ranking is possible, associated with an agent's perception. Generally, CV selection models aggregate opinions that dilute the idea of personalization. As a result, individual perception is not gathered adequately. To tackle the problem, an evidence-based approximation algorithm is put forward under GOFS, which promotes personalization and provides a holistic ranking based on the personalized rank vectors.
- The methods are integrated into a scientific framework, and the usefulness of the framework is testified by using a real case example of CV selection in an academic institution. Apart from this, a theoretical and statistical comparison is performed with existing models to understand the merits and limitations of the proposal.

Before presenting the methodology and case example in detail, it is essential to understand the decision process. Experts rate the CVs based on different criteria, and GOFS is used for rating. Similarly, experts share their opinion on each criterion in the GOFS fashion. By using the opinion vector, weights of criteria are determined. Some information relating to the importance of criteria is gathered as constraints to the model and solved using an optimization toolbox. Later, expert-driven ranking based on the personalized data from each expert for ranking CVs is put forward.

In this flow, the rest of the paper is structured as follows. A brief literature review on CV selection and GOFS-based decision models is provided in Sect. 8.2. Section 8.3 gave the core contributions of this paper by clearly describing the steps involved in the methods. A case example is demonstrated in Sect. 8.4 to showcase the usefulness of the proposed decision framework. A comparative study is presented in Sect. 8.5 that makes the application and method-oriented comparison with other models. Finally, in Sect. 8.6, conclusions along with future research directions are given.

8.2 Literature Review

A brief review of the CV selection models along with GOFS and the associated decision frameworks is presented in the section.

8.2.1 CV Selection Using Decision Models

Garg et al. [2] determined the cloud factors based on ISO standards and presented the AHP method for CV evaluation. Ghosh et al. [10] facilitated CV selection based on trust and competence measured using novel formulations. Liu et al. [11] gave an integrated model with TOPSIS and Delphi under the fuzzy context to evaluate CVs. Ranjan Kumar et al. [12] made a fuzzy-based AHP-TOPSIS method to assess CVs based on real-life data. Jatoth et al. [3] put forward an integrated model with grey numbers by adopting TOPSIS and AHP methods for CV selection. Kumar et al. [13] gave an AHP-TOPSIS combination for CV selection that aided in weight calculation and ranking CVs by using fuzzy context. Lang et al. [14] utilized the Delphi method for analyzing various factors associated with CV evaluation. Krishankumar et al. [15] gave a new CV evaluation model with intuitionistic fuzzy numbers by extending the VIKOR method. Recently, Mazdari and Khezri [16] prepared a detailed review on service selection based on different decision models and inferred that (i) fuzzy is a sophisticated tool for handling uncertainty and (ii) AHP and TOPSIS are popular methods for service selection. Al-Faifi et al. [17] gave a hybrid framework with K-means clustering and DEMATEL-ANP methods for CV assessment based on smart simulated data. Ramadass et al. [18, 19] made rational CV evaluation with the probabilistic linguistic term by developing new frameworks with COPRAS and PROMETHEE methods. Azadi et al. [20] extended data envelopment analysis in a networked structure for CV selection based on slack-based measures. Dahooie et al. [21] extended the CODAS method to a variant of intuitionistic fuzzy for selecting CVs for an academic sector. Sharma and Sehrawat [22] conducted a SWOT analysis for determining factors associated with cloud adoption based on fuzzy AHP with the DEMATEL approach and inferred controllable and uncontrollable elements for facilitating CV selection. Hussain et al. [23] developed a methodology for optimal service selection, which integrates best–worst and Copeland's approaches for evaluating service in the e-commerce sector by considering the quality of service and experience factors. Hussain et al. [24] put forward a framework called cloud service selection as a service by adopting the best–worst approach in fuzzy context for infrastructure-based CV selection.

8.2.2 GOFS-Based Decision Approaches

Yager [9] came up with a generic structure to provide flexibility to agents in preference elicitation. The structure allowed views from both the positive and negative perception in the form of an orthopair. Yager and Alajlan [25] showed the usage of GOFS in approximate reasoning. Attracted by these two works, researchers developed new mathematical frameworks under GOFS such as aggregation operators [26–36], distance measures [37, 38], entropy measures [39, 40], ranking methods [38–44], similarity measures [45–48], and variants of GOFS [49–53].

Briefly, we review the diverse ideas extended to GOFS that aid in rational decision-making. Further et al. [54] prepared a detailed review of various methods and models under the GOFS context and showcased its usage in the decision process, which motivated the authors to adopt the structure in this paper for CV selection.

Based on the review prepared above, it is observed that (i) GOFS is highly flexible for data/opinion sharing; (ii) most CV models adopt classical fuzzy data; (iii) popular ranking methods are AHP and TOPSIS for CV selection; (iv) partial information are either not gathered or unused during weight calculation. These inferences are framed as challenges in the study that the current research model addresses.

8.3 A New Scientific Framework for CV Selection

The core methodology is presented in this section with an overview of the basic idea that supports building the framework under the GOFS context.

8.3.1 Preliminaries

Let us review some theoretical concepts of GOFS.

Definition 8.1 [55] Fixed set BZ is considered with $DY \subset BZ$ also fixed. Then, $\overline{DY}$ is an intuitionistic fuzzy set in BZ that is given by,

$$\overline{DY} = \left\{bz, \mu_{\overline{DY}}(bz), \upsilon_{\overline{DY}}(bz) | bz \epsilon BZ\right\} \tag{8.1}$$

where $\mu_{\overline{DY}}(bz)$ and $\upsilon_{\overline{DY}}(bz)$ are the belongingness and non-belongingness grades, $\pi_{\overline{DY}}(bz) = 1 - \left(\mu_{\overline{DY}}(bz) + \upsilon_{\overline{DY}}(bz)\right)$ is the indeterminacy grade, $\mu_{\overline{DY}}(bz), \upsilon_{\overline{DY}}(bz), \pi_{\overline{DY}}(bz)$ are in 0 to 1 range and $\mu_{\overline{DY}}(bz) + \upsilon_{\overline{DY}}(bz) \leq 1$.

Definition 8.2 [9] BZ is as before and $bz \in BZ$. Then the GOFS GO on BZ is mathematically defined as,

$$GO = \{bz, \mu_{GO}(bz), \upsilon_{GO}(bz)|bz\epsilon BZ\} \tag{8.2}$$

where $\mu_{GO}(bz), \upsilon_{GO}(bz)$ are in 0 to 1 range and represents the grades of belongingness and non-belongingness, respectively. Moreover, $0 \leq (\mu_{GO}(bz))^q + (\upsilon_{GO}(bz))^q \leq 1$ with $q \geq 1$. At $q = 1$, we get IFS [55] and At $q = 2$, we get PFS [56].

Note 1 We represent $GO = (\mu_j, \upsilon_j)\forall j = 1, 2, \ldots, n$ as generalized orthopair fuzzy number (GOFN). GOFS contains such numbers.

Definition 8.3 [9] Consider two GOFNs, GO_1 and GO_2. Then, some operations are given by,

$$GO_1 \oplus GO_2 = \left(\left(1 - \left(1 - \mu_1^q\right)\left(1 - \mu_2^q\right)\right)^{1/q}, \upsilon_1\upsilon_2\right) \tag{8.3}$$

$$GO_2^\lambda = \left(\mu_2^\lambda, \left(1 - \left(1 - \upsilon_2^q\right)^\lambda\right)^{1/q}\right), \lambda > 0 \tag{8.4}$$

$$\lambda GO_2 = \left(\left(1 - \left(1 - \mu_2^q\right)^\lambda\right)^{1/q}, \upsilon_2^\lambda\right), \lambda > 0 \tag{8.5}$$

$$GO_1 \otimes GO_2 = \left(\mu_1\mu_2, \left(1 - \left(1 - \upsilon_1^q\right)\left(1 - \upsilon_2^q\right)\right)^{1/q}\right) \tag{8.6}$$

$$S(GO_1) = \mu_1^q - \upsilon_1^q \tag{8.7}$$

$$A(GO_1) = \mu_1^q + \upsilon_1^q \tag{8.8}$$

Operations such as addition, power operation, scalar multiplication, multiplication, score, and accuracy measures are provided in Eqs. (8.3–8.8).

8.3.2 *Mathematical Model with GOFS*

Service factors play a vital role in CV selection as they infer the potential and ability of the CV. These factors are diverse and heterogeneous, with conflicting trade-offs among them. Eventually, the relative importance of these factors varies from agent to agent. Common categorizations of weight assessment include (a) fully unknown category and (b) partially known category. The former category is useful when agents do not have any prior information on the weight values of the factors. Still, on the contrary, the latter category is used when agents share certain information on the weights of the factors, which are generally imprecise and vague.

Popular approaches in the former category are the analytical hierarchy process [13], variance measure [11], and entropy measure [16]. The primary limitation of these approaches is that they cannot utilize partial information from agents. To overcome the challenge, constrained optimization models are developed, forming an objective function and some constraints. Here, the constraints represent the partial information from agents, and by adopting optimization packages, weights are determined methodically.

Motivated by these claims, a new mathematical model is formulated with GOFS information, and the procedure is given below.

Step 1: Form an $e \times c$ matrix that collects opinions from agents on each service factor in the form of GOFS information. e agents give their views on c factors.

Step 2: Apply Eq. (8.5) to obtain a weighted accuracy measure by considering reliability values of agents that is a vector of order $1 \times e$. Further, the accuracy measure is determined by applying Eq. (8.8) for the opinion matrix that is also of the order $e \times c$

Step 3: Construct the model for determining the weights of the factors.

Model 1:

$$\text{Min}Z = \sum_{j=1}^{c} sc_j \sum_{k=1}^{e} \left(\left| A(GO_{kj}) - A\left(GO_j^+\right) \right| - \left| A(GO_{kj}) - A\left(GO_j^-\right) \right| \right)$$

Subject to

$$sc_j \geq 0; \sum_{j=1}^{c} sc_j = 1$$

$A\left(GO_j^+\right) = \max_{j \in B}(A(GO_i))\ or\ \min_{j \in C}(A(GO_i))$ and $A\left(GO_j^-\right) = \min_{j \in B}(A(GO_i))\ or\ \max_{j \in C}(A(GO_i))$. B denotes benefit, and C denotes cost.

Model 1 follows a distance norm that is close to human-centric decisions. Also, the idea of choosing points close to the ideal and away from nadir also closely resembles human cognition. These themes from the rationale behind the model. It must be noted that $B + C = c$ and since accuracy measure is calculated, the data points are no longer orthopair, but single values. Based on the equation above for positive and negative ideal values, it is clear that two vectors of order 1 by c are obtained.

8.3.3 Evidence-Based Ranking Algorithm with GOFS

Prioritization of CVs is an essential phase of the framework that aids in choosing CVs for the process. Existing frameworks for CV selection have directly considered the cumulative agents' views for prioritization, but each agent's individual ordering of CVs is not focused. Intuitively, pertaining to human cognition, personalized prioritization provides a sensible view of agents' opinions.

Driven by the intuitive idea, in this section, a new algorithm is presented that considers individual perception during prioritization and effectively handles uncertainty by adopting the concept of evidence theory [57]. Due to the hesitation/vagueness in the decision process, obtaining the complete information is not practical, and hence, an approximation is desirable. To achieve the same, Voorbraak [58] combined the concept of Bayesian approximation with evidence theory that allowed better management of uncertainty with partial information.

An algorithm for the prioritization of CVs based on agents' perception is proposed below.

Algorithm 1: Prioritization based on agents' perception.

Input: Data matrices from agents and factors' weights.

Output: Prioritization order of CVs.

Process:

Begin

(i) Calculate accuracy by using Eq. (8.8) for all data matrices.
(ii) Form weighted GOFS information and total weighted vectors for each data matrix by using Eqs. (8.9, 8.10).
(iii) Calculate Bayesian approximation matrix for each data matrix by using Eq. (8.11).
(iv) Aggregate the Bayesian values by using Eq. (8.12) to form cumulative approximations based on combination rule.
(v) Apply Eq. (8.13) to form the final prioritization order.

End

$$GO_{ij}^{k*} = \frac{1 - \left(1 - A\left(GO_{ij}^{k}\right)\right)^{wc_j}}{\sum_i 1 - \left(1 - A\left(GO_{ij}^{k}\right)\right)^{wc_j}} \tag{8.9}$$

$$TGO_i^k = 1 - \sum_j \left(wc_j . GO_{ij}^{k*}\right) \tag{8.10}$$

$$AB_{ij}^k = \begin{cases} \dfrac{GO_{ij}^{k*}}{TGO_i^k |v|} & if\ v\ is\ singleton \\ 0 & Otherwise \end{cases} \tag{8.11}$$

From Eq. (8.11), Bayesian values are determined, forming e matrices each of order $v \times c$. Cumulative Bayesian values are determined by using Eq. (8.12) that yields vectors of order $1 \times v$.

$$CAB_i^k = \frac{\prod_{j=1}^{c} AB_{ij}^k}{\sum_i \prod_{j=1}^{c} AB_{ij}^k} \tag{8.12}$$

where CAB_i^k is the cumulative Bayesian value of the ith CV from the kth agent.

Finally, Eq. (8.13) is applied to determine the final prioritization of CVs. Based on the individual ordering of CVs pertaining to the agents' opinions, a final order has been obtained that aids in the proper selection of CVs.

$$R_i = \left(1 - \prod_{k=1}^{e}\left(1 - \prod_{k=1}^{e}\left(CAB_{ij}^k\right)^{rp_k}\right)^{we_k}\right)^{\frac{1}{\sum_k (we_k)}} \tag{8.13}$$

where we_k is the weight of the agent and rp_k is the risk parameter of the agent.

R_i is the rank value of the ith CV, and by arranging in descending order, CVs are ordered. High value signifies high preference. Certain fundamental properties that hold for R_i are idempotency, commutativity, monotonicity, and boundedness.

The algorithm starts with calculating accuracy values that transform the GOFS into accuracy measures from each expert (Step i). Later, criteria weights calculated using the mathematical model are used to determine weighted accuracy measures and net weighted accuracy measures (Step ii). Following these measures on each decision matrix (from each expert), Bayesian approximation (Step iii) is calculated for each value in the decision matrices, and these are further aggregated (Step iv) to obtain a cumulative value for aiding in the ranking of CVs. CVs are ranked based on the data from each expert that are finally aggregated (Step v) to obtain net rank values and order of CVs. Through this algorithm, personalized and cumulative rankings of CVs have been obtained that aids in rational decision-making.

Example 1 Three air hostesses were rated based on their hospitality for appraisal purposes. Another factor is the facial charm. Expert chose GOFI for rating. We call three hostesses A, B, and C. A got a rating of (0.5, 0.5), B got a rating of (0.6, 0.5), and C got a rating of (0.4, 0.7) from the expert. For facial charm the expert gives values as (0.7, 0.2), (0.6, 0.4), and (0.6, 0.1). Weights for factors were 0.40 and 0.60. Accuracy measure was given by 0.25, 0.34, 0.41, respectively. Accuracy measure for facial charm is 0.35, 0.28, and 0.22, respectively. AB_{ij}^k for hospitality and facial charm were given by (0.34, 0.48, 0.50) and (0.59, 0.46, 0.30), respectively. CAB_i^k was given by (0.35, 0.38, 0.26) that decides the ranking of an air hostess.

The research model and its working are presented in Fig. 8.1a and b. The model of the proposed framework for CV assessment by developing integrated approaches is shown in Fig. 8.1a, and to support the model with its working, a flowchart is presented in Fig. 8.1b. Agents adopt GOFS as preference information. Agents construct their data matrices along with an opinion vector on each factor. Data matrices share the rating of CVs over each factor. Reliability values of agents are acquired from the officials and are used along with the opinion vector and partial information for determining the weights of the factors. Using this weight vector and the data matrix,

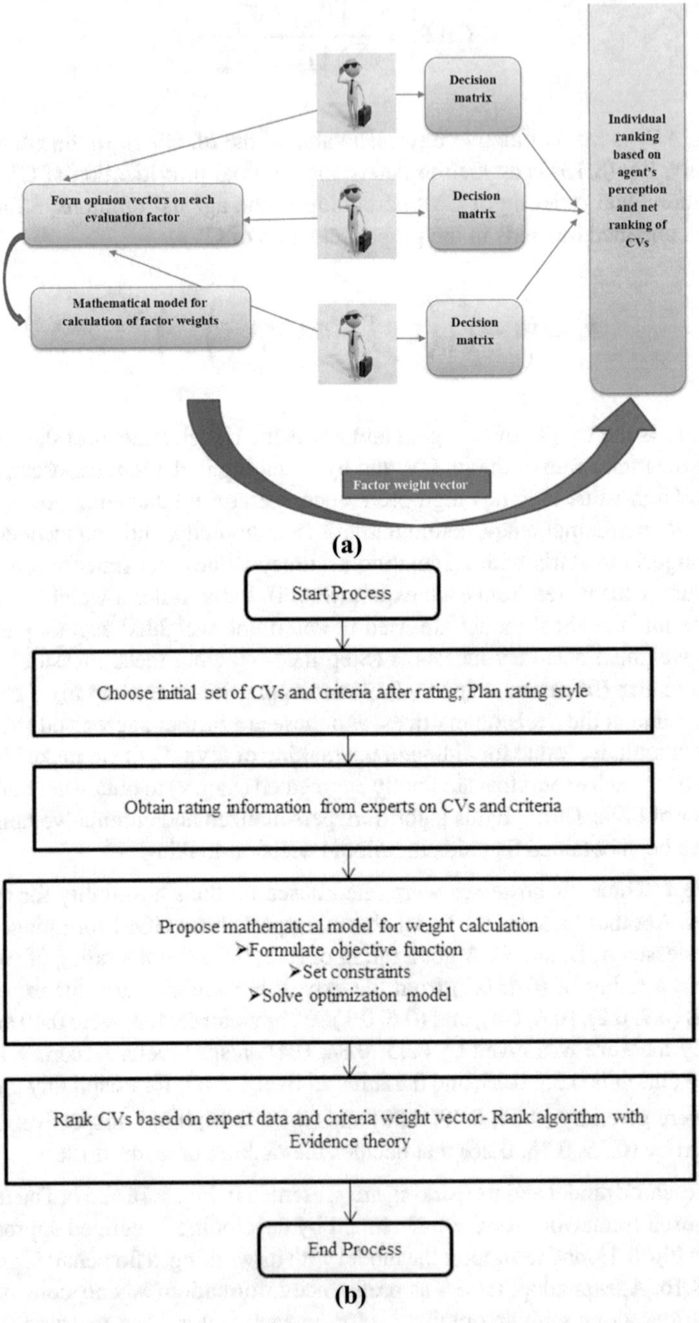

Fig. 8.1 **a** Proposed GOFS decision framework **b** Flowchart for the working of the framework

each agent determines the ranking values of CVs. A final net ranking order is also determined by fusing the individual rank values from agents. Intuitively, the model provides a bidirectional ranking from an individual's perception and group perception that adds to the novelty and usefulness of the developed framework.

8.4 Real Case Example—Selection of CVs

An example is provided in this section to showcase the usefulness of the model in decision-making. For this purpose, a case study of CV selection is developed under the context of an academic institute. Let us consider an institute, T2J (name anonymous), a leading academic institute in Tamil Nadu. It has seven departments with 400+ faculties and 10,000+ students. The institute stands in the frontline for offering quality education to students and uplifting faculty profiles. T2J makes a balanced contribution to teaching and research and encourages R&D activities by training students and faculties with industrial experts. T2J is popular for its workshops, conferences, and seminars globally and primarily focuses on outreach growth.

T2J recently prepared its annual report for 2020. The officials noted that in the 5-year plan, the enormous focus is made on data analytics and research related to discovering interesting patterns from the abundant source of data from the web. To achieve that goal, the institute plans for a new lab and targeted resource personnel. T2J also encouraged the faculties and students to participate in various research activities from their data-centric and analytics-driven departments. As the data grows, storage becomes crucial, and direct host storage using data centers or servers would exponentially increase the cost. With the view of cutting costs, T2J decides to opt for cloud technology. From an initial screening, it can be noted that there are diverse CVs available in the market, and a suitable selection of a CV is a typical decision problem involving multiple criteria. For this purpose, the officials decide to frame three experts, including a senior cloud architect, finance & legal officer, and senior professor from cloud computing. These members initially identify 17 CVs, and based on the SLA verification and peer interaction, 10 CVs are screened. Later, systematic screening is adopted via emails, phone calls, and virtual meets that helped the experts understand the CVs better. Finally, based on voting, four potential CVs are shortlisted. Based on the literature review and web discussion, it is evident that many such models adopt functional factors for CV assessment. Still, the technology, organization, and economics, popularly referred to as the TOE factors in software engineering [11, 15], are crucial for investigating service providers' performance. Additionally, two more factors, viz., the CV and customer relationship profile, are considered in the study to assess CVs. So the panel prepares a decision problem with four CVs that are rated based on five factors. For brevity, we refer the set of experts as $SE = (se_1, se_2, se_3)$ who provide her/his rating on each CV from the set $SV = (sv_1, sv_2, sv_3, sv_4)$ based on the factor set $SC = (sc_1, sc_2, sc_3, sc_4, sc_5)$. The panel decides to use GOFS for rating.

With this backdrop, an algorithm is put forward to rational select CVs by utilizing the proposed methods.

Algorithm 2: CV selection.

Input: Rating data of CVs and factors from experts with dimensions $m \times c$ and $e \times c$, respectively.

Output: Optimal CV from the set of CVs.

Process:

Begin

(i) Determine the weights of the factors by utilizing the $e \times c$ matrix.
(ii) Use the calculated weight vector and the data matrices of order $m \times c$ to rank CVs.
(iii) Perform sensitivity analysis of weights for robustness test.

End

Rating data from experts are provided in Table 8.1. GOFS is used for the rating with $q = 4$. Each agent shares their rating over a particular CV based on a factor. Preference and non-preference grades are provided as an orthopair by each agent that acts as preference information of a CV over an element. Ideally, an orthopair of (0.6, 0.5) infers that the preference is 60%, and the non-preference is 50%. Likewise, Table 8.1 is constructed.

Agents share their opinion (Table 8.2) on each evaluation factor used by the procedure presented in Sect. 8.3.2 for determining weights. These values are used for formulating the objective function based on ideal points and distance norms.

Table 8.1 Rating data from agents in the GOFS form

CVs	Factors for rating				
	sc_1	sc_2	sc_3	sc_4	sc_5
sv_1	(0.5, 0.5)	(0.6, 0.5)	(0.65, 0.55)	(0.6, 0.3)	(0.8, 0.5)
sv_2	(0.5, 0.45)	(0.66, 0.4)	(0.75, 0.55)	(0.6, 0.4)	(0.8, 0.6)
sv_3	(0.65, 0.6)	(0.8, 0.5)	(0.75, 0.5)	(0.7, 0.5)	(0.7, 0.5)
sv_4	(0.7, 0.6)	(0.8, 0.5)	(0.6, 0.7)	(0.8, 0.5)	(0.7, 0.5)
sv_1	(0.55, 0.55)	(0.85, 0.3)	(0.8, 0.4)	(0.85, 0.5)	(0.7, 0.5)
sv_2	(0.65, 0.7)	(0.65, 0.4)	(0.6, 0.5)	(0.6, 0.45)	(0.7, 0.4)
sv_3	(0.6, 0.7)	(0.7, 0.4)	(0.7, 0.5)	(0.5, 0.45)	(0.8, 0.5)
sv_4	(0.7, 0.45)	(0.7, 0.6)	(0.8, 0.4)	(0.55, 0.55)	(0.66, 0.55)
sv_1	(0.85, 0.5)	(0.55, 0.6)	(0.75, 0.35)	(0.55, 0.6)	(0.7, 0.65)
sv_2	(0.7, 0.5)	(0.65, 0.6)	(0.8, 0.45)	(0.65, 0.6)	(0.8, 0.6)
sv_3	(0.65, 0.6)	(0.7, 0.5)	(0.65, 0.45)	(0.7, 0.6)	(0.55, 0.5)
sv_4	(0.6, 0.5)	(0.75, 0.55)	(0.65, 0.6)	(0.6, 0.5)	(0.55, 0.7)

Table 8.2 Opinion of agents on evaluation factors

Agents	Factors for rating				
	sc_1	sc_2	sc_3	sc_4	sc_5
se_1	(0.6, 0.3)	(0.45, 0.5)	(0.55, 0.3)	(0.8, 0.4)	(0.6, 0.5)
se_2	(0.65, 0.45)	(0.65, 0.5)	(0.5, 0.4)	(0.7, 0.35)	(0.5, 0.5)
se_3	(0.7, 0.35)	(0.7, 0.4)	(0.5, 0.6)	(0.6, 0.4)	(0.7, 0.5)

Further, reliability values of agents are acquired from the officials as 0.3, 0.4, and 0.3, respectively, and they are used in the formulation. Figure 8.2 provides the positive and negative ideal points for each factor. Model 1 is constructed using the distance norm that forms objective functions pertaining to each agent's data and constraints. Objective function is given by $0.429sc_1+0.573sc_2+0.447sc_3+0.888sc_4+0.400sc_5$. $sc_1+sc_2+sc_3 \le 0.8$; $0.35 \le sc_3+sc_5 \le 0.4$, $0.3 \le sc_1+sc_4 \le 0.4$; $sc_2+sc_3+sc_5 \le 0.7$ are the constraints. By solving the model via the simplex solvers in the MATLAB® optimization toolbox, the weights are determined as 0.30, 0.20, 0.30, 0.10, 0.10, respectively.

Figure 8.3 shows the Bayesian values calculated using each agent's preference data and Eq. (8.11). In the x-axis, five factors are considered, and for each factor, Bayesian values associated with each CV are shown. These values are used to determine the rank values based on each agent's data, which is further used to determine the net ranking of CVs. Table 8.3 is formed by using Eqs. (8.12, 8.13). The evidence-based algorithm proposed in Sect. 8.3.3 aids in the rank calculation. From the perception of se_1, the rank order is given by $sv_4 \succ sv_3 \succ sv_2 \succ sv_1$. From the perception of se_2, the rank order is given by $sv_4 \succ sv_3 \succ sv_1 \succ sv_2$. Finally, from the perception of se_3, the rank order is given by $sv_2 \succ sv_4 \succ sv_1 \succ sv_3$. The last column of Table 8.3 shows the net rank values obtained from Eq. (8.13), and the order are given

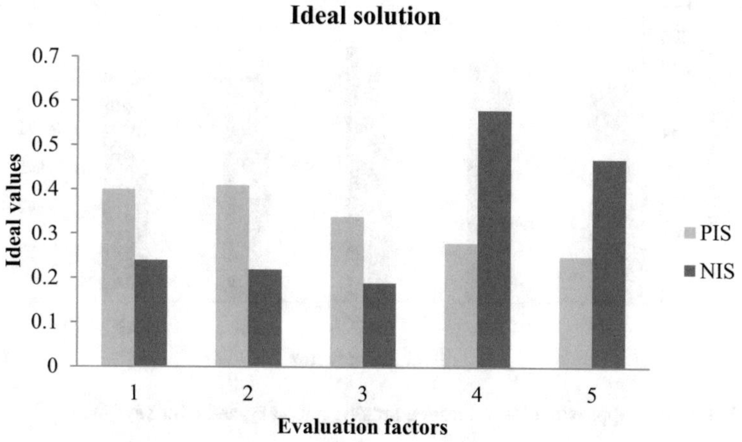

Fig. 8.2 Positive and negative ideal solution—factor wise

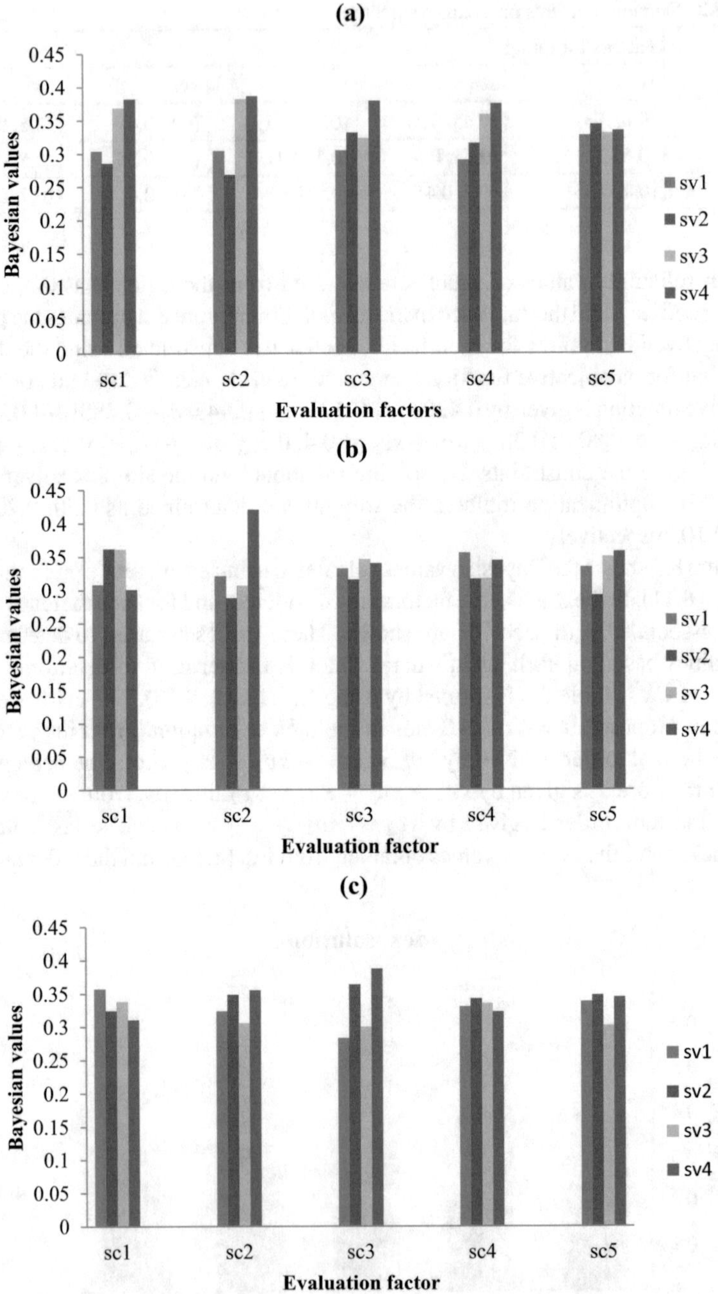

Fig. 8.3 Bayesian approximation values—**a** for se_1; **b** for se_2; and **c** for se_3

Table 8.3 Individual and net ranking of CVs

CVs	CAB_i^1	CAB_i^2	CAB_i^3	R_i
sv_1	0.14897	0.23684	0.22267	0.21904
sv_2	0.15299	0.18383	0.29806	0.24747
sv_3	0.29896	0.25801	0.19087	0.26505
sv_4	0.38895	0.32134	0.28968	0.344

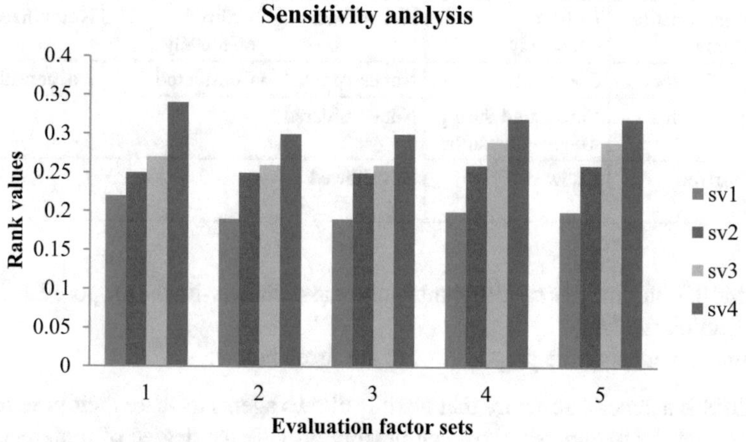

Fig. 8.4 Rank values for different weight sets

by $sv_4 \succ sv_3 \succ sv_2 \succ sv_1$, with sv_4 being the optimal option for the academic institute.

Weights of factors are shifted by using shift operation to understand the effects on the ranking order of CVs. This constitutes the sensitivity analysis process on criteria weights, and we form five test cases based on five factors that are shifted to form a new weight vector. It must be noted from Fig. 8.4 that the proposed framework is *robust and retains the ordering* for different weight sets. Further, it indicates that the model is unaffected by the alterations in weight values.

8.5 Comparative Analysis

The proposed framework is compared with other models for realizing the merits and shortcomings. Both application and method-oriented models are considered for comparison. From the application point of view, models such as Dahooie et al. [21], Sivagami et al. [19], and Hussain et al. [24] are considered. Further, from the method perspective, models such as Liu et al. [29], Krishankumar et al. [43], and Peng et al. [38] are compared with the proposed framework.

Table 8.4 Summary of diverse features—proposed versus other CV models

Features	Models—application			
	Proposed	[21]	[19]	[24]
Data	GOFS	IVIFS	PLTS	Fuzzy set
Subjective randomness	Handled effectively	Moderately handled	N/A	Moderately handled
Elicitation flexibility	Ensured adequately	Moderately ensured	N/A	Moderately ensured
Partial information on factors	Utilized effectively	Not utilized	Utilized effectively	Not utilized
Nature of factors	Considered	Not considered	Considered	Not considered
Agent's weights	Considered during weight calculation	Not considered		
Agent-driven ranking	Allowed	Not allowed		

Table 8.4 summarizes the different features associated with the proposed and other CV selection models.

Some novelties of the proposed work are listed below:

- GOFS is a generic structure that flexibly allows agents to share their preferences. By using the parameter q, we can flexibly provide the degree of preference and non-preference. Subjective randomness is also handled effectively by adopting GOFS.
- Partial information from agents is appropriately utilized and their reliability values to calculate the weights of evaluation factors.
- Consideration is given to the nature of factors during weight calculation. This helps in the rational estimation of weights by adopting distance norms.
- Unlike the other models, the proposed work considers an individual's data to rank based on their perception. Finally, net ranking is obtained by fusing these individual rank values.

Methodically, we experiment with proposed and other models (Liu et al. [29]; Krishankumar et al. [43]; Peng et al. [38]) by feeding the data matrix from agents. Data are fused using a weighted geometric operator for the existing models, and the resultant data are fed. Rank values are obtained for each model. From Fig. 8.5, it is clear that the proposed work produces a *unique ranking order* by effectively considering individual perception, which is lacking in earlier models.

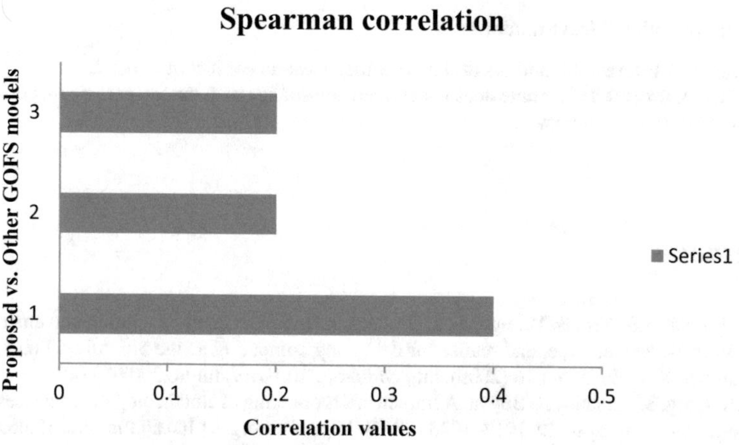

Fig. 8.5 Uniqueness of ranking order—correlation test

8.6 Conclusion

The framework proposed in this paper is a valuable addition to the CV selection models under uncertainty. GOFS is adopted as the preferred structure that promotes flexibility to agents and effectively shares their membership and non-membership grades. Partial information on factors is promptly utilized along with the reliability values of agents during weight estimation, and this framework also promotes individual perception-based ranking. Sensitivity analysis on factor weights reveals that the proposed framework is *robust* to weight alteration. The Spearman correlation infers that the proposed work produces a *unique ranking order* by considering individual agents' perceptions.

Some implications from the academic/institutional perception are (i) the tool is readily available for usage by agents as well as cloud users to assess CVs; (ii) the framework is bidirectional, in the sense that it helps both the vendors and the users; (iii) the framework can also be adopted by other venues where selection is substantial; and (iv) finally, some training must be given to the managers for promoting best usage of the tool. Some shortcomings of the framework are (i) reliability values of agents are directly obtained from officials that could enforce bias and (ii) objective weights are calculated, but subjective weights are not considered.

In the future, plans are made to address these shortcomings. Also, the interval variant of GOFS can be considered a data source, and integrated approaches can be developed to handle uncertainty in a generalized fashion. Further, plans are made to extend the framework for solving sustainable related issues such as electric vehicle charging system selection, carbon dioxide, hydrogen storage technology selection, waste treatment method selection, etc. Finally, learning concepts can be embedded with decision models for large-scale decision-making, such as large-scale selection of CVs by collecting linguistic data from online rating sources.

Compliance with Ethical Standards

- **Conflict of Interest** All authors declare that they have no conflict of interest.
- **Ethical Approval** This article does not contain any studies with human participants or animals performed by any authors.

References

1. R. Buyya, C.S. Yeo, S. Venugopal, J. Broberg, I. Brandic, Cloud computing and emerging IT platforms: vision, hype, and reality for delivering computing as the 5th utility. Futur. Gener. Comput. Syst. **25**, 599–616 (2009). https://doi.org/10.1016/j.future.2008.12.001
2. S.K. Garg, S. Versteeg, R. Buyya, A framework for ranking of cloud computing services. Futur. Gener. Comput. Syst. **29**, 1012–1023 (2013). https://doi.org/10.1016/j.future.2012.06.006
3. C. Jatoth, G.R. Gangadharan, U. Fiore, R. Buyya, SELCLOUD: a hybrid multi-criteria decision-making model for selection of cloud services. Soft Comput. 1–15 (2018). https://doi.org/10.1007/s00500-018-3120-2
4. G. Garrison, R.L. Wakefield, S. Kim, The effects of IT capabilities and delivery model on cloud computing success and firm performance for cloud supported processes and operations. Int. J. Inf. Manage. **35**, 377–393 (2015). https://doi.org/10.1016/j.ijinfomgt.2015.03.001
5. B. Martens, F. Teuteberg, Decision-making in cloud computing environments: a cost and risk-based approach. Inf. Syst. Front. **14**, 871–893 (2012). https://doi.org/10.1007/s10796-011-9317-x
6. S.C. Misra, A. Mondal, Identification of a company's suitability for the adoption of cloud computing and modelling its corresponding return on investment. Math. Comput. Model. **53**, 504–521 (2011). https://doi.org/10.1016/j.mcm.2010.03.037
7. M. Whaiduzzaman, A. Gani, N.B. Anuar, M. Shiraz, M.N. Haque, I.T. Haque, Cloud service selection using multicriteria decision analysis. Sci. World J. (2014). https://doi.org/10.1155/2014/459375
8. L. Sun, H. Dong, F.K. Hussain, O.K. Hussain, E. Chang, Cloud service selection: state-of-the-art and future research directions. J. Netw. Comput. Appl. **45**, 134–150 (2014). https://doi.org/10.1016/j.jnca.2014.07.019
9. R.R. Yager, Generalized orthopair fuzzy sets. IEEE Trans. Fuzzy Syst. **25**, 1222–1230 (2017). https://doi.org/10.1109/TFUZZ.2016.2604005
10. N. Ghosh, S.K. Ghosh, S.K. Das, SelCSP: a framework to facilitate selection of cloud service providers. IEEE Trans. Cloud Comput. **3**, 66–79 (2015). https://doi.org/10.1109/TCC.2014.2328578
11. S. Liu, F.T.S. Chan, W. Ran, Decision making for the selection of cloud vendor: aAn improved approach under group decision-making with integrated weights and objective/subjective attributes. Expert Syst. Appl. **55**, 37–47 (2016). https://doi.org/10.1016/j.eswa.2016.01.059
12. R.R. Kumar, S. Mishra, C. Kumar, Prioritizing the solution of cloud service selection using integrated MCDM methods under fuzzy environment. J. Supercomput. **73**, 4652–4682 (2017). https://doi.org/10.1007/s11227-017-2039-1
13. R.R. Kumar, C. Kumar, A multi criteria decision making method for cloud service selection and ranking. Int. J. Ambient Comput. Intell. **9**, 1–14 (2018). https://doi.org/10.4018/IJACI.2018070101
14. M. Lang, M. Wiesche, H. Krcmar, Criteria for selecting cloud service providers: a Delphi study of quality-of-service attributes. Inf. Manag. **55**, 746–758 (2018). https://doi.org/10.1016/j.im.2018.03.004
15. R. Krishankumar, K.S. Ravichandran, S.K. Tyagi, Solving cloud vendor selection problem using intuitionistic fuzzy decision framework. Neural Comput. Appl. **32**, 589–602 (2018). https://doi.org/10.1007/s00521-018-3648-1

16. M. Masdari, H. Khezri, Service selection using fuzzy multi-criteria decision making: a comprehensive review. Springer, Berlin Heidelberg **12**, 2803–2834 (2020). https://doi.org/10.1007/s12652-020-02441-w
17. A. Al-Faifi, B. Song, M.M. Hassan, A. Alamri, A. Gumaei, A hybrid multi-criteria decision method for cloud service selection from Smart data, Futur. Gener. Comput. Syst. **93**, 43–57 (2019). https://doi.org/10.1016/j.future.2018.10.023
18. S. Ramadass, R. Krishankumar, K.S. Ravichandran, H. Liao, S. Kar, E. Herrera-Viedma, Evaluation of cloud vendors from probabilistic linguistic information with unknown/partial weight values. Appl. Soft Comput. J. **97**, 106801 (2020). https://doi.org/10.1016/j.asoc.2020.106801
19. R. Sivagami, K.S. Ravichandran, R. Krishankumar, V. Sangeetha, S. Kar, X.Z. Gao, D. Pamucar, A scientific decision framework for cloud vendor prioritization under probabilistic linguistic term set context with unknown/partial weight information. Symmetry (Basel). **11**(5), 682 (2019). https://doi.org/10.3390/sym11050682
20. M. Azadi, A. Emrouznejad, F. Ramezani, F.K. Hussain, Efficiency measurement of cloud service providers using network data envelopment analysis. IEEE Trans. Cloud Comput. **32**, 1–12 (2019). https://doi.org/10.1109/TCC.2019.2927340
21. J.H. Dahooie, A.S. Vanaki, N. Mohammadi, Choosing the appropriate system for cloud computing implementation by using the interval-valued intuitionistic fuzzy CODAS multi-attribute decision-making method (case study: faculty of new sciences and technologies of Tehran university). IEEE Trans. Eng. Manag. **42**, 1–14 (2019). https://doi.org/10.1109/TEM.2018.2884866
22. M. Sharma, R. Sehrawat, Quantifying SWOT analysis for cloud adoption using FAHP-DEMATEL approach: evidence from the manufacturing sector. J. Enterp. Inf. Manag. **33**, 1111–1152 (2020). https://doi.org/10.1108/JEIM-09-2019-0276
23. A. Hussain, J. Chun, M. Khan, A novel customer-centric methodology for optimal service selection (MOSS) in a cloud environment. Futur. Gener. Comput. Syst. **105**, 562–580 (2020). https://doi.org/10.1016/j.future.2019.12.024
24. A. Hussain, J. Chun, M. Khan, A novel framework towards viable cloud service selection as a service (CSSaaS) under a fuzzy environment. Futur. Gener. Comput. Syst. **104**, 74–91 (2020). https://doi.org/10.1016/j.future.2019.09.043
25. R.R. Yager, N. Alajlan, Approximate reasoning with generalized orthopair fuzzy sets. Inf. Fusion. **38**, 65–73 (2017). https://doi.org/10.1016/j.inffus.2017.02.005
26. J. Wang, R. Zhang, X. Zhu, Z. Zhou, X. Shang, W. Li, Some q-rung orthopair fuzzy Muirhead means with their application to multiattribute group decision making. J. Intell. Fuzzy Syst. **36**, 1599–1614 (2019). https://doi.org/10.3233/JIFS-18607
27. J. Wang, G. Wei, J. Lu, F.E. Alsaadi, T. Hayat, C. Wei, Y. Zhang, Some q-rung orthopair fuzzy Hamy mean operators in multiple attribute decision-making and their application to enterprise resource planning systems selection. Int. J. Intell. Syst. **34**, 2429–2458 (2019). https://doi.org/10.1002/int.22155
28. M. Riaz, A. Razzaq, H. Kalsoom, D. Pamučar, H.M. Athar Farid, Y.M. Chu, q-Rung orthopair fuzzy geometric aggregation operators based on generalized and group-generalized parameters with application to water loss management. Symmetry (Basel). **12**, 1236 (2020). https://doi.org/10.3390/SYM12081236
29. P. Liu, S.M. Chen, P. Wang, The g-rung orthopair fuzzy power Maclaurin symmetric mean operators, in 2018 10th International Conference on Advanced Computational Intelligence (ICACI), vol. 10, pp. 156–161. (2018). https://doi.org/10.1109/ICACI.2018.8377599
30. P. Liu, J. Liu, Some q-rung orthopai fuzzy bonferroni mean operators and their application to multi-attribute group decision making. Int. J. Intell. Syst. **33**, 315–347 (2018). https://doi.org/10.1002/int.21933
31. H. Garg, A novel trigonometric operation-based q-rung orthopair fuzzy aggregation operator and its fundamental properties. Neural Comput. Appl. **32**, 15077–15099 (2020). https://doi.org/10.1007/s00521-020-04859-x
32. H. Garg, S.M. Chen, Multiattribute group decision making based on neutrality aggregation operators of q-rung orthopair fuzzy sets. Inf. Sci. (NY) **517**, 427–447 (2020). https://doi.org/10.1016/j.ins.2019.11.035

33. X. Peng, J. Dai, H. Garg, Exponential operation and aggregation operator for q-rung orthopair fuzzy set and their decision-making method with a new score function. Int. J. Intell. Syst. **33**, 2255–2282 (2018). https://doi.org/10.1002/int.22028
34. M. Riaz, H. Garg, H.M.A. Farid, M. Aslam, Novel q-rung orthopair fuzzy interaction aggregation operators and their application to low-carbon green supply chain management. J. Intell. Fuzzy Syst. **41**(2), 4109–4126 (2021). https://doi.org/10.3233/jifs-210506
35. Z. Yang, H. Garg, Interaction Power Partitioned Maclaurin symmetric mean operators under q-rung orthopair incertain linguistic information. Int. J. Fuzzy Syst. 40815 (2021). https://doi.org/10.1007/s40815-021-01062-5
36. H. Garg, New exponential operation laws and operators for interval-valued q-rung orthopair fuzzy sets in group decision making process. Neural Comput. Appl. **33**(20), 13937–13963 (2021). https://doi.org/10.1007/s00521-021-06036-0
37. W.S. Du, Minkowski-type distance measures for generalized orthopair fuzzy sets. Int. J. Intell. Syst. **33**, 802–817 (2018). https://doi.org/10.1002/int.21968
38. X. Peng, R. Krishankumar, K.S. Ravichandran, Generalized orthopair fuzzy weighted distance-based approximation (WDBA) algorithm in emergency decision-making. Int. J. Intell. Syst. **34**, 2364–2402 (2019). https://doi.org/10.1002/int.22140
39. T. Mahmood, Z. Ali, Entropy measure and TOPSIS method based on correlation coefficient using complex q-rung orthopair fuzzy information and its application to multi-attribute decision making. Soft Comput. **25**, 1249–1275 (2021). https://doi.org/10.1007/s00500-020-05218-7
40. L. Liu, J. Wu, G. Wei, C. Wei, J. Wang, Y. Wei, Entropy-based GLDS method for social capital selection of a PPP project with q-Rung orthopair fuzzy information. Entropy **22**, 414 (2020). https://doi.org/10.3390/E22040414
41. R. Krishankumar, S. Nimmagadda, A. Mishra, P. Rani, K.S. Ravichandran, A.H. Gandomi, Solving renewable energy source selection problems using a q-rung orthopair fuzzy-based integrated decision-making approach. J. Clean. Prod. **279**, 123329 (2020). https://doi.org/10.1016/j.ygyno.2016.04.081
42. R. Krishankumar, V. Sangeetha, P. Rani, K.S. Ravichandran, A.H. Gandomi, Selection of apt renewable energy source for smart cities using generalized orthopair fuzzy information, in: 2020 IEEE Symposium Series on Computational Intelligence Canberra Australia, vol. 42, pp. 2861–2868 (2020). https://doi.org/10.1109/ssci47803.2020.9308365
43. R. Krishankumar, Y. Gowtham, I. Ahmed, K.S. Ravichandran, S. Kar, Solving green supplier selection problem using q-rung orthopair fuzzy-based decision framework with unknown weight information. Appl. Soft Comput. J. **94**, 106431 (2020). https://doi.org/10.1016/j.asoc.2020.106431
44. R. Krishankumar, K.S. Ravichandran, S. Kar, F. Cavallaro, E.K. Zavadskas, A. Mardani, Scientific decision framework for evaluation of renewable energy sources under q-rung orthopair fuzzy set with partially known weight information. Sustain. **11**, 1–21 (2019). https://doi.org/10.3390/su11154202
45. Y. Donyatalab, E. Farrokhizadeh, S.A. Seyfi Shishavan, Similarity measures of q-rung orthopair fuzzy sets based on square root cosine similarity function. Adv. Intell. Syst. Comput. **1197**, 475–483 AISC (2021). https://doi.org/10.1007/978-3-030-51156-2_55
46. X. Peng, L. Liu, Information measures for q-rung orthopair fuzzy sets. Int. J. Intell. Syst. **34**, 1795–1834 (2019). https://doi.org/10.1002/int.22115
47. N. Jan, L. Zedam, T. Mahmood, E. Rak, Z. Ali, Generalized dice similarity measures for q-rung orthopair fuzzy sets with applications. Complex Intell. Syst. **6**, 545–558 (2020). https://doi.org/10.1007/s40747-020-00145-4
48. D. Liu, X. Chen, D. Peng, Some cosine similarity measures and distance measures between q-rung orthopair fuzzy sets. Int. J. Intell. Syst. **34**, 1572–1587 (2019). https://doi.org/10.1002/int.22108
49. P. Liu, T. Mahmood, Z. Ali, Complex q-rung orthopair fuzzy aggregation operators and their applications in multi-attribute group decision making. Information **11** (2020). https://doi.org/10.3390/info11010005

50. M. Lin, X. Li, L. Chen, Linguistic q-rung orthopair fuzzy sets and their interactional partitioned Heronian mean aggregation operators. Int. J. Intell. Syst. **35**, 217–249 (2020). https://doi.org/10.1002/int.22136
51. B.P. Joshi, A. Singh, P.K. Bhatt, K.S. Vaisala, Interval-valued q -rung orthopair fuzzy sets and their properties. J. Intell. Fuzzy Syst. **35**, 5225–5230 (2018). https://doi.org/10.3233/JIFS-169806
52. H. Garg, A new possibility degree measure for interval-valued q-rung orthopair fuzzy sets in decision-making. Int. J. Intell. Syst. **36**, 526–557 (2021). https://doi.org/10.1002/int.22308
53. H. Garg, CN-q-ROFS: connection number-based q-rung orthopair fuzzy set and their application to decision-making process. Int. J. Intell. Syst. **36**(7), 3106–3143 (2021)
54. X. Peng, Z. Luo, A review of q-rung orthopair fuzzy information: Bibliometrics and future directions. Springer, Netherlands **54**, 3361–3430 (2021). https://doi.org/10.1007/s10462-020-09926-2
55. K.T. Atanassov, Intuitionistic fuzzy sets. Fuzzy Sets Syst. **20**, 87–96 (1986). https://doi.org/10.1016/S0165-0114(86)80034-3
56. R.R. Yager, Pythagorean membership grades in multicriteria decision making. IEEE Trans. Fuzzy Syst. **22**, 958–965 (2014). https://doi.org/10.1109/TFUZZ.2013.2278989
57. K. Sentz, S. Ferson, Combination of evidence in Dempster-Shafer theory. (2002)
58. F. Voorbraak, A computationally efficient approximation of Dempster-Shafer theory. Int. J. Man. Mach. Stud. **30**, 525–536 (1989). https://doi.org/10.1016/S0020-7373(89)80032-X

Dr. R. Krishankumar is currently serving as a faculty in the Department of Computer Science and Engineering, Amrita School of Engineering, Amrita Vishwa Vidyapeetham. His research interest includes multi-criteria decision-making and soft computing. He has published around 50 articles in peer reviewed journals along with articles in conferences and chapters in books published by Elsevier and Springer. He served as reviewer for many top ranked journals and conferences. Also he is in the editorial team of journals encouraging mathematics and computer science themes. His current research focus is pertaining to design of uncertain models for decision-making.

Dr. Dragan Pamucar is an Associate Professor at the University of Defence in Belgrade, the Department of Logistics, Serbia. Dr. Dragan Pamucar obtained his M.Sc. at the Faculty of Transport and Traffic Engineering in Belgrade in 2009, and his Ph.D. degree in Applied Mathematics with specialization in Multi-Criteria Modelling and Soft Computing Techniques at University of Defence in Belgrade, Serbia in 2013. His research interest includes the fields of computational intelligence, multi-criteria decision-making problems, neuro-fuzzy systems, fuzzy, rough and intuitionistic fuzzy set theory, neutrosophic theory, with applications in a wide range of logistics problems. He has published 5 books and over 150 research papers in international SCI indexed journals.

Prof. K. S. Ravichandran is currently working in Rajiv Gandhi National Institute of Youth Development as Registrar. Prior to joining the institute, he served as Associate Dean Research in SASTRA Deemed University. He has 30 years of rigorous teaching and research experience. His area of interest is in computational intelligence, neural networks, decision-making under uncertainty, image processing, and software engineering. He has published more than 100 articles in international SCIE indexed journals. Some of his research contributions appear in IEEE transactions, Elsevier, and Springer venues. He has guided around 10 Ph.D. scholars and more than 50 PG scholars. He has worked on projects from government and private agencies as principal investigator. He recently completed a leadership programme hosted by MHRD—Government of India under the LEAP scheme. He has published articels in many conferences and book chapters.

Chapter 9
Supplier Selection Process Based on CODAS Method Using q-Rung Orthopair Fuzzy Information

Dynhora-Danheyda Ramírez-Ochoa, Luis Pérez-Domínguez, Erwin Adán Martínez-Gómez, Vianey Torres-Argüelles, Harish Garg, and Veronica Sansabas-Villapando

Abstract The process carried out for the selection of suppliers has been considered a critical activity within companies to maintain high competitiveness in the market. Likewise, a variety of factors that belong to the supply chain must be considered. In this way, to select a good supplier has become more complex related to different criteria involved. In this sense, this work presents a numerical case for the selection of providers, which validates the use of the Combinatorial Distance-Based Assessment (CODAS) method with fuzzy information of q-line orthopair in a multi-criteria environment. Similarly, the CODAS method achieves greater precision in the selection process, in addition, by employing fuzzy sets for decision-making in situations with uncertainty and multiple criteria. In this mode, the hybridization of the CODAS method and q-rung orthopair fuzzy set (q-ROFS) represent a proficient tool to deal with supplier appraisal.

D.-D. Ramírez-Ochoa · L. Pérez-Domínguez (✉) · E. A. Martínez-Gómez · V. Torres-Argüelles · V. Sansabas-Villapando
Department of Industrial Engineering and Manufacturing, Doctorate in Technology, Autonomous University of Ciudad Juarez, Ciudad Juarez, Chihuahua, Mexico
e-mail: luis.dominguez@uacj.mx

D.-D. Ramírez-Ochoa
e-mail: dynhora.ramirez@gmail.com; al206592@alumnos.uacj.mx

E. A. Martínez-Gómez
e-mail: emartine@uacj.mx

V. Torres-Argüelles
e-mail: vianey.torres@uacj.mx

V. Sansabas-Villapando
e-mail: al182970@alumnos.uacj.mx

H. Garg
School of Mathematics, Thapar Institute of Engineering and Technology, Deemed University, Patiala 147004, Punjab, India
e-mail: harishg58iitr@gmail.com; harish.garg@thapar.edu

H. Garg (ed.), *q-Rung Orthopair Fuzzy Sets*,
https://doi.org/10.1007/978-981-19-1449-2_9

Keywords q-Rung Orthopair · q-ROFS · CODAS · Green providers · Decision-making · Multi-criteria

9.1 Introduction

Currently, the economy is an issue of great importance for the government and society, in general, that impacts the growth of a country, but another equally important issue has emerged that affects the entire planet: the environmental issue.

Companies consider the environmental issue, not only within the supply chain but also in international environmental protection policies and regulations [14, 38]. That is why the selection of suppliers must consider both the traditional elements (cost, quality, service, delivery, performance history, production capacity and supplier risk management) as well as the management of sustainable products, carbon footprint, management of recycling, use of clean technologies. This means that traditional suppliers become ecological suppliers [5, 25], making it even more complex to make the best decision due to the multiple criteria that must be considered to select the best supplier [35, 38]. Among the elements that must be considered is the reputation and conduct of suppliers in environmental matters, where the situations cannot be determined with a true or false criterion, but with a wide range of answers [15].

These multi-criteria decision-making methods (MCDM) evaluate multiple conditions by means of algorithms and mathematical tools, obtaining the best alternative solution. These methods have the disadvantage that the people make the decisions subject to the weights and evaluate the complexity-uncertainty of the information [34, 35]. Due to this complexity, in 1965, Lotfi Zadeh started a mathematical and technological revolution by introducing a fuzzy logic that had an impact on science and technology [29]. This allowed the analysis of various situations of uncertainty or inaccuracy for the decision-making in situations, which I call fuzzy set (FS) [29, 31].

Since that year, people crated extensions and improvements to this topic, among which we can observe the fuzzy set with interval value (FIV) in which they do not use clear numbers for the degree of membership, but a sub-zero to one closed interval [31]. As well as the intuitive fuzzy set (IFS) that considers the function of degree of membership and that of non-member, that is, a neat number from zero to one [6, 16]. The IFS were with improved the bipolar fuzzy set (BFS) with the degree of positive and negative membership function [28], the Pythagorean fuzzy sets (PFS) for modeling the linguistic terms [27, 33].

Another important improvement is the q-rung orthopair fuzzy set (q-ROFS), which is based on IFS and PFN that presents degrees of membership, non-membership and indeterminacy of decision-makers [17, 27, 39]. Recent enhancements include the T-spherical fuzzy set (TSFS), among which are the evaluation of the performance of search and rescue robots by means of T-spherical Fuzzy Hamacher Aggregation Operators [31, 32]. The hybrid decision-making methods that employ multi-criteria fuzzy data such as the combination of Interval Analytical Hierarchy Process (IAHP) and

Combinative Distance-Based Assessment (CODAS) to prioritize alternative energy storage technologies [24].

Research methods within the commonly used MCDMs include the Technique for Order Preference by Similarity to an Ideal Solution (TOPSIS) that compares the distance of all alternatives with the best and worst solutions [4, 13]. Other method is the Multiple Criteria Optimization and Compromise Solution (VIKOR, VIseKriterijumska Optimizacija I Kompromisno Resenje) that seeks multi-criteria optimization by classifying a set of alternatives against various conflicts [4, 28]. On the other hand, there is the method multi-objective optimization on the basis of ratio analysis (MOORA) that evaluates the ranking of each alternative based on the analysis of reasons [20, 21]. While method combinative distance-based assessment (CODAS), uses the Euclidean Alternative Distance of the Negative Ideal and the Taxicab Distance [1]. In the same way, the Analytic Hierarchy Process (AHP) decomposes the elements in all hierarchies and ascertains the priority of the elements by quantitative judgment for integration and evaluation V [22, 23]. Likewise, a hybrid method has been generated with the advantages of two or more methods [4, 21]. Although the methods created to date present great characteristics that support decision-making, the selection of green suppliers is complicated because they must include uncertain criteria for evaluation and the experience of an expert.

Thus, the main contribution of this document is the development of the **CODAS method in a q-rung orthopair fuzzy set environment in a multi-criteria environment to improve the supplier selection process.**

9.2 q-Rung Orthopair Fuzzy Sets (q-ROFS)

In 2016, Yager introduced an enhancement to MCDM based on intuitive fuzzy set (IFS) and Pythagorean fuzzy sets (PFS), which he called q-rung orthopair fuzzy set (q-ROFS) [9, 12, 26]. Orthopair fuzzy sets are a membership classification of an element x into pairs of values in the unit interval $\langle A^{+}x, A^{-}x\rangle$, where one indicates compatibility with fuzzy set membership and the other provides support against membership [37]. The q-ROFS are a membership classification of an element x into pairs of values in the unit interval $\langle A^{+}x, A^{-}x\rangle$, where one indicates compatibility with fuzzy set membership and the other provides support against membership [10, 37]. Also, the q-ROFS establish the relationship of restrictions between the support for and opposition to the membership [11, 37]:

$$\left(A^{+}(x)\right)^{q} + \left(A^{-}(x)\right)^{q} \leq 1, \quad \text{with}, \quad q \geq 1$$

The q-ROFS are used to solve uncertain or inaccurate decision-making problems, showing in parallel way the degrees of membership, non-membership and indeterminate [17, 26, 27, 34, 38, 40]. Wang and Li [34] and Riaz et al. [26] present proposals for the selection of suppliers using q-ROFS [34], where they are exposed to the use of the weighted average prioritized operator (q-ROFWA) and weighted geometric operator (q-ROFPWG) [26]. And in another work by Riaz et al. [27], suggest a variety of methods are suggested to incorporate q-ROFS values, concluding that the prevailing q-ROFS established aggregation operators with expert decision-makers, so that in different situations the same results cannot be obtained [27]. Taking into account the results of the research carried out by Zhou and Chen [40] where they use the AHP-VIKOR-MRM method in a PFS scenario for the selection of suppliers, they see some limitations in their method and suggest the use of q-ROFS to make decision-making with greater precision [40]. While the work carried out by Garg [8] presents the case of the government of Momentum Jharkhand of the eastern state of India, in which it seeks the five food processing companies that it will create in rural areas to avoid emigration, to which establishes four important criteria to consider [8].

9.2.1 Algebraic Operations q-ROFS

This section defines some basic q-ROFS concepts that are used in the numerical case of this chapter. For convenience, the letter $(\widehat{a})$ represents the membership function and $(\widehat{b})$ represent to non-membership.

Definition 9.1 [8, 19]. Let be a fuzzy set $A = \left\{\left\langle x, \widehat{a}_A(x), \widehat{b}_A(x) | x \in X\right\rangle\right\}$ in a finite universe **X**, where $\widehat{a}_A : X \to [0, 1]$ shows the degree of membership and $\widehat{b}_A : X \to [0, 1]$ indicates the degree of non-membership of the element $x \in X$ to set A the condition $0 \leq (\widehat{a}_A(x)^q + \widehat{b}_A(x)^q) \leq 1, (q \geq 1)$.

The next example gives the degree of indeterminacy:

$$\pi_A(x) = (\widehat{a}_A(x)^q + \widehat{b}_A(x)^q - \widehat{a}_A(x)^q \widehat{b}_A(x)^q)^{\frac{1}{q}}$$

For ease of use, is named $\langle \widehat{a}_A(x), \widehat{b}_A(x) \rangle$ a q-ROFS denoted by $A = \langle \widehat{a}_A, \widehat{b}_A \rangle$. Given three q-ROFS $\alpha = \langle \widehat{a}, \widehat{b} \rangle$, $\alpha_1 = \langle \widehat{a}_1, \widehat{b}_1 \rangle$ and $\alpha_2 = \langle \widehat{a}_2, \widehat{b}_2 \rangle$ then the basic operations can be defined as follows:

$$\alpha_1 \oplus \alpha_2 = \left\langle \left(\widehat{a}_1^q + \widehat{a}_2^q - \widehat{a}_1^q \widehat{a}_2^q\right)^{\frac{1}{q}}, \widehat{b}_1 \widehat{b}_2 \right\rangle \tag{9.1}$$

$$\alpha_1 \otimes \alpha_2 = \langle \widehat{a}_1 \widehat{a}_2, (\widehat{b}_1^q + \widehat{b}_2^q - \widehat{b}_1^q \widehat{b}_2^q)^{\frac{1}{q}} \rangle \tag{9.2}$$

$$\lambda \alpha_1 = \langle (1 - (1 - \widehat{a}_1^q)^{\lambda})^{\frac{1}{q}}, \widehat{b}_1^{\lambda} \rangle \tag{9.3}$$

$$\alpha_1^{\lambda} = \langle \widehat{a}_1^{\lambda}, (1 - (1 - \widehat{b}_1^q)^{\lambda})^{\frac{1}{q}} \rangle \tag{9.4}$$

The subtraction and division operations between a and b are given by $a = \{a_1, a_2\}$ and $b = \{b_1, b_2\}$ for two q-ROFS [7].

$$a \ominus b = \left\langle \left(\frac{a_1^q - b_1^q}{1 - b_1^q} \right)^{\frac{1}{q}}, \frac{a_2}{b_2} \right\rangle, if 0 \leq \frac{a_2}{b_2} \leq \left(\frac{a_1^q - b_1^q}{1 - b_1^q} \right)^{\frac{1}{q}} \leq 1 \tag{9.5}$$

$$a \oslash b = \left\langle \frac{a_1}{b_1}, \left(\frac{a_2^q - b_2^q}{1 - b_2^q} \right)^{\frac{1}{q}} \right\rangle, if 0 \leq \frac{a_1}{b_1} \leq \left(\frac{1 - a_2^q}{1 - b_2^q} \right)^{\frac{1}{q}} \leq 1 \tag{9.6}$$

Definition 9.2 [8, 19] The q-ROFS $A = \langle \widehat{a}_A, \widehat{b}_A \rangle$ that defines a scoring function **S** of **A** where $S(A) \in [-1, 1]$ establishing that the higher the score **S(A)**, the higher the q-ROFS **A**, being as follows:

$$S(A) = \widehat{a}_A^q - \widehat{b}_A^q \tag{9.7}$$

Definition 9.3 [8, 19] A q-ROFS $A = \langle u_A, v_A \rangle$ defines a precision function **H** of **A**, where $H(A) \in [0, 1]$ establishing that the higher the degree of precision **H(A)**, the higher the **Ais**:

$$H(A) = \widehat{a}_A^q + \widehat{b}_A^q \tag{9.8}$$

Theorem 1 *[19] It is defined as a collection of q-ROFS, is defined where it is assumed* $A_k = \langle \widehat{a}_k, \widehat{b}_k \rangle (k = 1, 2, \ldots, n)$

$$q - ROFWA(\tilde{a}_1, \tilde{a}_2 \ldots, \tilde{a}_n) = \left\langle \left(1 - \prod_{k=1}^{n} (1 - \widehat{a}_k^q)^{w_k} \right)^{\frac{1}{q}}, \prod_{k=1}^{n} \widehat{b}_k^{w_k} \right\rangle \tag{9.9}$$

Definition 9.4 [18] Considering $X = \{x_1, x_2, \ldots, x_n\}$ there are two q-ROFS in X:

$$\alpha = \left\{ \langle x_i \mu_\alpha(x_i), \widehat{b}_\alpha(x_i) q \rangle \, | x_i \in X \right\}$$

$$\beta = \left\{ \langle x_i \mu_\beta(x_i), \widehat{b}_\beta(x_i) q \rangle \, | x_i \in X \right\}$$

defining the Euclidean distance measure as:

$$d_{q-ROFS}(\alpha, \beta) = \left(\frac{1}{2n} \sum_{x_i \in X} \left(\left| \mu_\alpha^q(x_i) - \mu_\beta^q(x_i) \right|^2 + \left| \widehat{b}_\alpha^q(x_i) - \widehat{b}_\beta^q(x_i) \right|^2 \right) \right)^{\frac{1}{2}} \tag{9.10}$$

Assuming that wi is the weight of $x \in X$ and $\Sigma_{i=1}^n \omega i = 1 (0 \leq 1, i = 1, 2, \ldots, n)$ the weighted Euclidean distance measure $d_{wqROF}(\alpha, \beta)$ between two q-ROFS αmu and βmu as follows:

$$d_{wqROF}(\alpha, \beta) = \left(\frac{1}{2n} \sum_{x_i \in X} \omega i \left(\left| \mu_\alpha^q(x_i) - \mu_\beta^q(x_i) \right|^2 + \left| \widehat{b}_\alpha^q(x_i) - \widehat{b}_\beta^q(x_i) \right|^2 \right) \right)^{\frac{1}{2}} \tag{9.11}$$

9.3 Combinative Distance-Based Assessment (CODAS)

The CODAS method takes into account characteristics that are not considered in the MCDM, being a combinatorial evaluation model based on distance, that uses the Euclidean distance measures of alternative of the negative ideal (main evaluation measure) and the Taxicab distance, used when the Euclidean distances are very close to each other [2, 3].

The best location of de water desalination plant in Libya has been found with the CODAS method [3]; to find the best supplier within the Libyan Iron and Steel Company (LISCO) in Libya that also include the modified BWM method (Best-Worst method) and MAIRCA (Multi-Attribute Ideal-Real Comparative Analysis), ideal-real comparative analysis of multiple attributes [1]; for the maintenance of process industries using CODAS in a diffuse environment with AHP [23]; evaluation of personnel to establish goals or assign new positions to the personnel of a textile company in Denizli using CRITIC (Criteria Importance Through Inter-criteria Correlation), PSI (Preference Selection Index) and CODAS, where they conclude the results of their investigations with problems and a solution to solve them with the use of fuzzy extensions by the CODAS method [30].

9.3.1 Steps for the CODAS Method

The CODAS method presents a series of steps for decision-making [3], these being the following:

Step 1: Develop the decision-making matrix:

$$X = [x_{ij}]_{nxm} = \begin{pmatrix} x_{11} & x_{12} & \cdots & x_{1m} \\ x_{21} & x_{22} & \cdots & x_{2m} \\ \vdots & \vdots & \ddots & \vdots \\ x_{n1} & x_{n2} & \cdots & x_{nm} \end{pmatrix} \tag{9.12}$$

where $x_{ij} (x_{ij} \geq 0)$ denotes the return value of alternative i in criterion j $(i \in \{1, 2, \ldots, n\})$, and, $(j \in \{1, 2, \ldots, m\})$.

Step 2: Calculate the normalized decision matrix:

Linear normalization of performance values used as given by the equation:

$$x_{ij} = \begin{cases} \frac{x_{ij}}{max x_i x_{ij}} & \text{if } j \in N_b \\ \frac{min x_i x_{ij}}{x_{ij}} & \text{if } j \in N_c \end{cases} \tag{9.13}$$

where N_b and N_c represent the sets of profit criteria, respectively.

Step 3: Calculate the weighted normalized decision matrix:

Equation (9.11) calculates the weighted normalized return values:

$$\beta_{ij} = w_j \widehat{n}_{ij} \tag{9.14}$$

where $wj(0 < w_j < 1)$ denotes the weight of the criterion j, and $\Sigma^m j = 1 w_j$

Step 4: Determine the ideal-negative solution (point), as seen i the equation:

$$\begin{aligned} ns &= [ns_j]_{1xm} \\ ns_j &= minn_i r_{ij} \end{aligned} \tag{9.15}$$

Step 5: Calculate the Euclidean and Taxicab distances from the alternatives of the ideal-negative solution:

$$\begin{aligned} E_j &= \sqrt{\sum_{j=1}^{m} (r_{ij} - ns_j)^2} \\ T_i &= \sum_{j=1}^{m} \left| r_{ij} - ns_j \right| \end{aligned} \tag{9.16}$$

Step 6: Construct the relative evaluation matrix, as seen in the equation:

$$\begin{aligned} R_\alpha &= [h_{ik}]_{nxm} \\ h_{ik} &= (E_i - E_k) + (\psi(E_j - E_k)x(T_i - T_k)) \end{aligned} \tag{9.17}$$

where $k \in \{1, 2, \ldots, n\}$ and ψ denote a threshold function to recognize Euclidean equality:

$$\begin{cases} \psi(x) = 1 \text{ if } & |x| \geq \tau \\ \\ 0\, if\, |x| \prec \tau \end{cases} \tag{9.18}$$

This function, τ is the threshold parameter that can be set by the decision-maker. It is suggested to set this parameter to a value between 0.01 and 0.05. If the difference between the Euclidean distances of two alternatives is less than τ, these two alternatives are also compared with the Taxicab distance.

Step 7: Calculate the evaluation score for each alternative:

$$H_i = \sum_{k=1}^{n} h_{ik} \tag{9.19}$$

Step 8: Rank the alternatives according to the decreasing values of the evaluation score (E), where the alternative with the highest E is the best option among the alternatives.

9.4 CODAS and q-Rung Orthopair Fuzzy Sets for the Supplier Selection Process

The proposed method comprises eight steps that combine the CODAS method with the q-rung orthopair fuzzy set (q-ROFS), as shown in Fig. 9.1.

The steps for the CODAS and q-rung orthopair method are as follows:

Step 1: Develop the decision-making matrix

$$Z = [z_{ij}]_{nxm} = \begin{pmatrix} y_{11} & y_{12} & \cdots & y_{1m} \\ y_{21} & y_{22} & \cdots & y_{2m} \\ \vdots & \vdots & \ddots & \vdots \\ y_{n1} & y_{n2} & \cdots & y_{nm} \end{pmatrix} \tag{9.20}$$

where $z_{ij}(z_{ij} \geq 0)$ denotes the return value of alternative i in criterion j $(i \in \{1, 2, \ldots, n\})$, and , $(j \in \{1, 2, \ldots, m\})$.

Step 2: Calculate the normalized decision matrix:

$$\widehat{n}_{ij} = \begin{cases} \frac{z_{ij}}{maxn_i z_{ij}} & \text{if} \quad j \in N_b \\ \frac{minn_i z_{ij}}{z_{ij}} & \text{if} \quad j \in N_c \end{cases} \tag{9.21}$$

When N_b and N_c represent the sets of profit criteria, respectively.

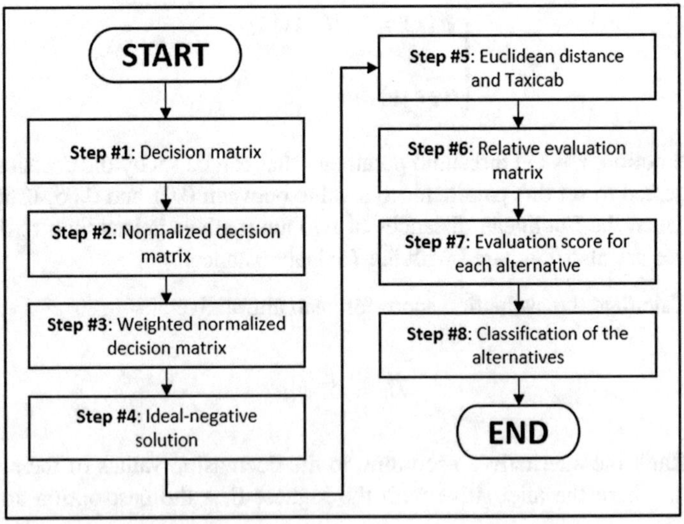

Fig. 9.1 Method CODAS and q-rung orthopair fuzzy set

Step 3: Calculate the weighted normalized decision matrix

The next equation calculates and gives the Weighted normalized performance values (9.28):

$$r_{ij} = w_j \widehat{n}_{ij} \tag{9.22}$$

Step 4: Determine the ideal-negative solution (point), as seen in the equation:

$$\widehat{ns} = [\widehat{ns}_j]_{1xm} \widehat{ns}_j = min_i r_{ij} \tag{9.23}$$

Step 5: Calculate the Euclidean and Taxicab distances from the alternatives of the ideal-negative solution:

$$\widehat{E}j = \left(\frac{1}{2n} \sum_{x_i \in X} \left(\left| \mu_\alpha^q(x_i) - \mu_\beta^q(x_i) \right|^2 + \left| \widehat{b}_\alpha^q(x_i) - \widehat{b}_\beta^q(x_i) \right|^2 \right) \right)^{\frac{1}{2}} \tag{9.24}$$

$$\widehat{T}_i j = \left(\sum_{x_i \in X} \left(\left| \mu_\alpha^q(x_i) - \mu_\beta^q(x_i) \right|^2 + \left| \widehat{b}_\alpha^q(x_i) - \widehat{b}_\beta^q(x_i) \right|^2 \right) \right) \tag{9.25}$$

Step 6: Construct the relative evaluation matrix, as seen in the equation:

$$\begin{gathered} \widehat{h}_{ik} = (\widehat{E_i - \widehat{E}_k}) + (\widehat{\psi(E_j - \widehat{E}_k)})x(\widehat{T}_i - \widehat{T}_k) \\ \widehat{R}_\alpha = [\widehat{h}_{ik}]_{nxm} \end{gathered} \tag{9.26}$$

where $k \in \{1, 2, \ldots, n\}$ and ψ denote a threshold function to recognize Euclidean equality:

$$\begin{cases} \psi(x) = 1 \text{ if } |x| \geq \tau \\ 0 \text{ if } |x| \prec \tau \end{cases} \tag{9.27}$$

Step 7: Calculate the evaluation score for each alternative:

$$\widehat{H}_i = \Sigma_{k=1}^n \widehat{h}_{ik} \tag{9.28}$$

Step 8: Rank the alternatives according to the decreasing values of the evaluation score (E), where the alternative with the highest E is the best option among the alternatives.

9.5 Case Numeric

To show the application of the CODAS and q-rung orthopair method, a selection of the best supplier in a supply chain activity with access to low-cost raw materials and effective delivery times, considering 11 potential suppliers $A = \{A_1, A_2, A_3, \ldots, A_{11}\}$. In which there are six criteria to evaluate, among which there is C_1: cost; C_2: delivery; C_3: quality, C_4: responsiveness; C_5: bonus; and C_6: price. The corresponding weight vector of the attribute is $\omega = (\omega_1, \omega_2, \omega_3, \omega_4, \omega_5, \omega_6) =$ (0.20, 0.20, 0.10, 0.15, 0.15, 0.20).

Step 1: Decision-making decision matrix.

The matrix established with the alternatives with the criteria, as shown in Table 9.1. The matrix established with the alternatives $A = \{A_1, A_2, A_3, \ldots, A_{11}\}$ with the Criteria $C = \{C_1, C_2, C_3, \ldots, C_6\}$, as shown in Table 9.1.

Step 2: Normalized decision matrix

Eq. (9.21) calculates the linear normalization of the yield values, see Table 9.2.

Step 3: Weighted normalized decision matrix.

Equation (9.22) calculates weighted normalized performance values, see Table 9.3, considering $q = 4$.

Step 4: Ideal-negative solution.

To calculate the ideal-negative solution, Eq. (9.23) is used, see Table 9.4.

Table 9.1 Decision matrix

Alternative	C_1		C_2		C_3		C_4		C_5		C_6	
A_1	0.5	0.2	0.3	0.3	0.4	0.3	0.3	0.3	0.4	0.2	0.4	0.2
A_2	0.6	0.3	0.5	0.2	0.4	0.5	0.4	0.5	0.3	0.4	0.5	0.3
A_3	0.3	0.4	0.2	0.5	0.3	0.4	0.3	0.4	0.3	0.2	0.4	0.4
A_4	0.4	0.4	0.2	0.6	0.4	0.4	0.4	0.5	0.2	0.6	0.4	0.5
A_5	0.2	0.6	0.3	0.4	0.4	0.2	0.4	0.3	0.3	0.4	0.5	0.4
A_6	0.5	0.2	0.4	0.2	0.5	0.4	0.5	0.4	0.3	0.3	0.7	0.1
A_7	0.7	0.2	0.4	0.3	0.4	0.3	0.4	0.3	0.4	0.4	0.6	0.1
A_8	0.3	0.5	0.2	0.5	0.5	0.2	0.5	0.2	0.5	0.3	0.6	0.2
A_9	0.4	0.3	0.4	0.4	0.6	0.3	0.6	0.3	0.5	0.4	0.5	0.3
A_{10}	0.2	0.4	0.5	0.4	0.3	0.4	0.3	0.4	0.3	0.4	0.2	0.4
A_{11}	0.4	0.3	0.4	0.3	0.1	0.4	0.4	0.4	0.5	0.3	0.3	0.2

Table 9.2 Normalized decision matrix

Alternative	C_1		C_2		C_3		C_4		C_5		C_6	
A_1	2.500	0.000	1.500	0.284	0.667	0.968	0.500	0.953	0.800	0.932	2.000	0.197
A_2	3.000	0.284	2.500	0.000	0.667	0.982	0.667	0.973	0.600	0.961	2.500	0.299
A_3	1.500	0.394	1.000	0.497	0.500	0.976	0.500	0.964	0.600	0.932	2.000	0.400
A_4	2.000	0.394	1.000	0.598	0.667	0.976	0.667	0.973	0.400	0.978	2.000	0.500
A_5	1.000	0.598	1.500	0.394	0.667	0.958	0.667	0.953	0.600	0.961	2.500	0.400
A_6	2.500	0.000	2.000	0.000	0.833	0.976	0.833	0.964	0.600	0.949	3.500	0.000
A_7	3.500	0.000	2.000	0.284	0.667	0.968	0.667	0.953	0.800	0.961	3.000	0.000
A_8	1.500	0.497	1.000	0.497	0.833	0.958	0.833	0.937	1.000	0.949	3.000	0.197
A_9	2.000	0.284	2.000	0.394	1.000	0.968	1.000	0.953	1.000	0.961	2.500	0.299
A_{10}	1.000	0.394	2.500	0.394	0.500	0.976	0.500	0.964	0.600	0.961	1.000	0.400
A_{11}	2.000	0.284	2.000	0.284	0.167	0.976	0.667	0.964	1.000	0.949	1.500	0.197

Table 9.3 Normalized decision matrix

Alternative	C_1		C_2		C_3		C_4		C_5		C_6	
A_1	1.324	0.000	1.235	0.007	0.384	0.879	0.313	0.824	0.525	0.754	1.284	0.002
A_2	1.358	0.007	1.324	0.000	0.384	0.929	0.424	0.895	0.379	0.852	1.324	0.008
A_3	1.235	0.024	1.000	0.061	0.283	0.907	0.313	0.863	0.379	0.754	1.284	0.026
A_4	1.284	0.024	1.000	0.128	0.384	0.907	0.424	0.895	0.250	0.915	1.284	0.062
A_5	1.000	0.128	1.235	0.024	0.384	0.841	0.424	0.824	0.379	0.852	1.324	0.026
A_6	1.324	0.000	1.284	0.000	0.502	0.907	0.554	0.863	0.379	0.810	1.389	0.000
A_7	1.389	0.000	1.284	0.007	0.384	0.879	0.424	0.824	0.525	0.852	1.358	0.000
A_8	1.235	0.061	1.000	0.061	0.502	0.841	0.554	0.771	1.000	0.810	1.358	0.002
A_9	1.284	0.007	1.284	0.024	1.000	0.879	1.000	0.824	1.000	0.852	1.324	0.008
A_{10}	1.000	0.024	1.324	0.024	0.283	0.907	0.313	0.863	0.379	0.852	1.000	0.026
A_{11}	1.284	0.007	1.284	0.007	0.094	0.907	0.424	0.863	1.000	0.810	1.235	0.002

Table 9.4 Ideal-negative solution

C_1		C_2		C_3		C_4		C_5		C_6	
1.389	0.128	1.324	0.128	0.094	0.841	0.313	0.771	0.250	0.754	1.389	0.062

Step 5: Euclidean and Taxicab Distance

Equation (9.24) calculates the Euclidean distance and Taxicab with Eq. (9.25) of the ideal-negative solution alternatives:

$$\widehat{E}_j = [0.335,\ 0.346,\ 0.358,\ 0.348,\ 0.328,\ 0.314,\ 0.330,\ 0.237,\ 0.152,\ 0.352,\ 0.339]$$

Table 9.5 Relative evaluation matrix

Ra	i_1	i_2	i_3	i_4	i_5	i_6	i_7	i_8	i_9	i_{10}	i_{11}
k_1	0.000	0.011	0.032	0.013	−0.011	−0.021	−0.008	−0.094	−0.213	0.016	0.005
k_2	−0.015	0.000	0.017	0.002	−0.026	−0.032	−0.023	−0.105	−0.225	0.005	−0.009
k_3	−0.032	−0.012	0.000	−0.010	−0.042	−0.044	−0.040	−0.116	−0.239	−0.006	−0.026
k_4	−0.018	−0.002	0.014	0.000	−0.028	−0.035	−0.025	−0.108	−0.226	0.003	−0.012
k_5	0.011	0.018	0.042	0.020	0.000	−0.014	0.003	−0.088	−0.203	0.023	0.015
k_6	0.030	0.033	0.062	0.035	0.019	0.000	0.023	−0.074	−0.186	0.037	0.034
k_7	0.008	0.016	0.040	0.018	−0.003	−0.016	0.000	−0.089	−0.206	0.021	0.012
k_8	0.127	0.115	0.156	0.116	0.116	0.080	0.120	0.000	−0.094	0.118	0.128
k_9	0.213	0.211	0.239	0.212	0.203	0.174	0.206	0.088	0.000	0.213	0.213
k_{10}	−0.022	−0.005	0.009	−0.003	−0.032	−0.038	−0.029	−0.112	−0.227	0.000	−0.016
k_{11}	−0.005	0.007	0.026	0.009	−0.015	−0.026	−0.012	−0.100	−0.213	0.012	0.000

$$\widehat{T_i} = [2.910,\ 2.895,\ 2.869,\ 2.747,\ 2.770,\ 2.720,\ 2.887,\ 1.976,\ 1.114,\ 2.538,\ 2.458]$$

Step 6: The relative evaluation matrix

Equation (9.26) calculates the relative evaluation matrix, considering $\psi = 0.05$, see Table 9.5.

Step 7: Evaluation score

Equation (9.28) calculates the evaluation score for each alternative.

$$\widehat{H_j}\ [-0.270,\ -0.410,\ -0.567,\ -0.437,\ -0.171,\ 0.012,\ -0.200,\ 0.984,\ 1.973,\ -0.476,\ -0.318]$$

Step 8: Ranking of the alternatives

Ranking of the alternatives based on step 7, see Table 9.6.

Ranking Results

$$A_9 \succ A_8 \succ A_6 \succ A_5 \succ A_7 \succ A_1 \succ A_{11} \succ A_2 \succ A_4 \succ A_{10} \succ A_3.$$

9.6 Discussions

To verify the viability and effectiveness of the proposed method, we apply other methods to calculate the same numerical example in this subsection, Table 9.8 shows the results, where their main difference lies in the normalization of the decision

Table 9.6 Ranking of the alternatives

Alternatives	$\widehat{H}_i$	Ranking
A_1	−0.270	6
A_2	−0.410	8
A_3	−0.567	11
A_4	−0.437	9
A_5	−0.171	4
A_6	0.012	3
A_7	−0.200	5
A_8	0.984	2
A_9	1.973	1
A_{10}	−0.476	10
A_{11}	−0.318	7

Table 9.7 Normalized decision matrix

Alternative	C_1		C_2		C_3		C_4		C_5		C_6	
A_1	0.2	0.5	0.3	0.3	0.4	0.3	0.3	0.3	0.4	0.2	0.2	0.4
A_2	0.3	0.6	0.2	0.5	0.4	0.5	0.4	0.5	0.3	0.4	0.3	0.5
A_3	0.4	0.3	0.5	0.2	0.3	0.4	0.3	0.4	0.3	0.2	0.4	0.4
A_4	0.4	0.4	0.4	0.3	0.4	0.4	0.5	0.3	0.5	0.6	0.4	0.2
A_5	0.6	0.2	0.4	0.3	0.4	0.2	0.4	0.3	0.3	0.4	0.4	0.5
A_6	0.2	0.5	0.2	0.4	0.5	0.4	0.5	0.4	0.3	0.3	0.1	0.7
A_7	0.2	0.7	0.3	0.4	0.4	0.3	0.4	0.3	0.4	0.4	0.1	0.6
A_8	0.5	0.3	0.5	0.2	0.5	0.2	0.5	0.2	0.5	0.3	0.2	0.6
A_9	0.3	0.4	0.4	0.4	0.6	0.3	0.6	0.3	0.5	0.4	0.3	0.5
A_{10}	0.4	0.2	0.4	0.5	0.3	0.4	0.3	0.4	0.3	0.4	0.4	0.2
A_{11}	0.3	0.4	0.3	0.4	0.1	0.4	0.4	0.4	0.5	0.3	0.2	0.3

matrix (step 2, Table 9.7) where the Eq. (9.29) is used for the cost and benefit criteria [18, 36].

$$\tilde{r}_{ij} = (\tilde{\widehat{a}}_{ij}, \tilde{\widehat{b}}_{ij}) = \begin{cases} (\widehat{a}_{ij}, \widehat{b}_{ij}) & \text{if for benefit criteria} \\ (\widehat{b}_{ij}, \widehat{a}_{ij}) & \text{if for cost criteria} \end{cases} \tag{9.29}$$

Ranking Results

$$A_1 \succ A_{10} \succ A_7 \succ A_{11} \succ A_6 \succ A_3 \succ A_5 \succ A_2 \succ A_4 \succ A_8 \succ A_9$$

Table 9.9 presents the comparison between the methods considering the following values $q = 4$, $\psi = 0.05$ y $\omega = (0.20, 0.20, 0.10, 0.15, 0.15, 0.20)$.

Table 9.8 Ranking of the alternatives

Alternatives	$\widehat{H}_i$	Ranking
A_1	0.302	1
A_2	−0.082	8
A_3	0.019	6
A_4	−0.202	9
A_5	−0.061	7
A_6	0.098	5
A_7	0.177	3
A_8	−0.245	10
A_9	−0.338	11
A_{10}	0.219	2
A_{11}	0.128	4

Table 9.9 Method comparison

Method	Ranking
Method based on [2, 19]	$A_1 \succ A_{10} \succ A_7 \succ A_{11} \succ A_6 \succ A_3 \succ A_5 \succ A_2 \succ A_4 \succ A_8 \succ A_9$
Proposed method	$A_9 \succ A_8 \succ A_6 \succ A_5 \succ A_7 \succ A_1 \succ A_{11} \succ A_2 \succ A_4 \succ A_{10} \succ A_3$

Considering other weight coefficients for the method based on the articles by Liu et al. [18] and Xu et al. [36], we have the results of Table 9.10 [18, 36].

With this method, Table 9.11 shows the decision results.

Table 9.12 presents the correlation matrix to visualize the degree of linear association that exists between the quantitative variables.

In addition, Fig. 9.2 shows a strong correlation in several of the data. For example, high correlation for q_6 to q_7, q_6 to q_8, q_7 to q_9, q_10 to q_11, q_{10} to q_{15}, finally q_{11} to q_{15}.

To check the validity and reliability of the data, the high correlation that exists between them is verified with Cronbach's alpha to 92.39%, which indicates that the items have a relatively high internal consistency and are highly correlated with each other, see Table 9.13.

9.7 Conclusions

This work presents a hybridization with CODAS and q-rung orthopair fuzzy set. According to numerical experiments, we can demonstrate that this hybridization has the potential to manipulate supplier selection problem in MCDM field. In the same time, our proposal is a guide in a systematic manner, robust and efficient tool for

Table 9.10 Decision results, considering different weights (q), based on [18, 36]

q	Rating
1	$A_{10} \succ A_1 \succ A_{11} \succ A_5 \succ A_3 \succ A_7 \succ A_4 \succ A_8 \succ A_6 \succ A_9 \succ A_2$
2	$A_1 \succ A_{10} \succ A_{11} \succ A_7 \succ A_3 \succ A_6 \succ A_5 \succ A_4 \succ A_8 \succ A_2 \succ A_9$
3	$A_1 \succ A_{10} \succ A_{11} \succ A_7 \succ A_6 \succ A_3 \succ A_5 \succ A_2 \succ A_4 \succ A_8 \succ A_9$
4	$A_1 \succ A_{10} \succ A_7 \succ A_{11} \succ A_6 \succ A_3 \succ A_5 \succ A_2 \succ A_4 \succ A_8 \succ A_9$
5	$A_1 \succ A_{10} \succ A_7 \succ A_6 \succ A_{11} \succ A_3 \succ A_2 \succ A_5 \succ A_4 \succ A_8 \succ A_9$
6	$A_1 \succ A_{10} \succ A_7 \succ A_6 \succ A_{11} \succ A_3 \succ A_2 \succ A_5 \succ A_4 \succ A_8 \succ A_9$
7	$A_1 \succ A_{10} \succ A_7 \succ A_6 \succ A_{11} \succ A_3 \succ A_2 \succ A_5 \succ A_4 \succ A_8 \succ A_9$
8	$A_1 \succ A_7 \succ A_{10} \succ A_6 \succ A_{11} \succ A_3 \succ A_2 \succ A_5 \succ A_4 \succ A_8 \succ A_9$
9	$A_1 \succ A_7 \succ A_{10} \succ A_6 \succ A_{11} \succ A_3 \succ A_2 \succ A_5 \succ A_4 \succ A_8 \succ A_9$
10	$A_1 \succ A_7 \succ A_{10} \succ A_6 \succ A_3 \succ A_{11} \succ A_2 \succ A_5 \succ A_4 \succ A_8 \succ A_9$
11	$A_1 \succ A_7 \succ A_{10} \succ A_6 \succ A_3 \succ A_{11} \succ A_2 \succ A_5 \succ A_4 \succ A_8 \succ A_9$
12	$A_1 \succ A_7 \succ A_{10} \succ A_6 \succ A_3 \succ A_{11} \succ A_2 \succ A_5 \succ A_4 \succ A_8 \succ A_9$
13	$A_1 \succ A_7 \succ A_{10} \succ A_6 \succ A_3 \succ A_2 \succ A_{11} \succ A_5 \succ A_4 \succ A_8 \succ A_9$
14	$A_1 \succ A_7 \succ A_{10} \succ A_6 \succ A_2 \succ A_3 \succ A_{11} \succ A_5 \succ A_4 \succ A_8 \succ A_9$
15	$A_1 \succ A_7 \succ A_{10} \succ A_6 \succ A_2 \succ A_3 \succ A_{11} \succ A_5 \succ A_4 \succ A_8 \succ A_9$

Table 9.11 Decision results of the proposed method, considering different weights

q	Rating
1	$A_9 \succ A_8 \succ A_{10} \succ A_{11} \succ A_3 \succ A_5 \succ A_4 \succ A_1 \succ A_6 \succ A_7 \succ A_2$
2	$A_9 \succ A_8 \succ A_{11} \succ A_{10} \succ A_5 \succ A_6 \succ A_1 \succ A_3 \succ A_7 \succ A_4 \succ A_2$
3	$A_9 \succ A_8 \succ A_6 \succ A_{11} \succ A_5 \succ A_7 \succ A_1 \succ A_{10} \succ A_4 \succ A_2 \succ A_3$
4	$A_9 \succ A_8 \succ A_6 \succ A_5 \succ A_7 \succ A_1 \succ A_{11} \succ A_2 \succ A_4 \succ A_{10} \succ A_3$
5	$A_9 \succ A_8 \succ A_6 \succ A_7 \succ A_5 \succ A_1 \succ A_2 \succ A_{11} \succ A_4 \succ A_{10} \succ A_3$
6	$A_9 \succ A_8 \succ A_6 \succ A_7 \succ A_5 \succ A_1 \succ A_2 \succ A_4 \succ A_{11} \succ A_{10} \succ A_3$
7	$A_9 \succ A_8 \succ A_6 \succ A_7 \succ A_5 \succ A_1 \succ A_2 \succ A_4 \succ A_{11} \succ A_{10} \succ A_3$
8	$A_9 \succ A_8 \succ A_6 \succ A_7 \succ A_5 \succ A_1 \succ A_2 \succ A_4 \succ A_{11} \succ A_{10} \succ A_3$
9	$A_9 \succ A_8 \succ A_6 \succ A_7 \succ A_5 \succ A_1 \succ A_2 \succ A_4 \succ A_{10} \succ A_{11} \succ A_3$
10	$A_9 \succ A_8 \succ A_6 \succ A_7 \succ A_5 \succ A_1 \succ A_2 \succ A_4 \succ A_{10} \succ A_3 \succ A_{11}$
11	$A_9 \succ A_8 \succ A_6 \succ A_7 \succ A_5 \succ A_1 \succ A_2 \succ A_4 \succ A_{10} \succ A_3 \succ A_{11}$
12	$A_9 \succ A_8 \succ A_6 \succ A_7 \succ A_5 \succ A_1 \succ A_2 \succ A_4 \succ A_{10} \succ A_3 \succ A_{11}$
13	$A_9 \succ A_8 \succ A_6 \succ A_7 \succ A_5 \succ A_1 \succ A_2 \succ A_4 \succ A_{10} \succ A_3 \succ A_{11}$
14	$A_9 \succ A_8 \succ A_6 \succ A_7 \succ A_5 \succ A_1 \succ A_2 \succ A_4 \succ A_{10} \succ A_3 \succ A_{11}$
15	$A_9 \succ A_8 \succ A_6 \succ A_7 \succ A_5 \succ A_1 \succ A_2 \succ A_4 \succ A_{10} \succ A_3 \succ A_{11}$

Table 9.12 Correlation matrix

q	q_1	q_2	q_3	q_4	q_5	q_6	q_7	q_8	q_9	q_{10}	q_{11}	q_{12}	q_{13}	q_{14}	q_{15}
q_1	1														
q_2	0.864														
q_3	0.345	0.591													
q_4	0.327	−0.027	−0.382												
q_5	0.227	0.227	0.464	0.227											
q_6	0.545	0.482	0.082	0.355	0.555										
q_7	0.545	0.482	0.082	0.355	0.555	1									
q_8	0.545	0.482	0.082	0.355	0.555	1	1								
q_9	0.555	0.455	0.064	0.409	0.609	0.991	0.991	0.991							
q_{10}	0.191	0.309	0.136	−0.1	0.1	0.482	0.482	0.482	0.418						
q_{11}	0.191	0.309	0.136	−0.1	0.1	0.482	0.482	0.482	0.418	1					
q_{12}	0.191	0.309	0.136	−0.1	0.1	0.482	0.482	0.482	0.418	1	1				
q_{13}	0.191	0.309	0.136	−0.1	0.1	0.482	0.482	0.482	0.418	1	1	1			
q_{14}	0.191	0.309	0.136	−0.1	0.1	0.482	0.482	0.482	0.418	1	1	1	1		
q_{15}	0.191	0.309	0.136	−0.1	0.1	0.482	0.482	0.482	0.418	1	1	1	1	1	1

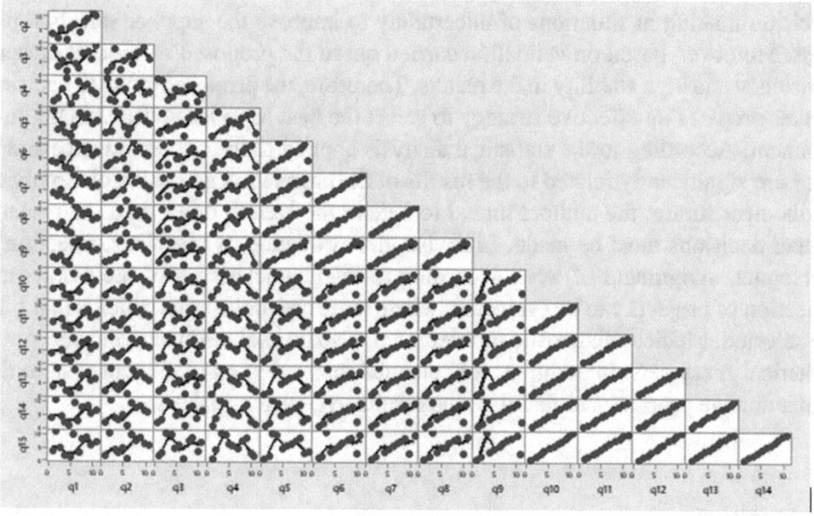

Fig. 9.2 Frequency graph

Table 9.13 Item and total statistics

Variable	Count	Mean	Std. dev.
q_1	11	6	3.317
q_2	11	6	3.317
q_3	11	6	3.317
q_4	11	6	3.317
q_5	11	6	3.317
q_6	11	6	3.317
q_7	11	6	3.317
q_8	11	6	3.317
q_9	11	6	3.317
q_{10}	11	6	3.317
q_{11}	11	6	3.317
q_{12}	11	6	3.317
q_{13}	11	6	3.317
q_{14}	11	6	3.317
q_{15}	11	6	3.317
Total	11	90	34.621
Cronbachs alpha =			0.9239

decision-making in situations of uncertainty to improve the supplier selection process. Moreover, based on validation carried out to the proposed method, it presents efficiency and high stability in the results. Therefore, the proposed guide to decision-maker prepares an effective strategy to select the best supplier under MCDM environment. According to the statistical analysis applied to the numerical experiment, they are significantly related to the results of the correlation and alpha of Cronbach. In the near future, the authors intend to extend the idea to other fields of research where decisions must be made. Likewise, the evaluation of machines, selection of personnel, assignment of work according to the characteristics of the worker and selection of projects can be evaluated, where there are multi-criteria fields that must be selected. Medical diagnostic problems can also be analyzed using the concept of spherical fuzzy sets. In addition, it is planned to develop an intelligent system that can automate processes to facilitate the results to decision-makers.

References

1. I. Badi, A.M. Abdulshahed, A. Shetwan, A case study of supplier selection for a steelmaking company in Libya by using the combinative distance-based assessment (codas) model. Decis. Mak.: Appl. Manag. Eng. **1**(1), 1–12 (2018). https://doi.org/10.31181/dmame180101b
2. I. Badi, M. Ballem, Supplier selection using the rough BWM-MAIRCA model: a case study in pharmaceutical supplying in Libya. Decis. Mak.: Appl. Manag. Eng. **1**(2), 16–33 (2018). https://doi.org/10.31181/dmame180101b
3. I. Badi, M. Ballem, A. Shetwan, Site selection of desalination plant in libya by using combinative distance-based assessment (codas) method. Int. J. Qual. Res. **12**(3) (2018). https://doi.org/10.18421/IJQR12.03-04
4. N. Banaeian, H. Mobli, B. Fahimnia, I.E. Nielsen, M. Omid, Green supplier selection using fuzzy group decision making methods: a case study from the agri-food industry. Comput. Oper. Res. **89**, 337–347 (2018). https://doi.org/10.1016/j.cor.2016.02.015
5. E.N. Bozgeyik, M.Türkay, A multi-objective optimization approach for sustainable supply chains incorporating business strategy, in 2019 International Symposium on Advanced Electrical and Communication Technologies (ISAECT) (IEEE, 2019), pp. 1–6. https://doi.org/10.1109/ISAECT47714.2019.9069702
6. K.H. Chang, C.H. Cheng, A risk assessment methodology using intuitionistic fuzzy set in FMEA. Int. J. Syst. Sci. **41**(12), 1457–1471 (2010). https://doi.org/10.1080/00207720903353633
7. W.S. Du, Research on arithmetic operations over generalized orthopair fuzzy sets. Int. J. Intell. Syst. **34**(5), 709–732 (2019). https://doi.org/10.1002/int.22073
8. H. Garg, A novel trigonometric operation-based q-rung orthopair fuzzy aggregation operator and its fundamental properties. Neural Comput. Appl. **32**(18), 15077–15099 (2020). https://doi.org/10.1007/s00521-020-04859-x
9. H. Garg, Cn-q-rofs: connection number-based q-rung orthopair fuzzy set and their application to decision-making process. Int. J. Int. Syst. **36**(7), 3106–3143 (2021). https://doi.org/10.1002/int.22406
10. H. Garg: New exponential operation laws and operators for interval-valued q-rung orthopair fuzzy sets in group decision making process. Neural Comput. Appl. 1–27 (2021). https://doi.org/10.1007/s00521-021-06036-0
11. H. Garg, A new possibility degree measure for interval-valued q-rung orthopair fuzzy sets in decision-making. Int. J. Intell. Syst. **36**(1), 526–557 (2021). https://doi.org/10.1002/int.22308

12. H. Garg, Z. Ali, Z. Yang, T. Mahmood, S. Aljahdali, Multi-criteria decision-making algorithm based on aggregation operators under the complex interval-valued q-rung orthopair uncertain linguistic information. J. Intell. Fuzzy Syst. Pre-press (Preprint), 1–30 (2021). https://doi.org/10.3233/JIFS-210442
13. M.K. Ghorabaee, M. Amiri, E.K. Zavadskas, R. Hooshmand, J. Antuchevičienė, Fuzzy extension of the codas method for multi-criteria market segment evaluation. J. Bus. Econ. Manag. **18**(1), 1–19 (2017). https://doi.org/10.3846/16111699.2016.1278559
14. A. Gustina, A.Y. Ridwan, M.D. Akbar, Multi-criteria decision making for green supplier selection and evaluation of textile industry using fuzzy axiomatic design (FAD) method, in *2019 5th International Conference on Science and Technology (ICST)*, vol. 1 (IEEE, 2019), pp. 1–6. https://doi.org/10.1109/ICST47872.2019.9166253
15. K. Kang, S. Gao, T. Gao, J. Zhang, Pricing and financing strategies for a green supply chain with a risk-averse supplier. IEEE Access **9**, 9250–9261 (2021). https://doi.org/10.1109/ACCESS.2021.3050130
16. M. Kaushal, R. Solanki, Q.D. Lohani, P.K. Muhuri, A novel intuitionistic fuzzy set generator with application to clustering, in *2018 IEEE International Conference on Fuzzy Systems (FUZZ-IEEE)* (2018), pp. 1–8. https://doi.org/10.1109/FUZZ-IEEE.2018.8491602
17. H. Liao, H. Zhang, C. Zhang, X. Wu, A. Mardani, A. Al-Barakati, A q-rung orthopair fuzzy glds method for investment evaluation of be angel capital in China. Technol. Econ. Dev. Econ. **26**(1), 103–134 (2020). https://doi.org/10.3846/tede.2020.11260
18. D. Liu, X. Chen, D. Peng, Some cosine similarity measures and distance measures between q-rung orthopair fuzzy sets. Int. J. Intell. Syst. **34**(7), 1572–1587 (2019). https://doi.org/10.1002/int.22108
19. P. Liu, P. Wang, Some q-rung orthopair fuzzy aggregation operators and their applications to multiple-attribute decision making. Int. J. Intell. Syst. **33**(2), 259–280 (2018). https://doi.org/10.1002/int.21927
20. A.I. Lubis, P. Sihombing, E.B. Nababan, Comparison saw and MOORA methods with attribute weighting using rank order centroid in decision making, in *2020 3rd International Conference on Mechanical, Electronics, Computer, and Industrial Technology (MECnIT)* (IEEE, 2020), pp. 127–131. https://doi.org/10.1109/MECnIT48290.2020.9166640
21. S.V.B. Manurung, F.G.N. Larosa, I.M.S. Simamora, A. Gea, E.R. Simarmata, A. Situmorang, Decision support system of best teacher selection using method moora and saw, in *2019 International Conference of Computer Science and Information Technology (ICoSNIKOM)* (IEEE, 2019), pp. 1–6. https://doi.org/10.1109/ICoSNIKOM48755.2019.9111550
22. M. Mesran, S. Suginam, D.P. Utomo, Implementation of AHP and WASPAS (weighted aggregated sum product assessment) methods in ranking teacher performance. IJISTECH (Int. J. Inform. Syst. Technol.) **3**(2), 173–182 (2020). https://doi.org/10.30645/IJISTECH.V3I2.43
23. D. Panchal, P. Chatterjee, R.K. Shukla, T. Choudhury, J. Tamosaitiene, Integrated fuzzy AHP-codas framework for maintenance decision in urea fertilizer industry. Econ. Comput. Econ. Cybern. Stud. Res. **51**(3) (2017). www.ipe.ro/RePEc/cys/ecocyb_pdf/ecocyb3_2017p179-196.pdf
24. J. Ren, Sustainability prioritization of energy storage technologies for promoting the development of renewable energy: a novel intuitionistic fuzzy combinative distance-based assessment approach. Renew. Energy **121**, 666–676 (2018). https://doi.org/10.1016/j.renene.2018.01.087
25. M. Riaz, H. Garg, H.M.A. Farid, M. Aslam, Novel q-rung orthopair fuzzy interaction aggregation operators and their application to low-carbon green supply chain management. J. Intell. Fuzzy Syst. Pre-press (Preprint), 1–18 (2021). https://doi.org/10.3233/JIFS-210506
26. M. Riaz, D. Pamucar, H.M. Athar Farid, M. Raza, Q-rung orthopair fuzzy prioritized aggregation operators and their application towards green supplier chain management. Symmetry **12**(6), 976 (2020). https://doi.org/10.3390/sym12060976
27. M. Riaz, A. Razzaq, H. Kalsoom, D. Pamučar, H.M. Athar Farid, Y.M. Chu, q-rung orthopair fuzzy geometric aggregation operators based on generalized and group-generalized parameters with application to water loss management. Symmetry **12**(8), 1236 (2020). https://doi.org/10.3390/sym12081236

28. M. Riaz, S.T. Tehrim, A robust extension of Vikor method for bipolar fuzzy sets using connection numbers of spa theory based metric spaces. Artif. Intell. Rev. **54**(1), 561–591 (2021). https://doi.org/10.1007/s10462-020-09859-w
29. M. Sugeno, The fuzzy theoretic turn, in *Fuzzy Approaches for Soft Computing and Approximate Reasoning: Theories and Applications* (Springer, 2021), pp. 1–4. https://doi.org/10.1007/978-3-030-54341-9_1
30. A. Tuş, E.A. Adalı, et al., Personnel assessment with codas and psi methods. Alphanum. J. **6**(2), 243–256 (2018). https://doi.org/10.17093/alphanumeric.432843
31. K. Ullah, N. Hassan, T. Mahmood, N. Jan, M. Hassan, Evaluation of investment policy based on multi-attribute decision-making using interval valued t-spherical fuzzy aggregation operators. Symmetry **11**(3), 357 (2019). https://doi.org/10.3390/sym11030357
32. K. Ullah, T. Mahmood, H. Garg, Evaluation of the performance of search and rescue robots using t-spherical fuzzy hamacher aggregation operators. Int. J. Fuzzy Syst. **22**(2), 570–582 (2020). https://doi.org/10.1007/s40815-020-0803-5
33. J. Wang, H. Gao, G. Wei, Y. Wei, Methods for multiple-attribute group decision making with q-rung interval-valued orthopair fuzzy information and their applications to the selection of green suppliers. Symmetry **11**(1), 56 (2019). https://doi.org/10.3390/sym11010056
34. R. Wang, Y. Li, A novel approach for green supplier selection under a q-rung orthopair fuzzy environment. Symmetry **10**(12), 687 (2018). https://doi.org/10.3390/sym10120687
35. M.Q. Wu, C.H. Zhang, X.N. Liu, J.P. Fan, Green supplier selection based on dea model in interval-valued pythagorean fuzzy environment. IEEE Access **7**, 108001–108013 (2019). https://doi.org/10.1109/ACCESS.2019.2932770
36. Y. Xu, X. Shang, J. Wang, W. Wu, H. Huang, Some q-rung dual hesitant fuzzy heronian mean operators with their application to multiple attribute group decision-making. Symmetry **10**(10), 472 (2018). https://doi.org/10.3390/sym10100472
37. R.R. Yager, Generalized orthopair fuzzy sets. IEEE Trans. Fuzzy Syst. **25**(5), 1222–1230 (2016). https://doi.org/10.1109/TFUZZ.2016.2604005
38. C.W. Yang, T.T. Lai, P.S. Chen, A survey of critical success factors in the implementation of reverse logistics in Taiwans optoelectronic industry. IEEE Access **8**, 193890–193897 (2020). https://doi.org/10.1109/ACCESS.2020.3030939
39. Z. Yang, H. Garg, Interaction power partitioned Maclaurin symmetric mean operators under q-rung orthopair uncertain linguistic information. Int. J. Fuzzy Syst, 1–19 (2021). https://doi.org/10.1007/s40815-021-01062-5
40. F. Zhou, T.Y. Chen, An integrated multicriteria group decision-making approach for green supplier selection under pythagorean fuzzy scenarios. IEEE Access **8**, 165216–165231 (2020). https://doi.org/10.1109/ACCESS.2020.3022377

Dynhora-Danheyda Ramírez-Ochoa was born in Hidalgo del Parral, Chihuahua, Mexico, where she completed her basic education. She moved to the City of Chihuahua to study engineering (2004) and a master's degree (2010) in Computer Systems, at the Autonomous University of Chihuahua (UACH). She is currently a student of the Doctorate in Technology at the Autonomous University of Ciudad Juarez, Chihuahua (starting in 2021). Furthermore, she began her professional-academic experience at UACH as a videoconference technician (2004–2008). Since 2006, she has been a professor at the Technological University of Chihuahua (UTCH) and is the representative of the academic body of Computer Security Technologies. Within the UTCH she also collaborates in various commissions that range from being responsible for the accreditation of educational programs, coordinator of special programs and head of the department of strategic programs. In addition to her professional experience in the private sector, she has worked as a data analyst and manager, as well as an independent consultant. Her research interests are data analysis, databases, decision-making, and project management.

Luis Pérez-Domínguez completed a B.Sc. in Industrial Engineering at Instituto Tecnologico de Villahermosa, Tabasco, Mexico in 2000 and M.Sc. degrees in Industrial Engineering from Instituto Tecnologico de Ciudad Juarez, Chihuahua, Mexico, in 2003 respectively. PhD. Science of Engineering, at the Autonomous University of Ciudad Juarez, Chihuahua, Mexico in 2016. Dr. Luis currently is professor-Research in the Universidad Autonoma de Ciudad Juarez. His research interests include multiple criteria decision-making, fuzzy sets applications and continuous improvement tools applied in the manufacturing field. Member of The Canadian Operational Research Society (CORS); Also, member of Society for Industrial and Applied Mathematics (SIAM). He is recognized as Research associated by Ministerio de Ciencia Tecnologia e Innovacion, Colombia (Ministry of Science Technology and Innovation in Colombia). He is member of Sistema Nacional de Investigadores recognized by CONACYT, Mexico. Dr. Perez also is a member of EURO Working Group on MCDA (EWG-MCDA). In addition, Dr. Perez is an advisory board member of Journals like: MUNDO FESC, RESPUESTAS. He is Member in GRUPO DE INVESTIGACIÓN EN SOFTWARE - GIS and GRINFESC registered in Miniciencias, Colombia.

Erwin Adán Martínez-Gómez was born in Tapachula, Chiapas, México in march 30 of 1973. His educational background is: Degree in industrial engineering, Tapachula Institute of Technology, Tapachula, Chiapas, 1999. Master of Science in industrial engineering, Orizaba Institute of Technology, Orizaba, Veracruz, 2002. PhD in strategic planning and direction of technology, Popular Autonomous University of Puebla, Puebla, Puebla, 2014. His professional experience as head of industrial maintenance in hospitals, coordinator of quality assurance in customer service, head of quality control in industry leather tannery and since 1999 works as a professor and researcher on Engineering and Technology Institute of the Autonomous University of Ciudad Juarez. Dr. Martínez, actual work is related with the determination of critical success factors for Six Sigma in the high technology industry. Other areas of specialty include strategic planning, quality improvement and simulation of manufacturing systems.

Vianey Torres-Argüelles received the Ph.D. degree in engineering from the Engineering Faculty, Universidad Autónoma de Querétaro. Her work is related to the analysis of complex systems. It links multidisciplinary areas, from natural to technological systems, with emphasis on the characterization of the essential attributes of the systems under study. Likewise, the research area is focused on sustainability and clean production.

Harish Garg working as an Associate Professor at Thapar Institute of Engineering & Technology, Deemed University, Patiala, Punjab, India. His research interests include Computational Intelligence, Reliability analysis, multi-criteria decision making, Evolutionary algorithms, Expert systems and decision support systems, Computing with words and Soft Computing. He has authored more than 365 papers (over 315 are SCI) published in refereed International Journals including Information Sciences, IEEE Transactions on Fuzzy Systems, Applied Intelligence, Expert Systems with Applications, Applied Soft Computing, IEEE Access, International Journal of Intelligent Systems, Computers and Industrial Engineering, Cognitive Computations, Soft Computing, Artificial Intelligence Review, IEEE/CAA Journal of Automatic Sinica, IEEE Transactions on Emerging Topics in Computational Intelligence, Computers & Operations Research, Measurement, Journal of Intelligent & Fuzzy Systems, International Journal of Uncertainty Fuzziness and Knowledge-Based Systems, and many more. He has also authored seven book chapters. His Google citations are over 14700+. He is the recipient of the Top-Cited Paper by India-based Author (2015–2019) from Elsevier Publisher. He is also an advisory board member of USERN. Dr. Garg is the Editor-in-Chief of Journal of Computational and Cognitive Engineering; Annals of Optimization Theory and Practice. He is also the Associate Editors for IEEE Transactions on Fuzzy Systems, Soft Computing, Alexandria Engineering Journal, Journal of Intelligent & Fuzzy Systems, Kybernetes, Complex and Intelligent Systems, Journal of Industrial & Management Optimization, Technological and Economic Development of Economy, International Journal of Computational Intelligence Systems, CAAI Transactions on Intelligence Technology, Mathematical Problems in Engineering, Complexity, etc.

Veronica Sansabas-Villapando is Master in International Business and Bachelor of Science in Accounting Currently working in education at Tecnologico Nacional de Mexico Campus Ciudad Juarez at Economic and administrative Sciences department, with experiences in Financial auditing in working until 1998–2003 at Delphi Energy-Sistemas Electricos y Conmutadores, S.A. de C.V.: Juarez, and 1990–1998 Chihuahua, MX and KPMG Cardenas Dosal. Currently Veronica is a PhD student at Tecnologia program in Industrial Engineering and Manufacturing Department of the Universidad Autonoma de Ciudad Juarez.

Chapter 10
Group Decision-Making Framework with Generalized Orthopair Fuzzy 2-Tuple Linguistic Information

Sumera Naz, Muhammad Akram, Feng Feng, and Abid Mahboob

Abstract Many decision-making problems in real-life scenarios depend on how to deal with uncertainty, which is typically a big challenge for decision-makers (DMs). Mathematical models are not common, but where the complexity is not usually probabilistic, various models emerged along with fuzzy logic and linguistic fuzzy approach. In the linguistic environment, multiple attribute group decision-making (MAGDM) is an essential part of modern decision-making science, and information aggregation operators play a crucial role in solving MAGDM problems. The notion of generalized orthopair fuzzy sets (GOFSs) (also known as q-rung orthopair fuzzy sets) serves as an extension of intuitionistic fuzzy sets ($q = 1$) and Pythagorean fuzzy sets ($q = 2$). The generalized orthopair fuzzy 2-tuple linguistic (GOFTL) set provides a better way to deal with uncertain and imprecise information in decision-making. The Maclaurin symmetric mean (MSM) aggregation operator is a useful tool to model the interrelationship between multi-input arguments. In this chapter, we generalize the traditional MSM to aggregate GOFTL information. Firstly, the GOFTL Maclaurin symmetric mean (GOFTLMSM) and the GOFTL weighted Maclaurin symmetric mean (GOFTLWMSM) operators are proposed along with desirable properties and some special cases. Furthermore, the GOFTL dual Maclaurin symmetric mean

S. Naz · A. Mahboob
Department of Mathematics, Division of Science and Technology, University of Education, Lahore, Pakistan
e-mail: sumera.naz@ue.edu.pk

A. Mahboob
e-mail: abid.mahboob@ue.edu.pk

M. Akram
Department of Mathematics, University of the Punjab, New Campus, Lahore 54590, Pakistan
e-mail: m.akram@pucit.edu.pk

F. Feng (✉)
Department of Applied Mathematics, School of Science, Xi'an University of Posts and Telecommunications, Xi'an 710121, China
e-mail: fengnix@hotmail.com

H. Garg (ed.), *q-Rung Orthopair Fuzzy Sets*,
https://doi.org/10.1007/978-981-19-1449-2_10

(GOFTLDMSM) and GOFTL weighted dual Maclaurin symmetric mean (GOFTL-WDMSM) operators with some properties and cases are presented. An efficient approach is developed to tackle the MAGDM problems within the GOFTL framework based on the GOFTLWMSM and GOFTLWDMSM operators. Finally, a numerical illustration regarding the selection of the most preferable supplier(s) in enterprise framework group (EFG) of companies is given to demonstrate the application of the proposed approach and exhibit its viability.

Keywords Generalized orthopair fuzzy 2-tuple linguistic set · Multiple attribute group decision-making · Maclaurin symmetric mean operator · Dual Maclaurin symmetric mean operator · Enterprise framework group of companies

10.1 Introduction

With the fast development of social and economic technology, market competition has become incredibly competitive. The importance of selecting the best supplier for a manufacturing organization is emphasized. In general, organizations must obtain information from all suppliers and utilize technology to evaluate which is the best supplier. The selection of a supplier is basically a multi-attribute decision-making (MADM) problem. Because of the uncertainty of today's decision-making issues, it is not possible for one decision-maker to interpret all of the information relevant to all decision-making strategies. As a consequence, many real-world decision-making difficulties necessitate group decision-making, i.e., MAGDM. The MAGDM is a tool that is based on the attribute values, rates the outcomes for the finite set of alternatives, and it can also select the appropriate supplier for the manufacturing organization. A significant issue in the real decision process is how to express the attribute values in a more effective and efficient way. Owing to the complexity of decision-making issues and the fuzziness of decision-making situations, communicating the attribute values of alternatives is inadequate in the modern world. Zadeh [1] introduced the fuzzy set (FS) theory as an extension of crisp sets to overcome the uncertainty of real-world problems, and the fuzzy value of FS lies between [0, 1]. The intuitionistic fuzzy set (IFS) was introduced by Atanassov [2] as a generalization of FS. IFS has two functions, membership degree (MD) and non-membership degree (NMD) with the condition that the sum of both the functions lies between [0, 1]. Maji et al. [3] combined intuitionistic fuzzy sets with soft sets to come up with the notion of intuitionistic fuzzy soft sets. Feng et al. [4] presented a new extension of the preference ranking organization method for enrichment evaluation (PROMETHEE), by taking advantage of intuitionistic fuzzy soft sets. Agarwal et al. [5] further proposed the notion of generalized intuitionistic fuzzy soft sets. Feng et al. [6] clarified and improved the concept of generalized intuitionistic fuzzy soft sets as a combination of an intuitionistic fuzzy soft set over the universe of discourse and an intuitionistic fuzzy set in the parameter set. IFS is unable to access all the information whenever the sum of both degrees is greater than 1. Yager [7] introduced the new set structure named as Pythagorean fuzzy set (PFS) to resolve the limitation of IFS. The square sum of both degrees lies between [0, 1] in PFS. Naz et al. [8] provided the theory of PF relations along with its application in the MADM environment. In PFS, the DMs have more space for assigning values to an object but due to some limitations, the DMs are not free to give any value. To enhance the expressiveness of PFS, Yager [9] presented the concept of the generalization of IFS and PFS, i.e., the GOFS, in which the MD and NMD satisfy the condition that the sum of their qth powers lies within the range of 0 to 1. The special cases of GOFSs with $q = 1$ and $q = 2$ are IFSs and PFSs, respectively. It has been widely used in medical analysis, design acknowledgment, cluster examination, etc. As far as GOFS is concerned, different aggregation operators have been introduced and applied, such as GOF weighted averaging and GOF weighted geometric operators [10], GOF power averaging and GOF power Maclaurin symmetric mean operators [11], GOF geometric Bonferroni mean and GOF weighted geometric Bonferroni mean operators [12], GOF weighted Heronian mean and GOF weighted dual Heronian mean operators [13]. Recently, Feng et al. [14] defined the novel score functions of generalized orthopair fuzzy membership grades.

The 2-tuple linguistic (2TL) representation model, firstly introduced by Herrera and Martinez [15], is one of the most crucial approaches to deal with linguistic decision-making problems. Several 2TL aggregation operators and decision-making approaches have been proposed. Deng et al. [16] expanded the generalized Heronian mean aggregation operators and their weighted form with 2TLPF numbers to propose the generalized 2TLPF Heronian mean aggregation operators. Wei and Gao [17] utilized the power average and power geometric operators with Pythagorean fuzzy 2-tuple linguistic information to develop some Pythagorean fuzzy 2-tuple linguistic power aggregation operators to solve the MAGDM problems. Wei et al. [35] proposed the concept of GOFTL sets to describe the MD and the NMD of an element to a 2-tuple linguistic variable with some of its operational laws and further developed some GOFTL Heronian mean aggregation operators. Ju et al. [18] proposed the GOFTL weighted averaging operator and the GOFTL weighted geometric operator to develop an approach to solve the MAGDM problem with GOFTL information. Further, the GOFTL Muirhead mean operator and the GOFTL dual Muirhead mean operator are also presented by them. Many scholars have recently developed a lot of decision-making strategies for fuzzy and generalized fuzzy scenarios [19–29, 36–47]. Information aggregation operator plays an important role in the process of decision-making, particularly in MAGDM. In practical MAGDM problems, the preferences of DMs change dynamically, and there always exist various interactions among different considered multi-attributes. The MSM operator, originally proposed by Maclaurin [30], and then developed by Detemple and Robertson [31], is a classical mean type aggregation operator, used to aggregate numerical values in modern information fusion theory. It has a prominent ability to capture the interrelationships among the multi-input arguments of the data. Inspired by the MSM operator, Qin and Liu [32] developed the dual Maclaurin symmetric mean (DMSM) operator. Wei et al. [33] incorporated MSM operator and DMSM operator to GOFSs and introduced the GOFMSM operator, GOFDMSM operator, GOFWMSM operator, and GOFWDMSM operator.

In comparison to different existing fuzzy set models, the GOFTLS is very versatile and can utilize the decision-making ideas of experts in a 2TL context. Further, the MSM as one of the efficient aggregation operators can take the interrelationships among any number of arguments into account and all the above aggregation operators, however, are inadequate to aggregate these GOFTL numbers (GOFTLNs) based on the traditional MSM and DMSM operators. So, an important subject is how to quantify such GOFTLNs on the basis of the MSM and DMSM operators. In this chapter, we extend MSM and DMSM to the GOFTL environment and design some new information aggregation operators such as GOFTLMSM operator, GOFTLDMSM operator, and its weighted forms. The key features of this chapter are identified as follows:

(1) An idea of GOFTLS is utilized to express the uncertainties in the data.
(2) We develop a family of GOFTLMSM operators, such as the GOFTLMSM, the GOFTLWMSM, the GOFTLDMSM, and the GOFTLWDMSM operators to fuse the GOFTLNs and their desirable properties and special cases are investigated.
(3) The newly developed GOFTLWMSM and GOFTLWDMSM operators are used to propose a novel MAGDM model, which can deal with the problems where the attributes have interrelationships.
(4) An illustrative example concerning the selection of the most suitable supplier from EFG of companies is presented to show the usefulness and effectiveness of our proposed MAGDM model. The main contribution of this work is that a new method for MAGDM is put forward based on the GOFTLWMSM and GOFTLWDMSM operators.
(5) The influence of parameters on MAGDM is interpreted. By comparing the proposed method to existing MAGDM approaches, the merits and advantages of the proposed approach are demonstrated. Further, the conclusions and future directions are also presented.

To achieve this goal, the structure of this chapter is arranged as follows: Sect. 10.2 briefly recalls some fundamental concepts relevant to the 2TL representation, the description of GOFTLS and MSM operator. In Sect. 10.3, a new information representation form, i.e., GOFTLMSM and GOFTLWMSM operators and their desirable axioms are briefly analyzed. In Sect. 10.4, the GOFTLDMSM and GOFTLWDMSM operators and some most preferable properties are discussed. In Sect. 10.5, a strategy for MAGDM is developed under the GOFTL environment based on GOFTLWMSM and GOFTLWDMSM operators. In Sect. 10.6, a numerical instance is given related to supplier selection from EFG of companies to illustrate the effectiveness and superiority of the developed method. Finally, Sect. 10.7 presents the conclusions and future directions. The structure of the chapter is given in Fig. 10.1.

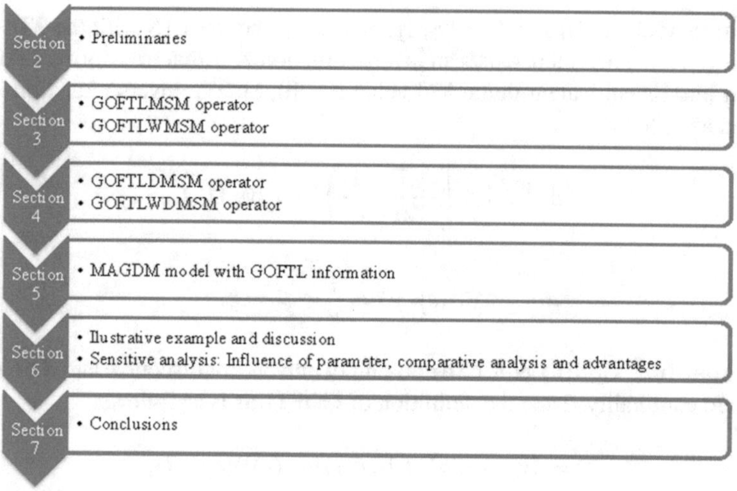

Fig. 10.1 Structure of this chapter

10.2 Preliminaries

The 2TL representation model, GOFS, and the MSM operator and its dual form are briefly discussed in this section.

10.2.1 The 2-Tuple Linguistic Representation Model

Tai and Chen [34] proposed the concept of 2TL representation model as follows:

Definition 10.1 ([34]) Let $S = \{s_0, s_1, s_2, \ldots, s_\tau\}$ be a linguistic term set (LTS) and $\xi \in [0, 1]$ be a number value that represents the aggregation result of linguistic symbolic. The function Δ is then defined to obtain the 2TL information equivalent to ξ as follows:

$$\Delta : [0, 1] \longrightarrow S \times \left[-\frac{1}{2\tau}, \frac{1}{2\tau}\right),$$

$$\Delta(\xi) = (s_j, \gamma) \text{ with } \begin{cases} s_j, & j = \text{round}(\xi \times \tau), \\ \gamma = \xi - \frac{j}{\tau}, & \gamma \in [-\frac{1}{2\tau}, \frac{1}{2\tau}) \end{cases}$$

where γ is the value of the symbolic translation and s_j has the closest index label to ξ.

Definition 10.2 ([34]) Let $S = \{s_0, s_1, s_2, \ldots, s_\tau\}$ be an LTS with the 2TL information (s_j, γ). Then there exists an inverse function Δ^{-1} that transforms 2TL information into its equivalent numerical value $\xi \in [0, 1]$. The inverse function Δ^{-1} is defined as follows:

$$\Delta^{-1} : S \times \left[-\frac{1}{2\tau}, \frac{1}{2\tau}\right) \longrightarrow [0, 1], f$$

$$\Delta^{-1}(s_j, \gamma) = \frac{j}{\tau} + \gamma = \xi.$$

Definition 10.3 ([35]) Let R be a non-empty universe of discourse and S be an LTS with odd cardinality. Then the definition of GOFTLSs is as follows:

$$D = \{\langle (s_{\psi(r)}, \gamma), (\vartheta_D(r), \varphi_D(r))\rangle | r \in R\},$$

where $s_{\psi(r)} \in S$, $\gamma \in [-\frac{1}{2\tau}, \frac{1}{2\tau})$, the numbers $\vartheta_D(r)$, $\varphi_D(r)$ represent the MD and NMD of r to the 2TL variable $(s_{\psi(r)}, \gamma)$, respectively, which satisfies the following conditions: $\vartheta_D(r) \in [0, 1]$, $\varphi_D(r) \in [0, 1]$, $0 \leq (\vartheta_D(r))^q + (\varphi_D(r))^q \leq 1$, for all $r \in R$ and $q \geq 1$.

For $r \in R$, $\pi_D(r) = (1 - (\vartheta_D(r))^q - (\varphi_D(r))^q)^{\frac{1}{q}}$ is degree of indeterminacy of the element r to 2TL variable $(s_{\psi(r)}, \gamma)$. For simplicity, $\wp = ((s_{\psi(\wp)}, \gamma), (\vartheta_\wp, \varphi_\wp))$ will be called a GOFTLN.

Definition 10.4 ([35]) Let $\wp_1 = ((s_1, \gamma_1), (\vartheta_1, \varphi_1))$ and $\wp_2 = ((s_2, \gamma_2), (\vartheta_2, \varphi_2))$ be two GOFTLNs. Then the operational laws of GOFTLNs are described as follows:

1. $\wp_1 \oplus \wp_2 = \Big(\Delta(\Delta^{-1}(s_1, \gamma_1) + \Delta^{-1}(s_2, \gamma_2)), (((\vartheta_1)^q + (\vartheta_2)^q - (\vartheta_1)^q \times (\vartheta_2)^q)^{\frac{1}{q}}, \varphi_1 \times \varphi_2)\Big)$;
2. $\wp_1 \otimes \wp_2 = \Big(\Delta(\Delta^{-1}(s_1, \gamma_1) \times \Delta^{-1}(s_2, \gamma_2)), (\vartheta_1 \times \vartheta_2, ((\varphi_1)^q + (\varphi_2)^q - (\varphi_1)^q \times (\varphi_2)^q)^{\frac{1}{q}})\Big)$;
3. $\lambda\wp_1 = \Big(\Delta(\lambda \times \Delta^{-1}(s_1, \gamma_1)), ((1 - (1 - (\vartheta_1)^q)^\lambda)^{\frac{1}{q}}, (\varphi_1)^\lambda)\Big)$, $\lambda > 0$;
4. $\wp_1^\lambda = \Big(\Delta((\Delta^{-1}(s_1, \gamma_1))^\lambda), ((\vartheta_1)^\lambda, (1 - (1 - (\varphi_1)^q)^\lambda)^{\frac{1}{q}})\Big)$, $\lambda > 0$.

Definition 10.5 ([18]) Let $\wp = ((s_{\psi(\wp)}, \gamma), (\vartheta_\wp, \varphi_\wp))$ be a GOFTLN. Then the score value $\mathcal{S}(\wp)$ and the accuracy value $\mathcal{H}(\wp)$ of GOFTLN $\wp$ are defined as:

$$\mathcal{S}(\wp) = \frac{\Delta^{-1}(s_{\psi(\wp)}, \gamma) \times (1 + \vartheta_\wp^q - \varphi_\wp^q)}{2}, \tag{10.1}$$

$$\mathcal{H}(\wp) = \Delta^{-1}(s_{\psi(\wp)}, \gamma) \times (\vartheta_\wp^q + \varphi_\wp^q). \tag{10.2}$$

Definition 10.6 ([18]) Let $\wp_1 = ((s_1, \gamma_1), (\vartheta_1, \varphi_1))$ and $\wp_2 = ((s_2, \gamma_2), (\vartheta_2, \varphi_2))$ be two GOFTLNs, then we can compare them using the rules below:

(1) If $\mathcal{S}(\wp_1) > \mathcal{S}(\wp_2)$, then $\wp_1 \succ \wp_2$;
(2) If $\mathcal{S}(\wp_1) = \mathcal{S}(\wp_2)$, then

- If $\mathcal{H}(\wp_1) > \mathcal{H}(\wp_2)$, then $\wp_1 \succ \wp_2$;
- If $\mathcal{H}(\wp_1) = \mathcal{H}(\wp_2)$, then $\wp_1 \sim \wp_2$.

10.2.2 *The MSM Operator and its Weighted Form*

Let $\ell_k (k = 1, 2, \ldots, n)$ be any set of non-negative numbers. Then the following operators are defined as:

1. MSM [30]: $\mathrm{MSM}^{(t)}(\ell_1, \ell_2, \ldots, \ell_n) = \left(\frac{\sum_{1 \le j_1 < \ldots < j_t \le n} \left(\prod_{k=1}^{t} \ell_{j_k} \right)}{C_n^t} \right)^{\frac{1}{t}}$;

2. Weighted MSM [30]: $\mathrm{WMSM}_{\varpi}^{(t)}(\ell_1, \ell_2, \ldots, \ell_n) = \left(\frac{\sum_{1 \le j_1 < \ldots < j_t \le n} \left(\prod_{k=1}^{t} (\ell_{j_k})^{\varpi_{j_k}} \right)}{C_n^t} \right)^{\frac{1}{t}}$;

3. Dual MSM [32]: $\mathrm{DMSM}^{(t)}(\ell_1, \ell_2, \ldots, \ell_n) = \frac{1}{t} \left(\prod_{1 \le j_1 < \ldots < j_t \le n} \left(\sum_{k=1}^{t} \ell_{j_k} \right)^{\frac{1}{C_n^t}} \right)$;

4. Weighted dual MSM [32]: $\mathrm{WDMSM}_{\varpi}^{(t)}(\ell_1, \ell_2, \ldots, \ell_n) = \frac{1}{t} \left(\prod_{1 \le j_1 < \ldots < j_t \le n} \left(\sum_{k=1}^{t} \left(\varpi_{j_k} \ell_{j_k} \right) \right)^{\frac{1}{C_n^t}} \right)$.

where $t = 1, 2, \ldots, n$, is a parameter, $j_1, j_2, \ldots, j_t$ are t integer values choose from the set $\{1, 2, \ldots, n\}$ of k integer values, C_n^t represents the binomial coefficient, and $C_n^t = n!/t!(n-t)!$. Nomenclature of proposed work is shown in Table 10.1.

10.3 The GOFTLMSM Aggregation Operator and its Weighted Form

In this section, we generalize the MSM operator and its weighted form with GOFTLSs to develop the GOFTLMSM operator and GOFTLWMSM operator for aggregating GOFTLNs and study its important characteristics and special cases. These operators can be used more broadly and effectively to fuse GOFTL information.

Table 10.1 Nomenclature of proposed work

Notation	Description
GOFS	Generalized orthopair fuzzy set
GOFTLS	Generalized orthopair fuzzy 2-tuple linguistic set
GOFTLMSM	GOFTL Maclaurin symmetric mean
GOFTLWMSM	GOFTL weighted Maclaurin symmetric mean
GOFTLDMSM	GOFTL dual Maclaurin symmetric mean
GOFTLWDMSM	GOFTL weighted dual Maclaurin symmetric mean
$\wp = \left((s_{\wp}, \gamma_{\wp}), (\vartheta_{\wp}, \varphi_{\wp})\right)$	GOFTL number
$(s_{\wp}, \gamma_{\wp})$	2-tuple linguistic term
$\vartheta_{\wp}$	Membership degree of GOFTLN
$\varphi_{\wp}$	Non-membership degree of GOFTLN
$\mathfrak{F}$	Alternative
ς	Attribute
$\mathcal{S}(\wp)$	Score function of GOFTLN
$\mathcal{H}(\wp)$	Accuracy function of GOFTLN
$\pi_D(r)$	Uncertainty of GOFTLN

10.3.1 The GOFTLMSM Operator

The GOFTLMSM operator is introduced in this subsection, along with its desirable characteristics and special cases.

Definition 10.7 Let $\wp_k = ((s_k, \gamma_k), (\vartheta_k, \varphi_k))$ $(k = 1, 2, \ldots, n)$ be any set of GOFTLNs. Then the GOFTLMSM operator is defined as follows:

$$\text{GOFTLMSM}^{(t)}(\wp_1, \wp_2, \ldots, \wp_n) = \left(\frac{\oplus_{1 \le j_1 < \ldots < j_t \le n} \left(\otimes_{k=1}^{t} \wp_{j_k} \right)}{C_n^t} \right)^{\frac{1}{t}}. \quad (10.3)$$

Theorem 10.1 can be deduced from the operational laws of GOFTLNs, as seen in Definition 10.4.

Theorem 10.1 *Let* $\wp_k = ((s_k, \gamma_k), (\vartheta_k, \varphi_k))$ $(k = 1, 2, \ldots, n)$ *be any set of GOFTLNs. Then the aggregated value by using GOFTLMSM operator is also a GOFTLN, and*

$$GOFTLMSM^{(t)}(\wp_1, \wp_2, \ldots, \wp_n) = \left(\begin{array}{c} \Delta\left(\left(\frac{\sum_{1\le j_1<\ldots<j_t\le n}\left(\prod_{k=1}^{t}\Delta^{-1}(s_{j_k},\gamma_{j_k})\right)}{C_n^t}\right)^{\frac{1}{t}}\right), \\ \left(\left(\sqrt[q]{1-\left(\prod_{1\le j_1<\ldots<j_t\le n}\left(1-\left(\prod_{k=1}^{t}\vartheta_{j_k}\right)^q\right)\right)^{\frac{1}{C_n^t}}}\right)^{\frac{1}{t}}, \right. \\ \left. \sqrt[q]{1-\left(1-\left(\prod_{1\le j_1<\ldots<j_t\le n}\left(1-\prod_{k=1}^{t}(1-(\varphi_{j_k})^q)\right)\right)^{\frac{1}{C_n^t}}\right)^{\frac{1}{t}}}\right) \end{array} \right). \tag{10.4}$$

Proof Depending on the rules of GOFTLNs, we have

$$\otimes_{k=1}^{t}\wp_{j_k} = \left(\Delta\left(\prod_{k=1}^{t}\Delta^{-1}\left(s_{j_k},\gamma_{j_k}\right)\right), \left(\prod_{k=1}^{t}\vartheta_{j_k}, \sqrt[q]{1-\prod_{k=1}^{t}\left(1-(\varphi_{j_k})^q\right)}\right)\right)$$

and

$$\oplus_{1\le j_1<\ldots<j_t\le n}\left(\otimes_{k=1}^{t}\wp_{j_k}\right) = \left(\begin{array}{c} \Delta\left(\sum_{1\le j_1<\ldots<j_t\le n}\left(\prod_{k=1}^{t}\Delta^{-1}\left(s_{j_k},\gamma_{j_k}\right)\right)\right), \\ \left(\sqrt[q]{1-\prod_{1\le j_1<\ldots<j_t\le n}\left(1-\left(\prod_{k=1}^{t}\vartheta_{j_k}\right)^q\right)}, \right. \\ \left. \prod_{1\le j_1<\ldots<j_t\le n}\sqrt[q]{1-\prod_{k=1}^{t}(1-(\varphi_{j_k})^q)} \right) \end{array} \right).$$

Thus we obtain

$$\frac{1}{C_n^t}\oplus_{1\le j_1<\ldots<j_t\le n}\left(\otimes_{k=1}^{t}\wp_{j_k}\right) = \left(\begin{array}{c} \Delta\left(\frac{1}{C_n^t}\sum_{1\le j_1<\ldots<j_t\le n}\left(\prod_{k=1}^{t}\Delta^{-1}\left(s_{j_k},\gamma_{j_k}\right)\right)\right), \\ \left(\sqrt[q]{1-\left(\prod_{1\le j_1<\ldots<j_t\le n}\left(1-\left(\prod_{k=1}^{t}\vartheta_{j_k}\right)^q\right)\right)^{\frac{1}{C_n^t}}}, \right. \\ \left. \left(\prod_{1\le j_1<\ldots<j_t\le n}\sqrt[q]{1-\prod_{k=1}^{t}(1-(\varphi_{j_k})^q)}\right)^{\frac{1}{C_n^t}} \right) \end{array} \right).$$

Accordingly,

$$\mathrm{GOFTLMSM}^{(t)}(\wp_1,\wp_2,\ldots,\wp_n) = \left(\begin{array}{c}\Delta\left(\left(\frac{\sum_{1\le j_1<\ldots<j_t\le n}\left(\prod_{k=1}^{t}\Delta^{-1}\left(s_{j_k},\gamma_{j_k}\right)\right)}{C_n^t}\right)^{\frac{1}{t}}\right),\\ \left(\left(\sqrt[q]{1-\left(\prod_{1\le j_1<\ldots<j_t\le n}\left(1-\left(\prod_{k=1}^{t}\vartheta_{j_k}\right)^q\right)\right)^{\frac{1}{C_n^t}}}\right)^{\frac{1}{t}},\\ \sqrt[q]{1-\left(1-\left(\prod_{1\le j_1<\ldots<j_t\le n}\left(1-\prod_{k=1}^{t}(1-(\varphi_{j_k})^q)\right)\right)^{\frac{1}{C_n^t}}\right)^{\frac{1}{t}}}\right)\end{array}\right).$$

Example 10.1 Take $\wp_1$=((s_2, 0.01), (0.3, 0.8)), $\wp_2$=((s_3, 0.05), (0.7, 0.4)), $\wp_3 =$ ((s_4, −0.02), (0.6, 0.8)) and $\wp_4 =$ ((s_5, 0.07), (0.5, 0.7)) are four GOFTLNs, and $S = \{s_0, s_1, s_2, s_3, s_4, s_5, s_6\}$ be the LTS, utilizing GOFTLMSM operator to aggregate these four GOFTLNs. Consider $t = 2$, $q = 4$; thus

$$\Delta\left(\left(\frac{1}{C_n^t}\sum_{1\le j_1<\ldots<j_t\le n}\left(\prod_{k=1}^{t}\Delta^{-1}\left(s_{j_k},\gamma_{j_k}\right)\right)\right)^{\frac{1}{t}}\right)$$

$$= \Delta\left(\left(\frac{1}{3!}\left(\begin{array}{l}\left(\Delta^{-1}(s_2,0.01)\times\Delta^{-1}(s_3,0.05)\right)+\left(\Delta^{-1}(s_2,0.01)\times\Delta^{-1}(s_4,-0.02)\right)\\ +\left(\Delta^{-1}(s_2,0.01)\times\Delta^{-1}(s_5,0.07)\right)+\left(\Delta^{-1}(s_3,0.05)\times\Delta^{-1}(s_4,-0.02)\right)+\\ \left(\Delta^{-1}(s_3,0.05)\times\Delta^{-1}(s_5,0.07)\right)+\left(\Delta^{-1}(s_4,-0.02)\times\Delta^{-1}(s_5,0.07)\right)\end{array}\right)\right)^{\frac{1}{2}}\right)$$

$$= \Delta(0.5997) = (s_4, -0.0670),$$

$$\left(\sqrt[q]{1-\left(\prod_{1\le j_1<\ldots<j_t\le n}\left(1-\left(\prod_{k=1}^{t}\vartheta_{j_k}\right)^q\right)\right)^{\frac{1}{C_n^t}}}\right)^{\frac{1}{t}}$$

$$= \left(\sqrt[4]{1-\left(\begin{array}{l}(1-(0.3\times 0.7)^4)\times(1-(0.3\times 0.6)^4)\times(1-(0.3\times 0.5)^4)\times(1-(0.7\times 0.6)^4)\\ \times(1-(0.7\times 0.5)^4)\times(1-(0.6\times 0.5)^4)\end{array}\right)^{\frac{1}{3!}}}\right)^{\frac{1}{2}}$$

$$= 0.5601,$$

$$\sqrt[q]{1-\left(1-\left(\prod_{1\le j_1<\ldots<j_t\le n}\left(1-\prod_{k=1}^{t}(1-(\varphi_{j_k})^q)\right)\right)^{\frac{1}{C_n^t}}\right)^{\frac{1}{t}}}$$

$$= \sqrt[4]{1-\left(1-\left(\begin{array}{l}(1-(1-(0.8)^4)\times(1-(0.4)^4))\times(1-(1-(0.8)^4)\times(1-(0.8)^4))\\ \times(1-(1-(0.8)^4)\times(1-(0.7)^4))\times(1-(1-(0.4)^4)\times(1-(0.8)^4))\\ \times(1-(1-(0.4)^4)\times(1-(0.7)^4))\times(1-(1-(0.8)^4)\times(1-(0.7)^4))\end{array}\right)^{\frac{1}{3!}}\right)^{\frac{1}{2}}}$$

$$= 0.7168.$$

Therefore, the aggregation value of the $\wp_1, \wp_2, \wp_3$ *and* $\wp_4$ by the GOFTLMSM operator mentioned below:

$$\mathrm{GOFTLMSM}^{(2)} = ((s_4, -0.0670), (0.5601, 0.7168)).$$

We can easily prove that the GOFTLMSM operator has the following properties:

Property 10.1 (Idempotency) If all $\wp_k$ $(k = 1, 2, \ldots, n)$ are equal, i.e., $\wp_k = \wp$ for all k, then

$$\mathrm{GOFTLMSM}^{(t)}(\wp_1, \wp_2, \ldots, \wp_n) = \wp.$$

Proof Because $\wp = ((s_\wp, \gamma_\wp), (\vartheta_\wp, \varphi_\wp))$, depending on the Theorem 10.1, we have

$$\mathrm{GOFTLMSM}^{(t)}(\wp, \wp, \ldots, \wp)$$

$$= \left(\begin{array}{l} \Delta\left(\left(\frac{1}{C_n^t}\sum\limits_{1\le j_1<\ldots<j_t\le n}\left(\prod\limits_{k=1}^{t}\left(\Delta^{-1}(s_\wp,\gamma_\wp)\right)\right)\right)^{\frac{1}{t}}\right), \\ \left(\left(\sqrt[q]{1-\left(\prod\limits_{1\le j_1<\ldots<j_t\le n}\left(1-\left(\prod\limits_{k=1}^{t}\vartheta_\wp\right)^q\right)\right)^{\frac{1}{C_n^t}}}\right)^{\frac{1}{t}}, \right. \\ \left.\sqrt[q]{1-\left(1-\left(\prod\limits_{1\le j_1<\ldots<j_t\le n}\left(1-\prod\limits_{k=1}^{t}\left(1-(\varphi_\wp)^q\right)\right)\right)^{\frac{1}{C_n^t}}\right)^{\frac{1}{t}}}\right) \end{array} \right).$$

$$= \left(\begin{array}{l} \Delta\left(\left(\frac{1}{C_n^t}\sum\limits_{1\le j_1<\ldots<j_t\le n}\left(\Delta^{-1}(s_\wp,\gamma_\wp)\right)^t\right)^{\frac{1}{t}}\right), \\ \left(\left(\sqrt[q]{1-\left(\prod\limits_{1\le j_1<\ldots<j_t\le n}\left(1-(\vartheta_\wp)^{qt}\right)\right)^{\frac{1}{C_n^t}}}\right)^{\frac{1}{t}}, \right. \\ \left.\sqrt[q]{1-\left(1-\left(\prod\limits_{1\le j_1<\ldots<j_t\le n}\left(1-(1-(\varphi_\wp)^q)^t\right)\right)^{\frac{1}{C_n^t}}\right)^{\frac{1}{t}}}\right) \end{array} \right).$$

$$= \left(\Delta\left(\left(\frac{1}{C_n^t}.(C_n^t)\left(\Delta^{-1}(s_\wp,\gamma_\wp)\right)^t\right)^{\frac{1}{t}}\right), \left(\left(\sqrt[q]{1-\left(\left(1-(\vartheta_\wp)^{qt}\right)^{C_n^t}\right)^{\frac{1}{C_n^t}}}\right)^{\frac{1}{t}}, \sqrt[q]{1-\left(1-\left(\left(1-(1-(\varphi_\wp)^q)^t\right)^{C_n^t}\right)^{\frac{1}{C_n^t}}\right)^{\frac{1}{t}}}\right)\right).$$

$$= \left(\Delta\left(\left(\Delta^{-1}(s_\wp,\gamma_\wp)\right)^t\right)^{\frac{1}{t}}, \left(\left(\sqrt[q]{1-(1-(\vartheta_\wp)^{qt})}\right)^{\frac{1}{t}}, \sqrt[q]{1-(1-(1-(1-(\varphi_\wp)^q)^t))^{\frac{1}{t}}}\right)\right).$$

$$= \left(\Delta\left(\Delta^{-1}(s_\wp,\gamma_\wp)\right), \left(\left(\sqrt[q]{(\vartheta_\wp)^{qt}}\right)^{\frac{1}{t}}, \sqrt[q]{(\varphi_\wp)^q}\right)\right) = ((s_\wp,\gamma_\wp),(\vartheta_\wp,\varphi_\wp)).$$

Property 10.2 (Commutativity) If $\wp_k$ $(k = 1, 2, \ldots, n)$ be any set of GOFTLNs, and $\wp_k'$ $(k = 1, 2, \ldots, n)$ be a permutation of $\wp_k$ $(k = 1, 2, \ldots, n)$, then

$$\mathrm{GOFTLMSM}^{(t)}(\wp_1, \wp_2, \ldots, \wp_n) = \mathrm{GOFTLMSM}^{(t)}(\wp_1', \wp_2', \ldots, \wp_n').$$

Proof As $\wp_k'$ $(k = 1, 2, \ldots, n)$ is the permutation of $\wp_k$ $(k = 1, 2, \ldots, n)$, depending on the definition of GOFTLMSM in Eq. 10.4, we have

$$\begin{aligned}\text{GOFTLMSM}^{(t)}(\wp_1, \wp_2, \ldots, \wp_n) &= \left(\frac{\oplus_{1\le j_1<\ldots<j_t\le n}(\otimes_{k=1}^t \wp_{j_k})}{C_n^t}\right)^{\frac{1}{t}} \\ &= \left(\frac{\oplus_{1\le j_1<\ldots<j_t\le n}(\otimes_{k=1}^t \wp_{j_k}')}{C_n^t}\right)^{\frac{1}{t}} \\ &= \text{GOFTLMSM}^{(t)}(\wp_1', \wp_2', \ldots, \wp_n').\end{aligned}$$

Property 10.3 (Monotonicity) Let $\wp_k$ and $\wp_k'$ $(k = 1, 2, \ldots, n)$ be two sets of GOFTLNs. If $(s_k, \gamma_k) \ge (s_k', \gamma_k')$, $\vartheta_k \ge \vartheta_k'$ and $\varphi_k \le \varphi_k'$, for all k, then

$$\text{GOFTLMSM}^{(t)}(\wp_1, \wp_2, \ldots, \wp_n) \ge \text{GOFTLMSM}^{(t)}(\wp_1', \wp_2', \ldots, \wp_n').$$

Proof Let $\wp_k = ((s_k, \gamma_k), (\vartheta_k, \varphi_k))$ and $\acute{\wp}_k = ((s_k', \gamma_k'), (\vartheta_k', \varphi_k'))$ be two GOFTLNs. Since

$$(s_k, \gamma_k) \ge (s_k', \gamma_k')\ (k = 1, 2, \ldots, n).$$

Thus we have

$$\Delta^{-1}(s_k, \gamma_k) \ge \Delta^{-1}(s_k', \gamma_k').$$

$$\prod_{k=1}^{t} \Delta^{-1}(s_k, \gamma_k) \ge \prod_{k=1}^{t} \Delta^{-1}(s_k', \gamma_k').$$

$$\frac{1}{C_n^t} \sum_{1\le j_1<\ldots<j_t\le n} \prod_{k=1}^{t} \Delta^{-1}(s_k, \gamma_k) \ge \frac{1}{C_n^t} \sum_{1\le j_1<\ldots<j_t\le n} \prod_{k=1}^{t} \Delta^{-1}(s_k', \gamma_k').$$

Because $t \ge 1$, $\vartheta_k \ge \vartheta_k' \ge 0$, $\varphi_k \le \varphi_k' \le 0$, we have $\vartheta_{j_k} \ge \vartheta_{j_k}' \ge 0$, $\varphi_{j_k} \le \varphi_{j_k}' \le 0$. Depending upon the above constraints, for all $j, k (j = 1, 2, \ldots, n;\ k = 1, 2, \ldots, t)$, we obtain

$$\prod_{k=1}^{t} \vartheta_{j_k} \ge \prod_{k=1}^{t} \vartheta_{j_k}' \Rightarrow 1 - \left(\prod_{k=1}^{t} \vartheta_{j_k}\right)^q \le 1 - \left(\prod_{k=1}^{t} \vartheta_{j_k}'\right)^q$$

$$\Rightarrow \prod_{1\le j_1<\ldots<j_t\le n} \left(1 - \left(\prod_{k=1}^{t} \vartheta_{j_k}\right)^q\right) \le \prod_{1\le j_1<\ldots<j_t\le n} \left(1 - \left(\prod_{k=1}^{t} \vartheta_{j_k}'\right)^q\right)$$

$$\Rightarrow \left(\sqrt[q]{1 - \left(\prod_{1\le j_1<\ldots<j_t\le n} \left(1 - \left(\prod_{k=1}^{t} \vartheta_{j_k}\right)^q\right)\right)^{\frac{1}{C_n^t}}}\right)^{\frac{1}{t}} \ge \left(\sqrt[q]{1 - \left(\prod_{1\le j_1<\ldots<j_t\le n} \left(1 - \left(\prod_{k=1}^{t} \vartheta_{j_k}'\right)^q\right)\right)^{\frac{1}{C_n^t}}}\right)^{\frac{1}{t}}.$$

Similarly, we have

$$\varphi_{j_k} \le \varphi'_{j_k} \Rightarrow 1-(\varphi_{j_k})^q \ge 1-(\varphi'_{j_k})^q \Rightarrow 1-\prod_{k=1}^{t}(1-(\varphi_{j_k})^q) \le 1-\prod_{k=1}^{t}(1-(\varphi'_{j_k})^q)$$

$$\Rightarrow \left(\prod_{1\le j_1<\ldots<j_t\le n}\left(1-\prod_{k=1}^{t}(1-(\varphi_{j_k})^q)\right)\right)^{\frac{1}{C_n^t}} \le \left(\prod_{1\le j_1<\ldots<j_t\le n}\left(1-\prod_{k=1}^{t}(1-(\varphi'_{j_k})^q)\right)\right)^{\frac{1}{C_n^t}}$$

$$\Rightarrow \left(1-\left(\prod_{1\le j_1<\ldots<j_t\le n}\left(1-\prod_{k=1}^{t}(1-(\varphi_{j_k})^q)\right)\right)^{\frac{1}{C_n^t}}\right)^{\frac{1}{t}} \ge \left(1-\left(\prod_{1\le j_1<\ldots<j_t\le n}\left(1-\prod_{k=1}^{t}(1-(\varphi'_{j_k})^q)\right)\right)^{\frac{1}{C_n^t}}\right)^{\frac{1}{t}}$$

$$\Rightarrow \sqrt[q]{1-\left(1-\left(\prod_{1\le j_1<\ldots<j_t\le n}\left(1-\prod_{k=1}^{t}(1-(\varphi_{j_k})^q)\right)\right)^{\frac{1}{C_n^t}}\right)^{\frac{1}{t}}}$$

$$\le \sqrt[q]{1-\left(1-\left(\prod_{1\le j_1<\ldots<j_t\le n}\left(1-\prod_{k=1}^{t}(1-(\varphi'_{j_k})^q)\right)\right)^{\frac{1}{C_n^t}}\right)^{\frac{1}{t}}}.$$

Property 10.4 (Boundedness) Suppose that $\wp_k (k = 1, 2, \ldots, n)$ is any set of GOFTLNs, $\wp^- = \min_k \wp_k = (\min_k (s_k, \gamma_k), (\min_k \vartheta_k, \max_k \varphi_k))$ and $\wp^+ = \max_k \wp_k = (\max_k (s_k, \gamma_k), (\max_k \vartheta_k, \min_k \varphi_k))$. Then

$$\wp^- \le \mathrm{GOFTLMSM}^{(t)}(\wp_1, \wp_2, \ldots, \wp_n) \le \wp^+.$$

Proof Utilizing Properties 10.1 and 10.3, we get

$$\mathrm{GOFTLMSM}^{(t)}(\wp_1, \wp_2, \ldots, \wp_n) \ge \mathrm{GOFTLMSM}^{(t)}(\wp^-, \wp^-, \ldots, \wp^-) = \wp^-$$
$$\mathrm{GOFTLMSM}^{(t)}(\wp_1, \wp_2, \ldots, \wp_n) \le \mathrm{GOFTLMSM}^{(t)}(\wp^+, \wp^+, \ldots, \wp^+) = \wp^+.$$

Theorem 10.2 *For given GOFTLNs $\wp_k (k = 1, 2, \ldots, n,\ t = 1, 2, \ldots, n)$, the GOFTLMSM is monotonically decreasing with respect to the parameter t.*

Now, we analyze some particular situations of the GOFTLMSM operator in terms of the parameter t.

Case 1. When $t = 1$, the GOFTLMSM operator converts into GOFTL averaging operator as follows:

$$\text{GOFTLMSM}^{(1)}(\wp_1,\wp_2,\ldots,\wp_n)$$

$$=\left(\begin{array}{c}\Delta\left(\frac{1}{C_n^1}\sum_{1\le j_1\le n}\left(\prod_{k=1}^{1}\Delta^{-1}\left(s_{j_k},\gamma_{j_k}\right)\right)\right)^{\frac{1}{1}},\\ \left(\left(\sqrt[q]{1-\left(\prod_{1\le j_1\le n}\left(1-\left(\prod_{k=1}^{1}\vartheta_{j_k}\right)^q\right)\right)^{\frac{1}{C_n^1}}}\right)^{\frac{1}{1}},\right.\\ \left.\sqrt[q]{1-\left(1-\left(\prod_{1\le j_1\le n}\left(1-\prod_{k=1}^{1}(1-(\varphi_{j_k})^q)\right)\right)^{\frac{1}{C_n^1}}\right)^{\frac{1}{1}}}\right)\end{array}\right).$$

$$=\left(\Delta\left(\frac{1}{n}\sum_{1\le j_1\le n}\Delta^{-1}\left(s_{j_1},\gamma_{j_1}\right)\right)^{\frac{1}{1}},\left(\sqrt[q]{1-\left(\prod_{1\le j_1\le n}(1-(\vartheta_{j_1})^q)\right)^{\frac{1}{n}}},\sqrt[q]{\left(\prod_{1\le j_1\le n}(1-(1-(\varphi_{j_1})^q))\right)^{\frac{1}{n}}}\right)\right).$$

$$=\left(\Delta\left(\frac{1}{n}\sum_{1\le j_1\le n}\Delta^{-1}\left(s_{j_1},\gamma_{j_1}\right)\right),\left(\sqrt[q]{1-\left(\prod_{1\le j_1\le n}(1-(\vartheta_{j_1})^q)\right)^{\frac{1}{n}}},\prod_{1\le j_1\le n}(\varphi_{j_1})^{\frac{1}{n}}\right)\right)(let\ j_1=j).$$

$$=\left(\Delta\left(\frac{1}{n}\sum_{j=1}^{n}\Delta^{-1}\left(s_j,\gamma_j\right)\right),\left(\sqrt[q]{1-\left(\prod_{j=1}^{n}(1-(\vartheta_j)^q)\right)^{\frac{1}{n}}},\prod_{j=1}^{n}(\varphi_j)^{\frac{1}{n}}\right)\right).$$

Case 2. When $t=2$, the GOFTLMSM operator converts into GOFTL Bonferoni mean operator as follows:

$$\text{GOFTLMSM}^{(2)}(\wp_1,\wp_2,\ldots,\wp_n)$$

$$=\left(\begin{array}{c}\Delta\left(\frac{1}{C_n^2}\sum_{1\le j_1<j_2\le n}\left(\prod_{k=1}^{2}\Delta^{-1}\left(s_{j_k},\gamma_{j_k}\right)\right)\right)^{\frac{1}{2}},\\ \left(\left(\sqrt[q]{1-\left(\prod_{1\le j_1<j_2\le n}\left(1-\left(\prod_{k=1}^{2}\vartheta_{j_k}\right)^q\right)\right)^{\frac{1}{C_n^2}}}\right)^{\frac{1}{2}},\right.\\ \left.\sqrt[q]{1-\left(1-\left(\prod_{1\le j_1<j_2\le n}\left(1-\prod_{k=1}^{2}(1-(\varphi_{j_k})^q)\right)\right)^{\frac{1}{C_n^2}}\right)^{\frac{1}{2}}}\right)\end{array}\right)$$

$$=\left(\begin{array}{c}\Delta\left(\frac{2}{n(n-1)}\sum_{1\le j_1<j_2\le n}\left(\Delta^{-1}\left(s_{j_1},\gamma_{j_1}\right)\Delta^{-1}\left(s_{j_2},\gamma_{j_2}\right)\right)\right)^{\frac{1}{2}},\\ \left(\left(\sqrt[q]{1-\left(\prod_{1\le j_1<j_2\le n}(1-(\vartheta_{j_1}\vartheta_{j_2})^q)\right)^{\frac{2}{n(n-1)}}}\right)^{\frac{1}{2}},\right.\\ \left.\sqrt[q]{1-\left(1-\left(\prod_{1\le j_1<j_2\le n}(1-(1-(\varphi_{j_1})^q)(1-(\varphi_{j_2})^q))\right)^{\frac{2}{n(n-1)}}\right)^{\frac{1}{2}}}\right)\end{array}\right).$$

$$\left(\begin{array}{c}\Delta\left(\frac{1}{n(n-1)}\sum_{j_1,j_2=1,j_1\neq j_2}^{n}\left(\Delta^{-1}\left(s_{j_1},\gamma_{j_1}\right)\Delta^{-1}\left(s_{j_2},\gamma_{j_2}\right)\right)\right)^{\frac{1}{2}},\\ \left(\begin{array}{c}\left(\sqrt[q]{1-\left(\prod_{j_1,j_2=1,j_1\neq j_2}^{n}\left(1-(\vartheta_{j_1}\vartheta_{j_2})^q\right)\right)^{\frac{1}{n(n-1)}}}\right)^{\frac{1}{2}},\\ \sqrt[q]{1-\left(1-\left(\prod_{j_1,j_2=1,j_1\neq j_2}^{n}\left(1-(1-(\varphi_{j_1})^q)(1-(\varphi_{j_2})^q)\right)\right)^{\frac{1}{n(n-1)}}\right)^{\frac{1}{2}}}\end{array}\right)\end{array}\right).$$

$$= \mathrm{GOFTLBM}^{(1,1)}(\wp_1,\wp_2,\ldots,\wp_n).$$

Case 3. When $t=n$, the GOFTLMSM operator converts into GOFTL geometric mean operator as follows:

$$\mathrm{GOFTLMSM}^{(n)}(\wp_1,\wp_2,\ldots,\wp_n)$$

$$= \left(\begin{array}{c}\Delta\left(\frac{1}{C_n^n}\sum_{1\le j_1\le\ldots<j_n\le n}\left(\prod_{k=1}^{n}\Delta^{-1}\left(s_{j_k},\gamma_{j_k}\right)\right)\right)^{\frac{1}{n}},\\ \left(\begin{array}{c}\left(\sqrt[q]{1-\left(\prod_{1\le j_1\le\ldots<j_n\le n}\left(1-\left(\prod_{k=1}^{n}\vartheta_{j_k}\right)^q\right)\right)^{\frac{1}{C_n^n}}}\right)^{\frac{1}{n}},\\ \sqrt[q]{1-\left(1-\left(\prod_{1\le j_1\le\ldots<j_n\le n}\left(1-\prod_{k=1}^{n}(1-(\varphi_{j_k})^q)\right)\right)^{\frac{1}{C_n^n}}\right)^{\frac{1}{n}}}\end{array}\right)\end{array}\right).$$

$$= \left(\Delta\left(\frac{1}{C_n^n}\cdot(C_n^n)\prod_{k=1}^{n}\Delta^{-1}\left(s_{j_k},\gamma_{j_k}\right)\right)^{\frac{1}{n}},\left(\begin{array}{c}\left(\sqrt[q]{1-\left(1-\left(\prod_{k=1}^{n}\vartheta_{j_k}\right)^q\right)^{(C_n^n)\cdot\frac{1}{C_n^n}}}\right)^{\frac{1}{n}},\\ \sqrt[q]{1-\left(1-\left(1-\prod_{k=1}^{n}(1-(\varphi_{j_k})^q)\right)^{(C_n^n)\cdot\frac{1}{C_n^n}}\right)^{\frac{1}{n}}}\end{array}\right)\right) \quad (\text{let } j_k=j).$$

$$= \left(\Delta\left(\prod_{j=1}^{n}\Delta^{-1}(s_j,\gamma_j)\right)^{\frac{1}{n}},\left(\left(\prod_{j=1}^{n}\vartheta_j\right)^{\frac{1}{n}},\sqrt[q]{1-\left(\prod_{j=1}^{n}(1-(\varphi_j)^q)^{\frac{1}{n}}\right)}\right)\right).$$

10.3.2 The GOFTLWMSM Operator

The significance of the aggregated arguments is not considered by the GOFTLMSM operator. However, the attribute weights play a crucial role in the aggregation process in many real-world circumstances, especially in MAGDM. To overcome the limitations of GOFTLMSM, we shall propose the GOFTLWMSM operator as follows:

Definition 10.8 Let $\wp_k = ((s_k,\gamma_k),(\vartheta_k,\varphi_k))$ $(k=1,2,\ldots,n)$ be any set of GOFTLNs, $\varpi=(\varpi_1,\varpi_2,\ldots,\varpi_n)^T$ be weight vector of $\wp_k$, where $\varpi_k>0$, $\Sigma_{k=1}^{n}$ $\varpi_k=1$. The GOFTLWMSM operator is defined as below:

$$\text{GOFTLWMSM}_{\varpi}^{(t)}(\wp_1, \wp_2, \ldots, \wp_n) = \left(\frac{\bigoplus_{1 \le j_1 < \ldots < j_t \le n} \left(\bigotimes_{k=1}^{t} \left(\wp_{j_k} \right)^{\varpi j_k} \right)}{C_n^t} \right)^{\frac{1}{t}}. \tag{10.5}$$

Theorem 10.3 *Let* $1 \le t \le n (t \in Z)$, $\wp_k = ((s_k, \gamma_k), (\vartheta_k, \varphi_k))$ $(k = 1, 2, \ldots, n)$ *be any set of GOFTLNs,* $\varpi = (\varpi_1, \varpi_2, \ldots, \varpi_n)^T$ *be weight vector of* $\wp_k$*, where* $\varpi_k > 0$, $\Sigma_{k=1}^{n} \varpi_k = 1$. *Then the aggregated value utilizing GOFTLWMSM operator is also a GOFTLN, and*

$$GOFTLWMSM_{\varpi}^{(t)}(\wp_1, \wp_2, \ldots, \wp_n)$$

$$= \left(\begin{array}{c} \Delta \left(\left(\frac{1}{C_n^t} \sum_{1 \le j_1 < \ldots < j_t \le n} \left(\prod_{k=1}^{t} \left(\Delta^{-1} \left(s_{j_k}, \gamma_{j_k} \right)^{\varpi j_k} \right) \right) \right)^{\frac{1}{t}} \right), \\ f \left(\begin{array}{c} \left(\sqrt[q]{1 - \left(\prod_{1 \le j_1 < \ldots < j_t \le n} \left(1 - \left(\prod_{k=1}^{t} \left(\vartheta_{j_k} \right)^{\varpi j_k} \right)^q \right) \right)^{\frac{1}{C_n^t}}} \right)^{\frac{1}{t}}, \\ \sqrt[q]{1 - \left(1 - \left(\prod_{1 \le j_1 < \ldots < j_t \le n} \left(1 - \prod_{k=1}^{t} (1 - (\varphi_{j_k})^q)^{\varpi j_k} \right) \right)^{\frac{1}{C_n^t}} \right)^{\frac{1}{t}}} \end{array} \right) \end{array} \right). \tag{10.6}$$

Proof The proof is similar to that of Theorem 10.1, it is not included here.

Example 10.2 Take $\wp_1 = ((s_2, 0.01), (0.3, 0.8))$, $\wp_2 = ((s_3, 0.05), (0.7, 0.4))$, $\wp_3 = ((s_4, -0.02), (0.6, 0.8))$ and $\wp_4 = ((s_5, 0.07), (0.5, 0.7))$ are four GOFTLSs and $S = \{s_0, s_1, s_2, s_3, s_4, s_5, s_6\}$ with weights $\varpi = (0.18, 0.25, 0.23, 0.34)^T$, utilizing GOFTLWMSM operator to combine these four GOFTLNs. Consider $t = 2$, $q = 4$; we have

$$\Delta \left(\left(\frac{1}{C_n^t} \sum_{1 \le j_1 < \ldots < j_t \le n} \left(\prod_{k=1}^{t} \Delta^{-1} \left(s_{j_k}, \gamma_{j_k} \right)^{\varpi j_k} \right) \right)^{\frac{1}{t}} \right)$$

$$= \Delta \left(\left(\frac{1}{3!} \left(\begin{array}{c} \left(\Delta^{-1} \left((s_2, 0.01)^{0.18} \right) \times \Delta^{-1} \left((s_3, 0.05)^{0.25} \right) \right) + \left(\Delta^{-1} \left((s_2, 0.01)^{0.18} \right) \times \Delta^{-1} \left((s_4, -0.02)^{0.23} \right) \right) + \\ \left(\Delta^{-1} \left((s_2, 0.01)^{0.18} \right) \times \Delta^{-1} \left((s_5, 0.07)^{0.34} \right) \right) + \left(\Delta^{-1} \left((s_3, 0.05)^{0.25} \right) \times \Delta^{-1} \left((s_4, -0.02)^{0.23} \right) \right) + \\ \left(\Delta^{-1} \left((s_3, 0.05)^{0.25} \right) \times \Delta^{-1} \left((s_5, 0.07)^{0.34} \right) \right) + \left(\Delta^{-1} \left((s_4, -0.02)^{0.23} \right) \times \Delta^{-1} \left((s_5, 0.07)^{0.34} \right) \right) \end{array} \right) \right)^{\frac{1}{2}} \right)$$

$$= \Delta (0.8887) = (s_5, 0.0553),$$

$$\left(\sqrt[q]{1 - \left(\prod_{1 \le j_1 < \ldots < j_t \le n} \left(1 - \left(\prod_{k=1}^{t} \left(\vartheta_{j_k} \right)^{\varpi j_k} \right)^q \right) \right)^{\frac{1}{C_n^t}}} \right)^{\frac{1}{t}}$$

$$= \left(\sqrt[4]{1 - \left(\begin{array}{c} (1 - ((0.3)^{0.18} \times (0.7)^{0.25})^4) \times (1 - ((0.3)^{0.18} \times (0.6)^{0.23})^4) \\ \times (1 - ((0.3)^{0.18} \times (0.5)^{0.34})^4) \times (1 - ((0.7)^{0.25} \times (0.6)^{0.23})^4) \\ \times (1 - ((0.7)^{0.25} \times (0.5)^{0.34})^4)(1 - ((0.6)^{0.23} \times (0.5)^{0.34})^4) \end{array} \right)^{\frac{1}{3!}}} \right)^{\frac{1}{2}}$$

$= 0.8544,$

$$\sqrt[q]{1-\left(1-\left(\prod_{1\le j_1<\ldots<j_2\le n}\left(1-\prod_{k=1}^{2}(1-(\varphi_{j_k})^q)^{\varpi_{j_k}}\right)\right)^{\frac{1}{C_n^2}}\right)^{\frac{1}{2}}}$$

$$= \sqrt[q]{1-\left(1-\begin{pmatrix}(1-(1-(0.8)^4)^{0.18}\times(1-(0.4)^4)^{0.25})\\ \times(1-(1-(0.8)^4)^{0.18}\times(1-(0.8)^4)^{0.23})\\ \times(1-(1-(0.8)^4)^{0.18}\times(1-(0.7)^4)^{0.34})\\ \times(1-(1-(0.4)^4)^{0.25}\times(1-(0.8)^4)^{0.23})\\ \times(1-(1-(0.4)^4)^{0.25}\times(1-(0.7)^4)^{0.34})\\ \times(1-(1-(0.8)^4)^{0.23}\times(1-(0.7)^4)^{0.34})\end{pmatrix}^{\frac{1}{3!}}\right)^{\frac{1}{2}}}$$

$= 0.5177.$

Therefore, the aggregation value of the $\wp_1, \wp_2, \wp_3$ *and* $\wp_4$ by the GOFTLWMSM operator is as follows:

$$\text{GOFTLWMSM}_{\varpi}^{(2)} = ((s_5, 0.0553), (0.8544, 0.5177)).$$

We can easily prove that the GOFTLWMSM operator has the following properties:

Property 10.5 (Commutativity) Let $\wp_k$ $(k = 1, 2, \ldots, n)$ be any set of GOFTLNs and $\wp_k'$ be any permutation of $\wp_k$. Then

$$\text{GOFTLWMSM}_{\varpi}^{(t)}(\wp_1, \wp_2, \ldots, \wp_n) = \text{GOFTLWMSM}_{\varpi}^{(t)}(\wp_1', \wp_2', \ldots, \wp_n').$$

Property 10.6 (Monotonicity) Let $\wp_k, \wp_k'$ $(k = 1, 2, \ldots, n)$ be two sets of GOFTLNs. If $(s_k, \gamma_k) \ge (s_k', \gamma_k')$, $\vartheta_k \ge \vartheta_k'$ and $\varphi_k \le \varphi_k'$, for all k, then

$$\text{GOFTLWMSM}_{\varpi}^{(t)}(\wp_1, \wp_2, \ldots, \wp_n) \ge \text{GOFTLWMSM}_{\varpi}^{(t)}(\wp_1', \wp_2', \ldots, \wp_n').$$

Property 10.7 (Boundedness) Suppose that $\wp_k (k = 1, 2, \ldots, n)$ is any set of GOFTLNs, $\wp^- = \min_k \wp_k = (\min_k(s_k, \gamma_k), (\min_k \vartheta_k, \max_k \varphi_k))$ and $\wp^+ = \max_k \wp_k = (\max_k(s_k, \gamma_k), (\max_k \vartheta_k, \min_k \varphi_k))$. Then

$$\wp^- \le \text{GOFTLWMSM}_{\varpi}^{(t)}(\wp_1, \wp_2, \ldots, \wp_n) \le \wp^+.$$

Now, we analyze some particular situations of the GOFTLWMSM operator in terms of the parameter t.

Case 1. When $t = 1$, the GOFTLWMSM operator converts into GOFTLW averaging (GOFTLWA) operator.

$$\text{GOFTLWA}_{\varpi}(\wp_1, \wp_2, \ldots, \wp_n) = \left(\Delta\left(\frac{1}{n}\sum_{k=1}^{n}\Delta^{-1}(s_k, \gamma_k)^{\varpi_k}\right), \left(\sqrt[q]{1-\left(\prod_{k=1}^{n}(1-(\vartheta_k)^{q\varpi_k})\right)^{\frac{1}{n}}}, \left(\prod_{k=1}^{n}\sqrt[q]{1-(1-(\varphi_k)^q)^{\varpi_k}}\right)^{\frac{1}{n}} \right) \right).$$

Case 2. When $t = 2$, the GOFTLWMSM operator converts into GOFTLW Bonferoni mean (GOFTLWBM) operator.

$$\text{GOFTLWBM}_{\varpi}^{(1,1)}(\wp_1, \wp_2, \ldots, \wp_n)$$
$$= \left(\begin{array}{c} \Delta\left(\frac{2}{n(n-1)}\sum_{1\le j_1<j_2\le n}\left(\left(\Delta^{-1}(s_{j_1}, \gamma_{j_1})\right)^{\varpi_{j_1}}\Delta^{-1}(s_{j_2}, \gamma_{j_2})^{\varpi_{j_2}}\right)\right)^{\frac{1}{2}}, \\ \left(\begin{array}{c} \sqrt[q]{1-\left(\prod_{1\le j_1<j_2\le n}\left(1-\left((\vartheta_{j_1})^{\varpi_{j_1}}(\vartheta_{j_2})^{\varpi_{j_2}}\right)^q\right)\right)^{\frac{2}{n(n-1)}}}, \\ \left(\prod_{1\le j_1<j_2\le n}\sqrt[q]{1-((1-(\varphi_{j_1})^q)^{\varpi_{j_1}}(1-(\varphi_{j_2})^q)^{\varpi_{j_2}})}\right)^{\frac{2}{n(n-1)}} \end{array} \right) \end{array} \right).$$

Case 3. When $t = n$, the GOFTLWMSM operator converts into the GOFW geometric (GOFTLWG) operator.

$$\text{GOFTLWGM}_{\varpi}(\wp_1, \wp_2, \ldots, \wp_n)$$
$$= \left(\Delta\left(\prod_{k=1}^{n}\Delta^{-1}(s_k, \gamma_k)^{\varpi_k}\right)^{\frac{1}{n}}, \left(\left(\prod_{k=1}^{n}(\vartheta_k)^{\varpi_k}\right)^{\frac{1}{n}}, \sqrt[q]{\left(1-\prod_{k=1}^{n}(1-(\varphi_k)^q)^{\varpi_k}\right)^{\frac{1}{n}}} \right) \right).$$

10.4 The GOFTLDMSM Aggregation Operator and its Weighted Form

In this section, we generalize the DMSM operator and its weighted form with GOFTLSs to develop the GOFTLDMSM operator and GOFTLWDMSM operator for aggregating GOFTLNs, and study its desirable properties and special cases.

10.4.1 The GOFTLDMSM Operator

The GOFTLDMSM operator is introduced in this subsection, along with its desirable characteristics and special cases.

Definition 10.9 Let $\wp_k = ((s_k, \gamma_k), (\vartheta_k, \varphi_k))$ $(k = 1, 2, \ldots, n)$ be any set of GOFTLNs. Then the GOFTLDMSM operator is defined as follows:

$$\mathrm{GOFTLDMSM}^{(t)}(\wp_1, \wp_2, \ldots, \wp_n) = \frac{1}{t}\left(\otimes_{1\le j_1<\ldots<j_t\le n}\left(\oplus_{k=1}^{t}\wp_{j_k}\right)^{\frac{1}{C_n^t}}\right). \tag{10.7}$$

Theorem 10.4 *Let $\wp_k = ((s_k, \gamma_k), (\vartheta_k, \varphi_k))$ $(k = 1, 2, \ldots, n)$ be any set of GOFTLNs. Then the aggregated value utilizing GOFTLDMSM operator is also a GOFTLN, and*

$$GOFTLDMSM^{(t)}(\wp_1, \wp_2, \ldots, \wp_n) = \left(\begin{array}{l} \Delta\left(\frac{1}{t}\prod_{1\le j_1<\ldots<j_t\le n}\left(\sum_{k=1}^{t}\left(\Delta^{-1}\left(s_{j_k}, \gamma_{j_k}\right)\right)\right)^{\frac{1}{C_n^t}}\right), \\ \left(\left(\sqrt[q]{1-\left(1-\prod_{1\le j_1<\ldots<j_t\le n}\left(1-\prod_{k=1}^{t}\left(1-(\vartheta_{j_k})^q\right)\right)^{\frac{1}{C_n^t}}\right)^{\frac{1}{t}}}, \right.\right. \\ \left.\left.\left(\sqrt[q]{1-\prod_{1\le j_1<\ldots<j_t\le n}\left(1-\left(\prod_{k=1}^{t}\varphi_{j_k}\right)^q\right)^{\frac{1}{C_n^t}}}\right)^{\frac{1}{t}}\right)\right) \end{array}\right). \tag{10.8}$$

Proof Depending on the operations of GOFTLNs, we have

$$\oplus_{k=1}^{t}\wp_{j_k} = \left(\Delta\left(\sum_{k=1}^{t}\Delta^{-1}\left(s_{j_k}, \gamma_{j_k}\right)\right), \left(\sqrt[q]{1-\prod_{k=1}^{t}(1-(\vartheta_{j_k})^q)}, \prod_{k=1}^{t}\varphi_{j_k}\right)\right).$$

and

$$\left(\oplus_{k=1}^{t}\wp_{j_k}\right)^{\frac{1}{C_n^t}} = \left(\Delta\left(\sum_{k=1}^{t}\Delta^{-1}\left(s_{j_k}, \gamma_{j_k}\right)\right)^{\frac{1}{C_n^t}}, \left(\left(\sqrt[q]{1-\prod_{k=1}^{t}(1-(\vartheta_{j_k})^q)}\right)^{\frac{1}{C_n^t}}, \sqrt[q]{1-\left(1-\left(\prod_{k=1}^{t}\varphi_{j_k}\right)^q\right)^{\frac{1}{C_n^t}}}\right)\right).$$

Thus we obtain

$$\otimes_{1\le j_1<\ldots<j_t\le n}\left(\oplus_{k=1}^{t}\wp_{j_k}\right)^{\frac{1}{C_n^t}} = \left(\begin{array}{l} \Delta\left(\prod_{1\le j_1<\ldots<j_t\le n}\left(\sum_{k=1}^{t}\Delta^{-1}\left(s_{j_k}, \gamma_{j_k}\right)\right)^{\frac{1}{C_n^t}}\right), \\ \left(\prod_{1\le j_1<\ldots<j_t\le n}\left(\sqrt[q]{1-\prod_{k=1}^{t}(1-(\vartheta_{j_k})^q)}\right)^{\frac{1}{C_n^t}}, \right. \\ \left.\sqrt[q]{1-\prod_{1\le j_1<\ldots<j_t\le n}\left(1-\left(\prod_{k=1}^{t}\varphi_{j_k}\right)^q\right)^{\frac{1}{C_n^t}}}\right) \end{array}\right).$$

Accordingly,

$$
\begin{aligned}
&\mathrm{GOFTLDMSM}^{(t)}(\wp_1, \wp_2, \ldots, \wp_n) \\
&= \left(\begin{array}{l}
\Delta\left(\frac{1}{t} \prod\limits_{1\le j_1<\ldots<j_t\le n} \left(\sum\limits_{k=1}^{t}\left(\Delta^{-1}\left(s_{j_k}, \gamma_{j_k}\right)\right)\right)^{\frac{1}{C_n^t}}\right), \\
\left(\sqrt[q]{1-\left(1-\prod\limits_{1\le j_1<\ldots<j_t\le n}\left(1-\prod\limits_{k=1}^{t}\left(1-(\vartheta_{j_k})^q\right)\right)^{\frac{1}{C_n^t}}\right)^{\frac{1}{t}}}, \\
\left(\sqrt[q]{1-\prod\limits_{1\le j_1<\ldots<j_t\le n}\left(1-\left(\prod\limits_{k=1}^{t}\varphi_{j_k}\right)^q\right)^{\frac{1}{C_n^t}}}\right)^{\frac{1}{t}}\right)
\end{array}\right).
\end{aligned}
$$

Example 10.3 Take $\wp_1 = ((s_2, 0.01), (0.3, 0.8))$, $\wp_2 = ((s_3, 0.05), (0.7, 0.4))$, $\wp_3 = ((s_4, -0.02), (0.6, 0.8))$ and $\wp_4 = ((s_5, 0.07), (0.5, 0.7))$ are four GOFTLNs, and $S = \{s_0, s_1, s_2, s_3, s_4, s_5, s_6\}$, utilizing GOFTLDMSM operator to aggregate these four GOFTLNs. Consider $t = 2$, $q = 4$; we have

$$
\begin{aligned}
&\Delta\left(\frac{1}{t} \prod_{1\le j_1<\ldots<j_t\le n} \left(\sum_{k=1}^{t}\left(\Delta^{-1}\left(s_{j_k}, \gamma_{j_k}\right)\right)\right)^{\frac{1}{C_n^t}}\right) \\
&= \Delta\left(\frac{1}{2}\left(\begin{array}{l}
\left(\Delta^{-1}(s_2, 0.01) + \Delta^{-1}(s_3, 0.05)\right)^{\frac{1}{3!}} \times \left(\Delta^{-1}(s_2, 0.01) + \Delta^{-1}(s_4, -0.02)\right)^{\frac{1}{3!}} \\
\times\left(\Delta^{-1}(s_2, 0.01) + \Delta^{-1}(s_5, 0.07)\right)^{\frac{1}{3!}} \times \left(\Delta^{-1}(s_3, 0.05) + \Delta^{-1}(s_4, -0.02)\right)^{\frac{1}{3!}} \\
\times\left(\Delta^{-1}(s_3, 0.05) + \Delta^{-1}(s_5, 0.07)\right)^{\frac{1}{3!}} \times \left(\Delta^{-1}(s_4, -0.02) + \Delta^{-1}(s_5, 0.07)\right)^{\frac{1}{3!}}
\end{array}\right)\right) \\
&= \Delta(0.5995) = (s_4, -0.0671),
\end{aligned}
$$

$$
\begin{aligned}
&\sqrt[q]{1-\left(1-\prod_{1\le j_1<\ldots<j_t\le n}\left(1-\prod_{k=1}^{t}\left(1-(\vartheta_{j_k})^q\right)\right)^{\frac{1}{C_n^t}}\right)^{\frac{1}{t}}} \\
&= \sqrt[4]{1-\left(1-\left(\begin{array}{l}
(1-(1-(0.3)^4)\times(1-(0.7)^4))\times(1-(1-(0.3)^4)\times(1-(0.6)^4)) \\
\times(1-(1-(0.3)^4)\times(1-(0.5)^4))\times(1-(1-(0.7)^4)\times(1-(0.6)^4)) \\
\times(1-(1-(0.7)^4)\times(1-(0.5)^4))\times(1-(1-(0.6)^4)\times(1-(0.5)^4))
\end{array}\right)^{\frac{1}{3!}}\right)^{\frac{1}{2}}} \\
&= 0.5597,
\end{aligned}
$$

$$
\begin{aligned}
&\left(\sqrt[q]{1-\prod_{1\le j_1<\ldots<j_t\le n}\left(1-\left(\prod_{k=1}^{t}\varphi_{j_k}\right)^q\right)^{\frac{1}{C_n^t}}}\right)^{\frac{1}{t}} \\
&= \left(\sqrt[4]{1-\left(\begin{array}{l}
(1-(0.8\times0.4)^4)\times(1-(0.8\times0.8)^4)\times(1-(0.8\times0.7)^4)\times(1-(0.4\times0.8)^4) \\
\times(1-(0.4\times0.7)^4)\times(1-(0.8\times0.7)^4)
\end{array}\right)^{\frac{1}{3!}}}\right)^{\frac{1}{2}} \\
&= 0.7137.
\end{aligned}
$$

Therefore, the aggregation value of the $\wp_1, \wp_2, \wp_3$ *and* $\wp_4$ by the GOFTLDMSM operator is as follows:

$$\text{GOFTLDMSM}^{(2)} = ((s_4, -0.0671), (0.5597, 0.7137)).$$

The GOFTLDMSM operator has the following properties:

Property 10.8 (Idempotency) If all $\wp_k$ $(k = 1, 2, \ldots, n)$ are equal, i.e., $\wp_k = \wp$ for all k, then

$$\text{GOFTLDMSM}^{(t)}(\wp_1, \wp_2, \ldots, \wp_n) = \wp.$$

Property 10.9 (Commutativity) Let $\wp_k$ $(k = 1, 2, \ldots, n)$ be any set of GOFTLNs and $\acute{\wp}_k$ $(k = 1, 2, \ldots, n)$ be any permutation of $\wp_k$ $(k = 1, 2, \ldots, n)$. Then

$$\text{GOFTLDMSM}^{(t)}(\wp_1, \wp_2, \ldots, \wp_n) = \text{GOFTLDMSM}^{(t)}(\wp_1', \wp_2', \ldots, \wp_n').$$

Property 10.10 (Monotonicity) Let $\wp_k, \acute{\wp}_k$ $(k = 1, 2, \ldots, n)$ be two sets of GOFTLNs. If $(s_k, \gamma_k) \geq (s_k', \gamma_k')$, $\vartheta_k \geq \vartheta_k'$ and $\varphi_k \leq \varphi_k'$, for all k, then

$$\text{GOFTLDMSM}^{(t)}(\wp_1, \wp_2, \ldots, \wp_n) \geq \text{GOFTLDMSM}^{(t)}(\wp_1', \wp_2', \ldots, \wp_n').$$

Property 10.11 (Boundedness) Suppose that $\wp_k$ $(k = 1, 2, \ldots, n)$ be any set of GOFTLNs, $\wp^- = \min_k \wp_k = (\min_k(s_k, \gamma_k), (\min_k \vartheta_k, \max_k \varphi_k))$ and $\wp^+ = \max_k \wp_k = (\max_k(s_k, \gamma_k), (\max_k \vartheta_k, \min_k \varphi_k))$. Then

$$\wp^- \leq \text{GOFTLDMSM}^{(t)}(\wp_1, \wp_2, \ldots, \wp_n) \leq \wp^+.$$

Now, we analyze some particular situations of the GOFTLDMSM operator in terms of the parameter t.

Case 1. When $t = 1$, the GOFTLDMSM operator converts into GOFTL geometric operator.

Case 2. When $t = 2$, the GOFTLDMSM operator converts into GOFTL geometric Bonferoni mean operator.

Case 3. When $t = n$, the GOFTLDMSM operator converts into GOFTL averaging operator.

10.4.2 The GOFTLWDMSM Operator

The significance of the aggregated arguments is not considered by the GOFTLDMSM operator. However, attribute weights play a crucial role in the aggregation process in many real-world circumstances, especially in MAGDM. To overcome the limitations of GOFTLDMSMs, we shall propose the GOFTLWDMSM operator as follows:

Definition 10.10 Suppose that $\wp_k = ((s_k, \gamma_k), (\vartheta_k, \varphi_k))$ $(k = 1, 2, \ldots, n)$ is any set of GOFTLNs and $\varpi = (\varpi_1, \varpi_2, \ldots, \varpi_n)^T$ is the weight vector of $\wp_k$, where $\varpi_k > 0$, $\sum_{k=1}^{n} \varpi_k = 1$. The GOFTLWDMSM operator is defined as follows:

$$\mathrm{GOFTLWDMSM}_{\varpi}^{(t)}(\wp_1, \wp_2, \ldots, \wp_n) = \frac{1}{t}\left(\otimes_{1 \le j_1 < \ldots < j_t \le n}\left(\oplus_{k=1}^{t}\left(\varpi_{j_k}\wp_{j_k}\right)\right)^{\frac{1}{C_n^t}}\right). \tag{10.9}$$

Theorem 10.5 *Let* $1 \le t \le n$ $(t \in Z)$, $\wp_k = ((s_k, \gamma_k), (\vartheta_k, \varphi_k))$ $(k = 1, 2, \ldots, n)$ *be any set of GOFTLNs and* $\varpi = (\varpi_1, \varpi_2, \ldots, \varpi_n)^T$ *be weight vector of* $\wp_k$, *where* $\varpi_k > 0$, $\sum_{k=1}^{t} \varpi_k = 1$. *Then the aggregated value by using GOFTLWDMSM operator is also a GOFTLN, and*

$$GOFTLWDMSM_{\varpi}^{(t)}(\wp_1, \wp_2, \ldots, \wp_n)$$
$$= \left(\begin{array}{l} \Delta\left(\frac{1}{t}\prod_{1 \le j_1 < \ldots < j_t \le n}\left(\sum_{k=1}^{t}\left(\Delta^{-1}\left(\varpi_{j_k} \times \left(s_{j_k}, \gamma_{j_k}\right)\right)\right)\right)^{\frac{1}{C_n^t}}\right), \\ \left(\left(\sqrt[q]{1-\left(1-\prod_{1 \le j_1 < \ldots < j_t \le n}\left(1-\prod_{k=1}^{t}\left(1-(\vartheta_{j_k})^q\right)^{\varpi_{j_k}}\right)^{\frac{1}{C_n^t}}\right)^{\frac{1}{t}}}\right), \left(\sqrt[q]{1-\prod_{1 \le j_1 < \ldots < j_t \le n}\left(1-\left(\prod_{k=1}^{t}(\varphi_{j_k})^{\varpi_{j_k}}\right)^q\right)^{\frac{1}{C_n^t}}}\right)^{\frac{1}{t}}\right) \end{array}\right). \tag{10.10}$$

Proof The proof is similar to that of Theorem 10.4, it is not included here.

Example 10.4 Take $\wp_1 = ((s_2, 0.01), (0.3, 0.8))$, $\wp_2 = ((s_3, 0.05), (0.7, 0.4))$, $\wp_3 = ((s_4, -0.02), (0.6, 0.8))$ and $\wp_4 = ((s_5, 0.07), (0.5, 0.7))$ are four GOFTLNs, and $S = \{s_0, s_1, s_2, s_3, s_4, s_5, s_6\}$, with weights $\varpi = (0.18, 0.25, 0.23, 0.34)$, utilizing GOFTLWDMSM operator to aggregate these four GOFTLNs. Consider $t = 2$, $q = 4$; thus

$$\Delta\left(\frac{1}{t}\prod_{1\le j_1<\ldots<j_t\le n}\left(\sum_{k=1}^{t}\left(\varpi_{j_k}\times\Delta^{-1}\left(s_{j_k},\gamma_{j_k}\right)\right)\right)^{\frac{1}{C_n^t}}\right)$$

$$=\Delta\left(\frac{1}{2}\left(\begin{array}{l}\left((0.18)\times\Delta^{-1}(s_2,0.01)+(0.25)\times\Delta^{-1}(s_3,0.05)\right)^{\frac{1}{3!}}\\ \times\left((0.18)\times\Delta^{-1}(s_2,0.01)+(0.23)\times\Delta^{-1}(s_4,-0.02)\right)^{\frac{1}{3!}}\\ \times\left((0.18)\times\Delta^{-1}(s_2,0.01)+(0.34)\times\Delta^{-1}(s_5,0.07)\right)^{\frac{1}{3!}}\\ \times\left((0.25)\times\Delta^{-1}(s_3,0.05)+(0.23)\times\Delta^{-1}(s_4,-0.02)\right)^{\frac{1}{3!}}\\ \times\left((0.25)\times\Delta^{-1}(s_3,0.05)+(0.34)\times\Delta^{-1}(s_5,0.07)\right)^{\frac{1}{3!}}\\ \times\left((0.23)\times\Delta^{-1}(s_4,-0.02)+(0.34)\times\Delta^{-1}(s_5,0.07)\right)^{\frac{1}{3!}}\end{array}\right)\right)$$

$$=\Delta(0.1553)=(s_1,-0.0114),$$

$$\sqrt[q]{1-\left(1-\prod_{1\le j_1<\ldots<j_t\le n}\left(1-\prod_{k=1}^{t}(1-(\vartheta_{j_k})^q)^{\varpi_{j_k}}\right)^{\frac{1}{C_n^t}}\right)^{\frac{1}{t}}}$$

$$=\sqrt[4]{1-\left(1-\left(\begin{array}{l}(1-(1-(0.3)^4)^{0.18}\times(1-(0.4)^4)^{0.25})\\ \times(1-(1-(0.3)^4)^{0.18}\times(1-(0.6)^4)^{0.23})\\ \times(1-(1-(0.3)^4)^{0.18}\times(1-(0.5)^4)^{0.34})\\ \times(1-(1-(0.4)^4)^{0.25}\times(1-(0.6)^4)^{0.23})\\ \times(1-(1-(0.4)^4)^{0.25}\times(1-(0.5)^4)^{0.34})\\ \times(1-(1-(0.6)^4)^{0.23}\times(1-(0.5)^4)^{0.34})\end{array}\right)^{\frac{1}{3!}}\right)^{\frac{1}{2}}}$$

$$=0.4050,$$

$$\left(\sqrt[q]{1-\prod_{1\le j_1<\ldots<j_t\le n}\left(1-\left(\prod_{k=1}^{t}(\varphi_{j_k})^{\varpi_{j_k}}\right)^q\right)^{\frac{1}{C_n^t}}}\right)^{\frac{1}{t}}$$

$$=\left(\sqrt[4]{1-\left(\begin{array}{l}(1-((0.1)^{0.18}\times(0.4)^{0.25})^4)\times(1-((0.1)^{0.18}\times(0.8)^{0.23})^4)\\ \times(1-((0.1)^{0.18}\times(0.3)^{0.34})^4)\times(1-((0.4)^{0.25}\times(0.8)^{0.23})^4)\\ \times(1-((0.4)^{0.25}\times(0.3)^{0.34})^4)\times(1-((0.8)^{0.23}\times(0.3)^{0.34})^4)\end{array}\right)^{\frac{1}{3!}}}\right)^{\frac{1}{2}}$$

$$=0.9079.$$

Therefore, the aggregation value of the $\wp_1, \wp_2, \wp_3$ *and* $\wp_4$ by the GOFTLWDMSM operator mentioned below:

$$\text{GOFTLWDMSM}_{\varpi}^{(2)} = ((s_1, -0.0114), (0.4050, 0.9079)).$$

The GOFTLWDMSM operator satisfies the following properties:

Property 10.12 (Commutativity) Let $\wp_k$ $(k = 1, 2, \ldots, n)$ be any set of GOFTLNs, and $\acute{\wp}_k$ be any permutation of $\wp_k$, then

$$\text{GOFTLWDMSM}_{\varpi}^{(t)}(\wp_1, \wp_2, \ldots, \wp_n) = \text{GOFTLWDMSM}_{\varpi}^{(t)}(\wp_1', \wp_2', \ldots, \wp_n').$$

Property 10.13 (Monotonicity) Let $\wp_k, \acute{\wp}_k$ $(k = 1, 2, \ldots, n)$ be two sets of GOFTLNs. If $(s_k, \gamma_k) \geq (s_k', \gamma_k')$, $\vartheta_k \geq \vartheta_k'$ and $\varphi_k \leq \varphi_k'$, for all k, then

$$\text{GOFTLWDMSM}_{\varpi}^{(t)}(\wp_1, \wp_2, \ldots, \wp_n) \geq \text{GOFTLWDMSM}_{\varpi}^{(t)}(\wp_1', \wp_2', \ldots, \wp_n').$$

Property 10.14 (Boundedness) Suppose that $\wp_k$ $(k = 1, 2, \ldots, n)$ is any set of GOFTLNs, $\wp^- = \min_k \wp_k = (\min_k(s_k, \gamma_k), (\min_k \vartheta_k, \max_k \varphi_k))$ and $\wp^+ = \max_k \wp_k = (\max_k(s_k, \gamma_k), (\max_k \vartheta_k, \min_k \varphi_k))$. Then

$$\wp^- \leq \text{GOFTLWDMSM}_{\varpi}^{(t)}(\wp_1, \wp_2, \ldots, \wp_n) \leq \wp^+.$$

Now, we analyze some special situations of the GOFTLWDMSM operator in terms of the parameter t.

Case 1. When $t = 1$, the GOFTLWDMSM operator converts into GOFTLWG operator.

Case 2. When $t = 2$, the GOFTLWDMSM operator converts into GOFTLW geometric Bonferoni mean (GOFTLWGBM) operator.

Case 3. When $t = n$, the GOFTLWDMSM operator converts into GOFTLWA operator.

10.5 An MAGDM Model with GOFTL Information

For MAGDM problems under GOFTL environment, let $\mathfrak{F} = \{\mathfrak{F}_1, \mathfrak{F}_2, \ldots, \mathfrak{F}_m\}$ be a set of alternatives, $\mathcal{D} = \{\mathcal{D}_1, \mathcal{D}_2, \ldots, \mathcal{D}_{\hat{n}}\}$ be the set of DMs, $\varsigma = \{\varsigma_1, \varsigma_2, \ldots, \varsigma_n\}$ be the set of attributes, and ϖ_k is the weight vector of attributes $\varsigma_k (k = 1, 2, \ldots, n)$, with $\varpi_k \in [0, 1]$, $\sum_{k=1}^{n} \varpi_k = 1$. Consider $N^l = (\wp_{jk}^l)_{m \times n}$ is the GOFTL decision matrix given by the DMs where $\wp_{jk}^l = ((s_{jk}^l, \gamma_{jk}^l), (\vartheta_{jk}^l, \varphi_{jk}^l))$ be a GOFTLN given by the DMs $\mathcal{D}_l$ $(l = 1, 2, \ldots, \hat{n})$ for the alternatives $\mathfrak{F}_j$ $(j = 1, 2, \ldots, m)$ with respect to the attributes ς_k $(k = 1, 2, \ldots, n)$, and $\vartheta_{jk}^l \in [0, 1]$, $\varphi_{jk}^l \in [0, 1]$, $(\vartheta_{jk}^l)^q + (\varphi_{jk}^l)^q \leq$

$1, q \geq 1$, and $s^l_{jk} \in S = \{s_0, s_1, s_2, \ldots, s_\tau\}$. To choose most preferable alternative(s), the developed operators are used to build MAGDM approach under GOFTL environment, consisting of the following steps in order to obtain the most preferable alternative(s):

Step 1. Utilize the GOFTLMSM operator in Eq. (10.4) or GOFTLDMSM operator in Eq. (10.8) to combine all individual GOFTL decision matrices $N^l = (\wp^l_{jk})_{m\times n}$ $(l = 1, 2, \ldots, \grave{n})$ into a collective GOFTL decision matrix $N = (\wp_{jk})_{m\times n}$.

$$\wp_{jk} = \text{GOFTLMSM}(\wp^1_{jk}, \wp^2_{jk}, \ldots, \wp^{\grave{n}}_{jk})$$

or

$$\wp_{jk} = \text{GOFTLDMSM}(\wp^1_{jk}, \wp^2_{jk}, \ldots, \wp^{\grave{n}}_{jk}).$$

Step 2. Utilize the GOFTLWMSM operator in Eq. (10.6) or GOFTLWDMSM operator in Eq. (10.10) to fuse the GOFTL assessment values $\wp_{jk} (k = 1, 2, \ldots, n)$ into overall assessment value $\wp_j$ of the alternative $\mathfrak{F}_j (j = 1, 2, \ldots, m)$.

$$\wp_j = \text{GOFTLWMSM}(\wp_{j1}, \wp_{j2}, \ldots, \wp_{j_n}), \quad (j = 1, 2, \ldots, m)$$

or

$$\wp_j = \text{GOFTLWDMSM}(\wp_{j1}, \wp_{j2}, \ldots, \wp_{j_n}), \quad (j = 1, 2, \ldots, m).$$

Step 3. Compute the score function $\mathcal{S}(\wp_j)$ and accuracy function $\mathcal{H}(\wp_j)$ of overall assessment value $\wp_j$ $(j = 1, 2, \ldots, m)$ by Eqs. (10.1, 10.2), respectively.

Step 4. Rank all the alternatives $\mathfrak{F}_j$ $(j = 1, 2, \ldots, m)$ according to Definition 10.6 and choose the most desirable alternative(s).

Step 5. End.

We give the following flowchart (Fig. 10.2) to better explain the steps of the MAGDM approach developed in this chapter.

10.6 Illustrative Example and Discussion

In this section, we will use the developed aggregation operators in the decision-making process for GOFTLSs to evaluate the approach with an illustrative example.

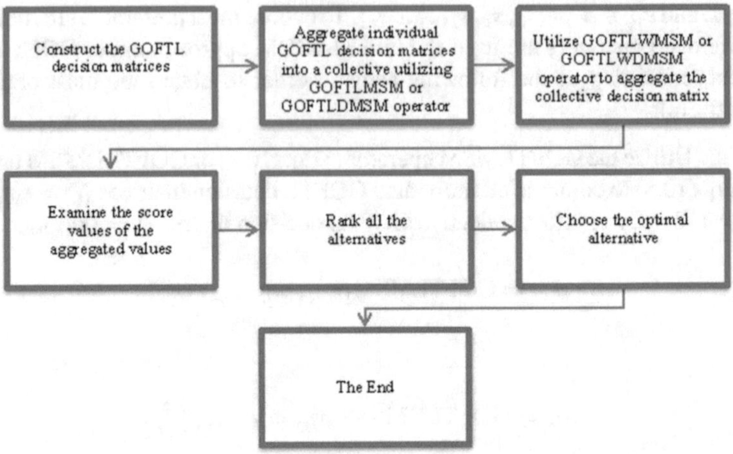

Fig. 10.2 The flowchart of the developed approach

10.6.1 Evaluation Process of the Proposed Method

In this subsection, we will present an illustrative example to demonstrate the evaluation of supplier selection from the EFG of companies with GOFTL information to explain the approach suggested in this study. An EFG of companies wants to choose a supplier and select the best global supplier for maximizing profit. Suppose that $\mathfrak{F} = \mathfrak{F}_j$ $(j = 1, 2, 3, 4, 5)$ be a group of five potential suppliers that are regarded as the suppliers, and the EFG of companies consider the following four attributes when evaluating the suppliers: (1) ς_1 represents functionality and technology; (2) ς_2 represents strategic fitness; (3) ς_3 represents supplier's ability; and (4) ς_4 represents supplier's reputation. Furthermore, $\varpi = (0.18, 0.25, 0.23, 0.34)^T$ be the weight vector of the four attributes and $\sum_{k=1}^{n} \varpi_k = 1$. Since the EFG of companies use 2TL environment i.e.,

$$S = \{s_0 = \text{extremely low}, s_1 = \text{very low}, s_2 = \text{poor}, s_3 = \text{fair}, s_4 = \text{good}, s_5 = \text{very good}, s_6 = \text{excellently}\}.$$

The individual linguistic decision matrices $N^l = (\wp^l_{jk})_{5\times4}$ $(l = 1, 2, 3, 4)$ are built according to the opinions of four experts, which are appeared in Tables 10.2, 10.3, 10.4 and 10.5. Since all attributes have an impact on others, and the attribute values take the form of GOFTL, in what follows, we will use the proposed approach to make a selection of the most preferable supplier(s). The hierarchical structure of the selected application is shown in Fig. 10.3.

Table 10.2 GOFTL decision framework given by $\mathcal{D}_1$

	ς_1	ς_2	ς_3	ς_4
$\mathfrak{F}_1$	$((s_6, 0), (0.4, 0.6))$	$((s_4, 0), (0.5, 0.2))$	$((s_5, 0), (0.3, 0.6))$	$((s_6, 0), (0.1, 0.5))$
$\mathfrak{F}_2$	$((s_4, 0), (0.2, 0.1))$	$((s_6, 0), (0.3, 0.2))$	$((s_4, 0), (0.8, 0.5))$	$((s_5, 0), (0.5, 0.2))$
$\mathfrak{F}_3$	$((s_5, 0), (0.2, 0.7))$	$((s_3, 0), (0.4, 0.6))$	$((s_6, 0), (0.1, 0.5))$	$((s_3, 0), (0.6, 0.8))$
$\mathfrak{F}_4$	$((s_2, 0), (0.7, 0.6))$	$((s_5, 0), (0.5, 0.4))$	$((s_3, 0), (0.3, 0.4))$	$((s_6, 0), (0.6, 0.4))$
$\mathfrak{F}_5$	$((s_3, 0), (0.3, 0.4))$	$((s_6, 0), (0.5, 0.1))$	$((s_3, 0), (0.6, 0.8))$	$((s_4, 0), (0.3, 0.7))$

Table 10.3 GOFTL decision framework given by $\mathcal{D}_2$

	ς_1	ς_2	ς_3	ς_4
$\mathfrak{F}_1$	$((s_5, 0), (0.3, 0.4))$	$((s_4, 0), (0.5, 0.2))$	$((s_3, 0), (0.2, 0.6))$	$((s_6, 0), (0.4, 0.3))$
$\mathfrak{F}_2$	$((s_4, 0), (0.4, 0.5))$	$((s_5, 0), (0.3, 0.8))$	$((s_4, 0), (0.5, 0.5))$	$((s_6, 0), (0.7, 0.4))$
$\mathfrak{F}_3$	$((s_5, 0), (0.6, 0.1))$	$((s_4, 0), (0.4, 0.6))$	$((s_3, 0), (0.8, 0.5))$	$((s_3, 0), (0.4, 0.7))$
$\mathfrak{F}_4$	$((s_3, 0), (0.6, 0.2))$	$((s_6, 0), (0.5, 0.4))$	$((s_5, 0), (0.2, 0.4))$	$((s_4, 0), (0.2, 0.4))$
$\mathfrak{F}_5$	$((s_5, 0), (0.6, 0.6))$	$((s_6, 0), (0.5, 0.6))$	$((s_4, 0), (0.1, 0.2))$	$((s_3, 0), (0.5, 0.2))$

Table 10.4 GOFTL decision framework given by $\mathcal{D}_3$

	ς_1	ς_2	ς_3	ς_4
$\mathfrak{F}_1$	$((s_4, 0), (0.1, 0.2))$	$((s_4, 0), (0.5, 0.5))$	$((s_3, 0), (0.3, 0.6))$	$((s_5, 0), (0.8, 0.3))$
$\mathfrak{F}_2$	$((s_4, 0), (0.7, 0.5))$	$((s_5, 0), (0.3, 0.2))$	$((s_5, 0), (0.6, 0.5))$	$((s_4, 0), (0.2, 0.2))$
$\mathfrak{F}_3$	$((s_5, 0), (0.2, 0.1))$	$((s_4, 0), (0.4, 0.4))$	$((s_6, 0), (0.2, 0.5))$	$((s_5, 0), (0.3, 0.6))$
$\mathfrak{F}_4$	$((s_4, 0), (0.1, 0.2))$	$((s_3, 0), (0.5, 0.3))$	$((s_5, 0), (0.7, 0.2))$	$((s_6, 0), (0.2, 0.1))$
$\mathfrak{F}_5$	$((s_5, 0), (0.5, 0.6))$	$((s_4, 0), (0.5, 0.1))$	$((s_4, 0), (0.2, 0.5))$	$((s_3, 0), (0.5, 0.6))$

Table 10.5 GOFTL decision framework given by $\mathcal{D}_4$

	ς_1	ς_2	ς_3	ς_4
$\mathfrak{F}_1$	$((s_2, 0), (0.5, 0.1))$	$((s_3, 0), (0.3, 0.5))$	$((s_4, 0), (0.3, 0.4))$	$((s_2, 0), (0.6, 0.3))$
$\mathfrak{F}_2$	$((s_3, 0), (0.7, 0.4))$	$((s_3, 0), (0.3, 0.1))$	$((s_5, 0), (0.6, 0.2))$	$((s_2, 0), (0.1, 0.2))$
$\mathfrak{F}_3$	$((s_3, 0), (0.4, 0.2))$	$((s_2, 0), (0.1, 0.3))$	$((s_2, 0), (0.2, 0.1))$	$((s_5, 0), (0.4, 0.6))$
$\mathfrak{F}_4$	$((s_2, 0), (0.1, 0.1))$	$((s_2, 0), (0.3, 0.1))$	$((s_2, 0), (0.7, 0.3))$	$((s_3, 0), (0.2, 0.1))$
$\mathfrak{F}_5$	$((s_1, 0), (0.2, 0.1))$	$((s_3, 0), (0.3, 0.2))$	$((s_4, 0), (0.2, 0.1))$	$((s_3, 0), (0.2, 0.6))$

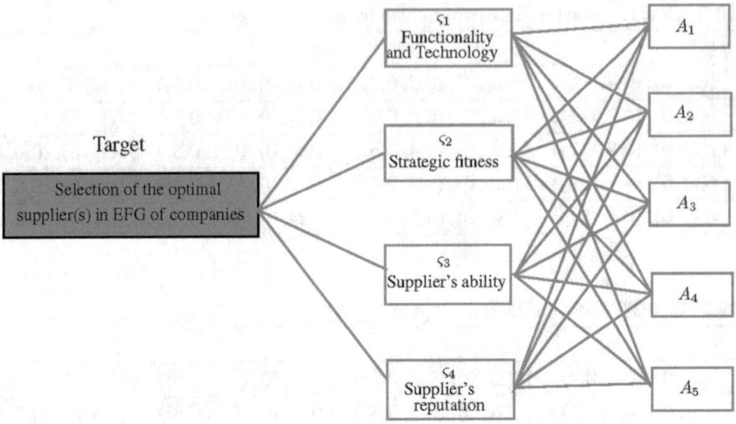

Fig. 10.3 Hierarchical structure of supplier selection

10.6.1.1 Decision-Making Procedure Using the GOFTLWMSM Operator

Step 1. Utilize the GOFTLMSM operator from Eq. (10.4) to combine all individual GOFTL decision matrices $N^l = (\wp^l_{jk})_{5\times 4}$ $(l = 1, 2, 3, 4)$ into a collective GOFTL decision framework $N = (\wp_{jk})_{5\times 4}$ shown in Table 10.6 (take $t = 3$ and $q = 4$).

Step 2. Utilize the GOFTLWMSM operator in Eq. (10.6) to combine the GOFTL estimated values $\wp_{jk}$ of supplier $\mathfrak{F}_j$ on all attributes $\varsigma_k (k = 1, 2, 3, 4)$ into the overall estimated value $\wp_j$ of the supplier $\mathfrak{F}_j$ $(j = 1, 2, 3, 4, 5)$ (take $q = 4$). The overall estimated values of suppliers $\mathfrak{F}_j$ $(j = 1, 2, 3, 4, 5)$ are given as:

$$\wp_1 = ((s_5, 0.0724), (0.7985, 0.3238)), \wp_2 = ((s_6, -0.0833), (0.8118, 0.2952)),$$
$$\wp_3 = ((s_5, 0.0599), (0.7742, 0.4057)), \wp_4 = ((s_5, 0.0563), (0.7866, 0.2451)),$$
$$\wp_5 = ((s_5, 0.0500), (0.7826, 0.3791)).$$

Step 3. Calculate the score function $\mathcal{S}(\wp_j)$ by using the Eq. (10.1) of overall estimated value.

$$\mathcal{S}(\wp_1) = 0.6320,\ \mathcal{S}(\wp_2) = 0.6539,\ \mathcal{S}(\wp_3) = 0.5949,\ \mathcal{S}(\wp_4) = 0.6135,\ \mathcal{S}(\wp_5) = 0.5982.$$

Step 4. Rank the suppliers depending on the score index, and get $\mathfrak{F}_2 \succ \mathfrak{F}_1 \succ \mathfrak{F}_4 \succ \mathfrak{F}_5 \succ \mathfrak{F}_3$, where the symbol $\succ$ means "the most preferable". Thus, the best supplier is $\mathfrak{F}_2$.

Step 5. End.

Table 10.6 The combined GOFTL decision framework utilizing GOFTLMSM operator

	ς_1	ς_2	ς_3	ς_4
$\mathfrak{F}_1$	$((s_4, 0.0103), (0.3493, 0.4259))$	$((s_4, -0.0446), (0.4578, 0.4189))$	$((s_4, -0.0516), (0.2778, 0.5673))$	$((s_5, -0.0782), (0.5142, 0.3746))$
$\mathfrak{F}_2$	$((s_4, -0.0446), (0.5207, 0.4385))$	$((s_5, -0.0565), (0.3000, 0.4548))$	$((s_4, 0.0802), (0.6263, 0.4653))$	$((s_4, 0.0103), (0.3691, 0.2801))$
$\mathfrak{F}_3$	$((s_4, 0.0733), (0.3441, 0.3944))$	$((s_3, 0.0291), (0.3567, 0.5183))$	$((s_4, -0.0035), (0.2856, 0.4643))$	$((s_4, -0.0142), (0.4263, 0.6924))$
$\mathfrak{F}_4$	$((s_3, -0.0553), (0.3281, 0.3650))$	$((s_4, -0.0367), (0.4578, 0.3467))$	$((s_4, -0.0684), (0.4777, 0.3492))$	$((s_5, -0.0623), (0.2817, 0.3323))$
$\mathfrak{F}_5$	$((s_3, 0.0319), (0.4063, 0.5144))$	$((s_5, -0.0623), (0.4578, 0.3508))$	$((s_4, -0.0446), (0.2595, 0.5613))$	$((s_3, 0.0386), (0.3827, 0.5948))$

10.6.1.2 Decision-Making Procedure Using the GOFTLWDMSM Operator

Step 1. Utilize the GOFTLDMSM operator from Eq. (10.8) to combine all individual GOFTL decision matrices $N^l = (\wp^l_{jk})_{5\times 4}$ $(l = 1, 2, 3, 4)$ into a collective GOFTL decision framework $N = (\wp_{jk})_{5\times 4}$ shown in Table 10.7 (suppose $t = 3$ and $q = 4$).

Step 2. Utilize the GOFTLWDMSM operator in Eq. (10.10) to aggregate the GOFTL estimated values $\wp_{jk}$ of supplier $\mathfrak{F}_j$ on all attributes ς_k $(k = 1, 2, 3, 4)$ into the global estimated value $\wp_j$ of the supplier $\mathfrak{F}_j$ $(j = 1, 2, 3, 4, 5)$ (suppose $q = 4$). The global estimated values of suppliers $\mathfrak{F}_j$ $(j = 1, 2, 3, 4, 5)$ are shown as follows:

$$\wp_1 = ((s_1, 0.0057), (0.3494, 0.7906)), \wp_2 = ((s_1, 0.0130), (0.3908, 0.7622)),$$
$$\wp_3 = ((s_1, -0.0029), (0.3073, 0.8268)), \wp_4 = ((s_1, -0.0037), (0.3563, 0.7343)),$$
$$\wp_5 = ((s_1, -0.0100), (0.3086, 0.7971)).$$

Step 3. Calculate the score function $\mathcal{S}(\wp_j)$ by using the Eq. (10.1) of overall estimated value.

$$\mathcal{S}(\wp_1) = 0.0538,\ \mathcal{S}(\wp_2) = 0.0616,\ \mathcal{S}(\wp_3) = 0.0443,\ \mathcal{S}(\wp_4) = 0.0591,\ \mathcal{S}(\wp_5) = 0.0474.$$

Step 4. Rank all results of suppliers depending on the score index, and get $\mathfrak{F}_2 \succ \mathfrak{F}_4 \succ \mathfrak{F}_1 \succ \mathfrak{F}_5 \succ \mathfrak{F}_3$, where the symbol $\succ$ means "the most preferable". Thus, the best supplier is $\mathfrak{F}_2$.

Step 5. End.

Tables 10.8, 10.9, 10.10, and 10.11 provide the overall evaluation values and ranking of alternatives.

10.6.2 Sensitivity Analysis

Our proposed technique consists of two critical parameters t and q. As GOFTLWMSM and GOFTLWDMSM operators have changeable parameters, various parameter values represent various risk preferential attitudes of DMs in solving MAGDM problems. The parameters q and t have a major effect on the ranking results. The ranking results are influenced by the parameters q and t. The GOFTLWMSM and GOFTLWDMSM operators are used to examine the effect of the parameters q and t on the score values and ranking outcomes.

To explain the effect of parameters t and q on the score values and ranking results, we discuss the impacts from the following two aspects:

(1) The value of a parameter t $(t = 3)$ is fixed and we analyze the effect of the q parameter on the score and ranking results of alternatives shown in Tables 10.12

Table 10.7 The combined GOFTL decision framework utilizing GOFTLDMSM operator

	ς_1	ς_2	ς_3	ς_4
$\mathfrak{F}_1$	$((s_4, 0.0370), (0.3876, 0.3255))$	$((s_4, -0.0421), (0.4695, 0.3485))$	$((s_4, -0.0434), (0.2830, 0.5557))$	$((s_5, -0.0466), (0.6067, 0.3485))$
$\mathfrak{F}_2$	$((s_4, -0.0421), (0.5971, 0.4138))$	$((s_5, -0.0439), (0.3000, 0.2856))$	$((s_4, 0.0828), (0.6544, 0.4482))$	$((s_4, 0.0370), (0.5099, 0.2464))$
$\mathfrak{F}_3$	$((s_4, 0.0819), (0.4279, 0.2281))$	$((s_3, 0.0398), (0.3711, 0.4814))$	$((s_4, 0.0348), (0.4548, 0.4456))$	$((s_4, -0.0023), (0.4581, 0.6760))$
$\mathfrak{F}_4$	$((s_3, -0.0441), (0.5466, 0.2595))$	$((s_4, -0.0058), (0.4695, 0.3243))$	$((s_4, -0.0458), (0.5917, 0.3314))$	$((s_5, -0.0449), (0.3743, 0.2379))$
$\mathfrak{F}_5$	$((s_3, 0.0763), (0.4676, 0.4672))$	$((s_5, -0.0449), (0.4695, 0.2167))$	$((s_4, -0.0421), (0.3650, 0.3859))$	$((s_3, 0.0411), (0.4250, 0.5642))$

Table 10.8 Overall estimated values by GOFTLWMSM and GOFTLWDMSM operators

	Results by GOFTLWMSM operator		Results by GOFTLWDMSM operator	
Alternative ($t = 1$)	Results	Score function	Results	Score function
$\mathfrak{F}_1$	$((s_5, 0.0725), (0.8017, 0.3114))$	0.6358	$((s_1, -0.0007), (0.2968, 0.8040))$	0.0489
$\mathfrak{F}_2$	$((s_6, -0.0831), (0.8363, 0.2830))$	0.6798	$((s_1, 0.0073), (0.3508, 0.7843))$	0.0554
$\mathfrak{F}_3$	$((s_5, 0.0604), (0.7776, 0.3588))$	0.6028	$((s_1, -0.0062), (0.3006, 0.8320))$	0.0424
$\mathfrak{F}_4$	$((s_5, 0.0564), (0.8002, 0.2449))$	0.6257	$((s_1, -0.0157), (0.3454, 0.7447))$	0.0533
$\mathfrak{F}_5$	$((s_5, 0.0508), (0.7927, 0.3506))$	0.6099	$((s_1, -0.0142), (0.3029, 0.8106))$	0.0440
Ranking	$\mathfrak{F}_2 \succ \mathfrak{F}_1 \succ \mathfrak{F}_4 \succ \mathfrak{F}_5 \succ \mathfrak{F}_3$		$\mathfrak{F}_2 \succ \mathfrak{F}_4 \succ \mathfrak{F}_1 \succ \mathfrak{F}_5 \succ \mathfrak{F}_3$	

Table 10.9 Overall estimated values by GOFTLWMSM and GOFTLWDMSM operators

	Results by GOFTLWMSM operator		Results by GOFTLWDMSM operator	
Alternative ($t = 2$)	Results	Score function	Results	Score function
$\mathfrak{F}_1$	$((s_5, 0.0725), (0.7997, 0.3194))$	0.6334	$((s_1, 0.0038), (0.3319, 0.7952))$	0.0522
$\mathfrak{F}_2$	$((s_6, -0.0832), (0.8209, 0.2930))$	0.6632	$((s_1, 0.0117), (0.3855, 0.7712))$	0.0596
$\mathfrak{F}_3$	$((s_5, 0.0601), (0.7752, 0.3893))$	0.5978	$((s_1, -0.0039), (0.3055, 0.8285))$	0.0438
$\mathfrak{F}_4$	$((s_5, 0.0564), (0.7925, 0.2451))$	0.6187	$((s_1, -0.0068), (0.3534, 0.7389))$	0.0573
$\mathfrak{F}_5$	$((s_5, 0.0504), (0.7861, 0.3725))$	0.6021	$((s_1, -0.0110), (0.3074, 0.8022))$	0.0463
Ranking	$\mathfrak{F}_2 \succ \mathfrak{F}_1 \succ \mathfrak{F}_4 \succ \mathfrak{F}_5 \succ \mathfrak{F}_3$		$\mathfrak{F}_2 \succ \mathfrak{F}_4 \succ \mathfrak{F}_1 \succ \mathfrak{F}_5 \succ \mathfrak{F}_3$	

Table 10.10 Overall estimated values by GOFTLWMSM and GOFTLWDMSM operators

	Results by GOFTLWMSM operator		Results by GOFTLWDMSM operator	
Alternative ($t = 3$)	Results	Score function	Results	Score function
$\mathfrak{F}_1$	$((s_5, 0.0724), (0.7985, 0.3238))$	0.6320	$((s_1, 0.0057), (0.3494, 0.7906))$	0.0538
$\mathfrak{F}_2$	$((s_6, -0.0833), (0.8118, 0.2952))$	0.6539	$((s_1, 0.0130), (0.3908, 0.7622))$	0.0616
$\mathfrak{F}_3$	$((s_5, 0.0599), (0.7742, 0.4057))$	0.5949	$((s_1, -0.0029), (0.3073, 0.8268))$	0.0443
$\mathfrak{F}_4$	$((s_5, 0.0563), (0.7866, 0.2451))$	0.6135	$((s_1, -0.0037), (0.3563, 0.7343))$	0.0591
$\mathfrak{F}_5$	$((s_5, 0.0500), (0.7826, 0.3791))$	0.5982	$((s_1, -0.0100), (0.3086, 0.7971))$	0.0474
Ranking	$\mathfrak{F}_2 \succ \mathfrak{F}_1 \succ \mathfrak{F}_4 \succ \mathfrak{F}_5 \succ \mathfrak{F}_3$		$\mathfrak{F}_2 \succ \mathfrak{F}_4 \succ \mathfrak{F}_1 \succ \mathfrak{F}_5 \succ \mathfrak{F}_3$	

Table 10.11 Overall assessment values by GOFTLWMSM and GOFTLWDMSM operators

	Results by GOFTLWMSM operator		Results by GOFTLWDMSM operator	
Alternative ($t = 4$)	Results	Score function	Results	Score function
$\mathfrak{F}_1$	$((s_5, 0.0724), (0.7974, 0.3266))$	0.6308	$((s_1, 0.0067), (0.3597, 0.7868))$	0.0549
$\mathfrak{F}_2$	$((s_5, 0.0833), (0.8055, 0.2961))$	0.6478	$((s_1, 0.0137), (0.3933, 0.7533))$	0.0633
$\mathfrak{F}_3$	$((s_5, 0.0596), (0.7734, 0.4170))$	0.5927	$((s_1, -0.0024), (0.3083, 0.8253))$	0.0448
$\mathfrak{F}_4$	$((s_5, 0.0563), (0.7795, 0.2452))$	0.6074	$((s_1, -0.0021), (0.3578, 0.7286))$	0.0604
$\mathfrak{F}_5$	$((s_5, 0.0496), (0.7799, 0.3823))$	0.5953	$((s_1, -0.0095), (0.3091, 0.7920))$	0.0484
Ranking	$\mathfrak{F}_2 \succ \mathfrak{F}_1 \succ \mathfrak{F}_4 \succ \mathfrak{F}_5 \succ \mathfrak{F}_3$		$\mathfrak{F}_2 \succ \mathfrak{F}_4 \succ \mathfrak{F}_1 \succ \mathfrak{F}_5 \succ \mathfrak{F}_3$	

Table 10.12 Score values by GOFTLWMSM and GOFTLWDMSM operators ($t = 3$)

	Score values by GOFTLWMSM					Score values by GOFTLWDMSM				
	$\mathcal{S}(\mathfrak{F}_1)$	$\mathcal{S}(\mathfrak{F}_2)$	$\mathcal{S}(\mathfrak{F}_3)$	$\mathcal{S}(\mathfrak{F}_4)$	$\mathcal{S}(\mathfrak{F}_5)$	$\mathcal{S}(\mathfrak{F}_1)$	$\mathcal{S}(\mathfrak{F}_2)$	$\mathcal{S}(\mathfrak{F}_3)$	$\mathcal{S}(\mathfrak{F}_4)$	$\mathcal{S}(\mathfrak{F}_5)$
$q = 2$	0.7168	0.7387	0.6748	0.7041	0.6778	0.0378	0.0453	0.0301	0.0433	0.0327
$q = 4$	0.6320	0.6539	0.5949	0.6135	0.5982	0.0538	0.0616	0.0443	0.0591	0.0474
$q = 6$	0.5695	0.5912	0.5387	0.5527	0.5413	0.0652	0.0724	0.0558	0.0688	0.0580
$q = 8$	0.5282	0.5487	0.5030	0.5148	0.5044	0.0727	0.0790	0.0638	0.0742	0.0650
$q = 10$	0.5014	0.5202	0.4809	0.4908	0.4811	0.0775	0.0830	0.0694	0.0773	0.0694
$q = 12$	0.4842	0.5009	0.4675	0.4755	0.4665	0.0805	0.0855	0.0732	0.0791	0.0724
$q = 14$	0.4732	0.4878	0.4593	0.4655	0.4573	0.0825	0.0871	0.0758	0.0801	0.0743
$q = 16$	0.4660	0.4788	0.4543	0.4588	0.4515	0.0838	0.0881	0.0777	0.0806	0.0756
$q = 18$	0.4614	0.4726	0.4513	0.4543	0.4479	0.0846	0.0887	0.0789	0.0810	0.0765
$q = 20$	0.4585	0.4682	0.4494	0.4513	0.4456	0.0851	0.0891	0.0798	0.0812	0.0770

Table 10.13 Ranking results by GOFTLWMSM and GOFTLWDMSM operators ($t = 3$)

	Ranking results by GOFTLWMSM	Ranking results by GOFTLWDMSM
$q = 2$	$\mathfrak{F}_2 \succ \mathfrak{F}_1 \succ \mathfrak{F}_4 \succ \mathfrak{F}_5 \succ \mathfrak{F}_3$	$\mathfrak{F}_2 \succ \mathfrak{F}_4 \succ \mathfrak{F}_1 \succ \mathfrak{F}_5 \succ \mathfrak{F}_3$
$q = 4$	$\mathfrak{F}_2 \succ \mathfrak{F}_1 \succ \mathfrak{F}_4 \succ \mathfrak{F}_5 \succ \mathfrak{F}_3$	$\mathfrak{F}_2 \succ \mathfrak{F}_4 \succ \mathfrak{F}_1 \succ \mathfrak{F}_5 \succ \mathfrak{F}_3$
$q = 6$	$\mathfrak{F}_2 \succ \mathfrak{F}_1 \succ \mathfrak{F}_4 \succ \mathfrak{F}_5 \succ \mathfrak{F}_3$	$\mathfrak{F}_2 \succ \mathfrak{F}_4 \succ \mathfrak{F}_1 \succ \mathfrak{F}_5 \succ \mathfrak{F}_3$
$q = 8$	$\mathfrak{F}_2 \succ \mathfrak{F}_1 \succ \mathfrak{F}_4 \succ \mathfrak{F}_5 \succ \mathfrak{F}_3$	$\mathfrak{F}_2 \succ \mathfrak{F}_4 \succ \mathfrak{F}_1 \succ \mathfrak{F}_5 \succ \mathfrak{F}_3$
$q = 10$	$\mathfrak{F}_2 \succ \mathfrak{F}_1 \succ \mathfrak{F}_4 \succ \mathfrak{F}_5 \succ \mathfrak{F}_3$	$\mathfrak{F}_2 \succ \mathfrak{F}_1 \succ \mathfrak{F}_4 \succ \mathfrak{F}_3 \succ \mathfrak{F}_5$
$q = 12$	$\mathfrak{F}_2 \succ \mathfrak{F}_1 \succ \mathfrak{F}_4 \succ \mathfrak{F}_3 \succ \mathfrak{F}_5$	$\mathfrak{F}_2 \succ \mathfrak{F}_1 \succ \mathfrak{F}_4 \succ \mathfrak{F}_3 \succ \mathfrak{F}_5$
$q = 14$	$\mathfrak{F}_2 \succ \mathfrak{F}_1 \succ \mathfrak{F}_4 \succ \mathfrak{F}_3 \succ \mathfrak{F}_5$	$\mathfrak{F}_2 \succ \mathfrak{F}_1 \succ \mathfrak{F}_4 \succ \mathfrak{F}_3 \succ \mathfrak{F}_5$
$q = 16$	$\mathfrak{F}_2 \succ \mathfrak{F}_1 \succ \mathfrak{F}_4 \succ \mathfrak{F}_3 \succ \mathfrak{F}_5$	$\mathfrak{F}_2 \succ \mathfrak{F}_1 \succ \mathfrak{F}_4 \succ \mathfrak{F}_3 \succ \mathfrak{F}_5$
$q = 18$	$\mathfrak{F}_2 \succ \mathfrak{F}_1 \succ \mathfrak{F}_4 \succ \mathfrak{F}_3 \succ \mathfrak{F}_5$	$\mathfrak{F}_2 \succ \mathfrak{F}_1 \succ \mathfrak{F}_4 \succ \mathfrak{F}_3 \succ \mathfrak{F}_5$
$q = 20$	$\mathfrak{F}_2 \succ \mathfrak{F}_1 \succ \mathfrak{F}_4 \succ \mathfrak{F}_3 \succ \mathfrak{F}_5$	$\mathfrak{F}_2 \succ \mathfrak{F}_1 \succ \mathfrak{F}_4 \succ \mathfrak{F}_3 \succ \mathfrak{F}_5$

and 10.13, respectively. As parameters have different values, we can obtain decision results for different score values, which means that the parameter q, which has strong flexibility control, and various parameter values can be perceived as evaluator risk preferences. According to the preferences of attitude, evaluators can pick favorable parameter q (see Figs. 10.4 and 10.5).

(2) The value of a q ($q = 4$) parameter is fixed and we investigate the influence of the t parameter on the score and ranking results and the decision-making outcomes for the different values of t parameter shown in Tables 10.14 and 10.15, respectively. But, the orders of alternatives are slightly different for the final results (see Figs. 10.6 and 10.7).

In practical decision-making problems, therefore, the value of t can be seen as the attitude of DMs towards optimism and pessimism. If DMs are cautious about

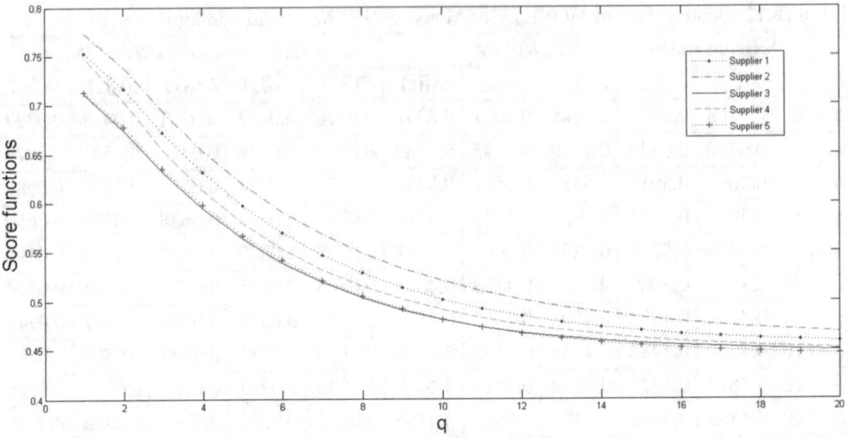

Fig. 10.4 Scores of alternatives based on the GOFTLWMSM operator ($t = 3$)

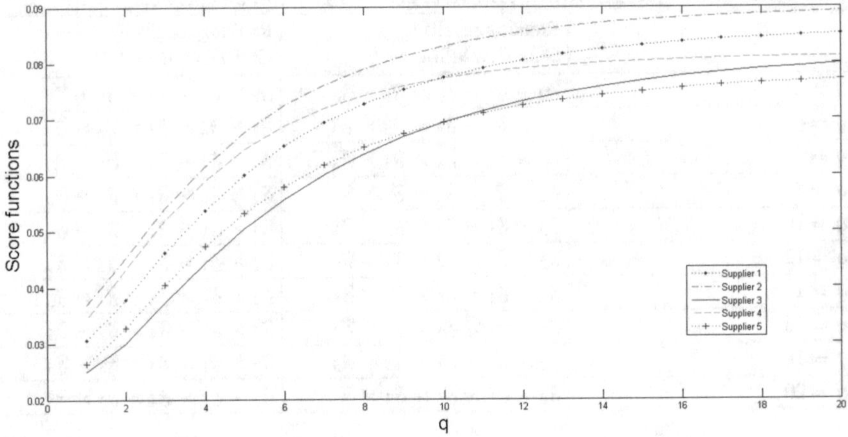

Fig. 10.5 Scores of alternatives based on the GOFTLWDMSM operator ($t = 3$)

Table 10.14 Score values by GOFTLWMSM and GOFTLWDMSM operators ($q = 4$ and $t = 1, 2, 3, 4$)

	Score values by GOFTLWMSM					Score values by GOFTLWDMSM				
	$\mathcal{S}(\mathfrak{F}_1)$	$\mathcal{S}(\mathfrak{F}_2)$	$\mathcal{S}(\mathfrak{F}_3)$	$\mathcal{S}(\mathfrak{F}_4)$	$\mathcal{S}(\mathfrak{F}_5)$	$\mathcal{S}(\mathfrak{F}_1)$	$\mathcal{S}(\mathfrak{F}_2)$	$\mathcal{S}(\mathfrak{F}_3)$	$\mathcal{S}(\mathfrak{F}_4)$	$\mathcal{S}(\mathfrak{F}_5)$
$t = 1$	0.6358	0.6798	0.6028	0.6257	0.6099	0.0489	0.0554	0.0424	0.0533	0.0440
$t = 2$	0.6334	0.6632	0.5978	0.6187	0.6021	0.0522	0.0596	0.0438	0.0573	0.0463
$t = 3$	0.6320	0.6539	0.5949	0.6135	0.5982	0.0538	0.0616	0.0443	0.0591	0.0474
$t = 4$	0.6308	0.6478	0.5927	0.6074	0.5953	0.0549	0.0633	0.0448	0.0604	0.0484

Table 10.15 Ranking results by GOFTLWMSM and GOFTLWDMSM operators ($q = 4$ and $t = 1, 2, 3, 4$)

	Ranking by GOFTLWMSM	Ranking by GOFTLWDMSM
$t = 1$	$\mathfrak{F}_2 \succ \mathfrak{F}_1 \succ \mathfrak{F}_4 \succ \mathfrak{F}_5 \succ \mathfrak{F}_3$	$\mathfrak{F}_2 \succ \mathfrak{F}_4 \succ \mathfrak{F}_1 \succ \mathfrak{F}_5 \succ \mathfrak{F}_3$
$t = 2$	$\mathfrak{F}_2 \succ \mathfrak{F}_1 \succ \mathfrak{F}_4 \succ \mathfrak{F}_5 \succ \mathfrak{F}_3$	$\mathfrak{F}_2 \succ \mathfrak{F}_4 \succ \mathfrak{F}_1 \succ \mathfrak{F}_5 \succ \mathfrak{F}_3$
$t = 3$	$\mathfrak{F}_2 \succ \mathfrak{F}_1 \succ \mathfrak{F}_4 \succ \mathfrak{F}_5 \succ \mathfrak{F}_3$	$\mathfrak{F}_2 \succ \mathfrak{F}_4 \succ \mathfrak{F}_1 \succ \mathfrak{F}_5 \succ \mathfrak{F}_3$
$t = 4$	$\mathfrak{F}_2 \succ \mathfrak{F}_1 \succ \mathfrak{F}_4 \succ \mathfrak{F}_5 \succ \mathfrak{F}_3$	$\mathfrak{F}_2 \succ \mathfrak{F}_4 \succ \mathfrak{F}_1 \succ \mathfrak{F}_5 \succ \mathfrak{F}_3$

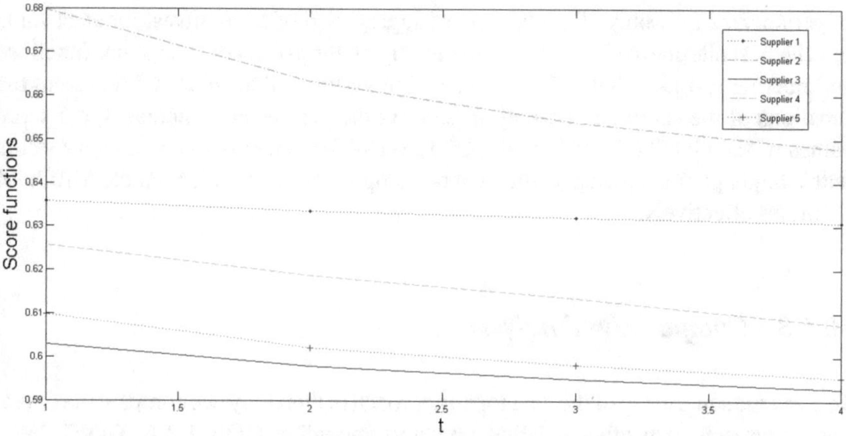

Fig. 10.6 Scores of alternatives based on the GOFTLWMSM operator ($q = 4$)

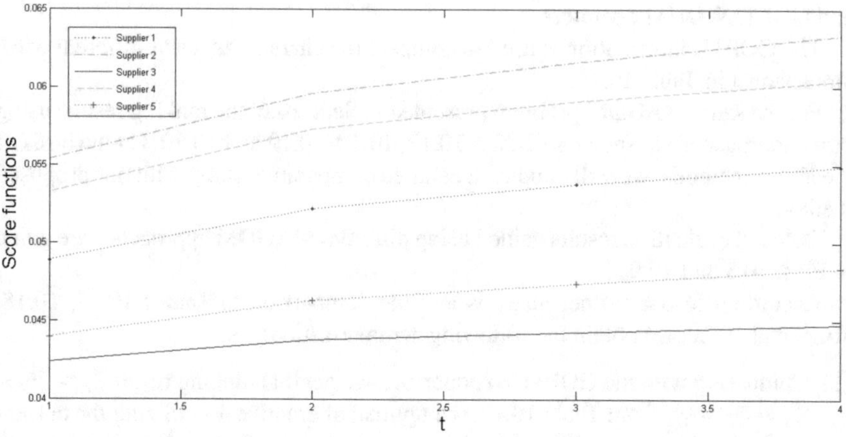

Fig. 10.7 Scores of alternatives based on the GOFTLWDMSM operator ($q = 4$)

their decisions then t should be given higher priority, while if DMs are gloomy about their decisions, they should be set lower value to t. This attribute demonstrates the versatility and efficiency of the approach proposed. As the value of t increases, the ranking effect of the five alternatives varies on the basis of such a factor that they find more attribute interdependencies. From Table 10.15, the parameter t can, therefore, be regarded as the evaluation attitude of the DMs. DMs should pick a suitable parameter value t based on their specific risk attitudes. For example, when a DM preferred for risk, he/she can choose a larger value of the parameter t that gives different results which are shown in Table 10.15 in which we observed that after increasing the value of parameter t they only affect the lowest aggregation of alternatives, for each value of t the best alternative is $\mathfrak{F}_2$. Now we investigate the effects of q on score functions and ranking results in Table 10.13. We assign various values to q and represent the functions of the score and ranking orders. As the parameter q increases, the score values of the GOFTLWMSM and GOFTLWDMSM operators become very stable with a larger parameter, suggesting that our proposed operators can handle MAGDM problems effectively.

10.6.3 Comparative Analysis

To show the superiority of the developed approach in this study, we compare the developed approach with other existing operators including GOFTLWA, GOFTLWG, GOFTLMM, GOFTLWMM, GOFTLDMM, GOFTLWDMM, GOFTLBM, GOFTLWBM, GOFTLGBM, GOFTLWGBM, GOFTLHM, GOFTLWHM, GOFTLDHM, and GOFTLWDHM operators.

The GOFTL information is used to compare the characteristics of different operators shown in Table 10.16.

For the same MAGDM problem presented in Sect. 10.6, the ranking results using other approaches are shown in Tables 10.17, 10.18, 10.19 and 10.20. On the basis of the above methods, we will conduct a detailed comparative study with our proposed method.

Detailed evaluation results gained using different MAGDM approaches are given in Figs. 10.8 and 10.9.

According to our further analysis and the comparison in Tables 10.17, 10.18, 10.19 and 10.20, we obtain the following summarization:

(1) Comparing with the GOFTLWA operator, we get this ranking result $\mathfrak{F}_2 \succ \mathfrak{F}_1 \succ \mathfrak{F}_4 \succ \mathfrak{F}_5 \succ \mathfrak{F}_3$ from Table 10.17, the optimal alternative $\mathfrak{F}_2$ utilizing the current method is the same as that of the operators suggested. This verifies the approach proposed in this study, which is logical and accurate.
(2) Comparing with the GOFTLWGM operator, we get this ranking result $\mathfrak{F}_2 \succ \mathfrak{F}_4 \succ \mathfrak{F}_1 \succ \mathfrak{F}_5 \succ \mathfrak{F}_3$ from Table 10.19, the optimal alternative $\mathfrak{F}_2$ utilizing the current method is the same as that of the operators suggested. This verifies the approach proposed in this study, which is logical and accurate.

Table 10.16 The characteristic comparison among different operators with GOFTLNs

Aggregation operators	Whether permits the sum of MD and NMD to be greater than one	Whether the operator can capture the interrelationship between the GOFTLNs	Whether parameters exist to manipulate ranking results
GOFTLWA	✓	×	×
GOFTLWG	✓	×	×
GOFTLMM	✓	✓	✓
GOFTLWMM	✓	✓	✓
GOFTLDMM	✓	✓	✓
GOFTLWDMM	✓	✓	✓
GOFTLBM	✓	✓	✓
GOFTLWBM	✓	✓	✓
GOFTLGBM	✓	✓	✓
GOFTLWGBM	✓	✓	✓
GOFTLHM	✓	✓	✓
GOFTLWHM	✓	✓	✓
GOFTLDHM	✓	✓	✓
GOFTLWDHM	✓	✓	✓
GOFTLMSM	✓	✓	✓
GOFTLWMSM	✓	✓	✓
GOFTLDMSM	✓	✓	✓
GOFTLWDMSM	✓	✓	✓

Table 10.17 Comparison of alternatives for different MAGDM methods

	Score and ranking value by GOFTLWA operator	Score and ranking value by GOFTLWBM operator	Score and ranking value by GOFTLWMSM operator
Alternative	Score function	Score function	Score function
$\mathfrak{F}_1$	0.6358	0.6334	0.6320
$\mathfrak{F}_2$	0.6798	0.6632	0.6539
$\mathfrak{F}_3$	0.6028	0.5978	0.5949
$\mathfrak{F}_4$	0.6257	0.6187	0.6135
$\mathfrak{F}_5$	0.6099	0.6021	0.5982
Ranking	$\mathfrak{F}_2 \succ \mathfrak{F}_1 \succ \mathfrak{F}_4 \succ \mathfrak{F}_5 \succ \mathfrak{F}_3$	$\mathfrak{F}_2 \succ \mathfrak{F}_1 \succ \mathfrak{F}_4 \succ \mathfrak{F}_5 \succ \mathfrak{F}_3$	$\mathfrak{F}_2 \succ \mathfrak{F}_1 \succ \mathfrak{F}_4 \succ \mathfrak{F}_5 \succ \mathfrak{F}_3$

Table 10.18 Comparison of alternatives for different MAGDM methods

Alternative	Score and ranking value by IFTLWMSM operator	Score and ranking value by PFTLWMSM operator	Score and ranking value by GOFTLWMSM operator
	Score function	Score function	Score
$\mathfrak{F}_1$	0.7526	0.7168	0.6320
$\mathfrak{F}_2$	0.7741	0.7387	0.6539
$\mathfrak{F}_3$	0.7117	0.6748	0.5949
$\mathfrak{F}_4$	0.7483	0.7041	0.6135
$\mathfrak{F}_5$	0.7135	0.6778	0.5982
Ranking	$\mathfrak{F}_2 \succ \mathfrak{F}_1 \succ \mathfrak{F}_4 \succ \mathfrak{F}_5 \succ \mathfrak{F}_3$	$\mathfrak{F}_2 \succ \mathfrak{F}_1 \succ \mathfrak{F}_4 \succ \mathfrak{F}_5 \succ \mathfrak{F}_3$	$\mathfrak{F}_2 \succ \mathfrak{F}_1 \succ \mathfrak{F}_4 \succ \mathfrak{F}_5 \succ \mathfrak{F}_3$

Table 10.19 Comparison of alternatives for different MAGDM methods

Alternative	Score and ranking value by GOFTLWGM operator	Score and ranking value by GOFTLWGBM operator	Score and ranking value by GOFTLWDMSM operator
	Score function	Score function	Score function
$\mathfrak{F}_1$	0.0489	0.0522	0.0538
$\mathfrak{F}_2$	0.0554	0.0596	0.0616
$\mathfrak{F}_3$	0.0424	0.0438	0.0443
$\mathfrak{F}_4$	0.0533	0.0573	0.0591
$\mathfrak{F}_5$	0.0440	0.0463	0.0474
Ranking	$\mathfrak{F}_2 \succ \mathfrak{F}_4 \succ \mathfrak{F}_1 \succ \mathfrak{F}_5 \succ \mathfrak{F}_3$	$\mathfrak{F}_2 \succ \mathfrak{F}_4 \succ \mathfrak{F}_1 \succ \mathfrak{F}_5 \succ \mathfrak{F}_3$	$\mathfrak{F}_2 \succ \mathfrak{F}_4 \succ \mathfrak{F}_1 \succ \mathfrak{F}_5 \succ \mathfrak{F}_3$

Table 10.20 Comparison of alternatives for different MAGDM methods

Alternative	Score and ranking value by IFTLWDMSM operator	Score and ranking value by PFTLWDMSM operator	Score and ranking value by GOFTLWDMSM operator
	Score function	Score function	Score function
$\mathfrak{F}_1$	0.0306	0.0378	0.0538
$\mathfrak{F}_2$	0.0369	0.0453	0.0616
$\mathfrak{F}_3$	0.0249	0.0301	0.0443
$\mathfrak{F}_4$	0.0344	0.0433	0.0591
$\mathfrak{F}_5$	0.0264	0.0327	0.0474
Ranking	$\mathfrak{F}_2 \succ \mathfrak{F}_4 \succ \mathfrak{F}_1 \succ \mathfrak{F}_5 \succ \mathfrak{F}_3$	$\mathfrak{F}_2 \succ \mathfrak{F}_4 \succ \mathfrak{F}_1 \succ \mathfrak{F}_5 \succ \mathfrak{F}_3$	$\mathfrak{F}_2 \succ \mathfrak{F}_4 \succ \mathfrak{F}_1 \succ \mathfrak{F}_5 \succ \mathfrak{F}_3$

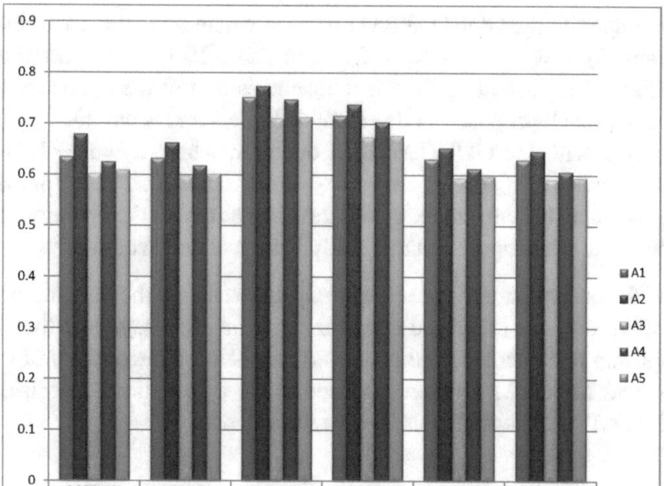

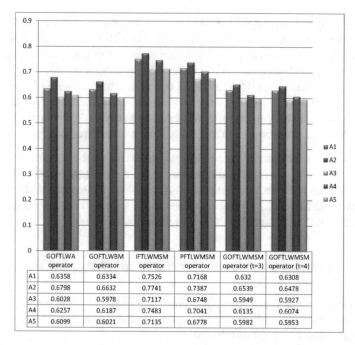

	GOFTLWA operator	GOFTLWBM operator	IFTLWMSM operator	PFTLWMSM operator	GOFTLWMSM operator (t=3)	GOFTLWMSM operator (t=4)
A1	0.6358	0.6334	0.7526	0.7168	0.632	0.6308
A2	0.6798	0.6632	0.7741	0.7387	0.6539	0.6478
A3	0.6028	0.5978	0.7117	0.6748	0.5949	0.5927
A4	0.6257	0.6187	0.7483	0.7041	0.6135	0.6074
A5	0.6099	0.6021	0.7135	0.6778	0.5982	0.5953

Fig. 10.8 The graphical representation for the comparison with different MAGDM approaches

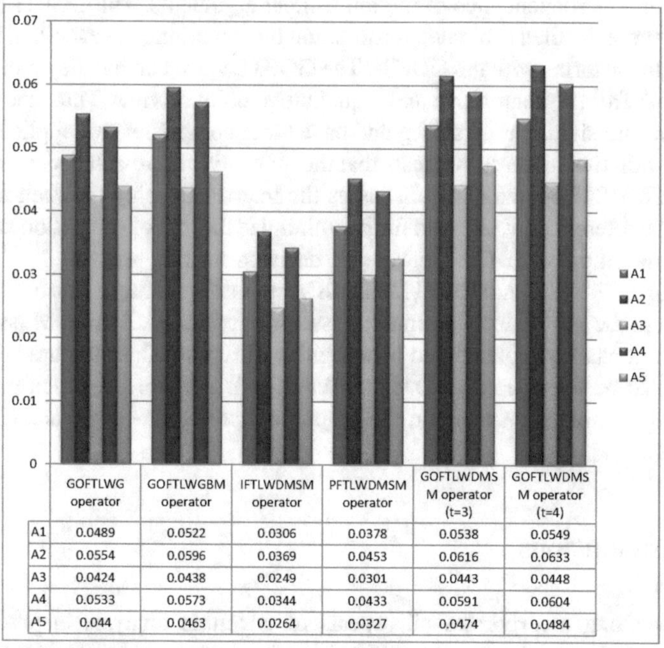

	GOFTLWG operator	GOFTLWGBM operator	IFTLWDMSM operator	PFTLWDMSM operator	GOFTLWDMSM operator (t=3)	GOFTLWDMSM operator (t=4)
A1	0.0489	0.0522	0.0306	0.0378	0.0538	0.0549
A2	0.0554	0.0596	0.0369	0.0453	0.0616	0.0633
A3	0.0424	0.0438	0.0249	0.0301	0.0443	0.0448
A4	0.0533	0.0573	0.0344	0.0433	0.0591	0.0604
A5	0.044	0.0463	0.0264	0.0327	0.0474	0.0484

Fig. 10.9 The graphical representation for the comparison with different MAGDM approaches

(3) Comparing with the GOFTLWBM operator, where $u = v = 1$ we get this ranking result $\mathfrak{F}_2 \succ \mathfrak{F}_1 \succ \mathfrak{F}_4 \succ \mathfrak{F}_5 \succ \mathfrak{F}_3$ from Table 10.17, the optimal alternative $\mathfrak{F}_2$ utilizing the current method is the same as that of the operators suggested. So, the approach proposed in this study is logical and accurate.
(4) Comparing with the GOFTLWGBM operator, where $u = v = 1$, we get this ranking result $\mathfrak{F}_2 \succ \mathfrak{F}_4 \succ \mathfrak{F}_1 \succ \mathfrak{F}_5 \succ \mathfrak{F}_3$ from Table 10.19, the optimal alternative $\mathfrak{F}_2$ utilizing the current method is the same as that of the operators suggested. So, the approach proposed in this study is logical and accurate.

We see that the order of alternatives is slightly different, but the best alternative is the same. Furthermore, our described operators have an advantage that they combined interrelationship between the arguments and consider the uncertainty of individuals in the practical MAGDM problems. Moreover, in a complicated decision-making process, the DM's risk attitude has its own importance.

10.6.4 Advantages and Superiorities of the Proposed Work

As analyzed and compared the above-mentioned summarization points, we try to describe how the proposed approach is superior in this subsection. The merits of our strategy have been outlined as follows:

- The proposed methods are more preferable than others because they effectively handle the interdependence of the multi-input arguments. The GOFTLMSM can convey more detailed evaluated information by combining the excellent feature of the linguistic terms with the GOFSs. The GOFTLMSM can handle practical problems both from a quantitative and a qualitative point of view. Thus, the proposed operators are significantly strong and have more comprehensive applications.
- The quantitative research suggests that the innovative approach depends upon the GOFTLWMSM operator that discusses the interrelationship between arguments of several references, which can further minimize the loss of evaluation knowledge and allow a more robust and reasonable decision-making process.
- The GOFTLWMSM and GOFTLWDMSM operators are more flexible and robust to address the group decision-making issues, making risk MAGDM issues more realistic. It can be implemented by assigning different values to parameter t, and it can also be interpreted as DM's different risk assessments based on relevant proposed solutions. As a result, the proposed method is realistic and causes less data loss.

10.7 Conclusions

Decision-making incorporates all aspects of uncertain analysis in practical and theoretical strategies. In the fuzzy linguistic framework, MAGDM is an integral part of modern decision-making research. Researchers take advantage of various non-classical sets to cope with problems involving group decision-making. Among these helpful mathematical tools, the 2TL sets and GOFSs are widely used due to linguistic terms and qth power of degrees. In the decision-making process, the GOFTLS provides a better technique to cope with vague and uncertain information. The MSM and DMSM operators are two important tools to deal with MAGDM issues, these two aggregation operators can remove the effects of DMs' inappropriate information and depict the interrelationship between any number of arguments. A novel approach based on MSM and DMSM operators is proposed in this chapter for addressing the MAGDM problems in the GOFTL framework. The crucial contributions of this research are classified into four categories. (1) The four aggregation operators under GOFTL information have been presented such as; the GOFTLMSM, GOFTLWMSM, GOFTLDMSM, and GOFTLWDMSM operators. These proposed aggregation operators' basic properties and special cases have also been carefully examined. (2) A novel approach has been presented based on the proposed operators to cope with the MAGDM issues in the GOFTL scenario to select the most preferable supplier(s). (3) The numerical analysis verified that our proposed algorithm is appropriate and more robust than some other existing MAGDM models. (4) The viability and reliability of the proposed method are more applicable and appropriate for dealing with realistic MAGDM techniques than some other existing techniques, and it also allows DMs to convey their fuzzy information in a better way. It is, therefore, easy to reflect the risk perception of DMs.

On the other hand, there are still some limitations in this research. Firstly, this work only deals with the aggregation of GOFNs. Secondly, this research is only applied to the evaluation of supplier(s) in the enterprise framework group (EFG) of companies. In fact, it has broad implications for assessment methods and related fields such as cognitive computation, engineering, natural and artificial cognitive systems, and management applications. To overcome discussed limitations, in the future, we will generalize the MSM and DMSM operators to other extended fuzzy scenarios to show their merits in information fusion domains and expand their applications in MAGDM.

Conflict of interest: The authors declare no conflict of interest.

Acknowledgements The authors are highly grateful to the anonymous referees for their valuable comments and suggestions. This work was partially supported by the National Natural Science Foundation of China [Grant Number 11301415], the Shaanxi Provincial Key Research and Development Program [Grant Number 2021SF-480], and the Natural Science Basic Research Plan in Shaanxi Province of China [Grant Number 2018JM1054].

References

1. L.A. Zadeh, Fuzzy sets. Inform. Control **8**(3), 338–353 (1965)
2. K.T. Atanassov, Intuitionistic fuzzy sets. Fuzzy Sets Syst. **20**(1), 87–96 (1986)
3. P.K. Maji, R. Biswas, A.R. Roy, Intuitionistic fuzzy soft sets. J. Fuzzy Math. **9**(3), 677–692 (2001)

4. F. Feng, Z. Xu, H. Fujita, M. Liang, Enhancing PROMETHEE method with intuitionistic fuzzy soft sets. Int. J. Intell. Syst. **35**, 1071–1104 (2020)
5. M. Agarwal, K.K. Biswas, M. Hanmandlu, Generalized intuitionistic fuzzy soft sets with applications in decision-making. Appl. Soft Comput. **13**, 3552–3566 (2013)
6. F. Feng, H. Fujita, M.I. Ali, R.R. Yager, X. Liu, Another view on generalized intuitionistic fuzzy soft sets and related multi-attribute decision making methods. IEEE Trans. Fuzzy Syst. **27**(3), 474–488 (2019)
7. R.R. Yager, Pythagorean membership grades in multi-criteria decision making. IEEE Trans. Fuzzy Syst. **22**(4), 958–965 (2014)
8. S. Naz, S. Ashraf, M. Akram, A novel approach to decision-making with Pythagorean fuzzy information. Mathematics **6**(6), 95 (2018)
9. R.R. Yager, Generalized orthopair fuzzy sets. IEEE Trans. Fuzzy Syst. **25**(5), 1222–1230 (2016)
10. P. Liu, P. Wang, Some q-rung orthopair fuzzy aggregation operators and their applications to multiple-attribute decision making. Int. J. Intell. Syst. **33**(2), 259–280 (2018)
11. G. Wei, C. Wei, J. Wang, H. Gao, Y. Wei, Some q-rung orthopair fuzzy Maclaurin symmetric mean operators and their applications to potential evaluation of emerging technology commercialization. Int. J. Intell. Syst. **34**(1), 50–81 (2019)
12. P. Liu, J. Liu, Some q-rung orthopair fuzzy Bonferroni mean operators and their application to multi-attribute group decision making. Int. J. Intell. Syst. **33**(2), 315–347 (2018)
13. Z. Liu, S. Wang, P. Liu, Multiple attribute group decision-making based on q-rung orthopair fuzzy Heronian mean operators. Int. J. Intell. Syst. **33**(12), 2341–2363 (2018)
14. F. Feng, Y. Zheng, B. Sun, M. Akram, Novel score functions of generalized orthopair fuzzy membership grades with application to multiple attribute decision making. Granular Comput., 1–17 (2021). https://doi.org/10.1007/s41066-021-00253-7
15. F. Herrera, L. Martinez, A 2-tuple fuzzy linguistic representation model for computing with words. IEEE Trans. Fuzzy Syst. **8**(6), 746–752 (2000)
16. X. Deng, J. Wang, G. Wei, Some 2-tuple linguistic Pythagorean Heronian mean operators and their application to multiple attribute decision-making. J. Exp. Theoret. Artif. Intell. **31**(4), 555–574 (2019)
17. G. Wei, H. Gao, Pythagorean 2-tuple linguistic power aggregation operators in multiple attribute decision making. Econ. Res. Ekonomska Istrazivanja **33**(1), 904–933 (2020)
18. Y. Ju, A. Wang, J. Ma, H. Gao, E.D. Santibanez Gonzalez, Some q-rung orthopair fuzzy 2-tuple linguistic Muirhead mean aggregation operators and their applications to multiple-attribute group decision making. Int. J. Intell. Syst. **35**(1), 184–213 (2020)
19. M. Akram, S. Naz, S.A. Edalatpanah, R. Mehreen, Group decision-making framework under linguistic q-rung orthopair fuzzy Einstein models. Soft Comput. **25**, 10309–10334 (2021)
20. M. Akram, G. Ali, Hybrid models for decision-making based on rough Pythagorean fuzzy bipolar soft information. Granular Comput. **5**(1), 1–15 (2020)
21. H. Garg, S.M. Chen, Multi-attribute group decision making based on neutrality aggregation operators of q-rung orthopair fuzzy sets. Inform. Sci. **517**, 427–447 (2020)
22. H. Garg, S. Naz, F. Ziaa, Z. Shoukatb, A ranking method based on Muirhead mean operator for group decision making with complex interval-valued q-rung orthopair fuzzy numbers. Soft Comput. **25**(22), 14001–140271-27 (2021). https://doi.org/10.1007/s00500-021-06231-0
23. P. Liu, S. Naz, M. Akram, M. Muzammal, Group decision-making analysis based on linguistic q-rung orthopair fuzzy generalized point weighted aggregation operators. Int. J. Mach. Learn. Cybern. (2021). https://doi.org/10.1007/s13042-021-01425-2
24. P. Liu, G. Shahzadi, M. Akram, Specific types of q-rung picture fuzzy Yager aggregation operators for decision-making. Int. J. Comput. Intell. Syst. **13**(1), 1072–1091 (2020)
25. M. Akram, S. Naz, S. Shahzadi, F. Ziaa, Geometric-arithmetic energy and atom bond connectivity energy of dual hesitant q-rung orthopair fuzzy graphs. J. Intell. Fuzzy Syst. **40**(1), 1287–1307 (2021)
26. S. Naz, M. Akram, S. Alsulamic, F. Ziaa, Decision-making analysis under interval-valued q-rung orthopair dual hesitant fuzzy environment. Int. J. Comput. Intell. Syst. **14**(1), 332–357 (2021)

27. S. Naz, M. Akram, Novel decision making approach based on hesitant fuzzy sets and graph theory. Computat. Appl. Math. **38**(1), 7 (2019)
28. M. Akram, S. Naz, A novel decision-making approach under complex Pythagorean fuzzy environment. Math. Computat. Appl. **24**(3), 73 (2019)
29. M. Akram, S. Naz, F. Smarandache, Generalization of maximizing deviation and TOPSIS method for MADM in simplified neutrosophic hesitant fuzzy environment. Symmetry **11**(8), 1058 (2019)
30. C. Maclaurin, A second letter to Martin Folkes, Esq concerning the roots of equations, with demonstration of other rules of algebra. Philos. Trans. Roy. Soc. Lond. Ser. A **36**, 59–96 (1729)
31. D.W. Detemple, J.M. Robertson, On generalized symmetric means of two variables, Publikacije Elektrotehnickog fakulteta. Serija Matematika i fizika (634/677), 236–238 (1979)
32. J. Qin, X. Liu, Approaches to uncertain linguistic multiple attribute decision making based on dual Maclaurin symmetric mean. J. Intell. Fuzzy Syst. **29**(1), 171–186 (2015)
33. G. Wei, C. Wei, J. Wang, H. Gao, Y. Wei, Some q-rung orthopair fuzzy Maclaurin symmetric mean operators and their applications to potential evaluation of technology commercialization. Int. J. Intell. Syst. **34**(1), 50–81 (2019)
34. W.S. Tai, C.T. Chen, A new evaluation model for intellectual capital based on computing with linguistic variable. Expert Syst. Appl. **36**(2), 3483–3488 (2009)
35. G. Wei, H. Gao, Y. Wei, Some q-rung orthopair fuzzy Heronian mean operators in multiple attribute decision making. Int. J. Intell. Syst. **33**(7), 1426–1458 (2018)
36. R. Krishankumar, K.S. Ravichandran, K.K. Murthy, A.B. Saeid, A scientific decision-making framework for supplier outsourcing using hesitant fuzzy information. Soft Comput. **22**, 7445–7461 (2018)
37. R. Krishankumar, L.S. Subrajaa, K.S. Ravichandran, S. Kar, A.B. Saeid, A framework for multi-attribute group decision-making using double hierarchy hesitant fuzzy linguistic term set. Int. J. Fuzzy Syst. **21**(4), 1130–1143 (2019)
38. M. Akram, A. Khan, A.B. Saeid, Complex Pythagorean Dombi fuzzy operators using aggregation operators and their decision-making. Expert Syst. **38**(2), e12626 (2021)
39. C. Jana, G. Muhiuddin, M. Pal, Multiple-attribute decision making problems based on SVTNH methods. J. Amb. Intelli. Human. Comput. **11**(9), 3717–3733 (2020)
40. C. Jana, G. Muhiuddin, M. Pal, Some dombi aggregation of q-rung orthopair fuzzy numbers in multiple-attribute decision making. Inte J. Intell. Syst. **34**(12), 3220–3240 (2019)
41. G. Shahzadi, G. Muhiuddin, M.A. Butt, A. Ashraf, Hamacher interactive hybrid weighted averaging operators under fermatean fuzzy numbers. J. Math., 1–17 (2021). https://doi.org/10.1155/2021/5556017
42. H. Garg, CN-q-ROFS: connection number-based q-rung orthopair fuzzy set and their application to decision-making process. Int. J. Intell. Syst. **36**(7), 3106–3143 (2021)
43. H. Garg, A new possibility degree measure for interval-valued q-rung orthopair fuzzy sets in decision-making. Int. J. Intell. Syst. **36**(1), 526–557 (2021)
44. H. Garg, New exponential operation laws and operators for interval-valued q-rung orthopair fuzzy sets in group decision making process. Neural Comput. Appl. **33**(20), 13937–13963 (2021). https://doi.org/10.1007/s00521-021-06036-0
45. M. Riaz, H. Garg, H.M.A. Farid, M. Aslam, Novel q-rung orthopair fuzzy interaction aggregation operators and their application to low-carbon green supply chain management. J. Intell. Fuzzy Syst. **41**(2), 4109–4126 (2021). https://doi.org/10.3233/JIFS-210506
46. Z. Yang, H. Garg, Interaction power partitioned Maclaurin symmetric mean operators under q-rung orthopair uncertain linguistic information. Int. J. Fuzzy Syst. 1–19 (2021). https://doi.org/10.1007/s40815-021-01062-5
47. H. Garg, A novel trigonometric operation-based q-rung orthopair fuzzy aggregation operator and its fundamental properties. Neural Comput. Appl. **32**(18), 15077–15099 (2020)
48. X. Peng, J. Dai, H. Garg, Exponential operation and aggregation operator for q-rung orthopair fuzzy set and their decision-making method with a new score function. Int. J. Intell. Syst. **33**(11), 2255–2282 (2018)

Sumera Naz completed her M.Sc., MPhil, and Ph.D. from the University of the Punjab, Lahore, Pakistan. She is serving as an Assistant Professor at the University of Education Lahore, Pakistan. Her research interests are in the field of extended fuzzy graph theory and decision analysis. She has diversity in her research work. She has published some work on generalized fuzzy graph theory, multi-attribute decision analysis, and some alternative theories of fuzzy Mathematics. She has published 25 research articles in international scientific journals. She has been acting as a referee for several international journals.

Muhammad Akram Muhammad Akram has received M.Sc. degrees in Mathematics and Computer Science, MPhil in (Computational) Mathematics and Ph.D. in (Fuzzy) Mathematics. He is currently a Professor in the Department of Mathematics at the University of the Punjab, Lahore. He has also served the Punjab University College of Information Technology as Assistant Professor and Associate Professor. Dr. Akram's research interests include fuzzy numerical methods, fuzzy graphs, fuzzy algebras, and fuzzy decision support systems. He has published 9 monographs and 385 research articles in international scientific journals. He has been an Editorial Board Member of several international academic journals, as well as a Reviewer/Referee for 140 International Journals including Mathematical Reviews (USA) and Zentralblatt MATH (Germany). Twelve students have successfully completed their Ph.D. research work under his supervision. Currently, he is supervising 5 Ph.D. students.

Feng Feng is currently a full professor and the director of the Department of Applied Mathematics, at Xi'an University of Posts and Telecommunications, China. He received the Ph.D. degree in Mathematics from Shaanxi Normal University. He was a visiting research faculty with the University of California, Los Angeles, USA from 2015 to 2016. He has authored over 60 papers (including 9 ESI Highly Cited Papers), with over 3200 citations in Web of Science. His research interests include soft computing, granular computing, fuzzy mathematics, decision analysis and data mining. He serves as an AE of Granular Computing, Journal of Mathematics and Annals of Communications in Mathematics. He has been an Editorial Board Member of 6 international academic journals, and has been a reviewer for over 60 SCIE journals. He was honored as the Chinese Highly Cited Researcher and the Young Sci-Tech New Star of Shaanxi Province.

Abid Mahboob did his M.Sc & M. Phil degree from Govt. College University, Lahore, Pakistan which takes pride in being the oldest educational institution of higher education in Pakistan, ranked 3rd in general category by Higher Education Commission of Pakistan. He got Chinese Council Scholarship in 2012 and completed his Ph.D. (in 2015) from USTC (University of science and Technology of China Hefei, China). In 2015, he joined University of Education Lahore as Assistant professor of Mathematics and till date serving as co-coordinator of mathematics at UE Vahari campus, Pakistan. He has been teaching experience of over ten years at different levels and supervised more than 15 MS/M.Phil students. He has published more than 25 research papers in international reputable journals.

Chapter 11
3PL Service Provider Selection with q-Rung Orthopair Fuzzy Based CODAS Method

Adem Pinar and Fatih Emre Boran

Abstract Since the 1990s, companies outsource their non-core logistic activities to third-party logistic (3PL) service providers. Selection of an appropriate 3PL service provider which gives a competitive advantage to the company can be considered as a multiple criteria group decision-making problem. 3PL service provider selection problem has several qualitative criteria which are graded with linguistic terms. A multicriteria decision-making (MCDM) method is required to consider vagueness and impreciseness evaluating the alternatives of 3PL company. Q-rung orthopair fuzzy (q-ROF) sets extend the intuitionistic and Pythagorean fuzzy sets by giving a wider area to decision-makers in expressing their evaluations. CODAS is a decision-making method based on calculating the distances between each alternative and the negative ideal solution which is the lowest value. At first, Euclidean distance is used, then Taxicab distance is used to exaggerate the discrimination. In this chapter, the q-Rung Orthopair Fuzzy CODAS method is proposed for the first time, by adapting the CODAS method to q-ROF sets. Information about the classical and fuzzy set applications of the CODAS method in the literature is given, and the third-party logistics service provider selection is made for a retail company with the proposed q-ROF CODAS method. In a case study, three decision-makers select the most suitable company with the help of seven criteria among six alternatives and the results were compared with another decision-making method.

Keywords Party logistics · q-rung orthopair fuzzy sets · CODAS method · MCDM

A. Pinar (✉)
Department of Logistics Management, Faculty of Business Administration, University of Turkish Aeronautical Association, 06790 Ankara, Turkey
e-mail: apinar@thk.edu.tr

F. E. Boran
Department of Energy Systems Engineering, Faculty of Technology, Gazi University, 06550 Ankara, Turkey
e-mail: emreboran@gazi.edu.tr

H. Garg (ed.), *q-Rung Orthopair Fuzzy Sets*,
https://doi.org/10.1007/978-981-19-1449-2_11

Nomenclature

μ_A	Membership degree in q-ROF set A
v_A	Non-membership degree in q-ROF set A
π_A	Hesitancy degree in q-ROF set A
λ_k	Importance degree of *k*th decision-maker
θ	The threshold parameter set by the decision-maker
P_i	Assessment score of *i*th alternative
w_j	Weight of the *j*th criteria
ns_j	The q-ROF Negative Ideal Solution (q-ROFNIS)

11.1 Introduction

The main purpose of supply chain management is to create competitive supply chains by integrating business processes among suppliers. In order to take part more effectively in this competition, companies aimed to focus on their core business by outsourcing some of their logistics activities to third-party logistics (3PL) providers.

3PL has many definitions in the literature; Berglund et al. [1] defined 3PL as, activities carried out by a logistics service provider, consisting of at least the management and execution of transportation and warehousing, in addition, value-added activities such as contract and inventory management, tracking, secondary assembly and installation of products. Bagchi and Virum [2] emphasize the logistic alliance perspective of 3PL and point out that a logistics alliance is a close and long-term relationship between a customer who sees each other as partners and a service provider, covering the fulfillment of a wide variety of logistics needs, where both partners are involved in the design and development of logistics solutions and he refers to a win–win agreement with the goal of participating in the development and measurement of performance.

The 3PL approach, which was introduced in the 1980s, has been accepted as an area of business since the 1990s and has been widely used by most companies since the 2000s. It is seen that outsourcing is widely used due to the benefits of reducing costs, increasing performance, and focusing on the main business, and 90% of Fortune 500 companies in the USA have some 3PL contracts and the market for 3 PL providers continues to grow [3, 4].

According to Singh et al. [5] selection of the most suitable 3PL service from many alternatives is of strategic importance for any company as it enhances its competitive advantage and strengthens the long-term relationship between the two companies. Besides, the increasing role of third-party logistic companies requires administrators that can link the processes, set up crucial relationships, focus on problems other than cost reduction, redesign the management processes to mitigate the risks from rapid changes in the market [4]. So, for most companies, increasing the efficiency in core businesses highly depends on the performance of the 3PL service providers which positively affects the companies focusing on its main function areas.

It is a multicriteria decision-making problem for companies to choose 3PL service providers according to the criteria they have determined. However, evaluation of 3PL services mostly depends on the satisfaction of the customer company where the criteria generally consist of qualitative issues like quality, customer relations, management philosophy, delivery performance, etc. During the decision-making process, these criteria should be quantified to get better results. To resolve this drawback fuzzy set-based decision-making methods are used. These MCDM methods, integrated with fuzzy numbers, quantify the ambiguity and uncertainty in human nature in the best way [6–9]. Since Zadeh's classical fuzzy approach, there are many fuzzy sets developed, like the intuitionistic fuzzy and Pythagorean fuzzy sets. However, q-ROF sets allow decision-makers to express their evaluations in a wider area than the aforementioned ones. In this study, a new approach is proposed for the 3PL service provider selection problems by integrating CODAS method with the q-ROF numbers recently proposed by Yager. Enhancing the classical CODAS method with q-ROF numbers, we closed the gap considering it an environment of vagueness and impreciseness.

In the second section, the literature applications of the CODAS method are examined, in the third section, the proposed q-ROF CODAS methodology is explained, in the fourth section, 3PL selection application is illustrated and compared with other methods and the advantages of the proposed method are revealed, finally, in the last part, the results are summarized.

11.2 Literature Survey

Ghorabaee et al. [10] proposed the fuzzy CODAS method to solve multicriteria group decision-making problems using linguistic variables and trapezoidal fuzzy numbers. They applied the market segment evaluation and selection problem under uncertainty to an example and made a comparison between fuzzy CODAS and two other MCDM methods (fuzzy EDAS and fuzzy TOPSIS) to validate the results. To demonstrate the validity of the results, a sensitivity analysis was performed with a randomly generated set of ten criteria weights.

Badi et al. [11] used the CODAS method to select the best desalination plant on the northwest coast of Libya. They minimized transportation costs with the objective function, chose the most suitable location with CODAS, and claimed that the study was efficient with their sensitivity analysis. Pamucar et al. [12] conducted a case study on choosing the optimum Power Generation Technology (PGT) in Libya using the CODAS method with Linguistic Neutrosophic Numbers. To test this integrated model, sensitivity analysis was performed by changing the weighting coefficients for 68 scenarios and the results were verified by comparing the four multicriteria decision-making models with the proposed model.

The CODAS method has been integrated with various fuzzy sets. Boltürk [13] extended the CODAS method to Pythagorean fuzzy CODAS to select the best supplier in an ambiguous environment. In this study, it is suggested that meaningful results were obtained by using the evaluations of the decision-makers for the most appropriate selection problem among eight different suppliers for a manufacturer.

Boltürk and Kahraman [14] extended the CODAS method to interval-valued intuitionistic fuzzy sets (IVIF) for the power plant selection problem. Dahooei et al. [15] proposed a new business intelligence assessment framework for enterprise systems with a similar IVIF-CODAS method. In this model, 34 criteria from the most important business intelligence indices were determined and these criteria were evaluated by five experts. The results obtained from the extended CODAS method were compared with three different ranking techniques and claimed that the proposed method is compatible with methods such as COPRAS-IVIF and MABAC-IVIF.

Yalçın and Pehlivan [16] used fuzzy CODAS and hesitant fuzzy linguistic term sets for blue-collar personnel selection in the manufacturing sector for uncertain environments where decision-makers hesitate and compared the results with four different fuzzy MCDM methods, they suggested that their methods are efficient and stable.

Deveci et al. [17] used the fuzzy CODAS method for the selection problem, which took into account 17 criteria for the selection of five alternative renewable energy sources. In their proposed IVIF-CODAS method, they developed a new aggregation operator for IVIF numbers. In addition, to eliminate the drawbacks of previous studies, they proposed a new distance measure for IVIF numbers instead of Euclidean and Taxicab distances, which in some cases were not distinctive.

Şeker [18] integrated the CODAS method with interval value trapezoidal fuzzy sets (IVITrFS) to better reflect the hesitancy and uncertainty inherent in the human nature of decision-making problems. Karaşan et al. [19] also studied the CODAS method and Spherical Fuzzy Sets together in their work on evaluating the life index in the suburbs in order to better quantify this hesitancy in human thought. Bolturk et al. and subsequently Aydoğmuş et al. [20] applied the CODAS method to Picture Fuzzy Sets consisting of four parameters as membership, non-membership, refusal, and abstention and stated that they got meaningful results for decision-making problems.

If we take a look at the recent studies with the CODAS method, Karagöz et al. [21] used the intuitionistic fuzzy CODAS method for the sorting center location problem in Istanbul and made a validity analysis by comparing it with WASPAS and TOPSIS. Ulutaş [22] integrated the fuzzy CODAS with the best–worst method (BWM) to determine the supplier selection of a furniture manufacturer, Şeker and Aydın [23] proposed the IVIF-AHP and CODAS hybrid method for the evaluation of different public transportation system alternatives in a crowded university campus.

There are also many studies on q-ROF decision-making methods based on q-ROF sets aggregation operators. Garg [24] developed a trigonometric operation-based q-ROF aggregation operator and applied it to a group decision-making technique. Garg and Chen [25] proposed a weighted averaging neutral aggregation operators (AOs) for q-ROF sets and a group decision-making process based on the proposed operator, Peng et al. [26] introduce a new score function and derive a weighted exponential aggregation operator for q-ROF numbers and applied to a multicriteria decision-making problem. Garg [27] introduced a novel concept of connection number-based q-ROF set (CN-q-ROFS) and used it in a method for solving the multiattribute group decision-making (MAGDM) problem. He et al. [28] proposed q-ROF cloud interaction weighted Maclaurin symmetric mean operator and a decision-making method related to this operator. Garg [29] proposed a q-connection number (q-CN) for interval-valued q-ROF set (IVq-ROFSs) and a multiple-attribute group decision-making method. Garg [30] also proposed a new possibility degree measure for interval-valued q-ROF sets and applied it to decision-making. Pinar and Boran [31, 32] studied q-ROF and q-RPF distance measures and applied them to decision-making.

11.3 q-ROF CODAS Method

In this section, after brief information about q-ROF sets is given, the methodology of decision-making including the steps of the q-ROF CODAS methodology (Fig. 11.1) will be explained.

11.3.1 q-Rung Orthopair Fuzzy Sets

Atanassov's [33] intuitionistic fuzzy set (IFS) concept has contributed to the non-membership function to Zadeh's classical fuzzy set theory. Later, Yager proposed the

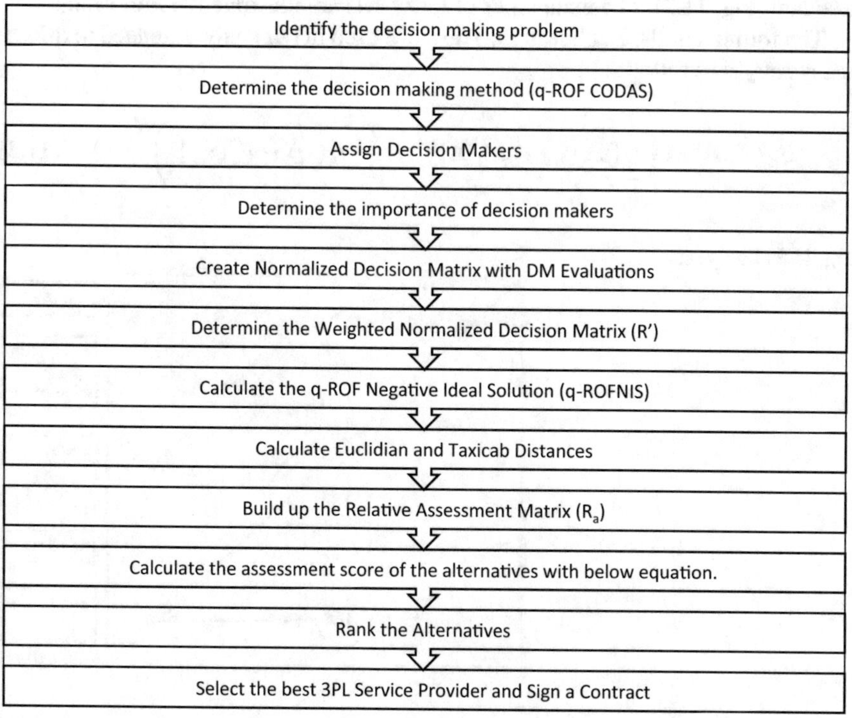

Fig. 11.1 Decision-making methodology with q-ROF CODAS

Pythagorean fuzzy sets (PFS), which is based on the sum of the squares of the membership and non-membership degrees and it covers a wider decision-making area than the IFS. For example, let's take (0.8; 0.6) values as membership degrees. If we try to express these membership degrees as IFS, it can easily be seen that IFS does not satisfy the condition $a + b \leq 1$ since $0.8 + 0.6$ equals 1.4. However, when we take these values as PFS, then $0.8^2 + 0.6^2 = 1$ and it fulfills the condition of PFS. After PFS, Yager [34] introduced the idea of q-ROF sets by generalizing IFS and PFS.

A q-ROF subset A of X has been shown with Eq. 11.1:

$$A = \{\langle x, \mu_A(x), v_A(x)\rangle |\ \ x \in \ X\} \tag{11.1}$$

Here, $\mu_A : X \to [0, 1]$ and $v_A : X \to [0, 1]$ show the membership and non-membership degrees, respectively, and satisfy the following conditions in the q-ROF sets:

$$(\mu_A(x))^q + (v_A(x))^q \leq 1 \tag{11.2}$$

Hesitation degree (π_A) is defined as: $\pi_A(x) = (1 - (\mu_A(x))^q - (v_A(x))^q)^{1/q}$.

Therefore, Yager [34] describes the Atanassovs [33] intuitionistic fuzzy set as the first q-level and the Pythagorean fuzzy sets as the second q-level q-ROF sets, as seen in Fig. 11.2. Based on the existing fuzzy sets, he states that *q-level* might vary from one to infinity and the illustration of a q-ROF set at the infinite q-level will be a square (Fig. 11.2). The main types of fuzzy sets are illustrated in Fig. 11.2.

The formula of the Euclidean (d_E) distance used in our study, extended to q-ROF fuzzy sets, is as follows:

$$d_E(A, B) = \left(\frac{1}{2}\left(\mu_A^q(x_i) - \mu_B^q(x_i)\right)^2 + \frac{1}{2}\left(v_A^q(x_i) - v_B^q(x_i)\right)^2\right)^{1/2} \tag{11.3}$$

Fig. 11.2 Fuzzy sets

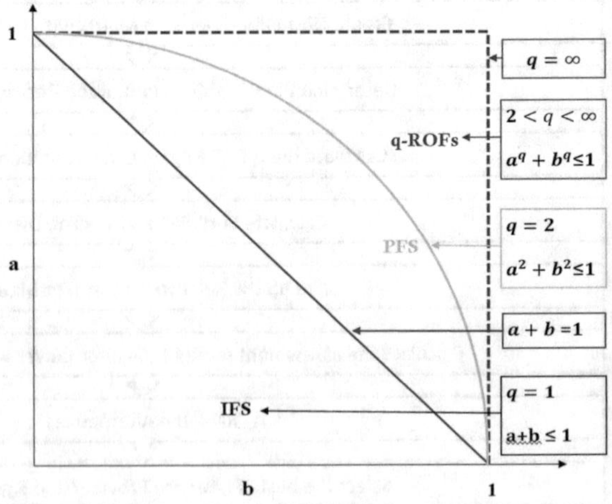

Similarly, the formula of the Taxicab (d_T) distances extended to q-ROF fuzzy sets, is as follows:

$$d_T(A, B) = \left\{\frac{1}{2}\left(\left|\mu_A^q(x_i) - \mu_B^q(x_i)\right| + \left|v_A^q(x_i) - v_B^q(x_i)\right|\right)\right\} \tag{11.4}$$

The q-ROFWA aggregation operator is presented as follows:

$$q - ROFWA(\alpha_1, \alpha_2, ..., \alpha_l) = \left\langle\left(1 - \prod_{k=1}^{l}(1-\mu_k^q)^{\lambda_k}\right)^{1/q}, \prod_{k=1}^{l} v_k^{\lambda_k}\right\rangle \tag{11.5}$$

11.3.2 q-ROF CODAS Methodology

In the q-ROF CODAS method, the set of alternatives is shown with A= $\{A_1, A_2, A_3, ..., A_m\}$ and the set of criteria is denoted with X=$\{X_1, X_2, X_3, ..., X_n\}$. The steps of the q-ROF CODAS methodology are proposed as follows:

Step 1. Determining the importance of decision-makers (DM)

In q-ROF CODAS, the calculation of decision-makers' importance levels is added to the first step of the classical CODAS method of Ghorabaee et al. [35]. Decision-makers are evaluated with the help of linguistic terms in Table 11.1 according to their expertise, then, these evaluations are converted to q-ROF numbers and DM weights are calculated so that the sum is 1. For example, to determine the importance level of the kth out of l decision-makers score function of the q-ROF number $D_k = [\mu_{k,} v_k, \pi_k]$ Eq. (11.5) is used in Eq. (11.6) [36, 37]

$$\lambda_k = \frac{(1 + \mu_k^q - v_k^q)}{\sum_{k=1}^{l}(1+\mu_k^q - v_k^q)} \text{ ve } \sum_{k=1}^{l}\lambda_k = 1 \tag{11.6}$$

Table 11.1 Linguistic terms for q-ROF numbers

Terms	μ	v
Extremely high (EH)	0.95	0.15
Very high (VH)	0.85	0.25
High (H)	0.75	0.35
Medium high (MH)	0.65	0.45
Medium (M)	0.55	0.55
Medium low (ML)	0.45	0.65
Low (L)	0.35	0.75
Very low (VL)	0.25	085
Extremely low (EL)	0.15	0.95

In this step, the criteria are also determined by DMs and the alternatives are evaluated using linguistic terms.

Step 2. Creation of Normalized Decision Matrix with DM Evaluations

Decision-makers' evaluations of alternatives are converted into q-ROF numbers. DMs weights (λ_k) that are determined in the first step and their evaluations in q-ROF numbers are usually shown as

$$\alpha_k = \langle \mu_k, v_k \rangle (k = 1, 2, 3, ..., l)$$

are aggregated and normalized in this step using the q-ROFWA aggregation operator [38].

An example of a q-ROF CODAS decision matrix is below:

$$\begin{bmatrix} \mu_{A_1}(x_1), v_{A_1}(x_1), \pi_{A_1}(x_1) & \mu_{A_1}(x_2), v_{A_1}(x_2), \pi_{A_1}(x_2) & \cdots & \mu_{A_1}(x_n), v_{A_1}(x_n), \pi_{A_1}(x_n) \\ \mu_{A_2}(x_1), v_{A_2}(x_1), \pi_{A_2}(x_1) & \mu_{A_2}(x_2), v_{A_2}(x_2), \pi_{A_2}(x_2) & \cdots & \mu_{A_2}(x_n), v_{A_2}(x_n), \pi_{A_2}(x_n) \\ \vdots & \vdots & \ddots & \vdots \\ \mu_{A_m}(x_1), v_{A_m}(x_1), \pi_{A_m}(x_1) & \mu_{A_m}(x_2), v_{A_m}(x_2), \pi_{A_m}(x_2) & \cdots & \mu_{A_m}(x_n), v_{A_m}(x_n), \pi_{A_m}(x_n) \end{bmatrix}$$

$\mathrm{R} = \left(\mathrm{r}_{ij}\right)_{mxn}$, where $\left(\mu_{A_i}(x_j), v_{A_i}(x_j), \pi_{A_i}(x_j)\right)$, (i = 1, 2, ..., m; j = 1, 2, ..., n).

Step 3. Determining Criteria Weights

DMs determine the importance of the criteria with linguistic terms given in Table 11.1, these evaluations are converted to q-ROF numbers, and the weight of each criterion is calculated using Eq. (11.7) based on the score function formula:

$$w_j = \frac{\sum_{k=1}^{l} \lambda_k (1 + \mu_{jk}^q - v_{jk}^q)}{\sum_{j=1}^{n} \sum_{k=1}^{l} \lambda_k (1 + \mu_{jk}^q - v_{jk}^q)} \tag{11.7}$$

where, $W = \{w_1, w_2, w_3, ..., w_n\}$, and $\sum_{j=1}^{n} w_j = 1$.

Step 4. Determine the Weighted Normalized Decision Matrix (R′)

The criteria weights are aggregated with the decision matrix created in the second step using Eq. (11.8) and the weighted normalized q-ROF decision matrix is obtained.

$$w_k \alpha_1 = \left\langle \left(1 - (1 - \mu_1^q)^{w_k}\right)^{1/q}, v_1^{w_k} \right\rangle, \pi_{A_i}(x_j) = \left(1 - \mu_{A_i}^q(x_j) - v_{A_i}^q(x_j)\right)^{1/q} \tag{11.8}$$

$r'_{ij} = (\mu'_{ij}, v'_{ij}, \pi'_{ij}) = \left(\mu_{A_i W}(x_j), v_{A_i W}(x_j), \pi_{A_i W}(x_j)\right)$ is an element of the matrix, where i = 1,2,3, ..., m; j = 1,2,3, ..., n. An example of matrix R' is as follows:

$$R' = \begin{bmatrix} \mu_{A_1W}(x_1), \upsilon_{A_1W}(x_1), \pi_{A_1W}(x_1) & \mu_{A_1W}(x_2), \upsilon_{A_1W}(x_2), \pi_{A_1W}(x_2) & \cdots & \mu_{A_1W}(x_n), \upsilon_{A_1W}(x_n), \pi_{A_1W}(x_n) \\ \mu_{A_2W}(x_1), \upsilon_{A_2W}(x_1), \pi_{A_2W}(x_1) & \mu_{A_2W}(x_2), \upsilon_{A_2W}(x_2), \pi_{A_2W}(x_2) & \cdots & \mu_{A_2W}(x_n), \upsilon_{A_2W}(x_n), \pi_{A_2W}(x_n) \\ \vdots & \vdots & \ddots & \vdots \\ \mu_{A_mW}(x_1), \upsilon_{A_mW}(x_1), \pi_{A_mW}(x_1) & \mu_{A_mW}(x_2), \upsilon_{A_mW}(x_2), \pi_{A_mW}(x_2) & \cdots & \mu_{A_mW}(x_n), \upsilon_{A_mW}(x_n), \pi_{A_mW}(x_n) \end{bmatrix}$$

Step 5. Calculate the q-ROF Negative Ideal Solution (q-ROFNIS)

q-ROFNIS maximizes cost while minimizing benefit. When J_1 and J_2 are benefit and cost criteria, respectively, q-ROFNIS (ns_j) can be expressed as:

$$ns_j = \left(\mu_{A^-W}(x_j), \upsilon_{A^-W}(x_j)\right) \tag{11.9}$$

$$\mu_{A^-W}(x_j) = \left(\left(\min_i \mu_{A_iW}(x_j) | j \in J_1\right), \left(\max_i \mu_{A_iW}(x_j) | j \in J_2\right)\right) \tag{11.10}$$

$$\upsilon_{A^-W}(x_j) = \left(\left(\max_i \upsilon_{A_iW}(x_j) | j \in J_1\right), \left(\min_i \upsilon_{A_iW}(x_j) | j \in J_2\right)\right) \tag{11.11}$$

Step 6. Calculate Euclidian and Taxicab (Manhattan) Distances

Euclidean (E_i) and Taxicab (T_i) distances are used to calculate the distance between the negative ideal solution and the values of each alternative denoted with q-ROF numbers. These distances are calculated with the following equations:

$$E_i = \sum_{j=1}^{m} d_E(r'_{ij} - ns_j) \tag{11.12}$$

$$T_i = \sum_{j=1}^{m} d_T(r'_{ij} - ns_j) \tag{11.13}$$

Step 7. Build up the Relative Assessment Matrix (R_a)

$$R_a = [p_{ik}]_{nxn}$$

$$p_{ik} = E_i - E_k + (t(E_i - E_k) \times (T_i - T_k))$$

where k ∈ {1, 2, 3, .. n} and t denotes a threshold function, is defined as follows:

$$t(x) = \begin{cases} 1 \text{ if } |x| \geq \theta \\ 0 \text{ if } |x| < \theta \end{cases} \tag{11.14}$$

Here θ is the threshold parameter set by the decision-maker. When the difference between Euclidian distances of the two alternatives is more than θ, Taxicab distance is also used for comparison. In our case study, we use $\theta = 0.02$ for the calculations.

Step 8. Calculate the assessment score of the alternatives with the below equation.

$$P_i = \sum_{k=1}^{n} p_{ik} \tag{11.15}$$

Step 9. Rank the Alternatives

Alternatives are ranked according to their evaluation scores from the largest to the smallest, that is, from the best to the worst.

11.4 Case Study

In this section, the most suitable 3PL service provider is selected among six alternatives for a retail company. The company has assigned three subject matter experts which determined seven criteria in line with the requirements of the company. The q-ROF CODAS method was applied with the following steps.

Step 1. Determining the Importance of Decision-Makers (DM)

At first, the decision-makers were evaluated with the help of the linguistic terms in Table 11.1 according to their expertise and DM weights were determined as [0.402; 0.299; 0.299], respectively. Then, DMs determined the eligibility criteria for an ideal 3PL service provider. Finally, decision-makers evaluated each alternative in linguistic terms according to the criteria, as presented in Table 11.2.

Step 2. Creation of Normalized Decision Matrix with DM Evaluations

As previously explained, the evaluation of the decision-makers in linguistic terms is converted to q-ROF numbers and the following decision matrix is obtained by aggregating these evaluations of the three decision-makers in q-ROF numbers:

	X1	X2	X3	X4	X5	X6	X7
A1	[0.850; 0.250; 0.718]	[0.850; 0.250; 0.718]	[0.361; 0.756; 0.805]	[0.780; 0.330; 0.788]	[0.787; 0.317; 0.783]	[0.826; 0.276; 0.745]	[0.826; 0.276; 0.745]
A2	[0.850; 0.250; 0.718]	[0.798; 0.306; 0.774]	[0.386; 0.719; 0.830]	[0.817; 0.286; 0.755]	[0.780; 0.330; 0.788]	[0.826; 0.276; 0.745]	[0.850; 0.250; 0.718]
A3	[0.850; 0.250; 0.718]	[0.787; 0.317; 0.783]	[0.416; 0.688; 0.844]	[0.780; 0.330; 0.788]	[0.826; 0.276; 0.745]	[0.798; 0.306; 0.774]	[0.716; 0.387; 0.832]

(continued)

(continued)

	X1	X2	X3	X4	X5	X6	X7
A4	[0.787; 0.317; 0.783]	[0.716; 0.387; 0.832]	[0.350; 0.750; 0.812]	[0.817; 0.286; 0.755]	[0.817; 0.286; 0.755]	[0.826; 0.276; 0.745]	[0.725; 0.377; 0.827]
A5	[0.780; 0.330; 0.788]	[0.787; 0.317; 0.783]	[0.434; 0.684; 0.843]	[0.780; 0.330; 0.788]	[0.708; 0.401; 0.834]	[0.696; 0.407; 0.841]	[0.759; 0.355; 0.803]
A6	[0.650; 0.450; 0.859]	[0.780; 0.330; 0.788]	[0.380; 0.735; 0.818]	[0.780; 0.330; 0.788]	[0.708; 0.401; 0.834]	[0.759; 0.355; 0.803]	[0.759; 0.355; 0.803]

Step 3. Determining Criteria Weights

DMs determine the importance of the criteria they choose based on their experience in linguistic terms, and the weights of the criteria translated into q-ROF numbers are calculated in Table 11.3.

Step 4. Determine the Weighted Normalized Decision Matrix (R′)

Using the criteria weights, weighted and normalized q-ROF CODAS decision matrix (R′) is obtained as follows:

	X1	X2	X3	X4	X5	X6	X7
A1	[0.509; 0.814; 0.690]	[0.508; 0.815; 0.689]	[0.199; 0.955; 0.496]	[0.423; 0.873; 0.638]	[0.436; 0.862; 0.651]	[0.495; 0.819; 0.691]	[0.469; 0.845; 0.665]
A2	[0.509; 0.814; 0.690]	[0.463; 0.840; 0.676]	[0.214; 0.947; 0.521]	[0.452; 0.858; 0.652]	[0.430; 0.867; 0.646]	[0.495; 0.819; 0.691]	[0.489; 0.834; 0.672]
A3	[0.509; 0.814; 0.690]	[0.455; 0.844; 0.673]	[0.231; 0.940; 0.539]	[0.423; 0.873; 0.638]	[0.467; 0.847; 0.662]	[0.471; 0.831; 0.684]	[0.387; 0.883; 0.633]
A4	[0.456; 0.843; 0.674]	[0.402; 0.869; 0.652]	[0.193; 0.954; 0.501]	[0.452; 0.858; 0.652]	[0.459; 0.851; 0.659]	[0.495; 0.819; 0.691]	[0.394; 0.880; 0.636]
A5	[0.450; 0.848; 0.669]	[0.455; 0.844; 0.673]	[0.241; 0.939; 0.541]	[0.423; 0.873; 0.638]	[0.380; 0.889; 0.624]	[0.396; 0.869; 0.655]	[0.417; 0.873; 0.640]
A6	[0.360; 0.888; 0.632]	[0.449; 0.849; 0.668]	[0.210; 0.950; 0.510]	[0.423; 0.873; 0.638]	[0.380; 0.889; 0.624]	[0.441; 0.851; 0.668]	[0.417; 0.873; 0.640]

Step 5. Calculate the q-ROF Negative Ideal Solution (q-ROFNIS):

Considering the six benefits and one cost criterion (X3), the negative ideal solution is calculated in Table 11.4.

Table 11.2 DM evaluations in linguistic terms

Criteria\Alt	A1	A2	A3	A4	A5	A6
DM-1						
X1	VH	VH	VH	H	VH	MH
X2	VH	VH	H	MH	H	VH
X3	VL	L	L	L	L	ML
X4	VH	H	VH	H	VH	VH
X5	H	VH	VH	H	H	H
X6	VH	VH	VH	VH	H	VH
X7	VH	VH	MH	H	VH	VH
DM-2						
X1	VH	VH	VH	H	H	MH
X2	VH	H	VH	H	VH	MH
X3	L	ML	ML	L	L	VL
X4	H	VH	MH	VH	H	H
X5	VH	MH	H	VH	H	M
X6	VH	H	H	H	MH	MH
X7	VH	VH	H	H	MH	MH
DM-3						
X1	VH	VH	VH	VH	MH	MH
X2	VH	H	H	H	H	H
X3	ML	L	ML	L	M	L
X4	MH	VH	H	VH	MH	MH
X5	H	H	VH	VH	M	H
X6	H	VH	H	VH	MH	MH
X7	H	VH	H	MH	MH	MH

Table 11.3 Criteria weights

Criteria		DM 1	DM 2	DM 3	Weight
X1	Quality	VH	VH	VH	0.1487
X2	Delivery	H	VH	EH	0.1476
X3	Cost	EH	EH	VH	0.1653
X4	Financial situation	H	H	MH	0.1228
X5	Customer relations	H	MH	VH	0.1289
X6	Reputation and position in the industry	VH	EH	VH	0.1558
X7	Management	VH	MH	H	0.1310

Table 11.4 Negative ideal solutions

Criteria	μ	ν	π
X1	0.360	0.888	0.632
X2	0.402	0.869	0.652
X3	0.241	0.939	0.541
X4	0.423	0.873	0.638
X5	0.380	0.889	0.624
X6	0.396	0.869	0.655
X7	0.387	0.883	0.633

Step 6. Calculate Euclidian and Taxicab (Manhattan) Distances

The distances of the alternatives from the negative ideal solution are calculated in Table 11.5.

Step 7. Build up the Relative Assessment Matrix (R_a)

At this stage, Taxicab distance was added to the values which are calculated above the threshold parameter ($\theta = 0.02$) of the Euclidean distances between the alternatives. The relative assessment matrix is presented in Table 11.6.

Step 8–9. Calculate the assessment score and rank the alternatives

Alternatives are ranked from best to worst as A1 > A2 > A3 > A4 > A5 > A6 according to their P_i values (Table 11.7).

Table 11.5 Euclidian and Taxicab (Manhattan) distances

Alternatives	Ei	Ti
A1	0.419	0.396
A2	0.406	0.383
A3	0.294	0.277
A4	0.279	0.261
A5	0.133	0.124
A6	0.115	0.106

Table 11.6 Relative assessment matrix (R_a)

Alternatives	A1	A2	A3	A4	A5	A6	P_i
A1	0.0000	0.0130	0.2451	0.2756	0.5586	0.5942	1.6865
A2	−0.0130	0.0000	0.2190	0.2495	0.5325	0.5682	1.5562
A3	−0.2451	−0.2190	0.0000	0.0150	0.3134	0.3491	0.2134
A4	−0.2756	−0.2495	−0.0150	0.0000	0.2830	0.3187	0.0617
A5	−0.5586	−0.5325	−0.3134	−0.2830	0.0000	0.0176	−1.6699
A6	−0.5942	−0.5682	−0.3491	−0.3187	−0.0176	0.0000	−1.8478

Table 11.7 Rankings

Alternatives	q-ROF TOPSIS	q-ROF CODAS (P_i)	Rankings
A1	0.8272	1.6865	1
A2	0.7744	1.5562	2
A3	0.6052	0.2134	3
A4	0.5121	0.0617	4
A5	0.3524	−1.6699	5
A6	0.2368	−1.8478	6

Comparison:

To demonstrate the validity of the q-ROF CODAS method, the results and rankings were compared with another q-ROF sets-based MCDM method, q-ROF TOPSIS [31]. The comparison of the two methods is presented in Fig. 11.3.

As illustrated in Fig. 11.3, the results of the proposed method were compared with the q-ROF version of the TOPSIS method, which is frequently used in the literature, and it was seen that both methods were compatible with each other. The q-ROF TOPSIS results were in the range of 0–1 as the nature of the method, q-ROF CODAS results were in the range of (+2, −2), but in the end, the rankings were the same in both methods from best to worst A1 > A2 > A3 > A4 > A5 > A6. Figure 11.3 shows that the intervals between some alternatives are very wide, such as the values of A2 and A3, A4 and A5. So, this feature makes q-ROF CODAS a more distinctive method than q-ROF TOPSIS.

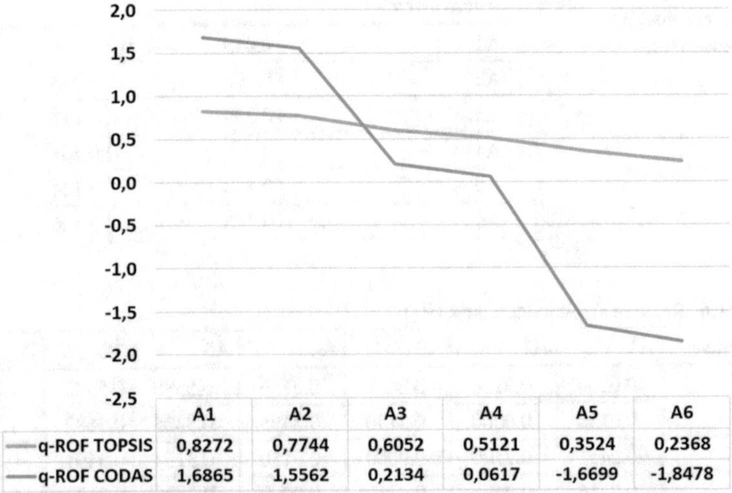

Fig. 11.3 Comparison of the methods

11.5 Conclusion

Today, many companies widely use third-party logistics service providers in their non-core business areas in order to compete more effectively with their competitors in the market. For the selection of the most suitable 3PL service provider, fuzzy MCDM methods, which successfully minimize the uncertainty in human decision-making behavior have been used. In our study, after the literature survey of classical and fuzzy decision-making methods is given, the CODAS method has been adapted to q-ROF sets for the first time and introduced as a new method. This is also the first study of CODAS method application to 3PL service provider selection. The proposed q-ROF CODAS method has been applied to the 3PL service provider evaluation problem. The results and analyses have shown that the method is consistent with other methods. Our method has these advantages; (a) quantifies the qualitative criteria better than classical methods, (b) models the uncertainty in human nature, (c) allows decision-makers a wider area to express their evaluations, (d) regarding results, besides its consistency, its more distinctive method than q-ROF TOPSIS.

For future research, CODAS method can be applied to other fuzzy sets, such as hesitant fuzzy, spherical fuzzy, q-rung picture fuzzy sets. Besides, q-ROF CODAS method can be applied to other decision-making problems like renewable energy, personnel, site, green supplier selection, etc., moreover, it can be integrated with other MCDM methods such as VIKOR, AHP, and DEA.

References

1. M. Berglund, Pv. Laarhoven, G. Sharman, S. Wandel, Third-party logistics: is there a future? Int. J. Logist. Manag. **10**(1), 59–70 (1999)
2. P.K. Bagchi, H. Virum, European logistics alliances: a management model. Int. J. Logist. Manag. **7**(1), 93–108 (1996)
3. A. Aguezzoul, Third-party logistics selection problem: a literature review on criteria and methods. Omega **49**, 69–78 (2014)
4. B.K. Sangka, S. Rahman, A. Yadlapalli, F. Jie, Managerial competencies of 3PL providers: a comparative analysis of Indonesian firms and multinational companies. Int. J. Logist. Manag. **30**(4), 1054–1077 (2019)
5. R.K. Singh, A. Gunasekaran, P. Kumar, Third party logistics (3PL) selection for cold chain management: a fuzzy AHP and fuzzy TOPSIS approach. Ann. Oper. Res. **267**(1), 531–553 (2018)
6. O. Torağay, M. Arıkan, Performance evaluation of faculty departments by a Delphi method based on 2-Tuple fuzzy Linguistic representation model and TOPSIS. Int. J. Basic Appl. Sci. **15**, 1–10 (2015)
7. A.R. Mishra, P. Rani, K. Pandey, Fermatean fuzzy CRITIC-EDAS approach for the selection of sustainable third-party reverse logistics providers using improved generalized score function. J. Ambient Intell. Humanized Comput. 1–17 (2021)
8. J. Fan, R. Guan, M. Wu, Z-MABAC method for the selection of third-party logistics suppliers in fuzzy environment. IEEE Access **8**, 199111–199119 (2020)
9. S. Jovčić, V. Simić, P. Průša, M. Dobrodolac, Picture fuzzy ARAS method for freight distribution concept selection. Symmetry **12**(7), 1062 (2020)

10. M.K. Ghorabaee, M. Amiri, E.K. Zavadskas, R. Hooshmand, J. Antuchevičienė, Fuzzy extension of the CODAS method for multi-criteria market segment evaluation. J. Bus. Econ. Manag. **18**(1), 1–19 (2017)
11. I. Badi, M. Ballem, A. Shetwan, Site selection of desalination plant in libya by using combinative distance-based assessment (CODAS) method. Int. J. Quality Res. **12**(3) (2018)
12. D. Pamučar, I. Badi, K. Sanja, R. Obradović, A novel approach for the selection of power-generation technology using a linguistic neutrosophic CODAS method: a case study in Libya. Energies **11**(9), 2489 (2018)
13. E. Bolturk, Pythagorean fuzzy CODAS and its application to supplier selection in a manufacturing firm. J. Enterp. Inf. Manag. **31**(4), 550–564 (2018)
14. E. Bolturk, C. Kahraman, Interval-valued intuitionistic fuzzy CODAS method and its application to wave energy facility location selection problem. J. Intell. Fuzzy Syst. **35**(4), 4865–4877 (2018)
15. J.H. Dahooei, E.K. Zavadskas, A.S. Vanaki, H.R. Firoozfar, M. Keshavarz-Ghorabaee, An evaluation model of business intelligence for enterprise systems with new extension of codas (CODAS-IVIF). **21**(3), 171–187s (2018)
16. N. Yalçın, P.N. Yapıcı, Application of the fuzzy CODAS method based on fuzzy envelopes for hesitant fuzzy linguistic term sets: a case study on a personnel selection problem. Symmetry **11**(4), 493 (2019)
17. K. Deveci, R. Cin, A. Kağızman, A modified interval valued intuitionistic fuzzy CODAS method and its application to multi-criteria selection among renewable energy alternatives in Turkey. Appl. Soft Comput. **96**(106660), 1568–4946 (2020)
18. S. Seker, A novel interval-valued intuitionistic trapezoidal fuzzy combinative distance-based assessment (CODAS) method. Soft. Comput. **24**(3), 2287–2300 (2020)
19. A. Karaşan, E. Boltürk, F.K. Gündoğdu, Assessment of livability indices of suburban places of Istanbul by using spherical fuzzy CODAS method, in *Decision Making with Spherical Fuzzy Sets* (Springer, 2021), pp. 277–93
20. H.Y. Aydoğmuş, E. Kamber, C. Kahraman, ERP selection using picture fuzzy CODAS method. J. Intell. Fuzzy Syst. **40**(6), 11363–11373
21. S. Karagoz, M. Deveci, V. Simic, N. Aydin, U. Bolukbas, A novel intuitionistic fuzzy MCDM-based CODAS approach for locating an authorized dismantling center: a case study of Istanbul. Waste Manag. Res. **38**(6), 660–672 (2020)
22. A. Ulutaş, *Supplier Evaluation with BWM and Fuzzy CODAS Methods. Handbook of Research on Recent Perspectives on Management, International Trade, and Logistics* (IGI Global, 2021), pp. 335–51
23. S. Seker, N. Aydin, Sustainable public transportation system evaluation: a novel two-stage hybrid method based on IVIF-AHP and CODAS. Int. J. Fuzzy Syst. **22**(1), 257–272 (2020)
24. H. Garg, A novel trigonometric operation-based q-rung orthopair fuzzy aggregation operator and its fundamental properties. Neural Comput. Appl. **32**(18), 15077–15099 (2020)
25. H. Garg, S.-M. Chen, Multiattribute group decision making based on neutrality aggregation operators of q-rung orthopair fuzzy sets. Inf. Sci. **517**, 427–447 (2020)
26. X. Peng, J. Dai, H. Garg, Exponential operation and aggregation operator for q-rung orthopair fuzzy set and their decision-making method with a new score function. Int. J. Intell. Syst. **33**(11), 2255–2282 (2018)
27. H. Garg, CN-q-ROFS: connection number-based q-rung orthopair fuzzy set and their application to decision-making process. Int. J. Intell. Syst. **36**(7), 3106–3143 (2021)
28. P. He, C. Li, H. Garg, J. Liu, Z. Yang, X. Guo, A q-rung orthopair cloud-based multi-attribute decision-making algorithm: considering the information error and multilayer heterogeneous relationship of attributes. IEEE Access. **9**, 132541–132557 (2021)
29. H. Garg, New exponential operation laws and operators for interval-valued q-rung orthopair fuzzy sets in group decision making process. Neural Comput. Appl. **33**(20), 13937–13963 (2021)
30. H. Garg, A new possibility degree measure for interval-valued q-rung orthopair fuzzy sets in decision-making. Int. J. Intell. Syst. **36**(1), 526–557 (2021)

31. A. Pinar, F. Boran, A q-rung orthopair fuzzy multi-criteria group decision making method for supplier selection based on a novel distance measure. Int. J. Mach. Learn. Cybern. **11**, 1749–1780 (2020)
32. A. Pinar, F.E. Boran, A novel distance measure on q-rung picture fuzzy sets and its application to decision making and classification problems. Artif. Intell. Rev. 1–34 (2021)
33. K.T. Atanassov, Intuitionistic fuzzy sets. Fuzzy Sets Syst. **20**(1), 87–96 (1986). https://doi.org/10.1016/S0165-0114(86)80034-3
34. R.R. Yager, Generalized orthopair fuzzy sets. IEEE Trans. Fuzzy Syst. **25**(5), 1222–1230 (2016)
35. M.K. Ghorabaee, E.K. Zavadskas, Z. Turskis, J. Antucheviciene, A new combinative distance-based assessment (CODAS) method for multi-criteria decision-making. Econ. Comput. Econ. Cybern. Stud. Res. **50**(3) (2016)
36. R. Wang, Y. Li, A novel approach for green supplier selection under a q-rung orthopair fuzzy environment. Symmetry **10**(12), 687 (2018). https://doi.org/10.3390/sym10120687
37. G. Wei, H. Gao, Y. Wei, Some q-rung orthopair fuzzy Heronian mean operators in multiple attribute decision making. Int. J. Intell. Syst. **33**(7), 1426–1458 (2018). https://doi.org/10.1002/int.21985
38. P.D. Liu, P. Wang, Some q-rung orthopair fuzzy aggregation operators and their applications to multiple-attribute decision making. Int. J. Intell. Syst. **33**(2), 259–280 (2018). https://doi.org/10.1002/int.21927

Adem Pinar received the B.Sc. degree in Systems Engineering from Turkish Military Academy in 1999, M.A. in International Relations from Trakya University in 2003, M.Sc. in Information Systems from Middle East Technical University in 2004, M.A. in Security Strategies and Leadership from Army War College in 2009, Ph.D. in Supply Chain and Logistics Management in Gazi University, in 2020 (all universities are in Turkey). As a 22-year logistician, he worked in many logistic units of armed forces in supply, acquisition, maintenance, inventory and warehouse management, logistics planning areas at tactical, operational, and strategic levels. He worked as a staff officer and logistics planner in national units and also worked in NATO Headquarters between 2011–2014 in Naples/Italy. He is currently an Assistant Professor in Logistics Department of Turkish Aeronautical Association University, Ankara. His research interests supply chain management, logistics planning, multiple-criteria decision making, q-rung orthopair fuzzy sets, management information systems, project management, distance measures and international relations.

Fatih Emre Boran received the B.Sc., M.Sc., and Ph.D. degrees in industrial engineering from Gazi University, Ankara, Turkey, in 2007, 2009, and 2013, respectively. He is currently a Professor with the Department of Energy Systems Engineering, Gazi University. His research interests include fuzzy sets and systems and decision making. He serves as an associate editor in IEEE Access.

Chapter 12
An Integrated Proximity Indexed Value and q-Rung Orthopair Fuzzy Decision-Making Model for Prioritization of Green Campus Transportation

Muhammet Deveci, Ilgin Gokasar, Dragan Pamucar, Sanjib Biswas, and Vladimir Simic

Abstract Global warming and air pollution are two of the most severe problems faced by humanity in recent years. Greenhouse gas (GHG) emissions caused by vehicles make up for a sizeable portion of the cause of these problems. To mitigate the effects of vehicles on air pollution and global warming, sustainable and environmentally friendly measures are to be taken. Therefore, authorities of many institutions have looked for alternative sustainable transportation systems to implement. In this study, four different alternatives are presented, which aim to provide sustainable and green university campuses by implementing alternative transportation systems. Presented alternatives are, namely, optimizing the performance of current shuttle operations, converting current shuttles into electric vehicles, implementing bicycle, scooter sharing, and implementing carpooling of the vehicles used in the campus. These alternatives are then assessed by the experts according to 10 criteria, which

M. Deveci (✉)
Department of Industrial Engineering, Turkish Naval Academy, National Defence University, 34940 Tuzla, Istanbul, Turkey
e-mail: muhammetdeveci@gmail.com

Royal School of Mines, Imperial College London, South Kensington Campus, London SW7 2AZ, UK

I. Gokasar
Department of Civil Engineering, Bogazici University, 34342 Bebek, Istanbul, Turkey
e-mail: ilgin.gokasar@boun.edu.tr

D. Pamucar
Department of Logistics, Military Academy, University of Defence in Belgrade, 11000 Belgrade, Serbia

S. Biswas
Decision Sciences and Operations Management, Calcutta Business School, Bishnupur, South 24 Parganas 743503, West Bengal, India
e-mail: sanjibb@acm.org

V. Simic
Faculty of Transport and Traffic Engineering, University of Belgrade, Vojvode Stepe 305, 11010 Belgrade, Serbia
e-mail: vsima@sf.bg.ac.rs

H. Garg (ed.), *q-Rung Orthopair Fuzzy Sets*,
https://doi.org/10.1007/978-981-19-1449-2_12

are grouped under 4 main criteria aspects, which are cost aspect, technical aspect, operational aspect, and health and environmental aspect. We aim to propose an extension of the Proximity Indexed Value (PIV) method in q-Rung Orthopair (q-ROF). q-ROF PIV method is combined with a new algorithm for determining the weight coefficients of the criteria, which is based on the application of a logarithmic additive function to define the relationship between the criteria. A case study can illustrate the formula and solution to the problem. As the assessments of the experts are gathered and analyzed, the advantage prioritization of the alternatives shows that converting current shuttles into electric vehicles is the most advantageous alternative.

Keywords Sustainability transport systems · Sharing system · Fuzzy sets · Multicriteria decision-making (MCDM) · q-Rung orthopair fuzzy

12.1 Introduction

A Campus is defined as "The ground and buildings of a university or college" in the Oxford dictionary. Buildings on a university campus may include lecture halls, libraries, residence halls, dining halls, parking structures, and even cinemas, theaters, and stadiums. From this definition, it can be said that university campuses function like a small town that has its planners, a governing body, and a distinctive population. University campuses have a vibrant academic and social life; as a result, there is a continuous movement between buildings, and campuses attract many trips during the day. Because of the demographically different population, continuous movement between buildings, and the difference between universities and other institutions in terms of the number of trips and distribution of trips during the day, transportation planning literature that focused on cities cannot fully address the transportation planning of university campuses [4, 51, 55].

Many old universities and newly founded urban universities have limited land because the land surrounding the campus is too expensive or urbanized. With the increase in car ownership over the second half of the last century [16], university campuses experienced car usage-related problems like parking shortage, land-use problems, and pedestrian mobility and security problems. Because of the problems that have been experienced and the lack of literature that fully addresses campus transportation problems, from the early 1990s, a campus transportation literature has grown. Many researchers and planners have been focused on transportation demand management strategies to address campus transportation problems. With the increase of the global awareness about sustainability and health in the twenty-first century, the goal of the planners and researchers has become to promote active transportation on university campuses [4, 51, 61].

In campus transportation literature, travel demand management strategies are widely studied and recommended as solutions for transportation problems. These strategies are aimed to reduce the utility of single-occupant vehicles while increasing the utility of active modes. One of the most common transportation problems faced by universities is parking capacity problems. Parking capacity problems are caused by either insufficient supply or underpriced or even free-to-use parking spaces. To decrease private car usage, public transportation can be motivated. Unlimited or discounted access programs are proven to be effective in decreasing private car usage and are recommended by many researchers [5, 10, 11]. Cycling is another alternative to the private car. As expected, increasing the modal share of cycling is more challenging than increasing the modal share of public transportation. A complete intervention package may be needed to increase the level of cycling [43]. A combination of facility improvements, educational campaigns, promotional campaigns, and infrastructure improvements are widely used strategies [9, 25, 43, 47]. Access to bicycles matters in the decision of cycling; hence, bicycle-sharing systems are widely adopted by many universities and cities [14, 17]. To decrease private vehicle usage and increase the level of cycling, affective and symbolic motives to use these modes also should be targeted. Cyclist image in society is found significant for the decision of cycling [15]. Also, effective and symbolic motives of driving are significant, especially for young people [54]. Because of that, strategies for demand management should not only focus on instrumental motives of mode choice, but also be focused on other motives of mode choice.

Many studies have been conducted in the literature using the fuzzy decision-making model such as [1, 2, 44]. The concept of a generalized fuzzy set called as q-rung orthopair fuzzy sets (q-ROFs) introduced by Yager [64], which is an extension of the existing intuitionistic fuzzy sets (IFSs) and Pythagorean fuzzy sets (PFSs). q-ROFs based multicriteria decision-making methods have been successfully implemented in various problems [19–22]. Riaz et al. [45] presented a novel q-rung orthopair fuzzy interaction aggregation operator for low-carbon green supply chain management problems.

Studies in the literature lack a comparative analysis regarding different sustainable and active transportation modes for university campuses. The presence of a comparative analysis between four different green transportation alternatives makes this study unique compared to other studies in the literature. This study also provides a guide for the decision-makers when going through a transition to more sustainable and greener transportation on the campuses. Even though the criteria, which are considered in the evaluation process of the alternatives, are subject to change according to different cultural, demographical, and geographical properties of the region, the methodology of this study applies to various criteria and alternatives. Therefore, one motivation of the study is to propose a guide for authorities in incentivizing the transition to more sustainable campus transportation.

This study is to propose a hybrid multicriteria framework based on the q-Rung Orthopair based on Proximity Indexed Value (q-ROF PIV) method and the Logarithm Methodology of Additive Weights (q-ROF PIV-LMAW). q-ROF PIV-LMAW methodology is based on defining the optimal alternative based on the absolute value of the dispersion in relation to the ideal reference point. The proposed methodology has several advantages, including: (i) The proposed multicriteria methodology enables the elimination of rank reversal problems, which is a significant limitation of numerous multicriteria techniques; (ii) The implementation of the LMAW methodology in q-ROF PIV enables the determination of the weight coefficients of the criteria based on the definition of the relationship between the criteria using a logarithmic additive function; and (iii) The presented multicriteria framework enables consistent, rational, and objective decision-making, taking into account the uncertainty in the information.

As of now, no prior work has presented a combination of fuzzy weighting using LMAW and qROF PIV-based ranking of alternatives. Hence, the present hybrid framework adds value to the growing literature by providing a new approach for multicriteria-based group decision-making.

12.2 Literature Review

Transportation is the transfer of an object or person from one place to another. Although transportation has been carried out by different means throughout history, today it is carried out mainly by automobiles, public transportation, ships, planes, trains, and so on. While these vehicles save time and make life easier, their disadvantages are an undeniable fact. The air, water, and soil pollution and global warming caused by greenhouse gases are the disadvantages of these vehicles. Making vehicles and transportation as green as possible will lead to sustainable and environmentally friendly transportation. Since transportation activities are high on university campuses, this study focuses on alternatives that have the potential in making the campuses green and sustainable. Although there are many ways to achieve a greener campus by means of transportation, evaluations will be made on four alternatives in this research. Optimizing the performance of shuttle operations, converting current shuttles into electric vehicles, implementing bicycle and scooter sharing, and implementing carpooling of the vehicles used in the campus are the four alternatives to be considered in terms of sustainability and green transportation.

Optimizing the performance of shuttle operations is an alternative solution, which is one of the easy and cheap ways to obtain a more sustainable and green vehicle over the current vehicles. Here, solutions such as providing less fuel consumption per kilometer without making a radical change over current vehicles, reducing the emission of exhaust gases into the air that will cause pollution, and reducing energy consumption by obtaining more feasible interior insulation are the first solutions to come to mind. In this context, looking at the studies in the literature, a study investigates the optimization and evaluation of the sustainable transportation system in a simulation-based study. Song et al. [53] propose a model, which works cooperatively with Simulation-Based Optimization (SBO) to find an optimal combination of transportation planning and operations strategy. Results of the study show that this method is feasible and applicable to optimize sustainable transportation systems and increase the effectiveness of the sustainability aspect of the transportation system. In another study, the aspects affecting the sustainability of public vehicles are investigated [62]. According to the study, speeds, signal delays, acceleration and deceleration trends, etc., are factors affecting the sustainability of public transportation services. This shows that as these factors are controlled, more sustainable transportation can be achieved. For example, if the vehicles move at a more regular speed rather than accelerating and decelerating constantly, GHG emissions produced by the vehicles reduce [52].

Converting current shuttles into electric vehicles is another alternative solution to achieve green campus transportation. This alternative provides a more sustainable and greener campus transport via decreasing fuel consumption and preventing exhaust gases. Within this framework, Casals et al. [13] conducted a study on the sustainability and environmental friendliness of electric vehicles. They compared the carbon emissions of vehicles, which have internal combustion engines and electric engines. Considering the research, results show that sustainability and green transport are more possible by the utilization of electric vehicles. Various studies are supporting the idea presented in this study, which is electric vehicles are a better option than vehicles running on fossil fuels in means of sustainable and green transportation [18, 26, 41].

Implementing bicycle and scooter sharing is another alternative that provides sustainable transportation on university campuses. Encouraging the use of bicycles and scooters, and facilitating efficient access to these transportation modes, will lead people to use bicycles and scooters more often. Bicycles and scooters use negligible amounts of energy. They are sustainable transportation modes, which is one of the major reasons that authorities are implementing bicycle and scooter sharing systems in many locations. In a study, Younes et al. [65] reviewed the sustainability and usage of bicycle and scooter sharing in Washington. Because of their research, they have shown that the utilization of these vehicles is one of the leading transportation mode alternatives in means of environmental friendliness. In a study, the characteristics of the bicycle-sharing system users are investigated [57]. Results of the data analysis made on the information of the users show that the main users of the bicycle-sharing systems are young people. This shows that implementing bicycle and scooter sharing systems to the university campuses is a feasible and sensible choice. In a different study, the cycling behavior of undergraduate students before and after implementing a public bicycle-sharing system is investigated in Spain [32]. Results of the questionnaires done for the study showed an increase of 14.6% in bicycle usage. This shows that implementing bicycle and scooter sharing systems on university campuses is a promising alternative since young students use bicycles and scooters for transportation. Zhu et al. [67] conducted a comparative analysis on bicycle and scooter sharing, considering the geographical conditions in Singapore. They explained how the use of bicycle and scooter sharing can be affected where geographical conditions do not allow it. If the conditions are not favorable, it has been observed that the use of these vehicles decreases. Implementing a sharing system within the university campus has an enormous potential in increasing the sustainability of the campus and providing a green campus.

Implementing carpooling of the vehicles used in the campus is the last alternative. Here, instead of using shuttles or other transportation modes, transportation users are expected to travel in other peoples' private vehicles. Here, increasing the carpooling of the vehicles on the campus will cause less vehicle circulation on the campuses, and thus, a more sustainable and green campus transport will be reached. In this framework, Bruck et al. [12] investigated that sustainable and carbon-minimizing transportation can be achieved by applying no optimization to vehicles or encouraging more environmentally friendly electric vehicles. The study states that traveling together with other individuals in a private vehicle rather than traveling alone will cause a decrease in the vehicle density in the traffic and provide more sustainable and greener transportation. According to Sustainable Commuting, public transport, walking, cycling, and carpooling are green transport modes [31]. Although carpooling is often seen as a preferred situation, sometimes people are hesitant about carpooling. Molina et al. [31] examined in which situations carpooling is not preferred, and because of the research, people refused to share their cars in case of epidemic diseases such as COVID-19. In addition, it was stated that if carpooling becomes widespread, time and economic savings can be achieved. Shaheen et al. [50] researched the benefits of carpooling, and because of this research, it is seen that carpooling is sustainable and economically feasible, as it is stated in other studies.

Our study fills a gap in the literature by providing an advantage prioritization method for alternatives, which aims to make university campus transportation more sustainable and greener. The assessed alternatives in this study are optimizing the performance of shuttle operations, converting current shuttles into electric vehicles, implementing bicycle, scooter sharing, and implementing carpooling of the vehicles used on the campus. These alternatives are then evaluated by the experts under 10 different criteria, which are grouped under 4 main criteria aspects. Results of this study create a consensus for the university authorities by providing an advantage prioritization of the stated alternatives so that authorities can use the study as a guide while going through a transition to a greener campus.

12.3 Case Study

Bogazici University, founded in 1863, is one of the oldest educational institutions in Turkey. The university has six campuses in Istanbul. South, North, Hisar, and Ucaksavar Campuses are located close to each other in the Hisarustu district of Istanbul. Saritepe Campus is in the northern part of Istanbul, and the Kandilli campus is on the Asia side of Istanbul. The educational and recreational center of the university is mainly South, North, Hisar, and Ucaksavar Campuses. The university's facilities are scattered among the aforementioned 4 campuses (http://boun.edu.tr/). For instance, the library is on the North campus, the student clubs are on South Campus, and the gym is on Hisar Campus. Even some departments have classrooms or laboratories on different campuses. Because of this decentralized structure of the university, many students make trips between campuses during the day. Primary transportation modes used for these trips are walking and shuttle buses. The distances between main campuses are favorable for cycling (Max distance is 1.3 km). However, the hilly topography of the area where the university is located discourages people to cycle between or within the campuses.

Trips that originated in South Campus suffer the most from this topographic disincentive because of the 350 m long ramp with a 9.6% average grade (14.5% maximum grade). Therefore, the students prefer to use shuttle buses for their trips originating from South Campus. This, as a result, is causing long queues at the shuttle bus stop in South Campus.

Because of the limited land area in North and South Campuses, there are limited dedicated parking areas available to the community. A considerable amount of the parking capacity comes from roadside parking. It is observed that most of the roadside parking areas were not properly marked. Because of that, illegal parking can be observed on campuses. There are 306 parking spots (57 of them are on the roadside) in South Campus and 224 parking spots (47 of them are on the roadside) in North Campus; however, 1586 parking permits were distributed in the last academic year. In consideration of limited South and North Campuses' parking capacity, there is an obvious and significant parking capacity shortage in these campuses.

In this study, four alternatives considering the advantage prioritization of green campus transportation, namely, (1) optimizing the performance of shuttle operations, (2) converting current shuttles into electric vehicles, (3) implementing bicycle, scooter sharing, and (4) implementing carpooling of the vehicles used in the campus are presented. Various criteria are identified under four main criteria topics, which are technical aspect, operational aspect, cost aspect, and environmental aspect to assess the advantages of each alternative. A case scenario is used to illustrate the formulation and solution of the problem.

12.3.1 Definition of Alternatives and Criteria

In this study, we defined four alternatives. These alternatives can be defined:

Alternative 1. Optimizing the performance of shuttle operations

This alternative includes the changes that can be made over the current shuttle operations. These operations include improvements, additions, or removals in shuttle operations to make them more sustainable and environmentally friendly, such as scheduling and demand management. As seen in the study by Song et al. [53], a more sustainable workspace can be created with good planning and efficient operations.

Shuttle optimization can be achieved by implementing various optimization techniques. For example, shuttle drivers usually start the engine early regardless of the number of commuters, which increases the operational costs and environmental effects by increasing GHG emissions. Therefore, briefing the drivers and educating them has the potential to increase the sustainability aspect of the transportation system.

For the case study specifically, the dynamic demand of the students and the staff is not available. However, with the average demand data obtained through the course and work schedules, the shuttle operation can still be optimized efficiently.

Alternative 2. Converting current shuttles into electric vehicles

Where there are sufficient funds and it is desired to have more sustainable and green shuttles, this conversion offers an alternative solution. GHG emissions caused by vehicles running on fossil fuels contribute to air pollution greatly. Therefore, converting these vehicles into electric vehicles reduces the GHG emission contribution of the vehicles [27], which makes this alternative a sustainable and green transportation alternative that can be considered by the authorities.

Alternative 3. Implementing bicycle, scooter sharing

The usage of bicycle and scooter sharing systems is increasing. Hence, the usage of vehicles running on fossil fuels decreases, which makes the GHG emission contribution of the university campus also decrease. The primary target group of bicycles and scooters is the young people [57]. Therefore, implementing bicycle and scooter sharing in the university campuses is an applicable alternative, which is sustainable and environmentally friendly.

To implement bicycle, scooter sharing, an amount of budget is required to fund the implementation [3]. Compared to the other alternatives, sharing implementation requires extra funding, which makes the implementation of this alternative more challenging. However, considering the preference of the users and the campus dynamics, in the long term, this alternative has the potential to have more users because of high accessibility and 7/24 availability.

Alternative 4. Implementing carpooling of the vehicles used on the campus

A possible carpooling scheme was proposed for the trips between four campuses of Bogazici University in Istanbul, Turkey. The four campuses are within walking distance of each other and have shuttle services between them. However, shuttle services can become very crowded at certain times of the day. Hence, to solve this problem, vehicles belonging to the university administration are suggested to be carpooled for the individuals in the university, as the administrative personnel makes trips between the campuses. It should be noted that this carpooling policy will not affect the campus traffic flow negatively. This is because the personnel vehicles are already present in the campus traffic, have dedicated parking slots, and will carry individuals of the university during their regular trips between campuses. Hence, they will not create additional traffic. It will have a positive effect because some individuals will probably prefer carpooling between campuses instead of using their car. Implementation of this alternative reduces the GHG emission contribution per person on the campus since fewer people may use their vehicles [12]. Therefore, implementing carpooling is a sustainable alternative that can be considered.

Definition of Criteria

The alternatives are evaluated in terms of four main criteria (main aspect) and ten subcriteria. The main aspects and subcriteria are listed as:

Technical Aspect

C11. Suitability of the terrain (grades of the roadway, nearby traffic, and so on.)-(Benefit)

The slope of the roadway and its closeness to the traffic is essential to the applicability and suitability of the alternatives. For example, when the average slope of the road is high, it is difficult to use bicycles and scooters [56]. Besides the grade levels of the roadways, closeness to the traffic is another example of the factors that must be considered while deciding on the suitability of the alternatives in the region.

C12. Maintenance team availability (Benefit)

Each vehicle will need a certain amount of maintenance because of constant and frequent use. The presence of the maintenance team at the designated places allows these vehicles to be maintained as soon as possible when they need maintenance. Otherwise, maintenance of these vehicles can take days or even weeks, which has the potential to distort the schedule of operations and slow down the provided transportation services.

C13. Parking and/or pick up locations availability (Benefit)

Vehicles need parking and pickup locations, and the availability of these locations on the university campuses is a must since they will serve the people on the university campus. Therefore, to provide efficient transportation, it is crucial to have space for parking and/or pickup locations for the vehicles. In the carpool alternative, it is unnecessary to provide parking and pickup locations since the university authority is not responsible for these vehicles.

Operational Aspect

C21. Operation speed (Benefit)

This criterion is related to the speed of the operations of the vehicles associated with each alternative. As the speed of the operation increases, the demand is managed better, and the waiting times of the users decrease since the number of trips of the vehicles can be increased. Also, as the operation speed increases, the travel time of the users decreases, which makes the users travel to their destination faster.

C22. Regularity (being based on a regular schedule) (Benefit)

Regularity is one of the most significant criteria for an efficient transportation network. People knowing the time that they can access the vehicles make the transportation network preferable and sustainable. A study is conducted in a city in India and the importance of regularity on the efficiency of the sustainable traffic network is investigated [42]. Results show that as the transportation system becomes more regular, fleet size can be minimized, which increases the sustainability of the transportation system because of fewer GHG emissions associated with a smaller number of vehicles.

C23. Punctuality (Benefit)

This criterion is related to the punctuality of the alternative sustainable transportation systems presented in the study. If the vehicles arrive at the station on time, it is said to be punctual. Since this study is related to university campuses, the punctuality of the vehicles is significantly important because the times of the classes, exams, etc. are exact and students must be punctual. Also, in a different study, the sustainability assessment of a bus system is conducted [46]. Results of the study show that the punctuality aspect is very important in means of providing sustainability to the transportation system.

Cost Aspect

C31. Available budget and funding (Benefit)

Implementing some alternatives presented in this study causes a certain cost. While the carpooling alternative does not cause any cost, other alternatives do. Therefore, the availability of the budget and funding associated with the alternative is essential. Since each alternative has different costs, the amount of needed budget is different.

C32. Operation and maintenance cost (Cost)

All vehicles that are used in transportation have operational and maintenance costs. These costs differ according to the type of vehicle. For example, the operational and maintenance cost of electric vehicles is higher than the costs of bicycles and scooters.

Environment and Health Aspect

C41. Emission levels (Cost)

This criterion is related to the levels of emission associated with the alternatives presented in this study. Since this study is mainly about making the university campuses green and sustainable, small emission levels are desired from the vehicles used in the alternatives. As it is stated in a study, carpooling and alternative fuels such as electricity are ways to reduce GHG emissions [49].

C42. The well-being of the users (Benefit)

Alongside the sustainability aspect regarding the GHG emissions, the well-being of the users is also another factor affecting the sustainability of the alternative. Not encountering any difficulties, comfort, reduced waiting times and travel times, regularity and punctuality are all important factors affecting the well-being of the users. The well-being of the users determines their preferences and, as they frequently use the transportation mode that they enjoy, sustainability is provided to the associated transportation system.

12.4 Preliminaries

Definition 12.1 The concept of Pythagorean Fuzzy Sets (PFS) was introduced by Yager [63]. The PFS is defined as

$$\tilde{A}^P = \{\langle x, \mu_{\tilde{A}^P}(x), \vartheta_{\tilde{A}^P}(x)\rangle : x \in U\}; \text{ U is the universe of discourse}$$

where, $\mu_{\tilde{A}^P}(x)$: U $\rightarrow$ [0,1] and $\vartheta_{\tilde{A}^P}(x)$: U $\rightarrow$ [0,1] are the degree of membership and degree of non-membership, respectively, where $0 \le (\mu_{\tilde{A}^P}(x))^2 + (\vartheta_{\tilde{A}^P}(x))^2 \le 1$; $\forall$ x $\in$ U; the degree of indeterminacy is being given by

$$\pi_{\tilde{A}^P}(x) = \sqrt{1 - (\mu_{\tilde{A}^P}(x))^2 - (\vartheta_{\tilde{A}^P}(x))^2}; \forall x \in U \tag{12.1}$$

Definition 12.2 As a generalization, the concept of q-Rung Orthopair Fuzzy Sets (q-ROFS) was proposed in [64]. A q-ROFS is defined as

$$\tilde{A}^Q = \left\{\langle x, \mu_{\tilde{A}^Q}(x), \vartheta_{\tilde{A}^Q}(x)\rangle : x \in U\right\}; \text{ U is the universe of discourse}$$

where, $\mu_{\tilde{A}^Q}(x)$: U $\rightarrow$ [0,1] and $\vartheta_{\tilde{A}^Q}(x)$: U $\rightarrow$ [0,1] are the degree of membership and degree of non-membership, respectively, where $0 \le \left(\mu_{\tilde{A}^Q}(x)\right)^q + \left(\vartheta_{\tilde{A}^Q}(x)\right)^q \le 1$; $\forall$ x $\in$ U; the degree of indeterminacy is being given by

$$\pi_{\tilde{A}^Q}(x) = \sqrt[q]{1 - \left(\mu_{\tilde{A}^Q}(x)\right)^q - \left(\vartheta_{\tilde{A}^Q}(x)\right)^q} \forall x \in U \tag{12.2}$$

Clearly, when q = 1, it $\tilde{A}^Q$ becomes an Atanassov's Intuitionistic Fuzzy Sets (IFS) and for q = 2 it represents PFS.

In this chapter, in the same way, as defined by expression (12.2), we shall use the notation $\mathbb{Q} = (\mu, \vartheta)$ for representing a q-Rung Orthopair Fuzzy Number (q-ROFN).

Definition 12.3 Basic operations on q-ROFN.

Let, $\mathbb{Q}_1 = (\mu_1, \vartheta_1)$, $\mathbb{Q}_2 = (\mu_2, \vartheta_2)$, and $\mathbb{Q} = (\mu, \vartheta)$ are the three q-ROFNs. The following definitions exhibit some of the basic operators [64].

$$\mathbb{Q}^c = (\vartheta, \mu) \tag{12.3}$$

$$\mathbb{Q}_1 \boxplus \mathbb{Q}_2 = \left(\sqrt[q]{\mu_1{}^q + \mu_2{}^q - \mu_1{}^q\mu_2{}^q}, \vartheta_1\vartheta_2\right) \tag{12.4}$$

$$\mathbb{Q}_1 \boxtimes \mathbb{Q}_2 = (\mu_1\mu_2, \sqrt[q]{\vartheta_1{}^q + \vartheta_2{}^q - \vartheta_1{}^q\ {}_2{}^q}) \tag{12.5}$$

$$\alpha\mathbb{Q} = \left(\sqrt[q]{1-(1-\mu^q)^\alpha}, \vartheta^\alpha\right); \alpha \text{ is a constant} \tag{12.6}$$

$$\mathbb{Q}^\alpha = \left(\mu^q, \sqrt[q]{1-(1-\vartheta^q)^\alpha}\right) \tag{12.7}$$

Definition 12.4 Score and Accuracy Function.

Simply, the Score Function (SF) is defined as [30]

$$\mathfrak{H} = \mu^q - \vartheta^q; \mathfrak{H} \in [-1, 1] \tag{12.8}$$

Extending the basic definition of SF, the researchers [58–60] put forth alternative definitions as given below.

$$\mathfrak{H}' = (1 + \mu^q - \vartheta^q)/2 \tag{12.9}$$

In this context, Peng et al. [39] proposed a novel definition of SF considering the degree of indeterminacy.

$$\mathfrak{H}'' = \mu^q - \vartheta^q + \left(\frac{e^{\mu^q-\vartheta^q}}{e^{\mu^q-\vartheta^q}+1} - \frac{1}{2}\right)\pi^q \tag{12.10}$$

However, in all these definitions, if $\mu = \vartheta$, the SF may not be useful to rank the q-ROFNs. Considering this fact, Peng and Dai [38] offered an improved definition of SF as given below

$$\mathfrak{H}^* = \frac{\mu^q - 2\vartheta^q - 1}{3} + \frac{\lambda}{3}(\mu^q + \vartheta^q + 2), \lambda \in [0,1] \tag{12.11}$$

The Accuracy Function (AF) is defined as [30]

$$\mathrm{H} = \mu^q + \vartheta^q, H \in [0,1] \tag{12.12}$$

If $\mathfrak{H}_1 > \mathfrak{H}_2$ then $\mathbb{Q}_1 > \mathbb{Q}_2$
If $\mathfrak{H}_1 < \mathfrak{H}_2$ then $\mathbb{Q}_1 < \mathbb{Q}_2$
If $\mathfrak{H}_1 = \mathfrak{H}_2$, then if $H_1 < H_2, \mathbb{Q}_1 < \mathbb{Q}_2$; $H_1 > H_2, \mathbb{Q}_1 > \mathbb{Q}_2$.

Definition 12.5 q-Rung Orthopair Fuzzy Weighted Averaging Operator (q-ROFWA) [30]

$$q-\text{ROFWA}\left(\mathbb{Q}_1,\mathbb{Q}_2,\ldots,\mathbb{Q}_r\right)=\left\langle\left(1-\prod_{k=1}^{r}\left(1-\mu_k^q\right)^{\alpha_k}\right)^{1/q},\prod_{k=1}^{r}\vartheta_k^{\alpha_k}\right\rangle \quad (12.13)$$

Here, α_k is the corresponding weight.

12.5 Proposed Methodologies

12.5.1 Proximity Indexed Value (PIV) Method

PIV method derives the best possible alternative among the choices subject to a list of criteria based on absolute dispersion value (i.e., proximity) concerning the ideal reference point [33]. In effect, PIV does not suffer from the common limitation of rank reversal as affected by the popular distance-based method, such as TOPSIS [7]. Rank reversal is one of the major issues affecting the performance of MCDM algorithms [8, 28, 37]. Due to its simplicity of calculation and inherent benefit, the PIV method has been used by many authors in solving various problems [7, 29, 48]. The procedural steps are described below.

12.5.2 Proposed q-ROF PIV Method

In this chapter for ranking the alternatives, we use an extended version of the classic PIV method in q-ROFS environment. The procedural steps (for the corresponding steps of the classical PIV method as depicted in Table 12.1) are given below.

Step 1. Normalization

For beneficial criteria, we use the same value of alternatives, while for the non-beneficial type of criteria we take the complement of the alternative values (using the expression (12.3)) for formulating the normalization matrix.

Step 2. Finding out the weighted normalized matrix

We apply the *Definition 12.3* (expression (12.6)) in the Eq. (12.15) to construct the weighted normalized matrix.

Step 3. Formulation of the WPI matrix

Here, for finding out v^+ and v^- we consider the following definition.

Table 12.1 Steps of the PIV method

Step 1. Normalization	Suppose, $R = [r_{ij}]_{m \times n}$ is the normalization matrix Where $r_{ij} = \frac{x_{ij}}{\sqrt{\sum_{i=1}^{m} x_{ij}^2}}$	(12.14)
Step 2. Constructing the weighted normalized matrix	$v_{ij} = w_j \times r_{ij}$ w_j is the weight of the jth criterion such that $\sum_{j=1}^{n} w_j = 1$	(12.15)
Step 3. Formation of the Weighted Proximity Index (WPI) matrix	$I_{ij} = \begin{cases} v^+ - v_{ij}; \text{ For beneficial criteria} \\ v_{ij} - v^-; \text{ For non-beneficial criteria} \end{cases}$ Where $v^+ = Max(v_{ij})$ and $v^- = Min(v_{ij})$	(12.16)
Step 4. Determination of Overall Proximity Value (OPV)	$\delta_i = \sum_{j=1}^{n} I_{ij}$	(12.17)
Decision Rule: Lower the value of δ_i, better proximity of the alternative to the ideal reference point and hence to be ranked first and so on		

$$v^+ = \begin{cases} \langle \max(\mu_{ij}), \min(\vartheta_{ij}) \rangle \text{ for beneficial criteria} \\ \langle \min(\mu_{ij}), \max(\vartheta_{ij}) \rangle \text{ for non-beneficial criteria} \end{cases} \quad (12.18)$$

$$v^- = \begin{cases} \langle \min(\mu_{ij}), \max(\vartheta_{ij}) \rangle \text{ for beneficial criteria} \\ \langle \max(\mu_{ij}), \min(\vartheta_{ij}) \rangle \text{ for non-beneficial criteria} \end{cases} \quad (12.19)$$

Further, we calculate the score values of v_{ij}, v^+, v^- using the expression (12.11) to derive I_{ij} values as defined in expression (12.16).

Step 4. Derive the OPV values for ranking of alternatives

We use the expression (12.17) in this regard and rank the alternatives in ascending order of the values, i.e., the alternative with the lowest OPV value is ranked first, and so on.

The proposed hybrid Fuzy LMAW-qROF PIV method provides the following advantages:

- Better stability in the result with respect to variations in the conditions, for example, changes in the alternative set. The basic mathematical framework remains unchanged despite variations in the criteria and/or alternative sets,
- Free from rank reversal problem,
- The model can handle both qualitative as well as quantitative information.

As of now, no prior work has presented a combination of fuzzy weighting using LMAW and qROF PIV-based ranking of alternatives. Hence, the present hybrid framework adds value to the growing literature by providing a new approach for multicriteria-based group decision-making.

12.6 Experimental Results

The weighting coefficients of the criteria were defined using the Logarithm Methodology of Additive Weights (LMAW) [37]. The LMAW methodology has been extended by applying fuzzy theory [66] to exploit uncertainty and subjectivism in expert assessments. Figure 12.1 presents the steps of the fuzzy LMAW-q ROF PIV methodology for determining the weighting coefficients of the criteria.

Five experts prioritized the criteria based on the fuzzy scale given in Table 12.2.

Based on expert assessments, five priority vectors have been defined, which are presented in Table 12.3.

Based on the priority vector (see Table 12.3) and absolute anti-ideal point (δ_{AIP}), relation vectors are defined. After averaging the relation vectors, fuzzy relation vectors were obtained for each level of criteria. Using the logarithmic additive function, the relation vectors were transformed into final fuzzy weighting coefficients of the criteria, Table 12.4.

Crisp values of the weighting coefficients of the criteria were used for the final prioritization of the alternatives.

In our study, five experts have taken part. The linguistic scale used in this study for a rating of the alternatives concerning the criteria is exhibited in Table 12.5. We refer to the work of Pinar and Boran [40].

The responses received from the experts are given in Table 12.6. We consider all experts equally important. To formulate the decision matrix (see Table 12.7), we use q-ROFWA by applying the expression (12.13). In our study, we consider $q = 5$. Next, we move to rank the alternatives by using the extended q-ROF PIV method as described in the previous section. Tables 12.8, 12.9, and 12.10 show the step-by-step process for ranking the alternatives by using the q-ROF PIV method.

Now we move to validate our results. For validation purposes, we compare the results obtained by using our q-ROF PIV method with that derived by using other established MCDM algorithms [24]. Accordingly, we calculate the score values for alternatives using the expression (12.11) from the decision matrix (please refer to Table 12.7) and apply the procedural steps of the classical EDAS method [23] to rank the alternatives.

EDAS method compares the alternatives in terms of their dispersion regarding the average solution point. In effect, EDAS provides a risk-averse solution other than considering extreme points as a baseline. The EDAS method, unlike TOPSIS, is free from the issue of rank reversal [23]. As the PIV method also works similarly to EDAS and is free from the rank reversal problem, we consider EDAS for validation.

Step 1: Defining a set of n criteria/subcriteria for evaluation

$\downarrow$

Step 2: Defining the priority vectors for experts
$$P^t = \left(\vartheta^t_{C1}, \vartheta^t_{C2}, .., \vartheta^t_{Cn} \right)$$

$\downarrow$

Step 3: Determining the absolute anti-ideal point
$$\delta_{AIP} < \min \left(\vartheta^t_{C1}, \vartheta^t_{C2}, .., \vartheta^t_{Cn} \right)$$

$\downarrow$

Step 4: Defining the relation vector
$$A^t = \left(\varphi^t_{C1}, \varphi^t_{C2}, .., \varphi^t_{Cn} \right)$$

$\downarrow$

Step 5: Determining the vector of the weight coefficients
$$w^t_j = \frac{\ln \left(\varphi^t_{Cn} \right)}{\ln \left(z^t \right)}$$

$\downarrow$

Step 6: Normalization of the decision matrix
$$\mathbb{Q}^c = (\vartheta, \mu)$$

$\downarrow$

Step 7: Multiply criteria weights with the normalized decision matrix
$$v_{ij} = w_j \times r_{ij}$$
Using
$$\alpha \mathbb{Q} = \left(\sqrt[q]{1 - (1 - \mu^q)^\alpha}, \vartheta^\alpha \right)$$

$\downarrow$

Step 8: Construction of the WPI matrix

$\downarrow$

Step 9: Calculation of OPV values & Ranking

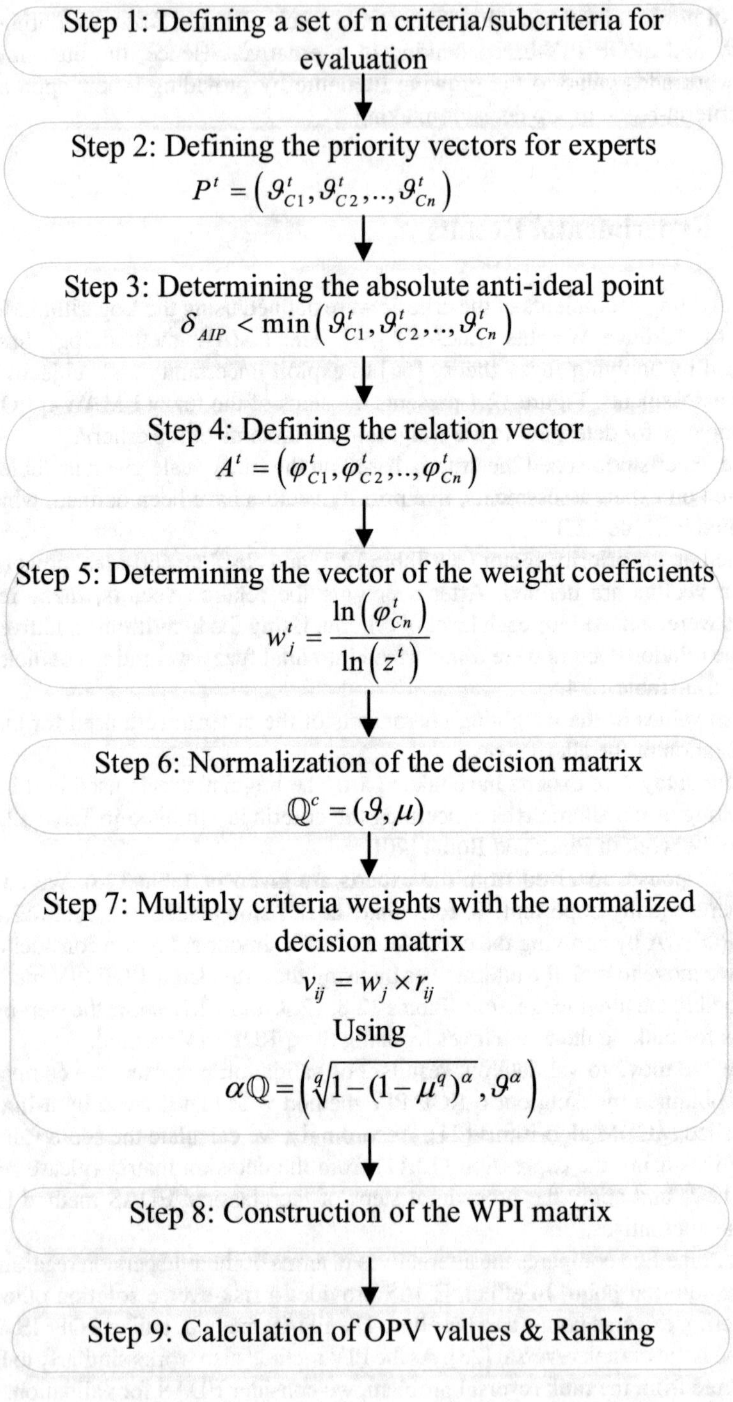

Fig. 12.1 Fuzzy LMAW-qROF PIV methodology

Table 12.2 Fuzzy linguistic scale for criteria evaluation

Linguistic terms	Membership function
Absolutely low (AL)	(1, 1, 1)
Very low (VL)	(1, 1.5, 2)
Low (L)	(1.5, 2, 2.5)
Medium low (ML)	(2, 2.5, 3)
Equal (E)	(2.5, 3, 3.5)
Medium high (MH)	(3, 3.5, 4)
High (H)	(3.5, 4, 4.5)
Very high (EH)	(4, 4.5, 5)
Absolutely high (AH)	(4.5, 5, 5)

Table 12.3 Priority vectors

Criteria	Expert 1	Expert 2	Expert 3	Expert 4	Expert 5
MC1	E	H	H	EH	H
MC2	AH	VH	EH	H	AH
MC3	H	AH	MH	AH	EH
MC4	MH	H	EH	MH	MH
MC1—Technical aspect					
C11	H	EH	H	EH	AH
C12	L	H	EH	EH	E
C13	EH	H	E	H	MH
MC2—Operational aspect					
C21	H	VH	AH	H	H
C22	AH	AH	EH	H	AH
C23	EH	AH	EH	EH	EH
MC3—Environment and health aspect					
C31	MH	EH	MH	AH	EH
C32	AH	AH	H	EH	AH
MC4—Environment and health aspect					
C33	EH	H	H	ML	EH
C34	H	E	MH	E	H

Table 12.11 confirms that the results obtained by using the q-ROF PIV algorithm (at q = 5) are completely consistent with that derived from score based EDAS method (at q = 5).

Next, we want to examine the stability of the results. Given variations in the set conditions, for example, changes in criteria weights or formation of the decision matrix, the outcome of the MCDM frameworks most often shows instability. Hence,

Table 12.4 Criteria vectors

Criteria	Fuzzy local weights	Crisp local weights	Crisp global weights
MC1	(0.177, 0.242, 0.332)	0.246094	–
C11	(0.261, 0.355, 0.485)	0.3612	0.0889
C12	(0.220, 0.314, 0.445)	0.3202	0.0788
C13	(0.237, 0.331, 0.463)	0.3372	0.0829
MC2	(0.197, 0.262, 0.348)	0.265903	–
C21	(0.246, 0.325, 0.432)	0.3297	0.0878
C22	(0.259, 0.339, 0.439)	0.3422	0.0910
C23	(0.256, 0.336, 0.443)	0.3406	0.0906
MC3	(0.191, 0.256, 0.342)	0.259634	–
C31	(0.365, 0.486, 0.651)	0.4933	0.1281
C32	(0.393, 0.514, 0.670)	0.5199	0.1349
MC4	(0.175, 0.239, 0.330)	0.243649	–
C33	(0.363, 0.511, 0.724)	0.5217	0.1271
C34	(0.341, 0.489, 0.702)	0.4999	0.1218

Table 12.5 q-ROF linguistic scale for ranking alternatives

Linguistic terms	μ	ν
Absolutely low (AL)	0.25	0.85
Very low (VL)	0.35	0.75
Low (L)	0.45	0.65
Medium (M)	0.55	0.55
Medium high (MH)	0.65	0.45
High (H)	0.75	0.35
Very high (EH)	0.25	0.85

to reduce the variations and thereby ensure the stability, reliability, and quality of the results, the sensitivity analysis is carried out [6, 34–36]. For this purpose, in our study, we vary the values of q from its original selected value of 5 (e.g., q = 1, 2, 3, 4, 7, 10) to generate various simulated cases and apply the proposed q-ROF PIV algorithm to rank the alternatives. Table 12.12 exhibits the summary of ranking results under different scenarios, which confirms that the q-ROF PIV method provides reasonable, consistent, and stable output. The pictorial representation of the results of the sensitivity analysis is shown in Fig. 12.2.

Further, to check the rank reversal possibility of the qROF PIV method, we conduct a simulated experiment. We delete the alternative A_3 and apply the qROF PIV model (while keeping the criteria set and the weights the same). We find the ranking order of the alternatives as follows.

Before deletion of A_3: $A_1 > A_2 > A_3 > A_4$.

After deletion of A_3: $A_1 > A_2 > A_4$.

Table 12.6 Responses of the experts

DM		C11		C12		C13		C21		C22	
EXP-1	A_1	0.35	0.75	0.25	0.85	0.75	0.35	0.65	0.45	0.75	0.35
	A_2	0.25	0.85	0.55	0.55	0.35	0.75	0.45	0.65	0.45	0.65
	A_3	0.85	0.25	0.55	0.55	0.35	0.75	0.35	0.75	0.35	0.75
	A_4	0.35	0.75	0.25	0.85	0.55	0.55	0.75	0.35	0.75	0.35
EXP 2		C11		C12		C13		C21		C22	
	A_1	0.85	0.25	0.85	0.25	0.85	0.25	0.75	0.35	0.55	0.55
	A_2	0.75	0.35	0.55	0.55	0.85	0.25	0.75	0.35	0.75	0.35
	A_3	0.65	0.45	0.55	0.55	0.55	0.55	0.55	0.55	0.85	0.25
	A_4	0.85	0.25	0.85	0.25	0.45	0.65	0.75	0.35	0.45	0.65
EXP 3		C11		C12		C13		C21		C22	
	A_1	0.25	0.85	0.45	0.65	0.45	0.65	0.85	0.25	0.85	0.25
	A_2	0.25	0.85	0.75	0.35	0.55	0.55	0.45	0.65	0.45	0.65
	A_3	0.85	0.25	0.55	0.55	0.75	0.35	0.55	0.55	0.55	0.55
	A_4	0.25	0.85	0.25	0.85	0.45	0.65	0.45	0.65	0.75	0.35
EXP 4		C11		C12		C13		C21		C22	
	A_1	0.75	0.35	0.65	0.45	0.85	0.25	0.85	0.25	0.85	0.25
	A_2	0.55	0.55	0.75	0.35	0.75	0.35	0.65	0.45	0.75	0.35
	A_3	0.85	0.25	0.55	0.55	0.75	0.35	0.55	0.55	0.25	0.85
	A_4	0.65	0.45	0.45	0.65	0.65	0.45	0.75	0.35	0.55	0.55
EXP 5		C11		C12		C13		C21		C22	
	A_1	0.35	0.75	0.55	0.55	0.65	0.45	0.55	0.55	0.85	0.25
	A_2	0.55	0.55	0.85	0.25	0.75	0.35	0.75	0.35	0.85	0.25
	A_3	0.85	0.25	0.85	0.25	0.85	0.25	0.65	0.45	0.35	0.75
	A_4	0.35	0.75	0.35	0.75	0.55	0.55	0.55	0.55	0.85	0.25

DM		C23		C31		C32		C33		C34	
EXP-1	A_1	0.85	0.25	0.55	0.55	0.35	0.75	0.75	0.35	0.75	0.35
	A_2	0.45	0.65	0.85	0.25	0.25	0.85	0.25	0.85	0.75	0.35
	A_3	0.35	0.75	0.65	0.45	0.35	0.75	0.25	0.85	0.75	0.35
	A_4	0.85	0.25	0.25	0.85	0.85	0.25	0.85	0.25	0.85	0.25
EXP 2		C23		C31		C32		C33		C34	
	A_1	0.85	0.25	0.85	0.25	0.35	0.75	0.65	0.45	0.45	0.65
	A_2	0.85	0.25	0.35	0.75	0.85	0.25	0.35	0.75	0.55	0.55
	A_3	0.85	0.25	0.55	0.55	0.65	0.45	0.35	0.75	0.75	0.35
	A_4	0.45	0.65	0.85	0.25	0.35	0.75	0.75	0.35	0.45	0.65
EXP 3		C23		C31		C32		C33		C34	
	A_1	0.85	0.25	0.65	0.45	0.65	0.45	0.35	0.75	0.75	0.35

(continued)

Table 12.6 (continued)

DM		C23		C31		C32		C33		C34	
	A_2	0.55	0.55	0.85	0.25	0.35	0.75	0.25	0.85	0.55	0.55
	A_3	0.25	0.85	0.65	0.45	0.75	0.35	0.25	0.85	0.55	0.55
	A_4	0.45	0.65	0.25	0.85	0.85	0.25	0.45	0.65	0.65	0.45
EXP 4		C23		C31		C32		C33		C34	
	A_1	0.85	0.25	0.85	0.25	0.55	0.55	0.55	0.55	0.45	0.65
	A_2	0.55	0.55	0.85	0.25	0.25	0.85	0.45	0.65	0.55	0.55
	A_3	0.25	0.85	0.75	0.35	0.45	0.65	0.25	0.85	0.85	0.25
	A_4	0.45	0.65	0.45	0.65	0.55	0.55	0.45	0.65	0.65	0.45
EXP 5		C23		C31		C32		C33		C34	
	A_1	0.55	0.55	0.45	0.65	0.55	0.55	0.35	0.75	0.45	0.65
	A_2	0.55	0.55	0.85	0.25	0.25	0.85	0.75	0.35	0.75	0.35
	A_3	0.65	0.45	0.65	0.45	0.45	0.65	0.85	0.25	0.85	0.25
	A_4	0.85	0.25	0.25	0.85	0.75	0.35	0.55	0.55	0.55	0.55

Therefore, we conclude that the proposed qROF PIV method is free from the rank reversal issue.

12.7 Discussion

In this study, 4 different alternatives are assessed considering 10 different criteria, which are listed under 4 main criteria aspects. To conduct the assessment, a questionnaire is prepared at which alternatives are asked to be scored according to each criterion. Then, this questionnaire is sent to the experts. After the results are gathered from the experts, alternatives are sorted in advantage order. According to the results of the proposed MCDM tool, optimizing the performance of shuttle operations is the most advantageous alternative. The second most advantageous alternative is converting current shuttles into electric vehicles. The third most advantageous is implementing bicycle scooter sharing and the least advantageous alternative is implementing carpooling of the vehicles used on the campus.

Implementing carpooling of the vehicles used in the campus is the least advantageous alternative. Carpooling vehicles is not a punctual and regular mode of transportation. Therefore, it is unlikely that the transportation needs of the residents of the campus will be fulfilled by this transportation method. Implementation of this alternative is not as sustainable as other alternatives, since vehicles running on fossil fuels are used mainly. Therefore, implementing carpooling of the vehicles used on the campus is not sustainable and impractical making this alternative the least advantageous one.

Table 12.7 Decision matrix for q-ROF PIV (q = 5)

	0.08889	0.08889	0.07880	0.07880	0.08299	0.08299	0.08768	0.08768	0.09100	0.09100
	C11		C12		C13		C21		C22	
A_1	0.6927	0.5300	0.6793	0.5091	0.7706	0.3641	0.7733	0.3521	0.8062	0.3131
A_2	0.5916	0.5980	0.7376	0.3921	0.7326	0.4172	0.6660	0.4714	0.7304	0.4192
A_3	0.8282	0.2812	0.6812	0.4698	0.7326	0.4172	0.5601	0.5622	0.6564	0.5802
A_4	0.6693	0.5574	0.6491	0.6151	0.5523	0.5649	0.6948	0.4336	0.7341	0.4054
	0.09058	0.09058	0.12807	0.12807	0.13498	0.13498	0.12712	0.12712	0.12180	0.12180
	C23		C31		C32		C33		C34	
A_1	0.8244	0.2927	0.7512	0.3985	0.5409	0.5982	0.6140	0.5465	0.6462	0.5074
A_2	0.6759	0.4857	0.8219	0.3114	0.6456	0.6490	0.5651	0.6579	0.6650	0.4590
A_3	0.6686	0.5715	0.6643	0.4455	0.6088	0.5491	0.6456	0.6490	0.7847	0.3349
A_4	0.7373	0.4435	0.6483	0.6307	0.7634	0.3900	0.7036	0.4588	0.6977	0.4483

Table 12.8 q-ROF normalization matrix.

	0.08889	0.08889	0.07880	0.07880	0.08299	0.08299	0.08768	0.08768	0.09100	0.09100
	C11		C12		C13		C21		C22	
A_1	0.6927	0.5300	0.6793	0.5091	0.7706	0.3641	0.7733	0.3521	0.8062	0.3131
A_2	0.5916	0.5980	0.7376	0.3921	0.7326	0.4172	0.6660	0.4714	0.7304	0.4192
A_3	0.8282	0.2812	0.6812	0.4698	0.7326	0.4172	0.5601	0.5622	0.6564	0.5802
A_4	0.6693	0.5574	0.6491	0.6151	0.5523	0.5649	0.6948	0.4336	0.7341	0.4054
	0.09058	0.09058	0.12807	0.12807	0.13498	0.13498	0.12712	0.12712	0.12180	0.12180
	C23		C31		C32		C33		C34	
A_1	0.8244	0.2927	0.7512	0.3985	0.5982	0.5409	0.5465	0.6140	0.6462	0.5074
A_2	0.6759	0.4857	0.8219	0.3114	0.6490	0.6456	0.6579	0.5651	0.6650	0.4590
A_3	0.6686	0.5715	0.6643	0.4455	0.5491	0.6088	0.6490	0.6456	0.7847	0.3349
A_4	0.7373	0.4435	0.6483	0.6307	0.3900	0.7634	0.4588	0.7036	0.6977	0.4483

Table 12.9 Score matrix

Alternatives	C11	C12	C13	C21	C22	C23	C31	C32	C33	C34
$S(A_1)$	−0.0925	−0.0992	−0.0474	−0.0363	−0.0135	−0.0038	−0.0013	−0.0578	−0.0896	−0.056
$S(A_2)$	−0.1143	−0.0651	−0.0666	−0.0803	−0.0567	−0.0802	0.04555	−0.0879	−0.0683	−0.0389
$S(A_3)$	−0.0018	−0.0896	−0.0666	−0.1078	−0.1052	−0.1027	−0.0278	−0.082	−0.0936	0.01982
$S(A_4)$	−0.1008	−0.1245	−0.113	−0.0681	−0.0523	−0.0636	−0.0885	−0.1326	−0.1183	−0.0323
S(V+)	−0.0018	−0.0651	−0.0474	−0.0363	−0.0135	−0.0038	0.04555	−0.1326	−0.1183	0.01982
S(V−)	−0.1143	−0.1245	−0.113	−0.1078	−0.1052	−0.1027	−0.0885	−0.0544	−0.0683	−0.056

Table 12.10 Final ranking

Alternatives	δ	Rank	Alternatives	δ	Rank
A_1	0.22281	1	A_3	0.32613	3
A_2	0.32044	2	A_4	0.4123	4

Table 12.11 Comparative analysis of results

Alternatives	q-ROF PIV	q-ROF score based EDAS
A_1	1	1
A_2	2	2
A_3	3	3
A_4	4	4

Table 12.12 Result of sensitivity analysis (q-ROF PIV)

Cases	Case 1	Case 2	Case 3	Case 4	Original	Case 5	Case 6
	q = 1	q = 2	q = 3	q = 4	q = 5	q = 7	q = 10
A1	1	1	1	1	1	1	1
A2	3	3	3	3	2	2	2
A3	2	2	2	2	3	3	3
A4	4	4	4	4	4	4	4

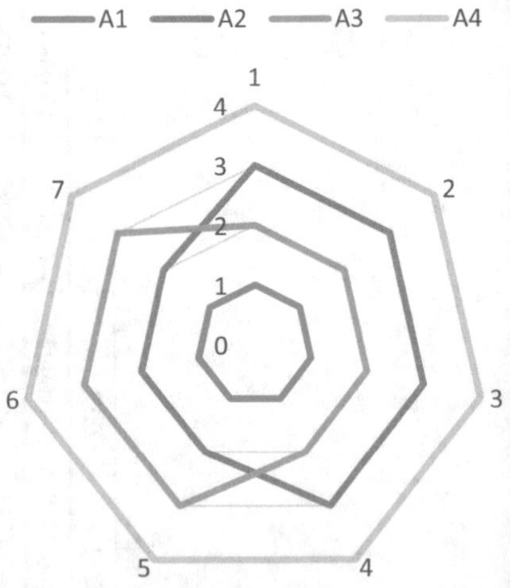

Fig. 12.2 Pictorial representation (sensitivity analysis)

According to the advantage prioritization results, implementing bicycle scooter sharing is the second least advantageous alternative. By implementing this alternative, fossil fuel consumption is reduced since the residents of the campus are expected to change their transportation usage habits from vehicles running on fossil fuels to bicycles and scooters. The utilization of such green vehicles improves the sustainability aspect of the campus greatly and is a big step toward reducing the GHG emissions associated with the campus and making the campus greener. However, the terrain of the campuses plays an important role in the success of this implementation.

The second most advantageous alternative is converting current shuttles into electric vehicles. This conversion reduces GHG emissions to a great extent. Hence, it maintains the punctuality and regularity of the transportation service and provides a comfortable transportation mode for the users. However, for its operation and maintenance, available budget and funding play an important role.

Optimizing the performance of shuttle operations is the most advantageous alternative. It improves the sustainability feature of the campus because fuel consumption is reduced. Also, it is more practical compared to other alternatives since this alternative provides a more regular and punctual form of transportation service.

12.8 Conclusion

The results of the advantage prioritization show that optimizing the performance of shuttle operations is the most advantageous alternative.

This paper fills a gap regarding the determination of the most advantageous alternative between the selected and assessed alternatives concerning a transition to a more sustainable and greener campus. Even if the alternatives are changed or adapted to the conditions and cultural aspects of the region of the campus, the methodology presented in this study can still prioritize the alternatives in advantage order.

Technology is developing rapidly and new technologies concerning fewer GHG emissions, and more sustainability features are emerging. Therefore, greener and more sustainable alternatives will inevitably be available shortly. Therefore, the criteria and the alternatives presented in this study are subject to change in the future.

There are also limitations to this study. For example, the suitability of the terrain of each campus is different. Grades of the roads present in the campuses vary, which affects the applicability of the alternatives. For example, implementing bicycle, scooter sharing is an alternative, which is highly affected by this aspect. Therefore, alternatives must be adapted to the conditions of the campus at which a transition to a green campus is conducted.

The proposed qROF PIV shows a hint of variations in the ranking orders of some alternatives when the value of q moves from a lower level to a higher level, although in our case the positions of the best and worst-performing alternatives remain unchanged. Therefore, the qROF PIV method may need to be tested for large criteria and alternative sets.

Nevertheless, the combined approach of fuzzy LMAW-qROF PIV is a new way of solving multi-criteria-based decision-making problems. The framework may further be extended using rough sets and/or a combination of fuzzy-rough sets for more granular level analysis.

References

1. Z. Ali, T. Mahmood, K. Ullah, Q. Khan, Einstein geometric aggregation operators using a novel complex interval-valued pythagorean fuzzy setting with application in green supplier chain management. Rep. Mech. Eng. **2**(1), 105–134 (2021). https://doi.org/10.31181/rme2001020105t
2. A. Alosta, O. Elmansuri, I. Badi, Resolving a location selection problem by means of an integrated AHP-RAFSI approach. Rep. Mech. Eng. **2**(1), 135–142 (2021). https://doi.org/10.31181/rme200102135a
3. A. Angelopoulos, D. Gavalas, C. Konstantopoulos, D. Kypriadis, G. Pantziou, An optimization model for the strategic design of a bicycle sharing system, in *Proceedings of the 20th Pan-Hellenic Conference on Informatics* (2016), pp. 1–6. https://doi.org/10.1145/3003733.3003784
4. C.J. Balsas, Sustainable transportation planning on college campuses. Transp. Policy **10**(1), 35–49 (2003)
5. E. Barata, L. Cruz, J.P. Ferreira, Parking at the UC campus: problems and solutions. Cities **28**(5), 406–413 (2011)
6. S. Biswas, Measuring the performance of healthcare supply chains in India: a comparative analysis of multi-criteria decision-making methods. Decis. Mak.: Appl. Manag. Eng. **3**(2), 162–189 (2020)
7. S. Biswas, O.P. Anand, Logistics competitiveness index-based comparison of BRICS and G7 countries: an integrated PSI-PIV approach. IUP J. Supply Chain Manag. **17**(2), 32–57 (2020)
8. S. Biswas, G. Bandyopadhyay, B. Guha, M. Bhattacharjee, An ensemble approach for portfolio selection in a multi-criteria decision-making framework. Decis. Mak.: Appl. Manag. Eng. **2**(2), 138–158 (2019)
9. H.R. Bowles, C. Rissel, A. Bauman, Mass community cycling events: who participates and is their behaviour influenced by participation? Int. J. Behav. Nutr. Phys. Act. **3**(1), 1–7 (2006)
10. J. Brown, D.B. Hess, D. Shoup, Unlimited access. Transportation **28**(3), 233–267 (2001)
11. J. Brown, D.B. Hess, D. Shoup, Fare-free public transit at universities: an evaluation. J. Plan. Educ. Res. **23**(1), 69–82 (2003)
12. B.P. Bruck, V. Incerti, M. Iori, M. Vignoli, Minimizing CO_2 emissions in a practical daily carpooling problem. Comput. Oper. Res. **81**, 40–50 (2017). https://doi.org/10.1016/j.cor.2016.12.003
13. L. Canals Casals, E. Martinez-Laserna, B. Amante García, N. Nieto, Sustainability analysis of the electric vehicle use in Europe for CO_2 emissions reduction. J. Clean. Prod. **127**, 425–437 (2016). https://doi.org/10.1016/j.jclepro.2016.03.120
14. R. Cervero, O.L. Sarmiento, E. Jacoby, L.F. Gomez, A. Neiman, Influences of built environments on walking and cycling: lessons from Bogota. Int. J. Sustain. Transp. **3**(4), 203–226 (2009)
15. M. Daley, C. Rissel, Perspectives and images of cycling as a barrier or facilitator of cycling. Transp. Policy **18**(1), 211–216 (2011)
16. J. Dargay, D. Gately, M. Sommer, vehicle ownership and income growth, worldwide: 1960–2030. Energy J. **28**(4), 143–170 (2007)
17. P. DeMaio, Bike-sharing: history, impacts, models of provision, and future. J. Public Transp. **12**(4), 41–56 (2009)
18. R. Faria, P. Moura, J. Delgado, A.T. de Almeida, A sustainability assessment of electric vehicles as a personal mobility system. Energy Convers. Manag. **61**, 19–30 (2012). https://doi.org/10.1016/j.enconman.2012.02.023
19. H. Garg, New exponential operation laws and operators for interval-valued q-rung orthopair fuzzy sets in group decision making process. Neural Comput. Appl. **33**(20), 13937–13963 (2021)
20. H. Garg, CN-q-ROFS: connection number-based q-rung orthopair fuzzy set and their application to decision-making process. Int. J. Intell. Syst. **36**(7), 3106–3143 (2021)
21. H. Garg, Z. Ali, T. Mahmood, Generalized dice similarity measures for complex q-Rung Orthopair fuzzy sets and its application. Complex Intell. Syst. **7**(2), 667–686 (2021)

22. H. Garg, A new possibility degree measure for interval-valued q-rung orthopair fuzzy sets in decision-making. Int. J. Intell. Syst. **36**(1), 526–557 (2021)
23. M.K. Ghorabaee, E.K. Zavadskas, L. Olfat, Z. Turskis, Multi-criteria inventory classification using a new method of evaluation based on distance from average solution (EDAS). Informatica **26**, 435–451 (2015). https://doi.org/10.15388/Informatica.2015.57
24. I. Ghosh, S. Biswas, A novel framework of ERP implementation in Indian SMEs: Kernel principal component analysis and intuitionistic Fuzzy TOPSIS driven approach. Accounting **3**(2), 107–118 (2017)
25. R. Greig, Cycling promotion in Western Australia. Health Promot. J. Austr. **12**(3), 250–253 (2001)
26. M. Hamurcu, T. Eren, Electric bus selection with multicriteria decision analysis for green transportation. Sustainability **12**(7), 2777 (2020). https://doi.org/10.3390/su12072777
27. A. Hoekstra, The underestimated potential of battery electric vehicles to reduce emissions. Joule **3**(6), 1412–1414 (2019). https://doi.org/10.1016/j.joule.2019.06.002
28. P. Karmakar, P. Dutta, S. Biswas, Assessment of mutual fund performance using distance-based multi-criteria decision-making techniques—an Indian perspective. Res. Bull. **44**(1), 17–38 (2018)
29. N.Z. Khan, T.S.A. Ansari, A.N. Siddiquee, Z.A. Khan, Selection of E-learning websites using a novel Proximity Indexed Value (PIV) MCDM method. J. Comput. Educ. **6**(2), 241–256 (2019)
30. P. Liu, P. Wang, Some q-rung orthopair fuzzy aggregation operators and their applications to multiple attribute decision making. Int. J. Intell. Syst. **33**(2), 259–280 (2018). https://doi.org/10.1002/int.21927
31. J.A. Molina, J.I. Giménez-Nadal, J. Velilla, Sustainable commuting: results from a social approach and international evidence on carpooling. Sustainability **12**(22), 9587 (2020). https://doi.org/10.3390/su12229587
32. J. Molina-Garcia, I. Castillo, A. Queralt, J.F. Sallis, Bicycling to university: evaluation of a bicycle-sharing program in Spain. Health Promot. Int. **30**(2), 350–358 (2013). https://doi.org/10.1093/heapro/dat045
33. S. Mufazzal, S.M. Muzakkir, A new multi-criterion decision-making (MCDM) method based on proximity indexed value for minimizing rank reversals. Comput. Ind. Eng. **119**, 427–438 (2018)
34. I. Mukhametzyanov, D. Pamucar, A sensitivity analysis in MCDM problems: a statistical approach. Decis. Mak.: Appl. Manag. Eng. **1**(2), 51–80 (2018)
35. D.S. Pamučar, D. Božanić, A. Ranđelović, Multi-criteria decision making: an example of sensitivity analysis. Serbian J. Manag. **12**(1), 01–27 (2017)
36. D. Pamučar, G. Ćirović, The selection of transport and handling resources in logistics centers using Multi-Attributive Border Approximation area Comparison (MABAC). Expert Syst. Appl. **42**(6), 3016–3028 (2015)
37. D. Pamucar, M. Žižović, S. Biswas, D. Božanić, A new logarithm methodology of additive weights (LMAW) for multi-criteria decision-making: application in logistics. Facta Univ. Ser.: Mech. Eng. (2021)
38. X. Peng, J. Dai, Research on the assessment of classroom teaching quality with q-rung orthopair fuzzy information based on multiparametric similarity measure and combinative distance-based assessment. Int. J. Intell. Syst. **34**(7), 1588–1630 (2019)
39. X. Peng, J. Dai, H. Garg, Exponential operation and aggregation operator for q-rung orthopair fuzzy set and their decision-making method with a new score function. Int. J. Intell. Syst. **33**(11), 2255–2282 (2018)
40. A. Pinar, F.E. Boran, A q-rung orthopair fuzzy multi-criteria group decision making method for supplier selection based on a novel distance measure. Int. J. Mach. Learn. Cybern. **11**(8), 1749–1780 (2020). https://doi.org/10.1007/s13042-020-01070-1
41. A. Poullikkas, Sustainable options for electric vehicle technologies. Renew. Sustain. Energy Rev. **41**, 1277–1287 (2015). https://doi.org/10.1016/j.rser.2014.09.016
42. C. Prathyusha, S. Singh, P. Shivananda, Strategies for sustainable, efficient, and economic integration of public transportation systems. Lect. Notes Civ. Eng. **121**, 157–169 (2021). https://doi.org/10.1007/978-981-33-4114-2_13

43. J. Pucher, J. Dill, S. Handy, Infrastructure, programs, and policies to increase bicycling: an international review. Prev. Med. **50**, 106–125 (2010)
44. K.R. Ramakrishnan, S. Chakraborty, A cloud topsis model for green supplier selection. Facta Univ. Ser.: Mech. Eng. **18**(3), 375–397 (2020)
45. M. Riaz, H. Garg, H.M.A. Farid, M. Aslam, Novel q-rung orthopair fuzzy interaction aggregation operators and their application to low-carbon green supply chain management. J. Intell. Fuzzy Syst. **41**(2), 4109–4126 (2021). https://doi.org/10.3233/jifs-210506
46. P. Ribeiro, F. Fonseca, P. Santos, Sustainability assessment of a bus system in a mid-sized municipality. J. Environ. Plan. Manag. **63**(2), 236–256 (2019). https://doi.org/10.1080/09640568.2019.1577224
47. G. Rose, H. Marfurt, Travel behaviour change impacts of a major ride to work day event. Transp. Res. Part a: Policy Pract. **41**(4), 351–364 (2007)
48. M. Seraj, S.M. Yahya, I.A. Badruddin, A.E. Anqi, M. Asjad, Z.A. Khan, Multi-response optimization of nanofluid-based IC engine cooling system using fuzzy PIV method. Processes **8**(1), 30 (2020)
49. S.A. Shaheen, T.E. Lipman, Reducing greenhouse emissions and fuel consumption. IATSS Res. **31**(1), 6–20 (2007). https://doi.org/10.1016/s0386-1112(14)60179-5
50. S. Shaheen, A. Cohen, A. Bayen, *The Benefits of Carpooling* (Transportation Sustainability Research Center, UC Berkeley, 2018). https://doi.org/10.7922/G2DZ06GF, https://escholarship.org/uc/item/7jx6z631
51. T. Shannon, B. Giles-Corti, T. Pikora, M. Bulsara, T. Shilton, F. Bull, Active commuting in a university setting: assessing commuting habits and potential for modal change. Transp. Policy **13**(3), 240–253 (2006)
52. P. Shridhar Bokare, A. Kumar Maurya, Study of the effect of speed, acceleration, and deceleration of small petrol car on its tailpipe emission. Int. J. Traffic Transp. Eng. **3**(4), 465–478 (2013). https://doi.org/10.7708/ijtte.2013.3(4).09
53. M. Song, M. Yin, X.M. Chen, L. Zhang, M. Li, A Simulation-based approach for sustainable transportation systems evaluation and optimization: theory, systematic framework and applications. Procedia. Soc. Behav. Sci. **96**, 2274–2286 (2013). https://doi.org/10.1016/j.sbspro.2013.08.257
54. L. Steg, Car use: lust and must. Instrumental, symbolic, and affective motives for car use. Transp. Res. Part A: Policy Pract. **39**(2), 147–162 (2005)
55. W. Toor, S. Havlick, *Transportation and Sustainable Campus Communities* (Island Press, Washington, DC, 2004)
56. S. Turner, C. Shafer, W. Stewart, Bicycle suitability criteria for state roadways in Texas (1997), https://www.researchgate.net/publication/228778749_Bicycle_suitability_criteria_for_state_roadways_in_Texas
57. M. Vogel, R. Hamon, G. Lozenguez, L. Merchez, P. Abry, J. Barnier, P. Borgnat, P. Flandrin, I. Mallon, C. Robardet, From bicycle sharing system movements to users: a typology of Vélo'v cyclists in Lyon based on a large-scale behavioural dataset. J. Transp. Geogr. **41**, 280–291 (2014). https://doi.org/10.1016/j.jtrangeo.2014.07.005
58. H. Wang, Y. Ju, P. Liu, Multi-attribute group decision-making methods based on q-rung orthopair fuzzy linguistic sets. Int. J. Intell. Syst. **34**(6), 1129–1157 (2019). https://doi.org/10.1002/int.22089
59. R. Wang, Y. Li, A novel approach for green supplier selection under a q-rung orthopair fuzzy environment. Symmetry **10**(12), 687 (2018). https://doi.org/10.3390/sym10120687
60. G. Wei, H. Gao, Y. Wei, Some q-rung orthopair fuzzy Heronian mean operators in multiple attribute decision making. Int. J. Intell. Syst. **33**(7), 1426–1458 (2018). https://doi.org/10.1002/int.21985
61. K.E. Whalen, A. Paez, J.A. Carrasco, Mode choice of university students commuting to school and the role of active travel, **31**, 132–142 (2013)
62. W. Wu, W. Ma, K. Long, H. Zhou, Y. Zhang, Designing sustainable public transportation: integrated optimization of bus speed and holding time in a connected vehicle environment. Sustainability **8**(11), 1170 (2016). https://doi.org/10.3390/su8111170

63. R.R. Yager, Pythagorean membership grades in multicriteria decision making. IEEE Trans. Fuzzy Syst. **22**(4), 958–965 (2013). https://doi.org/10.1109/Tfuzz.2013.2278989
64. R.R. Yager, Generalized orthopair fuzzy sets. IEEE Trans. Fuzzy Syst. **25**(5), 1222–1230 (2016). https://doi.org/10.1109/Tfuzz.2016.2604005
65. H. Younes, Z. Zou, J. Wu, G. Baiocchi, Comparing the temporal determinants of dockless scooter-share and station-based bike-share in Washington, D.C. Transp. Res. Part a: Policy Pract. **134**, 308–320 (2020). https://doi.org/10.1016/j.tra.2020.02.021
66. L.A. Zadeh, Fuzzy Sets. Inf. Control **8**(3), 338–353 (1965)
67. R. Zhu, X. Zhang, D. Kondor, P. Santi, C. Ratti, Understanding spatio-temporal heterogeneity of bike-sharing and scooter-sharing mobility. Comput. Environ. Urban Syst. **81**, 101483 (2020). https://doi.org/10.1016/j.compenvurbsys.2020.101483

Dr. Muhammet Deveci is currently a Visiting Professor at Royal School of Mines in the Imperial College London, London, UK. He is also an Associate Professor at the Departmentof Industrial Engineering in the Turkish Naval Academy, National Defence University, Istanbul, Turkey. Dr Deveci received his B.Sc. degree in Industrial Engineering from Cukurova University, Adana, Turkey in 2010, and his M.Sc.degree in Business Administration from Gazi University, Ankara, in 2012. He recieved his Ph.D. in Industrial Engineering at Yildiz Technical University,Istanbul, Turkey in 2017. He worked as a visiting researcher and postdoctoral researcher in 2014–2015 and 2018–2019, respectively, in the School of Computer Science at the University of Nottingham (UoN), UK. Dr Deveci published over 70 papers in journals indexed by SCI papers at reputable venues such as Elsevier, IEEE, and Springer, as well as more than 23 contributions in International Conferences related to his areas. He worked as a guest editor for many international journals such as IEEE Transactions on Fuzzy Systems, Applied Soft Computing, Sustainable Energy Technologies and Assessments, Journal of Petroleum Science and Engineering, and International of Journal of Hydrogen Energy. Dr Deveci is a member of editorial boards of some international journals. He has served as a reviewer for more than 100 journals. My research focuses on computational intelligence, handling of uncertainty, fuzzy sets and systems, fuzzy decision making, decision support systems, modelling and optimization, and their hybrids, applied to complex real-world problems such as climate change, renewable energy, sustainable transport, urban mobility, humanoid robots, autonomous vehicles, digitalization and circular economy, and so on.

Dr. Ilgin Gokasar received the B.S. degree in civil engineering from Bŏgaziçi University, Istanbul, Turkey, in 2000, and the M.S. and Ph.D. degrees in (intelligent transportation systems) civil engineering from Rutgers University, New Jersey, NJ, USA, in 2003 and 2006, respectively. She is currently an Associate Professor at the Department of Civil Engineering, Bogazici University. She is the Founder and Director of the BOUN-ITS Lab. Her research interests include intelligent transportation system, traffic safety, advanced public transportation systems, real-time traffic control, smart and sustainable transportation systems, and use of big data to address challenges in mobility, safety, sustainability, and resilience in multimodal transportation systems.

Dr. Dragan Pamucar is an Associate Professor at the University of Defence in Belgrade, the Department of Logistics, Serbia. Dr. Dragan Pamucar obtained his MSc at the Faculty of Transport and Traffic Engineering in Belgrade in 2009, and his Ph.D. degree in Applied Mathematics with specialization in Multi-Criteria Modelling and Soft Computing Techniques at University of Defence in Belgrade, Serbia in 2013. His research interest includes the fields of computational intelligence, multi-criteria decision making problems, neuro-fuzzy systems, fuzzy, rough and intuitionistic fuzzy set theory, neutrosophic theory, with applications in a wide range of logistics problems. He has published 10 books and over 170 research papers in international scholarly academic journals.

Dr. Sanjib Biswas is an Assistant Professor, Decision Sciences and Operations Management area at Calcutta Business School. He has a rich working experience of 17+ years in industry and academia. He has published and presented several research articles in several referred and Scopus/SCI indexed national and international journals and conferences of repute both in India and abroad. He is a member of the editorial board and reviewer of a number of Scopus/SCI/ABDC indexed journals of repute. His research interests include multi-criteria decision-making models and their applications in managerial decision making, multivariate analysis, logistics and supply chain management, sustainability, etc.

Dr. Vladimir Simic was born in Belgrade, Serbia in 1983. He received a Ph.D. degree in Transportation Engineering from the University of Belgrade, Faculty of Transport and Traffic Engineering, Serbia, in 2014. Since 2020, he is an Associate Professor of the Transport and Traffic Engineering Department at the University of Belgrade, Serbia. He has conducted intensive research on operations research applications in diverse fields of specialization, with a particular focus on developing advanced hybrid multi-criteria decision-making tools and real-life large-scale stochastic, fuzzy, interval, full- and semi-infinite programming optimization models. He published more than 80 papers, including 35 papers in journals from the JCR list. He is the second most influential author in the world in the end-of-life vehicle management research area (10.3390/en13215586). He regularly serves as an ad-hoc reviewer of many top-tier journals.

Chapter 13
Platform-Based Corporate Social Responsibility Evaluation with Three-Way Group Decisions Under Q-Rung Orthopair Fuzzy Environment

Decui Liang and Wen Cao

Abstract Platform-based enterprises are becoming an engine of new economic development. Although the rapid development of platform-based enterprises brings convenience to society, there exists some social responsibility problems, such as leakage of user privacy and lax supervision. Thus, the evaluation and classification of corporate social responsibility (CSR) for platform-based enterprises are conducive to the supervision of government agencies. In this paper, we propose an evaluation method to help government agencies effectively assess platform-based CSR, which is instrumental in the management of platform-based CSR. In our method, enterprises are evaluated with q-rung orthopair fuzzy sets (q-ROFSs) to reduce the uncertainties in evaluation process. The weights of criteria and experts are all learned from evaluation data. A new aggregation operator is proposed based on the compensative weighted averaging (CWA) operator. Moreover, considering the misclassification risk, we utilize TWDs to make a three classification for enterprises. Finally, we give an example to illustrate the application of our proposed method and validate the results via comparative experiment.

Keywords Platform-based enterprises · Corporate social responsibility · q-rung orthopair fuzzy sets · Multi-criteria decision-making · Three-way decisions

---

D. Liang (✉) · W. Cao
School of Management and Economics, University of Electronic Science and Technology of China, Chengdu 610054, China
e-mail: decuiliang@126.com

W. Cao
e-mail: caowenre@163.com

H. Garg (ed.), *q-Rung Orthopair Fuzzy Sets*,
https://doi.org/10.1007/978-981-19-1449-2_13

13.1 Introduction

The sharing platform, such as Airbnb and Uber, is gradually becoming a new business model with the development of Internet and information technology [19]. This kind of platform can efficiently integrate the information of both suppliers and demanders and provide great convenience for users. Compared with traditional entity enterprises, platform-based enterprises have some new characteristics. For example, platform-based enterprises often do not directly produce products, but act as a bridge connecting the production side and the demand side. However, due to these enterprises are particular and emerging, there exists some problems in actual operation. Some platform-based enterprises go against the original intention that provides convenience for users, but turn to monopolize market and disrupt the market economic order. Moreover, to avoid taking on the responsibility of employed users, some platform-based enterprises just play a platform role rather than a company role. In addition, the incidents that platform-based enterprises use their privileges to disclose and resell users' information are appearing in an endless stream. One reason for these problems is that platform-based enterprises do not pay attention to their social responsibilities. Corporate social responsibility (CSR) demands that enterprises should not focus solely on making profits, but take responsibility for consumers, communities, and the environment. In practice, the identification and evaluation of platform-based CSR are not be taken seriously.

Scholars' research on CSR is deepening with the development of society. The early studies mainly focus on qualitative researches for the CSR concept. Carroll [4] defined CSR should involve the economic, legal, ethical, and discretionary categories of business performance. Porter and Kramer [23] believed that enterprises should regard social responsibility as an opportunity rather than a cost or constraint, and they built a framework to link competitive advantage and CSR. In recent years, the CSR related literatures focus more on quantitative researches. Boccia et al. [3] designed a choice experiment and found that there is a positive correlation between social responsibility initiatives and consumers' attitudes towards the enterprise and its products. Li et al. [15] developed a theoretical model on the impact of internal CSR of platform-based enterprises on employee innovation performance with the aid of hierarchical linear model. Nikolaou et al. [20] introduced CSR and sustainable development into reverse logistics systems and proposed the performance indicators based on triple bottom line theory. Based on multiple regressions, Yoo and Lee [33] found that the higher the CSR support degree is, the more positive the consumer evaluation of enterprise. However, these studies mainly concentrate on the CSR of traditional entity enterprises. Owing to the fact that platform-based enterprises are developed in only recent years, there are little researches on CSR of these enterprises. Meanwhile, the researches on CSR evaluation, especially the evaluation of platform-based CSR, are rare. Considering that many platform-based enterprises do not take social responsibility well and there are few CSR evaluation mechanisms, it is necessary to construct a CSR evaluation system for platform enterprises, so as to regulate their behavior. Since platform-based enterprises are faced with more uncertainties and risks than entity enterprises, existing evaluation methods are not suitable to handle such situation. A new method to assess platform-based CSR is necessary.

Based on the above-mentioned, this paper aims to construct an evaluation method to help the government agencies or other institutions evaluate the CSR of platform-based enterprises. This evaluation model is a multi-criteria decision-making (MCDM) problem with a group of experts. Considering that in actual situations, experts tend to use fuzzy information to evaluate objects, we utilize the q-rung orthopair fuzzy sets (q-ROFSs) to assess platform-based enterprises [30]. This fuzzy set describes the relationship between membership and non-membership from power perspective, which can measure more information than Intuitionistic fuzzy sets [29] and Pythagorean fuzzy sets [26]. Q-ROFSs have been applied to some fields, such as project investment [8–11], medical science [12], and supply chain management [25]. Recently, scholars have expanded q-ROFSs from various aspects. Garg [9] proposed a new possibility degree measure for interval-valued q-ROFSs. Peng and Liu [22] developed a systematic transformation of information measures for q-ROFSs, which includes distance measure, similarity measure, entropy, and inclusion measure. Yang and Garg [31] combined power average operator and partitioned Maclaurin Symmetric mean operator together into the q-ROFSs environment. These studies further prove the applicability of q-ROFSs in decision-making problem. Hence, this paper also uses q-ROFSs to depict the evaluation information of platform-based enterprises.

To determine the weights of criteria and exeprts, we, respectively, utilize the Gini coefficient and the decision-making trial and evaluation laboratory (DEMATEL) method. These two methods can calculate the importance of object based on its data distribution. The trust matrix in DEMATEL is also computed based on evaluated results. Furthermore, we propose a new aggregation operator to fuse the information by combining q-ROFSs and compensative weighted averaging (CWA) operator [1]. After fusing the multi-experts' information, we use three-way decisions (TWDs) method to classify the enterprises. Developed from decision-theoretic rough sets, TWDs divide objects into three domains with least expected loss, which can be regarded as a three classification tool. Because the classification results always have low misclassification loss, TWDs method is suitable to solve the assessment problem of platform-based enterprises. In this paper, the conditional probability of TWDs is computed based on technique for order preference by similarity to an ideal solution (TOPSIS) method and regret theory. With the aid of TWDs, platform-based enterprises can be classified into the enterprises with high CSR, uncertain CSR, and low CSR.

The main purpose of this paper is to develop an MCDM method to assess the CSR of platform-based enterprises with q-ROFSs information. Our innovations are shown as below: (1) With the help of Gini coefficient and DEMATEL, we, respectively, calculate the weights of criteria and experts based on actual assessment results. These methods can give the corresponding weight objectively based on data density of evaluation object. Meanwhile, DEMATEL not only considers the distance among experts (data distribution) but also learns the trust relationships of experts, which is more in line with the actual interact scenarios. (2) We combine q-ROFSs with CWA operator and propose a q-ROFCWA operator to fuse experts' information. This operator can depict the compensation degree of experts' preferences in q-ROFSs environment. (3) Our method utilize regret theory to measure the experts' perception of regret degree between different alternatives. This theory is an irrational behavior theory and practical for real situations. (4) We use TWDs to measure the misclassification risk based on loss function matrix and make a three classification of platform-based enterprises with the minimum expected loss.

The remainder of this paper is arranged as follows: Sect. 13.2 introduces some basic conceptions of q-ROFSs and $TWDs$. In Sect. 13.3, we propose our decision-making method with the information aggregation part and the classification part. Section 13.4 gives an example of platform-based CSR evaluation to show the application of our method. Comparative experiment and sensitivity analysis are also implemented in this section. Section 13.5 summarizes the whole paper and gives some possible future works.

13.2 Preliminaries

In this section, we briefly introduce some basic conceptions of the evaluation information form, i.e., q-ROFSs and our main method, i.e., TWDs.

13.2.1 q-rung Orthopair Fuzzy Sets (q-ROFSs)

Proposed by Yager [30], q-ROFS is now extended into many decision-making problems [16]. According to Refs. [16, 30], the basic concepts of q-ROFS are briefly introduced as follows:

Definition 13.1 ([30]) Let $\mathbf{X}$ be a fixed set. A q-rung orthopair fuzzy set (q-ROFS) $\mathbf{Q}$ on $\mathbf{X}$ can be represented as the following mathematical symbol:

$$\mathbf{Q} = \{(x, u_Q(x), v_Q(x))|x \in \mathbf{X}\}, \tag{13.1}$$

where functions $u_Q(x) : \mathbf{X} \to [0, 1]$ and $v_Q(x) : \mathbf{X} \to [0, 1]$, respectively, denote the membership degree and non-membership degree of x to $\mathbf{Q}$ with the condition that $0 \leq u_Q(x)^q + v_Q(x)^q \leq 1$, $(q \geq 1)$. The indeterminacy degree is defined as: $\pi_Q(x) = (1 - (u_Q(x)^q + v_Q(x)^q))^{1/q}$.

To depict information easily, the q-rung orthopair number (q-ROFN) is proposed. A q-ROFN can be denoted by $\alpha = (u_Q, v_Q)$. According to Refs. [16, 21, 30], for any three q-ROFNs: $\alpha = (u, v)$, $\alpha_1 = (u_1, v_1)$, and $\alpha_2 = (u_2, v_2)$, their basic operations are defined as follows:
(1) $\overline{\alpha} = (v, u)$,
(2) $\alpha_1 \oplus \alpha_2 = ((u_1^q + u_2^q - u_1^q u_2^q)^{1/q}, v_1 v_2)$,
(3) $\alpha_1 \otimes \alpha_2 = (u_1 u_2, (v_1^q + v_2^q - v_1^q v_2^q)^{1/q})$,

(4) $\lambda\alpha = ((1-(1-u^q)^\lambda)^{1/q}, v^\lambda), \lambda > 0,$
(5) $\alpha^\lambda = (u^\lambda, (1-(1-v^q)^\lambda)^{1/q}), \lambda > 0,$
(6) $\lambda^\alpha = \begin{cases} ((\lambda^{1-u^q})^{1/q}, (1-\lambda^{v^q})^{1/q}), & \lambda \in (0,1), \\ ((\lambda^{u^q-1})^{1/q}, (1-\lambda^{-v^q})^{1/q}), & \lambda \in [1,+\infty). \end{cases}$

To compare q-ROFNs, the score function and accuracy function are proposed as follows:

Definition 13.2 ([17]) For a q-ROFN $\alpha = (u, v)$, its score function is defined as follows:

$$S(\alpha) = u^q - v^q, \tag{13.2}$$

where $S(\alpha) \in [-1, 1]$. The higher the score is, the more qualified the evaluation object. The accuracy function is defined as follows:

$$H(\alpha) = u^q + v^q, \tag{13.3}$$

where $H(\alpha) \in [0, 1]$. When the scores are the same, the higher the accuracy is, the more the evaluation object meets the requirements.

According to Ref. [16], score function and accuracy function can be used to compare q-ROFNs.

Definition 13.3 ([16]) For any two q-ROFNs $\alpha_1 = (u_1, v_1)$ and $\alpha_2 = (u_2, v_2)$, their score functions and accuracy functions are denoted as $S(\alpha_1), S(\alpha_2)$ and $H(\alpha_1), H(\alpha_2)$, respectively. To compare these two q-ROFNs, we have
(1) If $S(\alpha_1) > S(\alpha_2)$, then $\alpha_1 > \alpha_2$;
(2) If $S(\alpha_1) < S(\alpha_2)$, then $\alpha_1 < \alpha_2$;
(3) If $S(\alpha_1) = S(\alpha_2)$, then
  (3.1) If $H(\alpha_1) > H(\alpha_2)$, then $\alpha_1 > \alpha_2$;
  (3.2) If $H(\alpha_1) < H(\alpha_2)$, then $\alpha_1 < \alpha_2$;
  (3.3) If $H(\alpha_1) = H(\alpha_2)$, then $\alpha_1 = \alpha_2$.

13.2.2 Three Way Decisions (TWDs)

The core idea of TWDs method is that it will not make a decision when information is inadequate or indefinite [32]. Considering the misclassification risk, TWDs can divide evaluation objects into three regions: positive region (POS), boundary region (BND), and negative region (NEG), which respectively mean the acceptance decision, deferment decision, and rejection decision. This method can measure the risk loss of objects under different actions with the help of loss function, and make a decision with the minimum expected loss. For an evaluation object x, its loss function is given in Table 13.1.

In Table 13.1, a_P, a_B, and a_N, respectively, denote the acceptance, deferment, and rejection actions of x. $C(P)$ and $\neg C(N)$ are two opposite states of x. λ_{PP}, λ_{BP}, and

Table 13.1 The loss function matrix

	$C(P)$	$\neg C(N)$
a_P	λ_{PP}	λ_{PN}
a_B	λ_{BP}	λ_{BN}
a_N	λ_{NP}	λ_{NN}

λ_{NP} are the losses when taking the acceptance action, deferment action, and rejection action if x belongs to state C. Similarly, λ_{PN}, λ_{BN}, and λ_{NN} are the losses when taking the acceptance action, deferment action, and rejection action if x belongs to state $\neg C$.

13.3 CSR Evaluation Method Based on TWDs with q-ROFSs

To evaluate the social responsibility of platform-based enterprises, the evaluation information of multi-experts should be aggregated first. Then, we utilize TWDs to make a three classification with the minimum expected losses. For the sake of clarity, the framework of our CSR evaluation method is illustrated in Fig. 13.1.

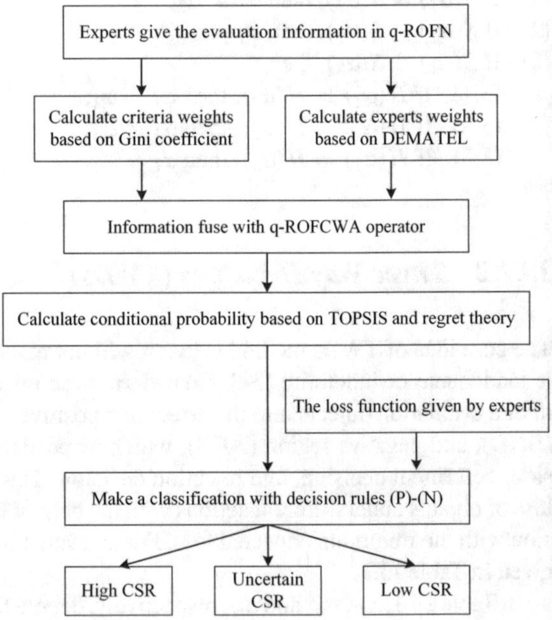

Fig. 13.1 The architecture of our CSR evaluation method

13.3.1 *Information Fusion Method*

13.3.1.1 Criteria Weights Calculation Based on Gini Coefficient

Let the set $A = \{a_1, a_2, \ldots, a_M\}$ denote M enterprises, and the set $C = \{c_1, c_2, \ldots, c_N\}$ denote N criteria. The weights of these criteria are denoted by the set $W = \{w_1, w_2, \ldots, w_N\}$. There are K experts invited to take part in the decision-making, and they are represented by the set $E = \{e_1, e_2, \ldots, e_K\}$. $X^k = (x_{ij}^k)_{M \times N}$ is the evaluated information of enterprise a_i by expert e_k under criterion c_j, where $i = 1, 2, \ldots, M,\ j = 1, 2, \ldots, N,\ k = 1, 2, \ldots, K$, and $x_{ij}^k = (u_{ij}^k, v_{ij}^k)$ is a q-ROFN.

For a criterion, if its evaluation data distribution is discrete, it means experts have great opinion divergence. In the case, this criterion has great influence on experts and should be assigned a large weight. Considering the Gini coefficient is used to describe the dispersion degree of data distribution, we use Gini coefficient to depict the opinion divergence and calculate the criteria weight. For criterion c_j, its Gini coefficient is calculated as follows:

$$Gini_j = \frac{\Sigma_{l=1}^{N} \Sigma_{k=1}^{K} \Sigma_{i=1}^{M} d(x_{il}^k, x_{ij}^k)}{2NKM \cdot (S(\overline{x_j}) + 1)}, \tag{13.4}$$

where $\overline{x_j}$ denotes the average value of criterion c_j under all enterprises and all experts. Based on the basic operations in Sect. 13.2, $\overline{x_j}$ can be calculated as follows:

$$\overline{x_j} = \frac{1}{MK} \mathop{\oplus}_{k=1}^{K} \mathop{\oplus}_{i=1}^{M} x_{ij}^k, \tag{13.5}$$

where $\overline{x_j}$ is also a q-ROFN. Moreover, in Eq. (13.4), $d(x_{il}^k, x_{ij}^k)$ is the Euclidean distance of two q-ROFNs, which can be calculated as follows [28]:

$$d(x_{il}^k, x_{ij}^k) = \sqrt{\frac{1}{2}(u_{il}^k - u_{ij}^k)^2 + \frac{1}{2}(v_{il}^k - v_{ij}^k)^2} \tag{13.6}$$

Based on Eqs. (13.4)–(13.6), the Gini coefficient can be obtained. Then, the weight of criterion c_j can be calculated as follows:

$$w_j = \frac{Gini_j}{\Sigma_{j=1}^{N} Gini_j}. \tag{13.7}$$

13.3.1.2 Determination of Experts Weights with DEMATEL Method

In social network, experts are usually influenced by the opinions of other experts they trust. If an expert is trusted by many experts, his or her opinions often have a great weight. Based on this idea, we utilize DEMATEL method to compute the weights of experts, which can calculate the importance of each individual with mutual influence degree. The trust of an expert to others can be measured by a trust matrix $TM^* = [tm^*_{kg}]_{K\times K}$, where tm^*_{kg} means the degree of e_g's trust in e_k. In this paper, we construct the trust matrix based on evaluation information. For experts, their evaluation results usually cannot reach a consensus. Some experts' opinions may be highly consistent, while others' evaluation results may vary greatly. If an expert's opinion is consistent with most experts, it may be because he or she has a high degree of influence, which makes other experts' opinions tend to him or her. Furthermore, expert who has a great disparity with others may have less influence, because very few experts share the same opinions with him or her. In this way, we calculate the distance of different experts and construct trust matrix. For an experts e_k, the distance between e_k and other experts can be computed as follows:

$$D_{e_k} = \sum_{g=1}^{K} \sum_{i=1}^{M} \sum_{j=1}^{N} d(x^k_{ij}, x^g_{ij}), \tag{13.8}$$

where D_{e_k} is the total distance between e_k and other experts, $k, g \in 1, 2, \ldots, K$. Based on total distance, the trust matrix can be constructed as $TM^* = [tm^*_{kg}]_{K\times K} = [D_{e_g}/D_{e_k}]_{K\times K}$. After standardizing TM^* into TM, the comprehensive impact matrix $IM = [im_{kg}]_{K\times K}$ can be obtained as follows:

$$IM = TM(I - TM)^{-1}. \tag{13.9}$$

Then, the affected degree $AD_k = \sum_{k=1}^{K} im_{kg}$ and impact degree $ID_k = \sum_{g=1}^{K} im_{kg}$ are computed by adding rows and columns, respectively. Based on the results of Ref. [2], the weights of experts can be computed as follows:

$$\omega^*_k = \sqrt{(ID_k + AD_k)^2 + (ID_k - AD_k)^2}, \tag{13.10}$$

where ω^*_k is the weight of expert e_k. We standardize this weight with the help of the following equation:

$$\omega_k = \frac{\omega^*_k}{\sum_{k=1}^{K} \omega^*_k}, \tag{13.11}$$

where ω_k is the standardized weight of expert e_k.

13.3.1.3 Information Fusion Process with CWA Operators

CWA operator has a special parameter to describe the compensation degree, including both non-compensatory minimum ("and") operator and fully compensatory maximum ("or") operator [1]. Based on the results of Refs. [1, 27], we introduce CWA operator into q-ROFSs and develop a new aggregation operator to fuse the multi-experts' information.

Definition 13.4 Let $x_i = (u_i, v_i)$ $(i = 1, 2, \ldots, m)$ be a collection of q-ROFNs. Their weight vector is $\boldsymbol{\omega} = (\omega_1, \omega_1, \ldots, \omega_m)$ with $\omega_i \in [0, 1]$ and $\sum_{i=1}^{m} \omega_i = 1$. Then, q-rung orthopair fuzzy compensative weighted averaging (q-ROFCWA) can be defined as follows:

$$q - ROFCWA(x_1, x_2, \ldots, x_m) = \left(\log_\lambda \left(\sum_{i=1}^{m} \omega_i \lambda^{u_i} \right), \left(1 - \log_\lambda \left(\sum_{i=1}^{m} \omega_i \lambda^{1-v_i^q} \right) \right)^{1/q} \right), \tag{13.12}$$

where $0 < \lambda$ and $\lambda \neq 1$. λ means the pessimism degree ($\lambda < 1$) or optimism degree ($\lambda > 1$) of the experts' perception to enterprise. Apparently, the aggregated value is also a q-ROFN.

According to Definition 4 and the basic operations of q-ROFNs, we can obtain the following properties:

Property 1 (Idempotency) Let $x_i = (u_i, v_i)$ $(i = 1, 2, \ldots, m)$ be a collection of q-ROFNs with a weight vector $\boldsymbol{\omega} = (\omega_1, \omega_1, \ldots, \omega_m)$ $(\sum_{i=1}^{m} \omega_i = 1)$. If $x_i = x = (u, v)$ $(i = 1, 2, \ldots, m)$, then $q - ROFCWA(x_1, x_2, \ldots, x_m) = x$.

Proof Since $x_i = x$ for all i and $\sum_{i=1}^{m} \omega_i = 1$, we have

$$q - ROFCWA(x, \ldots, x) = \left(\log_\lambda \left(\sum_{i=1}^{m} \omega_i \lambda^{u} \right), \left(1 - \log_\lambda \left(\sum_{i=1}^{m} \omega_i \lambda^{1-v^q} \right) \right)^{1/q} \right) = x.$$

Property 2 (Boundedness) Let $x_i = (u_i, v_i)$ $(i = 1, 2, \ldots, m)$ be a collection of q-ROFNs with a weight vector $\boldsymbol{\omega} = (\omega_1, \omega_1, \ldots, \omega_m)$ $(\sum_{i=1}^{m} \omega_i = 1)$. If $x^+ = (u^+, v^+) = \max_i S(x_i)$ and $x^- = (u^-, v^-) = \min_i S(x_i)$, then $x^- \leq q - ROFCWA(x_1, x_2, \ldots, x_m) \leq x^+$.

Proof Since $\sum_{i=1}^{m} \omega_i = 1$, we have

$$\begin{aligned} q - ROFCWA(x_1, x_2, \ldots, x_m) &= \left(\log_\lambda \left(\sum_{i=1}^{m} \omega_i \lambda^{u_i} \right), \left(1 - \log_\lambda \left(\sum_{i=1}^{m} \omega_i \lambda^{1-v_i^q} \right) \right)^{1/q} \right) \\ &\leq \left(\log_\lambda \left(\sum_{i=1}^{m} \omega_i \lambda^{u^+} \right), \left(1 - \log_\lambda \left(\sum_{i=1}^{m} \omega_i \lambda^{1-(v^+)^q} \right) \right)^{1/q} \right) \\ &\leq \left(\log_\lambda \left(\lambda^{u^+} \sum_{i=1}^{m} \omega_i \right), \left(1 - \log_\lambda \left(\lambda^{1-(v^+)^q} \sum_{i=1}^{m} \omega_i \right) \right)^{1/q} \right) \\ &= \left(\log_\lambda \left(\lambda^{u^+} \right), \left(1 - \log_\lambda \left(\lambda^{1-(v^+)^q} \right) \right)^{1/q} \right) \\ &= \left(u^+, v^+ \right). \end{aligned}$$

Similarly,

$$
\begin{aligned}
q-ROFCWA(x_1, x_2, \ldots, x_m) &= \left(\log_\lambda\left(\sum\nolimits_{i=1}^{m}\omega_i\lambda^{u_i}\right), \left(1-\log_\lambda\left(\sum\nolimits_{i=1}^{m}\omega_i\lambda^{1-v_i^q}\right)\right)^{1/q}\right)\\
&\geq \left(\log_\lambda\left(\sum\nolimits_{i=1}^{m}\omega_i\lambda^{u^-}\right), \left(1-\log_\lambda\left(\sum\nolimits_{i=1}^{m}\omega_i\lambda^{1-(v^-)^q}\right)\right)^{1/q}\right)\\
&\geq \left(\log_\lambda\left(\lambda^{u^-}\sum\nolimits_{i=1}^{m}\omega_i\right), \left(1-\log_\lambda\left(\lambda^{1-(v^-)^q}\sum\nolimits_{i=1}^{m}\omega_i\right)\right)^{1/q}\right)\\
&= \left(\log_\lambda\left(\lambda^{u^-}\right), \left(1-\log_\lambda\left(\lambda^{1-(v^-)^q}\right)\right)^{1/q}\right)\\
&= \left(u^-, v^-\right).
\end{aligned}
$$

Property 3 (Monotonicity) Let $x_i^1 = (u_i^1, v_i^1), x_i^2 = (u_i^2, v_i^2)$ $(i = 1, 2, \ldots, m)$ be two collections of q-ROFNs with weight vector $\boldsymbol{\omega} = (\omega_1, \omega_1, \ldots, \omega_m)$ $(\sum_{i=1}^{m}\omega_i = 1)$. If $S(x_i^1) \leq S(x_i^2)$ for all i, then $q-ROFCWA(x_1^1, x_2^1, \ldots, x_m^1) \leq q-ROFCWA(x_1^2, x_2^2, \ldots, x_m^2)$.

Proof Since $S(x_i^1) \leq S(x_i^2)$ for all i, i.e., $S\left((u_i^1, v_i^1)\right) \leq S\left((u_i^2, v_i^2)\right)$, we have

$$
\begin{aligned}
q-ROFCWA(x_1^1, x_2^1, \ldots, x_m^1) &= \left(\log_\lambda\left(\sum\nolimits_{i=1}^{m}\omega_i\lambda^{u_i^1}\right), \left(1-\log_\lambda\left(\sum\nolimits_{i=1}^{m}\omega_i\lambda^{1-(v_i^1)^q}\right)\right)^{1/q}\right)\\
&\leq \left(\log_\lambda\left(\sum\nolimits_{i=1}^{m}\omega_i\lambda^{u_i^2}\right), \left(1-\log_\lambda\left(\sum\nolimits_{i=1}^{m}\omega_i\lambda^{1-(v_i^2)^q}\right)\right)^{1/q}\right)\\
&= q-ROFCWA(x_1^2, x_2^2, \ldots, x_m^2).
\end{aligned}
$$

Based on q-ROFCWA operator, we can fuse the experts' information. For an expert $e_k, k \in (1, 2, \ldots, K)$, his or her evaluation information is defined by the matrix $X^k = (x_{ij}^k)_{M\times N}$, where $x_{ij}^k = (u_{ij}^k, v_{ij}^k)$ is a q-ROFN. The weight vector of experts is denoted as $\boldsymbol{\omega} = (\omega_1, \omega_2, \ldots, \omega_K)$ with $\sum_{k=1}^{K}\omega_k = 1$, which is computed using DEMATEL. Then, experts' information can be aggregated by using the following equation:

$$
x_{ij} = q-ROFCWA(x_{ij}^1, x_{ij}^2, \ldots, x_{ij}^k) = \left(\log_\lambda\left(\sum\nolimits_{k=1}^{K}\omega_k\lambda^{u_{ij}^k}\right), \left(1-\log_\lambda\left(\sum\nolimits_{k=1}^{K}\omega_k\lambda^{1-\left(v_{ij}^k\right)^q}\right)\right)^{1/q}\right). \tag{13.13}
$$

With the aid of q-ROFCWA operator, the final aggregated information is shown in Table 13.2.

Criteria information in Table 13.2 can evaluate enterprises from the perspective of objective indicators. However, an enterprise with a high criteria evaluation result may also have a high misclassification risk. Losses will occur if enterprises are classified only from the perspective of criteria evaluation results without considering the misclassification risks. Since CSR classification results of platform-based enterprises may directly affect their operation strategies and capital investment, the misclassification risks should not be ignored. To handle this problem, we use TWDs to depict the misclassification loss of enterprises in next subsection.

13.3.2 CSR Classification of Platform-Based Enterprises with TWDs

In TWDs, conditional probability means the probability that an evaluation object belongs to a certain state. In some traditional TWDs researches, the conditional probability is usually given by experts. These researches are based on the subjective perceptions of experts, ignoring the actual evaluation information. In this section, based on TOPSIS method and regret theory, we calculate conditional probability from objective evaluation information, i.e., the information in Table 13.2. TOPSIS method describes the closeness between evaluation object and ideal point, which can be regarded as the probability that the object belongs to a good one [14]. Meanwhile, considering that when experts evaluate objects, they always feel regret after choosing different alternatives. This psychological behavior can have an influence on decision-making [18]. Hence, to be in line with actual situations, we introduce regret theory when calculating the relative closeness in TOPSIS.

Let $a^+ = (a_1^+, a_2^+, \ldots, a_N^+)$ denote the positive ideal solution of information matrix X, which is a q-ROFN with $a_j^+ = (u_j^+, v_j^+) = \max_i a_{ij}$. Similarly, $a^- = (a_1^-, a_2^-, \ldots, a_N^-)$ denotes the negative ideal solution with $a_j^- = (u_j^-, v_j^-) = \min_i a_{ij}$, which is also a q-ROFN. Then, based on regret theory, the distances between evaluated enterprise and ideal solutions can be calculated as follows:

$$d^+(a_i, a^+) = \sum\nolimits_{j=1}^{N} \left(w_j |CI(x_{ij}) - S(a_j^+)| \right), \tag{13.14}$$

$$d^-(a_i, a^-) = \sum\nolimits_{j=1}^{N} \left(w_j |CI(x_{ij}) - S(a_j^-)| \right), \tag{13.15}$$

where $CI(x_{ij})$ is the comprehensive information of x_{ij} with regret theory. This information is obtained by comparing a_i with other enterprises under criterion c_j, which can be calculated as follows:

Table 13.2 The aggregated information table with q-ROFNs

Enterprises	c_1	c_2	…	c_N
a_1	$x_{11} = (u_{11}, v_{11})$	$x_{12} = (u_{12}, v_{12})$	…	$x_{1N} = (u_{1N}, v_{1N})$
a_2	$x_{21} = (u_{21}, v_{21})$	$x_{22} = (u_{22}, v_{22})$	…	$x_{2N} = (u_{2N}, v_{2N})$
⋮	⋮	⋮	⋱	⋮
a_M	$x_{M1} = (u_{M1}, v_{M1})$	$x_{M2} = (u_{M2}, v_{M2})$	…	$x_{MN} = (u_{MN}, v_{MN})$

$$CI(x_{ij}) = \frac{1}{M-1}\sum\nolimits_{l=1,l\neq i}^{M} mu(x_{ij}, x_{lj}), \tag{13.16}$$

where $mu(x_{ij}, x_{lj})$ is a collective utility that contains a direct utility and an indirect utility. Based on the results of Ref. [6], this collective utility can be calculated as follows:

$$mu(x_{ij}, x_{lj}) = U(x_{ij}) + R(U(x_{ij}) - U(x_{lj})). \tag{13.17}$$

In Eq. (13.17), $U(x_{ij})$ is the direct utility of x_{ij} and $R(U(x_{ij}) - U(x_{lj}))$ is the indirect utility that depicts the regret value between x_{ij} and x_{lj}. They can be calculated as follows:

$$U(x_{ij}) = \frac{1 - e^{-\theta \cdot S(x_{ij})}}{\theta}, \tag{13.18}$$

$$R(U(x_{ij}) - U(x_{lj})) = 1 - e^{-\delta(U(x_{ij}) - U(x_{lj}))}, \tag{13.19}$$

where $\theta \in (0, 1)$ means the risk aversion parameter of experts, $\delta \in (0, +\infty]$ is the regret aversion parameter of experts. Hence, based on Eqs. (13.14)–(13.19), the regret-based distance between an enterprise and ideal solutions can be obtained. Then, the relative closeness of enterprise a_i can be calculated as follows:

$$RC(a_i) = \frac{d^-(a_i, a^-)}{d^+(a_i, a^+) + d^-(a_i, a^-)}. \tag{13.20}$$

Actually, $RC(a_i)$ means the approximation degree between a_i and positive ideal solution, which can be also regarded as the conditional probability of a_i. In this case, we have $Pr(C|a_i) = RC(a_i)$. Then, the loss function of evaluated objects are given by experts, as shown in Table 13.3.

According to Bayesian decision procedure, the expected losses for a_i under different actions are defined as follows:

$$\begin{aligned} R(a_P|a_i) &= \lambda_{PP}^i Pr(C|a_i) + \lambda_{PN}^i Pr(\neg C|a_i), \\ R(a_B|a_i) &= \lambda_{BP}^i Pr(C|a_i) + \lambda_{BN}^i Pr(\neg C|a_i), \\ R(a_N|a_i) &= \lambda_{NP}^i Pr(C|a_i) + \lambda_{NN}^i Pr(\neg C|a_i). \end{aligned} \tag{13.21}$$

Table 13.3 The loss function matrix of all enterprises

Enterprises	λ_{PP}	λ_{BP}	λ_{NP}	λ_{PN}	λ_{BN}	λ_{NN}
a_1	λ_{PP}^1	λ_{BP}^1	λ_{NP}^1	λ_{PN}^1	λ_{BN}^1	λ_{NN}^1
a_2	λ_{PP}^2	λ_{BP}^2	λ_{NP}^2	λ_{PN}^2	λ_{BN}^2	λ_{NN}^2
$\vdots$	$\vdots$	$\vdots$	$\vdots$	$\vdots$	$\vdots$	$\vdots$
a_M	λ_{PP}^M	λ_{BP}^M	λ_{NP}^M	λ_{PN}^M	λ_{BN}^M	λ_{NN}^M

Considering that in TWDs, there always exists $Pr(C|a_i) + Pr(\neg C|a_i) = 1$. We can rewritten the above equations as follows:

$$\begin{aligned} R(a_P|a_i) &= \lambda^i_{PP} Pr(C|a_i) + \lambda^i_{PN}(1 - Pr(C|a_i)), \\ R(a_B|a_i) &= \lambda^i_{BP} Pr(C|a_i) + \lambda^i_{BN}(1 - Pr(C|a_i)), \\ R(a_N|a_i) &= \lambda^i_{NP} Pr(C|a_i) + \lambda^i_{NN}(1 - Pr(C|a_i)). \end{aligned} \tag{13.22}$$

Then, the action with minimum expected loss should be selected as the best classification result. We have the following decision rules:
(P) If $R(a_P|a_i) \leq R(a_B|a_i)$ and $R(a_P|a_i) \leq R(a_N|a_i)$, then, decide $a_i \in POS(C)$;
(B) If $R(a_B|a_i) \leq R(a_P|a_i)$ and $R(a_B|a_i) \leq R(a_N|a_i)$, then, decide $a_i \in BND(C)$;
(N) If $R(a_N|a_i) \leq R(a_P|a_i)$ and $R(a_N|a_i) \leq R(a_B|a_i)$, then, decide $a_i \in NEG(C)$.

In decision rules (P)-(N), rule (P) means enterprise a_i should be classified as a high social responsibility enterprise. Rule (B) means enterprise a_i needs more information and should not be classified at present. Rule (N) means enterprise a_i should be classified as a low social responsibility enterprise.

With the help of our method, platform-based enterprises can be effectively classified into three regions: high CSR, uncertain CSR, and low CSR. Compared with most MCDM methods, our method measures objects from a more extensive perspective. In addition to criteria information, we fully consider the misclassification loss of enterprises. Hence, our method can reduce decision risks in decision-making. Moreover, we introduce the psychological attitude measurement by using CWA operator and regret theory because experts' preference for risk may affect decision results. Our method is more realistic than usual evaluation method and suitable for evaluation of platform-based CSR.

13.4 An Illustrative Example

In China, platform-based enterprises are constructed rapidly in recent years. A large number of well-known platforms, such as Alibaba and Didi Travel, have developed relatively mature technologies. However, operations of some platform-based enterprises in China are not always successful [5]. For example, some platform-based enterprises pay much attention to their own enterprise construction, but ignore their responsibilities to society. Establishing an appropriate platform-based CSR evaluation system is conducive to promoting these enterprises to improve themselves and make due contributions to society. In this section, we show a CSR evaluation problem in China and apply our method to solve it.

13.4.1 Decision Analysis with Our Proposed Method

There are many researches on the evaluation index system of CSR. Carroll [4] defined that CSR should consist economic responsibility, legal responsibility, ethic responsibility, and beneficent responsibility. Based on triple bottom line theory, Elkington [7] constructed CSR from economic, social, and environmental perspectives. Pulido [24] divided CSR into seven aspects: organizational governance, human rights, labor practices, environment, fair operating practices, consumer issues, and community involvement and development. Based on the results of Ref. [24], we construct a platform-based CSR evaluation criteria system, as shown in Table 13.4.

Based on Table 13.4, 8 platform-based enterprises need to be evaluated. 5 experts are invited to assess enterprises with the form of q-ROFNs. Without loss of generality,

Table 13.4 Platform-based CSR evaluation criteria system

Evaluation index	Specific contents
Organizational governance (c_1)	Formulating strategies to embody the commitment of social responsibility; establishing the communication process of the organization and so on
Human rights (c_2)	Civil and political rights; economic, social and cultural rights; fundamental principles and rights at work and so on
Labor practices (c_3)	Conditions of work and social protection; health and safety at work; human development and training in the workplace and so on
Environment (c_4)	Prevention of pollution; sustainable resource use; climate change mitigation and adaptation; protection of the environment, biodiversity, and restoration of natural habitats
Fair operating practices (c_5)	Anticorruption; responsible political involvement; fair competition; promoting social responsibility in the value chain; respect for property rights
Consumer issues (c_6)	Fair marketing, factual and unbiased information and fair contractual practices; protecting consumers?? health and safety; consumer service, support, and complaint and dispute resolution; consumer data protection and privacy and so on
Community involvement and development (c_7)	Community involvement; education and culture; employment creation and skills development; technology development and access; wealth and income creation; health; social investment

Table 13.5 The evaluation results by expert e_1

Enterprises	c_1	c_2	c_3	c_4	c_5	c_6	c_7
a_1	(0.62,0.55)	(0.53,0.56)	(0.67,0.20)	(0.27,0.68)	(0.40,0.44)	(0.76,0.32)	(0.65,0.72)
a_2	(0.51,0.50)	(0.57,0.63)	(0.71,0.19)	(0.26,0.70)	(0.37,0.48)	(0.78,0.45)	(0.79,0.63)
a_3	(0.61,0.64)	(0.53,0.68)	(0.68,0.23)	(0.28,0.71)	(0.41,0.43)	(0.83,0.56)	(0.78,0.70)
a_4	(0.56,0.69)	(0.54,0.62)	(0.73,0.20)	(0.25,0.75)	(0.43,0.44)	(0.79,0.43)	(0.76,0.74)
a_5	(0.69,0.53)	(0.46,0.68)	(0.72,0.18)	(0.29,0.72)	(0.33,0.50)	(0.88,0.34)	(0.77,0.60)
a_6	(0.66,0.51)	(0.42,0.65)	(0.68,0.21)	(0.28,0.70)	(0.31,0.45)	(0.86,0.23)	(0.73,0.56)
a_7	(0.59,0.66)	(0.50,0.53)	(0.75,0.24)	(0.33,0.69)	(0.39,0.32)	(0.83,0.43)	(0.78,0.72)
a_8	(0.63,0.59)	(0.48,0.60)	(0.70,0.19)	(0.29,0.74)	(0.46,0.33)	(0.82,0.50)	(0.62,0.65)

Table 13.6 The evaluation results by expert e_2

Enterprises	c_1	c_2	c_3	c_4	c_5	c_6	c_7
a_1	(0.57,0.65)	(0.50,0.57)	(0.60,0.24)	(0.23,0.56)	(0.37,0.32)	(0.67,0.30)	(0.62,0.67)
a_2	(0.50,0.67)	(0.51,0.66)	(0.62,0.25)	(0.22,0.58)	(0.33,0.43)	(0.68,0.47)	(0.64,0.65)
a_3	(0.53,0.56)	(0.52,0.69)	(0.66,0.21)	(0.28,0.59)	(0.32,0.40)	(0.68,0.46)	(0.61,0.68)
a_4	(0.54,0.58)	(0.53,0.69)	(0.70,0.26)	(0.24,0.59)	(0.35,0.41)	(0.69,0.44)	(0.60,0.71)
a_5	(0.55,0.67)	(0.46,0.66)	(0.58,0.22)	(0.23,0.56)	(0.31,0.38)	(0.77,0.44)	(0.58,0.69)
a_6	(0.55,0.68)	(0.45,0.62)	(0.63,0.26)	(0.25,0.55)	(0.28,0.40)	(0.74,0.32)	(0.65,0.66)
a_7	(0.54,0.64)	(0.57,0.67)	(0.64,0.22)	(0.27,0.57)	(0.30,0.28)	(0.76,0.38)	(0.70,0.71)
a_8	(0.50,0.62)	(0.56,0.66)	(0.63,0.23)	(0.20,0.54)	(0.28,0.30)	(0.72,0.42)	(0.68,0.69)

Table 13.7 The evaluation results by expert e_3

Enterprises	c_1	c_2	c_3	c_4	c_5	c_6	c_7
a_1	(0.65,0.54)	(0.47,0.58)	(0.73,0.12)	(0.25,0.70)	(0.44,0.46)	(0.83,0.45)	(0.78,0.77)
a_2	(0.66,0.55)	(0.55,0.70)	(0.70,0.16)	(0.30,0.74)	(0.46,0.45)	(0.80,0.55)	(0.67,0.75)
a_3	(0.67,0.54)	(0.46,0.56)	(0.67,0.24)	(0.24,0.72)	(0.35,0.53)	(0.85,0.57)	(0.68,0.73)
a_4	(0.66,0.52)	(0.48,0.70)	(0.68,0.16)	(0.35,0.68)	(0.41,0.43)	(0.75,0.47)	(0.74,0.72)
a_5	(0.64,0.54)	(0.57,0.65)	(0.72,0.11)	(0.33,0.76)	(0.40,0.45)	(0.89,0.44)	(0.72,0.70)
a_6	(0.66,0.55)	(0.58,0.63)	(0.69,0.23)	(0.35,0.75)	(0.43,0.50)	(0.84,0.43)	(0.70,0.73)
a_7	(0.66,0.56)	(0.53,0.58)	(0.69,0.26)	(0.37,0.73)	(0.47,0.52)	(0.85,0.48)	(0.72,0.74)
a_8	(0.67,0.55)	(0.54,0.73)	(0.72,0.15)	(0.32,0.69)	(0.38,0.47)	(0.85,0.55)	(0.70,0.72)

we assume $q = 3$. The original assessment information is shown in Tables 13.5, 13.6, 13.7, 13.8 and 13.9.

With the help of Eqs. (13.4)–(13.7), the criteria weights based on Gini coefficient can be computed as follows (Tables 13.10, 13.11 and 13.12):

Then, by using Eq. (13.8), the standardized trust matrix can be obtained as follows:

Table 13.8 The evaluation results by expert e_4

Enterprises	c_1	c_2	c_3	c_4	c_5	c_6	c_7
a_1	(0.68,0.65)	(0.48,0.54)	(0.76,0.23)	(0.34,0.76)	(0.35,0.43)	(0.75,0.35)	(0.80,0.74)
a_2	(0.60,0.60)	(0.56,0.58)	(0.67,0.22)	(0.36,0.78)	(0.32,0.45)	(0.72,0.44)	(0.82,0.65)
a_3	(0.63,0.68)	(0.50,0.52)	(0.75,0.16)	(0.37,0.73)	(0.38,0.40)	(0.80,0.50)	(0.78,0.67)
a_4	(0.59,0.70)	(0.56,0.69)	(0.72,0.15)	(0.32,0.68)	(0.36,0.37)	(0.73,0.40)	(0.79,0.62)
a_5	(0.70,0.66)	(0.49,0.56)	(0.74,0.24)	(0.35,0.75)	(0.34,0.46)	(0.86,0.45)	(0.80,0.68)
a_6	(0.64,0.63)	(0.47,0.57)	(0.70,0.26)	(0.38,0.77)	(0.42,0.40)	(0.82,0.32)	(0.81,0.74)
a_7	(0.62,0.69)	(0.57,0.65)	(0.72,0.27)	(0.29,0.73)	(0.34,0.39)	(0.81,0.36)	(0.76,0.66)
a_8	(0.60,0.64)	(0.46,0.66)	(0.77,0.16)	(0.26,0.79)	(0.40,0.30)	(0.84,0.48)	(0.82,0.64)

Table 13.9 The evaluation results by expert e_5

Enterprises	c_1	c_2	c_3	c_4	c_5	c_6	c_7
a_1	(0.65,0.68)	(0.56,0.67)	(0.78,0.32)	(0.31,0.80)	(0.36,0.53)	(0.79,0.45)	(0.81,0.75)
a_2	(0.57,0.68)	(0.57,0.66)	(0.65,0.34)	(0.36,0.82)	(0.37,0.50)	(0.70,0.56)	(0.78,0.75)
a_3	(0.58,0.70)	(0.55,0.64)	(0.72,0.35)	(0.35,0.80)	(0.40,0.50)	(0.76,0.57)	(0.82,0.73)
a_4	(0.56,0.73)	(0.58,0.70)	(0.76,0.34)	(0.30,0.76)	(0.43,0.45)	(0.83,0.52)	(0.85,0.70)
a_5	(0.64,0.72)	(0.45,0.63)	(0.78,0.31)	(0.38,0.78)	(0.42,0.48)	(0.85,0.50)	(0.84,0.69)
a_6	(0.59,0.70)	(0.47,0.64)	(0.78,0.30)	(0.40,0.83)	(0.36,0.46)	(0.80,0.43)	(0.77,0.76)
a_7	(0.56,0.65)	(0.52,0.68)	(0.68,0.29)	(0.32,0.81)	(0.34,0.43)	(0.79,0.44)	(0.72,0.73)
a_8	(0.55,0.72)	(0.50,0.69)	(0.69,0.24)	(0.30,0.84)	(0.30,0.47)	(0.81,0.53)	(0.80,0.72)

Table 13.10 The criteria weights

	c_1	c_2	c_3	c_4	c_5	c_6	c_7
Weight	0.1116	0.1280	0.1313	0.2510	0.1489	0.1005	0.1287

Table 13.11 The trust matrix of experts

	e_1	e_2	e_3	e_4	e_5
e_1	0.1660	0.2147	0.1868	0.1764	0.2561
e_2	0.1283	0.1660	0.1443	0.1363	0.1979
e_3	0.1475	0.1908	0.1660	0.1568	0.2276
e_4	0.1561	0.2020	0.1757	0.1660	0.2409
e_5	0.1075	0.1391	0.1210	0.1143	0.1660

Table 13.12 The weights of experts

	e_1	e_2	e_3	e_4	e_5
Weight	0.2012	0.1967	0.1959	0.1978	0.2083

Table 13.13 The aggregated results of all experts

Enterprises	c_1	c_2	c_3	c_4	c_5	c_6	c_7
a_1	(0.63,0.63)	(0.50,0.60)	(0.69,0.24)	(0.27,0.74)	(0.38,0.46)	(0.75,0.39)	(0.70,0.74)
a_2	(0.55,0.63)	(0.55,0.66)	(0.66,0.25)	(0.29,0.77)	(0.36,0.47)	(0.73,0.51)	(0.72,0.71)
a_3	(0.59,0.65)	(0.51,0.64)	(0.69,0.26)	(0.29,0.74)	(0.37,0.47)	(0.77,0.54)	(0.71,0.71)
a_4	(0.57,0.68)	(0.53,0.69)	(0.72,0.25)	(0.28,0.72)	(0.39,0.42)	(0.75,0.46)	(0.71,0.71)
a_5	(0.63,0.66)	(0.48,0.65)	(0.68,0.23)	(0.30,0.74)	(0.35,0.46)	(0.84,0.45)	(0.70,0.68)
a_6	(0.61,0.65)	(0.47,0.63)	(0.69,0.26)	(0.32,0.77)	(0.34,0.45)	(0.80,0.37)	(0.72,0.72)
a_7	(0.59,0.65)	(0.53,0.64)	(0.69,0.26)	(0.31,0.75)	(0.35,0.42)	(0.80,0.43)	(0.73,0.72)
a_8	(0.57,0.65)	(0.50,0.68)	(0.69,0.20)	(0.27,0.78)	(0.34,0.40)	(0.80,0.51)	(0.70,0.69)

Table 13.14 The conditional probabilities of enterprises

	a_1	a_2	a_3	a_4	a_5	a_6	a_7	a_8
Conditional probability	0.5589	0.2927	0.4479	0.5713	0.6830	0.5238	0.6615	0.3404

With the aid of Eqs. (13.9)–(13.11), the weights of experts can be calculated, as shown below:

Next, we aggregate all evaluation values of experts based on q-ROFCWA operator. In this case, experts are assumed as pessimistic and strict. Hence, compensative parameter λ is equal to 0.0001. The results are shown in Table 13.13.

Then, we calculate the conditional probability based on Eqs. (13.14)–*(13.20). In regret theory, the risk aversion parameter θ and the regret aversion parameter δ are both equal to 0.02. The results are shown in Table 13.14.

To assess the risk loss, loss function matrix of enterprises is given by experts, as shown in Table 13.15.

Finally, with the help of decision rules (P)–(N), these platform-based enterprises can be classified into three classes, as shown below:

In Table 13.16, it can be seen that a_1, a_4, a_5 and a_7 are classified as the enterprises with high social responsibility, and a_2 and a_3 are divided into the enterprises with low social responsibility. a_6 and a_8 can not be classified as the enterprises with high or low social responsibility, they need further research by experts. For government agencies, they should ask enterprise a_2 and a_3 to carry out rectification, and implement the CSR in actual operation.

According to this example, our method can effectively evaluate platform-based enterprises with criteria information and risk loss. The enterprises that are classified

Table 13.15 The loss function of enterprises

Enterprises	λ_{PP}	λ_{BP}	λ_{NP}	λ_{PN}	λ_{BN}	λ_{NN}
a_1	0.30	0.54	0.66	0.92	0.71	0.54
a_2	0.32	0.63	0.69	0.90	0.80	0.58
a_3	0.37	0.60	0.68	0.95	0.76	0.63
a_4	0.30	0.58	0.77	0.89	0.75	0.61
a_5	0.26	0.55	0.69	0.91	0.83	0.59
a_6	0.32	0.52	0.76	0.96	0.70	0.60
a_7	0.34	0.65	0.71	0.93	0.82	0.69
a_8	0.38	0.50	0.70	0.85	0.78	0.70

Table 13.16 The classification resuls of enterprises

	High CSR	Uncertain CSR	Low CSR
Enterprises	$\{a_1, a_4, a_5, a_7\}$	$\{a_6, a_8\}$	$\{a_2, a_3\}$

Table 13.17 The results of comparative experiment

	High CSR	Uncertain CSR	Low CSR
Our method	$\{a_1, a_4, a_5, a_7\}$	$\{a_6, a_8\}$	$\{a_2, a_3\}$
Method 1	$\{a_1, a_4, a_5, a_6, a_7, a_8\}$	$\{\emptyset\}$	$\{a_2, a_3\}$
Method 2	$\{a_1, a_4, a_5, a_7, a_8\}$	$\{a_6\}$	$\{a_2, a_3\}$

as high CSR not only have good criteria values but also have low misclassification risk. Hence, our method is very suitable to solve the evaluation problem of platform-based CSR.

13.4.2 Comparative Experiment

In this section, we develop a comparative experiment to compare our method with the related studies. Considering Liu and Wang [17] (Method 1) and Garg and Chen [13] (Method 2) respectively proposed MCDM method of q-ROFSs, we use these methods to calculate the classification results based on our example in Sect. 13.4.1. The results are shown in Table 13.17.

In Table 13.17, the results of three methods are basically consistent, which can prove the effectiveness of our model. Besides, our method classifies a_6 and a_8 as uncertain CSR, while Method 1 and Method 2 respectively classify no objects and a_6 as uncertain CSR. This result shows that compared with Methods 1 and 2, our method is more cautious in classification because more enterprises are assigned to uncertain region. Then we calculate the loss of the results of three methods. Compared with Method 1 (4.3625) and Method 2 (3.8449), our method has the minimum loss (3.2709). The main difference between our method and comparison methods is that the aggregation operator in our method has an attitude factor λ to depict optimism degree or pessimism degree of experts. We can measure experts' attitude preference in aggregation process, while other two methods can not. In this way, our method is appropriate to deal with uncertain and high risk situation, such as the evaluation of platform-based CSR.

13.4.3 Sensitivity Analysis

To verify the influence of parameter value in our method on decision-making results, we conduct some sensitivity analyses. In Sect. 13.3.1.3, parameter λ in q-ROFCWA indicates the optimism degree or pessimism degree of experts. To explore the impact of λ, we change λ from 0.00001 to 10000 and make decisions using our method. The results are shown in Fig. 13.2.

In Fig. 13.2, with the increase of λ, more enterprises are classified into high CSR region, and enterprises in uncertain region are decreased. This may be because when λ increases, experts will become optimistic and have high tolerance for misclassification of uncertain enterprises. Then enterprises in uncertain region will be accepted as high CSR enterprises. This result shows parameter λ will affect the final decision results. Therefore, the value of λ needs to be taken seriously when using q-ROFCWA operator.

In Sect. 13.3.2, we utilize regret theory to depict information from the regret psychology perspective. In this theory, parameter θ denotes the risk aversion degree of experts, while parameter δ denotes the regret aversion degree of experts. To analyze the effect of θ, we change θ from 0.05 to 0.1 in steps of 0.05 and make decisions with our method. δ is fixed at 0.02 in this case. Furthermore, to explore the role of

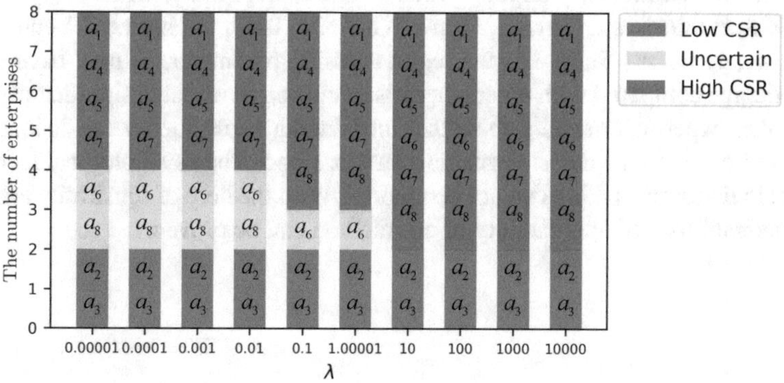

Fig. 13.2 Sensitivity analysis of the parameter λ in q-ROFCWA

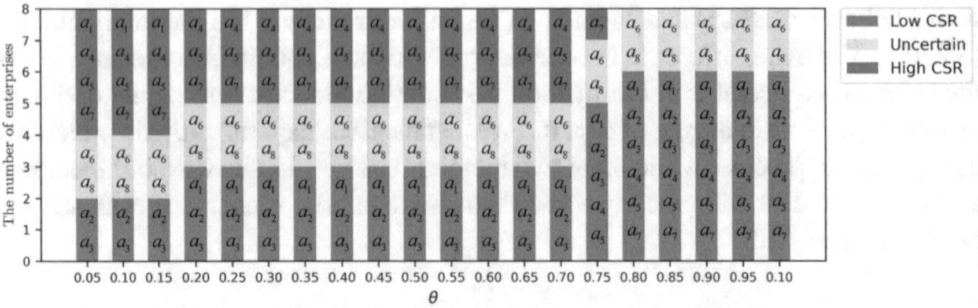

(a) The sensitivity analysis of θ

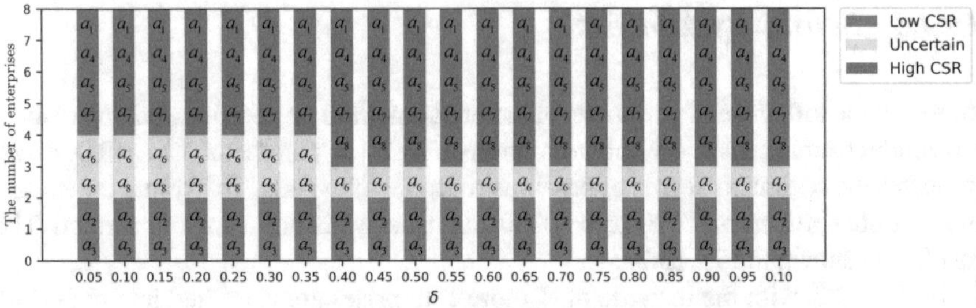

(b) The sensitivity analysis of δ

Fig. 13.3 Sensitivity analysis of the parameter θ and δ in regret theory

δ, we change δ from 0.05 to 0.1 in steps of 0.05 and also make decisions. θ is fixed at 0.02 in this case. The results are shown in Fig. 13.3.

In Fig. 13.3, with the change of parameters, the decision results will also change a lot. In subgraph (a), with the increase of θ, more enterprises move from high CSR region to low CSR region until there are no high CSR enterprises. This may be because the larger θ is, the more the experts attach importance to risk avoidance, and the worse the evaluation values of enterprises based on regret theory will be. As a result, more enterprises will be classified as low CSR. In subgraph (b), the results have changed only slightly when δ varies. Especially, when $\delta \leq 0.35$, the high SCR enterprises are a_1, a_4, a_5, a_7. When $\delta \geq 0.40$, the high SCR enterprises are a_1, a_4, a_5, a_7, a_8. Since δ is the regret aversion parameter, a_8 may have a high regret degree compared with other enterprises and can be affected significantly by δ. Sensitivity experiment results show that introducing regret theory can indeed have a certain impact on the decision-making results. Hence, the psychological factors of experts in decision-making can not be ignored. With the help of sensitivity analysis, the practicability and applicability of our method can be proved.

13.5 Conclusions

In this paper, we propose a new MCDM evaluation method to assess the CSR of platform-based enterprises. To contain more information, we utilize q-ROFSs to evaluate enterprises. We learn the criteria weights and experts' weights based on data distribution. Moreover, the trust relationships of experts are involved. Considering that CWA operator can depict the experts' attitude to objects, we introduce CWA operator into the q-ROFSs environment. With the aid of TWDs method, enterprises are classified into three classes with the minimum expected loss. Our method can evaluate platform-based enterprises with both the criteria evaluation information and the misclassification information, which can assess enterprises in a comprehensive perspective.

This paper provides a method guidance for government agencies or other evaluation institutions to assess platform-based CSR. On the one hand, our paper enriches the MCDM evaluation methods of platform-based CSR. On the other hand, for managers in related evaluation institutions, our method suggests that the psychological factors of experts in evaluation will affect the final decision-making results. When evaluating platform-based CSR, managers should fully consider this irrational behavior to make the final result more reasonable. It is necessary for managers to effectively identify experts' risk preference, optimism or pessimism attitude, and some other psychological factors.

In the future, we can further research on some more realistic scenarios. We can characterize a more complex social network such as the situation where experts' opinions will change after comparing with other experts. Due to the interaction between experts, their evaluation information may change after communication. How opinions change and how opinions reach agreement are worth studying. Meanwhile, we can use more complex evaluation form to assess enterprises, such as interval-valued q-rung orthopair.

Acknowledgements This work is partially supported by the Planning Fund for the Humanities and Social Sciences of the Ministry of Education of China (No. 19YJA630042), the National Natural Science Foundation of China (No. 72071030), the National Key R&D Program of China (No. 2020YFB1711900), and the Social Science Planning Project of the Sichuan Province (No. SC20C007).

References

1. M. Aggarwal, Compensative weighted averaging aggregation operators. Appl. Soft Comput. **28**, 368–378 (2015)
2. A. Baykasoğlu, İ Gölcük, Development of an interval type-2 fuzzy sets based hierarchical MADM model by combining DEMATEL and TOPSIS. Expert Syst. Appl. **70**, 37–51 (2017)
3. F. Boccia, R.M. Manzo, D. Covino, Consumer behavior and corporate social responsibility: an evaluation by a choice experiment. Corporate Soc. Respon. Environ. Manag. **26**(1), 97–105 (2019)

4. A.B. Carroll, A three-dimensional conceptual model of corporate performance. Acad. Manag. Rev. **4**(4), 497–505 (1979)
5. J.Y. Chen, The mirage and politics of participation in China's platform economy. J. Europ. Inst. Commun. Culture **27**(2), 154–170 (2020)
6. C.G. Chorus, Regret theory-based route choices and traffic equilibria. Transportmetrica **8**(4), 291–305 (2012)
7. J. Elkington, Towards the sustainable corporation: win-win-win business strategies for sustainable development. California Manag. Rev. **36**(2), 90–100 (1994)
8. H. Garg, A novel trigonometric operation-based q-rung orthopair fuzzy aggregation operator and its fundamental properties. Neural Comput. Appl. **32**, 15077–15099 (2020)
9. H. Garg, A new possibility degree measure for interval-valued q-rung orthopair fuzzy sets in decision-making. Int. J. Intell. Syst. **36**, 526–557 (2021)
10. H. Garg, CN-q-ROFS: connection number-based q-rung orthopair fuzzy set and their application to decision-making process. Int. J. Intell. Syst. **36**, 3106–3143 (2021)
11. H. Garg, New exponential operation laws and operators for interval-valued q-rung orthopair fuzzy sets in group decision making process. Neural Comput. Appl. (2021). https://doi.org/10.1007/s00521-021-06036-0
12. H. Garg, Z. Ali, Z.L. Yang, T. Mahmood, S. Aljahdali, Multi-criteria decision-making algorithm based on aggregation operators under the complex interval-valued q-rung orthopair uncertain linguistic information. J. Intelli. Fuzzy Syst. (2021). https://doi.org/10.3233/JIFS-210442
13. H. Garg, S.M. Chen, Multiattribute group decision making based on neutrality aggregation operators of q-rung orthopair fuzzy sets. Inform. Sci. **517**, 427–447 (2020)
14. C.L. Hwang, K.S. Yoon, *Multiple Attribute Decision Methods and Applications* (Springer, Berlin, Germany, 1981)
15. Y.B. Li, G.Q. Zhang, L.J. Liu, Platform corporate social responsibility and employee innovation performance: a cross-layer study mediated by employee intrapreneurship. SAGE Open **11**(2), 215824402110214 (2021)
16. P. Liu, J. Liu, Some q-rung orthopai fuzzy Bonferroni mean operators and their application to multi-attribute group decision making. Int. J. Intell. Syst. **33**(2), 315–347 (2018)
17. P.D. Liu, P. Wang, Some q-rung orthopair fuzzy aggregation operators and their applications to multiple-attribute decision making. Int. J. Intell. Syst. **33**, 259–280 (2018)
18. G. Loomes, R. Sugden, Regret theory: an alternative theory of rational choice under uncertainty. Econ. J. **92**(368), 805–824 (1982)
19. J. Mair, G. Reischauer, Capturing the dynamics of the sharing economy: institutional research on the plural forms and practices of sharing economy organizations. Technological Forecasting Soc. Change **125**, 11–20 (2017)
20. I.E. Nikolaou, K. Evangelinos, S. Allan, A reverse logistics social responsibility evaluation framework based on the triple bottom line approach. J. Cleaner Product. **56**, 173–184 (2013)
21. X.D. Peng, J.G. Dai, H. Garg, Exponential operation and aggregation operator for q-rung orthopair fuzzy set and their decision-making method with a new score function. Int. J. Intell. Syst. **33**(11), 2255–2282 (2018)
22. X.D. Peng, L. Liu, Information measures for q-rung orthopair fuzzy sets. Int. J. Intell. Syst. **34**(8), 1795–1834 (2019)
23. M.E. Porter, M.R. Kramer, Strategy and society: the link between competitive advantage and corporate social responsibility. Harv. Bus. Rev. **84**(12), 78–92, 163 (2006)
24. M.P. Pulido, Chapter 5-ISO 26000:2010 guidance on social responsibility: concept and practical application. Ethics Manag. Libraries Other Inform. Serv., 127–168 (2018)
25. M. Riaz, H. Garg, H.M.A. Farid, M. Aslam, Novel q-rung orthopair fuzzy interaction aggregation operators and their application to low-carbon green supply chain management. J. Intell. Fuzzy Syst. (2021). https://doi.org/10.3233/JIFS-210506
26. L. Wang, N. Li, Pythagorean fuzzy interaction power Bonferroni mean aggregation operators in multiple attribute decision making. Int. J. Intell. Syst. **35**(1), 150–183 (2020)
27. L.D. Wang, Y.J. Wang, A.K. Sangaiah, B.Q. Liao, Intuitionistic linguistic group decision-making methods based on generalized compensative weighted averaging aggregation operators. Soft Comput. **22**, 7605–7617 (2018)

28. S.D. Wen, Minkowski-type distance measures for generalized orthopair fuzzy sets. Int. J. Intell. Syst. **33**(4), 802–817 (2018)
29. Y.G. Xue, Y. Deng, H. Garg, Uncertain database retrieval with measure-Based belief function attribute values under intuitionistic fuzzy set. Inform. Sci. **546**(6), 436–447 (2021)
30. R.R. Yager, Generalized orthopair fuzzy sets. IEEE Trans. Fuzzy Syst. **25**(5), 1222–1230 (2017)
31. Z.L. Yang, H. Garg, Interaction power partitioned maclaurin symmetric mean operators under q-rung orthopair uncertain linguistic information. Int. J. Fuzzy Syst. (2021). https://doi.org/10.1007/s40815-021-01062-5
32. Y.Y. Yao, Three-way decisions with probabilistic rough sets. Inform. Sci. **180**(3), 341–353 (2010)
33. D. Yoo, J. Lee, The effects of corporate social responsibility (CSR) fit and CSR consistency on company evaluation: the role of CSR support. Sustainability **10**(8), 2956 (2018)

Decui Liang received the B.S. degree in information management and information system and the Ph.D. degree in management science and engineering from Southwest Jiaotong University, China, in 2008 and 2013, respectively. In 2012, he was a visiting Ph.D. student with the Department of Electrical and Computer Engineering, University of Alberta, Edmonton, AB, Canada. He is currently a Research Professor and doctoral supervisor with the School of Management and Economics, University of Electronic Science and Technology of China. His research interests include three-way decisions, multiple-attribute decision-making with uncertainty, risk analysis and management, social business analysis, text mining and so on.

Wen Cao received the B.S. degree from the Business School, Sichuan Agricultural University, Chengdu, China in 2018. She is currently pursuing the Ph.D. Degree with the School of Management and Economics, University of Electronic Science and Technology of China. Her current research interests include three-way decisions, group decisions and data analysis.

Chapter 14
MARCOS Technique by Using q-Rung Orthopair Fuzzy Sets for Evaluating the Performance of Insurance Companies in Terms of Healthcare Services

Tahir Mahmood and Zeeshan Ali

Abstract Evaluating and ranking confidential fitness assurance enterprises gives assurance organizations, assurance clients, and agencies a consistent mechanism for the coverage of decision-making methods. Additionally, since the planet's coverage region hurts from a gap of assessment of confidential fitness coverage enterprises through the COVID-19 epidemic, the necessity for a consistent, effective, and complete decision instrument is understandable. Appropriately, this manuscript wants to discover coverage corporations' important ranking in conditions of health care public services in Pakistan through the COVID-19 epidemic across a multi-criteria implementation assessment method. In this study, options are assessed and then ranked as per seven principles and evaluations of five specialists. Specialists' assessments and evaluations are completed of ambiguities, under the Measurement of Alternatives and Ranking according to the Compromise Solution (MARCOS) by using the q-rung orthopair fuzzy circumstances. Ultimately, a complete compassion assessment is presented to confirm the future method's permanence and efficiency. The announced methodology met the coverage evaluation dilemma through the COVID-19 pandemic extremely reasonable approach by using the compassion assessment conclusions.

Keywords q-Rung orthopair fuzzy sets · MARCOS method · Healthcare services · Evaluation of the performance of insurance

14.1 Introduction

MADM is an important strategy to choose the best tendency between a gathering of possible results over the variety of various principles. The rule of MADM has widely been utilized in different disciplines. Vulnerability is commonly happened, all things considered, issues due to the responsibility of different messed up hindrances,

T. Mahmood (✉) · Z. Ali
Department of Mathematics & Statistics, International Islamic University Islamabad, Islamabad, Pakistan
e-mail: tahirbakhat@iiu.edu.pk

H. Garg (ed.), *q-Rung Orthopair Fuzzy Sets*,
https://doi.org/10.1007/978-981-19-1449-2_14

detachment of information/information, and whim of the test. To manage this worry, the principle of fuzzy sets (FSs) was organized by Zadeh [1], as a talented strategy to deal with the concerns in the information, which works reality grade (TG) to explain the arrangement. Later their reality, particular specialists began to examine it extensively and genuinely to determine MADM inconveniences. For example, the complex FS (CFS) was developed by Ramot et al. [2]. Liu et al. [3] investigated the distance and cross-entropy measures by using CFS. Chiang and Lin [4] proposed correlation fuzzy sets. Brown [5] explored a note on FS. Torra [6] modified the FS is to elaborate the hesitant FS (HFS). Mahmood [7] utilized the idea of bipolar soft sets.

If somebody challenged such sort of information that covers the TG and FG, then the principle of FS has been failed. To resolve these sorts of troubles, the principle of intuitionistic FS (IFS), invented by Atanassov [8]. Further, the principle of interval-valued IFS (IV-IFS) was organized by Atanassov [9]. Xia et al. [10] elaborated the aggregation operators for IFSs. Garg and Kumar [11] initiated the TOPSIS method by using exponential distance measures based on IN-IFSs with connection numbers. The generalized improved score values by using the IV-IFSs were developed by Garg [12]. Garg and Rani [13] proposed similarity measures for IFSs. Ejegwa et al. [14] developed an improved correlation coefficient for IFSs. Sanam et al. [15] initiated the Einstein aggregation operators for IFSs. Kar et al. [16] investigated the trapezoidal intuitionistic types-2 information and their application in decision making. De [17] investigated the intuitionistic fuzzy metric distances. Fahmi et al. [18] organized the geometric operators by using the interval-valued intuitionistic neutrosophic fuzzy information.

Further, Yager [19] invented the Pythagorean FSs (PFSs) with $\mathbb{T}^2_{\mathcal{F}_{\text{pfs}}} + \mathbb{F}^2_{\mathcal{F}_{\text{pfs}}} \leq 1$. Later, their reality, particular agents began to examine it comprehensively and earnestly to determine MADM inconveniences. For example, Sarkar and Biswas [20] initiated the principle of the Pythagorean fuzzy AHP-TOPSIS method. Zulqarnain et al. [21] developed the TOPSIS method based on Pythagorean fuzzy hypersoft sets. Tang and Yang [22] initiated the interval-valued Pythagorean fuzzy multi-attribute group decision-making troubles. Click [23] determined the green supplier selection in the industry 4.0 era by using the Pythagorean fuzzy TOPSIS methods. Bakioglu and Atahan [24] elaborated the AHP, TOPSIS, and VIKOR methods by using the PFSs. Ejegwa et al. [25] investigated the correlation measures, variance, and deviation by using PFSs. Batool [26] explored the principle of Pythagorean probabilistic hesitant fuzzy information and its application in decision-making. Garg [27, 28] investigated the linguistic Pythagorean fuzzy sets and the novel correlation measures by using the PFSs. For simplicity, we have drawn the geometrical expressions of the IFSs and PFSs in the form of Fig. 14.1.

Yager [29] invented the q-rung orthopair FSs (q-ROFSs) with $\mathbb{T}^{\text{q}sc}_{\mathcal{F}_{qfs}} + \mathbb{F}^{\text{q}sc}_{\mathcal{F}_{qfs}} \leq 1$. Keeping the superiority of the q-ROFSs, Garg [30] initiated the principle of possible measures for q-ROFSs. Mahmood and Ali [31] developed the entropy measures, TOPSIS method, and correlation measures by using complex q-ROFSs. Riaz et al. [32] elaborated the principle of q-rung orthopair m-polar fuzzy information. Liu et al. [33] investigated the Banzhaf–Choquet-copula aggregation operators by using the

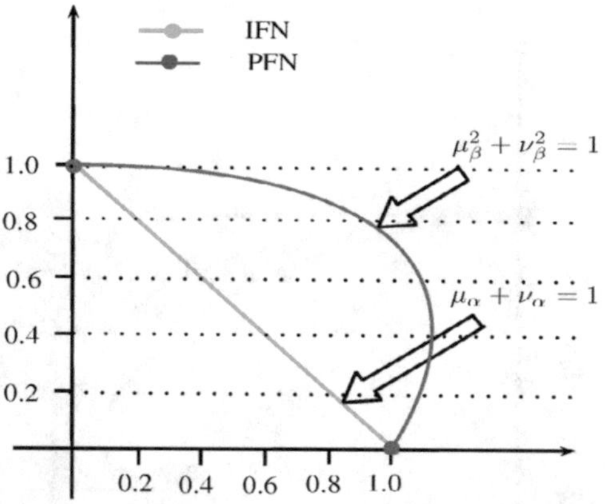

Fig. 14.1 Graphical expressions of the IFSs and PFSs

principle of q-ROFSs. Mahmood and Ali [34] investigated the principle of Hamacher aggregation operators by using the complex q-ROFSs. Pinar et al. [35] invented the q-rung orthopair fuzzy TOPSIS methods. Ali and Mahmood [36] maclurin symmetric mean operators by using the complex q-ROFSs. Jana et al. [37] diagnosed the Dombi aggregation operators by using the principle of q-ROFSs. The principle of the MARCOS method was developed by Stevic et al. [38]. Further, Bakir and Atalik [39] initiated the fuzzy AHP and fuzzy MARCOS techniques. Celik and Gul [40] utilized the MARCOS technique in the environment of type-2 fuzzy sets. Stankovic et al. [41] elaborated on the new fuzzy MARCOS techniques and their application in road traffic analysis. Ecer and Pamucar [42] introduced the principle of intuitionistic fuzzy MARCOS technique and their application in COVID-19. For simplicity, we have drawn the geometrical expressions of the IFSs, PFSs, and q-ROFSs in the form of Fig. 14.2.

Each association needs to contribute for it to develop, and speculations are made through projects. In this way, speculation the board is performed by applying project the executives' strategies. Distinctive venture the board programming programs are utilized to deal with numerous undertakings. There is a great deal of undertaking the board programming available, and four bits of the product were chosen and dissected. In this paper, the medical care administrations in the period of COVID-19 evaluated by the recipients of these undertakings in the United Arab Emirates are investigated. The examination needed for this review was led in the United Arab Emirates. The MARCOS technique was utilized to assess the sicknesses. The outcomes showed that Smartsheet had been appraised the risks by indications. This paper gives an outline of how multi-rules examination techniques can be utilized when positioning medical care administrations in the period of COVID-19. Consequently, researching the display of protection offices in an especially troublesome measure is an exceptionally basic issue. The graphical articulations of the started works are examined as Fig. 14.3.

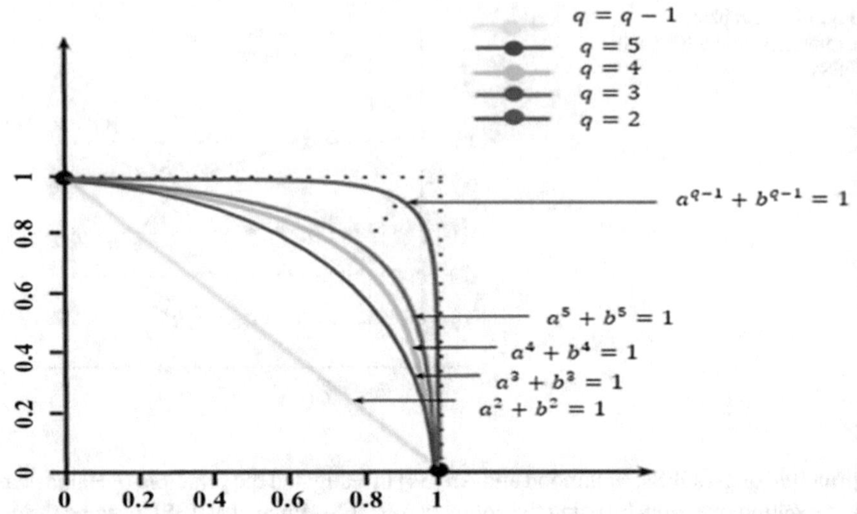

Fig. 14.2 Geometrical expressions of the IFSs, PFSs, and q-ROFSs

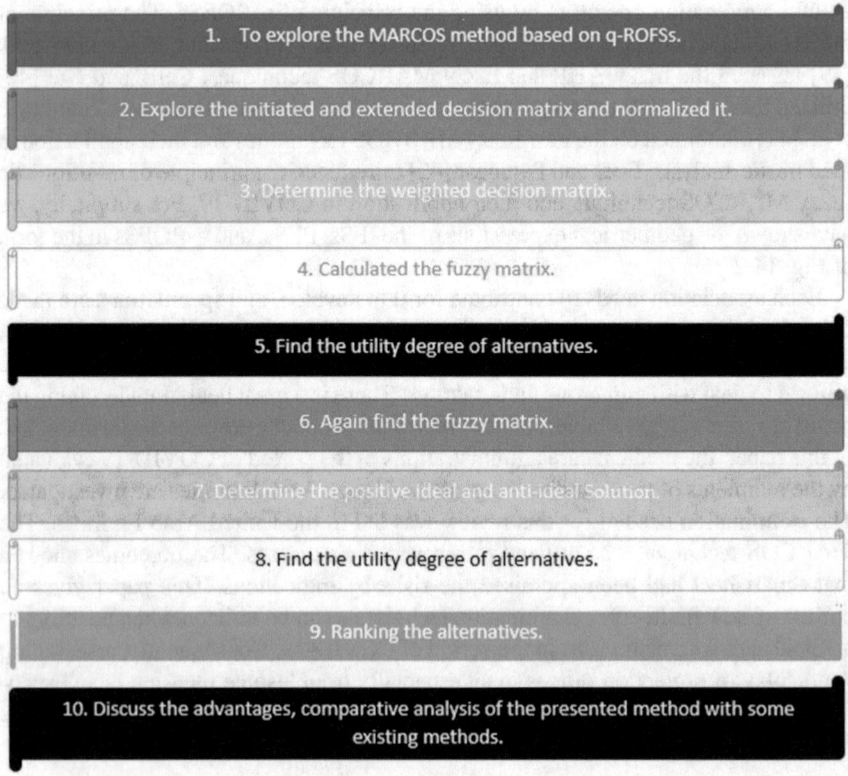

Fig. 14.3 Geometrical expression of the intuited works

Assessment of healthcare coverage enterprises, particularly concerning buyer-positioned considerations, has developed a crucial examination region of an abundance of coverage investigation facilities throughout the COVID-19 disease. This is predominantly due to the incontrovertible significance of health protection on the investment marketplace and the crucial responsibility of maintenance superiority in client happiness. This dilemma has remained one of the extremely crucial topics associated with the restitution industry, which has not been suitably investigated in the novel. In brief, this analysis is organized throughout the subsequent principal purposes:

1. To elaborate the MARCOS method by using the q-ROFSs and discussed their special cases.
2. By using the investigated q-rung orthopair fuzzy-MARCOS (q-ROF-MARCOS) method, we discuss some COVID-19 situations. Ultimately, a complete compassion assessment is presented to confirm the future method's permanence and efficiency. The announced methodology met the coverage evaluation dilemma through the COVID-19 pandemic extremely reasonable approach by using the compassion assessment conclusions.
3. To compare the elaborated approaches with some prevailing methods such as the TOPSIS method and with some measures.
4. To improve the quality of the elaborated methods, the geometrical expressions are also discussed in detail.

Main structure of analysis is of the shape: Sect. 14.2 talk about some prevailing studies. In Sect. 14.3, the q-ROF-MARCOS technique is produced in point. In Sect. 14.3, we also demonstrated the MADM technique based on the q-ROF-MARCOS technique. In Sect. 14.4, we invented the supremacy of the explored works. The finishing of the manuscript is discussed in Sect. 14.5.

14.2 Preliminaries

This study aims to briefly review some basic ideas such as IFSs, PFSs, q-ROFSs, and their useful laws. Additionally, in the overall manuscript the symbols $\mathbb{X}_{\text{uni}}$, stated the fixed set and the value of truth grade (TG) and falsity grade (FG) are denoted by $\mathbb{T}_{\mathcal{F}_{\text{fs}}}$ and $\mathbb{F}_{\mathcal{F}_{\text{fs}}}$. By using the above information, the principle of FS was developed by Zadeh [1], which covers the TG with a rule that is $\mathbb{T}_{\mathcal{F}_{\text{fs}}} \in [0, 1]$. But there were some troubles, if a person faced such sort of information that covers the TG and FG, then the principle of FS has been failed. To resolve these sorts of troubles, the principle of IFS was developed by Atanassov [8] is discussed below.

Definition 14.1 [8] An IFS $\mathcal{F}_{\text{ifs}}$ is stated by

$$\mathcal{F}_{ifs} = \left\{\left(\mathbb{T}_{\mathcal{F}_{ifs}}(\mathbb{x}_{el}), \mathbb{F}_{\mathcal{F}_{ifs}}(\mathbb{x}_{el})\right) : \mathbb{x}_{el} \in \mathbb{x}_{uni}\right\} \tag{14.1}$$

With a rule that is the sum of TG $\mathbb{T}_{\mathcal{F}_{\mathrm{ifs}}}$ and FG $\mathbb{F}_{\mathcal{F}_{\mathrm{ifs}}}$ is limited to [0,1], i.e., $0 \leq \mathbb{T}_{\mathcal{F}_{ifs}}(\mathbb{x}_{el}) + \mathbb{F}_{\mathcal{F}_{ifs}}(\mathbb{x}_{el}) \leq 1$. The refusal grade (RG) is stated by $\pi_{\mathcal{F}_{ifs}}(\mathbb{x}_{el}) = 1 - \left(\mathbb{T}_{\mathcal{F}_{ifs}}(\mathbb{x}_{el}) + \mathbb{F}_{\mathcal{F}_{ifs}}(\mathbb{x}_{el})\right)$. In the overall manuscript, the IFNs are stated by $\mathcal{F}_{ifs-i} = \left(\mathbb{T}_{\mathcal{F}_{ifs-i}}, \mathbb{F}_{\mathcal{F}_{ifs-i}}\right), i = 1, 2, \ldots, \mathbb{n}_{le}$.

If an expert obtained data in the shape $\mathbb{T}_{\mathcal{F}_{ifs}}(\mathbb{x}_{el}) + \mathbb{F}_{\mathcal{F}_{ifs}}(\mathbb{x}_{el}) > 1$, then the IFS has been unsuccessful. For example, $0.6 + 0.5 = 1.1 > 1$. To manage with these sorts of hurdles, the PFS, invented by Yager [19].

Definition 14.2 [19] A PFS $\mathcal{F}_{\mathrm{pfs}}$ is stated by

$$\mathcal{F}_{pfs} = \left\{\left(\mathbb{T}_{\mathcal{F}_{pfs}}(\mathbb{x}_{el}), \mathbb{F}_{\mathcal{F}_{pfs}}(\mathbb{x}_{el})\right) : \mathbb{x}_{el} \in \mathbb{x}_{uni}\right\} \tag{14.2}$$

With $0 \leq \mathbb{T}^2_{\mathcal{F}_{pfs}}(\mathbb{x}_{el}) + \mathbb{F}^2_{\mathcal{F}_{pfs}}(\mathbb{x}_{el}) \leq 1$. The refusal grade (RG) is stated by $\pi_{\mathcal{F}_{pfs}}(\mathbb{x}_{el}) = \left(1 - \left(\mathbb{T}^2_{\mathcal{F}_{pfs}}(\mathbb{x}_{el}) + \mathbb{F}^2_{\mathcal{F}_{pfs}}(\mathbb{x}_{el})\right)\right)^{\frac{1}{2}}$. In the overall manuscript, the PFNs are stated by $\mathcal{F}_{pfs-i} = \left(\mathbb{T}_{\mathcal{F}_{pfs-i}}, \mathbb{F}_{\mathcal{F}_{pfs-i}}\right), i = 1, 2, \ldots, \mathbb{n}_{le}$.

If an expert obtained data in the shape $\mathbb{T}^2_{\mathcal{F}_{pfs}}(\mathbb{x}_{el}) + \mathbb{F}^2_{\mathcal{F}_{pfs}}(\mathbb{x}_{el}) > 1$, then the IFS has been unsuccessful. For example, $0.9^2 + 0.8^2 = 0.81 + 0.64 = 1.45 > 1$. To manage with these sorts of hurdles, the q-ROFS, invented by Yager [29].

Definition 14.3 [29] A q-ROFS $\mathcal{F}_{\mathrm{qfs}}$ is stated by

$$\mathcal{F}_{qfs} = \left\{\left(\mathbb{T}_{\mathcal{F}_{qfs}}(\mathbb{x}_{el}), \mathbb{F}_{\mathcal{F}_{qfs}}(\mathbb{x}_{el})\right) : \mathbb{x}_{el} \in \mathbb{x}_{uni}\right\} \tag{14.3}$$

With $0 \leq \mathbb{T}^{\mathsf{q}_{SC}}_{\mathcal{F}_{qfs}}(\mathbb{x}_{el}) + \mathbb{F}^{\mathsf{q}_{SC}}_{\mathcal{F}_{qfs}}(\mathbb{x}_{el}) \leq 1, \mathsf{q}_{SC} \geq 1$. The refusal grade (RG) is stated by $\pi_{\mathcal{F}_{qfs}}(\mathbb{x}_{el}) = \left(1 - \left(\mathbb{T}^{\mathsf{q}_{SC}}_{\mathcal{F}_{qfs}}(\mathbb{x}_{el}) + \mathbb{F}^{\mathsf{q}_{SC}}_{\mathcal{F}_{qfs}}(\mathbb{x}_{el})\right)\right)^{\frac{1}{\mathsf{q}_{SC}}}$. In the overall manuscript, the q-ROFNs are stated by $\mathcal{F}_{qfs-i} = \left(\mathbb{T}_{\mathcal{F}_{qfs-i}}, \mathbb{F}_{\mathcal{F}_{qfs-i}}\right), i = 1, 2, \ldots, \mathbb{n}_{le}$.

For $\mathsf{q}_{SC} = 5$, we obtained easily, such that $0.9^5 + 0.8^5 = 0.59049 + 0.32768 = 0.91817 \leq 1$. For any two q-ROFNs $\mathcal{F}_{\mathrm{qfs}-1} = \left(\mathbb{T}_{\mathcal{F}_{\mathrm{qfs}-1}}, \mathbb{F}_{\mathcal{F}_{\mathrm{qfs}-1}}\right)$ and $\mathcal{F}_{\mathrm{qfs}-2} = \left(\mathbb{T}_{\mathcal{F}_{\mathrm{qfs}-2}}, \mathbb{F}_{\mathcal{F}_{\mathrm{qfs}-2}}\right)$, then

$$\mathcal{F}_{qfs-1} \oplus \mathcal{F}_{qfs-2} = \left(\left(\mathbb{T}^{\mathsf{q}_{SC}}_{\mathcal{F}_{qfs-1}} + \mathbb{T}^{\mathsf{q}_{SC}}_{\mathcal{F}_{qfs-2}} - \mathbb{T}^{\mathsf{q}_{SC}}_{\mathcal{F}_{qfs-1}}\mathbb{T}^{\mathsf{q}_{SC}}_{\mathcal{F}_{qfs-2}}\right)^{\frac{1}{\mathsf{q}_{SC}}}, \mathbb{F}_{\mathcal{F}_{qfs-1}}\mathbb{F}_{\mathcal{F}_{qfs-2}}\right) \tag{14.4}$$

$$\mathcal{F}_{qfs-1} \otimes \mathcal{F}_{qfs-2} = \left(\mathbb{T}_{\mathcal{F}_{qfs-1}}\mathbb{T}_{\mathcal{F}_{qfs-2}}, \left(\mathbb{F}^{\mathsf{q}_{SC}}_{\mathcal{F}_{qfs-1}} + \mathbb{F}^{\mathsf{q}_{SC}}_{\mathcal{F}_{qfs-2}} - \mathbb{F}^{\mathsf{q}_{SC}}_{\mathcal{F}_{qfs-1}}\mathbb{F}^{\mathsf{q}_{SC}}_{\mathcal{F}_{qfs-2}}\right)^{\frac{1}{\mathsf{q}_{SC}}}\right) \tag{14.5}$$

$$\psi_{SC}\mathcal{F}_{qfs-1} = \left(\left(1 - \left(1 - \mathbb{T}^{\mathsf{q}_{SC}}_{\mathcal{F}_{qfs-1}}\right)^{\psi_{SC}}\right)^{\frac{1}{\mathsf{q}_{SC}}}, \mathbb{F}^{\psi_{SC}}_{\mathcal{F}_{qfs-1}}\right) \tag{14.6}$$

$$\mathcal{F}_{qfs-1}^{\psi_{SC}} = \left(\mathbb{T}_{\mathcal{F}_{qfs-1}}^{\psi_{SC}}, \left(1 - \left(1 - \mathbb{F}_{\mathcal{F}_{qfs-1}}^{\mathsf{q}_{SC}}\right)^{\psi_{SC}}\right)^{\frac{1}{\mathsf{q}_{SC}}}\right) \tag{14.7}$$

Example 14.1 For any two q-ROFNs $\mathcal{F}_{\text{qfs}-1} = (0.9, 0.8)$ and $\mathcal{F}_{\text{qfs}-2} = (0.7, 0.6)$, then by using the Eqs. (14.4)–(14.7) for $\psi_{SC} = 2$, we get

$$\mathcal{F}_{\text{qfs}-1} \oplus \mathcal{F}_{\text{qfs}-2} = \left(\left(0.9^5 + 0.7^5 - 0.9^5 * 0.7^5\right)^{\frac{1}{5}}, 0.8 * 0.6\right) = (0.92007, 0.48)$$

$$\mathcal{F}_{\text{qfs}-1} \otimes \mathcal{F}_{\text{qfs}-2} = \left(0.9 * 0.7, \left(0.8^5 + 0.6^5 - 0.8^5 0.6^5\right)^{\frac{1}{5}}\right) = (0.63, 0.82404)$$

$$\psi_{SC}\mathcal{F}_{\text{qfs}-1} = 2 * \mathcal{F}_{\text{qfs}-1} = \left(\left(1 - \left(1 - 0.9^5\right)^2\right)^{\frac{1}{5}}, 0.8^5\right) = (0.96395, 0.64)$$

$$\mathcal{F}_{\text{qfs}-1}^{\psi_{SC}} = \mathcal{F}_{\text{qfs}-1}^{2} = \left(0.9^2, \left(1 - \left(1 - 0.8^5\right)^2\right)^{\frac{1}{5}}\right) = (0.81, 0.88665)$$

For $\mathsf{q}_{SC} = 1$, Eqs. (14.4)–(14.7) are reduced for IFSs [8], and similarly, for $\mathsf{q}_{SC} = 2$, Eqs. (14.4)–(14.7) is reduced for PFSs [19]. Moreover, we invented the score function (SF) and accuracy function (AF), such that

$$\mathbb{S}^{sf}\left(\mathcal{F}_{qfs-1}\right) = \mathbb{T}_{\mathcal{F}_{qfs-1}}^{\mathsf{q}_{SC}} - \mathbb{F}_{\mathcal{F}_{qfs-1}}^{\mathsf{q}_{SC}}, \mathbb{S}^{sf}\left(\mathcal{F}_{qfs-1}\right) \in [-1, 1] \tag{14.8}$$

$$\mathbb{H}^{af}\left(\mathcal{F}_{qfs-1}\right) = \mathbb{T}_{\mathcal{F}_{qfs-1}}^{\mathsf{q}_{SC}} + \mathbb{F}_{\mathcal{F}_{qfs-1}}^{\mathsf{q}_{SC}}, \mathbb{H}^{af}\left(\mathcal{F}_{qfs-1}\right) \in [0, 1] \tag{14.9}$$

For two q-ROFNs $\mathcal{F}_{\text{qfs}-1} = \left(\mathbb{T}_{\mathcal{F}_{\text{qfs}-1}}, \mathbb{F}_{\mathcal{F}_{\text{qfs}-1}}\right)$ and $\mathcal{F}_{\text{qfs}-2} = \left(\mathbb{T}_{\mathcal{F}_{\text{qfs}-2}}, \mathbb{F}_{\mathcal{F}_{\text{qfs}-2}}\right)$,

1. If $\mathbb{S}^{\text{sf}}\left(\mathcal{F}_{\text{qfs}-1}\right) > \mathbb{S}^{\text{sf}}\left(\mathcal{F}_{\text{qfs}-2}\right) \Rightarrow \mathcal{F}_{\text{qfs}-1} > \mathcal{F}_{\text{qfs}-2}$;
2. If $\mathbb{S}^{\text{sf}}\left(\mathcal{F}_{\text{qfs}-1}\right) < \mathbb{S}^{\text{sf}}\left(\mathcal{F}_{\text{qfs}-2}\right) \Rightarrow \mathcal{F}_{\text{qfs}-1} < \mathcal{F}_{\text{qfs}-2}$;
3. If $\mathbb{S}^{\text{sf}}\left(\mathcal{F}_{\text{qfs}-1}\right) = \mathbb{S}^{\text{sf}}\left(\mathcal{F}_{\text{qfs}-2}\right) \Rightarrow$

   (1) If $\mathbb{H}^{\text{af}}\left(\mathcal{F}_{\text{qfs}-1}\right) > \mathbb{H}^{\text{af}}\left(\mathcal{F}_{\text{qfs}-2}\right) \Rightarrow \mathcal{F}_{\text{qfs}-1} > \mathcal{F}_{\text{qfs}-2}$;
   (2) If $\mathbb{H}^{\text{af}}\left(\mathcal{F}_{\text{qfs}-1}\right) < \mathbb{H}^{\text{af}}\left(\mathcal{F}_{\text{qfs}-2}\right) \Rightarrow \mathcal{F}_{\text{qfs}-1} < \mathcal{F}_{\text{qfs}-2}$;
   (3) If $\mathbb{H}^{\text{af}}\left(\mathcal{F}_{\text{qfs}-1}\right) = \mathbb{H}^{\text{af}}\left(\mathcal{F}_{\text{qfs}-2}\right) \Rightarrow \mathcal{F}_{\text{qfs}-1} = \mathcal{F}_{\text{qfs}-2}$.

14.3 Q-ROF-MARCOS Method

The major influence of this study is to develop the q-ROF-MARCOS techniques. The principle of the MARCOS method was firstly developed by Stevic et al. [38]. Moreover, Bakir and Atalik [39] initiated the fuzzy AHP and fuzzy MARCOS techniques. Celik and Gul [40] utilized the MARCOS technique in the environment of type-2 fuzzy sets. Stankovic et al. [41] elaborated on the new fuzzy MARCOS techniques and their application in road traffic analysis. Ecer and Pamucar [42] introduced the principle of intuitionistic fuzzy MARCOS technique and their application in COVID-19. The procedure of the q-ROF-MARCOS technique is elaborated with the help of the following cases.

Case 1: Based on the family of alternative m and their criteria n, we developed the decision matrix with every entry in the form of q-ROFNs.

Case 2: To determine the ideal $\check{\mathfrak{A}}(\mathfrak{A}\mathfrak{I})$ and anti-ideal $\check{\mathfrak{A}}(\mathfrak{I}\mathfrak{D})$, we developed the modified decision matrix, such that

$$\check{\mathbb{X}}_{dm} = \begin{matrix} \check{\mathfrak{A}}(\mathfrak{A}\mathfrak{I}) \\ \check{\mathfrak{A}}_1 \\ \check{\mathfrak{A}}_2 \\ \dots \\ \check{\mathfrak{A}}_m \\ \check{\mathfrak{A}}(\mathfrak{I}\mathfrak{D}) \end{matrix} \overset{\begin{matrix} \check{\mathcal{C}}_1 & \quad\quad\quad\quad \check{\mathcal{C}}_2 & \quad\quad \dots\dots & \quad\quad \check{\mathcal{C}}_4 \end{matrix}}{\begin{bmatrix} \left(\mathbb{T}_{\mathcal{F}_{qfs-ai1}}, \mathbb{F}_{\mathcal{F}_{qfs-ai1}}\right) & \left(\mathbb{T}_{\mathcal{F}_{qfs-ai2}}, \mathbb{F}_{\mathcal{F}_{qfs-ai2}}\right) & \dots & \left(\mathbb{T}_{\mathcal{F}_{qfs-ain}}, \mathbb{F}_{\mathcal{F}_{qfs-ain}}\right) \\ \left(\mathbb{T}_{\mathcal{F}_{qfs-11}}, \mathbb{F}_{\mathcal{F}_{qfs-11}}\right) & \left(\mathbb{T}_{\mathcal{F}_{qfs-12}}, \mathbb{F}_{\mathcal{F}_{qfs-12}}\right) & \dots & \left(\mathbb{T}_{\mathcal{F}_{qfs-1n}}, \mathbb{F}_{\mathcal{F}_{qfs-1n}}\right) \\ \left(\mathbb{T}_{\mathcal{F}_{qfs-21}}, \mathbb{F}_{\mathcal{F}_{qfs-21}}\right) & \left(\mathbb{T}_{\mathcal{F}_{qfs-22}}, \mathbb{F}_{\mathcal{F}_{qfs-22}}\right) & \dots & \left(\mathbb{T}_{\mathcal{F}_{qfs-2n}}, \mathbb{F}_{\mathcal{F}_{qfs-2n}}\right) \\ \dots & \dots & \dots & \dots \\ \left(\mathbb{T}_{\mathcal{F}_{qfs-m1}}, \mathbb{F}_{\mathcal{F}_{qfs-m1}}\right) & \left(\mathbb{T}_{\mathcal{F}_{qfs-m2}}, \mathbb{F}_{\mathcal{F}_{qfs-m2}}\right) & \dots & \left(\mathbb{T}_{\mathcal{F}_{qfs-mn}}, \mathbb{F}_{\mathcal{F}_{qfs-mn}}\right) \\ \left(\mathbb{T}_{\mathcal{F}_{qfs-id1}}, \mathbb{F}_{\mathcal{F}_{qfs-id1}}\right) & \left(\mathbb{T}_{\mathcal{F}_{qfs-id2}}, \mathbb{F}_{\mathcal{F}_{qfs-id2}}\right) & \dots & \left(\mathbb{T}_{\mathcal{F}_{qfs-idn}}, \mathbb{F}_{\mathcal{F}_{qfs-idn}}\right) \end{bmatrix}} \tag{14.10}$$

where $\check{\mathfrak{A}}(\mathfrak{A}\mathfrak{I})$ expressed the cost alternative and $\check{\mathfrak{A}}(\mathfrak{I}\mathfrak{D})$ is expressed the benefit alternatives. By using the cost and benefit, $\check{\mathfrak{A}}(\mathfrak{A}\mathfrak{I})$ and $\check{\mathfrak{A}}(\mathfrak{I}\mathfrak{D})$ are obtained by using Eqs. (14.11) and (14.12), we have

$$\check{\mathfrak{A}}(\mathfrak{A}\mathfrak{I}) = \left(\max_i \mathbb{T}_{\mathcal{F}_{qfs-ij}}, \min_i \mathbb{F}_{\mathcal{F}_{qfs-ij}}\right) for\ cost\ types \tag{14.11}$$

$$\check{\mathfrak{A}}(\mathfrak{I}\mathfrak{D}) = \left(\min_i \mathbb{T}_{\mathcal{F}_{qfs-ij}}, \max_i \mathbb{F}_{\mathcal{F}_{qfs-ij}}\right) for\ benefit\ types \tag{14.12}$$

Case 3: By using Eqs. (14.13) and (14.14), we standardized the decision matrix $\overline{\overline{N}} = \left[\mathcal{F}_{\text{qfs - ij}}\right]$, we have

$$\mathcal{F}_{\text{qfs - ij}} = \left(\mathbb{T}_{\mathcal{F}_{\text{qfs - ij}}}, \mathbb{F}_{\mathcal{F}_{\text{qfs - ij}}}\right) for\ benefit\ types \tag{14.13}$$

$$\mathcal{F}_{\text{qfs - ij}} = \left(\mathbb{F}_{\mathcal{F}_{\text{qfs - ij}}}, \mathbb{T}_{\mathcal{F}_{\text{qfs - ij}}}\right) for\ cost\ types \tag{14.14}$$

Case 4: By using the information of the decision matrix $\overline{\overline{N}} = \left[\mathcal{F}_{\text{qfs - ij}}\right]$, we choose some weight vectors $\Xi = \{\Xi_1, \Xi_2, \dots, \Xi_n\}$ with a rule, that is, $\sum_{j=1}^{n} \Xi_j = 1, \Xi_j \in [0, 1]$ to invent the weighted from the decision matrix under the Eq. (14.15), such that

$$\Xi_j \mathcal{F}_{qfs-ij} = \left(\left(1 - \left(1 - \mathbb{T}^{q_{SC}}_{\mathcal{F}_{qfs-ij}}\right)^{\Xi_j}\right)^{\frac{1}{q_{SC}}}, \mathbb{F}^{\Xi_j}_{\mathcal{F}_{qfs-ij}}\right) \tag{14.15}$$

Case 5: By using Eqs. (14.16) and (14.17), we determine the positive and negative ideal solutions based on the information of the weighted decision matrix, such that

$$\mathcal{F}^{+}_{\text{qfs}} = \left\{ \begin{array}{l} \left(\max\limits_{i} \mathbb{T}_{\mathcal{F}_{\text{qfs - i1}}}, \min\limits_{i} \mathbb{F}_{\mathcal{F}_{\text{qfs - i1}}} \right), \left(\max\limits_{i} \mathbb{T}_{\mathcal{F}_{\text{qfs - i2}}}, \min\limits_{i} \mathbb{F}_{\mathcal{F}_{\text{qfs - i2}}} \right), \ldots, \\ \left(\max\limits_{i} \mathbb{T}_{\mathcal{F}_{\text{qfs - im}}}, \min\limits_{i} \mathbb{F}_{\mathcal{F}_{\text{qfs - im}}} \right) \end{array} \right\} \tag{14.16}$$

$$\mathcal{F}^{-}_{\text{qfs}} = \left\{ \begin{array}{l} \left(\min\limits_{i} \mathbb{T}_{\mathcal{F}_{\text{qfs - i1}}}, \max\limits_{i} \mathbb{F}_{\mathcal{F}_{\text{qfs - i1}}} \right), \left(\min\limits_{i} \mathbb{T}_{\mathcal{F}_{\text{qfs - i2}}}, \max\limits_{i} \mathbb{F}_{\mathcal{F}_{\text{qfs - i2}}} \right), \ldots, \\ \left(\min\limits_{i} \mathbb{T}_{\mathcal{F}_{\text{qfs - im}}}, \max\limits_{i} \mathbb{F}_{\mathcal{F}_{\text{qfs - im}}} \right) \end{array} \right\} \tag{14.17}$$

Case 6: By using the information in positive ideal solution $\mathcal{F}^{+}_{\text{qfs}}$ and negative ideal solution $\mathcal{F}^{-}_{\text{qfs}}$, we examine the fuzzy matrix $\overline{\overline{T}}_i$ based on the Eq. (14.18), such that

$$\begin{aligned} \overline{\overline{T}}_i &= \mathcal{F}^{+}_{qfs} \oplus \mathcal{F}^{-}_{qfs} = \left(\mathbb{T}^{+}_{\mathcal{F}_{qfs-ij}}, \mathbb{F}^{+}_{\mathcal{F}_{qfs-ij}} \right) \oplus \left(\mathbb{T}^{-}_{\mathcal{F}_{qfs-ij}}, \mathbb{F}^{-}_{\mathcal{F}_{qfs-ij}} \right) \\ &= \left(\left(\mathbb{T}^{+\mathrm{q}_{SC}}_{\mathcal{F}_{qfs-ij}} + \mathbb{T}^{-\mathrm{q}_{SC}}_{\mathcal{F}_{qfs-ij}} - \mathbb{T}^{+\mathrm{q}_{SC}}_{\mathcal{F}_{qfs-ij}} \mathbb{T}^{-\mathrm{q}_{SC}}_{\mathcal{F}_{qfs-ij}} \right)^{\frac{1}{\mathrm{q}_{SC}}}, \mathbb{F}^{+}_{\mathcal{F}_{qfs-ij}} \mathbb{F}^{-}_{\mathcal{F}_{qfs-ij}} \right) \end{aligned} \tag{14.18}$$

Case 7: By using the fuzzy decision matrix $\overline{\overline{T}}_i$, we invented the q-ROFNs $\mathcal{F}^{\text{num}}_{\text{qfs}}$ by using the Eq. (14.19), such that

$$\mathcal{F}^{\text{num}}_{\text{qfs}} = \max\left(\overline{\overline{T}}_{\mathrm{i}} \right) = \left(\max\limits_{\mathrm{i}} \mathbb{T}_{\mathcal{F}_{\overline{\overline{T}}_{\mathrm{i}}}}, \min\limits_{\mathrm{i}} \mathbb{F}_{\mathcal{F}_{\overline{\overline{T}}_{\mathrm{i}}}} \right) \tag{14.19}$$

Then by using $\mathcal{F}^{\text{num}}_{\text{qfs}}$, we determine the de-fuzzify function the above q-ROFN by using Eq. (14.20), such that

$$df_{dfn} = \frac{\mathbb{T}_{\mathcal{F}_{\overline{\overline{T}}_i}} + \mathbb{F}_{\mathcal{F}_{\overline{\overline{T}}_i}}}{2} \tag{14.20}$$

Case 8: By using the values of df_{dfn} and the values of positive ideal solution $\mathcal{F}^{+}_{\text{qfs - i}}$ and negative ideal solution $\mathcal{F}^{-}_{\text{qfs - i}}$, we determine the utility function of ideal $f\left(\mathcal{F}^{+}_{\text{qfs - i}}\right)$ and anti-ideal $f\left(\mathcal{F}^{-}_{\text{qfs - i}}\right)$ based on Eqs. (14.21) and (14.22), such that

$$f\left(\mathcal{F}^{+}_{\text{qfs - i}}\right) = \frac{\mathbb{S}^{\text{sf}}\left(\mathcal{F}^{-}_{\text{qfs - i}}\right)}{df_{dfn}} \tag{14.21}$$

$$f\left(\mathcal{F}^{-}_{\text{qfs - i}}\right)=\frac{\mathbb{S}^{\text{sf}}\left(\mathcal{F}^{+}_{\text{qfs - i}}\right)}{df_{dfn}} \tag{14.22}$$

To determine the value of $\mathcal{F}^{+}_{\text{qfs - i}}, \mathcal{F}^{-}_{\text{qfs - i}}$ and $f\left(\mathcal{F}^{+}_{\text{qfs - i}}\right), f\left(\mathcal{F}^{-}_{\text{qfs - i}}\right)$, we get the following case.

Case 9: By using the values of $\mathcal{F}^{+}_{\text{qfs - i}}, \mathcal{F}^{-}_{\text{qfs - i}}$ and $f\left(\mathcal{F}^{+}_{\text{qfs - i}}\right), f\left(\mathcal{F}^{-}_{\text{qfs - i}}\right)$, we determine the utility degree based on the Eq. (14.23), such that

$$\mathcal{F}_{\text{qfs - i}}=\frac{\mathbb{S}^{\text{sf}}\left(\mathcal{F}^{+}_{\text{qfs - i}}\right)+\mathbb{S}^{\text{sf}}\left(\mathcal{F}^{-}_{\text{qfs - i}}\right)}{1+\frac{1-f\left(\mathcal{F}^{+}_{\text{qfs}}\right)}{f\left(\mathcal{F}^{+}_{\text{qfs}}\right)}+\frac{1-f\left(\mathcal{F}^{-}_{\text{qfs}}\right)}{f\left(\mathcal{F}^{-}_{\text{qfs}}\right)}} \tag{14.23}$$

Case 10: Rank all alternatives to determine the best optimal.

As displayed above, we examined the MARCOS strategy by utilizing the q-ROFNs, which is the altered procedure of specific thoughts like IFSs and PFSs to decide the consistency and predominance of the expounded techniques. By utilizing the explained strategy, we foster the application in COVID-19 in the following review.

14.4 Analysis of Healthcare Services Under Q-ROFS-MARCOS Technique

Owing to its involvement in financial permanence and the natural world of confidence it appeared, the protection manufacturing has developed an indispensable role of the economic region in current years. As well, the health insurance (HI) marketplace, which is the focus of this survey, is a sizable marketplace and financial reserve. A certain scholar has employed it in the circumstances of separated areas [43–45]. Consequently, in this study, we suggest a q-ROF-MARCOS structure for choosing the most excellent individual HI theatre company including those having the biggest marketplace segment in Pakistan. Even Though HI includes sickness, personal fitness, fitness for outsiders, extra fitness, auxiliary fitness, and voyage fitness creations, this survey is constrained to simply personal HI. In this impression, ten sequestered HI corporations are analyzed for the current effort. Centered on the wide-ranging information assessment reviewed in Table 14.1, as perfectly as the views and understanding of five specialists in the HI care division, the extremely applicable standards are stipulated as supports:

As shown above information, we choose the troubles and resolved them by using the elaborated methods. For this, we choose the ten alternatives $\check{\mathfrak{A}}_j (j = 1, 2, \ldots, 10)$ and seven attributes $\check{C}_i (i = 1, 2, \ldots, 7)$ whose expressions are discussed in the form

Table 14.1 Expressions of the attributes and their representations

Expressions of the attributes	Their meanings
$\overset{\vee}{\mathcal{C}}_1(\boldsymbol{Effectiveness})$	It describes the efficacy of employed hospitals, sanatoria, and physicians
$\overset{\vee}{\mathcal{C}}_2(\boldsymbol{Responsibility})$	It describes accountability in compassion for COVID-19 affected role and their relatives
$\overset{\vee}{\mathcal{C}}_3(\boldsymbol{Network})$	It describes the amount of HI suppliers such as hospitals and infirmaries for affected roles with COVID-19
$\overset{\vee}{\mathcal{C}}_4(\boldsymbol{Support})$	It describes to client assistance delivering 24/7 assistance for troubles associated with COVID-19
$\overset{\vee}{\mathcal{C}}_5(\boldsymbol{Age})$	It describes growing older restrictions protected by COVID-19-associated medications
$\overset{\vee}{\mathcal{C}}_6(\boldsymbol{Payback\ Period})$	It requires immediate compensation for disease-associated practices (COVID-19 examinations, exhaustive treatment, etc.)
$\overset{\vee}{\mathcal{C}}_7(\boldsymbol{Premium\ Price})$	It describes the cost of the confidential HI strategy

of Table 14.1. The procedure of the q-ROF-MARCOS technique is elaborated with the help of the following cases.

Case 1: Based on the family of alternative m and their criteria n, we developed the decision matrix with every entry in the form of q-ROFNs which is discussed in the form of Table 14.2.

Case 2: To determine the ideal $\overset{\vee}{\mathfrak{A}}(\mathfrak{A}\mathfrak{I})$ and anti-ideal $\overset{\vee}{\mathfrak{A}}(\mathfrak{I}\mathfrak{D})$, by using the cost and benefit, $\overset{\vee}{\mathfrak{A}}(\mathfrak{A}\mathfrak{I})$ and $\overset{\vee}{\mathfrak{A}}(\mathfrak{I}\mathfrak{D})$ are obtained by using Eqs. (14.11) and (14.12); but, we choose all the benefit sorts of information, so we do not need to determine the anti-ideal $\overset{\vee}{\mathfrak{A}}(\mathfrak{I}\mathfrak{D})$. By using the information of Table 14.2, the ideal $\overset{\vee}{\mathfrak{A}}(\mathfrak{A}\mathfrak{I})$ is stated below:

$$\overset{\vee}{\mathfrak{A}}(\mathfrak{I}\mathfrak{D}) = \begin{Bmatrix} (0.5, 0.8), (0.51, 0.81), (0.52, 0.82), (0.53, 0.83), (0.54, 0.84), (0.55, 0.85), \\ (0.56, 0.86) \end{Bmatrix}$$

Case 3: By using Eqs. (14.13) and (14.14), we standardized the decision matrix $\overline{\overline{N}} = [\mathcal{F}_{\text{qfs - ij}}]$; but, we choose all the benefit sorts of information, so the information in Table 14.2 is not needed to be standardized.

Case 4: By using the information of the decision matrix $\overline{\overline{N}} = [\mathcal{F}_{\text{qfs - }ij}]$, we choose some weight vectors $\Xi = \{0.2, 0.1, 0.2, 0.1, 0.1, 0.1, 0.2\}$ with a rule, that is, $\sum_{j=1}^{n} \Xi_j = 1$, $\Xi_j \in [0, 1]$ to determine the weighted decision matrix by using Eq. (14.15) as discussed in the form of Table 14.3.

Table 14.2 Stated the decision matrix for $q_{SC} = 9$

	$\check{C}_1$	$\check{C}_2$	$\check{C}_3$	$\check{C}_4$
$\check{\mathfrak{A}}_1$	(0.9, 0.8)	(0.91, 0.81)	(0.92, 0.82)	(0.93, 0.83)
$\check{\mathfrak{A}}_2$	(0.8, 0.7)	(0.81, 0.71)	(0.82, 0.72)	(0.83, 0.73)
$\check{\mathfrak{A}}_3$	(0.7, 0.6)	(0.71, 0.61)	(0.72, 0.62)	(0.73, 0.63)
$\check{\mathfrak{A}}_4$	(0.6, 0.5)	(0.61, 0.51)	(0.62, 0.52)	(0.63, 0.53)
$\check{\mathfrak{A}}_5$	(0.9, 0.5)	(0.91, 0.51)	(0.92, 0.52)	(0.93, 0.53)
$\check{\mathfrak{A}}_6$	(0.9, 0.4)	(0.91, 0.41)	(0.92, 0.42)	(0.93, 0.43)
$\check{\mathfrak{A}}_7$	(0.8, 0.5)	(0.81, 0.51)	(0.82, 0.52)	(0.83, 0.53)
$\check{\mathfrak{A}}_8$	(0.5, 0.4)	(0.51, 0.41)	(0.52, 0.42)	(0.53, 0.43)
$\check{\mathfrak{A}}_9$	(0.9, 0.3)	(0.91, 0.31)	(0.92, 0.32)	(0.93, 0.33)
$\check{\mathfrak{A}}_{10}$	(0.9, 0.1)	(0.91, 0.11)	(0.92, 0.12)	(0.93, 0.13)
	$\check{C}_5$	$\check{C}_6$	$\check{C}_7$	
$\check{\mathfrak{A}}_1$	(0.94, 0.84)	(0.95, 0.85)	(0.96, 0.86)	
$\check{\mathfrak{A}}_2$	(0.84, 0.74)	(0.85, 0.75)	(0.86, 0.76)	
$\check{\mathfrak{A}}_3$	(0.74, 0.64)	(0.75, 0.65)	(0.76, 0.66)	
$\check{\mathfrak{A}}_4$	(0.64, 0.54)	(0.65, 0.55)	(0.66, 0.56)	
$\check{\mathfrak{A}}_5$	(0.94, 0.54)	(0.95, 0.55)	(0.96, 0.56)	
$\check{\mathfrak{A}}_6$	(0.94, 0.44)	(0.95, 0.45)	(0.96, 0.46)	
$\check{\mathfrak{A}}_7$	(0.84, 0.54)	(0.85, 0.55)	(0.86, 0.56)	
$\check{\mathfrak{A}}_8$	(0.54, 0.44)	(0.55, 0.45)	(0.56, 0.46)	
$\check{\mathfrak{A}}_9$	(0.94, 0.34)	(0.95, 0.35)	(0.96, 0.36)	
$\check{\mathfrak{A}}_{10}$	(0.94, 0.14)	(0.95, 0.15)	(0.96, 0.16)	

Case 5: By using Eqs. (14.16) and (14.17), we determine the positive and negative ideal solutions based on the information of the weighted decision matrix, such that

$$\mathcal{F}_{\mathrm{qfs}}^{+} = \left\{ \begin{array}{l} (0.8485, 0.9563), (0.7285, 0.9311), (0.6380, 0.9029), (0.5525, 0.8705), \\ (0.8408, 0.8705), (0.8408, 0.8325), (0.7284, 0.8705), (0.4684, 0.8325), \\ (0.8408, 0.7860), (0.8408, 0.6309) \end{array} \right\}$$

Table 14.3 Weighted decision matrix

	$\check{\mathcal{C}}_1$	$\check{\mathcal{C}}_2$	$\check{\mathcal{C}}_3$	$\check{\mathcal{C}}_4$
$\check{\mathfrak{A}}_1$	(0.7684, 0.9563)	(0.7235, 0.9791)	(0.7901, 0.9611)	(0.7452, 0.9815)
$\check{\mathfrak{A}}_2$	(0.6732, 0.9311)	(0.6322, 0.9663)	(0.6912, 0.9364)	(0.6492, 0.9690)
$\check{\mathfrak{A}}_3$	(0.5864, 0.9029)	(0.5510, 0.9518)	(0.6035, 0.9088)	(0.5669, 0.9548)
$\check{\mathfrak{A}}_4$	(0.5020, 0.8706)	(0.4726, 0.9349)	(0.5188, 0.8774)	(0.4882, 0.9385)
$\check{\mathfrak{A}}_5$	(0.7684, 0.8706)	(0.7235, 0.9349)	(0.7901, 0.8774)	(0.7452, 0.9385)
$\check{\mathfrak{A}}_6$	(0.7684, 0.8325)	(0.7235, 0.9147)	(0.7901, 0.8407)	(0.7452, 0.9191)
$\check{\mathfrak{A}}_7$	(0.6732, 0.8706)	(0.6322, 0.9349)	(0.6912, 0.8774)	(0.6492, 0.9385)
$\check{\mathfrak{A}}_8$	(0.4182, 0.8325)	(0.3949, 0.9147)	(0.4349, 0.8407)	(0.4104, 0.9191)
$\check{\mathfrak{A}}_9$	(0.7684, 0.7860)	(0.7235, 0.8895)	(0.7901, 0.7962)	(0.7452, 0.8951)
$\check{\mathfrak{A}}_{10}$	(0.7684, 0.6309)	(0.7235, 0.8019)	(0.7901, 0.6544)	(0.7452, 0.8154)
	$\check{\mathcal{C}}_5$	$\check{\mathcal{C}}_6$	$\check{\mathcal{C}}_7$	
$\check{\mathfrak{A}}_1$	(0.7569, 0.9827)	(0.7696, 0.9839)	(0.8408, 0.9703)	
$\check{\mathfrak{A}}_2$	(0.6578, 0.9703)	(0.6666, 0.9716)	(0.7285, 0.9466)	
$\check{\mathfrak{A}}_3$	(0.5749, 0.9563)	(0.5829, 0.9578)	(0.6380, 0.9202)	
$\check{\mathfrak{A}}_4$	(0.4960, 0.9402)	(0.5038, 0.9420)	(0.5525, 0.8905)	
$\check{\mathfrak{A}}_5$	(0.7569, 0.9402)	(0.7696, 0.9420)	(0.8408, 0.8905)	
$\check{\mathfrak{A}}_6$	(0.7569, 0.9212)	(0.7696, 0.9232)	(0.8408, 0.8561)	
$\check{\mathfrak{A}}_7$	(0.6578, 0.9402)	(0.6666, 0.9420)	(0.7285, 0.8905)	
$\check{\mathfrak{A}}_8$	(0.4182, 0.9212)	(0.4259, 0.9232)	(0.4684, 0.8561)	
$\check{\mathfrak{A}}_9$	(0.7569, 0.8977)	(0.7696, 0.9003)	(0.8408, 0.8152)	
$\check{\mathfrak{A}}_{10}$	(0.7569, 0.8215)	(0.7696, 0.8272)	(0.8408, 0.6931)	

$$\mathcal{F}_{\text{qfs}}^{-} = \left\{ \begin{array}{l} (0.7235, 0.9839), (0.6322, 0.9716), (0.5510, 0.9578), (0.4726, 0.9420), \\ (0.7235, 0.9420), (0.7235, 0.9232), (0.6322, 0.9420), (0.3949, 0.9232), \\ (0.7235, 0.9003), (0.7235, 0.8272) \end{array} \right\}$$

Case 6: By using the information in positive ideal solution $\mathcal{F}_{\text{qfs}}^{+}$ and negative ideal solution $\mathcal{F}_{\text{qfs}}^{-}$, we examine the fuzzy matrix $\overline{\overline{T}}_i$ based on Eq. (14.18), such that

$$\overline{\overline{T}}_i = \left\{ \begin{array}{l} (0.8584, 0.9409), (0.7476, 0.9047), (0.6548, 0.8648), (0.5661, 0.8200), \\ (0.8584, 0.8200), (0.8584, 0.7687), (0.7476, 0.8200), (0.4787, 0.7687), \\ (0.8584, 0.7077), (0.8584, 0.5219) \end{array} \right\}$$

Case 7: By using the fuzzy decision matrix $\overline{\overline{T}}_i$, we invented the q-ROFNs $\mathcal{F}_{\text{qfs}}^{\text{num}}$ using Eq. (14.19), such that

$$\mathcal{F}_{\text{qfs}}^{\text{num}} = (0.8584, 0.5219)$$

Then by using $\mathcal{F}_{\text{qfs}}^{\text{num}}$, we determine the de-fuzzify function the above q-ROFN by using Eq. (14.20), such that

$$df_{dfn} = 0.6902$$

Case 8: By using the values of df_{dfn} and the values of positive ideal solution $\mathcal{F}_{\text{qfs - i}}^{+}$ and negative ideal solution $\mathcal{F}_{\text{qfs - i}}^{-}$, we determine the utility function of ideal $f\left(\mathcal{F}_{\text{qfs - i}}^{+}\right)$ and anti-ideal $f\left(\mathcal{F}_{\text{qfs - i}}^{-}\right)$ based on Eqs. (14.21) and (14.22), such that

$$f\left(\mathcal{F}_{\text{qfs - i}}^{+}\right) = \left\{ \begin{array}{l} -1.1731, -1.0951, -0.9765, -0.8443, -0.7673, -0.6275, \\ -0.8226, -0.7059, -0.4846, -0.1840 \end{array} \right\}$$

$$f\left(\mathcal{F}_{\text{qfs - i}}^{-}\right) = \left\{ \begin{array}{l} -0.6652, -0.6788, -0.5523, -0.4091, -0.1116, 0.0260, \\ -0.3324, -0.2769, 0.1385, 0.2815 \end{array} \right\}$$

To determine the value of $\mathcal{F}_{\text{qfs - i}}^{+}, \mathcal{F}_{\text{qfs - i}}^{-}$ and $f\left(\mathcal{F}_{\text{qfs - i}}^{+}\right), f\left(\mathcal{F}_{\text{qfs - i}}^{-}\right)$, we get the following case.

Case 9: By using the values of $\mathcal{F}_{\text{qfs - i}}^{+}, \mathcal{F}_{\text{qfs - i}}^{-}$ and $f\left(\mathcal{F}_{\text{qfs - i}}^{+}\right), f\left(\mathcal{F}_{\text{qfs - i}}^{-}\right)$, we determine the utility degree based on the Eq. (14.23), such that

$$\begin{array}{l} \mathcal{F}_{\text{qfs}-1} = 0.378061, \mathcal{F}_{\text{qfs}-2} = 0.361509, \mathcal{F}_{\text{qfs}-3} = 0.275164, \mathcal{F}_{\text{qfs}-4} = 0.186903, \\ \mathcal{F}_{\text{qfs}-5} = 0.053872, \mathcal{F}_{\text{qfs}-6} = -0.01157, \mathcal{F}_{\text{qfs}-7} = 0.152596, \mathcal{F}_{\text{qfs}-8} = 0.112514 \\ \mathcal{F}_{\text{qfs}-9} = -0.05749, \mathcal{F}_{\text{qfs}-10} = -0.02334 \end{array}$$

Case 10: Rank all alternatives to determine the best optimal are discussed below:

$$\mathcal{F}_{\text{qfs}-1} \geq \mathcal{F}_{\text{qfs}-2} \geq \mathcal{F}_{\text{qfs}-3} \geq \mathcal{F}_{\text{qfs}-4} \geq \mathcal{F}_{\text{qfs}-7} \geq \mathcal{F}_{\text{qfs}-8} \geq \mathcal{F}_{\text{qfs}-5} \geq \mathcal{F}_{\text{qfs}-6} \geq \mathcal{F}_{\text{qfs}-10} \geq \mathcal{F}_{\text{qfs}}$$

As shown above, the best optimal is $\mathcal{F}_{\text{qfs}-1}$. Additionally, to modify further the elaborated approaches, we choose some existing methods and measures based on q-ROFSs, PFSs, and IFSs to advance the excellence of the research approaches.

14.5 Sensitivity Analysis

Certain researchers have used various kinds of speculations in the climate of MADM to decide the dependability and consistency of the explained approaches. In this review, we expounded the guideline of the MARCOS method by utilizing the q-ROFNs. Further, the consistency of the investigated MARCOS strategy for certain overall thoughts is decided by utilizing the data in Table 14.2. The data's above winning strategies are examined as follows: Zeng and Xiao [45] fostered the TOPSIS technique for IFSs, Sarkar and Biswas [20] started the TOPSIS strategy for PFSs, Pinar et al. [35] started the TOPSIS strategy for q-ROFSs, Ecer and Pamucar [42] explained the MARCOS technique for IFSs. By utilizing the data of Table 14.2, the near examination of the investigated and winning strategies is talked about as Table 14.4.

The graphical expression of the information in Table 14.4 is discussed in the form of Fig. 14.3.

Table 14.4 Stated the sensitive analysis

Methods	Score values	Ranking values
Zeng and Xiao [45] (TOPSIS method for IFSs)	*Cannot be calculated*	*Cannot be calculated*
Sarkar and Biswas [20] (TOPSIS method for PFSs)	*Cannot be calculated*	*Cannot be calculated*
Pinar et al. [35] (TOPSIS method for q-ROFSs)	$\mathcal{F}_{qfs-1} = 0.479071$, $\mathcal{F}_{qfs-2} = 0.462519$, $\mathcal{F}_{qfs-3} = 0.376174$, $\mathcal{F}_{qfs-4} = 0.287913$, $\mathcal{F}_{qfs-5} = 0.154882$, $\mathcal{F}_{qfs-6} = -0.11258$, $\mathcal{F}_{qfs-7} = 0.253606$, $\mathcal{F}_{qfs-8} = 0.213524$, $\mathcal{F}_{qfs-9} = -0.15850$, $\mathcal{F}_{qfs-10} = -0.12435$	$\mathcal{F}_{qfs-1} \geq \mathcal{F}_{qfs-2} \geq \mathcal{F}_{qfs-3} \geq \mathcal{F}_{qfs-4} \geq \mathcal{F}_{qfs-7} \geq \mathcal{F}_{qfs-8} \geq \mathcal{F}_{qfs-5} \geq \mathcal{F}_{qfs-6} \geq \mathcal{F}_{qfs-10} \geq \mathcal{F}_{qfs-9}$
Ecer and Pamucar [42] (MARCOS method for IFSs)	*Cannot be calculated*	*Cannot be calculated*
Proposed MARCOS method	$\mathcal{F}_{qfs-1} = 0.378061$, $\mathcal{F}_{qfs-2} = 0.361509$, $\mathcal{F}_{qfs-3} = 0.275164$, $\mathcal{F}_{qfs-4} = 0.186903$, $\mathcal{F}_{qfs-5} = 0.053872$, $\mathcal{F}_{qfs-6} = -0.01157$, $\mathcal{F}_{qfs-7} = 0.152596$, $\mathcal{F}_{qfs-8} = 0.112514$, $\mathcal{F}_{qfs-9} = -0.05749$, $\mathcal{F}_{qfs-10} = -0.02334$	$\mathcal{F}_{qfs-1} \geq \mathcal{F}_{qfs-2} \geq \mathcal{F}_{qfs-3} \geq \mathcal{F}_{qfs-4} \geq \mathcal{F}_{qfs-7} \geq \mathcal{F}_{qfs-8} \geq \mathcal{F}_{qfs-5} \geq \mathcal{F}_{qfs-6} \geq \mathcal{F}_{qfs-10} \geq \mathcal{F}_{qfs-9}$

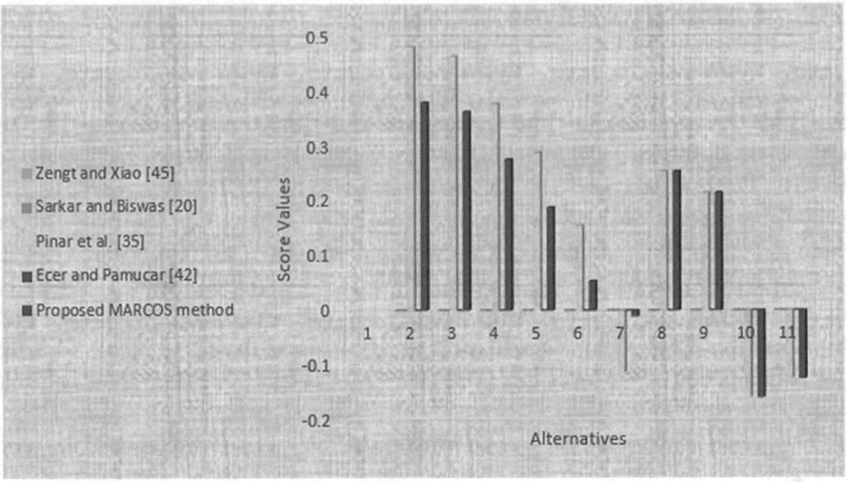

Fig. 14.4 Geometrical expressions of the information in Table 14.4

As shown above, the existing sorts of methods based on IFSs and PFSs are not able to resolve the elaborated sorts of information (q-rung orthopair fuzzy information's), but if we choose the existing sorts of information, then the elaborated sorts of methods based on q-ROFSs are more useful to resolve it very easily. Therefore, the elaborated approaches are more powerful and more suitable to manage awkward and inconsistent information easily (Fig. 14.4).

14.6 Conclusion

The key structure of article is brief below.

1. We initiated the principle of the MARCOS procedure by using the q-rung orthopair fuzzy natural environment to rank coverage enterprises.
2. The results produced ten coverage enterprises ranking in conditions of healthcare services in the era of COVID-19. The payback period, premium price, and network are established as extremely essential considerations.
3. A completed compassion assessment is presented to confirm the future method's permanence and efficiency. The announced methodology met the coverage evaluation dilemma through the COVID-19 pandemic extremely reasonable approach by using the compassion assessment conclusions.
4. The sensitive analysis of the initiated work is also discussed.

In the future, we will modify the principle of linear diophantine fuzzy sets [46–48], complex T-spherical fuzzy sets [49], complex neutrosophic sets [50–54], etc. [55–61], to improve the quality of the research works.

References

1. L.A. Zadeh, Fuzzy sets. Inf. Control **8**(3), 338–353 (1965)
2. D. Ramot, R. Milo, M. Friedman, A. Kandel, Complex fuzzy sets. IEEE Trans. Fuzzy Syst. **10**(2), 171–186 (2002)
3. P. Liu, Z. Ali, T. Mahmood, The distance measures and cross-entropy are based on complex fuzzy sets and their application in decision-making. J. Intell. Fuzzy Syst. **39**(3), 3351–3374 (2020)
4. D.A. Chiang, N.P. Lin, Correlation of fuzzy sets. Fuzzy Sets Syst. **102**(2), 221–226 (1999)
5. J.G. Brown, A note on fuzzy sets. Inf. Control **18**(1), 32–39 (1971)
6. V. Torra, Hesitant fuzzy sets. Int. J. Intell. Syst. **25**(6), 529–539 (2010)
7. T. Mahmood, A novel approach towards bipolar soft sets and their applications. J. Math. **2020**(4690808), 2020 (2020)
8. K. Atanassov, Intuitionistic fuzzy sets. Fuzzy Sets Syst. **20**(1), 87–96 (1986)
9. K.T. Atanassov, Interval-valued intuitionistic fuzzy sets. Fuzzy Sets Syst. **31**(3), 343–349 (1989)
10. M. Xia, Z. Xu, B. Zhu, Some issues on intuitionistic fuzzy aggregation operators based on Archimedean t-conorm and t-norm. Knowl.-Based Syst. **31**, 78–88 (2012)
11. H. Garg, K. Kumar, A novel exponential distance and its based TOPSIS method for interval-valued intuitionistic fuzzy sets using connection number of SPA theory. Artif. Intell. Rev. **53**(1), 595–624 (2020)
12. H. Garg, A new generalized improved score function of interval-valued intuitionistic fuzzy sets and applications in expert systems. Appl. Soft Comput. **38**, 988–999 (2016)
13. H. Garg, D. Rani, Novel similarity measure based on the transformed right-angled triangles between intuitionistic fuzzy sets and their applications. Cogn. Comput. **13**(2), 447–465 (2021)
14. P.A. Ejegwa, I.C. Onyeke, V. Adah, An algorithm for an improved intuitionistic fuzzy correlation measure with the medical diagnostic application. Ann. Optim. Theory Pract. **3**(3), 51–66 (2020)
15. A. Sanam, A. Saleem, Y. Muhammad, Some induced generalized Einstein aggregating operators and their application to group decision-making problems using intuitionistic fuzzy numbers. Ann. Optim. Theory Pract. **3**(3), 15–49 (2019)
16. R. Kar, A. Shaw, B. Das, An alternative approach to finding the optimal solution of assignment problem using Hungarian method by trapezoidal intuitionistic type-2 fuzzy data. Ann. Optim. Theory Pract. **3**(3), 155–173 (2020)
17. K.B.S.K. De, Decision-making under intuitionistic fuzzy metric distances. Ann. Optim. Theory Pract. **3**(2), 49–64 (2020)
18. A. Fahmi, F. Amin, S.B. Shah, Geometric operators are based on the linguistic interval-valued intuitionistic neutrosophic fuzzy number and their application in decision making. Ann. Optim. Theory Pract. **3**(1), 47–71 (2020)
19. R.R. Yager, Pythagorean membership grades in multicriteria decision making. IEEE Trans. Fuzzy Syst. **22**(4), 958–965 (2013)
20. B. Sarkar, A. Biswas, Pythagorean fuzzy AHP-TOPSIS integrated approach for transportation management through a new distance measure. Soft Comput. **25**(5), 4073–4089 (2021)
21. R.M. Zulqarnain, I. Siddique, F. Jarad, R. Ali, T. Abdeljawad, Development of TOPSIS technique under Pythagorean fuzzy hypersoft environment based on correlation coefficient and its application towards the selection of antivirus mask in COVID-19 pandemic. Complexity **2021**, 6634991 (2021)
22. Y. Tang, Y. Yang, Sustainable e-bike sharing recycling supplier selection: an interval-valued Pythagorean fuzzy MAGDM method based on preference information technology. J. Clean. Prod. **287**, 125530–125543 (2021)
23. A. Çalık, A novel Pythagorean fuzzy AHP and fuzzy TOPSIS methodology for green supplier selection in the Industry 4.0 era. Soft Comput. **25**(3), 2253–2265 (2021)

24. G. Bakioglu, A.O. Atahan, AHP integrated TOPSIS and VIKOR methods with Pythagorean fuzzy sets to prioritize risks in self-driving vehicles. Appl. Soft Comput. **99**, 106948–106961 (2021)
25. P. Ejegwa, S. Wen, Y. Feng, W. Zhang, N. Tang, Novel Pythagorean fuzzy correlation measures via Pythagorean fuzzy deviation, variance, and covariance with applications to pattern recognition and career placement. IEEE Trans. Fuzzy Syst. (2021). https://doi.org/10.1109/TFUZZ.2021.3063794
26. B. Batool, S.S. Abosuliman, S. Abdullah, S. Ashraf, EDAS method for decision support modeling under the Pythagorean probabilistic hesitant fuzzy aggregation information. J. Ambient Intell. Humaniz. Comput. 1–14 (2021). https://doi.org/10.1007/s12652-021-03181-1
27. H. Garg, Linguistic Pythagorean fuzzy sets and their applications in the multiattribute decision-making process. Int. J. Intell. Syst. **33**(6), 1234–1263 (2018)
28. H. Garg, Novel correlation coefficients between Pythagorean fuzzy sets and their applications to decision-making processes. Int. J. Intell. Syst. **31**(12), 1234–1252 (2016)
29. R.R. Yager, Generalized orthopair fuzzy sets. IEEE Trans. Fuzzy Syst. **25**(5), 1222–1230 (2016)
30. H. Garg, A new possibility degree measure for interval-valued q-rung orthopair fuzzy sets in decision-making. Int. J. Intell. Syst. **36**(1), 526–557 (2021)
31. T. Mahmood, Z. Ali, Entropy measure and TOPSIS method based on correlation coefficient using complex q-rung orthopair fuzzy information and its application to multi-attribute decision making. Soft Comput. **25**(2), 1249–1275 (2021)
32. M. Riaz, M.T. Hamid, D. Afzal, D. Pamucar, Y.M. Chu, Multi-criteria decision making in robotic agri-farming with q-rung orthopair m-polar fuzzy sets. PLoS ONE **16**(2), e0246485 (2021)
33. Y. Liu, G. Wei, S. Abdullah, J. Liu, L. Xu, H. Liu, Banzhaf–Choquet-copula-based aggregation operators for managing q-rung orthopair fuzzy information. Soft. Comput. **25**(10), 6891–6914 (2021)
34. T. Mahmood, Z. Ali, A novel approach of complex q-rung orthopair fuzzy Hamacher aggregation operators and their application for cleaner production assessment in gold mines. J. Ambient Intell. Humaniz. Comput. **12**, 8933–8959 (2021)
35. A. Pınar, B.D. Rouyendegh, Y.S. Özdemir, q-Rung orthopair fuzzy TOPSIS method for green supplier selection problem. Sustainability **13**, 985–1003 (2021)
36. Z. Ali, T. Mahmood, Maclaurin symmetric mean operators and their applications in the environment of complex q-rung orthopair fuzzy sets. Comput. Appl. Math. **39**, 1–27 (2020)
37. C. Jana, G. Muhiuddin, M. Pal, Some Dombi aggregation of Q-rung orthopair fuzzy numbers in multiple-attribute decision making. Int. J. Intell. Syst. **34**(12), 3220–3240 (2019)
38. Ž Stević, D. Pamučar, A. Puška, P. Chatterjee, Sustainable supplier selection in healthcare industries using a new MCDM method: Measurement of Alternatives and Ranking according to COmpromise Solution (MARCOS). Comput. Ind. Eng. **140**, 106231–106247 (2020)
39. M. Bakır, Ö. Atalık, Application of fuzzy AHP and fuzzy MARCOS approach for the evaluation of E-service quality in the airline industry. Decis. Mak. Appl. Manag. Eng. **4**(1), 127–152 (2021)
40. E. Celik, M. Gul, Hazard identification, risk assessment, and control for dam construction safety using an integrated BWM and MARCOS approach under interval type-2 fuzzy sets environment. Autom. Constr. **127**, 103699–103719 (2021)
41. M. Stanković, Ž Stević, D.K. Das, M. Subotić, D. Pamučar, A new fuzzy MARCOS method for road traffic risk analysis. Mathematics **8**(3), 457–471 (2020)
42. F. Ecer, D. Pamucar, MARCOS technique under intuitionistic fuzzy environment for determining the COVID-19 pandemic performance of insurance companies in terms of healthcare services. Appl. Soft Comput. **104**, 107199–107218 (2021)
43. T.P. Velavan, C.G. Meyer, The COVID-19 epidemic. Tropical Med. Int. Health **25**(3), 278–299 (2020)
44. M.S. Yang, Z. Ali, T. Mahmood, Complex q-rung orthopair uncertain linguistic partitioned Bonferroni mean operators with application in antivirus mask selection. Symmetry **13**(2), 249–268 (2021)

45. S. Zeng, Y. Xiao, TOPSIS method for intuitionistic fuzzy multiple-criteria decision making and its application to investment selection. Kybernetes **45**, 282–296 (2016)
46. M. Riaz, M.R. Hashmi, Linear Diophantine fuzzy set and its applications towards multi-attribute decision-making problems. J. Intell. Fuzzy Syst. **37**(4), 5417–5439 (2019)
47. M. Riaz, M.R. Hashmi, H. Kalsoom, D. Pamucar, Y.M. Chu, Linear Diophantine fuzzy soft rough sets for the selection of sustainable material handling equipment. Symmetry **12**(8), 1215–1237 (2020)
48. M. Riaz, M.R. Hashmi, D. Pamucar, Y.M. Chu, Spherical linear Diophantine fuzzy sets with modeling uncertainties in MCDM. Comput. Model. Eng. Sci. **126**(3), 1125–1164 (2021)
49. Z. Ali, T. Mahmood, M.S. Yang, TOPSIS method based on complex spherical fuzzy sets with Bonferroni mean operators. Mathematics **8**(10), 1739–1753 (2020)
50. M. Ali, F. Smarandache, Complex neutrosophic set. Neural Comput. Appl. **28**(7), 1817–1834 (2017)
51. L.Q. Dat, N.T. Thong, M. Ali, F. Smarandache, M. Abdel-Basset, H.V. Long, Linguistic approaches to interval complex neutrosophic sets in decision making. IEEE Access **7**, 38902–38917 (2019)
52. S.G. Quek, S. Broumi, G. Selvachandran, A. Bakali, M. Talea, F. Smarandache, Some results on the graph theory for complex neutrosophic sets. Symmetry **10**(6), 190–211 (2018)
53. C. Jana, G. Muhiuddin, M. Pal, Multiple-attribute decision-making problems based on SVTNH methods. J. Ambient. Intell. Humaniz. Comput. **11**(9), 3717–3733 (2020)
54. G. Shahzadi, G. Muhiuddin, M. Arif Butt, A. Ashraf, Hamacher interactive hybrid weighted averaging operators under Fermatean fuzzy numbers. J. Math. **2021**, 5556017 (2021)
55. H. Garg, CN-q-ROFS: connection number-based q-rung orthopair fuzzy set and their application to the decision-making process. Int. J. Intell. Syst. **36**(7), 3106–3143 (2021)
56. H. Garg, S.M. Chen, Multiattribute group decision-making based on neutrality aggregation operators of q-rung orthopair fuzzy sets. Inf. Sci. **517**, 427–447 (2020)
57. H. Garg, New exponential operation laws and operators for interval-valued q-rung orthopair fuzzy sets in the group decision-making process. Neural Comput. Appl. **33**(20), 13937–13963 (2021). https://doi.org/10.1007/s00521-021-06036-0
58. M. Riaz, H. Garg, H.M.A. Farid, M. Aslam, Novel q-rung orthopair fuzzy interaction aggregation operators and their application to low-carbon green supply chain management. J. Intell. Fuzzy Syst. **41**(2), 4109–4126. https://doi.org/10.3233/JIFS-210506
59. Z. Yang, H. Garg, Interaction power partitioned Maclaurin symmetric mean operators under q-rung orthopair uncertain linguistic information. Int. J. Fuzzy Syst. 1–19 (2021). https://doi.org/10.1007/s40815-021-01062-5
60. H. Garg, A novel trigonometric operation-based q-rung orthopair fuzzy aggregation operator and its fundamental properties. Neural Comput. Appl. **32**(18), 15077–15099 (2020)
61. X. Peng, J. Dai, H. Garg, Exponential operation and aggregation operator for q-rung orthopair fuzzy set and their decision-making method with a new score function. Int. J. Intell. Syst. **33**(11), 2255–2282 (2018)

Chapter 15
Interval Complex q-Rung Orthopair Fuzzy Aggregation Operators and Their Applications in Cite Selection of Electric Vehicle

Somen Debnath

Abstract In this chapter, the notion of interval complex q-rung orthopair fuzzy sets (IVC q-ROFSs) and its related interval complex q-rung orthopair fuzzy numbers (IVC q-ROFNs) are introduced to solve multi-attribute decision-making (MADM) problems. The IVC-q-ROFSs are formed by combining interval complex fuzzy sets (IVCFSs) and q-rung orthopair fuzzy sets (q-ROFSs) and these can be viewed as an extension of fuzzy sets (FSs), intuitionistic fuzzy sets (IFSs), interval-valued fuzzy sets (IVFSs), interval-valued intuitionistic fuzzy sets (IVIFSs), Pythagorean fuzzy sets (PFSs), q-ROFSs, complex Pythagorean fuzzy sets (CPFSs), complex q-ROFSs, etc. So, the IVC-q-ROFS is a hybrid structure that gives high flexibility and becomes more functional in various types of uncertain problems that are time-periodic by default. We also study some fundamental properties and aggregation operators based on IVC q-ROFNs. In some practical applications, due to the increase of complex uncertainty in fuzzy environments, there exist some issues where the decision-makers face challenges to express the incomplete knowledge precisely by a single complex-valued membership degree and a single complex-valued non-membership degree. Such problems arise because the data provided to the decision-makers need another superior environment than the existing environments where it is enabled to answer those questions where the complex-valued membership grade and the complex-valued non-membership grade are not precise, i.e., they represent uncertainty. So, to remove such uncertainty, there is a demand for another superior environment that is capable of accommodating such information. This leads to the invention of IVC q-ROFSs and, under this environment, we can solve MADM problems by an exact methodology. Moreover, we propose score function, accuracy function, and various aggregate operators by using IVC q-ROFSs. Based on aggregate operators, we propose weight operators on the set of attributes. Finally, we give a numerical example to illustrate the feasibility and validity of the proposed method in a practical scenario.

S. Debnath (✉)
Department of Mathematics, Umakanta Academy, Agartala 799001, Tripura, India
e-mail: somen008@rediffmail.com

H. Garg (ed.), *q-Rung Orthopair Fuzzy Sets*,
https://doi.org/10.1007/978-981-19-1449-2_15

Keywords Pythagorean fuzzy set · q-rung orthopair fuzzy set · Interval complex q-rung orthopair fuzzy set · MADM

15.1 Introduction

In 1965, Zadeh [63] defined fuzzy set (FS) as a class of objects with a continuum of grades of membership, i.e., an FS A is characterized by a membership function μ_A which associates each object of the universe X to a real number in the interval [0, 1]. The membership function can be viewed as an extension of the two-valued characteristic function. An FS is a more general framework than the ordinary set as it provides a much wider scope of applicability of different classes of imprecision found in human knowledge. After the invention of FS, it is enabled to capture the attention of researchers from various fields all over the world. In 2010, Zimmermann [65] published a review article on FS theory in which he discussed the applicability and usefulness of FS theory in the advancement of science and technology. Some other notable contributions related to FSs are proposed in [12, 15, 31, 49]. However, we are encountered with some real-world decision-making problems that involve hesitancy and the degree of hesitancy cannot be always measured by $1 - \mu_A(x)$. Therefore, the concept of FS is not capable enough to address the incomplete information present in the environment. To remove such an issue, Atanassov [5, 7] introduced an intuitionistic fuzzy set (IFS) as a generalization of FS. In IFS, every object to a universe has a membership degree ($\mu_A(x)$), a non-membership degree ($\gamma_A(x)$), and a hesitancy degree ($\pi_A(x)$) with the condition $\mu_A(x), \gamma_A(x) \in [0, 1]$ such that the sum of the membership and the non-membership degree is always less than or equal to 1. So, the invention of IFS provides a more general framework to the decision-maker with a more wide range of uncertain domains than FS. An IFS is sometimes treated as an FS when its non-membership degree becomes zero but the converse is not true. Later on, it has been observed that there exists another fuzzy environment where the membership degree is itself uncertain and IFS is not suitable for such an environment. This leads to the introduction of the interval-valued fuzzy set (IVFS) [11, 30]. In IVFS, the membership degree of every element of the universe is denoted by an interval that is a subset of [0, 1]. Due to the increase in popularity of fuzzy set among the researchers over the decades, it has been progressed rapidly and it leads to the development of new theories such as fuzzy logic and its practical application [50, 64], hesitant fuzzy set, and its application in decision-making problem [57, 67], picture fuzzy sets [14], bipolar fuzzy sets [66], dual hesitant fuzzy sets [71], linguistic fuzzy sets [4], interval-valued intuitionistic fuzzy sets [6], similarity measures for fuzzy sets [8], fuzzy sets in pattern recognition and machine intelligence [43], fuzzy sets for image processing and understanding[10], fuzzy optimization and decision-making [9], a new approach to fuzzy neural system modeling [37], GDM using fuzzy sets [34], and so on.

From the above discussion, it has been learned that every theory has its inherent limitation and it causes the invention of a new theory. To make it clear, we discuss the following: FS is capable enough to address the uncertain environment where the decision-maker provides only the membership degree and in the IFS, the decision-maker provides both the membership and the non-membership degree with its theoretical restriction. However, there exists some sort of uncertainty in human thinking that cannot be handled by either FS or IFS. For example, suppose in some uncertain environment, the decision-maker provides the membership and the non-membership values as 0.6 and 0.5, respectively, and their sum is 1.1 which is greater than 1. To handle such an issue, we need another superior tool that can accommodate this kind of environment. This leads to the introduction of the Pythagorean fuzzy set (PFS) proposed by Yager [58]. Properties and applications of PFSs are proposed in [59]. In PFS, the sum of the square of the membership grade and the square of the non-membership grade is less or equal to 1. So, PFS is more superior fuzzy environment than FS and IFS to cope with uncertain problems that are influenced by human thinking. Furthermore, Peng et al. [44] presented some results for PFSs, Zhang et al. [68] defined the extension of the TOPSIS method for MCDM using PFSs, Garg [17] presented linguistic PFSs and their application, Xiao et al. [55] defined the divergence measure of PFSs and its application in medical diagnosis, Wei et al. [54] proposed the similarity measures of PFSs based on the cosine function and their applications, Zhang et al. [70] defined the belief function of Pythagorean fuzzy rough approximation space and its applications, Chen et al. [13] proposed a new Chebyshev distance measures for PFSs with applications to multiple criteria decision analysis using an extended ELECTRE approach, Ejegwa [16] defined distance and similarity measures for PFSs, Ren et al. [48] presented the Pythagorean fuzzy TODIM approach to multi-criteria decision-making, etc.

We know that, when a theory failed to deliver to address a problem, then there is a necessity to discover another new theory to cope with such an issue. For example, when a decision-maker provides the membership degree = 0.8 and the non-membership degree = 0.7 to an uncertain problem, then it results $(0.8)^2 + (0.7)^2 = 0.64+0.49 = 1.13 > 1$. This type of problem cannot be handled by FS, IFS, and PFS. To control such an issue, Yager [60] introduced another mathematical tool known as q-rung orthopair fuzzy set (q-ROFS) and it is a more general and more flexible tool than FS, IFS, and PFS. In q-ROFS, the sum of the qth power of the membership degree and the qth power of the non-membership degree is less or equal to 1. So, due to the novelty of the q-ROFS, we remove the issue cited above. To get more insight into q-ROFS, we refer to the following: Krishankumar et al. [33] investigated the green supplier selection problem using a q-rung orthopair fuzzy-based decision framework with unknown weight information, Wang et al. [52] proposed the maximizing deviation method for multiple attribute decision-making under q-rung orthopair fuzzy environment, Rani et al. [47] investigated the multi-criteria weighted aggregated sum product assessment framework for fuel technology selection using q-ROFSs, Zhang et al. [69] proposed the group decision-making with incomplete q-rung orthopair fuzzy preference relations, Liu et al. [38] defined the multiple-attribute decision-making based on Archimedean Bonferroni operators of q-ROFNs, Joshi et al. [32] introduced the interval-valued q-ROFSs and their properties, new operators for group decision-making based on interval-valued q-ROFSs are developed in [26]. Wang et al.

[53] presented the similarity measures of q-ROFSs based on cosine function and their applications, Garg and Chen [19] defined the multi-attribute group decision-making based on neutrality aggregation operators of q-rung orthopair fuzzy sets. A new possibility degree measure for interval-valued q-ROFSs in decision-making was presented by Garg [22]. Li et al. [36] introduced some preference relations based on q-ROFSs, Garg [25] introduced the CN-q-ROFS for decision-making problems, Yang, et al. [61] investigated the power partitioned Maclaurin symmetric mean operators under q-rung orthopair linguistic domain, a novel trigonometric-based q-rung orthopair fuzzy aggregate operators and their properties are discussed in [27], etc.

So far in the above discussion, it has been observed that different researchers extended the notion of FS in different directions according to the problem domain is concerned. Moreover, in the context of another problem domain, while researching, mathematicians are eager to find the answer to a question, what will happen when the co-domain of FS is changed to the set of complex numbers instead of the closed unit interval [0, 1]. To answer this question, Ramot et al. [45] defined complex fuzzy set (CFS) as another extension of FS. The CFS is characterized by the complex-valued function denoted by $M_{\psi}(x)$ and it is defined as $M_{\psi}(x) = T_{\psi}(x).e^{i2\pi\omega_{T_{\psi}(x)}}$, as it looks like the form of $re^{i\theta}$, therefore $T_{\psi}(x)$ and $2\pi\omega_{T_{\psi}(x)}$ may be treated as an amplitude part and the phase part, respectively, and satisfied the condition $0 \leq T_{\psi}(x), \omega_{T_{\psi}(x)} \leq 1$. Some more works related to CFS are proposed in [35, 56, 62]. Although, there exist some fields of knowledge that cannot be presented by using only a complex-valued membership function. To eradicate such issues, Alkouri et al. [3] introduced complex intuitionistic fuzzy set (CIFS). The CIFS is characterized by a complex-valued membership function and complex-valued non-membership function under the restriction that the sum of the real part (similarly for the imaginary part) of the complex-valued membership grade and the real part (similarly imaginary part) of the complex-valued non-membership grade is less or equal to 1 and it is suitable to overcome both uncertainty and periodicity as the problem domain is concerned. Some notable contributions of CIFSs are given in [18, 20, 21, 46].

While working on the complex intuitionistic fuzzy environment, researchers found a problem where they assign $0.6e^{i2\pi 0.7}$ for membership grade and $0.7e^{i2\pi 0.4}$ non-membership grade and it cannot be described by IFSs as $0.6 + 0.7 = 1.3 > 1$ and $0.7 + 0.4 = 1.1 > 1$, though if any one of these two is greater than 1 then also difficulty arises. To encounter such types of problems, Ullah et al. [51] introduced complex Pythagorean fuzzy set (CPFS). In CPFS, the sum of the squares of the real part (similar for the imaginary part) of the membership grade and the real part (similar for an imaginary part) of the non-membership grade is less or equal to 1. Therefore, the above situation can be easily tackled by CPFS. Also, a novel decision-making approach under a complex Pythagorean fuzzy environment is proposed by Akram et al. [1]. Ma et al. [41] defined the group decision-making framework using complex Pythagorean fuzzy information, etc.

Uncertainties exist in real life in various forms depending on the problematic environment and no theory is ever been invented which can address all types of uncertainties in a different environment. Along with the advancement of science and technology, the CFS, CIFS, and CPFS have been applied successfully to various types of complex uncertain environment to some extent. As all these theoretical concepts have their inherent difficulties, so there are some issues where the decision-makers failed to answer the situation using CFS, CIFS, and CPFS. Such a situation arises because the complex-valued membership and non-membership grades are subjective, i.e., there may be an opportunity for a decision-maker to provide more than one complex-valued membership and non-membership function associated with an object of the universe. For example, suppose a decision-maker provides $0.9e^{i2\pi 0.7}$ for the membership grade and $0.7e^{i2\pi 0.8}$ for the non-membership grade. Such information cannot be fit for CIFS and CPFS. There is a demand for another powerful tool to address such problem and it leads to the invention of complex q-rung orthopair fuzzy sets (Cq-ROFSs) proposed by Liu et al. [40]. The Cq-ROFSs give more liberty to the decision-makers for their complex decision-making approaches. It is very much helpful while handling MAGDM and MCGDM problems under complex uncertain environments. In Cq-ROFSs, the sum of the qth power of the real part (same for an imaginary part) of the complex-valued membership grade and qth power of the real part (same for the imaginary part) of the complex-valued non-membership grade is less or equal to 1. Some more recent works on Cq-ROFSs and their extensions were proposed in [2, 23, 24, 28, 29, 39, 42].

In [24], Garg et al. proposed the algorithms for complex interval-valued q-rung orthopair fuzzy sets (CIVq-ROFSs) in decision-making based on aggregation operators, AHP, and TOPSIS. The aim of introducing CIVq-ROFSs is to generalize the notion of the interval-valued complex fuzzy set (IVCFS) and q-rung orthopair fuzzy set (q-ROFS). In this chapter, getting motivation from CIVq-ROFS, we introduce a similar structure called IVCq-ROFS as a generalization of CIFS, CPFS, and Cq-ROFS. The Cq-ROFS is characterized by the complex-valued membership grade and complex-valued non-membership grade in crisp form apart from the power of q. But due to the increase in the perception of human thinking as they encountered in their day-to-day practice, there may be another level of uncertainty and periodicity-related problems that cannot be handle by Cq-ROFS. Such a situation arises when the complex-valued membership and non-membership grades are both uncertain, i.e., the grades are subjective. To cope with such a situation, the IVCq-ROFS is introduced. The IVCq-ROFS is characterized by the interval complex-valued membership and interval complex-valued non-membership grade associated with each object of the universe. The restriction of the IVCq-ROFS is, the sum of the qth power of the real part (same for the complex part) of the complex-valued lower membership grade and the qth power of the real part (same for a complex part) of the complex-valued lower non-membership grade and the sum of the qth power of the real part (same for a complex part) of the complex-valued upper membership grade and the qth power of the real part (same for a complex part) of the complex-valued upper non-membership grade both is less or equal to 1. The IVCq-ROFS is a more generalized framework than q-ROFS to handle both the complex uncertainty and the periodicity-related problem. In IVCq-ROFS, if the upper limits of both the complex-valued membership and non-membership grades are zero, then the IVCq-ROFS reduces to Cq-ROFS. Similarly, in Cq-ROFS, if we take q = 2 then it reduces to CPFS. Furthermore, in Cq-ROFS, if we put q = 1 then it is reduced to CIFS. From this discussion, we claim that the IVCq-ROFS is an extension of CIFS, CPFS, Cq-ROFS. Researchers are always trying to develop a more generalized model as it accommodates the other existing models relevant to the environment. In this paper, we also study different laws and properties of IVCq-ROFSs, develop some aggregation operators, a new MADM problem-based model is developed by using the properties of IVCq-ROFSs. To check the validity and feasibility of the proposed model we cite a real-life-based problem and by solving it we show its reliability. The objectives of the proposed study are given in the form:

- To initiate the notion of IVCq-ROFSs that is analogous to CIVq-ROFSs, proposed in [24] and study their various properties.
- To introduce a new type of score and accuracy functions based on IVCq-ROFSs.
- To extend the notion of Cq-ROFSs.
- To develop new operators namely IVCq-ROFWA and IVCq-ROFGA and verify them.
- To introduce a new MADM model based on the new operators.
- To show the feasibility of the proposed method, a real-world-based example is undertaken.

The comparison of the restrictions of CIFS, CPFS, Cq-ROFS, and IVCq-ROFS is exhibited by Fig. 15.1.

The rest of the chapter is organized in the following manner: Sect. 15.2 includes a literature review where some basic definitions such as FS, IVFS, IFS, PFS, q-ROFS, CIFS, CPFS, and Cq-ROFS are introduced. In Sect. 15.3, IVCq-ROFSs and their related laws and operators are introduced. In Sect. 15.4, aggregate operator for MADM problems under IVCq-ROF environment has been introduced. In Sect. 15.5,

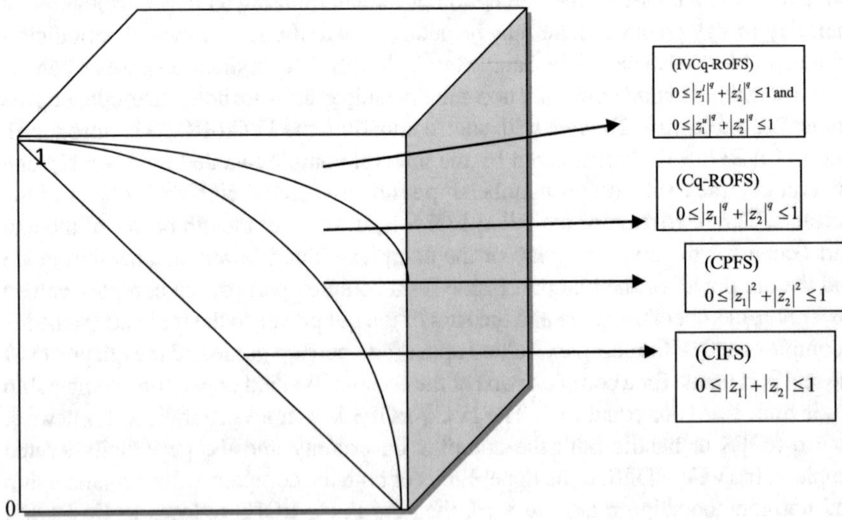

Fig. 15.1 Comparisons of the restrictions of CIFS, CPFS, Cq-ROFS, and IVCq-ROFS

a new MADM model based on IVCq-ROFWA and IVCq-ROFGA operators has been introduced. Section 15.6 includes a numerical example to validate the proposed model. In the last section, i.e., in Sect. 15.7, the conclusion of the chapter with future scope has been discussed.

15.2 Preliminaries

In this section, we give some basic definitions of FS, IFS, IVFS, PFS, q-ROFS, CFS, CIFS, CPFS, and Cq-ROFS which are useful for the subsequent sections of the chapter. Throughout our discussion, we consider U as the universe of discourse, μ, γ denote, respectively, the membership and the non-membership degrees, and $i \equiv \sqrt{-1}$.

Definition 15.2.1 [63] An FS X in U is defined as $\mathrm{X} = \{(u, \mu_X(u)) : u \in U\}$, where μ_X denotes the fuzzy membership function defined by $\mu_\mathrm{X} : U \to [0, 1]$. If μ_X takes only the values 0 or 1 then it reduces to a characteristic function.

Definition 15.2.2 [7] An IFS Υ in U is defined as $\Upsilon = \{(u, \mu_\Upsilon(u), \gamma_\Upsilon(u)) : u \in U\}$, where μ_Υ and γ_Υ denote, respectively, the membership function and the non-membership function defined by $\mu_\Upsilon, \gamma_\Upsilon : U \to [0, 1]$ such that $0 \leq \mu_\Upsilon(u) + \gamma_\Upsilon(u) \leq 1$ and $\Pi_\Upsilon(u) = 1 - \mu_\Upsilon(u) - \gamma_\Upsilon(u)$ is called the hesitancy degree. If $\gamma_\Upsilon(u) = 0$, then the IFS is reduced to FS.

Definition 15.2.3 [30] An IVFS Z in U is defined as $\mathrm{Z} = \{(u, \mu_\mathrm{Z}(u)) : u \in U\}$, where the membership function μ_Z is defined as $\mu_\mathrm{Z} : U \to Int([0, 1])$. Here, $Int([0, 1])$ represents the set of all subintervals within [0, 1]. Here, the membership degree is represented by an interval in the sense that the membership degree is itself uncertain.

Definition 15.2.4 [58] A PFS M in U is defined as $\mathrm{M} = \{(u, \mu_\mathrm{M}(u), \gamma_\mathrm{M}(u)) : u \in U\}$, where $\mu_\mathrm{M}, \gamma_\mathrm{M} : U \to [0, 1]$ such that $0 \leq \mu_\mathrm{M}^2(u) + \gamma_\mathrm{M}^2(u) \leq 1$. The hesitancy degree is defined as $\Pi_\mathrm{M}(u) = \left(1 - \left(\mu_\mathrm{M}^2(u) + \gamma_\mathrm{M}^2(u)\right)\right)^{1/2}$. Moreover, $\mathrm{M} = (\mu_\mathrm{M}, \gamma_\mathrm{M})$ is called a Pythagorean fuzzy number (PFN).

Definition 15.2.5 [53] A q-ROFS N in U is defined as $\mathrm{N} = \{(u, \mu_\mathrm{N}(u), \gamma_\mathrm{N}(u)) : u \in U\}$, where $\mu_\mathrm{N}, \gamma_\mathrm{N} : U \to [0, 1]$ such that $0 \leq \mu_\mathrm{N}^q(u) + \gamma_\mathrm{N}^q(u) \leq 1,\ q > 1$. The term $\Pi_\mathrm{N}(u) = \left(1 - \left(\mu_\mathrm{N}^q(u) + \gamma_\mathrm{N}^q(u)\right)\right)^{1/q}$ is called the hesitancy degree of u. Further, $\Psi = (\mu_\Psi, \gamma_\Psi)$ is called the q-rung orthopair fuzzy number (q-ROFN).

Definition 15.2.6 [62] A CFS P in U is defined as $\mathrm{P} = \{(u, \mu_\mathrm{P}(u)) : u \in U\}$, where $\mu_\mathrm{P} : U \to \{\zeta : \zeta \in \mathrm{P}, |\zeta| \leq 1\}$ such that $\mu_\mathrm{P}(u) = \zeta = x + iy$ provided $0 \leq |\zeta| \leq 1$. Otherwise, $\mu_\mathrm{P}(u) = \tau_\mathrm{P}(u).e^{i2\pi\omega_{\tau_\mathrm{P}}(u)}$ satisfying the conditions $0 \leq \tau_\mathrm{P}(u) \leq 1$ and $0 \leq \omega_{\tau_\mathrm{P}}(u) \leq 1$.

Definition 15.2.7 [3] A CIFS K in U is defined as $\mathrm{K} = \{(u, \mu_{\mathrm{K}}(u), \gamma_{\mathrm{K}}(u)) : u \in U\}$, where $\mu_{\mathrm{K}} : U \to \{\zeta_1 : \zeta \in \mathrm{K}, |\zeta_1| \le 1\}$, $\gamma_{\mathrm{K}} : U \to \{\zeta_2 : \zeta \in \mathrm{K}, |\zeta_2| \le 1\}$ such that $\mu_{\mathrm{K}}(u) = \zeta_1 = x_1 + iy_1$, $\gamma_{\mathrm{K}}(u) = \zeta_2 = x_2 + iy_2$ provided $0 \le |\zeta_1| + |\zeta_2| \le 1$.

Otherwise, $\mu_{\mathrm{K}}(u) = \tau_{\mathrm{K}}(u).e^{i2\pi\omega_{\tau_{\mathrm{K}}}(u)}$ and $\gamma_{\mathrm{K}}(u) = \sigma_{\mathrm{K}}(u).e^{i2\pi\omega_{\sigma_{\mathrm{K}}}(u)}$ satisfying the conditions $0 \le \tau_{\mathrm{K}}(u) + \sigma_{\mathrm{K}}(u) \le 1$ and $0 \le \omega_{\tau_{\mathrm{K}}}(u) + \omega_{\sigma_{\mathrm{K}}}(u) \le 1$.

Definition 15.2.8 [1] A CPFS H in U is defined as $\mathrm{H} = \{(u, \mu_{\mathrm{H}}(u), \gamma_{\mathrm{H}}(u)) : u \in U\}$, where $\mu_{\mathrm{H}} : U \to \{\zeta_1 : \zeta \in \mathrm{H}, |\zeta_1| \le 1\}$, $\gamma_{\mathrm{H}} : U \to \{\zeta_2 : \zeta \in \mathrm{H}, |\zeta_2| \le 1\}$ such that $\mu_{\mathrm{H}}(u) = \zeta_1 = x_1 + iy_1$, $\gamma_{\mathrm{H}}(u) = \zeta_2 = x_2 + iy_2$ provided $0 \le |\zeta_1|^2 + |\zeta_2|^2 \le 1$.

Otherwise, $\mu_{\mathrm{H}}(u) = \tau_{\mathrm{H}}(u).e^{i2\pi\omega_{\tau_{\mathrm{H}}}(u)}$ and $\gamma_{\mathrm{H}}(u) = \sigma_{\mathrm{H}}(u).e^{i2\pi\omega_{\sigma_{\mathrm{H}}}(u)}$ satisfying the conditions $0 \le \tau_{\mathrm{H}}^2(u) + \sigma_{\mathrm{H}}^2(u) \le 1$ and $0 \le \omega_{\tau_{\mathrm{H}}}^2(u) + \omega_{\sigma_{\mathrm{H}}}^2(u) \le 1$.

Furthermore, the term $\Pi_{\mathrm{H}}(u) = \kappa e^{i2\pi\omega_\kappa(u)}$ such that $\kappa = \left(1 - \left(\tau_{\mathrm{H}}^2(u) + \sigma_{\mathrm{H}}^2(u)\right)\right)^{1/2}$ and $\omega_\kappa(u) = \left(1 - \left(\omega_{\tau_{\mathrm{H}}}^2(u) + \omega_{\sigma_{\mathrm{H}}}^2(u)\right)\right)^{1/2}$ is called the complex hesitancy degree of u. Also, $\mathrm{H} = \left(\tau_{\mathrm{H}} e^{i2\pi\omega_{\tau_{\mathrm{H}}}}, \sigma_{\mathrm{H}} e^{i2\pi\omega_{\sigma_{\mathrm{H}}}}\right)$ is called the complex Pythagorean fuzzy number (CPFN).

Definition 15.2.9 [40] A Cq-ROFS T in U is defined as $\mathrm{T} = \{(u, \mu_{\mathrm{T}}(u), \gamma_{\mathrm{T}}(u)) : u \in U\}$ where $\mu_{\mathrm{T}} : U \to \{\zeta_1 : \zeta \in \mathrm{T}, |\zeta_1| \le 1\}$, $\gamma_{\mathrm{T}} : U \to \{\zeta_2 : \zeta \in \mathrm{T}, |\zeta_2| \le 1\}$ such that $\mu_{\mathrm{T}}(u) = \zeta_1 = x_1 + iy_1$, $\gamma_{\mathrm{T}}(u) = \zeta_2 = x_2 + iy_2$ provided $0 \le |\zeta_1|^q + |\zeta_2|^q \le 1$.

Otherwise, $\mu_{\mathrm{T}}(u) = \tau_{\mathrm{T}}(u).e^{i2\pi\omega_{\tau_{\mathrm{T}}}(u)}$ and $\gamma_{\mathrm{T}}(u) = \sigma_{\mathrm{T}}(u).e^{i2\pi\omega_{\sigma_{\mathrm{T}}}(u)}$ satisfying the condition $0 \le \tau_{\mathrm{T}}^q(u) + \sigma_{\mathrm{T}}^q(u) \le 1$ and $0 \le \omega_{\tau_{\mathrm{T}}}^q(u) + \omega_{\sigma_{\mathrm{T}}}^q(u) \le 1$.

Furthermore, the term $\Pi_{\mathrm{T}}(u) = \kappa e^{i2\pi\omega_\kappa(u)}$ such that $\kappa = \left(1 - \left(\tau_{\mathrm{T}}^q(u) + \sigma_{\mathrm{T}}^q(u)\right)\right)^{1/q}$ and $\omega_\kappa(u) = \left(1 - \left(\omega_{\tau_{\mathrm{T}}}^q(u) + \omega_{\sigma_{\mathrm{T}}}^q(u)\right)\right)^{1/q}$ is called the complex hesitancy degree of u. Also, $\mathrm{T} = \left(\tau_{\mathrm{T}} e^{i2\pi\omega_{\tau_{\mathrm{T}}}}, \sigma_{\mathrm{T}} e^{i2\pi\omega_{\sigma_{\mathrm{T}}}}\right)$ is called the complex q-rung orthopair fuzzy number (Cq-ROFN).

15.3 Interval Complex q-Rung Orthopair Fuzzy Sets

In this section, we first give the notion of interval complex q-rung orthopair fuzzy sets (IVCq-ROFSs), then we propose some basic operations on them.

Definition 15.3.1 [24] An IVCq-ROFS Ψ in X is defined as $\Psi = \{(x, \mu_\Psi(x), \gamma_\Psi(x)) : x \in X\}$, where $\mu_\Psi : X \to \{[\zeta_1^l, \zeta_1^u] : 0 \le |\zeta_1^l| \le |\zeta_1^u| \le 1\}$, $\gamma_\Psi : X \to \{[\zeta_2^l, \zeta_2^u] : 0 \le |\zeta_2^l| \le |\zeta_2^u| \le 1\}$ provided $0 \le |\zeta_1^l|^q + |\zeta_2^l|^q \le 1$ and $0 \le |\zeta_1^u|^q + |\zeta_2^2|^q \le 1$.

Otherwise, $\mu_\Psi(x) = \left[\tau_\Psi^l(x).e^{i2\pi\omega_{\tau_\Psi^l}(x)}, \tau_\Psi^u(x).e^{i2\pi\omega_{\tau_\Psi^u}(x)}\right]$ and $\gamma_\Psi(x) = \left[\sigma_\Psi^l(x).e^{i2\pi\omega_{\sigma_\Psi^l}(x)}, \sigma_\Psi^u(x).e^{i2\pi\omega_{\sigma_\Psi^u}(x)}\right]$ satisfying the conditions $0 \le (\tau_\Psi^l(x))^q + (\sigma_\Psi^l(x))^q \le 1$, $0 \le (\tau_\Psi^u(x))^q + (\sigma_\Psi^u(x))^q \le 1$ and $0 \le \left(\omega_{\tau_\Psi^l}(x)\right)^q + \left(\omega_{\sigma_\Psi^l}(x)\right)^q \le 1$ and $0 \le \left(\omega_{\tau_\Psi^u}(x)\right)^q + \left(\omega_{\sigma_\Psi^u}(x)\right)^q \le 1$.

Also, $0 \le \tau^l_\Psi(x) \le \tau^u_\Psi(x) \le 1,\ 0 \le \sigma^l_\Psi(x) \le \sigma^u_\Psi(x) \le 1$, and $0 \le \omega_{\tau^l_\Psi}(x) \le \omega_{\tau^u_\Psi}(x) \le 1,\ 0 \le \omega_{\sigma^l_\Psi}(x) \le \omega_{\sigma^u_\Psi}(x) \le 1$.

Furthermore, the term $\Pi_\Psi(x)=[\kappa^l e^{i2\pi\omega_{\kappa^l}(x)}, \kappa^u e^{i2\pi\omega_{\kappa^u}(x)}]$ such that $\kappa^l=\left(1-\left(\left(\tau^l_\Psi(x)\right)^q+\left(\sigma^l_\Psi(x)\right)^q\right)\right)^{1/q}$, $\kappa^u=(1-((\tau^u_\Psi(x))^q+(\sigma^u_\Psi(x))^q))^{1/q}$ and $\omega_{\kappa^l}(x)=\left(1-\left(\left(\omega_{\tau^l_\Psi}(x)\right)^q+\left(\omega_{\sigma^l_\Psi}(x)\right)^q\right)\right)^{1/q}$, $\omega_{\kappa^u}(x)=\left(1-\left(\left(\omega_{\tau^u_\Psi}(x)\right)^q+\left(\omega_{\sigma^u_\Psi}(x)\right)^q\right)\right)^{1/q}$ is called the interval complex hesitancy degree of u. Also, $\Psi=\left(\left[\tau^l e^{i2\pi\omega_{\tau^l}}, \tau^u e^{i2\pi\omega_{\tau^u}}\right], \left[\sigma^l e^{i2\pi\omega_{\sigma^l}}, \sigma^u e^{i2\pi\omega_{\sigma^u}}\right]\right)$ is called the interval complex q-rung orthopair fuzzy number (IVCq-ROFN).

For $\tau^l = \tau^u$ and $\sigma^l = \sigma^u$, the IVCq-ROFN reduces to Cq-ROFN.

Here, $\mu_\Psi(x)$ and $\gamma_\Psi(x)$ are IVCq-ROFNs in polar or Cartesian form.

We consider an example of IVCq-ROFN of the form $\left(\left[0.49e^{i(0.34)}, 0.65e^{i(0.56)}\right], \left[0.56e^{i(0.65)}, 0.75e^{i(0.7)}\right]\right)$. The advantage of using interval is that the decision-makers may provide more than one complex-valued membership and non-membership grade of an object of the universe within a stipulated time period, i.e., they are not certain about the grades. On the other hand in Cq-ROFS, the complex-valued membership and non-membership grades are precise. Therefore, IVCq-ROFS and Cq-ROFS models are applicable under different environments.

Definition 15.3.2 For any three IVCq-ROFNSs $\Psi_1 = \left\{\left[\tau^l_{\Psi_1} e^{i2\pi\omega_{\tau^l_{\Psi_1}}}, \tau^u_{\Psi_1} e^{i2\pi\omega_{\tau^u_{\Psi_1}}}\right], \left[\sigma^l_{\Psi_1} e^{i2\pi\omega_{\sigma^l_{\Psi_1}}}, \sigma^u_{\Psi_1} e^{i2\pi\omega_{\sigma^u_{\Psi_1}}}\right]\right\}$, $\Psi_2 = \left\{\left[\tau^l_{\Psi_2} e^{i2\pi\omega_{\tau^l_{\Psi_2}}}, \tau^u_{\Psi_2} e^{i2\pi\omega_{\tau^u_{\Psi_2}}}\right], \left[\sigma^l_{\Psi_2} e^{i2\pi\omega_{\sigma^l_{\Psi_2}}}, \sigma^u_{\Psi_2} e^{i2\pi\omega_{\sigma^u_{\Psi_2}}}\right]\right\}$, and $\Psi_3 = \left\{\left[\tau^l_{\Psi_3} e^{i2\pi\omega_{\tau^l_{\Psi_3}}}, \tau^u_{\Psi_3} e^{i2\pi\omega_{\tau^u_{\Psi_3}}}\right], \left[\sigma^l_{\Psi_3} e^{i2\pi\omega_{\sigma^l_{\Psi_3}}}, \sigma^u_{\Psi_3} e^{i2\pi\omega_{\sigma^u_{\Psi_3}}}\right]\right\}$,

(1) $\Psi_1 \subseteq \Psi_2$ iff $\tau^l_{\Psi_1} \le \tau^l_{\Psi_2}, \tau^u_{\Psi_1} \le \tau^u_{\Psi_2}, \omega_{\tau^l_{\Psi_1}} \le \omega_{\tau^l_{\Psi_2}}, \omega_{\tau^u_{\Psi_1}} \le \omega_{\tau^u_{\Psi_2}}$ and $\sigma^l_{\Psi_1} \ge \sigma^l_{\Psi_2}, \sigma^u_{\Psi_1} \ge \sigma^u_{\Psi_2}, \omega_{\sigma^l_{\Psi_1}} \ge \omega_{\sigma^l_{\Psi_2}}, \omega_{\sigma^u_{\Psi_1}} \ge \omega_{\sigma^u_{\Psi_2}}$.

(2) $\Psi_1 = \Psi_2$ iff $\tau^l_{\Psi_1} = \tau^l_{\Psi_2}, \tau^u_{\Psi_1} = \tau^u_{\Psi_2}, \omega_{\tau^l_{\Psi_1}} = \omega_{\tau^l_{\Psi_2}}, \omega_{\tau^u_{\Psi_1}} = \omega_{\tau^u_{\Psi_2}}$ and $\sigma^l_{\Psi_1} = \sigma^l_{\Psi_2}, \sigma^u_{\Psi_1} = \sigma^u_{\Psi_2}, \omega_{\sigma^l_{\Psi_1}} = \omega_{\sigma^l_{\Psi_2}}, \omega_{\sigma^u_{\Psi_1}} = \omega_{\sigma^u_{\Psi_2}}$.

(3) $\Psi_1^c = \left\{\left[\sigma^l_{\Psi_1} e^{i2\pi\omega_{\sigma^l_{\Psi_1}}}, \sigma^u_{\Psi_1} e^{i2\pi\omega_{\sigma^u_{\Psi_1}}}\right], \left[\tau^l_{\Psi_1} e^{i2\pi\omega_{\tau^l_{\Psi_1}}}, \tau^u_{\Psi_1} e^{i2\pi\omega_{\tau^u_{\Psi_1}}}\right]\right\}$

(4)

$$(a)\ \Psi_1 \vee \Psi_2 = \left\{\begin{array}{l} \left[\begin{array}{l} \max\left(\tau^l_{\Psi_1}, \tau^l_{\Psi_2}\right) e^{i2\pi \max\left(\omega_{\tau^l_{\Psi_1}}, \omega_{\tau^l_{\Psi_2}}\right)}, \\ \max\left(\tau^u_{\Psi_1}, \tau^u_{\Psi_2}\right) e^{i2\pi \max\left(\omega_{\tau^u_{\Psi_1}}, \omega_{\tau^u_{\Psi_2}}\right)} \end{array}\right], \\ \left[\begin{array}{l} \min\left(\sigma^l_{\Psi_1}, \sigma^l_{\Psi_2}\right) e^{i2\pi \min\left(\omega_{\sigma^l_{\Psi_1}}, \omega_{\sigma^l_{\Psi_2}}\right)}, \\ \min\left(\sigma^u_{\Psi_1}, \sigma^u_{\Psi_2}\right) e^{i2\pi \min\left(\omega_{\sigma^u_{\Psi_1}}, \omega_{\sigma^u_{\Psi_2}}\right)} \end{array}\right] \end{array}\right\}$$

(b) $\Psi_1 \wedge \Psi_2 = \left\{ \begin{array}{l} \left[\min\left(\tau^l_{\Psi_1}, \tau^l_{\Psi_2}\right) e^{i2\pi \min\left(\omega_{\tau^l_{\Psi_1}}, \omega_{\tau^l_{\Psi_2}}\right)}, \right. \\ \left. \min\left(\tau^u_{\Psi_1}, \tau^u_{\Psi_2}\right) e^{i2\pi \min\left(\omega_{\tau^u_{\Psi_1}}, \omega_{\tau^u_{\Psi_2}}\right)} \right], \\ \left[\max\left(\sigma^l_{\Psi_1}, \sigma^l_{\Psi_2}\right) e^{i2\pi \max\left(\omega_{\sigma^l_{\Psi_1}}, \omega_{\sigma^l_{\Psi_2}}\right)}, \right. \\ \left. \max\left(\sigma^u_{\Psi_1}, \sigma^u_{\Psi_2}\right) e^{i2\pi \max\left(\omega_{\sigma^u_{\Psi_1}}, \omega_{\sigma^u_{\Psi_2}}\right)} \right] \end{array} \right\}$

Theorem 15.3.3 *For any three IVCq-ROFNSs* $\Psi_1 = \left\{\left[\tau^l_{\Psi_1} e^{i2\pi\omega_{\tau^l_{\Psi_1}}}, \tau^u_{\Psi_1} e^{i2\pi\omega_{\tau^u_{\Psi_1}}}\right], \left[\sigma^l_{\Psi_1} e^{i2\pi\omega_{\sigma^l_{\Psi_1}}}, \sigma^u_{\Psi_1} e^{i2\pi\omega_{\sigma^u_{\Psi_1}}}\right]\right\}$, $\Psi_2 = \left\{\left[\tau^l_{\Psi_2} e^{i2\pi\omega_{\tau^l_{\Psi_2}}}, \tau^u_{\Psi_2} e^{i2\pi\omega_{\tau^u_{\Psi_2}}}\right], \left[\sigma^l_{\Psi_2} e^{i2\pi\omega_{\sigma^l_{\Psi_2}}}, \sigma^u_{\Psi_2} e^{i2\pi\omega_{\sigma^u_{\Psi_2}}}\right]\right\}$, *and* $\Psi_3 = \left\{\left[\tau^l_{\Psi_3} e^{i2\pi\omega_{\tau^l_{\Psi_3}}}, \tau^u_{\Psi_3} e^{i2\pi\omega_{\tau^u_{\Psi_3}}}\right], \left[\sigma^l_{\Psi_3} e^{i2\pi\omega_{\sigma^l_{\Psi_3}}}, \sigma^u_{\Psi_3} e^{i2\pi\omega_{\sigma^u_{\Psi_3}}}\right]\right\}$,

(a) $\Psi_1 \vee \Psi_2 = \Psi_2 \vee \Psi_1$ *and* $\Psi_1 \wedge \Psi_2 = \Psi_2 \wedge \Psi_1$ *(commutativity laws)*
(b) $\Psi_1 \vee (\Psi_2 \vee \Psi_3) = (\Psi_1 \vee \Psi_2) \vee \Psi_3$ *and* $\Psi_1 \wedge (\Psi_2 \wedge \Psi_3) = (\Psi_1 \wedge \Psi_2) \wedge \Psi_3$ *(associativity laws)*
(c) $\Psi_1 \vee (\Psi_2 \wedge \Psi_3) = (\Psi_1 \vee \Psi_2) \wedge (\Psi_1 \vee \Psi_3)$ *and* $\Psi_1 \wedge (\Psi_2 \vee \Psi_3) = (\Psi_1 \wedge \Psi_2) \vee (\Psi_1 \wedge \Psi_3)$ *(distributivity laws)*
(d) $\Psi_1 \vee (\Psi_1 \wedge \Psi_2) = \Psi_1$ *and* $\Psi_1 \wedge (\Psi_1 \vee \Psi_2) = \Psi_1$ *(absorption laws)*
(e) $(\Psi_1 \vee \Psi_2)^c = \Psi_1^c \wedge \Psi_2^c$ *and* $(\Psi_1 \wedge \Psi_2)^c = \Psi_1^c \vee \Psi_2^c$ *(De Morgan's laws)*

Proof By using the property (1) to (4), all proofs are straightforward.

Definition 15.3.4 For any two IVCq-ROFNSs $\Psi_1 = \left\{\left[\tau^l_{\Psi_1} e^{i2\pi\omega_{\tau^l_{\Psi_1}}}, \tau^u_{\Psi_1} e^{i2\pi\omega_{\tau^u_{\Psi_1}}}\right], \left[\sigma^l_{\Psi_1} e^{i2\pi\omega_{\sigma^l_{\Psi_1}}}, \sigma^u_{\Psi_1} e^{i2\pi\omega_{\sigma^u_{\Psi_1}}}\right]\right\}$ and $\Psi_2 = \left\{\left[\tau^l_{\Psi_2} e^{i2\pi\omega_{\tau^l_{\Psi_2}}}, \tau^u_{\Psi_2} e^{i2\pi\omega_{\tau^u_{\Psi_2}}}\right], \left[\sigma^l_{\Psi_2} e^{i2\pi\omega_{\sigma^l_{\Psi_2}}}, \sigma^u_{\Psi_2} e^{i2\pi\omega_{\sigma^u_{\Psi_2}}}\right]\right\}$,

$$\Psi_{1\cdot} \oplus \Psi_2$$

$$= \left\{ \begin{array}{l} \left[\begin{array}{l} \left(\left(\tau^l_{\Psi_{1\cdot}}\right)^q + \left(\tau^l_{\Psi_{2\cdot}}\right)^q - \left(\tau^l_{\Psi_{1\cdot}}\right)^q . \left(\tau^l_{\Psi_{2\cdot}}\right)^q\right)^{1/q} \\ .e^{2\pi\left(\left(\omega_{\tau^l_{\Psi_{1\cdot}}}\right)^q + \left(\omega_{\tau^l_{\Psi_{2\cdot}}}\right)^q - \left(\omega_{\tau^l_{\Psi_{1\cdot}}}\right)^q . \left(\omega_{\tau^l_{\Psi_2}}\right)^q\right)^{1/q}}, \\ \left(\left(\tau^u_{\Psi_{1\cdot}}\right)^q + \left(\tau^u_{\Psi_2}\right)^q - \left(\tau^u_{\Psi_{1\cdot}}\right)^q . \left(\tau^u_{\Psi_2}\right)^q\right)^{1/q} \\ .e^{2\pi\left(\left(\omega_{\tau^u_{\Psi_{1\cdot}}}\right)^q + \left(\omega_{\tau^u_{\Psi_2}}\right)^q - \left(\omega_{\tau^u_{\Psi_{1\cdot}}}\right)^q . \left(\omega_{\tau^u_{\Psi_2}}\right)^q\right)^{1/q}} \end{array} \right] \\ \left[\sigma^l_{\Psi_{1\cdot}} . \sigma^l_{\Psi_2} . e^{i2\pi\left(\omega_{\sigma^l_{\Psi_{1\cdot}}} . \omega_{\sigma^l_{\Psi_2}}\right)}, \sigma^u_{\Psi_{1\cdot}} . \sigma^u_{\Psi_2} . e^{i2\pi\left(\omega_{\sigma^u_{\Psi_{1\cdot}}} . \omega_{\sigma^u_{\Psi_2}}\right)} \right] \end{array} \right\}$$

Definition 15.3.5 For any two IVCq-ROFNSs $\Psi_1 = \left\{\left[\tau^l_{\Psi_1} e^{i2\pi\omega_{\tau^l_{\Psi_1}}}, \tau^u_{\Psi_1} e^{i2\pi\omega_{\tau^u_{\Psi_1}}}\right], \left[\sigma^l_{\Psi_1} e^{i2\pi\omega_{\sigma^l_{\Psi_1}}}, \sigma^u_{\Psi_1} e^{i2\pi\omega_{\sigma^u_{\Psi_1}}}\right]\right\}$ and $\Psi_2 = \left\{\left[\tau^l_{\Psi_2} e^{i2\pi\omega_{\tau^l_{\Psi_2}}}, \tau^u_{\Psi_2} e^{i2\pi\omega_{\tau^u_{\Psi_2}}}\right], \left[\sigma^l_{\Psi_2} e^{i2\pi\omega_{\sigma^l_{\Psi_2}}}, \sigma^u_{\Psi_2} e^{i2\pi\omega_{\sigma^u_{\Psi_2}}}\right]\right\}$,

$$\Psi_1 \otimes \Psi_2 = \left\{\begin{array}{l} \left[\tau^l_{\Psi_1}.\tau^l_{\Psi_2}.e^{i2\pi\left(\omega_{\tau^l_{\Psi_1}}.\omega_{\tau^l_{\Psi_2}}\right)}, \tau^u_{\Psi_1}.\tau^u_{\Psi_2}.e^{i2\pi\left(\omega_{\tau^u_{\Psi_1}}.\omega_{\tau^u_{\Psi_2}}\right)}\right], \\ \left[\begin{array}{l} \left(\left(\sigma^l_{\Psi_1}\right)^q + \left(\sigma^l_{\Psi_2}\right)^q - \left(\sigma^l_{\Psi_1}\right)^q.\left(\sigma^l_{\Psi_2}\right)^q\right)^{1/q} \\ .e^{2\pi\left(\left(\omega_{\sigma^l_{\Psi_1}}\right)^q + \left(\omega_{\sigma^l_{\Psi_2}}\right)^q - \left(\omega_{\sigma^l_{\Psi_1}}\right)^q.\left(\omega_{\sigma^l_{\Psi_2}}\right)^q\right)^{1/q}}, \\ \left(\left(\sigma^u_{\Psi_1}\right)^q + \left(\sigma^u_{\Psi_2}\right)^q - \left(\sigma^u_{\Psi_1}\right)^q.\left(\sigma^u_{\Psi_2}\right)^q\right)^{1/q} \\ .e^{2\pi\left(\left(\omega_{\sigma^u_{\Psi_1}}\right)^q + \left(\omega_{\sigma^u_{\Psi_2}}\right)^q - \left(\omega_{\sigma^u_{\Psi_1}}\right)^q.\left(\omega_{\sigma^u_{\Psi_2}}\right)^q\right)^{1/q}} \end{array}\right] \end{array}\right\}$$

Definition 15.3.6 For any real $\lambda>0$ and $\Psi_k = \left\{\left[\tau^l_k e^{i2\pi\omega_{\tau^l_k}}, \tau^u_k e^{i2\pi\omega_{\tau^u_k}}\right], \left[\sigma^l_k e^{i2\pi\omega_{\sigma^l_k}}, \sigma^u_k e^{i2\pi\omega_{\sigma^u_k}}\right]\right\}$, where $k = 1, 2$ be any two IVCq-ROFNs, then

(a) $$\lambda\Psi_k = \left\{\begin{array}{l} \left[\begin{array}{l} \left(1-\left(1-\left(\tau^l_k\right)^q\right)^\lambda\right)^{1/q}.e^{i2\pi\left(1-\left(1-\left(\omega_{\tau^l_k}\right)^q\right)^\lambda\right)^{1/q}}, \\ \left(1-\left(1-\left(\tau^u_k\right)^q\right)^\lambda\right)^{1/q}.e^{i2\pi\left(1-\left(1-\left(\omega_{\tau^u_k}\right)^q\right)^\lambda\right)^{1/q}} \end{array}\right], \\ \left[\left(\sigma^l_k\right)^\lambda.e^{i2\pi\left(\omega_{\sigma^l_k}\right)^\lambda}, \left(\sigma^u_k\right)^\lambda.e^{i2\pi\left(\omega_{\sigma^u_k}\right)^\lambda}\right] \end{array}\right\}, k = 1, 2$$

(b) $$\Psi_k^\lambda = \left\{\begin{array}{l} \left[\left(\tau^l_k\right)^\lambda.e^{i2\pi\left(\omega_{\tau^l_k}\right)^\lambda}, \left(\tau^u_k\right)^\lambda.e^{i2\pi\left(\omega_{\tau^u_k}\right)^\lambda}\right], \\ \left[\begin{array}{l} \left(1-\left(1-\left(\sigma^l_k\right)^q\right)^\lambda\right)^{1/q}.e^{i2\pi\left(1-\left(1-\left(\omega_{\sigma^l_k}\right)^q\right)^\lambda\right)^{1/q}}, \\ \left(1-\left(1-\left(\sigma^u_k\right)^q\right)^\lambda\right)^{1/q}.e^{i2\pi\left(1-\left(1-\left(\omega_{\sigma^u_k}\right)^q\right)^\lambda\right)^{1/q}} \end{array}\right], \end{array}\right\}, k = 1, 2$$

Theorem 15.3.7 *Let* $\Psi_j = \left\{\left[\tau^l_j e^{i2\pi\omega_{\tau^l_j}}, \tau^u_j e^{i2\pi\omega_{\tau^u_j}}\right], \left[\sigma^l_j e^{i2\pi\omega_{\sigma^l_j}}, \sigma^u_j e^{i2\pi\omega_{\sigma^u_j}}\right]\right\}, j = 1,$ *2 be any two IVCq-ROFNs and let* $\lambda_1, \lambda_2 > 0$ *be any real numbers, then we have*

(a) $\Psi_1 \oplus \Psi_2 = \Psi_2 \oplus \Psi_1$
(b) $\Psi_1 \otimes \Psi_2 = \Psi_2 \otimes \Psi_1$
(c) $\lambda_1(\Psi_1 \oplus \Psi_2) = \lambda_1\Psi_1 \oplus \lambda_1\Psi_2$
(d) $\lambda_1\Psi_1 \oplus \lambda_2\Psi_1 = (\lambda_1 + \lambda_2)\Psi_1$
(e) $\Psi_1^{\lambda_1} \otimes \Psi_1^{\lambda_2} = \Psi_1^{\lambda_1+\lambda_2}$
(f) $\Psi_1^{\lambda_1} \otimes \Psi_2^{\lambda_1} = (\Psi_1 \otimes \Psi_2)^{\lambda_1}$

Proof All proofs are obvious. So, left these as an exercise for the readers.

Definition 15.3.8 A score function S and an accuracy function A defined on an IVCq-ROFN $\Psi = \left(\left[\tau^l e^{i2\omega_{\tau^l}}, \tau^u e^{i2\omega_{\tau^u}}\right], \left[\sigma^l e^{i2\omega_{\sigma^l}}, \sigma^u e^{i2\omega_{\sigma^u}}\right]\right)$ is given by

$$S(\Psi) = \left|\left(\left(\tau^l\right)^q + (\tau^u)^q - \left(\sigma^l\right)^q - (\sigma^u)^q\right) + \left((\omega_{\tau^l})^q + (\omega_{\tau^u})^q - \left(\omega_{\sigma^l}\right)^q - (\omega_{\sigma^u})^q\right)\right|,$$

where $S(\Psi) \in [-1, 1]$.

$$A(\Psi) = \left|\left((\tau^u)^q - \left(\tau^l\right)^q + (\sigma^u)^q - \left(\sigma^l\right)^q\right) + \left((\omega_{\tau^u})^q - (\omega_{\tau^l})^q + (\omega_{\sigma^u})^q - (\omega_{\sigma^l})^q\right)\right|,$$

where $A(\Psi) \in [0, 1]$.

Definition 15.3.9 An order relation between two IVCq-ROFNs Ψ_1 and Ψ_2 can be defined as

1. If $S(\Psi_1) > S(\Psi_2)$ then $\Psi_1 > \Psi_2$.
2. If $S(\Psi_1) = S(\Psi_2)$ and
   (i) if $A(\Psi_1) > A(\Psi_2)$, then $\Psi_1 > \Psi_2$.
   (ii) if $A(\Psi_1) = A(\Psi_2)$, then $\Psi_1 = \Psi_2$

Theorem 15.3.10 *Let* $\Psi_j = \left\{\left[\tau_j^l e^{i2\pi\omega_{\tau_j^l}}, \tau_j^u e^{i2\pi\omega_{\tau_j^u}}\right], \left[\sigma_j^l e^{i2\pi\omega_{\sigma_j^l}}, \sigma_j^u e^{i2\pi\omega_{\sigma_j^u}}\right]\right\}$, $j = 1, 2$ *be any two IVCq-ROFNs and their associated complement be* $\Psi_j^c = \left\{\left[\sigma_j^l e^{i2\pi\omega_{\sigma_j^l}}, \sigma_j^u e^{i2\pi\omega_{\sigma_j^u}}\right], \left[\tau_j^l e^{i2\pi\omega_{\tau_j^l}}, \tau_j^u e^{i2\pi\omega_{\tau_j^u}}\right]\right\}$. *Then* $S(\Psi_1) \geq S(\Psi_2)$ *iff* $S\left(\Psi_1^c\right) \leq S\left(\Psi_2^c\right)$.

Proof By Definition 15.3.8, we have

$$S(\Psi_1) = \left|\left((\tau_1^l)^q + (\tau_1^u)^q - (\sigma_1^l)^q - (\sigma_1^u)^q\right) + \left(\left(\omega_{\tau_1^l}\right)^q + (\omega_{\tau_1^u})^q - \left(\omega_{\sigma_1^l}\right)^q - (\omega_{\sigma_1^u})^q\right)\right|$$

and

$$S(\Psi_2) = \left|\left((\tau_2^l)^q + (\tau_2^u)^q - (\sigma_2^l)^q - (\sigma_2^u)^q\right) + \left(\left(\omega_{\tau_2^l}\right)^q + (\omega_{\tau_2^u})^q - \left(\omega_{\sigma_2^l}\right)^q - (\omega_{\sigma_2^u})^q\right)\right|$$

Now,

$$S(\Psi_1) \geq S(\Psi_2)$$

$$\text{iff } \left|\left((\tau_1^l)^q + (\tau_1^u)^q - (\sigma_1^l)^q - (\sigma_1^u)^q\right) + \left(\left(\omega_{\tau_1^l}\right)^q + (\omega_{\tau_1^u})^q - \left(\omega_{\sigma_1^l}\right)^q - (\omega_{\sigma_1^u})^q\right)\right|$$
$$\geq \left|\left((\tau_2^l)^q + (\tau_2^u)^q - (\sigma_2^l)^q - (\sigma_2^u)^q\right) + \left(\left(\omega_{\tau_2^l}\right)^q + (\omega_{\tau_2^u})^q - \left(\omega_{\sigma_2^l}\right)^q - (\omega_{\sigma_2^u})^q\right)\right|$$
$$\text{Iff } \left|\left((\sigma_1^l)^q + (\sigma_1^u)^q - (\tau_1^l)^q - (\tau_1^u)^q\right) + \left(\left(\omega_{\sigma_1^l}\right)^q + (\omega_{\sigma_1^u})^q - \left(\omega_{\tau_1^l}\right)^q - (\omega_{\tau_1^u})^q\right)\right|$$
$$\leq \left|\left((\sigma_2^l)^q + (\sigma_2^u)^q - (\tau_2^l)^q - (\tau_2^u)^q\right) + \left(\left(\omega_{\sigma_2^l}\right)^q + (\omega_{\sigma_2^u})^q - \left(\omega_{\tau_2^l}\right)^q - (\omega_{\tau_2^u})^q\right)\right|$$

Thus, $S(\Psi_1) \geq S(\Psi_2)$ iff $S(\Psi_1^c) \leq S(\Psi_2^c)$.

15.4 Interval Complex q-Rung Orthopair Fuzzy Aggregate Operators for MADM Problems

In this section, we propose two types of aggregate operators such as interval complex q-rung orthopair fuzzy weight aggregate (IVCq-ROFWA) operator and the interval complex q-rung orthopair fuzzy group aggregate (IVCq-ROFGA) operator. Both the operators are based on IVCq-ROFNs. While defining these operators, we consider $W = (w_1, w_2, \ldots, w_k)^t$ as a weight vector, where $\sum_{j=1}^n w_j = 1$ and $w_j \in [0, 1], \forall j$. Also, $\Psi_j = \left(\left[\tau_j^l e^{i2\pi\omega_{\tau_j^l}}, \tau_j^u e^{i2\pi\omega_{\tau_j^u}}\right], \left[\sigma_j^l e^{i2\pi\omega_{\sigma_j^l}}, \sigma_j^u e^{i2\pi\omega_{\sigma_j^u}}\right] : j = 1, 2\ldots, n\right)$ represent the set of n IVCq-ROFNs.

Definition 15.4.1 The IVCq-ROFWA operator can be described by a function IVCq-ROFWA: $P^n \to P$ such that IVCq-ROFWA $(\Psi_1, \Psi_2, \ldots.., \Psi_n) = w_1\Psi_1 \oplus w_2\Psi_2 \oplus \ldots\ldots \oplus w_n\Psi_n = \oplus_{j=1}^n w_j\Psi_j$

Theorem 15.4.2 *For* $\Psi_j = \left(\left[\tau_j^l e^{i2\pi\omega_{\tau_j^l}}, \tau_j^u e^{i2\pi\omega_{\tau_j^u}}\right], \left[\sigma_j^l e^{i2\pi\omega_{\sigma_j^l}}, \sigma_j^u e^{i2\pi\omega_{\sigma_j^u}}\right] : j = 1, 2\ldots, n\right)$,

$$IVCq\text{-}ROFWA(\Psi_1, \Psi_2, \ldots.., \Psi_n) =$$
$$\left(\left[\left(1 - \prod_{j=1}^n \left(1 - (\tau_j^l)^q\right)^{w_j}\right)^{1/q} .e^{i2\pi\left(1-\prod_{j=1}^n\left(1-\omega_{\tau_j^l}^q\right)^{w_j}\right)^{1/q}}, \left(1 - \prod_{j=1}^n (1 - (\tau_j^u)^q)^{w_j}\right)^{1/q} .e^{i2\pi\left(1-\prod_{j=1}^n\left(1-\omega_{\tau_j^u}^q\right)^{w_j}\right)^{1/q}}\right], \left[\prod_{j=1}^n \left(\sigma_j^l\right)^{w_j} .e^{i2\pi\left(\prod_{j=1}^n \omega_{\sigma_j^l}^{w_j}\right)}, \prod_{j=1}^n \left(\sigma_j^u\right)^{w_j} .e^{i2\pi\left(\prod_{j=1}^n \omega_{\sigma_j^u}^{w_j}\right)}\right]\right)$$

Proof To prove this result, we use the concept of mathematical induction.

For n = 2,

IVCq-ROFWA $(\Psi_1, \Psi_2) = w_1\Psi_1 \oplus w_2\Psi_2$

$$= \left\{ \begin{array}{l} \left[\left(1-\left(1-(\tau_1^l)^q\right)^{w_1}\right)^{1/q} .e^{i2\pi\left(1-\left(1-\omega^q_{\tau_1^l}\right)^{w_1}\right)^{1/q}}, \right. \\ \left. \left(1-\left(1-(\tau_1^u)^q\right)^{w_1}\right)^{1/q} .e^{i2\pi\left(1-\left(1-\omega^q_{\tau_1^u}\right)^{w_1}\right)^{1/q}} \right], \\ \left[(\sigma_1^l)^{w_1}.e^{i2\pi\left(\omega_{\sigma_1^l}\right)^{w_1}}, (\sigma_1^u)^{w_1}.e^{i2\pi\left(\omega_{\sigma_1^u}\right)^{w_1}} \right] \end{array} \right\}$$

$$\oplus \left\{ \begin{array}{l} \left[\left(1-\left(1-(\tau_2^l)^q\right)^{w_2}\right)^{1/q} .e^{i2\pi\left(1-\left(1-\omega^q_{\tau_2^l}\right)^{w_2}\right)^{1/q}}, \right. \\ \left. \left(1-\left(1-(\tau_2^u)^q\right)^{w_2}\right)^{1/q} .e^{i2\pi\left(1-\left(1-\omega^q_{\tau_2^u}\right)^{w_2}\right)^{1/q}} \right], \\ \left[(\sigma_2^l)^{w_1}.e^{i2\pi\left(\omega_{\sigma_2^l}\right)^{w_2}}, (\sigma_2^u)^{w_1}.e^{i2\pi\left(\omega_{\sigma_2^u}\right)^{w_2}} \right] \end{array} \right\}$$

$$= \left\{ \begin{array}{l} \left[\begin{array}{l} \left(\begin{array}{l} \left(1-\left(1-(\tau_1^l)^q\right)^{w_1}\right) + \left(1-\left(1-(\tau_2^l)^q\right)^{w_2}\right) \\ -\left(1-\left(1-(\tau_1^l)^q\right)^{w_1}\right).\left(1-\left(1-(\tau_2^l)^q\right)^{w_2}\right) \end{array} \right)^{1/q} \\ .e^{i2\pi\left(\left(1-\left(1-\omega^q_{\tau_1^l}\right)^{w_1}\right)+\left(1-\left(1-\omega^q_{\tau_2^l}\right)^{w_2}\right)-\left(1-\left(1-\omega^q_{\tau_1^l}\right)^{w_1}\right).\left(1-\left(1-\omega^q_{\tau_2^l}\right)^{w_2}\right)\right)^{1/q}}, \\ \left(\begin{array}{l} \left(1-\left(1-(\tau_1^u)^q\right)^{w_1}\right) + \left(1-\left(1-(\tau_2^u)^q\right)^{w_2}\right) \\ -\left(1-\left(1-(\tau_1^u)^q\right)^{w_1}\right).\left(1-\left(1-(\tau_2^u)^q\right)^{w_2}\right) \end{array} \right)^{1/q} \\ .e^{i2\pi\left(\left(1-\left(1-\omega^q_{\tau_1^u}\right)^{w_1}\right)+\left(1-\left(1-\omega^q_{\tau_2^u}\right)^{w_2}\right)-\left(1-\left(1-\omega^q_{\tau_1^u}\right)^{w_1}\right).\left(1-\left(1-\omega^q_{\tau_2^u}\right)^{w_2}\right)\right)^{1/q}} \end{array} \right] \\ \left[(\sigma_1^l)^{w_1}.(\sigma_2^l)^{w_1} e^{i2\pi\left(\left(\omega_{\sigma_1^l}\right)^{w_1}.\left(\omega_{\sigma_2^l}\right)^{w_2}\right)}, (\sigma_1^u)^{w_1}.(\sigma_2^u)^{w_1} e^{i2\pi\left(\left(\omega_{\sigma_1^u}\right)^{w_1}.\left(\omega_{\sigma_2^u}\right)^{w_2}\right)} \right] \end{array} \right\}$$

After simplifying the terms in the above interval, we have

$$= \left\{ \begin{array}{l} \left(1-\left(1-(\tau_1^l)^q\right)^{w_1}.\left(1-(\tau_2^l)^q\right)^{w_2}\right)^{1/q} .e^{i2\pi\left(1-\left(1-\omega^q_{\tau_1^l}\right)^{w_1}.\left(1-\omega^q_{\tau_2^l}\right)^{w_2}\right)^{1/q}}, [\\ \left. \left(1-\left(1-(\tau_1^u)^q\right)^{w_1}.\left(1-(\tau_2^u)^q\right)^{w_2}\right)^{1/q} .e^{i2\pi\left(1-\left(1-\omega^q_{\tau_1^u}\right)^{w_1}.\left(1-\omega^q_{\tau_2^u}\right)^{w_2}\right)^{1/q}} \right], \\ \left[(\sigma_1^l)^{w_1}.(\sigma_2^l)^{w_1} e^{i2\pi\left(\left(\omega_{\sigma_1^l}\right)^{w_1}.\left(\omega_{\sigma_2^l}\right)^{w_2}\right)}, (\sigma_1^u)^{w_1}.(\sigma_2^u)^{w_1} e^{i2\pi\left(\left(\omega_{\sigma_1^u}\right)^{w_1}.\left(\omega_{\sigma_2^u}\right)^{w_2}\right)} \right] \end{array} \right\}$$

$$= \left\{ \begin{array}{l} \left[\left(1 - \coprod_{j=1}^{2}\left(1-(\tau_j^l)^q\right)^{w_j}\right)^{1/q} .e^{i2\pi\left(1-\coprod_{j=1}^{2}\left(1-\omega^q_{\tau_j^l}\right)^{w_j}\right)^{1/q}}, \right. \\ \left. \left(1 - \coprod_{j=1}^{2}(1-(\tau_j^u)^q)^{w_j}\right)^{1/q} .e^{i2\pi\left(1-\coprod_{j=1}^{2}\left(1-\omega^q_{\tau_j^u}\right)^{w_j}\right)^{1/q}} \right], \\ \left[\prod_{j=1}^{2}\left(\sigma_j^l\right)^{w_j} .e^{i2\pi\left(\prod_{j=1}^{2}\left(\omega_{\sigma_j^l}\right)^{w_j}\right)}, \prod_{j=1}^{2}\left(\sigma_j^u\right)^{w_j} .e^{i2\pi\left(\prod_{j=1}^{2}\left(\omega_{\sigma_j^u}\right)^{w_j}\right)} \right] \end{array} \right\}$$

For n = 2, the result is true.

Suppose the result is true for n = m. Then,

$$\text{IVCq-ROFWA}(\Psi_1, \Psi_2, \ldots, \Psi_m) = \left\{ \begin{array}{l} \left[\left(1 - \coprod_{j=1}^{m}\left(1-\left(\tau_j^l\right)^q\right)^{w_j}\right)^{1/q} .e^{i2\pi\left(1-\coprod_{j=1}^{m}\left(1-\omega^q_{\tau_j^l}\right)^{w_j}\right)^{1/q}}, \right. \\ \left. \left(1 - \coprod_{j=1}^{m}\left(1-\left(\tau_j^u\right)^q\right)^{w_j}\right)^{1/q} .e^{i2\pi\left(1-\coprod_{j=1}^{m}\left(1-\omega^q_{\tau_j^u}\right)^{w_j}\right)^{1/q}} \right], \\ \left[\prod_{j=1}^{m}\left(\sigma_j^l\right)^{w_j} .e^{i2\pi\left(\prod_{j=1}^{m}\left(\omega_{\sigma_j^l}\right)^{w_j}\right)}, \prod_{j=1}^{m}\left(\sigma_j^u\right)^{w_j} .e^{i2\pi\left(\prod_{j=1}^{m}\left(\omega_{\sigma_j^u}\right)^{w_j}\right)} \right] \end{array} \right\}$$

Now, we check for n = m + 1

$$\text{IVCq-ROFWA}(\Psi_1, \Psi_2, \ldots, \Psi_{m+1}) = \left\{ \begin{array}{l} \left[\left(1 - \coprod_{j=1}^{m}\left(1-(\tau_j^l)^q\right)^{w_j}\right)^{1/q} .e^{i2\pi\left(1-\coprod_{j=1}^{m}\left(1-\omega^q_{\tau_j^l}\right)^{w_j}\right)^{1/q}}, \right. \\ \left. \left(1 - \coprod_{j=1}^{m}(1-(\tau_j^u)^q)^{w_j}\right)^{1/q} .e^{i2\pi\left(1-\coprod_{j=1}^{m}\left(1-\omega^q_{\tau_j^u}\right)^{w_j}\right)^{1/q}} \right], \\ \left[\prod_{j=1}^{m}\left(\sigma_j^l\right)^{w_j} .e^{i2\pi\left(\prod_{j=1}^{m}\left(\omega_{\sigma_j^l}\right)^{w_j}\right)}, \prod_{j=1}^{m}\left(\sigma_j^u\right)^{w_j} .e^{i2\pi\left(\prod_{j=1}^{m}\left(\omega_{\sigma_j^u}\right)^{w_j}\right)} \right] \end{array} \right\} \oplus w_{m+1}\Psi_{m+1}$$

$$
= \left\{ \begin{array}{l} \left[\left(1 - \coprod_{j=1}^{m} \left(1 - (\tau_j^l)^q \right)^{w_j} \right)^{1/q} .e^{i2\pi \left(1 - \coprod_{j=1}^{m} \left(1 - \omega_{\tau_j^l}^q \right)^{w_j} \right)^{1/q}}, \right. \\ \left. \left(1 - \coprod_{j=1}^{m} (1 - (\tau_j^u)^q)^{w_j} \right)^{1/q} .e^{i2\pi \left(1 - \coprod_{j=1}^{m} \left(1 - \omega_{\tau_j^u}^q \right)^{w_j} \right)^{1/q}} \right], \\ \left[\prod_{j=1}^{m} \left(\sigma_j^l \right)^{w_j} .e^{i2\pi \left(\prod_{j=1}^{m} \left(\omega_{\sigma_j^l} \right)^{w_j} \right)}, \prod_{j=1}^{m} \left(\sigma_j^u \right)^{w_j} .e^{i2\pi \left(\prod_{j=1}^{m} \left(\omega_{\sigma_j^u} \right)^{w_j} \right)} \right] \end{array} \right\} \oplus
$$

$$
\left\{ \begin{array}{l} \left[\left(1 - \left(1 - (\tau_{m+1}^l)^q \right)^{w_{m+1}} \right)^{1/q} .e^{i2\pi \left(1 - \left(1 - \omega_{\tau_{m+1}^l}^q \right)^{w_{m+1}} \right)^{1/q}}, \right. \\ \left. \left(1 - (1 - (\tau_{m+1}^u)^q)^{w_{m+1}} \right)^{1/q} .e^{i2\pi \left(1 - \left(1 - \omega_{\tau_{m+1}^u}^q \right)^{w_{m+1}} \right)^{1/q}} \right], \\ \left[(\sigma_{m+1}^l)^{w_{m+1}} .e^{i2\pi \left(\omega_{\sigma_{m+1}^l} \right)^{w_{m+1}}}, (\sigma_{m+1}^u)^{w_{m+1}} .e^{i2\pi \left(\omega_{\sigma_{m+1}^u} \right)^{w_{m+1}}} \right] \end{array} \right\}
$$

$$
= \left\{ \begin{array}{l} \left[\left(1 - \coprod_{j=1}^{m+1} \left(1 - (\tau_j^l)^q \right)^{w_j} \right)^{1/q} .e^{i2\pi \left(1 - \coprod_{j=1}^{m+1} \left(1 - \omega_{\tau_j^l}^q \right)^{w_j} \right)^{1/q}}, \right. \\ \left. \left(1 - \coprod_{j=1}^{m+1} (1 - (\tau_j^u)^q)^{w_j} \right)^{1/q} .e^{i2\pi \left(1 - \coprod_{j=1}^{m+1} \left(1 - \omega_{\tau_j^u}^q \right)^{w_j} \right)^{1/q}} \right], \\ \left[\prod_{j=1}^{m+1} \left(\sigma_j^l \right)^{w_j} .e^{i2\pi \left(\prod_{j=1}^{m+1} \left(\omega_{\sigma_j^l} \right)^{w_j} \right)}, \prod_{j=1}^{m+1} \left(\sigma_j^u \right)^{w_j} .e^{i2\pi \left(\prod_{j=1}^{m+1} \left(\omega_{\sigma_j^u} \right)^{w_j} \right)} \right] \end{array} \right\}
$$

Thus, by the principle of mathematical induction, the result is true for all $n \in N$.

Example 15.4.3 Let $\Psi_1 = \left(\left[0.54e^{i2\pi(0.5)}, 0.84e^{i2\pi(0.6)} \right], \left[0.45e^{i2\pi(0.45)}, 0.56e^{i2\pi(0.55)} \right] \right)$, $\Psi_2 = \left(\left[0.6e^{i2\pi(0.36)}, 0.65e^{i2\pi(0.46)} \right], \left[0.7e^{i2\pi(0.6)}, 0.8e^{i2\pi(0.75)} \right] \right)$, and $\Psi_3 = \left(\left[0.55e^{i2\pi(0.6)}, 0.65e^{i2\pi(0.7)} \right], \left[0.64e^{i2\pi(0.75)}, 0.85e^{i2\pi(0.85)} \right] \right)$ be three IVCq-ROFNs and we consider the weight vector as $W = (0.3, 0.2, 0.5)^t$. Then, we can determine the IVCq-ROFWA for $q = 3$ in the following:

$$
IVCq\text{-}ROFWA(\Psi_1, \Psi_2, \Psi_3)
$$

$$= \left(\begin{array}{l} \left[(1-(0.949)(0.952)(0.913))^{1/3}e^{i2\pi(1-(0.960)(0.990)(0.885))^{1/3}}, \right. \\ \left. (1-(0.763)(0.937)(0.851))^{1/3}e^{i2\pi(1-(0.929)(0.979)(0.851))^{1/3}}\right], \\ \left[(0.091)(0.931)(0.8)e^{i2\pi(0.786)(0.902)(0.866)}, \right. \\ \left. (0.840)(0.956)(0.921)e^{i2\pi(0.835)(0.944)(0.921)}\right] \end{array} \right)$$

$$= \left(\left[0.559e^{i2\pi(0.541)}, 0.731e^{i2\pi(0.609)}\right], \left[0.067e^{i2\pi(0.613)}, 0.739e^{i2\pi(0.725)}\right]\right)$$

Definition 15.4.4 IVCq-ROFGA operator can be described by a function IVCq-ROFGA: $P^n \rightarrow P$ such that IVCq-ROFGA $(\Psi_1, \Psi_2, \ldots.., \Psi_n) = \Psi_1^{w_1} \otimes \Psi_2^{w_2} \otimes \ldots\ldots \otimes \Psi_n^{w_n} = \otimes_{j=1}^{n} \Psi_j^{w_j}$.

Theorem 15.4.5 *For* $\Psi_j = \left(\left[\tau_j^l e^{i2\pi\omega_{\tau_j^l}}, \tau_j^u e^{i2\pi\omega_{\tau_j^u}}\right], \left[\sigma_j^l e^{i2\pi\omega_{\sigma_j^l}}, \sigma_j^u e^{i2\pi\omega_{\sigma_j^u}}\right] : j = 1, 2 \ldots, n\right)$,

$$IVCq\text{-}ROFGA(\Psi_1, \Psi_2, \ldots.., \Psi_n)$$

$$= \left(\begin{array}{l} \left[\prod_{j=1}^{n} \left(\tau_j^l\right)^{w_j} . e^{i2\pi\left(\prod_{j=1}^{n} \omega_{\tau_j^l}^{w_j}\right)}, \prod_{j=1}^{n} \left(\tau_j^u\right)^{w_j} . e^{i2\pi\left(\prod_{j=1}^{n} \omega_{\tau_j^u}^{w_j}\right)} \right], \\ \left[\left(1 - \prod_{j=1}^{n}\left(1 - \left(\sigma_j^l\right)^q\right)^{w_j}\right)^{1/q} . e^{i2\pi\left(1-\prod_{j=1}^{n}\left(1-\omega_{\sigma_j^l}^q\right)^{w_j}\right)^{1/q}}, \right. \\ \left. \left(1 - \prod_{j=1}^{n}\left(1 - \left(\sigma_j^u\right)^q\right)^{w_j}\right)^{1/q} . e^{i2\pi\left(1-\prod_{j=1}^{n}\left(1-\omega_{\sigma_j^u}^q\right)^{w_j}\right)^{1/q}} \right] \end{array} \right)$$

Proof The proof is similar to Theorem 15.4.2. So, it is left as an exercise for the readers.

Example 15.4.6 Let $\Psi_1 = \left(\left[0.54e^{i2\pi(0.5)}, 0.84e^{i2\pi(0.6)}\right], \left[0.45e^{i2\pi(0.45)}, 0.56e^{i2\pi(0.55)}\right]\right)$, $\Psi_2 = \left(\left[0.6e^{i2\pi(0.36)}, 0.65e^{i2\pi(0.46)}\right], \left[0.7e^{i2\pi(0.6)}, 0.8e^{i2\pi(0.75)}\right]\right)$ and $\Psi_3 = \left(\left[0.55e^{i2\pi(0.6)}, 0.65e^{i2\pi(0.7)}\right], \left[0.64e^{i2\pi(0.75)}, 0.85e^{i2\pi(0.85)}\right]\right)$ be three IVCq-ROFNs and we consider the weight vector as $W = (0.3, 0.2, 0.5)^t$. Then, we can determine the IVCq-ROFGA for $q = 3$ in the following:

$$IVCq\text{-}ROFGA(\Psi_1, \Psi_2, \Psi_3)$$

$$= \left(\left[\begin{array}{l} \left[(0.831)(0.902)(0.741)e^{i2\pi(0.812)(0.815)(0.774)}, \right. \\ \left. (0.949)(0.917)(0.806)e^{i2\pi(0.857)(0.856)(0.836)}\right], \\ (1-(0.971)(0.919)(0.851))^{1/3}e^{i2\pi(1-(0.971)(0.952)(0.760))^{1/3}}, \\ (1-(0.943)(0.866)(0.621))^{1/3}e^{i2\pi(1-(0.946)(0.896)(0.621))^{1/3}} \end{array} \right] \right)$$

$$= \left(\left[0.555e^{i2\pi(0.512)}, 0.701e^{i2\pi(0.613)}\right], \left[0.621e^{i2\pi(0.667)}, 0.789e^{i2\pi(0.779)}\right]\right)$$

15.5 The MADM Model Based on IVCq-ROFWA and IVCq-ROFGA Operators

Recently, the MADM model is a very popular approach to solve complex decision-making problems involving selection from among a finite number of alternatives. The selection of the best alternative is done under certain criteria fixed by the decision-makers according to the environment domain. To conclude, the MADM method specifies how to attribute information is to be processed. In this section, we propose MADM model based on IVCq-ROFWA and IVCq-ROFGA operators given by

Let $\Theta = \{\Theta_1, \Theta_2, \Theta_3, \ldots., \Theta_m\}$ be the set of alternatives and $\Sigma = \{\Sigma_1, \Sigma_2, \Sigma_3, \ldots., \Sigma_n\}$ be the collection of criteria with a weighted vector $W = (w_1, w_2, \ldots.., w_n)^t$ such that $\sum_{i=1}^{n} w_i = 1$ and each $w_i \in [0, 1]$. For an alternative Θ_i under the criterion Σ_j, let the information provided by the decision-makers using IVCq-ROFNs is ψ_{ij}. With this information, we construct the decision-matrix $\Delta(\psi_{ij})_{m\times n}$. Furthermore, based on the IVCq-ROFWA and IVCq-ROFGA operators, we consider the following algorithm:

Algorithm: The steps of the algorithm are given by

Step 1. Start
Step 2. Normalized the decision-matrix $\Delta(\psi_{ij})_{m\times n}$ by using the following formula:

$$\Omega_{ij} = \left(\left[\tau_j^l e^{i2\pi\omega_{\tau_j^l}}, \tau_j^u e^{i2\pi\omega_{\tau_j^u}}\right], \left[\sigma_j^l e^{i2\pi\omega_{\tau_j^l}}, \sigma_j^u e^{i2\pi\omega_{\tau_j^u}}\right]\right), \ for\ profit$$
$$= \left(\left[\sigma_j^l e^{i2\pi\omega_{\tau_j^l}}, \sigma_j^u e^{i2\pi\omega_{\tau_j^u}}\right], \left[\tau_j^l e^{i2\pi\omega_{\tau_j^l}}, \tau_j^u e^{i2\pi\omega_{\tau_j^u}}\right]\right), \ for\ loss$$

Step 3. Aggregate all the attribute values by using IVCq-ROFWA or IVCq-ROFGA to get the comprehensive value of each alternative.
Step 4. Calculate the score or accuracy value of each aggregation.
Step 5. Rank all the alternatives and select the best place for investment.
Step 6. End

The diagrammatic representation of the proposed model can be shown with the help of the flow chart given by Fig. 15.2.

15.6 An Illustrative Example for the Validation of the Proposed MADM Model

To validate and usefulness of the proposed model, we consider the following example for practical application:

A company wants to invest money in selecting a place for setting an electric vehicle charging station. There are four alternatives called places and for each alternative,

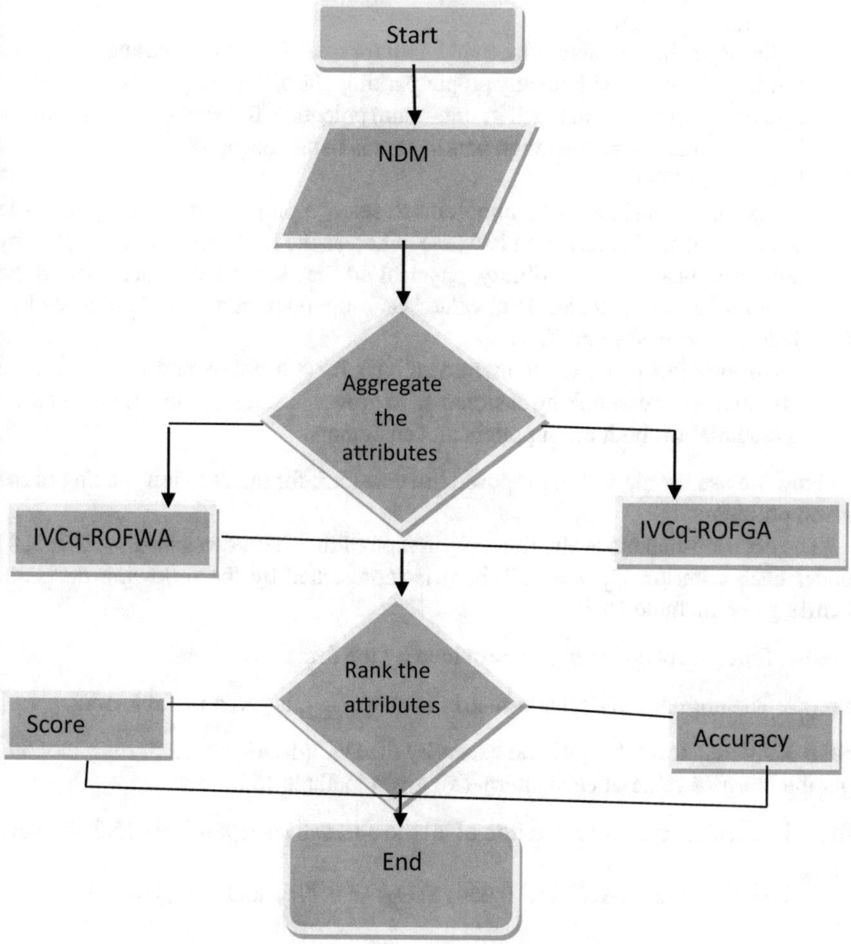

Fig. 15.2 Flowchart of the MADM model

there are four criteria. The set of alternatives is denoted by $\Theta = \{\Theta_1, \Theta_2, \Theta_3, \Theta_4\}$ and the set of criteria is denoted by $\Sigma = \{\Sigma_1, \Sigma_2, \Sigma_3, \Sigma_4\}$ and the criteria weight vector is $W = (0.2, 0.15, 0.35, 0.3)^t$. We consider the criteria as follows:

1. Population density (Σ_1):
   Population density is uneven across the globe. Physical factors like climate, water supply, natural resources, etc., affect the population density quite significantly. On the other hand, there are some social, political, and economic factors that affect the population density as well. By keeping these in mind, we find population density is an important criterion for the proposed study.

2. Parking area (Σ_2):
A parking area is another important factor for the safety of visitors and it prevents traffic jams. Also, without any proper parking place, it is not possible for setting up a charging station to reach its maximum potential. To avoid losing customers, it is important to have enough parking areas in any business.
3. Land value (Σ_3):
To execute a business, it is important to select a proper selection of land. The selection of land is based on its value, whereas the land value is determined by land use, location, accessibility, physical attributes, market value, price in the surrounding area, etc. So, land value is an important parameter for this study.
4. Transportation system (Σ_4):
Transportation is a vital economic activity for a business and it provides the essential service of linking customers. To grow a business, proper transportation is essential for both the suppliers and customers.

Now we use the algorithm proposed in Sect. 15.5, for the decision-making of the given problem:

The decision-makers evaluation information of the alternatives $\Theta_i (i = 1, 2, 3, 4)$ under each criterion $\Sigma_j (j = 1, 2, 3, 4)$ is represented by the following decision-matrix given in Table 15.1:

Step 1. Input the decision matrix provided by the decision makers.

Step 2. The normalized decision-matrix of $\Delta(\psi_{ij})_{4\times4}$ is given in Table 15.2.

Step 3. Aggregate all the attribute values by the IVCq-ROFA operator for $q = 4$ and get the ultimate value of each alternative given in Table 15.3.

Step 4. Calculate the score function of all the alternatives from Table 15.3, we have

$$S(\Theta_1) = 0.377,\ S(\Theta_2) = 0.054,\ S(\Theta_3) = 0.776, \text{ and } S(\Theta_4) = 0.295$$

Step 5. Ranking all the alternatives according to their score function, we have

$$\Theta_2 \prec \Theta_4 \prec \Theta_1 \prec \Theta_3$$

So, Θ_3 is the best place where the company may decide to set up an electric vehicle charging machine to earn maximum profit and minimum loss.

15.7 Conclusion and Future Work

In [40], Liu et al. developed the notion of Cq-ROFSs to handle those uncertain problems that cannot be described by the decision-makers under the q-ROFSs and CFSs. The main aim of the chapter is to introduce IVCq-ROFS as an extension of

Table 15.1 Decision-matrix $\Delta\left(\psi_{ij}\right)_{4\times 4}$

IVCq-ROFNs	Σ_1	Σ_2	Σ_3	Σ_4
Θ_1	$\left(\begin{matrix}[0.34e^{i2\pi(0.56)}, 0.64e^{i2\pi(0.75)}],\\ [0.55e^{i2\pi(0.65)}, 0.76e^{i2\pi(0.85)}]\end{matrix}\right)$	$\left(\begin{matrix}[0.64e^{i2\pi(0.42)}, 0.84e^{i2\pi(0.54)}],\\ [0.47e^{i2\pi(0.35)}, 0.56e^{i2\pi(0.55)}]\end{matrix}\right)$	$\left(\begin{matrix}[0.7e^{i2\pi(0.6)}, 0.8e^{i2\pi(0.74)}],\\ [0.54e^{i2\pi(0.45)}, 0.56e^{i2\pi(0.55)}]\end{matrix}\right)$	$\left(\begin{matrix}[0.43e^{i2\pi(0.56)}, 0.56e^{i2\pi(0.62)}],\\ [0.25e^{i2\pi(0.35)}, 0.46e^{i2\pi(0.45)}]\end{matrix}\right)$
Θ_2	$\left(\begin{matrix}[0.72e^{i2\pi(0.56)}, 0.77e^{i2\pi(0.6)}],\\ [0.81e^{i2\pi(0.65)}, 0.83e^{i2\pi(0.75)}]\end{matrix}\right)$	$\left(\begin{matrix}[0.45e^{i2\pi(0.6)}, 0.67e^{i2\pi(0.73)}],\\ [0.78e^{i2\pi(0.55)}, 0.87e^{i2\pi(0.57)}]\end{matrix}\right)$	$\left(\begin{matrix}[0.7e^{i2\pi(0.7)}, 0.8e^{i2\pi(0.74)}],\\ [0.87e^{i2\pi(0.45)}, 0.91e^{i2\pi(0.55)}]\end{matrix}\right)$	$\left(\begin{matrix}[0.57e^{i2\pi(0.37)}, 0.65e^{i2\pi(0.46)}],\\ [0.67e^{i2\pi(0.68)}, 0.78e^{i2\pi(0.73)}]\end{matrix}\right)$
Θ_3	$\left(\begin{matrix}[0.56e^{i2\pi(0.72)}, 0.76e^{i2\pi(0.74)}],\\ [0.7e^{i2\pi(0.6)}, 0.81e^{i2\pi(0.9)}]\end{matrix}\right)$	$\left(\begin{matrix}[0.57e^{i2\pi(0.53)}, 0.64e^{i2\pi(0.64)}],\\ [0.55e^{i2\pi(0.65)}, 0.59e^{i2\pi(0.75)}]\end{matrix}\right)$	$\left(\begin{matrix}[0.64e^{i2\pi(0.54)}, 0.84e^{i2\pi(0.64)}],\\ [0.75e^{i2\pi(0.77)}, 0.91e^{i2\pi(0.83)}]\end{matrix}\right)$	$\left(\begin{matrix}[0.57e^{i2\pi(0.56)}, 0.87e^{i2\pi(0.62)}],\\ [0.47e^{i2\pi(0.43)}, 0.55e^{i2\pi(0.56)}]\end{matrix}\right)$
Θ_4	$\left(\begin{matrix}[0.76e^{i2\pi(0.6)}, 0.84e^{i2\pi(0.7)}],\\ [0.55e^{i2\pi(0.65)}, 0.57e^{i2\pi(0.75)}]\end{matrix}\right)$	$\left(\begin{matrix}[0.44e^{i2\pi(0.54)}, 0.64e^{i2\pi(0.62)}],\\ [0.55e^{i2\pi(0.65)}, 0.66e^{i2\pi(0.75)}]\end{matrix}\right)$	$\left(\begin{matrix}[0.76e^{i2\pi(0.63)}, 0.87e^{i2\pi(0.7)}],\\ [0.5e^{i2\pi(0.35)}, 0.6e^{i2\pi(0.5)}]\end{matrix}\right)$	$\left(\begin{matrix}[0.64e^{i2\pi(0.65)}, 0.74e^{i2\pi(0.7)}],\\ [0.55e^{i2\pi(0.65)}, 0.66e^{i2\pi(0.57)}]\end{matrix}\right)$

Table 15.2 Normalized decision-matrix

IVCq-ROFNs	Σ_1	Σ_2	Σ_3	Σ_4
Θ_1	$\left(\begin{array}{l}[0.34e^{i2\pi(0.56)}, 0.64e^{i2\pi(0.75)}],\\ [0.55e^{i2\pi(0.65)}, 0.76e^{i2\pi(0.85)}]\end{array}\right)$	$\left(\begin{array}{l}[0.47e^{i2\pi(0.35)}, 0.56e^{i2\pi(0.55)}],\\ [0.64e^{i2\pi(0.42)}, 0.84e^{i2\pi(0.54)}]\end{array}\right)$	$\left(\begin{array}{l}[0.7e^{i2\pi(0.6)}, 0.8e^{i2\pi(0.74)}],\\ [0.54e^{i2\pi(0.45)}, 0.56e^{i2\pi(0.55)}]\end{array}\right)$	$\left(\begin{array}{l}[0.43e^{i2\pi(0.56)}, 0.56e^{i2\pi(0.62)}],\\ [0.25e^{i2\pi(0.35)}, 0.46e^{i2\pi(0.45)}]\end{array}\right)$
Θ_2	$\left(\begin{array}{l}[0.72e^{i2\pi(0.56)}, 0.77e^{i2\pi(0.6)}],\\ [0.81e^{i2\pi(0.65)}, 0.83e^{i2\pi(0.75)}]\end{array}\right)$	$\left(\begin{array}{l}[0.78e^{i2\pi(0.55)}, 0.87e^{i2\pi(0.57)}],\\ [0.45e^{i2\pi(0.6)}, 0.67e^{i2\pi(0.73)}]\end{array}\right)$	$\left(\begin{array}{l}[0.87e^{i2\pi(0.45)}, 0.91e^{i2\pi(0.55)}],\\ [0.7e^{i2\pi(0.7)}, 0.8e^{i2\pi(0.74)}]\end{array}\right)$	$\left(\begin{array}{l}[0.57e^{i2\pi(0.37)}, 0.65e^{i2\pi(0.46)}],\\ [0.67e^{i2\pi(0.68)}, 0.78e^{i2\pi(0.73)}]\end{array}\right)$
Θ_3	$\left(\begin{array}{l}[0.7e^{i2\pi(0.6)}, 0.81e^{i2\pi(0.9)}],\\ [0.56e^{i2\pi(0.72)}, 0.76e^{i2\pi(0.74)}]\end{array}\right)$	$\left(\begin{array}{l}[0.57e^{i2\pi(0.53)}, 0.64e^{i2\pi(0.64)}],\\ [0.55e^{i2\pi(0.65)}, 0.59e^{i2\pi(0.75)}]\end{array}\right)$	$\left(\begin{array}{l}[0.75e^{i2\pi(0.77)}, 0.91e^{i2\pi(0.83)}],\\ [0.64e^{i2\pi(0.54)}, 0.84e^{i2\pi(0.64)}]\end{array}\right)$	$\left(\begin{array}{l}[0.57e^{i2\pi(0.56)}, 0.87e^{i2\pi(0.62)}],\\ [0.47e^{i2\pi(0.43)}, 0.55e^{i2\pi(0.56)}]\end{array}\right)$
Θ_4	$\left(\begin{array}{l}[0.55e^{i2\pi(0.65)}, 0.57e^{i2\pi(0.75)}],\\ [0.76e^{i2\pi(0.6)}, 0.84e^{i2\pi(0.7)}]\end{array}\right)$	$\left(\begin{array}{l}[0.44e^{i2\pi(0.54)}, 0.64e^{i2\pi(0.62)}],\\ [0.55e^{i2\pi(0.65)}, 0.66e^{i2\pi(0.75)}]\end{array}\right)$	$\left(\begin{array}{l}[0.76e^{i2\pi(0.63)}, 0.87e^{i2\pi(0.7)}],\\ [0.5e^{i2\pi(0.35)}, 0.6e^{i2\pi(0.5)}]\end{array}\right)$	$\left(\begin{array}{l}[0.55e^{i2\pi(0.65)}, 0.66e^{i2\pi(0.57)}],\\ [0.64e^{i2\pi(0.65)}, 0.74e^{i2\pi(0.7)}]\end{array}\right)$

Table 15.3 The ultimate value of each alternative under IVCq-ROFA operator

Alternatives	IVCq-ROFA
Θ_1	$([0.576e^{i2\pi(0.558)}, 0.696e^{i2\pi(0.693)}], [0.441e^{i2\pi(0.444)}, 0.596e^{i2\pi(0.563)}])$
Θ_2	$([[0.780e^{i2\pi(0.481)}, 0.837e^{i2\pi(0.544)}]], [0.665e^{i2\pi(0.668)}, 0.778e^{i2\pi(0.737)}])$
Θ_3	$([[0.678e^{i2\pi(0.668)}, 0.861e^{i2\pi(0.796)}]], [0.555e^{i2\pi(0.549)}, 0.687e^{i2\pi(0.648)}])$
Θ_4	$([[0.649e^{i2\pi(0.630)}, 0.761e^{i2\pi(0.673)}]], [0.593e^{i2\pi(0.515)}, 0.693e^{i2\pi(0.628)}])$

Cq-ROFS by replacing the complex-valued membership degree by interval complex-valued membership degree and complex-valued non-membership degree by interval complex-valued non-membership degree with the restriction that the sum of the qth power of the supremum of the complex membership interval grade and the qth power of the supremum of the complex non-membership interval grade should be less equal to one. This restriction makes IVCq-ROFS flexible to the Cq-ROFS in the sense that it enables to procure more complex uncertain information, i.e., its information domain space becomes wider. In this work, we have introduced a new theory which is known as IVCq-ROF set theory and its associated IVCq-ROF number. We also investigated fundamental properties and laws based on IVCq-ROFNs. Then, we define the score function and the accuracy function based on IVCq-ROFNs. The IVCq-ROFSs are superior to the PFSs, q-ROFSs, CIFSs, CPFSs, Cq-ROFSs to describe the uncertain complex environment in the form of an interval. Based on IVCq-ROFNs two new aggregate operators called IVCq-ROFWA and IVCq-ROGA are introduced which are very much useful in multi-criteria decision-making problems. Moreover, with the help of these operators, a new methodology has been introduced to solve MADM problems. Finally, for a practical application of the new MADM model, a suitable example is adopted which justifies the usefulness and the feasibility of the proposed method.

In the future, the proposed study can be handy to introduce some new theories such as linguistic interval complex q-rung orthopair fuzzy set (LIVCq-ROFS), similarity measures between IVCq-ROFSs, information measures for IVCq-ROFSs, generalized IVCq-ROFSs, Einstein IVCq-ROFS operators, and many more. In the end, the present study can also be extended and applicable in different decision-making process given in [1, 2, 19, 20, 25, 29, 33, 42, 48].

Conflict of Interest The author declares no conflict of interest regarding the publication of the chapter.

References

1. M. Akram, S. Naz, A novel decision-making approach under complex Pythagorean fuzzy environment. Math. Comput. Appl. **24**, 73 (2019). https://doi.org/10.3390/mca24030073
2. Z. Ali, T. Mahmood, Maclaurin symmetric mean operators and their applications in the environment of complex q-rung orthopair fuzzy sets. Comput. Appl. Math. **39**, 1–27 (2020)
3. A. Alkouri, A.R. Salleh, Complex intuitionistic fuzzy sets. AIP Conf. Proc. **1482**, 464–470 (2012). https://doi.org/10.1063/1.4757515
4. B. Arfi, Fuzzy decision making in politics: a linguistic fuzzy-set approach (LFSA). Polit. Anal. **13**, 23–56 (2005). https://doi.org/10.1093/pan/mpi002
5. K.T. Atanassov, Intuitionistic fuzzy sets. Fuzzy Sets Syst. **20**, 87–96 (1986)
6. K. Atanassov, G. Gargov, Interval valued intuitionistic fuzzy sets. Fuzzy Sets Syst. **31**, 343–349 (1989)
7. K. Atanassov, Intuitionistic fuzzy sets. Int. J. Bioautom. **20**, 1–6 (2016)
8. I. Beg, S. Ashraf, Similarity measures for fuzzy sets. Appl. Comput. Math. **8**(2), 192–202 (2009)
9. D.C. Bisht, P.K. Srivastava, Fuzzy optimization and decision making. Adv. Fuzzy Logic Approach. Eng. Sci. (IGI Global), 310–326 (2019). https://doi.org/10.4018/978-1-5225-5709-8.ch014
10. I. Bloch, Fuzzy sets for image processing and understanding. Fuzzy Sets Syst. **281**, 280–291 (2015)
11. H. Bustince, Interval-valued fuzzy sets in soft computing. Int. J. Comput. Intell. Syst. **3**, 215–222 (2010). https://doi.org/10.1080/18756891.2010.9727692
12. Y.C. Chen, An application of fuzzy set theory to the external performance evaluation of distribution centers in logistics. Soft. Comput. **6**, 64–70 (2002)
13. T.Y. Chen, New Chebyshev distance measures for Pythagorean fuzzy sets with applications to multiple criteria decision analysis using an extended ELECTRE approach. Expert Syst. Appl. **147** (2019). https://doi.org/10.1016/j.eswa.2019.113164
14. B.C. Cuong, V. Kreinovich, Picture fuzzy sets—a new concept for computational intelligence problems, in *2013 Third World Congress on Information and Communication Technologies (WICT 2013)* (2013). https://doi.org/10.1109/wict.2013.7113099.
15. Y. Egusa, H. Akahori, A. Morimura, N. Wakami, An application of fuzzy set theory for an electronic video camera image stabilizer. IEEE Trans. Fuzzy Syst. **3**, 351–356 (1995)
16. P.A. Ejegwa, Distance and similarity measures for Pythagorean fuzzy sets. Granul. Comput. **5**, 225–238 (2020)
17. H. Garg, Linguistic Pythagorean fuzzy sets and its applications in multiattribute decision-making process. Int. J. Intell. Syst. **33**, 1234–1263 (2018)
18. H. Garg, D. Rani, Some results on information measures for complex intuitionistic fuzzy sets. Int. J. Intell. Syst. **34**, 2319–2363 (2019)
19. H. Garg, S.M. Chen, Multiattribute group decision making based on neutrality aggregation operators of q-rung orthopair fuzzy sets. Inf. Sci. **517**, 427–447 (2020)
20. H. Garg, D. Rani, New generalised Bonferroni mean aggregation operators of complex intuitionistic fuzzy information based on Archimedean t-norm and t-conorm. J. Exp. Theor. Artif. Intell. **32**, 81–109 (2020)
21. H. Garg, D. Rani, Robust averaging–geometric aggregation operators for complex intuitionistic fuzzy sets and their applications to MCDM process. Arab. J. Sci. Eng. **45**, 2017–2033 (2020)
22. H. Garg, A new possibility degree measure for interval-valued q-rung orthopair fuzzy sets in decision-making. Int. J. Intell. Syst. **36**, 526–557 (2021)
23. H. Garg, Z. Ali, T. Mahmood, Generalized dice similarity measures for complex q-rung orthopair fuzzy sets and its application. Complex Intell. Syst. **7**, 667–686 (2021)
24. H. Garg, Z. Ali, T. Mahmood, Algorithms for complex interval-valued q-rung orthopair fuzzy sets in decision making based on aggregation operators, AHP, and TOPSIS. Expert. Syst. **38**(1), e12609 (2021). https://doi.org/10.1111/exsy.12609

25. H. Garg, CN-q-ROFS: connection number-based q-rung orthopair fuzzy set and their application to decision-making process. Int. J. Intell. Syst. **36**, 3106–3143 (2021)
26. H. Garg, New exponential operation laws and operators for interval-valued q-rung orthopair fuzzy sets in group decision making process. Neural Comput. Appl. **33**(20), 13937–13963 (2021). https://doi.org/10.1007/s00521-021-06036-0
27. H. Garg, A novel trigonometric operation-based q-rung orthopair fuzzy aggregation operator and its fundamental properties. Neural Comput. Appl. **32**, 15077–15099 (2020). https://doi.org/10.1007/s00521-020-04859-x
28. H. Garg, Z. Ali, Z. Yang, T. Mahmood, S. Aljahdali, Multi-criteria decision-making algorithm based on aggregation operators under the complex interval-valued q-rung orthopair uncertain linguistic information. J. Intell. Fuzzy Syst. **41**(1), 1627–1656 (2021). https://doi.org/10.3233/JIFS-210442
29. H. Garg, J. Gwak, T. Mahmood, Z. Ali, Power aggregation operators and VIKOR methods for complex q-rung orthopair fuzzy sets and their applications. Mathematics **8**, 538 (2020). https://doi.org/10.3390/math8040538
30. M.B. Gorzałczany, An interval-valued fuzzy inference method-some basic properties. Fuzzy Sets Syst. **31**, 243–251 (1989)
31. J.K. Hamidi, K. Shahriar, B. Rezai, H. Bejari, Application of fuzzy set theory to rock engineering classification systems: an illustration of the rock mass excavability index. Rock Mech. Rock Eng. **43**, 335–350 (2010)
32. B.P. Joshi, A. Singh, P.K. Bhatt, K.S. Vaisla, Interval valued q-rung orthopair fuzzy sets and their properties. J. Intell. Fuzzy Syst. **35**, 5225–5230 (2018)
33. R. Krishankumar, Y. Gowtham, I. Ahmed, K.S. Ravichandran, S. Kar, Solving green supplier selection problem using q-rung orthopair fuzzy-based decision framework with unknown weight information. Appl. Soft Comput. **94** (2020). https://doi.org/10.1016/j.asoc.2020.106431
34. H.M. Lee, Group decision making using fuzzy sets theory for evaluating the rate of aggregative risk in software development. Fuzzy Sets Syst. **80**, 261–271 (1996)
35. C. Li, T.W. Chiang, Complex fuzzy computing to time series prediction—a multi-swarm PSO learning approach, in *Asian Conference on Intelligent Information and Database Systems*, vol. 6592 (Springer, Berlin, Heidelberg, 2011), pp. 242–251. https://doi.org/10.1007/978-3-642-20042-7_25
36. H. Li, S. Yin, Y. Yang, Some preference relations based on q-rung orthopair fuzzy sets. Int. J. Intell. Syst. **34**, 2920–2936 (2019)
37. Y. Lin, G.A. Cunningham, A new approach to fuzzy-neural system modeling. IEEE Trans. Fuzzy Syst. **3**, 190–198 (1995)
38. P. Liu, P. Wang, Multiple-attribute decision-making based on Archimedean Bonferroni operators of q-rung orthopair fuzzy numbers. IEEE Trans. Fuzzy Syst. **27**, 834–848 (2018)
39. P. Liu, Z. Ali, T. Mahmood, N. Hassan, Group decision-making using complex q-rung orthopair fuzzy Bonferroni mean. Int. J. Comput. Intell. Syst. **13**, 822–851 (2020)
40. P. Liu, T. Mahmood, Z. Ali, Complex q-rung orthopair fuzzy aggregation operators and their applications in multi-attribute group decision making. Information **11**, 5 (2020). https://doi.org/10.3390/info11010005
41. X. Ma, M. Akram, K. Zahid, J.C.R. Alcantud, Group decision-making framework using complex Pythagorean fuzzy information. Neural Comput. Appl. **33**, 2085–2105 (2021)
42. T. Mahmood, Z. Ali, Entropy measure and TOPSIS method based on correlation coefficient using complex q-rung orthopair fuzzy information and its application to multi-attribute decision making. Soft. Comput. **25**, 1249–1275 (2021)
43. S. Mitra, S.K. Pal, Fuzzy sets in pattern recognition and machine intelligence. Fuzzy Sets Syst. **156**, 381–386 (2005)

44. X. Peng, Y. Yang, Some results for Pythagorean fuzzy sets. Int. J. Intell. Syst. **30**, 1133–1160 (2015)
45. D. Ramot, R. Milo, M. Friedman, A. Kandel, Complex fuzzy sets. IEEE Trans. Fuzzy Syst. **10**, 171–186 (2002)
46. D. Rani, H. Garg, Distance measures between the complex intuitionistic fuzzy sets and their applications to the decision-making process. Int. J. Uncertain. Quantif. **7**, 423–439 (2017). https://doi.org/10.1615/Int.J.UncertaintyQuantification.2017020356
47. P. Rani, A.R. Mishra, Multi-criteria weighted aggregated sum product assessment framework for fuel technology selection using q-rung orthopair fuzzy sets. Sustain. Prod. Consum. **24**, 90–104 (2020). https://doi.org/10.1016/j.spc.2020.06.015
48. P. Ren, Z. Xu, X. Gou, Pythagorean fuzzy TODIM approach to multi-criteria decision making. Appl. Soft Comput. **42**, 246–259 (2016)
49. G.C. Sousa, B.K. Bose, A fuzzy set theory based control of a phase-controlled converter DC machine drive. IEEE Trans. Ind. Appl. **30**, 34–44 (1994)
50. K. Tanaka, B. Werners, An Introduction to fuzzy logic for practical applications. Math. Methods Oper. Res.-ZOR **46**, 435 (1997)
51. K. Ullah, T. Mahmood, Z. Ali, N. Jan, On some distance measures of complex Pythagorean fuzzy sets and their applications in pattern recognition. Complex Intell. Syst. **6**, 15–27 (2020)
52. J. Wang, G.W. Wei, C. Wei, J. Wu, Maximizing deviation method for multiple attribute decision making under q-rung orthopair fuzzy environment. Def. Technol. **16**, 1073–1087 (2020)
53. P. Wang, J. Wang, G. Wei, C. Wei, Similarity measures of q-rung orthopair fuzzy sets based on cosine function and their applications. Mathematics **7**, 340 (2019). https://doi.org/10.3390/math7040340
54. G. Wei, Y. Wei, Similarity measures of Pythagorean fuzzy sets based on the cosine function and their applications. Int. J. Intell. Syst. **33**, 634–652 (2018)
55. F. Xiao, W. Ding, Divergence measure of Pythagorean fuzzy sets and its application in medical diagnosis. Appl. Soft Comput. **79**, 254–267 (2019)
56. D.E. Tamir, N.D. Rishe, A. Kandel, Complex fuzzy sets and complex fuzzy logic an overview of theory and applications. Fifty years of fuzzy logic and its applications. Stud. Fuzziness Soft Comput. **326**, 661–681 (2015). https://doi.org/10.1007/978-3-319-19683-1_31
57. V. Torra, Hesitant fuzzy sets. Int. J. Intell. Syst. **25**, 529–539 (2010)
58. R.R. Yager, Pythagorean fuzzy subsets, in *2013 Joint IFSA World Congress and NAFIPS Annual Meeting (IFSA/NAFIPS)* (2013), pp. 57–61. https://doi.org/10.1109/IFSA-NAFIPS.2013.6608375
59. R.R. Yager, Properties and applications of Pythagorean fuzzy sets, in *Imprecision and Uncertainty in Information Representation and Processing, Studies in Fuzziness and Soft Computing*, vol. 332. (Springer, Cham, 2016), pp. 119–136. https://doi.org/10.1007/978-3-319-26302-1_9
60. R.R. Yager, Generalized orthopair fuzzy sets. IEEE Trans. Fuzzy Syst. **25**, 1222–1230 (2016)
61. Z. Yang, H. Garg, Interaction power partitioned Maclaurin symmetric mean operators under q-rung orthopair uncertain linguistic information. Int. J. Fuzzy Syst. 1–19 (2021). https://doi.org/10.1007/s40815-021-01062-5
62. O. Yazdanbakhsh, S. Dick, A systematic review of complex fuzzy sets and logic. Fuzzy Sets Syst. **338**, 1–22 (2018)
63. L.A. Zadeh, Fuzzy sets. Inf. Control **8**, 338–353 (1965)
64. L.A. Zadeh, Fuzzy logic. Computer **21**, 83–93 (1988)
65. H.J. Zimmermann, Fuzzy set theory. Wiley Interdiscip. Rev.: Comput. Stat. **2**, 317–332 (2010)
66. W.R. Zhang, Yin Yang bipolar fuzzy sets, in *1998 IEEE International Conference on Fuzzy Systems Proceedings. IEEE World Congress on Computational Intelligence*, vol. 1 (1998), pp. 835–840. https://doi.org/10.1109/FUZZY.1998.687599
67. N. Zhang, G. Wei, Extension of VIKOR method for decision making problem based on hesitant fuzzy set. Appl. Math. Model. **37**, 4938–4947 (2013)
68. X. Zhang, Z. Xu, Extension of TOPSIS to multiple criteria decision making with Pythagorean fuzzy sets. Int. J. Intell. Syst. **29**, 1061–1078 (2014)

69. Z. Zhang, S.M. Chen, Group decision making with incomplete q-rung orthopair fuzzy preference relations. Inf. Sci. **553**, 376–396 (2020). https://doi.org/10.1016/j.ins.2020.10.015
70. S.P. Zhang, P. Sun, J.S. Mi, T. Feng, Belief function of Pythagorean fuzzy rough approximation space and its applications. Int. J. Approx. Reason. **119**, 58–80 (2020)
71. B. Zhu, Z. Xu, M. Xia, Dual hesitant fuzzy sets. J. Appl. Math. **2012**, 1–13 (2012). https://doi.org/10.1155/2012/879629

Dr. Somen Debnath has earned his Ph.D. degree in Mathematics from Tripura University (A central university), Suryamaninagar, Agartala, Tripura 799022, India. At present, he is working as a Post Graduate Teacher(in Mathematics) at Umakanta Academy, Agartala-799001, Tripura, India. His research interests include fuzzy sets, rough sets, soft sets, hypersoft sets, neutrosophic sets, Pythagorean fuzzy sets, matrix representation of interval-valued intuitionistic fuzzy sets, game theory, multi-criteria decision making, etc. He has over 10 years of teaching and over 6 years of research experience. He has published 20^{+} research articles in national and international peer-reviewed journals so far.

Chapter 16
A Novel Fermatean Fuzzy Analytic Hierarchy Process Proposition and Its Usage for Supplier Selection Problem in Industry 4.0 Transition

Alper Camci, Muharrem Eray Ertürk, and Sait Gül

Abstract In multi-attribute decision-making (MADM) field, fuzzy sets provide differing representation manners in modeling the expert's preferences. Fermatean fuzzy set (FFS) is a special concept of q-rung orthopair fuzzy set when we set $q =$ 3. Thus, FFS has an extensive domain from a geometric perspective. In the study, a well-known MADM tool, Analytic Hierarchy Process (AHP), is extended into FFS environment with the aim of providing more freedom to the experts in stating their thoughts and expertise. The applicability of the proposition is demonstrated in a real supplier selection problem in Industry 4.0 transition. In today's environment, supplier selection is of strategic importance to companies. Further advances in computer, communication, and automation technologies transformed the corporate landscape. The need for integration of suppliers and customers adds further challenge to the supplier selection. This paper aims to explore the supplier selection and develop a set of criteria in tune with Industry 4.0.

Keywords q-Rung orthopair fuzzy set · Fermatean fuzzy set · AHP method · Supplier selection · Industry 4.0

16.1 Introduction

As one of the most attractive fields of managerial decision analysis, multiple attribute decision making (MADM) aims at searching for the most convenient alternatives by concerning some attributes having significant effects on the decision process. In a

A. Camci · S. Gül (✉)
Faculty of Engineering and Natural Sciences, Management Engineering Department, Bahçeşehir University, Beşiktaş, İstanbul 34353, Turkey
e-mail: sait.gul@eng.bau.edu.tr

A. Camci
e-mail: alper.camci@eng.bau.edu.tr

M. E. Ertürk
Graduate School of Business, Bahçeşehir University, Beşiktaş, İstanbul 34353, Turkey
e-mail: eray.erturk@mefaendustri.com

H. Garg (ed.), *q-Rung Orthopair Fuzzy Sets*,
https://doi.org/10.1007/978-981-19-1449-2_16

decision matrix, alternatives and attributes are represented on rows and columns, respectively where the performance scores of the alternatives are shown in its cells. These scores may be determined objectively or subjectively. While objective judgments are directly measured such as cost in currency, distance in mile, etc., subjective judgments are gathered from the real experts. In the traditional MADM, the subjective judgments are crisp integer numbers as representatives of linguistic evaluations. However, in today's continuously changing business environment, uncertainty and vagueness are very catchy issues in each decision analysis operation. In order to cope with the representation issue of uncertainty, fuzzy set concepts have been developed in the literature.

From a subjective judgment perspective, linguistic evaluation scales which are mostly based on fuzzy concepts are used today in specifying the preferences and expertise of the experts. Zadeh [95] defined and conceptualized the fuzzy set to symbolize the uncertain elements in expert evaluations by developing the membership degree (a) ranging in a closed unit interval. The membership degree measures the agreement level in a judgment. Since then, many fuzzy concepts have been developed and new elements have been added to the fuzzy set definitions. As one of the most important developments in this field, Atanassov [4] developed the intuitionistic fuzzy set (IFS) via introducing a new independent element: non-membership degree (b). With this element, the experts have a chance of declaring their disagreement level. The only rule that must be satisfied by an IFS is $0 \leq \boldsymbol{a} + \boldsymbol{b} \leq 1$. Furthermore, Atanassov [4] introduced another element in the fuzzy set definition about the experts' hesitancy levels: $\boldsymbol{c} = 1 - \boldsymbol{a} - \boldsymbol{b}$. Herewith, IFS has become the first three-dimensional fuzzy set definition due to its capability of handling three different types of expert judgments: membership, non-membership, and hesitancy.

Following IFS, several extensions, e.g., Pythagorean fuzzy sets (***PFS***: $0 \leq \boldsymbol{a}^2 + \boldsymbol{b}^2 \leq 1$) by Yager [92], q-rung orthopair fuzzy sets (q-ROFS: $0 \leq \boldsymbol{a}^q + \boldsymbol{b}^q \leq 1$) by Yager [93], spherical fuzzy sets ($0 \leq \boldsymbol{a}^2 + \boldsymbol{b}^2 + \boldsymbol{c}^2 \leq 1$) by Kutlu Gündoğdu and Kahraman [45], picture fuzzy sets ($0 \leq \boldsymbol{a} + \boldsymbol{b} + \boldsymbol{c} \leq 1$) by Cuong and Kreinovich [14], Fermatean fuzzy sets (***FFS***: $0 \leq \boldsymbol{a}^3 + \boldsymbol{b}^3 \leq 1$) by Senapati and Yager [84], etc. have been developed in decades. As Liu and Wang [54] stated, the MADM tools should be improved by considering these novel fuzzy concepts in order to become more appropriate and comprehensive for the continuous complications of today's society. Considering this need, this study proposes a Fermatean fuzzy version of the Analytic Hierarchy Process (FF-AHP).

Yager [93] initiated the theory of q-rung orthopair fuzzy sets (q-ROFS) requiring the sum of the qth power of the membership degree, and the qth power of the non-membership degree is bounded by 1. q-ROFS is a generalization of IFS and PFS since these two are special cases of q-ROFS. As depicted in Fig. 16.1 and detailed in Sect. 16.3, it is seen that as the power increases, the space of acceptable orthopairs increases, and more orthopairs satisfy the restricting constraint. Thus, the decision analyst can adjust the value of the parameter q with the aim of obtaining the expert judgment representation field [54].

After the first appearance of q-ROFS in the literature, many scholars from different areas have found this concept interesting and it has been highly appreciated since then.

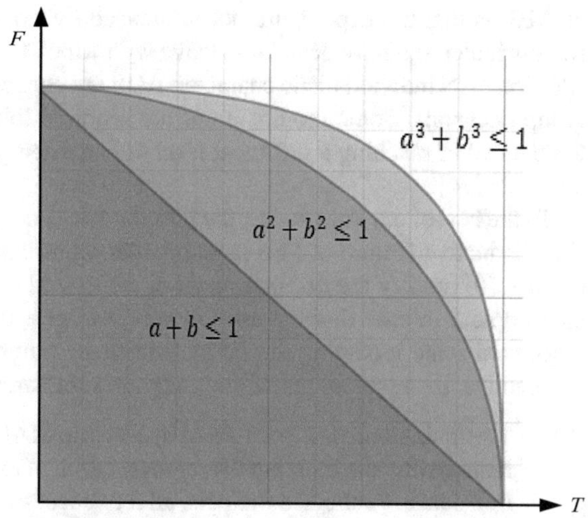

Fig. 16.1 Geometric representation of IFS, PFS, and FFS

For example, different versions of q-ROFS have been established, e.g., connection number-based q-ROFS by Garg [22] and IVq-ROFS by Ju et al. [40] and Garg [23]. Also, many aggregation operators (AO) have been developed, e.g., trigonometric operation-based AO for q-ROFS by Garg [20], power AOs for complex q-ROFS by Garg et al. [25], exponential operation laws, and AO for interval-valued q-ROFS (IVq-ROFS) by Garg [21], interaction power partitioned Maclaurin symmetric mean operators by Yang and Garg [94], AOs for complex IVq-ROFS by Garg et al. [26], and interaction AOs for q-ROFS by Riaz et al. [69].

As a special type of q-ROFS, Senapati and Yager [84] proposed the concept of FFS. Setting $q = 3$, the novel q-ROFS concept is named as Fermatean fuzzy sets (FFS). Under it, an expert has gained more freedom in specifying his/her judgment due to the chance of expressing agreeing (membership) and/or disagreeing (non-membership) opinions about the state of a subject. For illustration purposes, an election can represent the FFS environment. Suppose a voter thinks a specific candidate satisfies the general expectations with a possibility of 0.60 where the same candidate dissatisfies the general expectations with a possibility of 0.90. This judgment of the voter cannot be modeled as an IFS because $0.60 + 0.90 = 1.50 > 1$, and also as a PFS because $0.60^2 + 0.90^2 = 1.17 > 1$. However, since $0.60^3 + 0.90^3 = 0.945 < 1$, a FFS is capable of modeling this judgment.

To reveal the importance of attributes that are considered in a MADM application, attribute prioritization is needed. AHP is one of the most cited attribute weighting methodologies in the MADM literature. This method can find the weights via processing the subjective judgments of the experts. The experts' subjectivity originates from gathering the expert's comparative judgments regarding the importance of attributes [31]. AHP which was developed by Thomas L. Saaty in the 1970s relies on the prioritization of decision elements. It is based on the pairwise comparisons of the experts focusing on only two attributes at a time [9]. Moreover, another advantage of AHP is that the experts are not influenced by exterior factors while comparing two attributes and have deep knowledge with which to compare the two attributes. In addition to its individualistic character, AHP is a convenient MADM tool for making group decisions. Some modifications are required to use it in group decisions due to the need of reaching a sufficient level of consistency between integrated decision elements [74].

To the best of our knowledge, the literature does not have a FFS extension of AHP. The distinctive feature of FFS is its broader representation domain provided to the experts. To modify the original method, we extend AHP into the Fermatean fuzzy environment in case that the experts accept to give their comparisons by utilizing a nominal scale including the Fermatean fuzzy numbers (FFNs). From a MADM perspective, the motivations of the study are summarized as follows:

(i) The main improvement offered by FFNs in MADM applications is its comprehensive domain relaxing the representation of expert judgments. If the evaluation data are collected in FFNs, all the successive operations can be performed subsequently as FFNs.
(ii) If the evaluation (pairwise comparison in AHP cases) collection process can be fuzzified under FFS, the AHP should be modified accordingly under FFS.
(iii) Thanks to FFS' broader geometric area yielding more opportunities to the decision-makers, FFS-based extension of AHP is needed.
(iv) Also, the properly and extensively arranged supplier selection problems requiring the imprecise evaluations of the experts should be coped with in a broader context of FF-AHP to obtain efficient and respectable solutions.

In literature, some MADM tools have been extended to the FFS environment until now: TOPSIS by Senapati and Yager [84], WPM by Senapati and Yager [82], WASPAS by Keshavarz-Ghorabaee et al. [43], EDAS by Mishra et al. [60], and VIKOR, SAW, and ARAS by Gül [30]. As seen, all these MADM tools are appropriate for selecting the best alternative or ranking the alternatives. Any method that could be used for weighting the decision attributes has not been developed yet. In order to fulfill this gap in the literature, this study aims to modify AHP with FFS conditions. Thus, we attempt to contribute as follows:

(i) The originality of the paper arises from the proposition of an AHP version using FFNs to prioritize the attributes in the decision problem.
(ii) To model the judgment of the expert in the proposed FF-AHP, we develop a new pairwise comparison scale including FFNs and their correspondences as linguistic evaluations. Using this scale, the experts will have more flexibility in expressing their expertise.
(iii) The uncertainty and ambiguity in human judgments are modeled in an extensive context which FFS provides.
(iv) The FF-AHP method is applied to a supplier selection problem in Industry 4.0 transition period. Three experts were consulted for the case study. As a beginning point, we aimed to determine the most important attributes to guide the companies in this managerial decision problem. The evaluation of the

alternatives and selection process are not considered here. The attributes that are found as important ones can be accepted and assessed as key performance indicators (KPIs).

The study has been organized as follows: the details of the supplier selection problem during the Industry 4.0 transition period is studied and the attributes are defined in Sect. 16.2; FFS and its operations, as well as a brief literature review on FFS-based MADM, are detailed in Sect. 16.3; the steps of novel FF-AHP are given in Sect. 16.4; Sect. 16.5 covers the implementation of FF-AHP in a real supplier selection problem; and finally, Sect. 16.6 discusses the findings and concludes the study.

16.2 Supplier Selection in Industry 4.0 Transition

Industry 4.0 or the 4th Industrial Revolution has become one of the most popular topics of academic discussion across a wide spectrum of fields ever since its first appearance at the Hannover industrial fair in Germany in 2011 [64]. The main idea of the Industry 4.0 is the integration of systems and components (human to human, human to machine, or machine to machine) within a supply chain, sharing digitized data and information using cyber-physical systems, the internet of things (IoT), and the internet of services (IoS) [70]. The goal of such an integrated system is to respond to customer demand, which is the primary data point in a supply chain, quickly and in a cost-effective manner regardless of the order size, thus staying profitable even with a lot size of one [10].

According to Lasi et al. [47], the fourth industrial revolution stems from the application pull and technology push triggers in the environment.

Application push triggers are as follows [47]:

- Short development periods (time to market)
- Individualization on demand (buyer's market leads to an increasing individualization of products)
- Higher flexibility in product development and production
- Decentralization or leaner organizational hierarchies to ensure faster decision-making
- Economic and ecological increase in resource efficiency

Technology pull triggers are as follows [47]:

- Ever-increasing mechanization and automation
- Digitalization and networking
- Miniaturization
- Lean and agile management philosophy.

With Industry 4.0 the role of the supplier has become very important as they are expected to be a part of an integrated, data-driven, digital and customer-focused supply chain. Thus, supplier selection has become an even more important task for the supply management function. In literature, the terms vendor and supplier are used interchangeably especially in the selection problems. In one of the earliest studies on vendor selection Dickson [17] defined the problem as selecting one vendor among several possible alternatives and proposed 23 attributes for selection. The supplier selection problem has become one of the most popular topics in decision-making literature, especially in MADM. De Boer et al. [15] and Sachdeva et al. [75] offer extensive literature reviews on the field utilizing different MADM tools. The most recurring main selection attributes in this field are cost (price), quality, delivery (lead time), and service level as their basic [11, 18, 19, 33, 42].

Industry 4.0 is considered to be a revolutionary concept of integrating business functions, suppliers, and customers within a supply chain using data, sensors, and automation technologies in order to cater to customer need regardless of the size of the need. Suppliers play a much more crucial role in Industry 4.0 supply chains as integration with the supply chain partners is one of the prerequisites. Sachdeva et al. [75] add technological capabilities and relationship between the supplier and the customer as the main Industry 4.0 related attributes in addition to the more traditional attributes of cost/price, rejection rate, and delivery delay in their supplier selection study where they use TOPSIS as the MADM methodology. Hasan et al. [36] on the other hand take a more technology and systems-focused approach for the same problem using TOPSIS and Multi-Choice Goal Programming (MCGP) methods stressing both the complexity and uncertainty related to the supply chain as well as the technology and digitalization level of the suppliers. Using type 2 fuzzy AHP and COPRAS-G methodologies Kayapinar Kaya and Aycin [42] primarily focus on the smart and digital technology attributes in addition to the more classical attributes of cost, quality, delivery, and capacity. Our opinion is that the limited number of studies pertaining to Industry 4.0 supplier selection fall short in developing a comprehensive attribute set. Using the literature resources as well as the expert opinions, a new set of attributes with sub-attributes was developed as shown in Table 16.1.

16.3 Preliminaries: Fermatean Fuzzy Sets

To cope with vague and uncertain situations, Zadeh [95] first developed the fuzzy set. Following him, fuzzy sets have been implemented in different areas such as decision-making (e.g., [88, 90, 91]), medical diagnosis (e.g., [1, 76]), and pattern recognition (e.g., [2]). Until today, many extensions have been developed like rough sets [68], intuitionistic fuzzy sets [4], soft sets [63], bipolar valued fuzzy sets [49], hesitant fuzzy sets [89], dense fuzzy sets [16], neutrosophic sets [86], etc. Although all these extensions are benefitted by the researchers, the intuitionistic fuzzy set concept [4] has gained much more attention in the literature.

In IFS, imprecision can be modeled more extensively because it includes both agreement (membership) and disagreement (non-membership) levels. The boundary

Table 16.1 Supplier selection attributes and sub-attributes

Main attribute	Sub attribute	Description	Industry 4.0 References
Economic	Unit cost	Total expenditure by the customer to procure one unit of a particular product or service	Santana et al. [80], Zhong et al. [96], Tang et al. [87], Manavalan and Jayakrishna [58], Gilchrist [28]
	Payment method	Method of monetary transfer between customer and the supplier after the sale transaction	
	Logistics costs	All of the expenses incurred moving and handling products from supplier to customer	
	Financial stability	Ability to withstand an adverse financial situation, such as sales decline, financing problems, economic downturns	
	After-sales service cost	Cost of any support provided to the customers by the supplier after the purchase of the product	
Strategic	Strategic alignment	Alignment between its competitive strategies and the customer needs (priorities) that the supplier aims to meet with its products	Salam [77], Bär et al. [7], Benton and Maloni [8]
	Strategic contribution	Supplier's contribution to the customer's strategic performance drivers that will help achieve the strategic goals	
	Competitive advantage	Factors that enable a company to produce a comparable good or service at a lower price or better than its competitors	
	Balance of power	State of stability and equilibrium between customers and suppliers	

(continued)

Table 16.1 (continued)

Main attribute	Sub attribute	Description	Industry 4.0 References
Human Resources	Human resource quality	Skills, training, education, experience, and knowledge levels of employees in an organization required to perform tasks to achieve organizational goals	Hasan et al. [37], Cisneros-Cabrera et al. [12], Nguyen et al. [66], Mohelska and Sokolova [61]
	Agile teams	A highly skilled and motivated group of people with diverse backgrounds came together to produce solutions incrementally	
	Compatibility of cultures	To be able to understand, communicate, respect, and collaborate with the cultures of other organizations	
	Innovation culture	A dynamic culture that supports creativity takes on new challenges and encourages risk-taking	
Quality	Product quality	Supplier's products' conformance to customer requirements at an acceptable price	Gunasekaran et al. [29], Illés et al. [38]
	Process quality	Supplier's ability to produce quality products consistently	
	Quality certification-assurance	Accreditation by an accepted agency showing that supplier's quality management processes are documented to ensure product and process quality, such as ISO 9001	
Lean management	Continuous improvement	Continuing practice to improve products, services, and processes	Sanders et al. [79], Rosin et al. [72]
	Lean operations	Providing customer value by eliminating waste	

(continued)

Table 16.1 (continued)

Main attribute	Sub attribute	Description	Industry 4.0 References
	Cost reduction ability	Ability to reducing the cost of products through lean principles	
Delivery	Lead time	Time between the receiving of a customer order request and the delivery of the order to the customer	Ghadge et al. [27], Cruz-Mejia et al. [13], Long et al. [55]
	On-time delivery	Delivering the order at the time specified by the customer	
	Order fullfilment rate	Percentage of the successfully completed among the total number of orders received	
	Geographical location	Location of supplier's facilities in relation to the customer's	
	Flexibility	Supplier's ability to modify the timing of delivery of its products	
Technology	Production facilities	Capabilities and technologies of production facilities	Muscio and Ciffolilli [65], Schumacher et al. [81]
	Technology level	Technology level of supplier's operations	
	R&D capabilities	Organization, skills, and strategies to conduct R&D activities in-house and in collaboration with the customer	
	Digitalization level	Maturity and readiness level a company reach during its digital transformation	
Information technologies	Integration ability	Supplier's ability to connect its IT systems with the customer's systems to work in a cohesive and coordinated fashion	Rojko [71], Lu [56], Sanchez et al. [78], Lezzi et al. [50], Kim [44]
	ERP system	Integrated business management system to help manage all aspects of a business	

(continued)

Table 16.1 (continued)

Main attribute	Sub attribute	Description	Industry 4.0 References
	Digital data	Data stored and analyzed in computer systems	
	IT security risk	Probability of external actors' illegally getting access, attacking or harming the organization through IT connections	
	Cloud usage	Using third-party, external servers for storage and retrieval of data as well as other IT applications	
Sustainability	Environmental impact	Potential effects of supplier's operations and products on the environment	Bai et al. [6], Oláh et al. [67], Jena et al. [39], Liboni et al. [51], Luthra et al. [57]
	Waste	Non-value-bearing by-products of a supplier's operations that need to be discarded	
	Carbon footprint—energy use	Equivalent greenhouse gas emission due to energy usage	
	Fair working conditions	Socially acceptable working conditions that favor the wellbeing, rights, and security of the employees	

of IFS is that the sum of these degrees should be between 0 and 1. To smooth this boundary, Yager [92] enlarged the preference domain of IFS by setting the squared sum of the degrees within the unit interval and name it PFS. PFS also has a similar restriction since both IFS and PFS are not able to model some expert judgments. By generalizing PFS, Yager [93] developed a more embracive theory of q-ROFS such that the sum of qth power of degrees is restricted by 1. In 2019, Senapati and Yager [84] proposed a special case of q-ROFS by fixing $q = 3$ and rename it as FFS. Here, the sum of cubes is defined in an interval. As shown in Fig. 16.1, the preference domain of human judgments in FFS is broader because it covers both IFS and PFS.

From a geometric perspective displayed in Fig. 16.1, it is obvious that FFS is a terser case of q-ROFS and brings a broader preference depiction area to the experts than IFS and PFS do. It is noticed that IFS represents all the points beneath the line of $a+b \leq 1$, PFS represents all the points beneath the curve of $a^2+b^2 \leq 1$, and FFS represents all the points beneath the curve of $a^3+b^3 \leq 1$. Thus, the area represented by FFS is larger than the area covered by IFS and PFS.

Definitions of FFS and important operations defined on FFS is given as follows [82–.82].

Definition 16.1 Let X be a universal set. Then a FFS A in X is defined as follows:

$$A = \{(x, a(x), b(x)) | x \in X\} \tag{16.1}$$

where a, b are mappings from X to [0, 1]. For all $x \in X, a(x)$ is positive membership degree of $x \in A$ and $b(x)$ is negative membership degree of $x \in A$ such that

$$0 \leq (a(x))^3 + (b(x))^3 \leq 1, \forall x \in X \tag{16.2}$$

and $c(x) = \sqrt[3]{1 - (a(x))^3 - (b(x))^3}$ is the indeterminacy (hesitancy) degree of x in A [84].

Definition 16.2 Let $X = (a_1, b_1), Y = (a_2, b_2)$ and $F = (a, b)$ be FFNs, then their operations are as follows:

$$X \oplus Y = \left(\sqrt[3]{(a_1^3 + a_2^3 - a_1^3 a_2^3}, b_1 b_2\right) \tag{16.3}$$

$$X \otimes Y = \left(a_1 a_2, \sqrt[3]{(b_1^3 + b_2^3 - a_1^3 b_2^3}\right) \tag{16.4}$$

$$X \ominus Y = \left(\sqrt[3]{\frac{a_1^3 - a_2^3}{1 - a_2^3}}, \frac{b_1}{b_2}\right) if\ a_1 \geq a_2\ and\ b_1 \leq \min\left\{b_2, \frac{b_2 c_1}{c_2}\right\} \tag{16.5}$$

$$X \oslash Y = \left(\frac{a_1}{a_2}, \sqrt[3]{\frac{b_1^3 - b_2^3}{1 - b_2^3}}\right) if\ b_1 \geq b_2\ and\ a_1 \leq \min\left\{a_2, \frac{a_2 c_1}{c_2}\right\} \tag{16.6}$$

$$\tau F = \left(\sqrt[3]{(1 - (1 - a^3)^\tau}, b^\tau\right) \tag{16.7}$$

$$F^\tau = \left(a^\tau, \sqrt[3]{(1 - (1 - b^3)^\tau}\right) \tag{16.8}$$

$$F^C = (b, a) \tag{16.9}$$

Definition 16.3 Let $F = (a, b)$ be a FFN, then the score and accuracy functions are as follows:

$$sc(F) = a^3 - b^3 \tag{16.10}$$

$$acc(F) = a^3 + b^3 \tag{16.11}$$

Definition 16.4 Let $X = (a_1, b_1)$ and $Y = (a_2, b_2)$ be two FFNs. The ranking rules are defined as follows:

i .If $sc(X) < sc(Y)$, then $X < Y$
ii .If $sc(X) > sc(Y)$, then $X > Y$
iii .If $sc(X) = sc(Y)$, then

a .If $acc(X) < acc(Y)$, then $X < Y$
b .If $acc(X) > acc(Y)$, then $X > Y$
c .If $acc(X) = acc(Y)$, then $X \approx Y$.

Definition 16.5 Let $X = (a_1, b_1)$ and $Y = (a_2, b_2)$ be two FFNs, then the Euclidean distance between them is defined as follows:

$$d_{euc}(X, Y) = \sqrt{\frac{1}{2}\left[\left(a_1^3 - a_2^3\right)^2 + \left(b_1^3 - b_2^3\right)^2 + \left(c_1^3 - c_2^3\right)^2\right]} \tag{16.12}$$

Definition 16.6 Let $X = (a_1, b_1)$ and $Y = (a_2, b_2)$ be FFNs, then the average (arithmetic mean) operator is given as follows:

$$aver(F_1, F_2) = \left(\frac{a_1^3 + a_2^3}{2}, \frac{b_1^3 + b_2^3}{2}\right) \tag{16.13}$$

FFS literature today is enriching day by day. Particularly, various aggregation operators have been developed by Senapati and Yager [83], Aydemir and Yilmaz Gunduz [5], Garg et al. [24], Akram et al. [3], Hadi et al. [34], and Shahzadi and Aksam [85]. Also, some papers modifying MADM tools for FFS have appeared in the literature. Senapati and Yager [84] proposed a FFS-based TOPSIS and show its applicability on a location selection problem for a house. Senapati and Yager [82] extended WPM (weighted product method) under FFS environment and applied it in a bridge construction method selection problem. Liu et al. [52] proposed FF linguistic set and integrated it with TOPSIS. Liu et al. [53] proposed some novel distance measures for FF linguistic sets and integrated them with modified TOPSIS and TODIM (Tomada de Decisão Interativa Multicritério) approaches. The last two studies demonstrated their methodology by analyzing the house selection problem which was previously introduced by Senapati and Yager [84]. Keshavarz-Ghorabaee et al. [43] extended WASPAS (weighted aggregated sum product assessment) into FFS and implemented it in a green construction supplier evaluation problem. Mishra et al. [60] developed an improved generalized score function and used it in a novel FFS extension of CRITIC-EDAS approach. The method is applied in the selection of sustainable third-party reverse logistics providers. As seen from the literature review regarding FFS-based MADM, there is no proposition of any AHP extension. In order to fulfill this gap, we propose a FF-AHP version as explained in the next section in detail.

16.4 A Novel Fermatean Fuzzy AHP Extension

The algorithm of FF-AHP method is shown in Fig. 16.2. The steps of FF-AHP (Fermatean fuzzy version of the AHP) are detailed as follows:

Step 1. Decision Hierarchy.

As the name of the method indicates, the modeling approach of AHP is based on building a hierarchy of the decision problem. A hierarchy is built by exposing the goal at Level 0, the main attributes at Level 1, the sub-attribute, and the sub-sub-attribute of the related main attributes or sub-attributes at consecutive Level 2, 3, etc., and the alternatives at the bottom level. The attributes are represented by A_i where i is the index for the attribute number. If an attribute has sub-attributes, the index will be properly determined. For instance, let $A_i = \{A_1, A_2, \ldots, A_n\}$ show the main attributes and suppose that A_i has k sub-attributes. So, these sub-attributes will be shown by $\{A_{i1}, A_{i2}, \ldots, A_{ik}\}$. In this fashion, the alternatives (where j is the index for alternative numbers) are listed at the bottom level: $S_j = \{S_1, S_2, \ldots, S_m\}$.

Step 2. Individual Comparison Matrices from the Experts.

The dataset processed by AHP covers the comparison judgments of the experts regarding the importance differences between elements at each level. To compare elements, the experts use a proper scale that is developed for the current decision-making environment. Saaty [73] developed a 9-point scale in which 1 shows equal importance between compared elements while 9 shows the perfect importance of an element against the other.

As a novelty, we propose a new linguistic scale for FF-AHP by implementing the procedure proposed by Kutlu Gündoğdu and Kahraman [46] for spherical fuzzy version of AHP. We established the FFN-based scale (Table 16.2) as correspondences to the evaluations from 1 to 9 and their reciprocals via setting the score index values of FFNs equal to the crisp evaluations. Equation (16.14) gives the score index used in establishing FFN-based AHP evaluation scale.

$$SI(F) = \left|10\left(a^3 - b^3\right)\right| \tag{16.14}$$

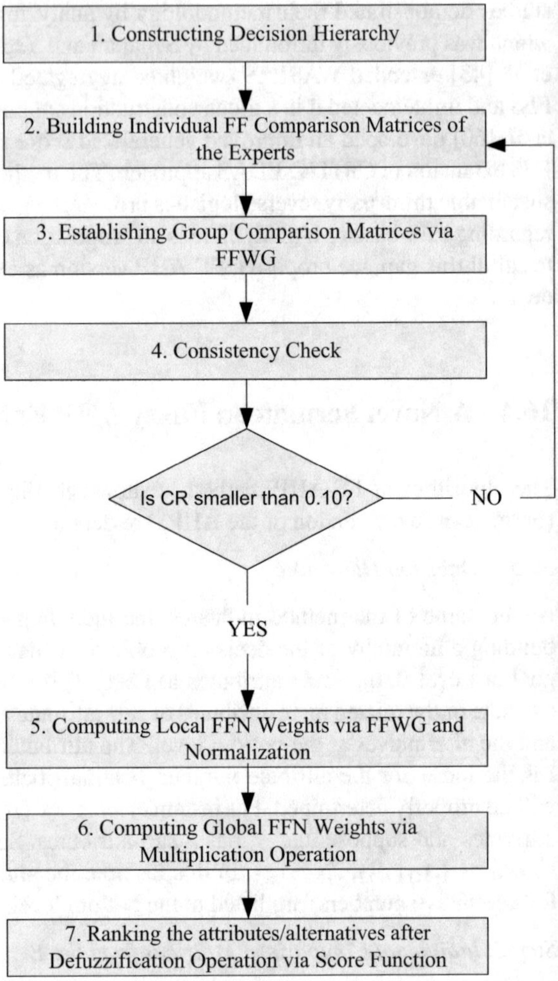

Fig. 16.2 The flow chart of FF-AHP

Table 16.2 The evaluation scale for FF-AHP

SI	*Meaning*	*FFN*	*SI*	*FFN*
9	Absolutely more importance (AMI)	(0.970, 0.233)	8	(0.929, 0.121)
7	Very high importance (VHI)	(0.900, 0.307)	6	(0.844, 0.107)
5	High importance (HI)	(0.794, 0.083)	4	(0.737, 0.068)
3	Slightly high importance (SHI)	(0.670, 0.091)	2	(0.585, 0.059)
1	Equal importance (EI)	(0.465, 0.082)	1/2	(0.369, 0.062)
1/3	Slightly low importance (SLI)	(0.322, 0.038)	1/4	(0.293, 0.054)
1/5	Low importance (LI)	(0.272, 0.050)	1/6	(0.256, 0.048)
1/7	Very low importance (VLI)	(0.243, 0.040)	1/8	(0.233, 0.053)
1/9	Absolutely low importance (ALI)	(0.224, 0.050)		

For instance, (0.465, 0.082) will be the FFN correspondence to the evaluation of 5 because $SI(0.465, 0.082) = \left|10\left(0.465^3 - 0.082^3\right)\right| = 5$.

Step 3. Group comparison matrices.

The pairwise comparison matrices for each level are constructed by the experts who use FFN scale given in Table 16.2. In a group environment, the aggregation of the individual comparisons is handled via Fermatean fuzzy weighted geometric mean (FFGW) operator (Eq. 16.15) proposed by Senapati and Yager [83] where k is the number of the experts consulted. FF-AHP operate this aggregation process by considering the expert's weights $\omega_e = \{\omega_1, \ldots, \omega_k\}$ representing their expertise level. As an alternative to the aggregation of individual expressions, group consensus approaches, e.g., Borda rule, can be used for reaching the group decision [35, 48] after finding the attribute weights individualistically.

$$FFWG(F_1, \ldots, F_k) = \left(\prod_{e=1}^{k} a_{F_e}^{\omega_e}, \prod_{e=1}^{k} b_{F_e}^{\omega_e}\right) \tag{16.15}$$

Step 4. Consistency check.

One of the most important features of AHP is its consistency analysis part. AHP computes a consistency ratio to reveal the inconsistency which is hidden in the expert evaluations. In this manner, AHP monitors the decision process that cannot be totally consistent all the time. Thus, an acceptable inconsistency is needed for each matrix (Gül and Topçu, [32]. As Saaty [73] proposed, the consistency ratio must not exceed 10%. So, the closer the consistency ratio is to zero, the greater the consistency. If an acceptable consistency ratio is not be obtained for a single matrix, the expert should modify his/her evaluations. To conduct consistency analysis, all the matrices are defuzzified via score function given in Eq. (16.10). Then, the classical consistency calculation defined for AHP is applied. A potential future study should focus on developing a special consistency analysis for FF-AHP.

Step 5. Local FFN Weights.

The local weights of the main attributes, sub-attributes determined with respect to the related main attribute, and alternatives with respect to each attribute or sub-attribute if the main attribute has, are obtained in this step. In each comparison matrix, FFGW operator (Eq. 16.15) is performed for each row in order to compute its local FFN weight. Here, it is assumed that all the w values in the equation are equal to the average where their sum is equal to 1. For example, if a comparison matrix has a size of 5×5, all these weights will be equal to 1/5.

Step 6. Global FFN Weights.

The final meaningful information is the rankings of the attributes and/or alternatives in MADM applications. In AHP, the ultimate rankings can be obtained via generating global scores. To find them, the local weights are combined by following the hierarchical sequence. The local FFN weights which are determined at each level are combined so that the sequence of the computations is started from the top-level (main criteria) and finalized in the bottom level (alternatives). The global weight of each child element that belongs to a parent element is computed through the distribution of the parent element's weight into its child elements. The distribution is based on the multiplication (Eq. 16.4) of the related local FFN weights. For instance, suppose A_i attribute has $\{A_{i1}, A_{i2}, \ldots, A_{ik}\}$ sub-attributes; let local FFN weight of A_i be (a_{A_i}, b_{A_i}) and let local FFN weight of A_{ik} be $(a_{A_{ik}}, b_{A_{ik}})$. The global weight of A_{ik} can be found as follows:

$$A_i \otimes A_{ik} = \left(a_{A_i} * a_{A_{ik}}, \sqrt[3]{\left(b_{A_i}^3 + b_{A_{ik}}^3 - b_{A_i}^3 * b_{A_{ik}}^3\right)}\right.$$

The multiplication operations continue until the determination of global priorities of the alternatives.

Step 7. Ranking of the alternatives/attributes.

In the final step, AHP can rank the alternatives and/or attributes in descending order of defuzzified global weights. For this purpose, Eq. (16.10) is performed. The highest final global priority will show the most important attribute or sub-attribute, or the most convenient alternative. Weights can be determined via normalizing defuzzified priorities.

16.5 An Application in Turkey

In this study, a novel FF-AHP approach has been proposed for handling supplier selection problems in Industry 4.0 era. In this manner, we aim to show the applicability of FF-AHP approach in a real managerial decision-making case and the solution strength of the FFS extension about the representation of uncertainty. In the application, we have limited the study with obtaining the attribute weights and guide the companies that have to cope with a similar supplier selection problem. The attribute weights represent the importance of the attributes so that a company interested can focus on the most important ones if they have limited resources that can be reserved for this operation.

In Step 1, the supplier selection attribute set was determined. The hierarchical structure shown in Fig. 16.3 is used while aggregating the local weights. The hierarchy has 4 levels: the goal of the problem: "Selection of the best supplier for Industry 4.0" is shown at level 0, levels 1 and 2 expose the main attributes and sub-attributes, and level 3 shows the alternatives. As mentioned before, we excluded the evaluation of the supplier alternatives in the case study because we limited our aim by weighting the attributes.

Step 2 is the process for gathering the comparison judgments from the experts who utilized the expressions given in Table 16.2. For this purpose, three experts were

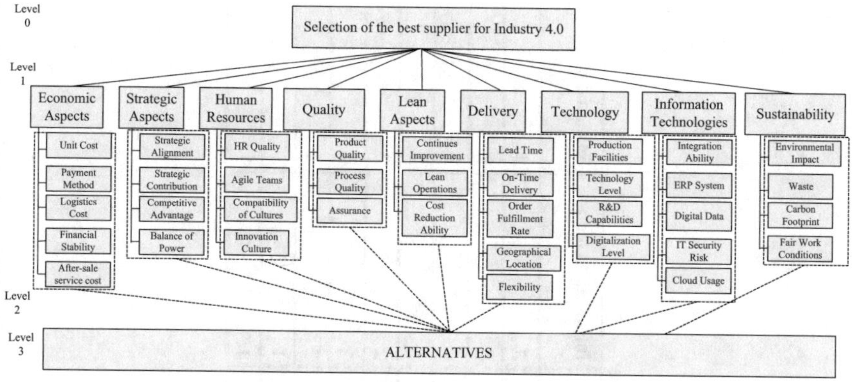

Fig. 16.3 Hierarchy of supplier selection problem in industry 4.0 transition

consulted. Expert-1 is a purchasing manager who is an industrial engineer having expertise of 16 years. Expert-2 who is a factory manager has an expertise of 18 years and served 2 other companies in the manager position, previously. Expert-3 is a mechanical engineer who has an expertise of 12 years.

In Step 3 which focuses on building the group comparison matrices, these experts have formed a decision team and gave consensus results so that Eq. (16.15) was not operated. For an instance, the pairwise comparison matrix which was built by the team for main attributes is given in Table 16.3. Table 16.4 shows the FFN correspondences of the comparisons. Since there are nine main attributes, nine more comparison matrices were constructed in order to get the local priorities of the sub-attributes.

Step 4 computes the inconsistencies of matrices and checks their acceptability against the rule forcing CR to be lower than or equal to 0.10. The calculation details of the original consistency analysis can be found in Gül and Topçu [32]. CR values are 9.2% for the main attributes, 9.8% for economic aspects, 5.7% for strategic aspects, 6.1% for human resources, 4.6% for quality aspects, 0.8% for lean aspects, 9.8% for delivery aspects, 6.1% for technology aspects, 8.3% for information technologies, and 2.1% for sustainability. Thus, all the matrices are found as acceptable for further operations since each had inconsistency which is lower than 10%.

In Step 5, local FFN weights are obtained by performing FFWG operator (Eq. 16.15). This operation is done for each comparison matrix. Table 16.5 exposes all the local and global weights. For instance, the local FFN weight of Economic Aspects main attribute is shown as follows:

$$FFWG(C_1, \ldots, C_9) = \left(\prod_{e=1}^{9} a_{C_e}^{1/9}, \prod_{e=1}^{9} b_{C_e}^{1/9} \right)$$

where

Table 16.3 Pairwise comparison matrix for main attributes (evaluations)

	Economic	Strategic	Human resources	Quality	Lean	Delivery	Technology	Information tech	Sustainability
Economic	1	2	3	3	2	3	2	3	5
Strategic	1/2	1	3	4	5	2	4	5	3
Human resources	1/3	1/3	1	2	2	4	3	4	3
Quality	1/3	1/4	½	1	2	3	2	1	2
Lean	1/2	1/5	½	1/2	1	2	2	2	2
Delivery	1/3	1/2	¼	1/3	1/2	1	2	2	4
Technology	1/2	1/4	1/3	1/2	1/2	1/2	1	2	2
Information tech	1/3	1/5	¼	1	1/2	1/2	1/2	1	3
Sustainability	1/5	1/3	1/3	1/2	1/2	1/4	1/2	1/3	1

Table 16.4 Pairwise comparison matrix for main attributes (FFN correspondences)

	Economic	Strategic	Human resources	Quality	Lean	Delivery	Technology	Information tech	Sustainability
Economic	(0.465, 0.082)	(0.585, 0.059)	(0.670, 0.091)	(0.670, 0.091)	(0.585, 0.059)	(0.670, 0.091)	(0.585, 0.059)	(0.670, 0.091)	(0.794, 0.083)
Strategic	(0.369, 0.062)	(0.465, 0.082)	(0.670, 0.091)	(0.737, 0.068)	(0.794, 0.083)	(0.585, 0.059)	(0.737, 0.068)	(0.794, 0.083)	(0.670, 0.091)
Human resources	(0.322, 0.038)	(0.322, 0.038)	(0.465, 0.082)	(0.585, 0.059)	(0.585, 0.059)	(0.737, 0.068)	(0.670, 0.091)	(0.737, 0.068)	(0.670, 0.091)
Quality	(0.322, 0.038)	(0.293, 0.054)	(0.369, 0.062)	(0.465, 0.082)	(0.585, 0.059)	(0.670, 0.091)	(0.585, 0.059)	(0.465, 0.082)	(0.585, 0.059)
Lean	(0.369, 0.062)	(0.272, 0.050)	(0.369, 0.062)	(0.369, 0.062)	(0.465, 0.082)	(0.585, 0.059)	(0.585, 0.059)	(0.585, 0.059)	(0.585, 0.059)
Delivery	(0.322, 0.038)	(0.369, 0.062)	(0.293, 0.054)	(0.322, 0.038)	(0.369, 0.062)	(0.465, 0.082)	(0.585, 0.059)	(0.585, 0.059)	(0.737, 0.068)
Technology	(0.369, 0.062)	(0.293, 0.054)	(0.322, 0.038)	(0.369, 0.062)	(0.369, 0.062)	(0.369, 0.062)	(0.465, 0.082)	(0.585, 0.059)	(0.585, 0.059)
Information tech	(0.322, 0.038)	(0.272, 0.050)	(0.293, 0.054)	(0.465, 0.082)	(0.369, 0.062)	(0.369, 0.062)	(0.369, 0.062)	(0.465, 0.082)	(0.670, 0.091)
Sustainability	(0.272, 0.050)	(0.322, 0.038)	(0.322, 0.038)	(0.369, 0.062)	(0.369, 0.062)	(0.293, 0.054)	(0.369, 0.062)	(0.322, 0.038)	(0.465, 0.082)

Table 16.5 Local and global weights and rankings of attributes and sub-attributes

Main attribute	Sub-attribute	Local weights		Global weights		SC	Local ranking	w	Global ranking
Economic		0.627	0.077						
	Unit cost	0.365	0.051	0.229	0.084	0.011	3	0.0203	18
	Payment method	0.299	0.054	0.187	0.085	0.006	5	0.0106	25
	Logistics costs	0.434	0.070	0.272	0.093	0.019	4	0.0344	7
	Financial stability	0.730	0.118	0.457	0.128	0.093	1	0.1661	1
	After-sales service cost	0.630	0.099	0.395	0.112	0.060	2	0.1069	2
Strategic		0.629	0.075						
	Strategic alignment	0.605	0.074	0.381	0.094	0.054	1	0.0967	3
	Strategic contribution	0.509	0.072	0.320	0.093	0.032	2	0.0570	4
	Competitive advantage	0.400	0.056	0.252	0.084	0.015	3	0.0273	10
	Balance of power	0.378	0.059	0.238	0.086	0.013	4	0.0227	14
Human resources		0.542	0.063						
	Human resource quality	0.571	0.071	0.310	0.085	0.029	1	0.0516	5
	Agile teams	0.509	0.072	0.276	0.086	0.020	2	0.0362	6
	Compatibility of cultures	0.424	0.058	0.230	0.076	0.012	3	0.0208	16
	Innovation culture	0.378	0.059	0.205	0.077	0.008	4	0.0144	23
Quality		0.453	0.061						
	Product quality	0.542	0.065	0.245	0.080	0.014	1	0.0254	12
	Process quality	0.399	0.068	0.180	0.082	0.005	3	0.0095	27
	Quality certification-assurance	0.465	0.067	0.210	0.081	0.009	2	0.0156	21
Lean		0.449	0.061						

(continued)

Table 16.5 (continued)

Main attribute	Sub-attribute	Local weights		Global weights		SC	Local ranking	w	Global ranking
	Continuous improvement	0.567	0.076	0.255	0.087	0.016	1	0.0282	9
	Lean operations	0.381	0.058	0.171	0.075	0.005	3	0.0082	30
	Cost reduction ability	0.465	0.067	0.209	0.081	0.009	2	0.0152	22
Delivery		0.428	0.056						
	Lead time	0.618	0.077	0.265	0.086	0.018	1	0.0318	8
	On time delivery	0.548	0.071	0.235	0.082	0.012	2	0.0221	15
	Order fulfillment rate	0.465	0.063	0.199	0.076	0.007	3	0.0132	24
	Geographical location	0.412	0.059	0.177	0.072	0.005	4	0.0091	29
	Flexibility	0.334	0.055	0.143	0.070	0.003	5	0.0046	33
Technology		0.403	0.059						
	Production facilities	0.617	0.078	0.248	0.088	0.015	1	0.0260	11
	Technology level	0.540	0.075	0.217	0.086	0.010	2	0.0171	19
	R&D capabilities	0.393	0.055	0.158	0.072	0.004	3	0.0064	32
	Digitalization level	0.357	0.057	0.144	0.073	0.003	4	0.0046	34
Information tech		0.395	0.060						
	Integration ability	0.618	0.077	0.244	0.088	0.014	1	0.0247	13
	ERP System	0.583	0.077	0.230	0.087	0.012	2	0.0205	17
	Digital data	0.452	0.057	0.179	0.074	0.005	3	0.0094	28
	IT Security risk	0.419	0.061	0.165	0.077	0.004	4	0.0072	31

(continued)

Table 16.5 (continued)

Main attribute	Sub-attribute	Local weights		Global weights		SC	Local ranking	w	Global ranking
	Cloud usage	0.318	0.053	0.126	0.072	0.002	5	0.0029	36
Sustainability		0.341	0.052						
	Environmental impact	0.632	0.072	0.215	0.080	0.009	1	0.0168	20
	Waste	0.550	0.079	0.187	0.086	0.006	2	0.0106	26
	Carbon footprint	0.400	0.056	0.136	0.068	0.002	3	0.0040	35
	Fair working conditions	0.336	0.060	0.114	0.071	0.001	4	0.0020	37

$$\prod_{e=1}^{9} a_{C_e}^{1/9} = 0.465^{1/9} * 0.585^{\frac{1}{9}} * 0.670^{\frac{1}{9}} * 0.670^{\frac{1}{9}} * \cdots * 0.794^{\frac{1}{9}} = 0.627$$

$$\prod_{e=1}^{9} b_{C_e}^{1/9} = 0.082^{1/9} * 0.059^{\frac{1}{9}} * 0.091^{\frac{1}{9}} * 0.059^{\frac{1}{9}} * \cdots * 0.083^{\frac{1}{9}} = 0.077$$

After calculating the local weights of each main attribute and their sub-attributes, the global weights of the sub-attributes are found in Step 6 via operating the FFN multiplication operation. The global weights are also given in Table 16.5. As an example, the global weight of Unit Cost sub-attribute of Economic Aspects main attribute was computed as follows:

$$C_1 \otimes C_{11} = \left(a_{C_1} * a_{C_{11}}, \sqrt[3]{(b_{C_1}^3 + b_{C_{11}}^3 - b_{C_1}^3 * b_{C_{11}}^3)}\right)$$

$$(0.627, 0.077) \otimes (0.365, 0.051) = \left(0.627 * .365, \sqrt[3]{(0.077^3 + 0.051^3 - 0.077^3 * 0.051^3)}\right)$$

$$= (0.229, 0.084)$$

In Step 7, the global weights were defuzzied via Eq. (16.10) and then the attributes were ranked in descending order of the defuzzified values. Global and local rankings of the attributes and sub-attributes are shown in Table 16.5. The defuzzified score of global FFN weight of Unit Cost and its weight is shown as follows:$sc(w_{11}) = sc(0.229, 0.084) = 0.229^3 - 0.084^3 = 0.011$

By considering the score function values, the sub-attributes are locally ranked. After normalizing them, the global weights and also the global ranks of sub-attributes are obtained. The outputs of FF-AHP are summarized in Table 16.5.

The findings outlined in Table 16.5 show that Strategic, Economic, and Human Resource attributes dominate the Industry 4.0 supplier selection decisions. Also, the top 7 attributes come from these three main attributes. After these three main attributes related to the overall business of the suppliers, next three are related to the operations, quality, lean, and delivery. Only after these, come the technology and the information technology, strengthening the notion that industry 4.0 is not just about technology [62]. For a lasting and mutually beneficial relationship, companies primarily would like to pair up with suppliers, which are financially stable. This is a sign of the long-term health and robustness of a supplier. The second most important sub-attribute, after-sales service cost shows the supplier's commitment to its relationship with the customer in the life cycle of product. Following these two economic sub-attributes, the two strategic sub-attributes alignment and contribution shows how well the suppliers will be able to work with their customers in the long term and contribute to their overall business strategies. This is an important aspect of integration in the Industry 4.0 supply chains. Even though Industry 4.0 transformation is based on technological advances, human resource-related attributes rank high in our list, as human resource quality and agile teams are seen as important factors for developing customer solutions quickly. Coming at the 7th and 8th spot, logistics costs and lead time sub-attributes show the need for an efficient and responsive supply chain with low uncertainty.

Based on our findings, when selecting suppliers for Industry 4.0 supply chains it is prudent to assess the financial, organizational, and human resource attributes following by the operational attributes like quality, lean management, and delivery. These findings also should caution the supplier firms in their quest for Industry 4.0 transition as they should prepare their organizations and operations first for the transition before even investing heavily in Industry 4.0 related technologies.

16.6 Discussion and Concluding Remarks

Mentzer et al. [59] define a supply chain as a set of three or more entities (companies or individuals) directly involved in the upstream (supply-side) and downstream (distribution side) flows of products, services, finances, and/or information from a source to a final customer. In a supply chain, each entity acts as supplier or customer to each adjacent entity, so there is always a supplier-customer dyad at each intersection of the supply chain. In such a relationship, materials or services flow from supplier to the customer, and the funds flow from customer to supplier. As most companies are focusing on their core businesses and competencies, suppliers, and vendors are crucial for providing inputs for manufacturing processes. The role of the supplier has become very important as they are expected to be a part of integrated, data-driven, digital, and customer-focused supply chains in today's Industry 4.0 environments. Thus, supplier selection has become an even more crucial decision for companies. The first contribution of this study has been a comprehensive attribute set for supplier selection in Industry 4.0 transition. Results show that rather than technology-related attributes, financial, strategic, and human resource attributes play a significant role in these selection decisions. Technology attributes rank even lower than the operational attributes of quality, lean management, and delivery. These findings suggest supplier companies go through an organizational transition to robust, strategically positioned organizations having a highly skilled and agile workforce and efficient operations before the technology transformations.

MADM asks the experts regarding their expertise and knowledge related to a specific decision problem. The fuzzy set literature provides several assessment tools for easing the data collection process. As mentioned before, IFS and PFS are good candidates to achieve this aim, but they have a limited capability than FFS. Their mathematical limits are expanded in FFS by considering the sum of the cubes of the membership and non-membership degrees rather than directly taking their sum (like IFS) or their squared sum (like PFS). This extension presents a broader preference domain to the experts. From a technical perspective, this study is the first attempt aiming to extend the well-known AHP MADM tool into a FFS environment to benefit from its extensive representation domain. For data collection, a new FFN-based linguistic evaluation scale is also proposed as a contribution to the literature.

The proposed method may be utilized in any MADM problem having attributes that could be compared in a pairwise manner with the purpose of reaching a selection or ranking decision. For example, web service selection problem can be modeled by FF-AHP and the most appropriate web service can be chosen if the experts are ready to share their opinions about the attributes and alternatives [41].

In order to justify the importance of the use of FF-AHP method, we also conducted a crisp AHP application in the same supplier selection problem and summarize the local and global importance ranking of the attributes. The results are shown in Table 16.6. As seen in the table, although there is only one differing rank pair in terms of local rankings of the sub-attributes (unit cost and logistics cost have changed their rankings; please see the underlined attributes), most of the global ranks have changed (please see the italic attributes). Only seven over 37 attributes have kept their rankings: financial stability, strategic alignment, human resource quality, product facilities, technology level, cloud usage, and fair working conditions. From the comparison, it is clear that usage of FF-AHP gave significantly different results than the crisp version because the uncertainty and vagueness in human judgments cannot be modeled by crisp numbers but the FFS version is capable of doing it, extensively. So, it can be inferred that the results are more confident and reliable than the crisp version of AHP.

In the future, some limitations of the study can be addressed. Firstly, the FF-AHP method needs a dedicated and original consistency analysis that can work with FFNs. Also, an entropy-based tool can be integrated into FF-AHP with the aim of reducing the subjectivity level. Finally, a full application of FF-AHP that will include the evaluations of the alternatives should be conducted and the results should be compared with different contemporary fuzzy AHP extensions such as IF-AHP and PF-AHP.

Table 16.6 Comparison results of crisp and FF-AHP applications

Main attribute	Sub-attribute	FF-AHP			Crisp AHP			
		Local ranking	Global weights	Global ranking	Local weights	Local ranking	Global weights	Global ranking
Economic					0.227			
	Unit cost	3	0.020	18	0.061	4	0.014	24
	Payment method	5	0.011	25	0.034	5	0.008	30
	Logistics costs	4	0.034	7	0.103	3	0.023	16
	Financial stability	1	0.166	*1*	0.488	1	0.111	*1*
	After-sales service cost	2	0.107	2	0.314	2	0.071	3
Strategic					0.229			
	Strategic alignment	1	0.097	3	0.471	1	0.108	2
	Strategic contribution	2	0.057	*4*	0.280	2	0.064	*4*
	Competitive advantage	3	0.027	10	0.136	3	0.031	9
	Balance of power	4	0.023	14	0.114	4	0.026	13
Human resources					0.147			
	Human resource quality	1	0.052	5	0.416	1	0.061	5
	Agile teams	2	0.036	6	0.294	2	0.043	8
	Compatibility of cultures	3	0.021	16	0.170	3	0.025	14
	Innovation culture	4	0.014	23	0.120	4	0.018	21
Quality					0.092			
	Product quality	1	0.025	12	0.493	1	0.045	6
	Process quality	3	0.009	27	0.196	3	0.018	20
	Quality certification-assurance	2	0.016	21	0.311	2	0.029	10

(continued)

Table 16.6 (continued)

Main attribute	Sub-attribute	FF-AHP			Crisp AHP			
		Local ranking	Global weights	Global ranking	Local weights	Local ranking	Global weights	Global ranking
Lean					0.083			
	Continuous improvement	1	0.028	9	0.540	1	0.045	7
	Lean operations	3	0.008	30	0.163	3	0.014	25
	Cost reduction ability	2	0.015	22	0.297	2	0.025	15
Delivery					0.072			
	Lead time	1	0.032	8	0.388	1	0.028	12
	On time delivery	2	0.022	15	0.271	2	0.020	18
	Order fulfillment rate	3	0.013	24	0.165	3	0.012	26
	Geographical location	4	0.009	29	0.115	4	0.008	28
	Flexibility	5	0.005	33	0.061	5	0.004	35
Technology					0.060			
	Production facilities	1	0.026	*11*	0.472	1	0.028	*11*
	Technology level	2	0.017	*19*	0.316	2	0.019	*19*
	R&D capabilities	3	0.006	32	0.122	3	0.007	31
	Digitalization level	4	0.005	34	0.091	4	0.005	33
Information tech					0.053			
	Integration ability	1	0.025	13	0.374	1	0.020	17
	ERP system	2	0.021	17	0.314	2	0.017	23
	Digital data	3	0.009	28	0.146	3	0.008	29

(continued)

Table 16.6 (continued)

Main attribute	Sub-attribute	FF-AHP			Crisp AHP			
		Local ranking	Global weights	Global ranking	Local weights	Local ranking	Global weights	Global ranking
	IT security risk	4	0.007	31	0.116	4	0.006	32
	Cloud usage	5	0.003	*36*	0.051	5	0.003	*36*
Sustainability					0.036			
	Environmental impact	1	0.017	20	0.485	1	0.017	22
	Waste	2	0.011	26	0.319	2	0.011	27
	Carbon footprint	3	0.004	35	0.123	3	0.004	34
	Fair working conditions	4	0.002	*37*	0.073	4	0.003	*37*

References

1. M.A.M. Abushariah, A.A.M. Alqudah, O.Y. Adwan, R.M.M. Yousef, Automatic heart disease diagnosis system based on artificial neural network (ANN) and adaptive neuro-fuzzy inference systems (ANFIS) approaches. J. Softw. Eng. Appl. **7**, 1055–1064 (2014)
2. A.B. Ajiboye, R.F. Weir, A Heuristic fuzzy logic approach to EMG pattern recognition for multifunctional prosthesis control. IEEE Trans. Neural Syst. Rehabil. Eng. **13**(3), 280–291 (2005)
3. M. Akram, G. Shahzadi, A.A.H. Ahmadini, Decision-making framework for an effective sanitizer to reduce COVID-19 under fermatean fuzzy environment. J. Math. **2020**, 3263407 (2020)
4. K.T. Atanassov, Intuitionistic fuzzy sets. Fuzzy Sets Syst. **20**, 87–96 (1986)
5. S.B. Aydemir, S. Yilmaz Gunduz, Fermatean fuzzy TOPSIS method with Dombi aggregation operators and its application in multi-criteria decision making. J. Intell. Fuzzy Syst. **39**(1), 851–869 (2020)
6. C. Bai, P. Dallasega, G. Orzes, J. Sarkis, Industry 4.0 technologies assessment: a sustainability perspective. Int. J. Prod. Econ. **229**, 107776 (2020)
7. K. Bär, Z.N.L. Herbert-Hansen, W. Khalid, Considering industry 4.0 aspects in the supply chain for an SME. Prod. Eng. **12**, 747–758 (2018)
8. W.C. Benton, M. Maloni, The influence of power driven buyer/seller relationships on supply chain satisfaction. J. Oper. Manag. **23**(1), 1–22 (2005)
9. F.T. Bozbura, A. Beskese, C. Kahraman, Prioritization of human capital measurement indicators using fuzzy AHP. Expert Syst. Appl. **32**(4), 1100–1112 (2007)
10. M. Brettel, N. Friederichsen, M. Keller, M. Rosenberg, How virtualization, decentralization and network building change the manufacturing landscape: an industry 4.0 perspective. Int. J. Mech. Ind. Sci. Eng. **8**(1), 37–44 (2014)
11. Y. Chen, Structured methodology for supplier selection and evaluation in a supply chain. Inf. Sci. **181**(9), 1651–1670 (2011)
12. S. Cisneros-Cabrera, G. Pishchulov, P. Sampaio, N. Mehandjiev, Z. Liu, K. Kununka, An approach and decision support tool for forming industry 4.0 supply chain collaborations. Comput. Ind. **125**, 103391 (2021). https://doi.org/10.1016/j.compind.2020.103391
13. O. Cruz-Mejia, J. Marmolejo, P. Vasant, Lead time performance in a internet product delivery supply chain with automatic consolidation. J. Ambient. Intell. Humaniz. Comput. **9**, 867–874 (2018)
14. B.C. Cuong, V. Kreinovich, Picture fuzzy sets—a new concept for computational intelligence problems, in Proceeding of the 3rd World Congress on Information and Communication Technologies, December 15–18 (Hanoi, Vietnam, 2013), pp. 1–6
15. L. De Boer, E. Labro, P. Morlacchi, A review of methods supporting supplier selection. Eur. J. Purch. Supply Manag. **7**(2), 75–89 (2001)
16. S.K. De, I. Beg, Triangular dense fuzzy sets and new defuzzification methods. J. Intell. Fuzzy Syst. **31**(1), 469–477 (2016)
17. G.W. Dickson, An analysis of vendor selection systems and decisions. J. Purchas. **2**(1), 5–17 (1966)
18. F. Dweiri, S. Kumar, S. Khan, V. Jain, Designing an integrated AHP based decision support system for supplier selection in automotive industry. Expert Syst. Appl. **62**, 273–283 (2016)
19. M.R. Galankashi, S.A. Helmi, P. Hashemzahi, Supplier selection in automobile industry: a mixed balanced scorecard–fuzzy AHP approach. Alex. Eng. J. **55**(1), 93–100 (2016)
20. H. Garg, A novel trigonometric operation-based q-rung orthopair fuzzy aggregation operator and its fundamental properties. Neural Comput. Appl. **32**, 15077–15099 (2020)
21. H. Garg, New exponential operation laws and operators for interval-valued q-rung orthopair fuzzy sets in group decision making process. Neural Comput. Appl. **33**(20), 13937–13963 (2021a). https://doi.org/10.1007/s00521-021-06036-0
22. H. Garg, CN-q-ROFS: Connection number-based q-rung orthopair fuzzy set and their application to decision-making process. Int. J. Intell. Syst. **36**, 3106–3143 (2021b)

23. H. Garg, A new possibility degree measure for interval-valued q-rung orthopair fuzzy sets in decision-making. Int. J. Intell. Syst. **36**, 526–557 (2021c)
24. H. Garg, G. Shahzadi, M. Akram, Decision-making analysis based on fermatean fuzzy Yager aggregation operators with application in COVID-19 testing facility. Math. Problem. Eng. **2020a**, 7279027 (2020a)
25. H. Garg, J. Gwak, T. Mahmood, Z. Ali, Power aggregation operators and VIKOR methods for complex q-rung orthopair fuzzy sets and their applications. Mathematics **8**, 538 (2020b)
26. H. Garg, Z. Ali, Z. Yang, T. Mahmood, S. Aljahdali, Multi-criteria decision-making algorithm based on aggregation operators under the complex interval-valued q-rung orthopair uncertain linguistic information. J. Intell. Fuzzy Syst. **41**(1), 1627–1656 (2021). https://doi.org/10.3233/JIFS-210442
27. A. Ghadge, M. Er Kara, H. Moradlou, M. Goswami, The impact of industry 4.0 implementation on supply chains. J. Manuf. Technol. Manag. **31**(4), 669–686 (2020)
28. A. Gilchrist, Introducing industry 4.0. Industry 4.0. Apress, Berkeley, CA (2016)
29. A. Gunasekaran, N. Subramanian, W.T.E. Ngai, Quality management in the 21st century enterprises: research pathway towards industry 4.0. Int. J. Prod. Econ. **207**, 125–129 (2018)
30. S. Gül, Fermatean fuzzy set extensions of SAW, ARAS, and VIKOR with applications in COVID-19 testing laboratory selection problem. Expert Syst. Early View e12769 (2021). https://doi.org/10.1111/exsy.12769
31. S. Gül, Ö. Kabak, Y.I. Topçu, A multiple criteria credit rating approach utilizing social media data. Data Knowl. Eng. **116**, 80–99 (2018)
32. S. Gül, Y.İ Topçu, A multi-attribute decision support model for the selection of touristic activities. Int. J. Anal. Hierarchy Process **7**(3), 560–588 (2015)
33. S.H. Ha, R. Krishnan, A hybrid approach to supplier selection for the maintenance of a competitive supply chain. Expert Syst. Appl. **34**(2), 1303–1311 (2008)
34. A. Hadi, W. Khan, A. Khan, A novel approach to MADM problems using Fermatean fuzzy Hamacher aggregation operators. Int. J. Intell. Syst. **36**(7), 3464–3499 (2021)
35. A. Hafezalkotob, A. Hafezalkotob, H. Liao, F. Herrera, Interval multimoora method integrating interval borda rule and interval best–worst-method-based weighting model: case study on hybrid vehicle engine selection. IEEE Trans. Cybern. **50**(3), 1157–1169 (2020)
36. M.M. Hasan, D. Jiang, S. Ullah, M. Noor-E-Alam, Resilient supplier selection in logistics 4.0 with heterogeneous information. Expert Syst. Appl. **139**, 112799 (2020). https://doi.org/10.1016/j.eswa.2019.07.016
37. M.Z. Hasan, A. Mallik, J.C. Tsou, Learning method design for engineering students to be prepared for industry 4.0: a Kaizen approach. High. Educ Skill. Work-Based Learn. **11**(1), 182–198 (2021)
38. B. Illés, P. Tamás, P. Dobos, R. Skapinyecz, New challenges for quality assurance of manufacturing processes in industry 4.0. Solid State Phenom. **261**, 481–486 (2017)
39. M.C. Jena, S.K. Mishra, H.S. Moharana, Application of industry 4.0 to enhance sustainable manufacturing. Environ. Prog. Sustain. Energy **39**(1), 13360 (2020)
40. Y. Ju, C. Luo, J. Ma, H. Gao, E.D.S. Gonzalez, Some interval-valued q-rung orthopair weighted averaging operators and their applications to multiple-attribute decision making. Int. J. Intell. Sys. **34**(10), 2584–2606 (2019)
41. K.R. Kalantari, A. Ebrahimnejad, H. Motameni, Presenting a new fuzzy system for web service selection aimed at dynamic software rejuvenation. Complex Intell. Syst. **6**, 697–710 (2020)
42. S. Kayapinar Kaya, E. Aycin, An integrated interval type 2 fuzzy AHP and COPRAS-G methodologies for supplier selection in the era of industry 4.0. Neural Comput. Appl. **33**, 10515–10535 (2021)
43. M. Keshavarz-Ghorabaee, M. Amiri, T. Hashemi-Tabatabaei, E.K. Zavadskas, A. Kaklauskas, A new decision-making approach based on fermatean fuzzy sets and WASPAS for green construction supplier evaluation. Mathematics **8**, 2202 (2020)
44. J.H. Kim, A review of cyber-physical system research relevant to the emerging IT trends: industry 4.0, IoT, big data, and cloud computing. J. Ind. Integr. Manag. **02**(03), 1750011 (2017)

45. F. Kutlu Gündoğdu, C. Kahraman, Spherical fuzzy sets and spherical fuzzy TOPSIS method. J. Intell. Fuzzy Syst. **36**(1), 337–352 (2019)
46. F. Kutlu Gündoğdu, C. Kahraman, A novel spherical fuzzy analytic hierarchy process and its renewable energy application. Soft. Comput. **24**, 4607–4621 (2020)
47. H. Lasi, P. Fettke, H.G. Kemper, T. Feld, M. Hoffmann, industry 4.0. Bus. Info. Syst. Eng. **6**(4), 239–242 (2014)
48. G. Laska, Wind energy and multi-criteria analysis in making decisions on the location of wind farms. Proced. Eng. **182**, 418–424 (2017)
49. K.M. Lee, Bipolar-valued fuzzy sets and their operations. Proc. Int. Conf. on Intelligent Technologies. (Bangkok, Thailand, 2000), pp. 307–312
50. M. Lezzi, M. Lazoi, A. Corallo, Cybersecurity for industry 4.0 in the current literature: a reference framework. Comput. Ind. **103**, 97–110 (2018)
51. L.B. Liboni, L.O. Cezarino, C.J.C. Jabbour, B.G. Oliveira, N.O. Stefanelli, Smart industry and the pathways to HRM 4.0: implications for SCM. Supply Chain Manag. **24**(1), 124–146 (2019)
52. D. Liu, Y. Liu, X. Chen, Fermatean fuzzy linguistic set and its application in multicriteria decision making. Int. J. Intell. Syst. **34**(5), 878–894 (2019a)
53. D. Liu, Y. Liu, L. Wang, Distance measure for Fermatean fuzzy linguistic term sets based on linguistic scale function: an illustration of the TODIM and TOPSIS methods. Int. J. Intell. Syst. **34**(11), 2807–2834 (2019b)
54. P. Liu, P. Wang, Some q-rung orthopair fuzzy aggregation operators and their applications to multiple-attribute decision making. Int. J. Intell. Syst. **33**(2), 259–280 (2018)
55. F. Long, P. Zeiler, B. Bertsche, Modelling the flexibility of production systems in industry 4.0 for analysing their productivity and availability with high-level petri nets. IFAC-PapersOnLine **50**(1), 5680–5687 (2017)
56. Y. Lu, Industry 4.0: a survey on technologies, applications and open research issues. J. Ind. Inf. Integration **6**, 1–10 (2017)
57. S. Luthra, K. Govindan, D. Kannan, S. Mangla, C. Garg, An integrated framework for sustainable supplier selection and evaluation in supply chains. J. Clean. Prod. **140**, 1686–1698 (2017)
58. E. Manavalan, K. Jayakrishna, A review of internet of things (IoT) embedded sustainable supply chain for industry 4.0 requirements. Comput. Ind. Eng. **127**, 925–953 (2019)
59. J. Mentzer, W.D. Witt, J. Keebler, S. Min, N. Nix, D. Smith, Z. Zacharia, Defining supply chain (SC) management. J. Bus. Logist. **22**(2), 1–25 (2001)
60. A.R. Mishra, P. Rani, K. Pandey, Fermatean fuzzy CRITIC-EDAS approach for the selection of sustainable third-party reverse logistics providers using improved generalized score function. J. Ambient. Intell. Humaniz. Comput. (2021). https://doi.org/10.1007/s12652-021-02902-w
61. H. Mohelska, M. Sokolova, Management approaches for industry 4.0—the organizational culture perspective. Technol. Econ. Dev. Econ. **24**(6), 2225–2240 (2018)
62. S. Moica, J. Ganzarain, D. Ibarra, F. Peti, Change made in shop floor management to transform a conventional production system into an "industry 4.0" case studies in SME automotive production manufacturing, in 7th International Conference on Industrial Technology and Management (ICITM), 7–9 March (Oxford UK, 2018)
63. D. Molodtsov, Soft set theory—first results. Comput. Math. Appl. **37**, 19–31 (1999)
64. F. Mosconi, *The new European industrial policy: global competitiveness and the manufacturing renaissance* (Routledge, London, England, 2015)
65. A. Muscio, A. Ciffolilli, What drives the capacity to integrate industry 4.0 technologies? evidence from European R&D projects. Econ. Innov. New Technol. **29**(2), 169–183 (2020)
66. H. Nguyen, G. Onofrei, D. Truong, Supply chain communication and cultural compatibility: performance implications in the global manufacturing industry. Bus. Process. Manag. J. **27**(1), 253–274 (2021)
67. J. Oláh, N. Aburumman, J. Popp, M.A. Khan, H. Haddad, N. Kitukutha, Impact of industry 4.0 on environmental sustainability. Sustain. **12**(11), 4674 (2020)
68. Z. Pawlak, Rough sets. Int. J. Comput. Inform. Sci. **11**(5), 341–356 (1982)

69. M. Riaz, H. Garg, H.M.A. Farid, M. Aslam, Novel q-rung orthopair fuzzy interaction aggregation operators and their application to low-carbon green supply chain management. J. Intell. Fuzzy Syst. **41**(2), 4109–4126 (2021). https://doi.org/10.3233/JIFS-210506
70. V. Roblek, M. Meško, A. Krapež, A complex view of industry 4.0. SAGE Open **6**(2), 1–11 (2016)
71. A.Rojko, Industry 4.0 concept: background and overview. Int. J. Interact. Mobile Technol. **11**(5), 77–90 (2017)
72. F. Rosin, P. Forget, S. Lamouri, R. Pellerin, Impacts of industry 4.0 technologies on lean principles. Int. J. Prod. Research 58(6), 1644–1661 (2020).
73. T.L. Saaty, *The Analytic Hierarchy Process* (RWS Publications, Pittsburgh, USA, 1980)
74. T.L. Saaty, How to make a decision: the analytic hierarchy process. Eur. J. Oper. Res. **48**(1), 9–26 (1990)
75. N. Sachdeva, A. Shrivastava, A. Chauhan, Modeling supplier selection in the era of industry 4.0. Benchmark. Inte. J. **28**(5), 1809–1836 (2021)
76. A. Saibene, M. Assale, M. Giltri, Expert systems: definitions, advantages and issues in medical field applications. Expert Syst. Appl. **177**, 114900 (2021)
77. M.A. Salam, Analyzing manufacturing strategies and industry 4.0 supplier performance relationships from a resource-based perspective. Benchmark. Int. J. **28**(5), 1697–1716 (2021)
78. M. Sanchez, E. Exposito, J. Aguilar, Industry 4.0: survey from a system integration perspective. Int. J. Comput. Integr. Manuf. **331**(10–11), 1017–1041 (2020)
79. A. Sanders, C. Elangeswaran, J.P. Wulfsberg, Industry 4.0 implies lean manufacturing: research activities in industry 4.0 function as enablers for lean manufacturing. J. Ind. Eng. Manag. **9**(3), 811–833 (2016)
80. A. Santana, P. Afonso, A. Zanin, R. Wernke, Costing models for capacity optimization in industry 4.0: trade-off between used capacity and operational efficiency. Proced. Manuf. **13**, 1183–1190 (2017)
81. A. Schumacher, S. Erol, W. Sihn, A maturity model for assessing industry 4.0 readiness and maturity of manufacturing enterprises. Proced. CIRP **52**, 161–166 (2016)
82. T. Senapati, R.R. Yager, Some new operations over fermatean fuzzy numbers and application of fermatean fuzzy WPM in multiple criteria decision making. Informatica **30**(2), 391–412 (2019a)
83. T. Senapati, R.R. Yager, Fermatean fuzzy weighted averaging/geometric operators and its application in multi-criteria decision-making methods. Eng. Appl. Artif. Intell. **85**, 112–121 (2019b)
84. T. Senapati, R.R. Yager, Fermatean fuzzy sets. J. Ambient. Intell. Humaniz. Comput. **11**, 663–674 (2020)
85. G. Shahzadi, M. Aksam, Group decision-making for the selection of an antivirus mask under fermatean fuzzy soft information. J. Intell. Fuzzy Syst. **40**(1), 1401–1416 (2021)
86. F. Smarandache, *A Unifying Field In Logics Neutrosophy: Neutrosophic Probability, Set and Logic* (American Research Press, Rehoboth, USA, 1999)
87. C.S. Tang, P. Lucas, L.P. Veelenturf, The strategic role of logistics in the industry 4.0 era. Transp. Res. Part E Logist. Transp. Rev. **129**, 1–11 (2019)
88. H.M. Tornyeviadzi, F.A. Neba, H. Mohammed, R. Seidu, Nodal vulnerability assessment of water distribution networks: an integrated fuzzy AHP-TOPSIS approach. Int. J. Crit. Infrastruct. Prot. **34**, 100434 (2021). https://doi.org/10.1016/j.ijcip.2021.100434
89. V. Torra, Hesitant fuzzy sets. Int. J. Intell. Syst. **25**(6), 529–539 (2010)
90. S.H. Tsaur, T.Y. Chang, C.H. Yen, The evaluation of airline service quality by fuzzy MCDM. Tour. Manag. **23**, 107–115 (2002)
91. B. Ünver, İ Altın, S. Gürgen, Risk ranking of maintenance activities in a two-stroke marine diesel engine via fuzzy AHP method. Appl. Ocean Res. **111**, 102648 (2021). https://doi.org/10.1016/j.apor.2021.102648
92. R.R. Yager, Pythagorean membership grades in multicriteria decision making. IEEE Trans. Fuzzy Syst. **22**(4), 958–965 (2014)
93. R.R. Yager, Generalized orthopair fuzzy sets. IEEE Trans. Fuzzy Syst. **25**(5), 1222–1230 (2017)

94. Z. Yang, H. Garg, Interaction power partitioned maclaurin symmetric mean operators under q-rung orthopair uncertain linguistic information. Int. J. Fuzzy Syst. (2021). https://doi.org/10.1007/s40815-021-01062-5
95. L.A. Zadeh, Fuzzy sets. Inf. Control **8**, 338–353 (1965)
96. Y. Zhong, O. Segu, H.C. Moon, Service transformation under industry 4.0: investigating acceptance of facial recognition payment through an extended technology acceptance model. Technol. Soc. **64**, 101515 (2021)

Dr. Alper Camci serves as a faculty member in the Management Engineering Department of Bahçeşehir University's Faculty of Engineering and Natural Sciences in Istanbul. Prior to joining Bahçeşehir University, Dr. Camci served as an academic director in Arcadia University and faculty member for Okan University in Istanbul and Franklin University in Columbus. Dr. Camci completed his Ph.D. at University of Central Florida in Industrial Engineering and Management Systems, MS at Wayne State University and undergraduate studies at Istanbul Technical University in Mechanical Engineering. Dr Camci's academic interests include project and program management, operations management, complexity in organizations, technology and innovation management, supply chain management, organizational transformation and change management.

Muharrem Eray Ertürk works as a production engineer in Mefa Group which one of the leading original equipment manufacturers for white goods industry with five production facilities across Turkey and more than 2000 employees. Prior to this position, Mr. Ertürk also worked as a project/process engineer and a quality engineer with the same company. Mr. Ertürk completed his MBA degree with Bahçeşehir University in Istanbul and B.Sc. degree in Mechanical Engineering with Ondokuz Mayıs University in Samsun, Turkey.

Dr. Sait Gül is an assistant professor at Management Engineering Department of Bahçeşehir University Faculty of Engineering and Natural Sciences. He received his B.Sc. degrees in Maritime Transport and Management Engineering (2007) and Industrial Engineering (2008) from Istanbul University and holds his MS in Engineering Management (2012) and Ph.D. in Industrial Engineering (2017), both from Istanbul Technical University. His research interests include operations research, multiple criteria decision analysis, fuzzy theory applications, knowledge management and big data analysis.

Chapter 17
Pentagonal q-Rung Orthopair Numbers and Their Applications

Irfan Deli

Abstract In the chapter, Pentagonal q-rung orthopair numbers(Pq-RO-numbers) that are generalization of the fuzzy numbers and intuitionistic fuzzy numbers are defined on real number R. Then, normal Pq-RO-numbers defined in [0, 1] and by using the concept of s-norm and t-norm their laws of operations are proposed including their properties. Also, to compare any two Pq-RO-numbers, 1. and 2. rank value of Pq-RO-numbers are proposed. Furthermore, some operators of qth rung orthopair fuzzy numbers such as; qth rung orthopair fuzzy number weighted aggregation mean operator and qth rung orthopair fuzzy number weighted geometric mean operator are developed. Finally, by using the Pq-RO-numbers and related concepts, a multi-attribute decision-making method is developed and a real example is initiated to illustrate the proposed method.

17.1 Introduction

Multi-criteria decision-making is a problem that can give the ranking of desired alternatives based on the some related criteria. Also, Liu and Wang [30] said that "In real decision process, an important problem is how to express the attribute value more efficiently and accurately. In the real world, because of the complexity of decision-making problems and the fuzziness of decision-making environments, it is not enough to express attribute values of alternatives by exact values." Since information of the real problems cannot always express by crisp numbers, fuzzy set theory proposed by Zadeh [50]. Then, several extensions and application fuzzy sets including linear Diophantine fuzzy sets and linear Diophantine fuzzy algebraic structures have been studied in Riaz et al. [34] and Kamaci [26]. After fuzzy set theory, intuitionistic fuzzy set theory proposed by Atanassov [1] and then the theory has achieved a great success in dealing with contain imprecision and uncertainty in various complex decision-making problems. In this theory, an element(x) of universe have a membership degree $\mu(x)$, a non-membership degree $\nu(x)$ and

I. Deli (✉)
Kilis 7 Aralık University, 79000 Kilis, Turkey
e-mail: irfandeli@kilis.edu.tr

H. Garg (ed.), *q-Rung Orthopair Fuzzy Sets*,
https://doi.org/10.1007/978-981-19-1449-2_17

hesitant membership degree $\pi(x) = 1 - \mu(x) - \nu(x)$ such that $0 \leq \mu(x) + \nu(x) \leq 1$. To tackle more complex problems with vague information in the real world, several extensions and applications of intuitionistic fuzzy sets have been studied in Xu [45], Xu and Wang [46], Xu and Yager [47] and the theory generalized by many researchers to Pythagorean fuzzy sets in Yager and Abbasov [54] and Yager [55] such that $0 \leq \mu^2(x) + \nu^2(x) \leq 1$, to Fermatean fuzzy sets in Senapati and Yager [37] such that $0 \leq \mu^3(x) + \nu^3(x) \leq 1$ and to Spherical fuzzy sets in Gündoğdu and Kahraman [23] such that $0 \leq \mu^2(x) + \nu^2(x) + \pi^2(x) \leq 1$.

Recently, q-rung orthopair fuzzy sets (or generalized orthopair fuzzy sets) proposed by Yager [51] and Yager and Alajlan [52] are new an ideas to express fuzzy data in complex decision-making problems with generalized orthopair fuzzy sets. In the theory an element(x) of universe have a membership degree $\mu(x)$, a non-membership degree $\nu(x)$ such that $0 \leq \mu^q(x) + \nu^q(x) \leq 1$ $(q \geq 1)$. Also, Liu and Wang [30] developed two new methods by introducing the q-rung orthopair fuzzy weighted averaging operator and the q-rung orthopair fuzzy weighted geometric operator. Riaz et al. [35] studied on novel q-rung orthopair fuzzy interaction aggregation operators based on low-carbon green supply chain management. Yang and Garg [53] developed interaction power partitioned maclaurin symmetric mean operators under q-rung orthopair uncertain linguistic information. Peng et al. [33] proposed exponential operation and aggregation operator including score function for q-rung orthopair fuzzy sets. Presently q-rung orthopair fuzzy sets are being studied and used in different fields of science such as on integrals in Ai et al. [2], on operators in Akram et al. [3], Ali and Mahmood [5], Aydemir and Gunduz [6, 7], Darko and Liang [9], Garg and Chen [18], Riaz et al. [36], Xing et al. [48], on decision making method in Akram et al. [4] Cheng et al. [8], Hussain et al. [24], Li et al. [29], Liao et al. [31], Wang et al. [40–42], Tian et al. [38], on distance measures in Du [15], Zeng et al. [56], on correlation and correlation coefficient in Du [16], on similarity measures in Farhadinia et al. [17], Jan et al. [25], Wang et al. [39], on preference relations in Li et al. [27], on information measures in Peng and Liu [32], on exponential operation and aggregation operator including score function in Peng et al. [33], on connection number-based q-rung orthopair fuzzy set in Garg [21]. Also, Garg et al. [22] initiated aggregation operators under the complex interval-valued q-rung orthopair uncertain linguistic information. And then, Garg [19, 20] introduced possibility degree measure for interval-valued qth rung orthopair fuzzy sets, and new exponential operation laws and operators for interval-valued q-rung orthopair fuzzy sets, respectively.

The concept of a fuzzy number and intuitionistic fuzzy number is of the most important for modeling an ill-known quantity. To handle some uncertain information, which contain an ill-known quantity, many investigations drawn much attentions in theory and practice on real number R in Deli and Şubaş [10, 11], Deli [12, 13], Deli and Çağman [14], Li [28], Wang et al. [43], Wei [44], Uthra et al. [49], Zimmermann [57]. As far as we know, there is no study on Pentagonal q-rung orthopair numbers (Pq-RO-numbers) based on generalized trapezoidal hesitant fuzzy numbers in the literature. Therefore, in the chapter we first proposed Pentagonal q-rung orthopair numbers (Pq-RO-numbers) that are generalization of the fuzzy numbers and intuitionistic fuzzy numbers on real number R. We second defined normal

Pq-RO-numbers by using the concept of s-norm and t-norm, and proposed their laws of operations including their properties. We third introduced 1. and 2. rank value of Pq-RO-numbers to compare any two Pq-RO-numbers. We fourth developed some operators of qth rung orthopair fuzzy numbers such as; qth rung orthopair fuzzy numbers weighted aggregation mean operator and qth rung orthopair fuzzy number weighted geometric mean operator. Finally, we initiated a multi-attribute decision-making method with a real example under the Pq-RO-numbers and related concepts.

In the chapter, some of the definitions, operations and properties are quoted or inspired or generalized by Yager and Abbasov [54], and Yager [55], Senapati and Yager [37], Liu and Wang [30], Deli and Şubaş [10, 11], Deli [12, 13], Deli and Çağman [14], Li [28], Wang et al. [43], Wei [44], Uthra et al. [49], Zimmermann [57].

17.2 Preliminary

In this section, we recall some basic notions of the fuzzy sets, intuitionistic fuzzy sets, qth rung orthopair fuzzy sets, single valued neutrosophic sets, single-valued trapezoidal neutrosophic numbers and intuitionistic pentagonal fuzzy numbers.

Definition 17.1 (Zadeh [50]) Let E be a universe. Then, a fuzzy set X over E is defined by

$$X = \{(\mu_X(x)/x) : x \in E\}$$

where μ_X is called membership function of X and defined by $\mu_X : E \to [0.1]$. For each $x \in E$, the value $\mu_X(x)$ represents the degree of x belonging to the fuzzy set X.

Definition 17.2 (Atanassov [1]) Let E be a universe. Then, an intuitionistic fuzzy set K over E is defined by

$$K = \{< x, \mu_K(x), \upsilon_K(x) >: \ x \in E\}$$

where $\mu_K : E \to [0, 1]$ and $\upsilon_K : E \to [0, 1]$ such that $0 \leq \mu_K(x) + \upsilon_K(x) \leq 1$ for any $x \in E$. For each $x \in E$, the values $\mu_K(x)$ and $\upsilon_K(x)$ are the degree of membership and degree of non-membership of x, respectively.

Definition 17.3 (Yager [51]) Let E be a universe. Then, a qth rung orthopair fuzzy set A^q over E is defined by

$$A^q = \{< x, A^\mu(x), A^\nu(x) >: x \in E\}.$$

where $q \geq 1$, $A^{\mu}(x)$ indicates support for membership of x in A^q and $A^{\nu}(x)$ indicates support against membership of x in A^q. They are respectively defined by

$$A^{\mu} : E \to [0, 1], \quad A^{\nu} : E \to [0, 1]$$

such that $0 \leq A^{\mu}(x))^q + (A^{\nu}(x)^q \leq 1$.

Definition 17.4 (Liu and Wang [30]) Let $A^q = \{< x, A^{\mu}(x), A^{\nu}(x) >: x \in E\}$, $A_1^q = \{< x, A_1^{\mu}(x), A_1^{\nu}(x) >: x \in E\}$ and $A_2^q = \{< x, A_2^{\mu}(x), A_2^{\nu}(x) >: x \in E\}$ be three qth rung orthopair fuzzy sets and $\gamma > 0$ be any real number. Then,

1. $A_1^q \oplus A_2^q = \{< x, \sqrt[q]{(A_1^{\mu}(x))^q + (A_2^{\mu}(x))^q - (A_1^{\mu}(x))^q.(A_2^{\mu}(x))^q}, A_1^{\nu}(x).A_2^{\nu}(x) >: x \in E\}$
2. $A_1^q \otimes A_2^q = \{< x, A_1^{\mu}(x).A_2^{\mu}(x), \sqrt[q]{(A_1^{\nu}(x))^q + (A_2^{\nu}(x))^q - (A_1^{\nu}(x))^q.(A_2^{\nu}(x))^q} >: x \in E\}$
3. $(A^q)^{\gamma} = \{< x, A^{\mu}(x), \sqrt[q]{(1 - (1 - (A^{\nu}(x))^q)^{\gamma}} >: x \in E\}$
4. $\gamma(A^q) = \{< x, \sqrt[q]{(1 - (1 - (A^{\mu}(x))^q)^{\gamma}}, A^{\nu}(x) >: x \in E\}$.

Theorem 17.1 (Liu and Wang [30]) *Let* $A^q = \{< x, A^{\mu}(x), A^{\nu}(x) >: x \in E\}$, $A_1^q = \{< x, A_1^{\mu}(x), A_1^{\nu}(x) >: x \in E\}$ *and* $A_2^q = \{< x, A_2^{\mu}(x), A_2^{\nu}(x) >: x \in E\}$ *be three qth rung orthopair fuzzy sets and* $\gamma, \gamma_1, \gamma_2 > 0$ *be any real number. Then, Then, the following are valid;*

1. $A_1^q \oplus A_2^q = A_2^q \oplus A_1^q$
2. $A_1^q \otimes A_2^q = A_2^q \otimes A_1^q$
3. $\gamma(A_1^q \oplus A_2^q) = \gamma A_1^q \oplus \gamma A_2^q$
4. $\gamma_1 A_1^q \oplus \gamma_2 A_1^q = (\gamma_1 + \gamma_2) A_1^q$
5. $(A_1^q \otimes A_2^q)^{\gamma} = (A_1^q)^{\gamma} \otimes (A_2^q)^{\gamma}$
6. $(A^q)^{\gamma_1} \otimes (A^q)^{\gamma_2} = (A^q)^{\gamma_1 + \gamma_2}$.

For convenience, if universe set the only element we use $A^q =< A^{\mu}, A^{\nu} >$ instead of $A^q = \{< x, A^{\mu}(x), A^{\nu}(x) >: x \in E\}$.

Definition 17.5 (Liu and Wang [30]) Let $A^q =< A^{\mu}, A^{\nu} >$ be a qth rung orthopair fuzzy set. Then,

1. score function of A^q, is denoted $S(A^q)$, is defined as follows:

$$S(A^q) = (A^{\mu}(x))^q - (A^{\nu}(x))^q$$

2. an accuracy function H of A^q, is denoted $A(A^q)$, is defined as follows:

$$A(A^q) = (A^{\mu}(x))^q + (A^{\nu}(x))^q$$

Obviously, $S(A^q) \in [-1, 1]$ and $A(A^q) \in [0, 1]$.

If $A_1^q =< A_1^{\mu}, A_1^{\nu} >$ and $A_2^q =< A_2^{\mu}, A_2^{\nu} >$ be two qth rung orthopair fuzzy sets. Based on the above score function and the accuracy function, a comparison method is proposed as

1. If $S(A_1^q) > S(A_2^q)$, then $A_1^q > A_2^q$
2. If $S(A_1^q) = S(A_2^q)$, then
   (a) If $A(A_1^q) > A(A_2^q)$, then $A_2^q > A_1^q$
   (b) If $A(A_1^q) = A(A_2^q)$, then $A_1^q = A_2^q$.

Definition 17.6 (Liu and Wang [30]) Let $A_j^q =< A_j^\mu, A_j^\nu >$, $j = 1, 2, \ldots, n$ are collection of some qth rung orthopair fuzzy sets. Then

1. qth rung orthopair fuzzy weighted aggregation mean operator, denoted by $q - OFW_{ao}$, is defined as

$$q - OFW_{ao}(A_1^q, A_2^q, \ldots, A_n^q) = \bigoplus_{j=1}^n w_j A_j^q$$
$$= \left\langle (1 - \prod_{j=1}^n (1 - (A_j^\mu)^q)^{w_j})^{\frac{1}{q}}, \prod_{j=1}^n (A_j^\nu)^{w_j} \right\rangle$$

2. qth rung orthopair fuzzy weighted geometric operator, denoted by $q - OFW_{go}$, is defined as

$$q - OFW_{go}(A_1^q, A_2^q, \ldots, A_n^q) = \bigotimes_{j=1}^n (A_j^q)^{w_j}$$
$$= \left\langle \prod_{j=1}^n (A_j^\mu)^{w_j}, (1 - \prod_{j=1}^n (1 - (A_j^\nu)^q)^{w_j})^{\frac{1}{q}} \right\rangle$$

where $w = (w_1, w_2, \ldots, w_n)^T$ is a weight vector for every $j \in I_n$ such that, $w_j \in [0, 1]$ and $\sum_{j=1}^n w_j = 1$.

Also Liu and Wang [30] show that the operators hold some properties such as

1. (Idempotency)if $A_j^q = A^q =< A^\mu, A^\nu >$, $j = 1, 2, \ldots, n$, then

$$q - OFW_{ao}(A_1^q, A_2^q, \ldots, A_n^q) = A^q \text{ and } q - OFW_{go}(A_1^q, A_2^q, \ldots, A_n^q) = A^q$$

2. (Monotonicity) Suppose that $A_j^q =< A_j^\mu, A_j^\nu >$ and $B_j^q =< B_j^\mu, B_j^\nu >$, $j = 1, 2, \ldots, n$ are two collection of some qth rung orthopair fuzzy sets. If $A_j^\mu \geq B_j^\mu$ $A_j^\nu \leq B_j^\nu$ for all j, then

$$q - OFW_{ao}(A_1^q, A_2^q, \ldots, A_n^q) \geq q - OFW_{ao}(B_1^q, B_2^q, \ldots, B_n^q)$$

and

$$q - OFW_{go}(A_1^q, A_2^q, \ldots, A_n^q) \geq q - OFW_{go}(B_1^q, B_2^q, \ldots, B_n^q)$$

3. (Boundedness) If $A^- =< \min_{1 \leq j \leq n} A_j^\mu), \max_{1 \leq j \leq n} A_j^\nu >$ and $A^+ =< \max_{1 \leq j \leq n} A_j^\mu), \min_{1 \leq j \leq n} A_j^\nu >$, then

$$A^- \leq q - OFW_{ao}(A_1^q, A_2^q, \ldots, A_n^q) \leq A^+ \text{ and } A^- \leq q - OFW_{go}(A_1^q, A_2^q, \ldots, A_n^q) \leq A^+$$

Definition 17.7 (Zimmermann [57]) t-norms are associative, monotonic and commutative two valued functions t that map from $[0, 1] \times [0, 1]$ into $[0, 1]$. For fuzzy sets X_1 and X_2, these properties are formulated with the following conditions:

1. $t(0, 0) = 0$ and $t(\mu_{X_1}(x), 1) = t(1, \mu_{X_1}(x)) = \mu_{X_1}(x), \;\; x \in E$
2. If $\mu_{X_1}(x) \leq \mu_{X_3}(x)$ and $\mu_{X_2}(x) \leq \mu_{X_4}(x)$, then $t(\mu_{X_1}(x), \mu_{X_2}(x)) \leq t(\mu_{X_3}x), \mu_{X_4}(x))$
3. $t(\mu_{X_1}(x), \mu_{X_2}(x)) = t(\mu_{X_2}(x), \mu_{X_1}(x))$
4. $t(\mu_{X_1}(x), t(\mu_{X_2}(x), \mu_{X_3}(x))) = t(t(\mu_{X_1}(x), \mu_{X_2})(x), \mu_{X_3}(x))$.

Also, s-norm are associative, monotonic, and commutative two placed functions s which map from $[0, 1] \times [0, 1]$ into $[0, 1]$. These properties are formulated with the following conditions:

1. $s(1, 1) = 1$ and $s(\mu_{X_1}(x), 0) = s(0, \mu_{X_1}(x)) = \mu_{X_1}(x), x \in E$
2. if $\mu_{X_1}(x) \leq \mu_{X_3}(x)$ and $\mu_{X_2}(x) \leq \mu_{X_4}(x)$, then $s(\mu_{X_1}(x), \mu_{X_2}(x)) \leq s(\mu_{X_3}(x), \mu_{X_4}(x))$
3. $s(\mu_{X_1}(x), \mu_{X_2}(x)) = s(\mu_{X_2}(x), \mu_{X_1}(x))$
4. $s(\mu_{X_1}(x), s(\mu_{X_2}(x), \mu_{X_3}(x))) = s(s(\mu_{X_1}(x), \mu_{X_2})(x), \mu_{X_3}(x))$.

t-norms and s-norms are related in a sense of logical duality. Typical dual pairs of non-parametrized t-norm and s-norms are complied below:

1. Drastic product:

$$t_w(\mu_{X_1}(x), \mu_{X_2}(x)) = \begin{cases} \min\{\mu_{X_1}(x), \mu_{X_2}(x)\}, & \max\{\mu_{X_1}(x)\mu_{X_2}(x)\} = 1 \\ 0, & \text{otherwise} \end{cases} \tag{17.1}$$

2. Drastic sum:

$$s_w(\mu_{X_1}(x), \mu_{X_2}(x)) = \begin{cases} \max\{\mu_{X_1}(x), \mu_{X_2}(x)\}, & \min\{\mu_{X_1}(x)\mu_{X_2}(x)\} = 0 \\ 1, & \text{otherwise} \end{cases} \tag{17.2}$$

3. Bounded product:

$$t_1(\mu_{X_1}(x), \mu_{X_2}(x)) = \max\{0, \mu_{X_1}(x) + \mu_{X_2}(x) - 1\} \tag{17.3}$$

4. Bounded sum:

$$s_1(\mu_{X_1}(x), \mu_{X_2}(x)) = \min\{1, \mu_{X_1}(x) + \mu_{X_2}(x)\} \tag{17.4}$$

5. Einstein product:

$$t_{1.5}(\mu_{X_1}(x), \mu_{X_2}(x)) = \frac{\mu_{X_1}(x) \cdot \mu_{X_2}(x)}{2 - [\mu_{X_1}(x) + \mu_{X_2}(x) - \mu_{X_1}(x) \cdot \mu_{X_2}(x)]} \tag{17.5}$$

6. Einstein sum:

$$s_{1.5}(\mu_{X_1}(x), \mu_{X_2}(x)) = \frac{\mu_{X_1}(x) + \mu_{X_2}(x)}{1 + \mu_{X_1}(x) \cdot \mu_{X_2}(x)} \tag{17.6}$$

7. Algebraic product:

$$t_2(\mu_{X_1}(x), \mu_{X_2}(x)) = \mu_{X_1}(x) \cdot \mu_{X_2}(x) \tag{17.7}$$

8. Algebraic sum:

$$s_2(\mu_{X_1}(x), \mu_{X_2}(x)) = \mu_{X_1}(x) + \mu_{X_2}(x) - \mu_{X_1}(x) \cdot \mu_{X_2}(x) \tag{17.8}$$

9. Hamacher product:

$$t_{2.5}(\mu_{X_1}(x), \mu_{X_2}(x)) = \frac{\mu_{X_1}(x) \cdot \mu_{X_2}(x)}{\mu_{X_1}(x) + \mu_{X_2}(x) - \mu_{X_1}(x) \cdot \mu_{X_2}(x)} \tag{17.9}$$

10. Hamacher sum:

$$s_{2.5}(\mu_{X_1}(x), \mu_{X_2}(x)) = \frac{\mu_{X_1}(x) + \mu_{X_2}(x) - 2 \cdot \mu_{X_1}(x) \cdot \mu_{X_2}(x)}{1 - \mu_{X_1}(x) \cdot \mu_{X_2}(x)} \tag{17.10}$$

11. Minimum:

$$t_3(\mu_{X_1}(x), \mu_{X_2}(x)) = \min\{\mu_{X_1}(x), \mu_{X_2}(x)\} \tag{17.11}$$

12. Maximum:

$$s_3(\mu_{X_1}(x), \mu_{X_2}(x)) = \max\{\mu_{X_1}(x), \mu_{X_2}(x)\} \tag{17.12}$$

Definition 17.8 (Uthra et al. [49]) Let $w_A, u_A \in [0, 1]$ such that $w_A + u_A \in [0, 1]$, $\hat{w_A} \in [0, w_A]$, $\hat{u_A} \in [u_A, 1]$ and $a_i, b_i \in R(i = 1, 2, \ldots, 5)$ such that $b_1 \leq a_1 \leq b_2 \leq a_2 \leq b_3 \leq a_3 \leq a_4 \leq b_4 \leq a_5 \leq b_5$. Then, an intitutionistic pentagonal fuzzy number

$$A = \langle((a_1, a_2, a_3, a_4, a_5), (b_1, b_2, b_3, b_4, b_5); w_A, u_A\rangle$$

is a special intitutionistic set on the real on R, whose membership function μ_A and non-membership function ν_A are given as follows

$$\mu_A(x) = \begin{cases} \hat{w_A} - (x - a_2)\hat{w_A}/(a_1 - a_2) & (a_1 \leq x < a_2) \\ w_A + (x - a_3)(\hat{w_A} - w_A)/(a_2 - a_3) & (a_2 \leq x < a_3) \\ w_A + (x - a_3)(\hat{w_A} - w_A)/(a_4 - a_3) & (a_3 \leq x \leq a_4) \\ \hat{w_A} - (x - a_4)\hat{w_A}/(a_5 - a_4) & (a_4 < x \leq a_5) \\ 0 & \text{otherwise,} \end{cases}$$

and

$$\nu_A(x) = \begin{cases} 1 + (\hat{u_A} - 1)(x - b_1)/(b_2 - b_1) & (b_1 \le x < b_2) \\ \hat{u_A} + (u_A - \hat{u_A})(x - b_2))/(b_3 - b_2) & (b_2 \le x < b_3) \\ u_A + (x - b_3)(\hat{u_A} - u_A)/(b_4 - b_3) & (b_3 \le x \le b_4) \\ \hat{u_A} - (x - b_4)(1 - \hat{u_A})/(b_5 - b_4) & (b_4 < x \le b_5) \\ 1 & \text{otherwise,} \end{cases}$$

respectively.

Definition 17.9 (Uthra et al. [49]) Let $A = \langle((a_1, a_2, a_3, a_4, a_5), (b_1, b_2, b_3, b_4, b_5); w_A, u_A\rangle$ be an intitutionistic pentagonal fuzzy number and $\hat{w_A} \in [0, w_A]$, $\hat{u_A} \in [u_A, 1]$. Then, rank of A, is denoted by $R(A)$ is defined as

$$R(A) = \frac{w_A \cdot x_\mu \cdot y_\mu + u_A x_\nu \cdot y_\nu}{w_A + u_A} \tag{17.13}$$

where $x_\mu = \frac{a_1+2a_2+3a_3+2a_4+a_5}{9}$, $y_\mu = \frac{3\hat{w_A}+w_A}{9}$, $x_\nu = \frac{b_1+2b_2+3b_3+2b_4+b_5}{9}$ and $y_\nu = \frac{3\hat{u_A}+5+u_A}{9}$.

17.3 Pentagonal q-Rung Orthopair Numbers

In this section, we define Pentagonal q-rung orthopair numbers with their properties. Some of the definitions are quoted or inspired or generalized by Deli [13], Deli and Çağman [14], Liu and Wang [30], Li [28], Uthra et al. [49], Yager [55].

Definition 17.10 Let $q \ge 1$, $w_{\overline{A}}, u_{\overline{A}} \in [0, 1]$ such that $w_{\overline{A}}^q + u_{\overline{A}}^q$, $\hat{w}_{\overline{A}}^q + \hat{u}_{\overline{A}}^q \in [0, 1]$, $\hat{w}_{\overline{A}} \in [0, w_{\overline{A}}]$, $\hat{u}_{\overline{A}} \in [u_{\overline{A}}, 1]$ and $a_i \in R (i = 1, 2, \ldots, 5)$ such that $a_1 \le a_2 \le a_3 \le a_4 \le a_5$. Then, a Pentagonal q-rung orthopair number (Pq-RO-number)

$$\overline{A} = \langle a_1, a_2, a_3, a_4, a_5; (\hat{w}_{\overline{A}}, w_{\overline{A}}), (u_{\overline{A}}, \hat{u}_{\overline{A}})\rangle$$

is a special q-rung orthopair set on the real on R, whose membership function A^μ and non-membership function A^ν are given as follows

$$A^\mu(x) = \begin{cases} (x - a_1)\hat{w}_{\overline{A}}/(a_2 - a_1) & (a_1 \le x < a_2) \\ w_{\overline{A}} + (x - a_3)(\hat{w}_{\overline{A}} - w_{\overline{A}})/(a_2 - a_3) & (a_2 \le x < a_3) \\ w_{\overline{A}} + (x - a_3)(\hat{w}_{\overline{A}} - w_{\overline{A}})/(a_4 - a_3) & (a_3 \le x \le a_4) \\ (x - a_5)\hat{w}_{\overline{A}}/(a_4 - a_5) & (a_4 < x \le a_5) \\ 0 & \text{otherwise,} \end{cases}$$

and

$$A^{\nu}(x) = \begin{cases} (a_2 - x + \hat{u}_{\overline{A}}(x - a_1)/(a_2 - a_1) & (a_1 \leq x < a_2) \\ \hat{u}_{\overline{A}} + (u_{\overline{A}} - \hat{u}_{\overline{A}})(x - a_2))/(a_3 - a_2) & (a_2 \leq x < a_3) \\ u_{\overline{A}} + (x - a_3)(\hat{u}_{\overline{A}} - u_{\overline{A}})/(a_4 - a_3) & (a_3 \leq x \leq a_4) \\ (x - a_4) + \hat{u}_{\overline{A}}(a_5 - x))/(a_5 - a_4) & (a_4 < x \leq a_5) \\ 1 & \text{otherwise,} \end{cases}$$

respectively.

Example 17.1 Assume that $\overline{A} = \langle 1, 2, 3, 4, 5; (0,5,0,6), (0.7, 0.8)\rangle$ be a Pq-RO-number ($q > 2$). The meanings of $\overline{A}$ is interpreted as follows: For example; the membership degree of the element $2.5 \in R$ belonging to A is 0.55 whereas the non-membership degree is 0.75, i.e., $A^{\mu}(2.5) = 0.55$, $A^{\nu}(2.5) = 0.75$.

Note 1 If $a_1, a_2, a_3, a_4, a_5 \in [0, 1]$, then $\overline{A} = \langle a_1, a_2, a_3, a_4, a_5; (\hat{w}_{\overline{A}}, w_{\overline{A}}), (u_{\overline{A}}, \hat{u}_{\overline{A}})\rangle$ is a normal Pq-RO-number.

Note that the set of all normal Pq-RO-numbers on R will be denoted by Δ.

Definition 17.11 Let $\overline{A}_1 = \langle a_1, a_2, a_3, a_4, a_5; (\hat{w}_{\overline{A}_1}, w_{\overline{A}_1}), (u_{\overline{A}_1}, \hat{u}_{\overline{A}_1})\rangle$ and $\overline{A}_2 = \langle b_1, b_2, b_3, b_4, b_5; (\hat{w}_{\overline{A}_2}, w_{\overline{A}_2}), (u_{\overline{A}_2}, \hat{u}_{\overline{A}_2})\rangle \in \Delta$ and $\gamma > 0$, then

1.

$$\overline{A}_1 \underset{s_k, t_k}{\widetilde{\oplus}} \overline{A}_2 = \langle \sqrt[q]{s_k(a_1^q, b_1^q)}, \sqrt[q]{s_k(a_2^q, b_2^q)}, \sqrt[q]{s_k(a_3^q, b_3^q)}, \sqrt[q]{s_k(a_4^q, b_4^q)}, \sqrt[q]{s_k(a_5^q, b_5^q)}; (\sqrt[q]{s_k(\hat{w}^q_{\overline{A}_1}, \hat{w}^q_{\overline{A}_2})}, \sqrt[q]{s_k(w^q_{\overline{A}_1}, w^q_{\overline{A}_2})}), (\sqrt[q]{t_k(u^q_{\overline{A}_1}, u^q_{\overline{A}_2})}, \sqrt[q]{t_k(\hat{u}^q_{\overline{A}_1}, \hat{u}^q_{\overline{A}_2})})\rangle$$

where s_k is a s-norm and t_k is a t-norm given in Definition 17.7.

2.

$$\overline{A}_1 \underset{s_k, t_k}{\widetilde{\otimes}} \overline{A}_2 = \begin{cases} \langle \sqrt[q]{t_k(a_1^q, b_1^q)}, \sqrt[q]{t_k(a_2^q, b_2^q)}, \sqrt[q]{t_k(a_3^q, b_3^q)}, \sqrt[q]{t_k(a_4^q, b_4^q)}, \sqrt[q]{t_k(a_5^q, b_5^q)}; \\ (\sqrt[q]{t_k(\hat{w}^q_{\overline{A}_1}, \hat{w}^q_{\overline{A}_2})}, \sqrt[q]{t_k(w^q_{\overline{A}_1}, w^q_{\overline{A}_2})}), (\sqrt[q]{s_k(u^q_{\overline{A}_1}, u^q_{\overline{A}_2})}, \sqrt[q]{s_k(\hat{u}^q_{\overline{A}_1}, \hat{u}^q_{\overline{A}_2})})\rangle, & (d_1 > 0, d_2 > 0) \\ \langle \sqrt[q]{t_k(a_1^q, b_5^q)}, \sqrt[q]{t_k(a_2^q, b_4^q)}, \sqrt[q]{t_k(a_3^q, b_3^q)}, \sqrt[q]{t_k(a_4^q, b_2^q)}, \sqrt[q]{t_k(a_5^q, b_1^q)}; \\ (\sqrt[q]{t_k(\hat{w}^q_{\overline{A}_1}, \hat{w}^q_{\overline{A}_2})}, \sqrt[q]{t_k(w^q_{\overline{A}_1}, w^q_{\overline{A}_2})}), (\sqrt[q]{s_k(u^q_{\overline{A}_1}, u^q_{\overline{A}_2})}, \sqrt[q]{s_k(\hat{u}^q_{\overline{A}_1}, \hat{u}^q_{\overline{A}_2})})\rangle, & (d_1 < 0, d_2 > 0) \\ \langle \sqrt[q]{t_k(a_5^q, b_5^q)}, \sqrt[q]{t_k(a_4^q, b_4^q)}, \sqrt[q]{t_k(a_3^q, b_3^q)}, \sqrt[q]{t_k(a_2^q, b_2^q)}, \sqrt[q]{t_k(a_1^q, b_1^q)}; \\ (\sqrt[q]{t_k(\hat{w}^q_{\overline{A}_1}, \hat{w}^q_{\overline{A}_2})}, \sqrt[q]{t_k(w^q_{\overline{A}_1}, w^q_{\overline{A}_2})}), (\sqrt[q]{s_k(u^q_{\overline{A}_1}, u^q_{\overline{A}_2})}, \sqrt[q]{s_k(\hat{u}^q_{\overline{A}_1}, \hat{u}^q_{\overline{A}_2})})\rangle, & (d_1 < 0, d_2 < 0) \end{cases}$$

where s_k is a s-norm and t_k is a t-norm given in Definition 17.7.

3.

$$\gamma.\overline{A}_1 = \langle \sqrt[q]{1 - (1 - a_1^q)^{\gamma}}, \sqrt[q]{1 - (1 - a_2^q)^{\gamma}}, \sqrt[q]{1 - (1 - a_3^q)^{\gamma}}, \sqrt[q]{1 - (1 - a_4^q)^{\gamma}}, \sqrt[q]{1 - (1 - a_5^q)^{\gamma}}; (\sqrt[q]{1 - (1 - \hat{w}^q_{\overline{A}_1})^{\gamma}}, \sqrt[q]{1 - (1 - w^q{}_{\overline{A}_1})^{\gamma}}), (u^{\gamma}_{\overline{A}_1}, \hat{u}^{\gamma}_{\overline{A}_1})\rangle$$

4.

$$\overline{A_1}^{\gamma} = \langle a_1^{\gamma}, a_2^{\gamma}, a_3^{\gamma}, a_4^{\gamma}, a_5^{\gamma}; (\hat{w}^{\gamma}_{\overline{A_1}}, w^{\gamma}_{\overline{A_1}}), (\sqrt[q]{1-(1-u^{q}{}_{\overline{A_1}})^{\gamma}}, \sqrt[q]{1-(1-\hat{u}^{q}_{\overline{A_1}})^{\gamma}})\rangle$$

In this study, we will use the algebraic sum as the s-norm and algebraic product as the t-norm. Now let's give the operations $\overline{A_1} \overbrace{\oplus}_{s_2,t_2} \overline{A_2}$ and $\overline{A_1} \overbrace{\oplus}_{s_2,t_2} \overline{A_2}$ that depend on algebraic sum and algebraic product as follow;

5.

$$\begin{aligned}\overline{A_1} \overbrace{\oplus}_{s_2,t_2} \overline{A_2} = \langle & \sqrt[q]{a_1^q + b_1^q - a_1^q \cdot b_1^q}, \sqrt[q]{a_2^q + b_2^q - a_2^q \cdot b_2^q}, \sqrt[q]{a_3^q + b_3^q - a_3^q \cdot b_3^q}, \\ & \sqrt[q]{a_4^q + b_4^q - a_4^q \cdot b_4^q}, \sqrt[q]{a_5^q + b_5^q - a_5^q \cdot b_5^q}; (\sqrt[q]{\hat{w}^q_{\overline{A_1}} + \hat{w}^q_{\overline{A_2}} - \hat{w}^q_{\overline{A_1}} \cdot \hat{w}^q_{\overline{A_2}}}, \\ & \sqrt[q]{w^q_{\overline{A_1}} + w^q_{\overline{A_2}} - w^q_{\overline{A_1}} \cdot w^q_{\overline{A_2}}}), (u_{\overline{A_1}} \cdot u_{\overline{A_2}}, \hat{u}_{\overline{A_1}} \cdot \hat{u}_{\overline{A_2}})\rangle\end{aligned}$$

6.

$$\overline{A_1} \overbrace{\otimes}_{s_2,t_2} \overline{A_2} = \begin{cases} \langle a_1 \cdot b_1, a_2 \cdot b_2, a_3 \cdot b_3, a_4 \cdot b_4, a_5 \cdot b_5; (\hat{w}_{\overline{A_1}} \cdot \hat{w}_{\overline{A_2}}, w_{\overline{A_1}} \cdot w_{\overline{A_2}}), \\ (\sqrt[q]{u^q_{\overline{A_1}} + u^q_{\overline{A_2}} - u^q_{\overline{A_1}} \cdot u^q_{\overline{A_2}}}, \sqrt[q]{\hat{u}^q_{\overline{A_1}} + \hat{u}^q_{\overline{A_2}} - \hat{u}^q_{\overline{A_1}} \cdot \hat{u}^q_{\overline{A_2}}})\rangle, \ (a_1 > 0, a_2 > 0) \\ \langle a_1 \cdot b_5, a_2 \cdot b_4, a_3 \cdot b_3, a_4 \cdot b_2, a_5 \cdot b_1; (\hat{w}_{\overline{A_1}} \cdot \hat{w}_{\overline{A_2}}, w_{\overline{A_1}} \cdot w_{\overline{A_2}}), \\ (\sqrt[q]{u^q_{\overline{A_1}} + u^q_{\overline{A_2}} - u^q_{\overline{A_1}} \cdot u^q_{\overline{A_2}}}, \sqrt[q]{\hat{u}^q_{\overline{A_1}} + \hat{u}^q_{\overline{A_2}} - \hat{u}^q_{\overline{A_1}} \cdot \hat{u}^q_{\overline{A_2}}})\rangle, \ (d_1 < 0, a_2 > 0) \\ \langle a_5 \cdot b_5, a_4 \cdot b_4, a_3 \cdot b_3, a_2 \cdot b_2, a_1 \cdot b_1; (\hat{w}_{\overline{A_1}} \cdot \hat{w}_{\overline{A_2}}, w_{\overline{A_1}} \cdot w_{\overline{A_2}}), \\ (\sqrt[q]{u^q_{\overline{A_1}} + u^q_{\overline{A_2}} - u^q_{\overline{A_1}} \cdot u^q_{\overline{A_2}}}, \sqrt[q]{\hat{u}^q_{\overline{A_1}} + \hat{u}^q_{\overline{A_2}} - \hat{u}^q_{\overline{A_1}} \cdot \hat{u}^q_{\overline{A_2}}})\rangle, \ (d_1 < 0, d_2 < 0) \end{cases}$$

Example 17.2 Assume that $\overline{A_1} = \langle 0.1, 0.2, 0.3, 0.4, 0.5; (0, 5, 0, 6), (0.7, 0.8)\rangle$ and $\overline{A_2} = \langle 0.1, 0.3, 0.5, 0.7, 0.9; (0, 6, 0, 7), (0.5, 0.8)\rangle$ be two Pq-RO-numbers ($q = 3$). Then, we have

1.

$$\overline{A_1} \overbrace{\oplus}_{s_3,t_3} \overline{A_2} = \langle 0.1, 0.3, 0.5, 0.7, 0.9; (0.6, 0.7), (0.5, 0.8)\rangle$$

where s_3 is a s-norm and t_3 is a t-norm given in Definition 17.7.

2.

$$\overline{A_1} \overbrace{\otimes}_{s_3,t_3} \overline{A_2} = \langle 0.1, 0.2, 0.3, 0.4, 0.5; (0.5, 0.6), (0.7, 0.8)\rangle$$

where s_3 is a s-norm and t_3 is a t-norm given in Definition 17.7.

3.

$$\overline{A_1} \overbrace{\oplus}_{s_2,t_2} \overline{A_2} = \langle 0.1260, 0.3264, 0.5297, 0.7275, 0.9137; (0.6797, 0.7856), (0.35, 0.64)\rangle$$

where s_2 is a s-norm and t_2 is a t-norm given in Definition 17.7.

4.

$$\overline{A}_1 \overbrace{\underset{s_2,t_2}{\otimes}} \overline{A}_2 = \langle 0.01, 0.06, 0.15, 0.28, 0.45; (0.3, 0.42), (0.7519, 0.9133)\rangle$$

where s_2 is a s-norm and t_2 is a t-norm given in Definition 17.7.

5. $5.\overline{A}_1 = \langle 0.3464, 0.5236, 0.6628, 0.7761, 0.8660; (0.8660, 0.9322), (0.1681, 0.3277)\rangle$
6. $\overline{A}_1^5 = \langle 0.0000, 0.0003, 0.0024, 0.0102, 0.0313; (0.0313, 0.0778), (0.9742, 0.9944)\rangle$.

Theorem 17.2 *Let $\overline{A}_1$ and $\overline{A}_2$ be two normal Pq-RO-numbers and $\gamma > 0$. Then $\overline{A}_1 \overbrace{\underset{s_k,t_k}{\oplus}} \overline{A}_2$, $\overline{A}_1 \overbrace{\underset{s_k,t_k}{\otimes}} \overline{A}_2$, $\gamma.\overline{A}_1$ and $\overline{A}_1^{\gamma}$ are normal Pq-RO-numbers.*

Proof It is clear from Definitions 17.7–17.11.

Proposition 17.1 *Let $\overline{A}_1 = \langle a_1, a_2, a_3, a_4, a_5; (\hat{w}_{\overline{A}_1}, w_{\overline{A}_1}), (u_{\overline{A}_1}, \hat{u}_{\overline{A}_1})\rangle$ and $\overline{A}_2 = \langle b_1, b_2, b_3, b_4, b_5; (\hat{w}_{\overline{A}_2}, w_{\overline{A}_2}), (u_{\overline{A}_2}, \hat{u}_{\overline{A}_2})\rangle \in \Delta$ and $\lambda, \lambda_1, \lambda_2 > 0$ be any real number. Then, the following are valid*

1. $\overline{A}_1 \overbrace{\underset{s_k,t_k}{\oplus}} \overline{A}_2 = \overline{A}_2 \overbrace{\underset{s_k,t_k}{\oplus}} \overline{A}_1$
2. $\overline{A}_1 \overbrace{\underset{s_k,t_k}{\otimes}} \overline{A}_2 = \overline{A}_2 \overbrace{\underset{s_k,t_k}{\otimes}} \overline{A}_1$
3. $\lambda(\overline{A}_1 \overbrace{\underset{s_k,t_k}{\oplus}} \overline{A}_2) = \lambda\overline{A}_1 \overbrace{\underset{s_k,t_k}{\oplus}} \lambda\overline{A}_2$
4. $\lambda_1\overline{A}_1 \overbrace{\underset{s_k,t_k}{\oplus}} \lambda_2\overline{A}_1 = (\lambda_1 + \lambda_2)\overline{A}_1$
5. $(\overline{A}_1 \overbrace{\underset{s_k,t_k}{\otimes}} \overline{A}_2)^{\gamma} = \overline{A}_1^{\gamma} \overbrace{\underset{s_k,t_k}{\otimes}} \overline{A}_2^{\gamma}$
6. $\overline{A}_1^{\lambda_1} \overbrace{\underset{s_k,t_k}{\otimes}} \overline{A}_1^{\lambda_2} = \overline{A}_1^{\lambda_1+\lambda_2}$.

Corollary 17.1 *Let $\overline{A}_1 = \langle a_1, a_2, a_3, a_4, a_5; (\hat{w}_{\overline{A}_1}, w_{\overline{A}_1}), (u_{\overline{A}_1}, \hat{u}_{\overline{A}_1})\rangle$ and $\overline{A}_2 = \langle b_1, b_2, b_3, b_4, b_5; (\hat{w}_{\overline{A}_2}, w_{\overline{A}_2}), (u_{\overline{A}_2}, \hat{u}_{\overline{A}_2})\rangle \in \Delta$ and $\lambda, \lambda_1, \lambda_2 > 0$ be any real number. Then, the following are valid*

1. $\overline{A}_1 \overbrace{\underset{s_2,t_2}{\oplus}} \overline{A}_2 = \overline{A}_2 \overbrace{\underset{s_2,t_2}{\oplus}} \overline{A}_1$
2. $\overline{A}_1 \overbrace{\underset{s_2,t_2}{\otimes}} \overline{A}_2 = \overline{A}_2 \overbrace{\underset{s_2,t_2}{\otimes}} \overline{A}_1$
3. $\lambda(\overline{A}_1 \overbrace{\underset{s_2,t_2}{\oplus}} \overline{A}_2) = \lambda\overline{A}_1 \overbrace{\underset{s_2,t_2}{\oplus}} \lambda\overline{A}_2$
4. $\lambda_1\overline{A}_1 \overbrace{\underset{s_2,t_2}{\oplus}} \lambda_2\overline{A}_1 = (\lambda_1 + \lambda_2)\overline{A}_1$
5. $(\overline{A}_1 \overbrace{\underset{s_2,t_2}{\otimes}} \overline{A}_2)^{\gamma} = \overline{A}_1^{\gamma} \overbrace{\underset{s_2,t_2}{\otimes}} \overline{A}_2^{\gamma}$

6. $\overline{A}_1^{\lambda_1} \overset{\frown}{\underset{s_2,t_2}{\otimes}} \overline{A}_1^{\lambda_2} = \overline{A}_1^{\lambda_1+\lambda_2}$.

Proof Suppose that $\overline{A}_1 = \langle a_1, a_2, a_3, a_4, a_5; (\hat{w}_{\overline{A}_1}, w_{\overline{A}_1}), (u_{\overline{A}_1}, \hat{u}_{\overline{A}_1})\rangle$ and $\overline{A}_2 = \langle b_1, b_2, b_3, b_4, b_5; (\hat{w}_{\overline{A}_2}, w_{\overline{A}_2}), (u_{\overline{A}_2}, \hat{u}_{\overline{A}_2})\rangle \in \Delta$ and $\lambda, \lambda_1, \lambda_2 > 0$ be any real number. Then,

1.

$$\begin{aligned}
\overline{A}_1 \overset{\frown}{\underset{s_2,t_2}{\oplus}} \overline{A}_2 = &\langle \sqrt[q]{a_1^q + b_1^q - a_1^q \cdot b_1^q}, \sqrt[q]{a_2^q + b_2^q - a_2^q \cdot b_2^q}, \sqrt[q]{a_3^q + b_3^q - a_3^q \cdot b_3^q},\\
&\sqrt[q]{a_4^q + b_4^q - a_4^q \cdot b_4^q}, \sqrt[q]{a_5^q + b_5^q - a_5^q \cdot b_5^q}; (\sqrt[q]{\hat{w}^q_{\overline{A}_1} + \hat{w}^q_{\overline{A}_2} - \hat{w}^q_{\overline{A}_1} \cdot \hat{w}^q_{\overline{A}_2}},\\
&\sqrt[q]{w^q_{\overline{A}_1} + w^q_{\overline{A}_2} - w^q_{\overline{A}_1} \cdot w^q_{\overline{A}_2}}), (u_{\overline{A}_1} \cdot u_{\overline{A}_2}, \hat{u}_{\overline{A}_1} \cdot \hat{u}_{\overline{A}_2})\rangle\\
= &\langle \sqrt[q]{b_1^q + a_1^q - b_1^q \cdot a_1^q}, \sqrt[q]{b_2^q + a_2^q - b_2^q \cdot a_2^q}, \sqrt[q]{b_3^q + a_3^q - b_3^q \cdot a_3^q},\\
&\sqrt[q]{b_4^q + a_4^q - b_4^q \cdot a_4^q}, \sqrt[q]{b_5^q + a_5^q - b_5^q \cdot a_5^q}; (\sqrt[q]{\hat{w}^q_{\overline{A}_2} + \hat{w}^q_{\overline{A}_1} - \hat{w}^q_{\overline{A}_2} \cdot \hat{w}^q_{\overline{A}_1}},\\
&\sqrt[q]{w^q_{\overline{A}_2} + w^q_{\overline{A}_1} - w^q_{\overline{A}_2} \cdot w^q_{\overline{A}_1}}), (u_{\overline{A}_2} \cdot u_{\overline{A}_1}, \hat{u}_{\overline{A}_2} \cdot \hat{u}_{\overline{A}_1})\rangle\\
= &\overline{A}_2 \overset{\frown}{\underset{s_2,t_2}{\oplus}} \overline{A}_1
\end{aligned}$$

2. The proof can be made similar to the proof of the 1.
3. Since

$$\begin{aligned}
\lambda(\overline{A}_1 \overset{\frown}{\underset{s_2,t_2}{\oplus}} \overline{A}_2) = &\lambda\langle \sqrt[q]{a_1^q + b_1^q - a_1^q \cdot b_1^q}, \sqrt[q]{a_2^q + b_2^q - a_2^q \cdot b_2^q}, \sqrt[q]{a_3^q + b_3^q - a_3^q \cdot b_3^q},\\
&\sqrt[q]{a_4^q + b_4^q - a_4^q \cdot b_4^q}, \sqrt[q]{a_5^q + b_5^q - a_5^q \cdot b_5^q}; (\sqrt[q]{\hat{w}^q_{\overline{A}_1} + \hat{w}^q_{\overline{A}_2} - \hat{w}^q_{\overline{A}_1} \cdot \hat{w}^q_{\overline{A}_2}},\\
&\sqrt[q]{w^q_{\overline{A}_1} + w^q_{\overline{A}_2} - w^q_{\overline{A}_1} \cdot w^q_{\overline{A}_2}}), (u_{\overline{A}_1} \cdot u_{\overline{A}_2}, \hat{u}_{\overline{A}_1} \cdot \hat{u}_{\overline{A}_2})\rangle\\
= &\langle \sqrt[q]{1 - (1 - (a_1^q + b_1^q - a_1^q \cdot b_1^q))^\lambda}, \sqrt[q]{1 - (1 - (a_2^q + b_2^q - a_2^q \cdot b_2^q))^\lambda},\\
&\sqrt[q]{1 - (1 - (a_3^q + b_3^q - a_3^q \cdot b_3^q))^\lambda} \sqrt[q]{1 - (1 - (a_4^q + b_4^q - a_4^q \cdot b_4^q))^\lambda},\\
&\sqrt[q]{1 - (1 - (a_5^q + b_5^q - a_5^q \cdot b_5^q))^\lambda}; (\sqrt[q]{1 - (1 - (\hat{w}^q_{\overline{A}_1} + \hat{w}^q_{\overline{A}_2} - \hat{w}^q_{\overline{A}_1} \cdot \hat{w}^q_{\overline{A}_2}))^\lambda},\\
&\sqrt[q]{1 - (1 - (w^q_{\overline{A}_1} + w^q_{\overline{A}_2} - w^q_{\overline{A}_1} \cdot w^q_{\overline{A}_2}))^\lambda}), ((u_{\overline{A}_1} \cdot u_{\overline{A}_2})^\lambda, (\hat{u}_{\overline{A}_1} \cdot \hat{u}_{\overline{A}_2})^\lambda)\rangle
\end{aligned}$$

and

$$\begin{aligned}
\lambda\overline{A}_1 \overset{\frown}{\underset{s_2,t_2}{\oplus}} \lambda\overline{A}_2 = &\langle \sqrt[q]{1 - (1 - a_1^q)^\gamma}, \sqrt[q]{1 - (1 - a_2^q)^\gamma}, \sqrt[q]{1 - (1 - a_3^q)^\gamma}, \sqrt[q]{1 - (1 - a_4^q)^\gamma}, \sqrt[q]{1 - (1 - a_5^q)^\gamma};\\
&(\sqrt[q]{1 - (1 - \hat{w}^q_{\overline{A}_1})^\gamma}, \sqrt[q]{1 - (1 - w^q{}_{\overline{A}_1})^\gamma}), (u^\gamma_{\overline{A}_1}, \hat{u}^\gamma_{\overline{A}_1})\rangle\\
&+\langle \sqrt[q]{1 - (1 - b_1^q)^\gamma}, \sqrt[q]{1 - (1 - b_2^q)^\gamma}, \sqrt[q]{1 - (1 - b_3^q)^\gamma}, \sqrt[q]{1 - (1 - b_4^q)^\gamma}, \sqrt[q]{1 - (1 - b_5^q)^\gamma};\\
&(\sqrt[q]{1 - (1 - \hat{w}^q_{\overline{A}_2})^\gamma}, \sqrt[q]{1 - (1 - w^q{}_{\overline{A}_2})^\gamma}), (u^\gamma_{\overline{A}_2}, \hat{u}^\gamma_{\overline{A}_2})\rangle
\end{aligned}$$

$$= \langle \sqrt[q]{\sqrt[q]{(1-(1-a_1^q)^\gamma)^q} + \sqrt[q]{(1-(1-b_1^q)^\gamma)^q} - \sqrt[q]{(1-(1-a_1^q)^\gamma)^q} \cdot \sqrt[q]{(1-(1-b_1^q)^\gamma)^q}},$$
$$\sqrt[q]{\sqrt[q]{(1-(1-a_2^q)^\gamma)^q} + \sqrt[q]{(1-(1-b_2^q)^\gamma)^q} - \sqrt[q]{(1-(1-a_2^q)^\gamma)^q} \cdot \sqrt[q]{(1-(1-b_2^q)^\gamma)^q}},$$
$$\sqrt[q]{\sqrt[q]{(1-(1-a_3^q)^\gamma)^q} + \sqrt[q]{(1-(1-b_3^q)^\gamma)^q} - \sqrt[q]{(1-(1-a_3^q)^\gamma)^q} \cdot \sqrt[q]{(1-(1-b_3^q)^\gamma)^q}},$$
$$\sqrt[q]{\sqrt[q]{(1-(1-a_4^q)^\gamma)^q} + \sqrt[q]{(1-(1-b_4^q)^\gamma)^q} - \sqrt[q]{(1-(1-a_4^q)^\gamma)^q} \cdot \sqrt[q]{(1-(1-b_4^q)^\gamma)^q}},$$
$$\sqrt[q]{\sqrt[q]{(1-(1-a_5^q)^\gamma)^q} + \sqrt[q]{(1-(1-b_5^q)^\gamma)^q} - \sqrt[q]{(1-(1-a_5^q)^\gamma)^q} \cdot \sqrt[q]{(1-(1-b_5^q)^\gamma)^q}};$$
$$(\sqrt[q]{\sqrt[q]{(1-(1-\hat{w}^q_{\overline{A}_1})^\gamma)^q} + \sqrt[q]{(1-(1-\hat{w}^q_{\overline{A}_2})^\gamma)^q} - \sqrt[q]{(1-(1-\hat{w}^q_{\overline{A}_1})^\gamma)^q} \cdot \sqrt[q]{(1-(1-\hat{w}^q_{\overline{A}_2})^\gamma)^q}},$$
$$\sqrt[q]{\sqrt[q]{(1-(1-w^q{}_{\overline{A}_1})^\gamma)^q} + \sqrt[q]{(1-(1-w^q{}_{\overline{A}_2})^\gamma)^q} - \sqrt[q]{(1-(1-w^q{}_{\overline{A}_1})^\gamma)^q} \cdot \sqrt[q]{(1-(1-w^q{}_{\overline{A}_2})^\gamma)^q}}),$$
$$((u_{\overline{A}_1} \cdot u_{\overline{A}_2})^\lambda, (\hat{u}_{\overline{A}_1} \cdot \hat{u}_{\overline{A}_2})^\lambda)$$

$$= \langle \sqrt[q]{(1-(1-a_1^q)^\gamma)^q} + \sqrt[q]{(1-(1-b_1^q)^\gamma)^q} - \sqrt[q]{(1-(1-a_1^q)^\gamma)^q} \cdot \sqrt[q]{(1-(1-b_1^q)^\gamma)},$$
$$\sqrt[q]{(1-(1-a_2^q)^\gamma)^q} + \sqrt[q]{(1-(1-b_2^q)^\gamma)^q} - \sqrt[q]{(1-(1-a_2^q)^\gamma)^q} \cdot \sqrt[q]{(1-(1-b_2^q)^\gamma)},$$
$$\sqrt[q]{(1-(1-a_3^q)^\gamma)^q} + \sqrt[q]{(1-(1-b_3^q)^\gamma)^q} - \sqrt[q]{(1-(1-a_3^q)^\gamma)^q} \cdot \sqrt[q]{(1-(1-b_3^q)^\gamma)},$$
$$\sqrt[q]{(1-(1-a_4^q)^\gamma)^q} + \sqrt[q]{(1-(1-b_4^q)^\gamma)^q} - \sqrt[q]{(1-(1-a_4^q)^\gamma)^q} \cdot \sqrt[q]{(1-(1-b_4^q)^\gamma)},$$
$$\sqrt[q]{(1-(1-a_5^q)^\gamma)^q} + \sqrt[q]{(1-(1-b_5^q)^\gamma)^q} - \sqrt[q]{(1-(1-a_5^q)^\gamma)^q} \cdot \sqrt[q]{(1-(1-b_5^q)^\gamma)};$$
$$(\sqrt[q]{(1-(1-\hat{w}^q_{\overline{A}_1})^\gamma)^q} + \sqrt[q]{(1-(1-\hat{w}^q_{\overline{A}_2})^\gamma)^q} - \sqrt[q]{(1-(1-\hat{w}^q_{\overline{A}_1})^\gamma)^q} \cdot \sqrt[q]{(1-(1-\hat{w}^q_{\overline{A}_2})^\gamma)},$$
$$\sqrt[q]{(1-(1-w^q{}_{\overline{A}_1})^\gamma)^q} + \sqrt[q]{(1-(1-w^q{}_{\overline{A}_2})^\gamma)^q} - \sqrt[q]{(1-(1-w^q{}_{\overline{A}_1})^\gamma)^q} \cdot \sqrt[q]{(1-(1-w^q{}_{\overline{A}_2})^\gamma)}),$$
$$((u_{\overline{A}_1} \cdot u_{\overline{A}_2})^\lambda, (\hat{u}_{\overline{A}_1} \cdot \hat{u}_{\overline{A}_2})^\lambda)$$

$$= \langle \sqrt[q]{1-(1-(a_1^q+b_1^q-a_1^q \cdot b_1^q))^\lambda}, \sqrt[q]{1-(1-(a_2^q+b_2^q-a_2^q \cdot b_2^q))^\lambda},$$
$$\sqrt[q]{1-(1-(a_3^q+b_3^q-a_3^q \cdot b_3^q))^\lambda} \sqrt[q]{1-(1-(a_4^q+b_4^q-a_4^q \cdot b_4^q))^\lambda},$$
$$\sqrt[q]{1-(1-(a_5^q+b_5^q-a_5^q \cdot b_5^q))^\lambda}; (\sqrt[q]{1-(1-(\hat{w}^q_{\overline{A}_1}+\hat{w}^q_{\overline{A}_2}-\hat{w}^q_{\overline{A}_1} \cdot \hat{w}^q_{\overline{A}_2}))^\lambda},$$
$$\sqrt[q]{1-(1-(w^q_{\overline{A}_1}+w^q_{\overline{A}_2}-w^q_{\overline{A}_1} \cdot w^q_{\overline{A}_2}))^\lambda}), ((u_{\overline{A}_1} \cdot u_{\overline{A}_2})^\lambda, (\hat{u}_{\overline{A}_1} \cdot \hat{u}_{\overline{A}_2})^\lambda))$$

we have $\lambda(\overline{A}_1 \underset{s_2,t_2}{\widehat{\oplus}} \overline{A}_2) = \lambda\overline{A}_1 \underset{s_2,t_2}{\widehat{\oplus}} \lambda\overline{A}_2$.

Proofs of 4, 5, and 6 can be made similar to the proof of the 1–3 of the theorem.

Now we give a compare method based on Definitions 17.5, 17.7 and 17.9 as follows.

Definition 17.12 Let $\overline{A} = \langle a_1, a_2, a_3, a_4, a_5; (\hat{w}_{\overline{A}}, w_{\overline{A}}), (u_{\overline{A}}, \hat{u}_{\overline{A}})\rangle$ be a Pq-RO-number, s_k be a s-norm and t_k be a t-norm. Then,

1. 1. rank value of $\overline{A}$, is denoted by $R_1^k(A)$ is defined as,

$$R_1^k(A) = \frac{w_A.s_k(x_\mu.y_\mu) + u_A.t_k(x_\nu.y_\nu)}{w_A + u_A} \tag{17.14}$$

where $x_\mu = \frac{a_1+2a_2+3a_3+2a_4+a_5}{9}$, $y_\mu = \frac{3\hat{w_A}+w_A}{9}$, $x_\nu = \frac{b_1+2b_2+3b_3+2b_4+b_5}{9}$ and $y_\nu = \frac{3\hat{u_A}+5+u_A}{9}$.

2. 2. rank value of $\overline{A}$, is denoted by $R_2^k(A)$ is defined as

$$R_2^k(A) = \frac{w_A.s_k(x_\mu.y_\mu) - u_A.t_k(x_\nu.y_\nu)}{w_A - u_A} \tag{17.15}$$

where $x_\mu = \frac{a_1+2a_2+3a_3+2a_4+a_5}{9}$, $y_\mu = \frac{3\hat{w_A}+w_A}{9}$, $x_\nu = \frac{b_1+2b_2+3b_3+2b_4+b_5}{9}$ and $y_\nu = \frac{3\hat{u_A}+5+u_A}{9}$.

If $\overline{A}_1 = \langle a_1, a_2, a_3, a_4, a_5; (\hat{w}_{\overline{A}_1}, w_{\overline{A}_1}), (u_{\overline{A}_1}, \hat{u}_{\overline{A}_1})\rangle$ and $\overline{A}_2 = \langle b_1, b_2, b_3, b_4, b_5; (\hat{w}_{\overline{A}_2}, w_{\overline{A}_2}), (u_{\overline{A}_2}, \hat{u}_{\overline{A}_2})\rangle \in \Delta$. Based on 1. rank value and 2. rank value, a comparison method is proposed as

1. If $R_1^k(A_1) > R_1^k(A_1)$, then $A_1 > A_2$
2. (a) If $R_1^k(A_1) = R_1^k(A_1)$ and $R_2^k(A_1) > R_2^k(A_1)$ then $A_1 < A_2$
   (b) If $R_1^k(A_1) = R_1^k(A_1)$, $R_2^k(A_1) = R_2^k(A_1)$ then $A_1 = A_2$

Example 17.3 Assume that

$$\overline{A_1} = \langle 0.7476, 0.8205, 0.9269, 0.9652, 1.0000; (0.7734, 0.9149), (0.3842, 0.4459)\rangle$$
$$\overline{A_2} = \langle 0.9200, 0.9546, 0.9609, 1.0000, 1.0000; (1.0000, 0.9867), (0.0000, 0.0000)\rangle$$
$$\overline{A_3} = \langle 0.6117, 0.6971, 0.7667, 0.8245, 0.8727; (0.9791, 0.9724), (0.2766, 0.4119)\rangle$$

and

$$\overline{A_4} = \langle 0.8235, 0.8897, 0.9069, 1.0000, 1.0000; (1.0000, 0.9849), (0.0000, 0.0000)\rangle$$

be four Pq-RO-numbers. Then, we have

$$\begin{array}{ll} R_1^2(\overline{A_1}) = 0.9760 & R_2^2(\overline{A_1}) = 0.9760 \\ R_1^2(\overline{A_2}) = 0.8579 & R_2^2(\overline{A_2}) = 1.1269 \\ R_1^2(\overline{A_3}) = 0.9822 & R_2^2(\overline{A_3}) = 0.9822 \\ R_1^2(\overline{A_4}) = 0.7938 & R_2^2(\overline{A_4}) = 0.9885 \end{array}$$

In here,

$$R_1^2(\overline{A_3}) > R_1^2(\overline{A_1}) > R_1^2(\overline{A_2}) > R_1^2(\overline{A_4}),$$

Therefore we have

$$\overline{A_3} > \overline{A_1} > \overline{A_2} > \overline{A_4}.$$

Definition 17.13 Let $\overline{A}_j = \langle a_j, b_j, c_j, d_j, e_j; (\hat{w}_{\overline{A}_j}, w_{\overline{A}_j}), (u_{\overline{A}_j}, \hat{u}_{\overline{A}_j})\rangle$, $j = 1, 2, \ldots, n$ are collection of some be normal Pq-RO-numbers. Then, qth rung orthopair fuzzy number weighted aggregation mean operator, denoted by $Pq - RONW_{ao}$, is defined as

$$Pq - RONW_{ao}(\overline{A}_1, \overline{A}_2, \ldots, \overline{A}_n) = \overbrace{\underset{s_k,t_k}{\oplus}}^{n}_{j=1} w_j \overline{A}_j$$

where $w = (w_1, w_2, \ldots, w_n)^T$ is a weight vector for every $j \in I_n$ such that, $w_j \in [0, 1]$ and $\sum_{j=1}^{n} w_j = 1$.

Example 17.4 Assume that

$$\overline{A_1} = \langle 0.90, 0.95, 0.0.95, 1.0, 1.0; (1.0, 0.9), (0.0, 0.0)\rangle$$
$$\overline{A_2} = \langle 0.25, 0.35, 0.45, 0.55, 0, 65; (03, 0.4), (0.4, 0.3)\rangle$$
$$\overline{A_3} = \langle 0.40, 0.50, 0.80, 0.90, 1.00; (0.5, 0.7), (0.4, 0.6)\rangle$$

and

$$\overline{A_4} = \langle 0.40, 0.50, 0.80, 0.90, 1.00; (0.5, 0.7), (0.4, 0.6)\rangle$$

be four Pq-RO-numbers and $w = (0.14, 0.35, 0.21, 0.30)^T$ be weight vector of the $\overline{A_p}(p = 1, 2, 3, 4)$. Then, we have

$$Pq - RONW_{ao}(\overline{A}_1, \overline{A}_2, \overline{A}_3, \overline{A}_4) = \langle 0.8632, 0.9178, 0.9721, 1.0000, 1.0000; (1.0000, 0.9447), (0.0000, 0.0000)\rangle$$

Theorem 17.3 *Suppose that* $\overline{A}_j = \langle a_j, b_j, c_j, d_j, e_j; (\hat{w}_{\overline{A}_j}, w_{\overline{A}_j}), (u_{\overline{A}_j}, \hat{u}_{\overline{A}_j})\rangle$, $j = 1, 2, \ldots, n$ *are collection of some be normal Pq-RO-numbers. Then, for s-norm* s_2 *and t-norm* t_2*, qth rung orthopair fuzzy number weighted aggregation mean operator, denoted by* $Pq - RONW_{ao}$*, is defined as*

$$\begin{aligned}Pq - RONW_{ao}(\overline{A}_1, \overline{A}_2, \ldots, \overline{A}_n) &= \overbrace{\underset{s_2,t_2}{\oplus}}^{n}_{j=1} w_j \overline{A}_j \\ &= \langle (\sqrt[q]{1 - \prod_{j=1}^{n}(1-(a_j)^q)^{w_j}}, \sqrt[q]{1 - \prod_{j=1}^{n}(1-(b_j)^q)^{w_j}}, \\ &\sqrt[q]{1 - \prod_{j=1}^{n}(1-(c_j)^q)^{w_j}}, \sqrt[q]{1 - \prod_{j=1}^{n}(1-(d_j)^q)^{w_j}}, \\ &\sqrt[q]{1 - \prod_{j=1}^{n}(1-(e_j)^q)^{w_j}}; (\sqrt[q]{1 - \prod_{j=1}^{n}(1-(\hat{w}_{\overline{A}_j})^q)^{w_j}}, \\ &\sqrt[q]{1 - \prod_{j=1}^{n}(1-(w_{\overline{A}_j})^q)^{w_j}}), (\prod_{j=1}^{n}(u_{\overline{A}_j})^{w_j}, \prod_{j=1}^{n}(\hat{u}_{\overline{A}_j})^{w_j})\rangle\end{aligned}$$

where $w = (w_1, w_2, \ldots, w_n)^T$ *is a weight vector for every* $j \in I_n$ *such that* $w_j \in [0, 1]$ *and* $\sum_{j=1}^{n} w_j = 1$.

Proposition 17.2 *Let* $\overline{A}_j = \langle a_j, b_j, c_j, d_j, e_j; (\hat{w}_{\overline{A}_j}, w_{\overline{A}_j}), (u_{\overline{A}_j}, \hat{u}_{\overline{A}_j})\rangle$, $j = 1, 2, \ldots, n$ *are collection of normal Pq-RO-numbers. Then, the following properties is hold.*

1. *(Idempotency) if* $\overline{A}_j = \overline{A} = \langle a, b, c, d, e; (\hat{w}_{\overline{A}}, w_{\overline{A}}), (u_{\overline{A}}, \hat{u}_{\overline{A}})\rangle$, $j = 1, 2, \ldots, n$, *then*

$$Pq - RONW_{ao}(\overline{A}_1, \overline{A}_2, \ldots, \overline{A}_n) = \overline{A}$$

2. *(Monotonicity) Suppose that* $\overline{A}_j = \langle a_j, b_j, c_j, d_j, e_j; (\hat{w}_{\overline{A}_j}, w_{\overline{A}_j}), (u_{\overline{A}_j}, \hat{u}_{\overline{A}_j})\rangle$ *and* $\overline{B}_j = \langle \acute{a}_j, \acute{b}_j, \acute{c}_j, \acute{d}_j, \acute{e}_j; (\acute{\hat{w}}_{\overline{B}_j}, \acute{w}_{\overline{B}_j}), (\acute{u}_{\overline{B}_j}, \acute{\hat{u}}_{\overline{B}_j})\rangle$, $j = 1, 2, \ldots, n$ *are two collection of some Pq-RO-numbers. If* $a_j \geq \acute{a}_j$, $b_j \geq \acute{b}_j$, $c_j \geq \acute{c}_j$, $d_j \geq \acute{d}_j$, $e_j \geq \acute{e}_j$, $\hat{w}_{\overline{A}_j} \geq \acute{\hat{w}}_{\overline{B}_j}$, $w_{\overline{A}_j} \geq \acute{w}_{\overline{B}_j}$, $u_{\overline{A}_j} \leq \acute{u}_{\overline{B}_j}$, $\hat{u}_{\overline{A}_j} \leq \acute{\hat{u}}_{\overline{B}_j}$ *for all j, then*

$$Pq - RONW_{ao}(\overline{A}_1, \overline{A}_2, \ldots, \overline{A}_n) \geq Pq - RONW_{ao}(\overline{B}_1, \overline{B}_2, \ldots, \overline{B}_n)$$

3. *(Boundedness) If*

$$\overline{A^-}_j = \langle \min_{1\leq j\leq n}\{a_j\}, \min_{1\leq j\leq n}\{b_j\}, \min_{1\leq j\leq n}\{c_j\}, \min_{1\leq j\leq n}\{d_j\}, \min_{1\leq j\leq n}\{e_j\}; (\min_{1\leq j\leq n}\{\hat{w}_{\overline{A}_j}\}, \min_{1\leq j\leq n}\{w_{\overline{A}_j}\}), (\max_{1\leq j\leq n}\{u_{\overline{A}_j}\}, (\max_{1\leq j\leq n}\{\hat{u}_{\overline{A}_j}\})\rangle$$

and

$$\overline{A^+}_j = \langle \max_{1\leq j\leq n} a_j, \max_{1\leq j\leq n} b_j, \max_{1\leq j\leq n} c_j, \max_{1\leq j\leq n} d_j, \max_{1\leq j\leq n} e_j; (\max_{1\leq j\leq n} \hat{w}_{\overline{A}_j}, \max_{1\leq j\leq n} w_{\overline{A}_j}), (\min_{1\leq j\leq n} u_{\overline{A}_j}, \min_{1\leq j\leq n} \hat{u}_{\overline{A}_j})\rangle,$$

then

$$\overline{A^-}_j \leq Pq - RONW_{ao}(\overline{A}_1, \overline{A}_2, \ldots, \overline{A}_n) \leq \overline{A^+}_j$$

Definition 17.14 Let $\overline{A}_j = \langle a_j, b_j, c_j, d_j, e_j; (\hat{w}_{\overline{A}_j}, w_{\overline{A}_j}), (u_{\overline{A}_j}, \hat{u}_{\overline{A}_j})\rangle$, $j = 1, 2, \ldots, n$ are collection of some be normal Pq-RO-numbers. Then, Pq-RO-number weighted geometric operator, denoted by $Pq - RONW_{go}$, is defined as

$$Pq - RONW_{go}(\overline{A}_1, \overline{A}_2, \ldots, \overline{A}_n) = \overbrace{\bigotimes_{s_k,t_k}}^{n}{}_{j=1} \overline{A}_j^{\,w_j}$$

where $w = (w_1, w_2, \ldots, w_n)^T$ is a weight vector for every $j \in I_n$ such that $w_j \in [0, 1]$ and $\sum_{j=1}^{n} w_j = 1$.

Example 17.5 Assume that

$$\overline{A_1} = \langle 0.45, 0.50, 0.55, 0.60, 0.65; (0.2, 0.8), (0.3, 0.9)\rangle$$
$$\overline{A_2} = \langle 0.40, 0.50, 0.80, 0.90, 1.00; (0.5, 0.7), (0.4, 0.6)\rangle$$
$$\overline{A_3} = \langle 0.25, 0.35, 0.45, 0.55, 0, 65; (03, 0.4), (0.4, 0.3)\rangle$$

and

$$\overline{A_4} = \langle 0.25, 0.35, 0.45, 0.55, 0, 65; (03, 0.4), (0.4, 0.3)\rangle$$

be four Pq-RO-numbers and $w = (0.14, 0.35, 0.21, 0.30)^T$ be weight vector of the $\overline{A_p}(p = 1, 2, 3, 4)$. Then, we have

$$Pq - RONW_{go}(\overline{A}_1, \overline{A}_2, \overline{A}_3, \overline{A}_4) = \langle 0.3200, 0.4168, 0.5661, 0.6615, 0.7558; (0.3389, 0.5361), (0.3880, 0.5949)\rangle$$

Theorem 17.4 *Suppose that* $\overline{A}_j = \langle a_j, b_j, c_j, d_j, e_j; (\hat{w}_{\overline{A}_j}, w_{\overline{A}_j}), (u_{\overline{A}_j}, \hat{u}_{\overline{A}_j})\rangle$, $j = 1, 2, \ldots, n$ *are collection of some be normal Pq-RO-numbers. Then, for s-norm* s_2 *and t-norm* t_2*, Pq-RO-number weighted geometric operator, denoted by* $Pq - RONW_{go}$*, is defined as*

$$\begin{aligned} Pq - RONW_{go}(\overline{A}_1, \overline{A}_2, \ldots, \overline{A}_n) &= \overbrace{\underset{s_2,t_2\, j=1}{\otimes}}^{n} \overline{A}_j{}^{w_j} \\ &= \langle (\textstyle\prod_{j=1}^n (a_j)^{w_j}, \prod_{j=1}^n (b_j)^{w_j}, \prod_{j=1}^n (c_j)^{w_j}, \prod_{j=1}^n (d_j)^{w_j}, \\ &\textstyle\prod_{j=1}^n (e_j)^{w_j}; (\prod_{j=1}^n (\hat{w}_{\overline{A}_j})^{w_j}, \prod_{j=1}^n (w_{\overline{A}_j})^{w_j}), \\ &(\sqrt[q]{1 - \textstyle\prod_{j=1}^n (1 - (u_{\overline{A}_j})^q)^{w_j}}, \sqrt[q]{1 - \textstyle\prod_{j=1}^n (1 - (\hat{w}_{\overline{A}_j})^q)^{w_j}})\rangle \end{aligned}$$

where $w = (w_1, w_2, \ldots, w_n)^T$ *is a weight vector for every* $j \in I_n$ *such that,* $w_j \in [0, 1]$ *and* $\sum_{j=1}^n w_j = 1$.

Proposition 17.3 *Let* $\overline{A}_j = \langle a_j, b_j, c_j, d_j, e_j; (\hat{w}_{\overline{A}_j}, w_{\overline{A}_j}), (u_{\overline{A}_j}, \hat{u}_{\overline{A}_j})\rangle$, $j = 1, 2, \ldots, n$ *are collection of normal Pq-RO-numbers. Then, the following properties is hold.*

1. *(Idempotency) if* $\overline{A}_j = \overline{A} = \langle a, b, c, d, e; (\hat{w}_{\overline{A}}, w_{\overline{A}}), (u_{\overline{A}}, \hat{u}_{\overline{A}})\rangle$, $j = 1, 2, \ldots, n$, *then*

$$Pq - RONW_{go}(\overline{A}_1, \overline{A}_2, \ldots, \overline{A}_n) = \overline{A}$$

2. *(Monotonicity) Suppose that* $\overline{A}_j = \langle a_j, b_j, c_j, d_j, e_j; (\hat{w}_{\overline{A}_j}, w_{\overline{A}_j}), (u_{\overline{A}_j}, \hat{u}_{\overline{A}_j})\rangle$ *and* $\overline{B}_j = \langle \acute{a}_j, \acute{b}_j, \acute{c}_j, \acute{d}_j, \acute{e}_j; (\hat{\acute{w}}_{\overline{B}_j}, \acute{w}_{\overline{B}_j}), (\acute{u}_{\overline{B}_j}, \hat{\acute{u}}_{\overline{B}_j})\rangle$, $j = 1, 2, \ldots, n$ *are two collection of some Pq-RO-numbers. If* $a_j \geq \acute{a}_j$, $b_j \geq \acute{b}_j$, $c_j \geq \acute{c}_j$, $d_j \geq \acute{d}_j$, $e_j \geq \acute{e}_j$, $\hat{w}_{\overline{A}_j} \geq \hat{\acute{w}}_{\overline{B}_j}$, $\acute{w}_{\overline{A}_j} \geq \acute{w}_{\overline{B}_j}$, $u_{\overline{A}_j} \leq \acute{u}_{\overline{B}_j}$, $\hat{u}_{\overline{A}_j} \leq \hat{\acute{u}}_{\overline{B}_j}$ *for all j, then*

$$Pq - RONW_{go}(\overline{A}_1, \overline{A}_2, \ldots, \overline{A}_n) \geq Pq - RONW_{go}(\overline{B}_1, \overline{B}_2, \ldots, \overline{B}_n)$$

3. *(Boundedness) If*

$$\begin{aligned} \overline{A^-}_j = \langle &\min_{1\leq j\leq n}\{a_j\}, \min_{1\leq j\leq n}\{b_j\}, \min_{1\leq j\leq n}\{c_j\}, \min_{1\leq j\leq n}\{d_j\}, \min_{1\leq j\leq n}\{e_j\}; \\ &(\min_{1\leq j\leq n}\{\hat{w}_{\overline{A}_j}\}, \min_{1\leq j\leq n}\{w_{\overline{A}_j}\}), (\max_{1\leq j\leq n}\{u_{\overline{A}_j}\}, (\max_{1\leq j\leq n}\{\hat{u}_{\overline{A}_j}\})\rangle \end{aligned}$$

and

$$\overline{A^+}_j = \langle \max_{1\le j\le n} a_j, \max_{1\le j\le n} b_j, \max_{1\le j\le n} c_j, \max_{1\le j\le n} d_j, \max_{1\le j\le n} e_j; \\ (\max_{1\le j\le n} \hat{w}_{\overline{A}_j}, \max_{1\le j\le n} w_{\overline{A}_j}), (\min_{1\le j\le n} u_{\overline{A}_j}, \min_{1\le j\le n} \hat{u}_{\overline{A}_j})\rangle,$$

then

$$\overline{A^-}_j \le Pq - RONW_{go}(\overline{A}_1, \overline{A}_2, \ldots, \overline{A}_n) \le \overline{A^+}_j$$

17.4 Multi-criteria Decision-Making Method Based on Pq-RO-Numbers

In this section, we developed a multi-criteria decision-making method based on Pq-RO-numbers. Some of the definitions are quoted or inspired by Gündoğdu and Kahraman [23], Deli and Şubaş [10, 11], Deli [12, 13], Deli (2021b), Deli and Çağman [14], Li [28], Uthra et al. [49].

Definition 17.15 Let $X = (x_1, x_2, \ldots, x_m)$ be a set of alternatives, $U = (u_1, u_2, \ldots, u_n)$ be the set of criterions. If $\overline{A}_{ij} = \langle a_{ij}, b_{ij}, c_{ij}, d_{ij}, e_{ij}; (\hat{w}_{\overline{A}_{ij}}, w_{\overline{A}_{ij}}), (u_{\overline{A}_{ij}}, \hat{u}_{\overline{A}_{ij}})\rangle \in \triangle$, then

1.

$$[\overline{A}_{ij}]_{m\times n} = \begin{array}{c} \\ x_1 \\ x_2 \\ \vdots \\ x_m \end{array} \begin{array}{c} \begin{array}{cccc} u_1 & u_2 & \cdots & u_n \end{array} \\ \begin{bmatrix} \overline{A}_{11} & \overline{A}_{12} & \cdots & \tilde{a}_{1n} \\ \overline{A}_{21} & \overline{A}_{22} & \cdots & \\ \vdots & \vdots & \vdots & \vdots \\ \overline{A}_{m1} & \overline{A}_{m2} & \cdots & A_{mn} \end{bmatrix} \end{array}$$

is called a Pq-RON multi-criteria decision-making matrix of the decision maker.

2.

$$[\overline{A}_i]_{1\times n} = \begin{array}{cc} & \begin{array}{cccc} u_1 & u_2 & \cdots & u_n \end{array} \\ x_i & \begin{pmatrix} \overline{A}_{i1} & \overline{A}_{i2} & \cdots & \tilde{a}_{in} \end{pmatrix} \end{array}$$

is called a Pq-RON multi-criteria decision-making sub-matrix of the decision maker based on the alternatives x_i $(i = 1, 2, \ldots, m)$.

Now, we can give an algorithm of the Pq-RON multi-criteria decision-making method as follows:

Flow card of the Pq-RON multi-criteria decision-making method as in Fig. 17.1.

Algorithm:
[***Step 1***.] Construct the Pq-RON multi-criteria decision-making matrix $[\overline{A}_{ij}]_{m\times n}$ for $i = 1, 2, \ldots, m$ and $j = 1, 2, \ldots, n$;
[***Step 2***.] Insert the weights of the criterions $w = (w_1, w_2, \ldots, w_n)$;

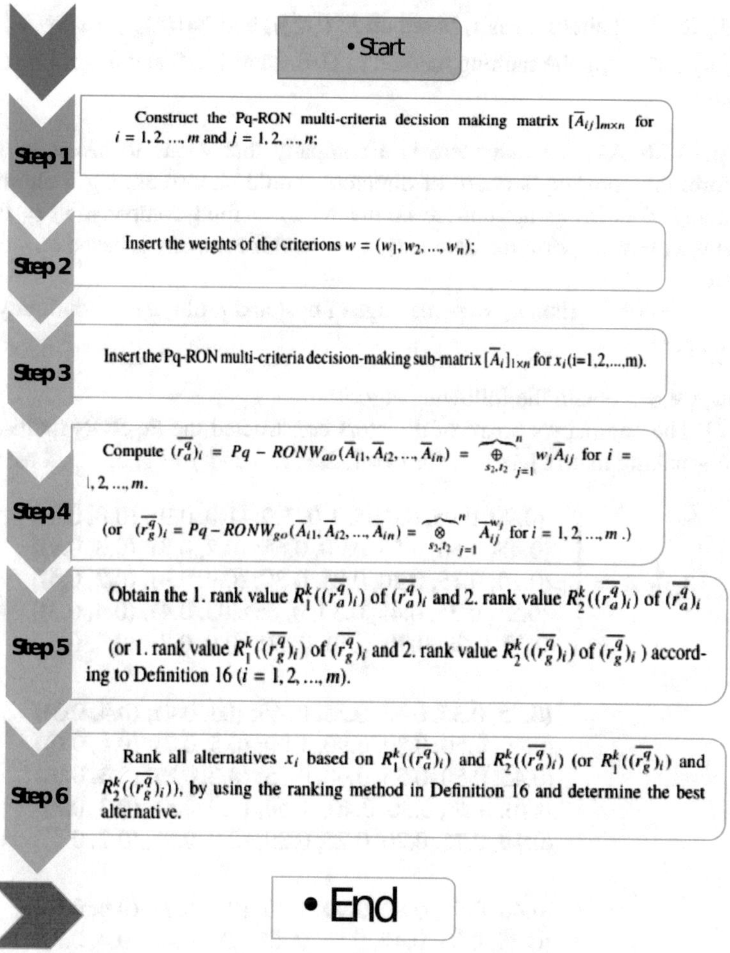

Fig. 17.1 Flow card of the Pq-RON multi-criteria decision-making method

[***Step** 3*.] Insert the Pq-RON multi-criteria decision-making sub-matrix $[\overline{A}_i]_{1\times n}$ for x_i $(i = 1, 2, \ldots, m)$.

[***Step** 4*.] Compute $(\overline{r_a^q})_i = Pq - RONW_{ao}(\overline{A}_{i1}, \overline{A}_{i2}, \ldots, \overline{A}_{in}) = \overset{n}{\underset{s_2,t_2\ j=1}{\overbrace{\oplus}}} w_j\overline{A}_{ij}$ for $i = 1, 2, \ldots, m$.

(or $(\overline{r_g^q})_i = Pq - RONW_{go}(\overline{A}_{i1}, \overline{A}_{i2}, \ldots, \overline{A}_{in}) = \overset{n}{\underset{s_2,t_2\ j=1}{\overbrace{\otimes}}} \overline{A}_{ij}^{w_j}$ for $i = 1, 2, \ldots, m$.)

[***Step** 5*.] Obtain the 1. rank value $R_1^k((\overline{r_a^q})_i)$ of $(\overline{r_a^q})_i$ and 2. rank value $R_2^k((\overline{r_a^q})_i)$ of $(\overline{r_a^q})_i$ (or 1. rank value $R_1^k((\overline{r_g^q})_i)$ of $(\overline{r_g^q})_i$ and 2. rank value $R_2^k((\overline{r_g^q})_i)$ of $(\overline{r_g^q})_i$) according to Definition 17.12 $(i = 1, 2, \ldots, m)$.

[**Step 6.**] Rank all alternatives x_i based on $R_1^k((\overline{r_a^q})_i)$ and $R_2^k((\overline{r_a^q})_i)$ (or $R_1^k((\overline{r_g^q})_i)$ and $R_2^k((\overline{r_g^q})_i)$), by using the ranking method in Definition 17.12 and determine the best alternative.

Example 17.6 Assume that there is a company that wants to invest in Turkey. Therefore, the company's board of directors should choose among 5 alternatives, by denoted $X = \{x_1 =$ agriculture company, $x_2 =$ food company, $x_3 =$ farming company, $x_4 =$ transportation company, $x_5 =$ health company $\}$, based on 4 criteria, denoted

by $\quad U = \{u_1 =$ risk, $u_2 =$ percentage of cost and profit, $u_3 =$ continuity, $u_4 =$ performance$\}$.

Now, we can obtain the following algorithm:
[**Step 1**]. The company's board of directors constructed the Pq-RON multi-criteria decision-making matrix $[\overline{A}_{ij}]_{5\times 4}$ for $i = 1, 2, \ldots, 5$ and $j = 1, 2, \ldots, 4$ as;

$$[\overline{A}_{ij}]_{5\times 4} = \begin{pmatrix} \langle 0.90, 0.95, 0.0.95, 1.0, 1.0; (1.0, 0.9), (0.0, 0.0)\rangle \\ \langle 0.45, 0.50, 0.55, 0.60, 0.65; (0.2, 0.8), (0.3, 0.9)\rangle \\ \langle 0.10, 0.15, 0.20, 0.25, 0.30; (0.9, 0.8), (0.2, 0.4)\rangle \\ \langle 0.25, 0.35, 0.45, 0.55, 0, 65; (03, 0.4), (0.4, 0.3)\rangle \\ \langle 0.10, 0.20, 0.30, 0.40, 0, 50; (0.9, 0.8), (0.3, 0.3)\rangle \end{pmatrix}$$

$$\langle 0.25, 0.35, 0.45, 0.55, 0, 65; (03, 0.4), (0.4, 0.3)\rangle$$
$$\langle 0.40, 0.50, 0.80, 0.90, 1.00; (0.5, 0.7), (0.4, 0.6)\rangle$$
$$\langle 0.45, 0.50, 0.55, 0.60, 0.65; (0.2, 0.8), (0.3, 0.9)\rangle$$
$$\langle 0.10, 0.20, 0.30, 0.40, 0, 50; (0.9, 0.8), (0.3, 0.3)\rangle$$
$$\langle 0.10, 0.15, 0.20, 0.25, 0.30; (0.9, 0.8), (0.2, 0.4)\rangle$$

$$\langle 0.40, 0.50, 0.80, 0.90, 1.00; (0.5, 0.7), (0.4, 0.6)\rangle$$
$$\langle 0.25, 0.35, 0.45, 0.55, 0, 65; (03, 0.4), (0.4, 0.3)\rangle$$
$$\langle 0.10, 0.20, 0.30, 0.40, 0, 50; (0.9, 0.8), (0.3, 0.3)\rangle$$
$$\langle 0.45, 0.50, 0.55, 0.60, 0.65; (0.2, 0.8), (0.3, 0.9)\rangle$$
$$\langle 0.90, 0.95, 0.0.95, 1.0, 1.0; (1.0, 0.9), (0.0, 0.0)\rangle$$

$$\langle 0.40, 0.50, 0.80, 0.90, 1.00; (0.5, 0.7), (0.4, 0.6)\rangle$$
$$\langle 0.25, 0.35, 0.45, 0.55, 0, 65; (03, 0.4), (0.4, 0.3)\rangle$$
$$\langle 0.90, 0.95, 0.0.95, 1.0, 1.0; (1.0, 0.9), (0.0, 0.0)\rangle$$
$$\langle 0.10, 0.15, 0.20, 0.25, 0.30; (0.9, 0.8), (0.2, 0.4)\rangle$$
$$\langle 0.10, 0.20, 0.30, 0.40, 0, 50; (0.9, 0.8), (0.3, 0.3)\rangle$$

[**Step 2.**] We inserted the weights of the criterions as

$$w = (0.14, 0.35, 0.21, 0.30)^T$$

[***Step 3***.] We inserted the Pq-RON multi-criteria decision-making sub-matrix as

$$[\overline{A}_1]_{1\times 4} = \begin{pmatrix} \langle 0.90, 0.95, 0.0.95, 1.0, 1.0; (1.0, 0.9), (0.0, 0.0)\rangle \\ \langle 0.25, 0.35, 0.45, 0.55, 0, 65; (03, 0.4), (0.4, 0.3)\rangle \\ \langle 0.40, 0.50, 0.80, 0.90, 1.00; (0.5, 0.7), (0.4, 0.6)\rangle \\ \langle 0.40, 0.50, 0.80, 0.90, 1.00; (0.5, 0.7), (0.4, 0.6)\rangle \end{pmatrix}$$

for x_1,

$$[\overline{A}_2]_{1\times 4} = \begin{pmatrix} \langle 0.45, 0.50, 0.55, 0.60, 0.65; (0.2, 0.8), (0.3, 0.9)\rangle \\ \langle 0.40, 0.50, 0.80, 0.90, 1.00; (0.5, 0.7), (0.4, 0.6)\rangle \\ \langle 0.25, 0.35, 0.45, 0.55, 0, 65; (03, 0.4), (0.4, 0.3)\rangle \\ \langle 0.25, 0.35, 0.45, 0.55, 0, 65; (03, 0.4), (0.4, 0.3)\rangle \end{pmatrix}$$

for x_2,

$$[\overline{A}_3]_{1\times 4} = \begin{pmatrix} \langle 0.10, 0.15, 0.20, 0.25, 0.30; (0.9, 0.8), (0.2, 0.4)\rangle \\ \langle 0.45, 0.50, 0.55, 0.60, 0.65; (0.2, 0.8), (0.3, 0.9)\rangle \\ \langle 0.10, 0.20, 0.30, 0.40, 0, 50; (0.9, 0.8), (0.3, 0.3)\rangle \\ \langle 0.90, 0.95, 0.0.95, 1.0, 1.0; (1.0, 0.9), (0.0, 0.0)\rangle \end{pmatrix}$$

for x_3,

$$[\overline{A}_4]_{1\times 4} = \begin{pmatrix} \langle 0.25, 0.35, 0.45, 0.55, 0, 65; (03, 0.4), (0.4, 0.3)\rangle \\ \langle 0.10, 0.20, 0.30, 0.40, 0, 50; (0.9, 0.8), (0.3, 0.3)\rangle \\ \langle 0.45, 0.50, 0.55, 0.60, 0.65; (0.2, 0.8), (0.3, 0.9)\rangle \\ \langle 0.10, 0.15, 0.20, 0.25, 0.30; (0.9, 0.8), (0.2, 0.4)\rangle \end{pmatrix}$$

for x_4,

$$[\overline{A}_5]_{1\times 4} = \begin{pmatrix} \langle 0.10, 0.20, 0.30, 0.40, 0, 50; (0.9, 0.8), (0.3, 0.3)\rangle \\ \langle 0.10, 0.15, 0.20, 0.25, 0.30; (0.9, 0.8), (0.2, 0.4)\rangle \\ \langle 0.90, 0.95, 0.0.95, 1.0, 1.0; (1.0, 0.9), (0.0, 0.0)\rangle \\ \langle 0.10, 0.20, 0.30, 0.40, 0, 50; (0.9, 0.8), (0.3, 0.3)\rangle \end{pmatrix}$$

for x_5.

[***Step 4***.] We computed $(\overline{r_a^2})_i$ (and $(\overline{r_g^2})_i$) ($i = 1, 2, 3, 4, 5.$) as

$$(\overline{r_a^2})_1 = \langle 0.8632, 0.9178, 0.9721, 1.0000, 1.0000; (1.0000, 0.9447), (0.0000, 0.0000)\rangle$$
$$(\overline{r_a^2})_2 = \langle 0.7476, 0.8205, 0.9269, 0.9652, 1.0000; (0.7734, 0.9149), (0.3842, 0.4459)\rangle$$
$$(\overline{r_a^2})_3 = \langle 0.9200, 0.9546, 0.9609, 1.0000, 1.0000; (1.0000, 0.9867), (0.0000, 0.0000)\rangle$$
$$(\overline{r_a^2})_4 = \langle 0.6117, 0.6971, 0.7667, 0.8245, 0.8727; (0.9791, 0.9724), (0.2766, 0.4119)\rangle$$
$$(\overline{r_a^2})_5 = \langle 0.8235, 0.8897, 0.9069, 1.0000, 1.0000; (1.0000, 0.9849), (0.0000, 0.0000)\rangle$$

or

$(\overline{r_g^2})_1 = \langle 0.3801, 0.4828, 0.6700, 0.7688, 0.8600; (0.4608, 0.5961), (0.3732, 0.4790)\rangle$
$(\overline{r_g^2})_2 = \langle 0.3200, 0.4168, 0.5661, 0.6615, 0.7558; (0.3389, 0.5361), (0.3880, 0.5949)\rangle$
$(\overline{r_g^2})_3 = \langle 0.3273, 0.4225, 0.4952, 0.5682, 0.6282; (0.5487, 0.8288), (0.2384, 0.6819)\rangle$
$(\overline{r_g^2})_4 = \langle 0.1559, 0.2405, 0.3193, 0.3955, 0.4702; (0.5627, 0.7260), (0.2925, 0.6005)\rangle$
$(\overline{r_g^2})_5 = \langle 0.1586, 0.2508, 0.3316, 0.4113, 0.4837; (0.9201, 0.8200), (0.2329, 0.3122)\rangle$

[***Step 5.***] We obtained the 1. rank value $R_1^2((\overline{r_a^q})_i)$ of $(\overline{r_a^q})_i$ and 2. rank value $R_2^2((\overline{r_a^q})_i)$ of $(\overline{r_a^q})_i$ (or 1. rank value $R_1^2((\overline{r_g^q})_i)$ of $(\overline{r_g^q})_i$ and 2. rank value $R_2^2((\overline{r_g^q})_i)$ of $(\overline{r_g^q})_i$) according to Definition 17.12 ($i = 1, 2, \ldots, 5$).

Obtain the 1. and 2. rank value of $\overline{r}_i$, is denoted by $R_k^2(\overline{r}_i)$ $(k = 1, 2)$, according to Definition 17.12

$$\begin{array}{ll}
R_1^2((\overline{r_a^2})_1) = 0.9760 & R_2^2((\overline{r_a^2})_1) = 0.9760 \\
R_1^2((\overline{r_a^2})_2) = 0.8579 & R_2^2((\overline{r_a^2})_2) = 1.1269 \\
R_1^2((\overline{r_a^2})_3) = 0.9822 & R_2^2((\overline{r_a^2})_3) = 0.9822 \\
R_1^2((\overline{r_a^2})_4) = 0.7938 & R_2^2((\overline{r_a^2})_4) = 0.9885 \\
R_1^2((\overline{r_a^2})_5) = 0.9581 & R_2^2((\overline{r_a^2})_5) = 0.9581
\end{array}$$

or

$$\begin{array}{ll}
R_1^2((\overline{r_g^2})_1) = 0.6281 & R_2^2((\overline{r_g^2})_1) = 1.1116 \\
R_1^2((\overline{r_g^2})_2) = 0.5464 & R_2^2((\overline{r_g^2})_2) = 1.1215 \\
R_1^2((\overline{r_g^2})_3) = 0.5791 & R_2^2((\overline{r_g^2})_3) = 0.7256 \\
R_1^2((\overline{r_g^2})_4)) = 0.4286 & R_2^2((\overline{r_g^2})_4) = 0.6694 \\
R_1^2((\overline{r_g^2})_5) = 0.5140 & R_2^2((\overline{r_g^2})_5) = 0.7429
\end{array}$$

[***Step 6.***] Since

$$R_1^2((\overline{r_a^2})_3) > R_1^2((\overline{r_a^2})_1) > R_1^2((\overline{r_a^2})_5) > R_1^2((\overline{r_a^2})_2) > R_1^2((\overline{r_a^2})_4),$$

we ranked all alternatives x_i based on $R_1^2((\overline{r_a^2})_i)$ by using the ranking method in Definition 17.12 as

$$x_3 > x_1 > x_5 > x_2 > x_4.$$

Finally the best alternative is x_3 and the worst alternative is x_4.
or
Since

$$R_1^2((\overline{r_g^2})_1) > R_1^2((\overline{r_g^2})_3) > R_1^2((\overline{r_g^2})_2) > R_1^2((\overline{r_g^2})_5) > R_1^2((\overline{r_g^2})_4),$$

we ranked all alternatives x_i based on $R_1^2((\overline{r_g^2})_i)$ by using the ranking method in Definition 17.12 as

$$x_1 > x_3 > x_2 > x_5 > x_4.$$

Finally the best alternative is x_1 and the worst alternative is x_4.

17.5 Conclusion

In this chapter, a new multi-criteria decision-making method to solve Pentagonal q-rung orthopair numbers(Pq-RO-numbers) that are generalization of the fuzzy numbers and intuitionistic fuzzy numbers on real number R are developed. In summary, the following conclusions can be made:

1. Pentagonal q-rung orthopair numbers(Pq-RO-numbers) and their operations is defined.
2. Normal Pq-RO-numbers by using the concept of s-norm and t-norm and their laws of operations including their properties are initiated.
3. The concept of 1. and 2. rank value of Pq-RO-numbers to compare any two Pq-RO-numbers are introduced.
4. Some operators of qth rung orthopair fuzzy numbers such as qth rung orthopair fuzzy numbers weighted aggregation mean operator and qth rung orthopair fuzzy number weighted geometric mean operator are proposed.
5. Moreover, a new decision-making approach by using the Pq-RO-numbers and related concepts is developed, and then a real example to illustrate the development.

In future research, more attention may be focused on developing different types of aggregation operators such as bonferroni mean operators, muirhead mean operators and maclaurin symmetric mean operator and also distance measures for Pq-RO-numbers. Also, the subject will be applied to other fields, such as supply chain management, risk management, pattern recognition, game theory, society, economics, management, military, engineering technology, and so on.

References

1. K. Atanassov, Intuitionistic fuzzy sets. Fuzzy Sets Syst. **20**, 87–96 (1986)
2. Z. Ai, Z. Xu, R.R. Yager, J. Ye, q-Rung orthopair fuzzy integrals in the frame of continuous archimedean t-norms and t-conorms and their application. IEEE Trans. Fuzzy Syst. **29**(5), 996–1007 (2021)
3. M. Akram, G. Shahzadi, X. Peng, Extension of Einstein geometric operators to multi-attribute decision making under q-rung orthopair fuzzy information. Granular Computing (2020). https://doi.org/10.1007/s41066-020-00233-3

4. M. Akram, G. Shahzadi, A hybrid decision-making model under q-rung orthopair fuzzy Yager aggregation operators. Granul. Comput. (2020). https://doi.org/10.1007/s41066-020-00229-z
5. Z. Ali, T. Mahmood, Maclaurin symmetricmean operators and their applications in the environment of complex q-rung orthopair fuzzy sets. Comput. Appl. Math. **39**(161), 1–27 (2020)
6. S.B. Aydemir, S.Y. Gunduz, A novel approach to multi-attribute group decision making based on power neutrality aggregation operator for q-rung orthopair fuzzy sets. Int. J. Intell. Syst. **36**, 1454–1481 (2021)
7. S.B. Aydemir, S.Y. Gunduz, Extension of multi-Moora method with some q-rung orthopair fuzzy Dombi prioritized weighted aggregation operators for multi-attribute decision making. Soft. Comput. **24**, 18545–18563 (2020)
8. S. Cheng, S. Jianfu, M. Alrasheedi, P. Saeidi, A.R. Mishra, P. Ran, A new extended VIKOR approach using q-rung orthopair fuzzy sets for sustainable enterprise risk management assessment in manufacturing small and medium-sized enterprises. Int. J. Fuzzy Syst. **23**, 1347–1369 (2021)
9. A.P. Darko, D. Liang, Some q-rung orthopair fuzzy Hamacher aggregation operators and their application to multiple attribute group decision making with modified EDAS method. Eng. Appl. Artif. Intell. **87**, 1–17 (2020)
10. I. Deli, Y. Şubaş, A ranking method of single valued neutrosophic numbers and its applications to multi-attribute decision making problems. Int. J. Mach. Learn. Cybern. **8**(4), 1309–1322 (2017)
11. I. Deli, Y. Şubaş, Some weighted geometric operators with SVTrN-numbers and their application to multi-criteria decision making problems. J. Intell. Fuzzy Syst. **32**(1), 291–301 (2017)
12. I. Deli, Operators on single valued trapezoidal neutrosophic numbers and SVTN-group decision making. Neutrosophic Sets Syst. **22**, 130–150 (2018)
13. I. Deli, Theory of single valued trapezoidal neutrosophic numbers and their applications to multi robot systems, in *Toward Humanoid Robots: The Role of Fuzzy Sets, A Handbook on Theory and Applications*. Studies in Systems, Decision and Control, Vol. 344 (Springer Nature Switzerland AG, 2021), pp. 255–276
14. İ. Deli, N. Çağman, Spherical fuzzy numbers and multi-criteria decision-making, in *Decision Making with Spherical Fuzzy Sets: Theory and Applications*. Studies in Fuzziness and Soft Computing, Vol. 392 (Springer Nature Switzerland AG, 2021), pp. 53–84
15. W.S. Du, Minkowski-type distance measures for generalized orthopair fuzzy sets. Int. J. Intell. Syst. **33**, 802–817 (2018)
16. W.S. Du, Correlation and correlation coefficient of generalized orthopair fuzzy sets. Int. J. Intell. Syst. **34**, 564–583 (2019)
17. B. Farhadinia, S. Effati, F. Chiclana, A family of similarity measures for q-rung orthopair fuzzy sets and their applications to multiple criteria decision making. Int. J. Intell. Syst. **36**, 1535–1559 (2021)
18. H. Garg, S.M. Chen, Multiattribute group decision making based on neutrality aggregation operators of q-rung orthopair fuzzy sets. Inf. Sci. **517**, 427–447 (2020)
19. H. Garg, A. Zeeshan, Y. Zaoli, M. Tahir, A. Sultan, Multi-criteria decision-making algorithm based on aggregation operators under the complex interval-valued q-rung orthopair uncertain linguistic information. J. Intell. Fuzzy Syst. **41**(1), 1627–1656 (2021). https://doi.org/10.3233/JIFS-210442
20. H. Garg, A new possibility degree measure for interval-valued q-rung orthopair fuzzy sets in decision-making. Int. J. Intell. Syst. **36**, 526–557 (2021)
21. H. Garg, CN-q-ROFS: connection number-based q-rung orthopair fuzzy set and their application to decision-making process. Int. J. Intell. Syst. **36**(7), 3106–3143 (2021). https://doi.org/10.1002/int.22406
22. H. Garg, New exponential operation laws and operators for interval-valued q-rung orthopair fuzzy sets in group decision making process. Neural Comput. Appl. **33**(20), 13937–13963 (2021). https://doi.org/10.1007/s00521-021-06036-0
23. F.K. Gündoğdu, C. Kahraman, Spherical fuzzy sets and spherical fuzzy TOPSIS method. J. Intell. Fuzzy Syst. **36**(1), 337–352 (2019)

24. A. Hussain, M.I. Ali, T. Mahmood, Covering based q-rung orthopair fuzzy rough set model hybrid with TOPSIS for multi-attribute decision making. J. Intell. Fuzzy Syst. **37**, 981–993 (2019)
25. N. Jan, L. Zedam, T. Mahmood, E. Rak, Z. Ali, Generalized dice similarity measures for q-rung orthopair fuzzy sets with applications. Complex Intell. Syst. **6**, 545–558 (2020)
26. H. Kamacı, Linear Diophantine fuzzy algebraic structures. J. Ambient Intell. Humaniz. Comput. (2021). https://doi.org/10.1007/s12652-020-02826-x
27. H. Li, S. Yin, Y. Yang, Some preference relations based on q-rung orthopair fuzzy sets. Int. J. Intell. Syst. **34**(11), 2920–2936 (2019)
28. D.F. Li, *Decision and Game Theory in Management With Intuitionistic Fuzzy Sets*. Studies in Fuzziness and Soft Computing, Vol. 308 (Springer, 2014)
29. Z. Li, G. Wei, R. Wang, J. Wu, C. Wei, Y. Wei, EDAS method for multiple attribute group decision making under q-rung orthopair fuzzy environment. Technol. Econ. Dev. Econ. **26**(1), 86–102 (2020)
30. P. Liu, P. Wang, Some q-rung orthopair fuzzy aggregation operators and their applications to multiple-attribute decision making. Int. J. Intell. Syst. **33**, 259–280 (2018)
31. H. Liao, H. Zhang, C. Zhang, X. Wu, A. Mardani, A. Al-Barakati, q-rung orthopair fuzzy GLDS method for investment evaluation of be Angel capital in China. Technol. Econ. Dev. Econ. **26**(1), 103–134 (2020)
32. X. Peng, L. Liu, Information measures for q-rung orthopair fuzzy sets. Int. J. Intell. Syst. **34**(8), 1795–1834 (2019)
33. X. Peng, J. Dai, H. Garg, Exponential operation and aggregation operator for q-rung orthopair fuzzy set and their decision-making method with a new score function. Int. J. Intell. Syst. **33**, 2255–2282 (2018)
34. M. Riaz, M.R. Hashmi, Linear Diophantine fuzzy set and its applications towards multi-attribute decision making problems. J. Intell. Fuzzy Syst. **37**(4), 5417–5439 (2019)
35. M. Riaz, H. Garg, H.M.A. Farid, M. Aslam, Novel q-rung orthopair fuzzy interaction aggregation operators and their application to low-carbon green supply chain management. J. Intell. Fuzzy Syst. **41**(2), 4109–4126 (2021). https://doi.org/10.3233/JIFS-210506
36. M. Riaz, W. Salabun, H.M.A. Farid, N. Ali, J. Trobski, A robust q-rung orthopair fuzzy information aggregation using Einstein operations with application to sustainable energy planning decision management. Energies **13**(9), 1–39 (2020)
37. T. Senapati, R.R. Yager, Fermatean fuzzy weighted averaging/geometric operators and its application in multicriteria decision making methods. Eng. Appl. Artif. Intell. **85**, 112–121 (2014)
38. X. Tian, M. Niu, W. Zhang, L. Li, E.H. Viedma, A novel TODIM based on prospect theory to select Green supplier with q-rung orthopair fuzzy set. Technol. Econ. Dev. Econ. **27**(2), 284–310 (2021)
39. P. Wang, J. Wang, G. Wei, C. Wei, Similarity measures of q-rung orthopair fuzzy sets based on cosine function and their applications. Mathematics **07**, 1–29 (2019)
40. J. Wang, G. Wei, C. Wei, Y. Wei, MABAC method for multiple attribute group decision making under qrung orthopair fuzzy environment. Def. Techno. **16**, 208–216 (2020)
41. Y.M. Wang, J.B. Yang, D.L. Xu, K.S. Chin, On the centroids of fuzzy numbers. Fuzzy Sets Syst. **157**, 919–926 (2006)
42. J. Wang, G. Wei, C. Wei, J. Wu, Maximizing deviation method for multiple attribute decision making under q-rung orthopair fuzzy environment. Def. Technol. **16**, 1073–1087 (2020)
43. H. Wang, F. Smarandache, Q. Zhang, R. Sunderraman, Single valued neutrosophic sets. Multispace Multistruct. **4**, 410–413 (2010)
44. G. Wei, Some arithmetic aggregation operators with intuitionistic trapezoidal fuzzy numbers and their application to group decision making. J. Comput. **5**(3), 345–351 (2010)
45. Z.S. Xu, Intuitionistic fuzzy aggregation operators. IEEE Trans. Fuzzy Syst. **15**(6), 1179–1187 (2007)
46. Y. Xu, H. Wang, The induced generalized aggregation operators for intuitionistic fuzzy sets and their application in group decision making. Appl. Soft Comput. **12**, 1168–1179 (2012)

47. Z. Xu, R.R. Yager, Some geometric aggregation operators based on intuitionistic fuzzy sets. Int. J. Gen Syst **35**(4), 417–433 (2006)
48. Y. Xing, R. Zhang, Z. Zhou, J. Wang, Some q-rung orthopair fuzzy point weighted aggregation operators for multi-attribute decision making. Soft. Comput. **23**, 11627–11649 (2019)
49. G. Uthra, K. Thangavelu, S. Shunmugapriya, Ranking generalized intuitionistic fuzzy numbers. Int. J. Math. Trends Technol. **56**(7), 530–538 (2018)
50. L.A. Zadeh, Fuzzy Sets. Inf. Control **8**, 338–353 (1965)
51. R.R. Yager, Generalized orthopair fuzzy sets. IEEE Trans. Fuzzy Syst. **25**(5), 1222–1230 (2017)
52. R.R. Yager, N. Alajlan, Approximate reasoning with generalized orthopair fuzzy sets. Inf. Fusion **38**, 65–73 (2017)
53. Z. Yang, H. Garg, Interaction power partitioned Maclaurin symmetric mean operators under q-rung orthopair uncertain linguistic information. Int. J. Fuzzy Syst. 1–19 (2021). https://doi.org/10.1007/s40815-021-01062-5
54. R.R. Yager, A.M. Abbasov, Pythagorean membership grades, complex numbers, and decision making. Int. J. Intell. Syst. **28**, 436–452 (2013)
55. R.R. Yager, Pythagorean membership grades in multicriteria decision making. IEEE Trans. Fuzzy Syst. **22**, 958–965 (2014)
56. S. Zeng, Y. Hu, X. Xie, Q-rung orthopair fuzzy weighted induced logarithmic distance measures and their application in multiple attribute decision making. Eng. Appl. Artif. Intell. **100**(104167), 1–7 (2021)
57. H.-J. Zimmermann, *Fuzzy Set Theory and Its Applications* (Kluwer Academic Publishers, 1993)

Irfan Deli was born in 1986 Denizli, Turkey. He received his Ph.D. degree from the Department of Mathematics, Faculty of Arts and Sciences, Gaziosmanpaşa University, Tokat, Turkey in 2013. At present he is working as Assistant Professor in University of 7 Aralık, Kilis, Turkey. His main interest areas include soft sets, fuzzy sets, intuitionistic fuzzy sets, neutrosophic sets, game theory, decision making, optimization, and so on.

Chapter 18
q-Rung Orthopair Fuzzy Soft Set-Based Multi-criteria Decision-Making

Muhammad Riaz and Hafiz Muhammad Athar Farid

Abstract In this chapter, we develop a hybrid structure named as "q-rung orthopair fuzzy soft sets" (q-ROFSSs) by combining the features of Yager's "q-rung orthopair fuzzy sets" (q-ROFSs) and Molodtsov's soft sets. Certain new concepts of q-ROFSSs theory including algebraic features on these sets are proposed. The significance of linguistic variables in q-ROFSS information is discussed and extended towards real-life circumstances. Mathematical models for "multi-criteria decision-making" (MCDM) problems based on q-ROFSS information are developed by using four different techniques including, "Technique for Order of Preference by Similarity to Ideal Solution" (TOPSIS), "Vlse Kriterijumska Optimizacija Kompromisno Resenje" (VIKOR), "choice value method", and new similarity measures (SMs). Additionally, a practical application of proposed MCDM approaches is presented related to appropriate persons for key ministries in a government, selection of agriculture land and COVID-19.

Keywords TOPSIS · VIKOR · COVID-19 · Similarity measures · Medical diagnosis · Agriculture land selection · q-rung orthopair fuzzy soft information

18.1 Introduction

Modern set theory provides rational, fair, balanced, and encompassing solutions to many of the problems that humans are facing in daily life. The bulk of commonly faced difficulties in everyday life is characterized by uncertainty, imprecision, and ambiguity. Zadeh's "fuzzy sets" (FSs) and Molodstov's "soft sets" (SSs) are essential mathematical frameworks for coping with these uncertainties. The reason behind these problems is lack of knowledge, and vague and uncertain information. With a reasonably qualitative storey, the fuzzy methodology delivers promising results in a wide number of fields. The use of words or sentences rather than numbers is prompted by the fact that philosophical descriptions are often not accurate than mathematical descriptions in terms of numbers, sets, and models. Zadeh [1] has been pioneered

M. Riaz (✉) · H. M. Athar Farid
Department of Mathematics, University of the Punjab, Lahore, Pakistan
e-mail: mriaz.math@pu.edu.pk

H. Garg (ed.), *q-Rung Orthopair Fuzzy Sets*,
https://doi.org/10.1007/978-981-19-1449-2_18

the idea of FS theory. Atanassov [2] initiated the novel idea of "intuitionistic fuzzy set" (IFS), Yager [3–5] introduced "Pythagorean fuzzy set" (PFS) as an extension of IFS. The concepts of IFS and PFS are further generalized by Yager [6], and the novel concept of q-ROFS is developed. Molodtsov [7] pioneered the perception of a revolutionary type of model for resolving ambiguities, known traditionally as a SSs.

FSs, SSs, and their further expansions are robust computational models for tack a wide range of real-world issues. To cope with legitimate difficulties, the researchers devised a number of mathematical frameworks. Çağman et al. [8] initiated the idea of fuzzy SSs (FSSs). Feng et al. [9] offered a dynamic approach to decision-making based on FSSs. Feng et al. [11] merged SSs, FSs, and rough sets. Majumdar and Samanta [10] gave the notion of generalized FSSs. Davvaz and Sadrabadi [12] proposed the applications of IFSs in medical. Maji et al. [13] proposed "intuitionistic fuzzy soft sets" (IFSSs). Feng et al. [14], Li and Cui [15], Osmanoglu and Tokat [16], Garg and Arora [17] developed different ideas related to IFSSs. Guleria and Bajaj [18] developed matrices for "Pythagorean fuzzy soft sets" (PFSSs). Naz et al. [19], Peng and Yang [20], Peng et al. [21], and Peng et al. [22] gave the different contributions for PFSSs. Peng and Selvachandran encapsulated the concepts of PFSs [23]. Riaz and Naeem [24, 25], Fei et al. [26], and Fei and Deng [27] gave the notions of soft σ-algebra, soft likelihood functions for PFSSs.

Numerous scholars have pondered on TOPSIS and VIKOR decision-making strategies including Adeel et al. [29], Hwang and Yoon [28], Eraslan and Karaaslan [30], Liu et al. [32], Naeem et al. [31], Kumar and Garg [33], Riaz et al. [34, 35], Opricovic and Tzeng [37, 38], Li and Nan [36], Naeem et al. [40], Mohd and Abdullah [39], and Kalkan et al. [41].

The TOPSIS method's pros include its ease of application, universality, and assessment of distances to an ideal solution; significant downsides are its high subjectivity. TOPSIS has a straightforward procedure that is simple to use and programme; the number of steps remains constant regardless of the number of attributes. When comparing to other approaches, the outcomes are acquired relatively quickly. TOPSIS is frequently used in fields such as logistical, factory automation and architecture, sales department, resource stewardship, finance, technology, personnel management, and irrigation. Its use of Euclidean distance ignores attribute connection, making it tricky to weight and maintain consistency of evaluation. A significant departure of one indicator from the optimal answer has a significant impact on the findings. When the criteria of alternatives do not vary significantly, the technique is appropriate.

Xu et al. [42–44] developed many AOs for IFS. Mahmood et al. [45], Jose and Kuriaskose [47], Wei et al. [46], Hashmi et al. [48], Zhang et al. [50], Wang and Liu [49], Zhao et al. [51], Wang et al. [53], Mu et al. [52], Garg [54] and Rahman et al.et al. [55] proposed many AOs. Zeng et al. [56], Zhang et al.et al. [57], Zeng et al. [58], Riaz et al.et al. [60] and Sitara et al. [59] developed many approaches for MCDM. Many AOs are developed related to q-ROFS by different researchers, like Einstein [61], prioritized [62] Einstein prioritized [63], Bonferroni mean [67], Heronian mean [68], confidence-based AOs [69], and group-generalized AOs [70]. The concept of "linear Diophantine fuzzy Set" (LDFS) was given by Riaz and Hashmi[64]. Riaz et al. introduced Einstein AOs [66] and prioritized AOs [65] related to LDFSs.

In practically every field of sciences, the idea of SM is essential. It is usually faked to see if an object, scenario, or document is genuine. The SM is a useful tool for determining the degree of similarity between two or more data sets. The SMs established by means of the notions of FSs, SSs, IFSs, PFSs, and q-ROFSs are broadly. Kharal [72], Hong and Kim [71], Kamaci [73], Hyung et al. [75], Hung and Yang [74], Chen [76–78], Wang et al. [79], and Muthukumar and Krishnan [80] explored SMs on many models. In recent times, Peng and Garg [81] made multiparametric SMs on PFSs. Garg introduced the idea of connection number-based q-ROFS and possibility degree measures for q-ROFSs [83]. Garg developed a series of AOs related to q-ROFSs, namely, trigonometric operation-based AOs [86], interaction power partitioned Maclaurin symmetric mean AOs [85] and exponential based AOs [84].

The following is how the chapter is structured: The fundamental properties of q-ROFSSs are described in Sect. 18.2. We devote Sect. 18.3 to a "multi-criteria decision-making" (MCDM) application based on q-ROFS matrices. In Sect. 18.4, we proposed the q-ROFS-TOPSIS algorithm and demonstrated its application in selecting appropriate individuals for important ministries in a cabinet. In Sect. 18.5, we propose q-ROFS-VIKOR and apply it to agricultural land selection. Section 18.6 develops an SM and a weighted SM for q-ROFSSs. In COVID-19, we describe a bioscience application based on this SM. Finally, in Sect. 18.7, we summarize our study.

Reader is suggested to see [1–8, 13] for preliminary notions.

18.2 q-ROFSSs

We assume that $\widetilde{\mathscr{Y}}^{\triangledown}$ be the "universal set" and $\breve{\urcorner}$ the "set of parameters", and $\breve{\xi}^{\beth}, \breve{\xi}^{\beth}_1, \breve{\xi}^{\beth}_2, \breve{\xi}^{\beth}_3$ be subsets of $\breve{\urcorner}$.

Definition 18.2.1 A q-ROFSS on $\widetilde{\mathscr{Y}}$ is an object of the form

$$
\begin{aligned}
(\beth^{\urcorner}, \breve{\xi}^{\beth}) &= \left\{ \left(\breve{\hbar}^{\blacklozenge}, \left\{\breve{\aleph}^{\eth}, (\mu^{\gamma}_{\beth\urcorner}(\breve{\aleph}^{\eth}), \nu^{\gamma}{}_{\beth\urcorner}(\breve{\aleph}^{\eth}))\right\} : \breve{\hbar}^{\blacklozenge} \in \breve{\xi}^{\beth}, \breve{\aleph}^{\eth} \in \widetilde{\mathscr{Y}}^{\triangledown} \right\} \\
&= \left\{ \left(\breve{\hbar}^{\blacklozenge}, \left\{ \frac{\breve{\aleph}^{\eth}}{(\mu^{\gamma}_{\beth\urcorner}(\breve{\aleph}^{\eth}), \nu^{\gamma}{}_{\beth\urcorner}(\breve{\aleph}^{\eth}))} \right\} \right) : \breve{\hbar}^{\blacklozenge} \in \breve{\xi}^{\beth}, \breve{\aleph}^{\eth} \in \widetilde{\mathscr{Y}}^{\triangledown} \right\} \\
&= \left\{ \left(\breve{\hbar}^{\blacklozenge}, \left\{ \frac{(\mu^{\gamma}_{\beth\urcorner}(\breve{\aleph}^{\eth}), \nu^{\gamma}{}_{\beth\urcorner}(\breve{\aleph}^{\eth}))}{\breve{\aleph}^{\eth}} \right\} \right) : \breve{\hbar}^{\blacklozenge} \in \breve{\xi}^{\beth}, \breve{\aleph}^{\eth} \in \widetilde{\mathscr{Y}}^{\triangledown} \right\}
\end{aligned}
$$

where $\mu^{\gamma}_{\beth\urcorner}, \nu^{\gamma}{}_{\beth\urcorner} : \widetilde{\mathscr{Y}}^{\triangledown} \to [0, 1]$ are the "membership degree (MSD) and non-membership degree (NMSD) mappings", s.t each $\breve{\aleph}^{\eth} \in \widetilde{\mathscr{Y}}^{\triangledown}$ satisfies the property

$$0 \leq \mu^{\gamma q}{}_{\beth\urcorner}(\breve{\aleph}^{\eth}) + \nu^{\gamma q}{}_{\beth\urcorner}(\breve{\aleph}^{\eth}) \leq 1$$

Table 18.1 Tabulatory delineation of q-ROFSS $(\beth^{\urcorner}, \breve{\xi}^{\gimel})$

$(\beth^{\urcorner}, \breve{\xi}^{\gimel})$	$\breve{\hbar}^{\blacklozenge}{}_1$	$\breve{\hbar}^{\blacklozenge}{}_2$	$\cdots$	$\breve{\hbar}^{\blacklozenge}{}_n$
$\breve{\aleph}^{\eth}_1$	$(\mu^{\gamma}_{11}, \nu^{\gamma}{}_{11})$	$(\mu^{\gamma}_{12}, \nu^{\gamma}{}_{12})$	$\cdots$	$(\mu^{\gamma}_{1n}, \nu^{\gamma}{}_{1n})$
$\breve{\aleph}^{\eth}_2$	$(\mu^{\gamma}_{21}, \nu^{\gamma}{}_{21})$	$(\mu^{\gamma}_{22}, \nu^{\gamma}{}_{22})$	$\cdots$	$(\mu^{\gamma}_{2n}, \nu^{\gamma}{}_{2n})$
$\vdots$	$\vdots$	$\vdots$	$\ddots$	$\vdots$
$\breve{\aleph}^{\eth}_m$	$(\mu^{\gamma}_{m1}, \nu^{\gamma}{}_{m1})$	$(\mu^{\gamma}_{m2}, \nu^{\gamma}{}_{m2})$	$\cdots$	$(\mu^{\gamma}_{mn}, \nu^{\gamma}{}_{mn})$

Table 18.2 Tabular delineation of $(\beth^{\urcorner}, \breve{\xi}^{\gimel})$

$(\beth^{\urcorner}, \breve{\xi}^{\gimel})$	$\breve{\hbar}^{\blacklozenge}{}_1$	$\breve{\hbar}^{\blacklozenge}{}_2$	$\breve{\hbar}^{\blacklozenge}{}_3$	$\breve{\hbar}^{\blacklozenge}{}_4$	$\breve{\hbar}^{\blacklozenge}{}_5$
$\breve{\aleph}^{\eth}_1$	(0.000,1.000)	(0.230,0.810)	(0.000,1.000)	(0.000,1.000)	(0.560,0.180)
$\breve{\aleph}^{\eth}_2$	(0.000,1.000)	(0.000,1.000)	(0.000,1.000)	(0.000,1.000)	(0.410,0.610)
$\breve{\aleph}^{\eth}_3$	(0.000,1.000)	(0.750,0.420)	(0.000,1.000)	(0.000,1.000)	(0.000,1.000)
$\breve{\aleph}^{\eth}_4$	(0.000,1.000)	(0.540,0.340)	(0.000,1.000)	(0.000,1.000)	(0.000,1.000)

with $q \geq 1$. If we write $\mu^{\gamma}_{ij} = \mu^{\gamma}_{\beth^{\urcorner}}(e_j)(\breve{\aleph}^{\eth}_i)$ and $\nu^{\gamma}{}_{ij} = \nu^{\gamma}{}_{\beth^{\urcorner}}(e_j)(\breve{\aleph}^{\eth}_i), i = 1, \cdots, m;$ $j = 1, \cdots, n$, then the q-ROFSS $(\beth^{\urcorner}, \breve{\xi}^{\gimel})$ also expressed as in Table 18.1.

The matrix representing q-ROFSS $(\beth^{\urcorner}, \breve{\xi}^{\gimel})$ is termed as q-ROFS matrix and has form

$$(\beth^{\urcorner}, \breve{\xi}^{\gimel}) = [(\mu^{\gamma}_{ij}, \nu^{\gamma}{}_{ij})]_{m\times n}$$
$$= \begin{pmatrix} (\mu^{\gamma}_{11}, \nu^{\gamma}{}_{11}) & (\mu^{\gamma}_{12}, \nu^{\gamma}{}_{12}) & \cdots & (\mu^{\gamma}_{1n}, \nu^{\gamma}{}_{1n}) \\ (\mu^{\gamma}_{21}, \nu^{\gamma}{}_{21}) & (\mu^{\gamma}_{22}, \nu^{\gamma}{}_{22}) & \cdots & (\mu^{\gamma}_{2n}, \nu^{\gamma}{}_{2n}) \\ \vdots & \vdots & \ddots & \vdots \\ (\mu^{\gamma}_{m1}, \nu^{\gamma}{}_{m1}) & (\mu^{\gamma}_{m2}, \nu^{\gamma}{}_{m2}) & \cdots & (\mu^{\gamma}_{mn}, \nu^{\gamma}{}_{mn}) \end{pmatrix}$$

Example 18.2.2 Let $\mathscr{Y}^{\nabla} = \{\breve{\aleph}^{\eth}_i : i = 1, \cdots, 4\}$ and $\breve{\daleth} = \{\breve{\hbar}^{\blacklozenge}{}_i : i = 1, 2, \cdots, 5\}$. Take $\breve{\xi}^{\gimel} = \{\breve{\hbar}^{\blacklozenge}{}_2, \breve{\hbar}^{\blacklozenge}{}_5\}$. Then, $(\beth^{\urcorner}, \breve{\xi}^{\gimel}) = \Big\{\Big(\breve{\hbar}^{\blacklozenge}{}_2, \Big\{\Big(\frac{\breve{\aleph}^{\eth}_1}{(0.230,0.810)}\Big), \Big(\frac{\breve{\aleph}^{\eth}_3}{(0.750,0.420)}\Big), \Big(\frac{\breve{\aleph}^{\eth}_4}{(0.540,0.340)}\Big)\Big\}\Big),$ $\Big(\breve{\hbar}^{\blacklozenge}{}_5, \Big\{\Big(\frac{\breve{\aleph}^{\eth}_1}{(0.560,0.180)}\Big), \Big(\frac{\breve{\aleph}^{\eth}_2}{(0.410,0.610)}\Big)\Big\}\Big)\Big\}$ is a q-ROFSS over $\mathscr{Y}^{\nabla}$ and take $q = 3$. The tabular delineation of $(\beth^{\urcorner}, \breve{\xi}^{\gimel})$ is given in Table 18.2.

The corresponding q-ROFS matrix is

$$(\beth^{\urcorner}, \breve{\xi}^{\gimel}) = [\mu^{\gamma}_{ij}, \nu^{\gamma}{}_{ij}]_{4\times 5}$$
$$= \begin{pmatrix} (0.000, 1.000) & (0.230, 0.810) & (0.000, 1.000) & (0.000, 1.000) & (0.560, 0.180) \\ (0.000, 1.000) & (0.000, 1.000) & (0.000, 1.000) & (0.000, 1.000) & (0.410, 0.610) \\ (0.000, 1.000) & (0.750, 0.420) & (0.000, 1.000) & (0.000, 1.000) & (0.000, 1.000) \\ (0.000, 1.000) & (0.540, 0.340) & (0.000, 1.000) & (0.000, 1.000) & (0.000, 1.000) \end{pmatrix}$$

Definition 18.2.3 A q-ROFSS $\beth^{\daleth}_{\breve{\xi}^{\daleth}_1}$ is called *q-ROFS subset* of $\beth^{\daleth}_{\breve{\xi}^{\daleth}_2}$ i.e. $\beth^{\daleth}_{\breve{\xi}^{\daleth}_1} \widetilde{\subseteq} \beth^{\daleth}_{\breve{\xi}^{\daleth}_2}$, if

(i) $\breve{\xi}^{\daleth}_1 \subseteq \breve{\xi}^{\daleth}_2$, and
(ii) $\beth^{\daleth}_{\breve{\xi}^{\daleth}_1}(\hbar^{\blacklozenge})$ is q-ROFS subset of $\beth^{\daleth}_{\breve{\xi}^{\daleth}_2}(\hbar^{\blacklozenge})$, for all $e \in \breve{\xi}^{\daleth}_1$.

Definition 18.2.4 The *union* of two q-ROFSSs $\left(\beth^{\daleth}_1, \breve{\xi}^{\daleth}_1\right)$ and $\left(\beth^{\daleth}_2, \breve{\xi}^{\daleth}_2\right)$ defined over $\widetilde{\mathscr{Y}}^{\nabla}$ is given as $\left(\beth^{\daleth}, \breve{\xi}^{\daleth}_1 \cup \breve{\xi}^{\daleth}_2\right) = \left(\beth^{\daleth}_1, \breve{\xi}^{\daleth}_1\right) \widetilde{\cup} \left(\beth^{\daleth}_2, \breve{\xi}^{\daleth}_2\right)$, and for all $\hbar^{\blacklozenge} \in \breve{\xi}^{\daleth}$,

$$
\beth^{\daleth}(\hbar^{\blacklozenge}) = \begin{cases}
\beth^{\daleth}_1(\hbar^{\blacklozenge}), & \text{if } \hbar^{\blacklozenge} \in \breve{\xi}^{\daleth}_1 \ \& \ \hbar^{\blacklozenge} \notin \breve{\xi}^{\daleth}_2 \\
\beth^{\daleth}_2(\hbar^{\blacklozenge}), & \text{if } \hbar^{\blacklozenge} \in \breve{\xi}^{\daleth}_2 \ \& \ \hbar^{\blacklozenge} \notin \breve{\xi}^{\daleth}_1 \\
\beth^{\daleth}_1(\hbar^{\blacklozenge}) \cup \beth^{\daleth}_2(\hbar^{\blacklozenge}), & \text{if } \hbar^{\blacklozenge} \in \breve{\xi}^{\daleth}_1 \cap \breve{\xi}^{\daleth}_2
\end{cases}
$$

where $\beth^{\daleth}_1(\hbar^{\blacklozenge}) \cup \beth^{\daleth}_2(\hbar^{\blacklozenge})$ is the union of two q-ROFSSs.

Definition 18.2.5 The *intersection* of two q-ROFSSs $\left(\beth^{\daleth}_1, \breve{\xi}^{\daleth}_1\right)$ and $\left(\beth^{\daleth}_2, \breve{\xi}^{\daleth}_2\right)$ is another q-ROFSS $\left(\beth^{\daleth}, \breve{\xi}^{\daleth}_1 \cap \breve{\xi}^{\daleth}_2\right) = \left(\beth^{\daleth}_1, \breve{\xi}^{\daleth}_1\right) \widetilde{\cap} \left(\beth^{\daleth}_2, \breve{\xi}^{\daleth}_2\right)$, where $\beth^{\daleth}(\hbar^{\blacklozenge}) = \beth^{\daleth}_1(\hbar^{\blacklozenge}) \cap \beth^{\daleth}_2(\hbar^{\blacklozenge})$ for all $\hbar^{\blacklozenge} \in \breve{\xi}^{\daleth}_1 \cap \breve{\xi}^{\daleth}_2$.

Definition 18.2.6 The *difference* of two q-ROFSSs $(\beth^{\daleth}_1, \breve{\xi}^{\daleth}_1)$ and $(\beth^{\daleth}_2, \breve{\xi}^{\daleth}_2)$ over $\widetilde{\mathscr{Y}}^{\nabla}$ is defined as

$$(\beth^{\daleth}_1, \breve{\xi}^{\daleth}_1) \widetilde{\setminus} (\beth^{\daleth}_2, \breve{\xi}^{\daleth}_2)$$
$$= \left\{ \left(e, \{\breve{\aleph}^{\eth}, \min\{\mu^{\gamma}_{\beth^{\daleth}_1(\hbar^{\blacklozenge})}(\breve{\aleph}^{\eth}), \nu^{\gamma}_{\beth^{\daleth}_2(\hbar^{\blacklozenge})}(\breve{\aleph}^{\eth})\}, \max\{\nu^{\gamma}_{\beth^{\daleth}_1(\hbar^{\blacklozenge})}(\breve{\aleph}^{\eth}), \mu^{\gamma}_{\beth^{\daleth}_2(\hbar^{\blacklozenge})}(\breve{\aleph}^{\eth})\}\}\right) : \right.$$
$$\left. \breve{\aleph}^{\eth} \in \widetilde{\mathscr{Y}}^{\nabla}, \hbar^{\blacklozenge} \in \daleth \right\}$$

Definition 18.2.7 The *complement* of a q-ROFSS $(\beth^{\daleth}, \breve{\xi}^{\daleth})$ is a mapping $\beth^{\daleth c} : \breve{\xi}^{\daleth} \to q - ROF^{\widetilde{\mathscr{Y}}^{\nabla}}$ given by $\beth^{\daleth c}(\hbar^{\blacklozenge}) = [\beth^{\daleth}(\hbar^{\blacklozenge})]^c$, for all $\hbar^{\blacklozenge} \in \breve{\xi}^{\daleth}$. It is represented as $(\beth^{\daleth}, \breve{\xi}^{\daleth})^c$ or $(\beth^{\daleth c}, \breve{\xi}^{\daleth})$. Thus, if

$$\beth^{\daleth}(\hbar^{\blacklozenge}) = \{(\breve{\aleph}^{\eth}, \mu^{\gamma}_{\beth^{\daleth}(\hbar^{\blacklozenge})}(\breve{\aleph}^{\eth}), \nu^{\gamma}_{\beth^{\daleth}(\hbar^{\blacklozenge})}(\breve{\aleph}^{\eth})) : \breve{\aleph}^{\eth} \in \widetilde{\mathscr{Y}}^{\nabla}\}$$

then

$$\beth^{\daleth c}(\hbar^{\blacklozenge}) = \{(\breve{\aleph}^{\eth}, \nu^{\gamma}_{\beth^{\daleth}(\hbar^{\blacklozenge})}(\breve{\aleph}^{\eth}), \mu^{\gamma}_{\beth^{\daleth}(\hbar^{\blacklozenge})}(\breve{\aleph}^{\eth})) : \breve{\aleph}^{\eth} \in \widetilde{\mathscr{Y}}^{\nabla}\}$$

for all $\hbar^{\blacklozenge} \in \breve{\xi}^{\daleth}$.

Definition 18.2.8 A q-ROFSS given on $\widetilde{\mathscr{Y}}^{\nabla}$ is known as *null q-ROFSS* if it is given as

$$\Phi = \left\{ \left(\hbar^{\blacklozenge}, \left\{\frac{\breve{\aleph}^{\eth}}{(0, 1)}\right\}\right) : \hbar^{\blacklozenge} \in \daleth, \breve{\aleph}^{\eth} \in \widetilde{\mathscr{Y}}^{\nabla} \right\}$$

Definition 18.2.9 A q-ROFSS given on $\widetilde{\mathscr{Y}}^{\nabla}$ is known as *absolute q-ROFSS* if it is given as

$$\breve{\widetilde{\mathscr{Y}}}^{\nabla} = \left\{\left(\check{\hbar}^{\blacklozenge}, \left\{\frac{\breve{\aleph}^{\eth}}{(1,0)}\right\}\right) : \check{\hbar}^{\blacklozenge} \in \check{\daleth}, \breve{\aleph}^{\eth} \in \widetilde{\mathscr{Y}}^{\nabla}\right\}$$

Definition 18.2.10 If

$$\beth\daleth^{(1)}_{\breve{\xi}^{\beth}_1} = \left\{\left(\check{\hbar}^{\blacklozenge}, \left\{\frac{\breve{\aleph}^{\eth}}{(\mu^{\gamma}_{\beth\daleth^{(1)}_{\breve{\xi}^{\beth}_1}}(\breve{\aleph}^{\eth}), \nu^{\gamma}_{\beth\daleth^{(1)}_{\breve{\xi}^{\beth}_1}}(\breve{\aleph}^{\eth}))}\right\}\right) : \check{\hbar}^{\blacklozenge} \in \breve{\xi}^{\beth}_1, \breve{\aleph}^{\eth} \in \widetilde{\mathscr{Y}}^{\nabla}\right\}$$

and

$$\beth\daleth^{(2)}_{\breve{\xi}^{\beth}_2} = \left\{\left(\check{\hbar}^{\blacklozenge}, \left\{\frac{\breve{\aleph}^{\eth}}{(\mu^{\gamma}_{\beth\daleth^{(2)}_{\breve{\xi}^{\beth}_2}}(\breve{\aleph}^{\eth}), \nu^{\gamma}_{\beth\daleth^{(2)}_{\breve{\xi}^{\beth}_2}}(\breve{\aleph}^{\eth}))}\right\}\right) : \check{\hbar}^{\blacklozenge} \in \breve{\xi}^{\beth}_2, \breve{\aleph}^{\eth} \in \widetilde{\mathscr{Y}}^{\nabla}\right\}$$

are two q-ROFSSs, then

$$\beth\daleth^{(1)}_{\breve{\xi}^{\beth}_1} \widetilde{\oplus} \beth\daleth^{(2)}_{\breve{\xi}^{\beth}_2}$$
$$= \left\{\left(\check{\hbar}^{\blacklozenge}, \left\{\frac{\breve{\aleph}^{\eth}}{\left(\sqrt[q]{(\mu^{\gamma}_{\beth\daleth^{(1)}_{\breve{\xi}^{\beth}_1}}(\breve{\aleph}^{\eth}))^q + (\mu^{\gamma}_{\beth\daleth^{(2)}_{\breve{\xi}^{\beth}_2}}(\breve{\aleph}^{\eth}))^q - (\mu^{\gamma}_{\beth\daleth^{(1)}_{\breve{\xi}^{\beth}_1}}(\breve{\aleph}^{\eth})\mu^{\gamma}_{\beth\daleth^{(2)}_{\breve{\xi}^{\beth}_2}}(\breve{\aleph}^{\eth}))^q}, \nu^{\gamma}_{\beth\daleth^{(1)}_{\breve{\xi}^{\beth}_1}}(\breve{\aleph}^{\eth})\nu^{\gamma}_{\beth\daleth^{(2)}_{\breve{\xi}^{\beth}_2}}(\breve{\aleph}^{\eth})\right)}\right\}\right)$$
$$: \check{\hbar}^{\blacklozenge} \in \check{\daleth}, \breve{\aleph}^{\eth} \in \widetilde{\mathscr{Y}}^{\nabla}\Bigg\}$$

and

$$\beth\daleth^{(1)}_{\breve{\xi}^{\beth}_1} \widetilde{\otimes} \beth\daleth^{(2)}_{\breve{\xi}^{\beth}_2}$$
$$= \left\{\left(\check{\hbar}^{\blacklozenge}, \left\{\frac{\breve{\aleph}^{\eth}}{\left(\mu^{\gamma}_{\beth\daleth^{(1)}_{\breve{\xi}^{\beth}_1}}(\breve{\aleph}^{\eth})\mu^{\gamma}_{\beth\daleth^{(2)}_{\breve{\xi}^{\beth}_2}}(\breve{\aleph}^{\eth}), \sqrt[q]{(\nu^{\gamma}_{\beth\daleth^{(1)}_{\breve{\xi}^{\beth}_1}}(\breve{\aleph}^{\eth}))^q + (\nu^{\gamma}_{\beth\daleth^{(2)}_{\breve{\xi}^{\beth}_2}}(\breve{\aleph}^{\eth}))^q - (\nu^{\gamma}_{\beth\daleth^{(1)}_{\breve{\xi}^{\beth}_1}}(\breve{\aleph}^{\eth})\nu^{\gamma}_{\beth\daleth^{(2)}_{\breve{\xi}^{\beth}_2}}(\breve{\aleph}^{\eth}))^q}\right)}\right\}\right)$$
$$: \check{\hbar}^{\blacklozenge} \in \check{\daleth}, \breve{\aleph}^{\eth} \in \widetilde{\mathscr{Y}}^{\nabla}\Bigg\}$$

Definition 18.2.11 For q-ROFSS the *necessity operator* is

$$(\beth\daleth, \breve{\xi}^{\beth}) = \left\{\left(\check{\hbar}^{\blacklozenge}, \left\{\frac{\breve{\aleph}^{\eth}}{(\mu^{\gamma}_{\beth\daleth}(\breve{\aleph}^{\eth}), \nu^{\gamma}_{\beth\daleth}(\breve{\aleph}^{\eth}))}\right\}\right) : \check{\hbar}^{\blacklozenge} \in \breve{\xi}^{\beth}, \breve{\aleph}^{\eth} \in \widetilde{\mathscr{Y}}^{\nabla}\right\}$$

is defined as

$$\widetilde{\square}(\beth\daleth, \breve{\xi}^{\beth}) = \left\{\left(\check{\hbar}^{\blacklozenge}, \left\{\frac{\breve{\aleph}^{\eth}}{(\mu^{\gamma}_{\beth\daleth}(\breve{\aleph}^{\eth}), \sqrt[q]{1 - \breve{\zeta}^{q}_{(\beth\daleth, \breve{\xi}^{\beth})}(\breve{\aleph}^{\eth})})}\right\}\right) : \check{\hbar}^{\blacklozenge} \in \breve{\xi}^{\beth}, \breve{\aleph}^{\eth} \in \widetilde{\mathscr{Y}}^{\nabla}\right\}$$

Definition 18.2.12 For q-ROFSS the *possibility operator* is

$$(\gimel\daleth, \breve{\xi}^{\gimel}) = \left\{\left(\breve{\hbar}^{\blacklozenge}, \left\{\frac{\breve{\aleph}^{\eth}}{(\mu^{\gamma}_{\gimel\daleth}(\breve{\aleph}^{\eth}), \nu^{\gamma}{}_{\gimel\daleth}(\breve{\aleph}^{\eth}))}\right\}\right) : \breve{\hbar}^{\blacklozenge} \in \breve{\xi}^{\gimel}, \breve{\aleph}^{\eth} \in \widetilde{\mathscr{Y}}^{\triangledown}\right\}$$

is defined as

$$\widetilde{\diamond}(\gimel\daleth, \breve{\xi}^{\gimel}) = \left\{\left(\breve{\hbar}^{\blacklozenge}, \left\{\frac{\breve{\aleph}^{\eth}}{(\sqrt[q]{1 - \varrho^{q}_{(\gimel\daleth, \breve{\xi}^{\gimel})}(\breve{\aleph}^{\eth})}, \nu^{\gamma}{}_{\gimel\daleth}(\breve{\aleph}^{\eth}))}\right\}\right) : \breve{\hbar}^{\blacklozenge} \in \breve{\xi}^{\gimel}, \breve{\aleph}^{\eth} \in \widetilde{\mathscr{Y}}^{\triangledown}\right\}$$

Remark The above-given operators transformed any given q-ROFSS into the FSS.

Example 18.2.13 Take $\widetilde{\mathscr{Y}}^{\triangledown} = \{\breve{\aleph}^{\eth}_1, \cdots, \breve{\aleph}^{\eth}_4\}$ and $\breve{\daleth} = \{\breve{\hbar}^{\blacklozenge}{}_1, \breve{\hbar}^{\blacklozenge}{}_2, \cdots, \breve{\hbar}^{\blacklozenge}{}_6\}$. Assume that $\breve{\xi}^{\gimel}_1 = \{\breve{\hbar}^{\blacklozenge}{}_2, \breve{\hbar}^{\blacklozenge}{}_4\}, \breve{\xi}^{\gimel}_2 = \{\breve{\hbar}^{\blacklozenge}{}_1, \breve{\hbar}^{\blacklozenge}{}_4, \breve{\hbar}^{\blacklozenge}{}_5\}$ and $\breve{\xi}^{\gimel}_3 = \{\breve{\hbar}^{\blacklozenge}{}_2, \breve{\hbar}^{\blacklozenge}{}_4, \breve{\hbar}^{\blacklozenge}{}_6\}$. Consider the q-ROFSSs by taking $q = 3$,

$$\gimel\daleth^{(1)}_{\breve{\xi}^{\gimel}_1} = \begin{pmatrix} (0.000, 1.000) & (0.270, 0.780) & (0.000, 1.000) & (0.390, 0.480) & (0.000, 1.000) & (0.000, 1.000) \\ (0.000, 1.000) & (0.110, 0.040) & (0.000, 1.000) & (0.730, 0.540) & (0.000, 1.000) & (0.000, 1.000) \\ (0.000, 1.000) & (0.560, 0.600) & (0.000, 1.000) & (0.590, 0.510) & (0.000, 1.000) & (0.000, 1.000) \\ (0.000, 1.000) & (0.620, 0.620) & (0.000, 1.000) & (0.370, 0.560) & (0.000, 1.000) & (0.000, 1.000) \end{pmatrix},$$

$$\gimel\daleth^{(2)}_{\breve{\xi}^{\gimel}_2} = \begin{pmatrix} (0.560, 0.270) & (0.000, 1.000) & (0.000, 1.000) & (0.450, 0.580) & (0.330, 0.780) & (0.000, 1.000) \\ (0.110, 0.850) & (0.000, 1.000) & (0.000, 1.000) & (0.090, 0.280) & (0.420, 0.510) & (0.000, 1.000) \\ (0.760, 0.490) & (0.000, 1.000) & (0.000, 1.000) & (0.620, 0.670) & (0.920, 0.210) & (0.000, 1.000) \\ (0.540, 0.710) & (0.000, 1.000) & (0.000, 1.000) & (0.540, 0.820) & (0.870, 0.480) & (0.000, 1.000) \end{pmatrix}$$

and

$$\gimel\daleth^{(3)}_{\breve{\xi}^{\gimel}_3} = \begin{pmatrix} (0.000, 1.000) & (0.310, 0.540) & (0.000, 1.000) & (0.390, 0.010) & (0.000, 1.000) & (0.220, 0.870) \\ (0.000, 1.000) & (0.250, 0.040) & (0.000, 1.000) & (0.760, 0.210) & (0.000, 1.000) & (0.530, 0.160) \\ (0.000, 1.000) & (0.490, 0.320) & (0.000, 1.000) & (0.620, 0.370) & (0.000, 1.000) & (0.420, 0.190) \\ (0.000, 1.000) & (0.630, 0.450) & (0.000, 1.000) & (0.460, 0.540) & (0.000, 1.000) & (0.880, 0.320) \end{pmatrix}.$$

It may be observed that $\gimel\daleth^{(1)}_{\breve{\xi}^{\gimel}_1} \widetilde{\subseteq} \gimel\daleth^{(3)}_{\breve{\xi}^{\gimel}_3}$, whereas neither $\gimel\daleth^{(1)}_{\breve{\xi}^{\gimel}_1} \widetilde{\subseteq} \gimel\daleth^{(2)}_{\breve{\xi}^{\gimel}_2}$ nor $\gimel\daleth^{(2)}_{\breve{\xi}^{\gimel}_2} \widetilde{\subseteq} \gimel\daleth^{(3)}_{\breve{\xi}^{\gimel}_3}$. Moreover,

$$\gimel\daleth\beth_{\breve{\xi}_1}^{(1)} \tilde{\cup} \gimel\daleth\beth_{\breve{\xi}_2}^{(2)} = \begin{pmatrix} (0.560, 0.270) & (0.270, 0.780) & (0.000, 1.000) & (0.450, 0.480) & (0.330, 0.780) & (0.000, 1.000) \\ (0.110, 0.850) & (0.110, 0.040) & (0.000, 1.000) & (0.730, 0.280) & (0.420, 0.510) & (0.000, 1.000) \\ (0.760, 0.490) & (0.560, 0.600) & (0.000, 1.000) & (0.620, 0.510) & (0.920, 0.210) & (0.000, 1.000) \\ (0.540, 0.710) & (0.620, 0.620) & (0.000, 1.000) & (0.540, 0.560) & (0.870, 0.480) & (0.000, 1.000) \end{pmatrix},$$

$$\gimel\daleth\beth_{\breve{\xi}_1}^{(1)} \tilde{\cap} \gimel\daleth\beth_{\breve{\xi}_2}^{(2)} = \begin{pmatrix} (0.000, 1.000) & (0.000, 1.000) & (0.000, 1.000) & (0.390, 0.580) & (0.000, 1.000) & (0.000, 1.000) \\ (0.000, 1.000) & (0.000, 1.000) & (0.000, 1.000) & (0.090, 0.540) & (0.000, 1.000) & (0.000, 1.000) \\ (0.000, 1.000) & (0.000, 1.000) & (0.000, 1.000) & (0.590, 0.670) & (0.000, 1.000) & (0.000, 1.000) \\ (0.000, 1.000) & (0.000, 1.000) & (0.000, 1.000) & (0.370, 0.820) & (0.000, 1.000) & (0.000, 1.000) \end{pmatrix},$$

$$(\gimel\daleth\beth_{\breve{\xi}_1}^{(1)})^c = \begin{pmatrix} (1.000, 0.000) & (0.780, 0.270) & (1.000, 0.000) & (0.480, 0.390) & (1.000, 0.000) & (1.000, 0.000) \\ (1.000, 0.000) & (0.040, 0.110) & (1.000, 0.000) & (0.540, 0.730) & (1.000, 0.000) & (1.000, 0.000) \\ (1.000, 0.000) & (0.600, 0.560) & (1.000, 0.000) & (0.510, 0.590) & (1.000, 0.000) & (1.000, 0.000) \\ (1.000, 0.000) & (0.620, 0.620) & (1.000, 0.000) & (0.560, 0.370) & (1.000, 0.000) & (1.000, 0.000) \end{pmatrix},$$

$$\gimel\daleth\beth_{\breve{\xi}_1}^{(1)} \tilde{\setminus} \gimel\daleth\beth_{\breve{\xi}_2}^{(2)} = \begin{pmatrix} (0.000, 1.000) & (0.000, 1.000) & (0.000, 1.000) & (0.390, 0.480) & (0.000, 1.000) & (0.000, 1.000) \\ (0.000, 1.000) & (0.000, 1.000) & (0.000, 1.000) & (0.280, 0.540) & (0.000, 1.000) & (0.000, 1.000) \\ (0.000, 1.000) & (0.000, 1.000) & (0.000, 1.000) & (0.590, 0.620) & (0.000, 1.000) & (0.000, 1.000) \\ (0.000, 1.000) & (0.000, 1.000) & (0.000, 1.000) & (0.370, 0.560) & (0.000, 1.000) & (0.000, 1.000) \end{pmatrix},$$

$$\tilde{\Box}\gimel\daleth\beth_{\breve{\xi}_1}^{(1)} = \begin{pmatrix} (0.000, 1.000) & (0.270, 0.960) & (0.000, 1.000) & (0.390, 0.920) & (0.000, 1.000) & (0.000, 1.000) \\ (0.000, 1.000) & (0.110, 0.990) & (0.000, 1.000) & (0.730, 0.680) & (0.000, 1.000) & (0.000, 1.000) \\ (0.000, 1.000) & (0.560, 0.830) & (0.000, 1.000) & (0.590, 0.810) & (0.000, 1.000) & (0.000, 1.000) \\ (0.000, 1.000) & (0.620, 0.780) & (0.000, 1.000) & (0.370, 0.930) & (0.000, 1.000) & (0.000, 1.000) \end{pmatrix},$$

$$\tilde{\Diamond}\gimel\daleth\beth_{\breve{\xi}_1}^{(1)} = \begin{pmatrix} (0.000, 1.000) & (0.620, 0.780) & (0.000, 1.000) & (0.880, 0.480) & (0.000, 1.000) & (0.000, 1.000) \\ (0.000, 1.000) & (0.990, 0.040) & (0.000, 1.000) & (0.840, 0.540) & (0.000, 1.000) & (0.000, 1.000) \\ (0.000, 1.000) & (0.800, 0.600) & (0.000, 1.000) & (0.860, 0.510) & (0.000, 1.000) & (0.000, 1.000) \\ (0.000, 1.000) & (0.780, 0.620) & (0.000, 1.000) & (0.830, 0.560) & (0.000, 1.000) & (0.000, 1.000) \end{pmatrix},$$

$$\beth^{(1)}_{\breve{\xi}_1} \tilde{\oplus} \beth^{(2)}_{\breve{\xi}_2}$$
$$= \begin{pmatrix} (0.560, 0.270) & (0.270, 0.780) & (0.000, 1.000) & (0.570, 0.280) & (0.330, 0.780) & (0.000, 1.000) \\ (0.110, 0.850) & (0.110, 0.040) & (0.000, 1.000) & (0.730, 0.150) & (0.420, 0.510) & (0.000, 1.000) \\ (0.760, 0.490) & (0.560, 0.600) & (0.000, 1.000) & (0.770, 0.340) & (0.920, 0.210) & (0.000, 1.000) \\ (0.540, 0.710) & (0.620, 0.620) & (0.000, 1.000) & (0.620, 0.460) & (0.870, 0.480) & (0.000, 1.000) \end{pmatrix}$$

and

$$\beth^{(1)}_{\breve{\xi}_1} \tilde{\otimes} \beth^{(2)}_{\breve{\xi}_2}$$
$$= \begin{pmatrix} (0.000, 1.000) & (0.000, 1.000) & (0.000, 1.000) & (0.180, 0.700) & (0.000, 1.000) & (0.000, 1.000) \\ (0.000, 1.000) & (0.000, 1.000) & (0.000, 1.000) & (0.060, 0.590) & (0.000, 1.000) & (0.000, 1.000) \\ (0.000, 1.000) & (0.000, 1.000) & (0.000, 1.000) & (0.360, 0.770) & (0.000, 1.000) & (0.000, 1.000) \\ (0.000, 1.000) & (0.000, 1.000) & (0.000, 1.000) & (0.200, 0.880) & (0.000, 1.000) & (0.000, 1.000) \end{pmatrix}.$$

Proposition 18.2.14 *Every q-ROFSS* $(\beth, \breve{\xi}^{\beth})$ *may be sandwiched between* Φ *and* $\breve{\tilde{\mathscr{Y}}}^{\nabla}$, *i.e.* $\Phi \tilde{\subseteq} \beth_{\breve{\xi}^{\beth}} \tilde{\subseteq} \breve{\tilde{\mathscr{Y}}}^{\nabla}$.

Proposition 18.2.15 *If* $\beth^{(1)}_{\breve{\xi}_1}$, $\beth^{(2)}_{\breve{\xi}_2}$ *and* $\beth^{(3)}_{\breve{\xi}_3}$ *are three q-ROFSSs over* $\tilde{\mathscr{Y}}^{\nabla}$, *then*

(i) $\beth^{(1)}_{\breve{\xi}_1} \tilde{\cap} \beth^{(1)}_{\breve{\xi}_1} = \beth^{(1)}_{\breve{\xi}_1}$.
(ii) $\beth^{(1)}_{\breve{\xi}_1} \tilde{\cup} \beth^{(1)}_{\breve{\xi}_1} = \beth^{(1)}_{\breve{\xi}_1}$.
(iii) $\beth^{(1)}_{\breve{\xi}_1} \tilde{\cap} \beth^{(2)}_{\breve{\xi}_2} = \beth^{(2)}_{\breve{\xi}_2} \tilde{\cap} \beth^{(1)}_{\breve{\xi}_1}$.
(iv) $\beth^{(1)}_{\breve{\xi}_1} \tilde{\cup} \beth^{(2)}_{\breve{\xi}_2} = \beth^{(2)}_{\breve{\xi}_2} \tilde{\cup} \beth^{(1)}_{\breve{\xi}_1}$.
(v) $\beth^{(1)}_{\breve{\xi}_1} \tilde{\cap} (\beth^{(2)}_{\breve{\xi}_2} \tilde{\cap} \beth^{(3)}_{\breve{\xi}_3}) = (\beth^{(1)}_{\breve{\xi}_1} \tilde{\cap} \beth^{(2)}_{\breve{\xi}_2}) \tilde{\cap} \beth^{(3)}_{\breve{\xi}_3}$.
(vi) $\beth^{(1)}_{\breve{\xi}_1} \tilde{\cup} (\beth^{(2)}_{\breve{\xi}_2} \tilde{\cup} \beth^{(3)}_{\breve{\xi}_3}) = (\beth^{(1)}_{\breve{\xi}_1} \tilde{\cup} \beth^{(2)}_{\breve{\xi}_2}) \tilde{\cup} \beth^{(3)}_{\breve{\xi}_3}$.
(vii) $\beth^{(1)}_{\breve{\xi}_1} \tilde{\cup} (\beth^{(2)}_{\breve{\xi}_2} \tilde{\cap} \beth^{(3)}_{\breve{\xi}_3}) = (\beth^{(1)}_{\breve{\xi}_1} \tilde{\cup} \beth^{(2)}_{\breve{\xi}_2}) \tilde{\cap} (\beth^{(1)}_{\breve{\xi}_1} \tilde{\cup} \beth^{(3)}_{\breve{\xi}_3})$.
(viii) $\beth^{(1)}_{\breve{\xi}_1} \tilde{\cap} (\beth^{(2)}_{\breve{\xi}_2} \tilde{\cup} \beth^{(3)}_{\breve{\xi}_3}) = (\beth^{(1)}_{\breve{\xi}_1} \tilde{\cap} \beth^{(2)}_{\breve{\xi}_2}) \tilde{\cup} (\beth^{(1)}_{\breve{\xi}_1} \tilde{\cap} \beth^{(3)}_{\breve{\xi}_3})$.

Proposition 18.2.16 *If* $\beth^{(1)}_{\breve{\xi}_1}$ *and* $\beth^{(2)}_{\breve{\xi}_2}$ *are two q-ROFSSs over* $\tilde{\mathscr{Y}}^{\nabla}$, *then*

(i) $\beth^{(1)}_{\breve{\xi}_1} \tilde{\cap} \beth^{(2)}_{\breve{\xi}_2} \tilde{\subseteq} \beth^{(1)}_{\breve{\xi}_1} \tilde{\subseteq} \beth^{(1)}_{\breve{\xi}_1} \tilde{\cup} \beth^{(2)}_{\breve{\xi}_2}$
(ii) $\beth^{(1)}_{\breve{\xi}_1} \tilde{\cap} \beth^{(2)}_{\breve{\xi}_2} \tilde{\subseteq} \beth^{(2)}_{\breve{\xi}_2} \tilde{\subseteq} \beth^{(1)}_{\breve{\xi}_1} \tilde{\cup} \beth^{(2)}_{\breve{\xi}_2}$.

The above propositions are easy consequences of definition.

Remark Consider the q-ROFSSs $\beth^{(1)}_{\breve{\xi}_1}$ and $\beth^{(2)}_{\breve{\xi}_2}$ given in Example 18.2.13. We have

$$(\gimel\daleth^{(1)}_{\breve{\xi}_1^{\gimel}} \tilde{\cup} \gimel\daleth^{(2)}_{\breve{\xi}_2^{\gimel}})^c$$
$$= \begin{pmatrix} (0.270, 0.560) & (0.780, 0.270) & (1.000, 0.000) & (0.480, 0.450) & (0.780, 0.330) & (1.000, 0.000) \\ (0.850, 0.110) & (0.040, 0.110) & (1.000, 0.000) & (0.280, 0.730) & (0.510, 0.420) & (1.000, 0.000) \\ (0.490, 0.760) & (0.600, 0.560) & (1.000, 0.000) & (0.510, 0.620) & (0.210, 0.920) & (1.000, 0.000) \\ (0.710, 0.540) & (0.620, 0.620) & (1.000, 0.000) & (0.560, 0.540) & (0.480, 0.870) & (1.000, 0.000) \end{pmatrix} \quad (18.1)$$

$$(\gimel\daleth^{(1)}_{\breve{\xi}_1^{\gimel}})^c$$
$$= \begin{pmatrix} (1.000, 0.000) & (0.780, 0.270) & (1.000, 0.000) & (0.480, 0.390) & (1.000, 0.000) & (1.000, 0.000) \\ (1.000, 0.000) & (0.040, 0.110) & (1.000, 0.000) & (0.540, 0.730) & (1.000, 0.000) & (1.000, 0.000) \\ (1.000, 0.000) & (0.600, 0.560) & (1.000, 0.000) & (0.510, 0.590) & (1.000, 0.000) & (1.000, 0.000) \\ (1.000, 0.000) & (0.620, 0.620) & (1.000, 0.000) & (0.560, 0.370) & (1.000, 0.000) & (1.000, 0.000) \end{pmatrix},$$

$$(\gimel\daleth^{(2)}_{\breve{\xi}_2^{\gimel}})^c$$
$$= \begin{pmatrix} (0.270, 0.560) & (1.000, 0.000) & (1.000, 0.000) & (0.580, 0.450) & (0.780, 0.330) & (1.000, 0.000) \\ (0.850, 0.110) & (1.000, 0.000) & (1.000, 0.000) & (0.280, 0.090) & (0.510, 0.420) & (1.000, 0.000) \\ (0.490, 0.760) & (1.000, 0.000) & (1.000, 0.000) & (0.670, 0.620) & (0.210, 0.920) & (1.000, 0.000) \\ (0.710, 0.540) & (1.000, 0.000) & (1.000, 0.000) & (0.820, 0.540) & (0.480, 0.870) & (1.000, 0.000) \end{pmatrix}$$

Since $\breve{\xi}_1^{\gimel} \cap \breve{\xi}_2^{\gimel} = \{\hbar^{\check{\blacklozenge}}{}_4\}$, so

$$(\gimel\daleth^{(1)}_{\breve{\xi}_1^{\gimel}})^c \tilde{\cap} (\gimel\daleth^{(2)}_{\breve{\xi}_2^{\gimel}})^c$$
$$= \begin{pmatrix} (0.000, 1.000) & (0.000, 1.000) & (0.000, 1.000) & (0.480, 0.450) & (0.000, 1.000) & (0.000, 1.000) \\ (0.000, 1.000) & (0.000, 1.000) & (0.000, 1.000) & (0.280, 0.730) & (0.000, 1.000) & (0.000, 1.000) \\ (0.000, 1.000) & (0.000, 1.000) & (0.000, 1.000) & (0.510, 0.620) & (0.000, 1.000) & (0.000, 1.000) \\ (0.000, 1.000) & (0.000, 1.000) & (0.000, 1.000) & (0.560, 0.540) & (0.000, 1.000) & (0.000, 1.000) \end{pmatrix} \quad (18.2)$$

From (18.1) and (18.2), we show that "De Morgan's law" does not hold in q-ROFSSs.

Theorem 18.2.17 *If* $(\gimel\daleth_1, \breve{\xi}_1^{\gimel})$ *and* $(\gimel\daleth_2, \breve{\xi}_2^{\gimel})$ *are two q-ROFSSs over* $\widetilde{\mathscr{Y}}^{\triangledown}$, *then*

(a) $\left((\gimel\daleth_1, \breve{\xi}_1^{\gimel}) \tilde{\cup} (\gimel\daleth_2, \breve{\xi}_2^{\gimel})\right)^c \neq (\gimel\daleth_1, \breve{\xi}_1^{\gimel})^c \tilde{\cap} (\gimel\daleth_2, \breve{\xi}_2^{\gimel})^c$, *and*

(b) $\left((\gimel\daleth_1, \breve{\xi}_1^{\gimel}) \tilde{\cap} (\gimel\daleth_2, \breve{\xi}_2^{\gimel})\right)^c \neq (\gimel\daleth_1, \breve{\xi}_1^{\gimel})^c \tilde{\cup} (\gimel\daleth_2, \breve{\xi}_2^{\gimel})^c$.

Remark Consider again the q-ROFSS $\gimel\daleth^{(1)}_{\breve{\xi}_1^{\gimel}}$ given in Example 18.2.13. We have

$$\gimel\daleth^{(1)}_{\breve{\xi}_1^{\gimel}}$$
$$= \begin{pmatrix} (0.000, 1.000) & (0.270, 0.780) & (0.000, 1.000) & (0.390, 0.480) & (0.000, 1.000) & (0.000, 1.000) \\ (0.000, 1.000) & (0.110, 0.040) & (0.000, 1.000) & (0.730, 0.540) & (0.000, 1.000) & (0.000, 1.000) \\ (0.000, 1.000) & (0.560, 0.600) & (0.000, 1.000) & (0.590, 0.510) & (0.000, 1.000) & (0.000, 1.000) \\ (0.000, 1.000) & (0.620, 0.620) & (0.000, 1.000) & (0.370, 0.560) & (0.000, 1.000) & (0.000, 1.000) \end{pmatrix}$$

$$\therefore (\beth\urcorner^{(1)}_{\breve{\xi}_1 \beth})^c$$

$$= \begin{pmatrix} (1.000, 0.000) & (0.780, 0.270) & (1.000, 0.000) & (0.480, 0.390) & (1.000, 0.000) & (1.000, 0.000) \\ (1.000, 0.000) & (0.040, 0.110) & (1.000, 0.000) & (0.540, 0.730) & (1.000, 0.000) & (1.000, 0.000) \\ (1.000, 0.000) & (0.600, 0.560) & (1.000, 0.000) & (0.510, 0.590) & (1.000, 0.000) & (1.000, 0.000) \\ (1.000, 0.000) & (0.620, 0.620) & (1.000, 0.000) & (0.560, 0.370) & (1.000, 0.000) & (1.000, 0.000) \end{pmatrix}.$$

Now,

$$\beth\urcorner^{(1)}_{\breve{\xi}_1 \beth} \widetilde{\cup} (\beth\urcorner^{(1)}_{\breve{\xi}_1 \beth})^c$$

$$= \begin{pmatrix} (1.000, 0.000) & (0.780, 0.270) & (1.000, 0.000) & (0.480, 0.390) & (1.000, 0.000) & (1.000, 0.000) \\ (1.000, 0.000) & (0.110, 0.040) & (1.000, 0.000) & (0.730, 0.540) & (1.000, 0.000) & (1.000, 0.000) \\ (1.000, 0.000) & (0.600, 0.560) & (1.000, 0.000) & (0.590, 0.510) & (1.000, 0.000) & (1.000, 0.000) \\ (1.000, 0.000) & (0.620, 0.620) & (1.000, 0.000) & (0.560, 0.370) & (1.000, 0.000) & (1.000, 0.000) \end{pmatrix}$$

$$\neq \widetilde{\breve{\mathscr{Y}}}^{\nabla}$$

and

$$\beth\urcorner^{(1)}_{\breve{\xi}_1 \beth} \widetilde{\cap} (\beth\urcorner^{(1)}_{\breve{\xi}_1 \beth})^c$$

$$= \begin{pmatrix} (0.000, 1.000) & (0.270, 0.780) & (0.000, 1.000) & (0.390, 0.480) & (0.000, 1.000) & (0.000, 1.000) \\ (0.000, 1.000) & (0.040, 0.110) & (0.000, 1.000) & (0.540, 0.730) & (0.000, 1.000) & (0.000, 1.000) \\ (0.000, 1.000) & (0.560, 0.600) & (0.000, 1.000) & (0.510, 0.590) & (0.000, 1.000) & (0.000, 1.000) \\ (0.000, 1.000) & (0.620, 0.620) & (0.000, 1.000) & (0.370, 0.560) & (0.000, 1.000) & (0.000, 1.000) \end{pmatrix}$$

$$\neq \Phi$$

Such findings relate to the underlying statement.

Theorem 18.2.18 *If* $(\beth\urcorner, \breve{\xi}^{\beth})$ *is any PF SS over* $\widetilde{\breve{\mathscr{Y}}}^{\nabla}$, *then*

(1) $(\beth\urcorner, \breve{\xi}^{\beth}) \widetilde{\cup} \beth\urcorner^{c}_{\breve{\xi}\beth} \neq \widetilde{\breve{\mathscr{Y}}}^{\nabla}$, *and*
(2) $(\beth\urcorner, \breve{\xi}^{\beth}) \widetilde{\cap} \beth\urcorner^{c}_{\breve{\xi}\beth} \neq \Phi$.

Definition 18.2.19 We know that

$$\beth\urcorner^{(1)}_{\breve{\xi}_1 \beth} \widetilde{\otimes} \beth\urcorner^{(2)}_{\breve{\xi}_2 \beth}$$

$$= \left\{ \left(\hbar^{\blacklozenge}, \left\{ \frac{\breve{\aleph}^{\eth}}{(\mu^{\gamma}_{\beth\urcorner^{(1)}_{\breve{\xi}_1 \beth}}(\breve{\aleph}^{\eth}) \mu^{\gamma}_{\beth\urcorner^{(2)}_{\breve{\xi}_2 \beth}}(\breve{\aleph}^{\eth}), \sqrt[q]{(\nu^{\gamma}_{\beth\urcorner^{(1)}_{\breve{\xi}_1 \beth}}(\breve{\aleph}^{\eth}))^q + (\nu^{\gamma}_{\beth\urcorner^{(2)}_{\breve{\xi}_2 \beth}}(\breve{\aleph}^{\eth}))^q - (\nu^{\gamma}_{\beth\urcorner^{(1)}_{\breve{\xi}_1 \beth}}(\breve{\aleph}^{\eth}) \nu^{\gamma}_{\beth\urcorner^{(2)}_{\breve{\xi}_2 \beth}}(\breve{\aleph}^{\eth}))^q})} \right\} \right) \right.$$

$$\left. : \hbar^{\blacklozenge} \in \breve{\daleth}, \breve{\aleph}^{\eth} \in \widetilde{\breve{\mathscr{Y}}}^{\nabla} \right\}$$

If we substitute $\beth\urcorner^{(1)}_{\breve{\xi}_1 \beth} = \beth\urcorner^{(2)}_{\breve{\xi}_2 \beth} = (\beth\urcorner, \breve{\xi}^{\beth})$, then

$$(\beth\urcorner, \breve{\xi}^{\beth}) \widetilde{\otimes} (\beth\urcorner, \breve{\xi}^{\beth})$$

$$= \left\{ \left(\hbar^{\breve{\blacklozenge}}, \left\{ \frac{\breve{\aleph}^{\eth}}{(\breve{\zeta}^{2}_{(\beth\urcorner, \breve{\xi}^{\beth})}(\breve{\aleph}^{\eth}), \sqrt[q]{2\varrho^{q}_{(\beth\urcorner, \breve{\xi}^{\beth})}(\breve{\aleph}^{\eth}) - \varrho^{2q}_{(\beth\urcorner, \breve{\xi}^{\beth})}(\breve{\aleph}^{\eth})})} \right\} \right) : \hbar^{\blacklozenge} \in \breve{\daleth}, \breve{\aleph}^{\eth} \in \widetilde{\breve{\mathscr{Y}}}^{\nabla} \right\}$$

That is,

$$((\beth^{\urcorner}, \xi^{\beth}))^2 = \left\{ \left(\hbar^{\blacklozenge}, \left\{ \frac{\breve{\aleph}^{\eth}}{\left(\breve{\zeta}^{2}_{(\beth^{\urcorner},\xi^{\beth})}(\breve{\aleph}^{\eth}), \sqrt[q]{1-(1-\varrho^{q}_{(\beth^{\urcorner},\xi^{\beth})})^{2}}\right)} \right\} \right) : \hbar^{\blacklozenge} \in \breve{\urcorner}, \breve{\aleph}^{\eth} \in \tilde{\mathscr{Y}}^{\nabla} \right\}$$

In general, if k is any non-negative real number, then

$$((\beth^{\urcorner}, \xi^{\beth}))^k = \left\{ \left(\hbar^{\blacklozenge}, \left\{ \frac{\breve{\aleph}^{\eth}}{\left(\breve{\zeta}^{k}_{(\beth^{\urcorner},\xi^{\beth})}(\breve{\aleph}^{\eth}), \sqrt[q]{1-(1-\varrho^{q}_{(\beth^{\urcorner},\xi^{\beth})})^{k}}\right)} \right\} \right) : \hbar^{\blacklozenge} \in \breve{\urcorner}, \breve{\aleph}^{\eth} \in \tilde{\mathscr{Y}}^{\nabla} \right\}$$

In particular, for $k = \frac{1}{2}$, we have

$$((\beth^{\urcorner}, \breve{\xi}^{\beth}))^{\frac{1}{2}} = \left\{ \left(\hbar^{\blacklozenge}, \left\{ \frac{\breve{\aleph}^{\eth}}{\left(\sqrt{\mu^{\gamma}_{\beth^{\urcorner}}(\breve{\aleph}^{\eth})}, \sqrt[q]{1-\sqrt{1-\varrho^{q}_{(\beth^{\urcorner},\xi^{\beth})}}}\right)} \right\} \right) : \hbar^{\blacklozenge} \in \breve{\urcorner}, \breve{\aleph}^{\eth} \in \tilde{\mathscr{Y}}^{\nabla} \right\}$$

$((\beth^{\urcorner}, \xi^{\beth}))^2$ is known as "*concentration*" of $(\beth^{\urcorner}, \breve{\xi}^{\beth})$ and express as $con((\beth^{\urcorner}, \breve{\xi}^{\beth}))$ whereas $((\beth^{\urcorner}, \breve{\xi}^{\beth}))^{\frac{1}{2}}$ is known as "*dilation*" of $(\beth^{\urcorner}, \xi^{\beth})$ and express as $dil((\beth^{\urcorner}, \breve{\xi}^{\beth}))$.

Example 18.2.20 For q-ROFSS $\beth^{\urcorner(1)}_{\breve{\xi}^{\beth}_1} = (\beth^{\urcorner}, \breve{\xi}^{\beth})$ given in Example 18.2.13, concentration and dilation are

$$con((\beth^{\urcorner}, \xi^{\beth})) = \begin{pmatrix} (0.000, 1.000) & (0.0729, 0.8978) & (0.000, 1.000) & (0.1521, 0.5934) & (0.000, 1.000) & (0.000, 1.000) \\ (0.000, 1.000) & (0.0121, 0.0503) & (0.000, 1.000) & (0.5329, 0.6620) & (0.000, 1.000) & (0.000, 1.000) \\ (0.000, 1.000) & (0.3136, 0.7277) & (0.000, 1.000) & (0.3481, 0.6280) & (0.000, 1.000) & (0.000, 1.000) \\ (0.000, 1.000) & (0.3844, 0.7488) & (0.000, 1.000) & (0.1369, 0.6843) & (0.000, 1.000) & (0.000, 1.000) \end{pmatrix}$$

and

$$dil((\beth^{\urcorner}, \xi^{\beth})) = \begin{pmatrix} (0.000, 1.000) & (0.5196, 0.6503) & (0.000, 1.000) & (0.6245, 0.3847) & (0.000, 1.000) & (0.000, 1.000) \\ (0.000, 1.000) & (0.3317, 0.0317) & (0.000, 1.000) & (0.8544, 0.4346) & (0.000, 1.000) & (0.000, 1.000) \\ (0.000, 1.000) & (0.7483, 0.4856) & (0.000, 1.000) & (0.7681, 0.4095) & (0.000, 1.000) & (0.000, 1.000) \\ (0.000, 1.000) & (0.6200, 0.6200) & (0.000, 1.000) & (0.6083, 0.4515) & (0.000, 1.000) & (0.000, 1.000) \end{pmatrix}$$

respectively.

We see that as the q-ROFSS concentration is increased, the MSD value decreases but the NMSD value increases. In the situation of q-ROFSS dilation, the value of the MSD surpasses and the value of the NMSD decreases when compared to the corresponding actual numbers.

Taking this into account, we can associate phonetic phrases such "*very*", "*moderate*", "*highly*", and "*not*" with the q-ROFSS $(\beth^{\urcorner}, \xi^{\beth})$ by assigning various positive real numbers to £. For Example,

$$£ = \frac{1}{2} \Rightarrow \text{“}very\text{''}$$
$$£ = \frac{3}{4} \Rightarrow \text{“}moderate\text{''}$$
$$£ = \frac{1}{5} \Rightarrow \text{“}highly\text{''}$$
$$£ = 4 \Rightarrow \text{“}not\text{''}$$

Examine the given illustration for successfully understanding these concepts.

Example 18.2.21 Choose $\widetilde{\mathscr{Y}}^{\nabla} = \{\text{Ali, Smad, Akbar, Palaj}\}$ are the students and $\breve{\urcorner} = \{\breve{\hbar}^{\blacklozenge}{}_1, \cdots, \breve{\hbar}^{\blacklozenge}{}_5\}$ are the assemblage of parameters, where

$$\breve{\hbar}^{\blacklozenge}{}_1 = \text{Regular}$$
$$\breve{\hbar}^{\blacklozenge}{}_2 = \text{Hard working}$$
$$\breve{\hbar}^{\blacklozenge}{}_3 = \text{Fashionable}$$
$$\breve{\hbar}^{\blacklozenge}{}_4 = \text{Obedient}$$
$$\breve{\hbar}^{\blacklozenge}{}_5 = \text{Good in indoor games}$$

Consider the q-ROFSS representing members of $\widetilde{\mathscr{Y}}^{\nabla}$ and the value of trait $\breve{\hbar}^{\blacklozenge}{}_j$ in the form of q-ROFNs by taking $q = 3$ is

$$(\beth^{\urcorner}, \breve{\xi}^{\beth})$$
$$= \begin{pmatrix} (0.830, 0.280) & (0.540, 0.210) & (0.370, 0.640) & (0.590, 0.160) & (0.860, 0.110) \\ (0.310, 0.260) & (0.560, 0.570) & (0.430, 0.320) & (0.740, 0.250) & (0.130, 0.050) \\ (0.520, 0.270) & (0.640, 0.120) & (0.450, 0.570) & (0.610, 0.350) & (0.290, 0.510) \\ (0.480, 0.590) & (0.350, 0.210) & (0.570, 0.130) & (0.210, 0.210) & (0.880, 0.410) \end{pmatrix}$$

The element at (1, 1) spot, i.e. (0.830, 0.280), demonstrates that Ali has an 83
Now,

$$very((\beth^{\urcorner}, \breve{\xi}^{\beth}))$$
$$= \begin{pmatrix} (0.910, 0.200) & (0.730, 0.150) & (0.610, 0.480) & (0.770, 0.110) & (0.930, 0.080) \\ (0.560, 0.190) & (0.750, 0.420) & (0.660, 0.230) & (0.860, 0.180) & (0.360, 0.040) \\ (0.720, 0.190) & (0.800, 0.090) & (0.670, 0.420) & (0.780, 0.250) & (0.540, 0.370) \\ (0.690, 0.440) & (0.590, 0.150) & (0.750, 0.090) & (0.460, 0.150) & (0.940, 0.300) \end{pmatrix},$$

$$moderate((\beth^{\urcorner}, \breve{\xi}^{\beth}))$$
$$= \begin{pmatrix} (0.870, 0.240) & (0.630, 0.180) & (0.470, 0.570) & (0.670, 0.140) & (0.890, 0.100) \\ (0.420, 0.230) & (0.650, 0.510) & (0.530, 0.280) & (0.800, 0.220) & (0.220, 0.040) \\ (0.610, 0.230) & (0.720, 0.100) & (0.550, 0.510) & (0.690, 0.310) & (0.400, 0.450) \\ (0.580, 0.520) & (0.460, 0.180) & (0.660, 0.110) & (0.310, 0.180) & (0.910, 0.360) \end{pmatrix},$$

$$highly((\gimel^{\daleth}, \breve{\xi}^{\gimel}))$$
$$= \begin{pmatrix} (0.960, 0.130) & (0.880, 0.090) & (0.820, 0.320) & (0.900, 0.070) & (0.970, 0.050) \\ (0.790, 0.120) & (0.890, 0.270) & (0.840, 0.150) & (0.940, 0.110) & (0.660, 0.020) \\ (0.880, 0.120) & (0.910, 0.050) & (0.850, 0.270) & (0.910, 0.160) & (0.780, 0.240) \\ (0.860, 0.290) & (0.810, 0.090) & (0.890, 0.060) & (0.730, 0.090) & (0.970, 0.190) \end{pmatrix}$$

and

$$not((\gimel^{\daleth}, \breve{\xi}^{\gimel}))$$
$$= \begin{pmatrix} (0.470, 0.530) & (0.090, 0.410) & (0.020, 0.940) & (0.120, 0.310) & (0.550, 0.220) \\ (0.010, 0.490) & (0.100, 0.890) & (0.030, 0.590) & (0.300, 0.480) & (0.000, 0.100) \\ (0.070, 0.510) & (0.170, 0.240) & (0.040, 0.890) & (0.140, 0.640) & (0.010, 0.840) \\ (0.050, 0.910) & (0.020, 0.410) & (0.110, 0.260) & (0.000, 0.410) & (0.600, 0.720) \end{pmatrix}.$$

Definition 18.2.22 A q-ROFSS $(\gimel^{\daleth}, \breve{\xi}^{\gimel})$ is termed as a "*q-rung orthopair fuzzy soft point*" (q-ROFS point), denoted as $\aleph^{\hbar}{}_{\gimel^{\daleth}}$, if for the element $\aleph^{\hbar} \in \breve{\xi}^{\gimel}$ we have

(i) $\gimel^{\daleth}(\aleph^{\hbar}) \neq \Phi$, and
(ii) $\gimel^{\daleth}(\aleph^{\hbar'}) = \breve{\tilde{\mathscr{Y}}}^{\nabla}$, for all $\aleph^{\hbar'} \in \breve{\xi}^{\gimel} - \{\aleph^{\hbar}\}$.

Definition 18.2.23 A q-ROFS point $\aleph^{\hbar}{}_{\gimel^{\daleth}} \widetilde{\in} (\gimel^{\daleth}, \breve{\xi}^{\gimel})$ is said to be in q-ROFSS $(\gimel^{\daleth}{}_1, \breve{\xi}^{\gimel}_1)$, i.e. $\aleph^{\hbar}{}_{\gimel^{\daleth}} \widetilde{\in} (\gimel^{\daleth}{}_1, \breve{\xi}^{\gimel}_1)$ if $\aleph^{\hbar} \in \breve{\xi}^{\gimel}_1 \Rightarrow \gimel^{\daleth}(\aleph^{\hbar}) \subseteq \gimel^{\daleth}{}_1(\aleph^{\hbar})$.

Example 18.2.24 Let $\tilde{\mathscr{Y}}^{\nabla} = \{i, n, k\}$ and $\breve{\daleth} = \{\aleph^{\hbar}{}_1, \aleph^{\hbar}{}_2\}$, then

$$\aleph^{\hbar}{}_{\gimel^{\daleth}{}_1} = \{(\aleph^{\hbar}{}_1, \{(i, 0.420, 0.570), (k, 0.430, 0.420)\})\},$$

and

$$\aleph^{\hbar}{}_{\gimel^{\daleth}{}_2} = \{(\aleph^{\hbar}{}_2, \{(n, 0.370, 0.560), (k, 0.680, 0.290)\})\}$$

are two distinct q-ROFS points contained in the q-ROFSS

$$\gimel^{\daleth}{}_{\breve{\daleth}} = \{(\aleph^{\hbar}{}_1, \{(i, 0.420, 0.570), (k, 0.430, 0.420)\}), (\aleph^{\hbar}{}_2, \{(n, 0.370, 0.560), (k, 0.680, 0.290)\})\}$$

Notice that $\gimel^{\daleth}{}_{\breve{\daleth}} = \aleph^{\hbar}{}_{\gimel^{\daleth}{}_1} \widetilde{\cup} \aleph^{\hbar}{}_{\gimel^{\daleth}{}_2}$, i.e. a q-ROFSS is union of its q-ROFS points.

We design a dynamic SM for q-ROFSSs based on Frobenius inner product of matrices and cosine similarity measure.

Definition 18.2.25 Let $\tilde{\mathscr{Y}}^{\nabla} = \{\breve{\aleph}^{\eth}_i : i = 1, \cdots, m\}$ be a crisp set and $\breve{\daleth} = \{\breve{\hbar}^{\blacklozenge}{}_j : j = 1, \cdots, n\}$ be the aggregate of attributes. If

$$\gimel^{\daleth}{}_1 = \begin{pmatrix} (\mu^{\gamma}_{11}, \nu^{\gamma}{}_{11})_{\gimel^{\daleth}{}_1} & (\mu^{\gamma}_{12}, \nu^{\gamma}{}_{12})_{\gimel^{\daleth}{}_1} & \cdots & (\mu^{\gamma}_{1n}, \nu^{\gamma}{}_{1n})_{\gimel^{\daleth}{}_1} \\ (\mu^{\gamma}_{21}, \nu^{\gamma}{}_{21})_{\gimel^{\daleth}{}_1} & (\mu^{\gamma}_{22}, \nu^{\gamma}{}_{22})_{\gimel^{\daleth}{}_1} & \cdots & (\mu^{\gamma}_{2n}, \nu^{\gamma}{}_{2n})_{\gimel^{\daleth}{}_1} \\ \vdots & \vdots & \ddots & \vdots \\ (\mu^{\gamma}_{m1}, \nu^{\gamma}{}_{m1})_{\gimel^{\daleth}{}_1} & (\mu^{\gamma}_{m2}, \nu^{\gamma}{}_{m2})_{\gimel^{\daleth}{}_1} & \cdots & (\mu^{\gamma}_{mn}, \nu^{\gamma}{}_{mn})_{\gimel^{\daleth}{}_1} \end{pmatrix}$$

and

$$\beth_2 = \begin{pmatrix} (\mu^{\gamma}_{11}, \nu^{\gamma}{}_{11})_{\beth_2} & (\mu^{\gamma}_{12}, \nu^{\gamma}{}_{12})_{\beth_2} & \cdots & (\mu^{\gamma}_{1n}, \nu^{\gamma}{}_{1n})_{\beth_2} \\ (\mu^{\gamma}_{21}, \nu^{\gamma}{}_{21})_{\beth_2} & (\mu^{\gamma}_{22}, \nu^{\gamma}{}_{22})_{\beth_2} & \cdots & (\mu^{\gamma}_{2n}, \nu^{\gamma}{}_{2n})_{\beth_2} \\ \vdots & \vdots & \ddots & \vdots \\ (\mu^{\gamma}_{m1}, \nu^{\gamma}{}_{m1})_{\beth_2} & (\mu^{\gamma}_{m2}, \nu^{\gamma}{}_{m2})_{\beth_2} & \cdots & (\mu^{\gamma}_{mn}, \nu^{\gamma}{}_{mn})_{\beth_2} \end{pmatrix}$$

are q-ROFS matrices of q-ROFSSs $(\beth_1, \breve{\daleth})$ and $(\beth_2, \breve{\daleth})$, then *SM* between $(\beth_1, \breve{\daleth})$ and $(\beth_2, \breve{\daleth})$ is given as

$$Sim(\beth_1, \beth_2) = \frac{< \beth_1, \beth_2 >}{\|\beth_1\| \|\beth_2\|}$$

where

$$< \beth_1, \beth_2 > = tr(\beth_1^T \beth_2)$$
$$\|\beth_1\| = \sqrt{< \beth_1, \beth_1 >}$$

Here, $tr(\beth_1^T \beth_2)$ ("*trace* of the matrix $\beth_1^T \beth_2$") shows the sum of elements at principal diagonal of the matrix $\beth_1^T \beth_2$. The above definition holds good if hesitation margin ε_{ij} is also taken into account. Moreover, this SM satisfies the following:

(1) $0 \le Sim(\beth_1, \beth_2) \le 1$.
(2) $Sim(\beth_1, \beth_2) = 1 \Leftrightarrow \beth_1 = \beth_2$.
(3) $Sim(\beth_1, \beth_2) = Sim(\beth_2, \beth_1)$.
(4) $Sim(\beth, \beth^c) = 1$ iff $\beth$ is a crisp set.
(5) If $(\beth_1, \breve{\daleth}) \tilde{\sqsubseteq} (\beth_2, \breve{\daleth}) \tilde{\sqsubseteq} (\beth_3, \breve{\daleth})$, then $Sim(\beth_1, \beth_3) \le Sim(\beth_2, \beth_3)$.

Example 18.2.26 Let $\mathscr{Y}^{\nabla} = \{\breve{\aleph}^{\eth}_1, \cdots, \breve{\aleph}^{\eth}_4\}$ be the universe and $\breve{\daleth} = \{\check{\hbar}^{\blacklozenge}{}_i \mid i = 1, 2, 3\}$ be the aggregate of attributes. Consider the q-ROFS matrices by taking $q = 3$,

$$\beth_1 = \begin{pmatrix} (0.950, 0.210) & (0.730, 0.460) & (0.530, 0.710) \\ (0.380, 0.820) & (1.000, 0.000) & (0.670, 0.520) \\ (0.280, 0.570) & (0.580, 0.310) & (0.620, 0.790) \\ (0.000, 1.000) & (0.910, 0.190) & (0.630, 0.740) \end{pmatrix}$$

and

$$\beth_2 = \begin{pmatrix} (0.540, 0.290) & (0.610, 0.670) & (0.760, 0.020) \\ (0.070, 0.530) & (0.560, 0.110) & (0.390, 0.790) \\ (0.580, 0.170) & (0.360, 0.340) & (0.170, 0.580) \\ (0.210, 0.830) & (0.490, 0.480) & (0.210, 0.870) \end{pmatrix}$$

representing q-ROFSSs $(\beth_1, \breve{\daleth})$ and $(\beth_2, \breve{\daleth})$, respectively. Now,

$$
\begin{aligned}
<\beth^{\urcorner}_1, \beth^{\urcorner}_2> &= (0.95, 0.21).(0.54, 0.29) + (0.73, 0.46).(0.61, 0.67) + \cdots \\
&\quad + (0.63, 0.74).(0.21, 0.87) \\
&= 6.7180, \\
\|\beth^{\urcorner}_1\| &= \sqrt{0.95^2 + 0.21^2 + 0.73^2 + \cdots + 0.74^2} \\
&= 3.1089, \\
\|\beth^{\urcorner}_2\| &= \sqrt{0.54^2 + 0.29^2 + 0.61^2 + \cdots + 0.87^2} \\
&= 2.4784. \\
\therefore Sim(\beth^{\urcorner}_1, \beth^{\urcorner}_2) &= \frac{<\beth^{\urcorner}_1, \beth^{\urcorner}_2>}{\|\beth^{\urcorner}_1\|\|\beth^{\urcorner}_2\|} \\
&= \frac{6.7180}{3.1089 \times 2.4784} \\
&= 0.8719.
\end{aligned}
$$

Example 18.2.27 Let $\widetilde{\mathscr{Y}}^{\triangledown} = \{\breve{\aleph}^{\eth}_1, \breve{\aleph}^{\eth}_2, \breve{\aleph}^{\eth}_3\}$ and $\breve{\urcorner} = \{\breve{\hbar}^{\blacklozenge}_1, \breve{\hbar}^{\blacklozenge}_2, \breve{\hbar}^{\blacklozenge}_3\}$. Let

$$
\beth^{\urcorner}_1 = \begin{pmatrix}
(0.520, 0.730, 0.440) & (0.890, 0.150, 0.430) & (0.620, 0.590, 0.520) \\
(0.460, 0.730, 0.500) & (1.000, 0, 000, 0.000) & (0.510, 0.510, 0.690) \\
(0.320, 0.190, 0.930) & (0.640, 0.270, 0.720) & (0.870, 0.030, 0.490)
\end{pmatrix}
$$

$$
\beth^{\urcorner}_2 = \begin{pmatrix}
(0.680, 0.520, 0.520) & (0.310, 0.690, 0.650) & (0.440, 0.020, 0.900) \\
(0.610, 0.500, 0.610) & (0.330, 0.570, 0.750) & (0.810, 0.160, 0.560) \\
(0.520, 0.280, 0.810) & (0.290, 0.220, 0.930) & (0.210, 0.390, 0.900)
\end{pmatrix}
$$

be the q-ROFS matrices representing the q-ROFSSs $(\beth^{\urcorner}_1, \breve{\urcorner})$ and $(\beth^{\urcorner}_2, \breve{\urcorner})$, respectively.
Now,

$$
\begin{aligned}
<\beth^{\urcorner}_1, \beth^{\urcorner}_2> &= (0.52, 0.73, 0.44).(0.68, 0.52, 0.52) + \cdots \\
&\quad +(0.74, 0.63, 0.49).(0.35, 0.54, 0.90) \\
&= 7.0581, \\
\|\beth^{\urcorner}_1\| &= \sqrt{0.52^2 + 0.73^2 + 0.44^2 + \cdots + 0.49^2} \\
&= 2.9987, \\
\|\beth^{\urcorner}_2\| &= \sqrt{0.68^2 + 0.52^2 + 0.31^2 + \cdots + 0.54^2} \\
&= 2.9994 \\
\therefore Sim(\beth^{\urcorner}_1, \beth^{\urcorner}_2) &= \frac{<\beth^{\urcorner}_1, \beth^{\urcorner}_2>}{\|\beth^{\urcorner}_1\|\|\beth^{\urcorner}_2\|} \\
&= \frac{7.0581}{2.9987 \times 2.9994} \\
&= 0.7847
\end{aligned}
$$

Example 18.2.28 Let $\widetilde{\mathscr{Y}}^{\nabla} = \{\breve{\aleph}_1^{ð}, \breve{\aleph}_2^{ð}, \breve{\aleph}_3^{ð}\}$ and $\breve{\daleth} = \{\breve{\hbar}^{\blacklozenge}{}_1, \breve{\hbar}^{\blacklozenge}{}_2, \breve{\hbar}^{\blacklozenge}{}_3\}$. Consider the q-ROFS matrices

$$(\gimel_1, \breve{\daleth}) = \begin{pmatrix} (0.270, 0.390) & (0.420, 0.510) & (0.610, 0.430) \\ (0.250, 0.560) & (0.580, 0.490) & (0.920, 0.360) \\ (0.760, 0.230) & (0.460, 0.480) & (0.540, 0.210) \end{pmatrix}$$

$$(\gimel_2, \breve{\daleth}) = \begin{pmatrix} (0.450, 0.210) & (0.260, 0.890) & (0.540, 0.390) \\ (0.290, 0.280) & (0.460, 0.440) & (0.640, 0.310) \\ (0.270, 0.540) & (0.280, 0.330) & (0.890, 0.160) \end{pmatrix}$$

$$(\gimel_3, \breve{\daleth}) = \begin{pmatrix} (0.930, 0.150) & (0.450, 0.590) & (0.330, 0.140) \\ (0.390, 0.280) & (0.510, 0.550) & (0.640, 0.270) \\ (0.710, 0.320) & (0.330, 0.180) & (0.090, 0.560) \end{pmatrix}$$

Then $Sim(\gimel_2, \gimel_1) = 0.8925 > 0.82$ and $Sim(\gimel_1, \gimel_3) = 0.8491 > 0.82$ but $Sim(\gimel_2, \gimel_3) = 0.8027 \not> 0.82$. This advocates that the relation of being similar is not transitive.

Definition 18.2.29 Two q-ROFSSs $(\gimel_1, \breve{\daleth}_1)$ and $(\gimel_2, \breve{\daleth}_2)$ defined over $(\widetilde{\mathscr{Y}}^{\nabla}, \breve{\daleth})$ are called λ *-similar*, denoted as $(\gimel_1, \breve{\daleth}_1) \approx^{\lambda} (\gimel_2, \breve{\daleth}_2)$, if $Sim(\gimel_1, \gimel_2) \geq \lambda$ for some $0 < \lambda < 1$.

18.2.1 Weighted SM for q-ROFSSs

In this part, we introduce "weighted SM" among two q-ROFSSs and discuss almost all of its unique properties.

Definition 18.2.30 Let $\gimel_1$ and $\gimel_2$ be as given in Definition 18.2.25. Let the weight of $\breve{\hbar}^{\blacklozenge}{}_j$ is $\breve{\coprod}_j^{\top} \in [0, 1]$ for $j = 1, 2, \cdots, n$. The "*weighted SM*" betwixt $\gimel_1$ and $\gimel_2$ is given as

$$Sim_W(\gimel_1, \gimel_2) = \frac{< \gimel_1, \gimel_2 >}{\|\gimel_1\| \|\gimel_2\|}$$

where

$$< \gimel_1, \gimel_2 > = \frac{\Sigma_{i,j} \breve{\coprod}_j^{\top} (\mu_{ij}^{\gamma}, \nu^{\gamma}{}_{ij})_{\gimel_1} . (\mu_{ij}^{\gamma}, \nu^{\gamma}{}_{ij})_{\gimel_2}}{\Sigma_j \breve{\coprod}_j^{\top}}$$

$$\|\gimel_1\| = \sqrt{< \gimel_1, \gimel_1 >}$$

This weighted SM satisfies the same properties as given in Definition 18.2.25.

Example 18.2.31 Consider the q-ROFSSs given by the q-ROFS matrices

$$\gimel^{\urcorner}{}_1 = \begin{pmatrix} (0.520, 0.730) & (0.890, 0.150) & (0.620, 0.590) \\ (0.460, 0.730) & (1.000, 0.000) & (0.510, 0.510) \\ (0.320, 0.190) & (0.640, 0.270) & (0.870, 0.030) \end{pmatrix}$$

$$\gimel^{\urcorner}{}_2 = \begin{pmatrix} (0.680, 0.520) & (0.310, 0.690) & (0.440, 0.020) \\ (0.610, 0.500) & (0.330, 0.570) & (0.810, 0.160) \\ (0.520, 0.280) & (0.290, 0.220) & (0.210, 0.390) \end{pmatrix}$$

Assume that the weights of the attributes $\hbar^{\check{\blacklozenge}}{}_1$, $\hbar^{\check{\blacklozenge}}{}_2$ and $\hbar^{\check{\blacklozenge}}{}_3$ are $\breve{\coprod}_1^{\intercal} = 0.52$, $\breve{\coprod}_2^{\intercal} = 0.31$ and $\breve{\coprod}_3^{\intercal} = 0.47$ respectively. Then,

$$\begin{aligned} < \gimel^{\urcorner}{}_1, \gimel^{\urcorner}{}_2 > &= 1.7672 \\ \|\gimel^{\urcorner}{}_1\| &= 1.7020 \\ \|\gimel^{\urcorner}{}_2\| &= 1.3479 \\ \therefore Sim_W(\gimel^{\urcorner}{}_1, \gimel^{\urcorner}{}_2) &= 0.7703 \end{aligned}$$

18.3 MCDM Using q-Rung Orthopair Fuzzy Soft Information

There are numerous statements that we commonly use in ordinary living that contain fuzziness aspect. We frequently employ numerical or, occasionally, verbal terms to illustrate an event, reference towards something, evaluate someone else's competence, and in a variety of other circumstances involving fuzziness. It is common practice to utilize lingual phrases. When determining on a scenario or explaining an incident, these statements rarely indicate certainty. For instance, the terms "poor", "middle class", "lower middle class", "higher middle class", "upper class", and "rich" are used to describe a person's income or possessions. When riding at 80 km/h on a difficult road, we say "fast," yet when traveling on a highway, we say "slow." These cases indicate how well the human mind actually does work and makes decisions in confusing and unclear circumstances, as well as how it perceives, evaluates, and regulates processes.

With such introduction of FSs, scientific and technological have advanced tremendously. The logic of FSs has found widespread use in both theory and practice studies spanning from biological studies to machine learning, and from thermodynamics to industry and literature. In ordinary living, we frequently encounter challenges that are neither specific and unambiguous. This problem brings about a lot of judgement systems. Using these strategies, we attempt to make a faultless and intellectual conclusion. As a result, new scientific models and approaches for mitigating risk and ambiguities are critical. Smart decision-making is an active component of trading, finance, and other significant challenges. It address every day moo reasonable terms and evaluations at low-ranking administration level and long-term sustainable planning by decision experts of business. Conclusions provided at any level can have legitimate or negative consequences, but is there an unambiguous framework that decision-makers should embrace in order to guarantee triumph, or should they substitute the conventional plans of treating a problem?

Before making a united and consistent judgement, decision-makers should evaluate a number of aspects. As a result, it is vital to ensure that all of these factors are considered before the assurance is completed. It is unavoidable in parliamentary law to organize the decision-making process in an orderly fashion in order to ensure that all relevant realities and numbers are evaluated. Aside from other enormous implications, the science of mathematics also aids us in reaching judgements based on logical reasoning. IFSs, PFSs, and SSs are all covered by q-ROFSSs. In everyday decision-making situations, q-ROFSSs have a larger participation space than the IFSS and PFSS. As a result, q-ROFSSs can model discrepancy and unpredictability in decision-making situations better than IFSSs and PFSSs.

Now, underneath the banner of q-ROFSSs, we propose a methodology for dealing with MCDM problems utilizing the choice value technique, accompanied by an illustration.

Algorithm 18.1:

Step 1: Consider $\breve{\aleph}_i^{\eth}$ as the collection of m alternatives and $\breve{\daleth}_i$ as the assemblage of n parameters.

Step 2: Obtain the q-ROFS matrix from the DMs.

Step 3: Evaluate "relative importance" i.e. weight $\breve{\coprod}_i^{\top}$ of all parameters s.t $\Sigma_{i=1}^{n}$ $\breve{\coprod}_i^{\top} = 1$.

Step 4: By using $C = \beth^{\daleth}{}_{\breve{\daleth}} \times \breve{\Upsilon}^{t}$, get the "matrix of choice values".

Step 5: Calculate the score value s for each alternative utilizing $s_j = n_{\mu_j^{\gamma}} - n_{\nu^{\gamma}{}_j}$, where $n_{\mu_j^{\gamma}}$ denotes number of times μ_j^{γ} goes beyond or equals other values of $\mu_k^{\gamma}, k \neq j$.

Step 6: The choice with the highest score value seems to be the mandatory option.

Pictorial view of Algorithm 18.1 is given in Fig. 18.1.

Example 18.3.1 Assume an entrepreneur wants to contrive a fresh firm with the least amount of risk. Let $\mathscr{Y}^{\nabla} = \{\breve{\aleph}_i^{\eth} : i = 1, \cdots, 7\}$ be the assemblage of possible choices for business. To reduce and prevent aspect, he intends to put his funds in the top 2 rated enterprises in a 3:2 ratio. He picks the set of characteristics after consulting with his five investment planners as $\breve{\daleth} = \{\beth_i^{\beta} : i = 1, \cdots, 5\}$, where

$$\beth^{\beta}{}_1 = \text{Investment safety}$$
$$\beth^{\beta}{}_2 = \text{Product viability}$$
$$\beth^{\beta}{}_3 = \text{Prospects}$$
$$\beth^{\beta}{}_4 = \text{Impact on market}$$
$$\beth^{\beta}{}_5 = \text{Standing reputation of the business.}$$

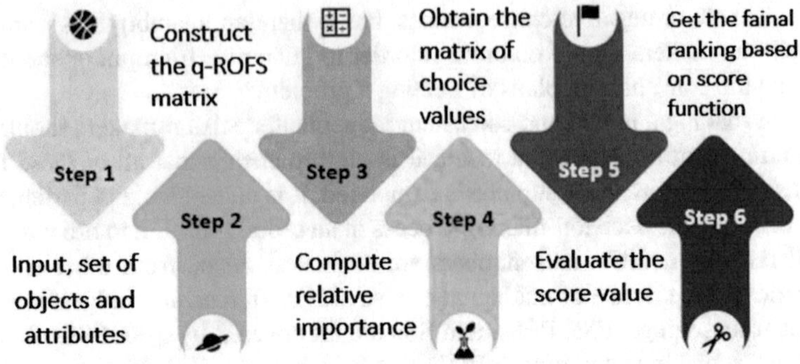

Fig. 18.1 Pictorial view of Algorithm 18.1

Table 18.3 Tabular delineation of $\gimel\daleth_{\daleth}$

$\gimel\daleth_{\daleth}$	$\gimel^{\beta}{}_1$	$\gimel^{\beta}{}_2$	$\gimel^{\beta}{}_3$	$\gimel^{\beta}{}_4$	$\gimel^{\beta}{}_5$
$\aleph^{\eth}_1$	(0.420,0.560)	(0.370,0.540)	(0.590,0.110)	(0.230,0.590)	(0.110,0.920)
$\aleph^{\eth}_2$	(0.340,0.130)	(0.520,0.410)	(0.540,0.110)	(0.330,0.020)	(0.220,0.140)
$\aleph^{\eth}_3$	(0.890,0.240)	(0.770,0.310)	(0.560,0.150)	(0.500,0.130)	(0.280,0.130)
$\aleph^{\eth}_4$	(0.430,0.440)	(0.560,0.670)	(0.830,0.290)	(0.470,0.580)	(0.370,0.090)
$\aleph^{\eth}_5$	(0.560,0.670)	(0.490,0.520)	(0.570,0.380)	(0.210,0.340)	(0.380,0.360)
$\aleph^{\eth}_6$	(0.790,0.340)	(0.440,0.430)	(0.560,0.580)	(0.910,0.390)	(0.330,0.390)
$\aleph^{\eth}_7$	(0.540,0.240)	(0.510,0.420)	(0.550,0.550)	(0.110,0.090)	(0.390,0.560)

Studying the history and trends of these businesses, the DMs gathered data in the form of Table 18.3 of the q-ROFS-set $\gimel\daleth_{\daleth}$.

These data can be represented as a q-ROFS matrix,

$$\gimel\daleth_{\daleth} = \begin{pmatrix} (0.420, 0.560) & (0.370, 0.540) & (0.590, 0.110) & (0.230, 0.590) & (0.110, 0.920) \\ (0.340, 0.130) & (0.520, 0.410) & (0.540, 0.110) & (0.330, 0.020) & (0.220, 0.140) \\ (0.890, 0.240) & (0.770, 0.310) & (0.560, 0.150) & (0.500, 0.130) & (0.280, 0.130) \\ (0.430, 0.440) & (0.560, 0.670) & (0.830, 0.290) & (0.470, 0.580) & (0.370, 0.090) \\ (0.560, 0.670) & (0.490, 0.520) & (0.570, 0.380) & (0.210, 0.340) & (0.380, 0.360) \\ (0.790, 0.340) & (0.440, 0.430) & (0.560, 0.580) & (0.910, 0.390) & (0.330, 0.390) \\ (0.540, 0.240) & (0.510, 0.420) & (0.550, 0.550) & (0.110, 0.090) & (0.390, 0.560) \end{pmatrix}$$

Consider that the four investment managers disclose the “relative relevance”, i.e. fuzzified weights, to each characteristic and that this information is given in the format of the corresponding matrix:

$$M = \begin{pmatrix} 0.540 & 0.380 & 0.590 & 0.890 & 0.760 \\ 0.370 & 0.470 & 0.480 & 0.940 & 0.880 \\ 0.820 & 0.460 & 0.760 & 0.230 & 0.790 \\ 0.180 & 0.320 & 0.570 & 0.460 & 0.690 \end{pmatrix}$$

After normalizing the entries of M, the normalized matrix appears to be

$$\widehat{M} = \begin{pmatrix} 0.5070 & 0.4610 & 0.4850 & 0.6390 & 0.4850 \\ 0.3480 & 0.5700 & 0.3940 & 0.6750 & 0.5620 \\ 0.7700 & 0.5580 & 0.6250 & 0.1650 & 0.5040 \\ 0.1690 & 0.3880 & 0.4680 & 0.3300 & 0.4410 \end{pmatrix}$$

Thus, the weighted values for the attributes are

$$\check{\Upsilon}(\gimel^{\beta}{}_1) = 0.1880,\ \check{\Upsilon}(\gimel^{\beta}{}_2) = 0.2070,\ \check{\Upsilon}(\gimel^{\beta}{}_3) = 0.2070,\ \check{\Upsilon}(\gimel^{\beta}{}_4) = 0.1900,\ \check{\Upsilon}(\gimel^{\beta}{}_5) = 0.2090$$

Hence, the WV is

$$\check{\Upsilon} = \begin{pmatrix} 0.1880 & 0.2070 & 0.2070 & 0.1900 & 0.2090 \end{pmatrix}$$

Thus, the q-ROF-matrix for choice values is

$$C = \gimel^{\urcorner}{}_{\daleth} \times \check{\Upsilon}^{t}$$

$$= \begin{pmatrix} (0.420, 0.560) & (0.370, 0.540) & (0.590, 0.110) & (0.230, 0.590) & (0.110, 0.920) \\ (0.340, 0.130) & (0.520, 0.410) & (0.540, 0.110) & (0.330, 0.020) & (0.220, 0.140) \\ (0.890, 0.240) & (0.770, 0.310) & (0.560, 0.150) & (0.500, 0.130) & (0.280, 0.130) \\ (0.430, 0.440) & (0.560, 0.670) & (0.830, 0.290) & (0.470, 0.580) & (0.370, 0.090) \\ (0.560, 0.670) & (0.490, 0.520) & (0.570, 0.380) & (0.210, 0.340) & (0.380, 0.360) \\ (0.790, 0.340) & (0.440, 0.430) & (0.560, 0.580) & (0.910, 0.390) & (0.330, 0.390) \\ (0.540, 0.240) & (0.510, 0.420) & (0.550, 0.550) & (0.110, 0.090) & (0.390, 0.560) \end{pmatrix} \begin{pmatrix} 0.1880 \\ 0.2070 \\ 0.2070 \\ 0.1900 \\ 0.2090 \end{pmatrix}$$

$$= \begin{pmatrix} (0.34440, 0.54420) \\ (0.39200, 0.16510) \\ (0.59620, 0.19220) \\ (0.53520, 0.41050) \\ (0.44400, 0.45210) \\ (0.59740, 0.42860) \\ (0.42340, 0.38010) \end{pmatrix}$$

The values of the score function along with ranking are given in Table 18.4.

Table 18.4 demonstrates that

$$\breve{\aleph}^{\eth}_3 \succ \breve{\aleph}^{\eth}_6 \succ \breve{\aleph}^{\eth}_2 = \breve{\aleph}^{\eth}_4 \succ \breve{\aleph}^{\eth}_7 \succ \breve{\aleph}^{\eth}_5 \succ \breve{\aleph}^{\eth}_1$$

Investor should invest 60% of the capital on $\breve{\aleph}^{\eth}_3$, and 40% on $\breve{\aleph}^{\eth}_6$.

Table 18.4 Values of score function and ranking

$\widetilde{\mathscr{Y}}^{\nabla}$	s	Ranking
$\aleph^{\eth}_1$	$0-6=-6$	6
$\aleph^{\eth}_2$	$1-0=1$	3
$\aleph^{\eth}_3$	$5-1=4$	1
$\aleph^{\eth}_4$	$4-3=1$	3
$\aleph^{\eth}_5$	$3-5=-2$	5
$\aleph^{\eth}_6$	$6-4=2$	2
$\aleph^{\eth}_7$	$2-2=0$	4

Table 18.5 Linguistic terms and fuzzy weights for assessing alternatives

Linguistic terms	Fuzzy weights
Very Good (VG)	(0.800, 1.000]
Good (G)	(0.500, 0.800)
Fair (F)	(0.200, 0.500)
Bad (B)	(0.100, 0.200)
Very Bad (VB)	[0.000, 0.100)

18.4 MCDM with TOPSIS Approach Based on q-ROFSSs

In this section, we take advantages of q-ROFSS information to introduce a robust MCDM approach. For this purpose, we develop an extension of TOPSIS approach for uncertain q-ROFSS information, which is proposed in Algorithm 18.2. Following that, we will look into the issue of selecting competent candidates for government ministries in a democracy.

Algorithm 18.2:

Step 1: Consider $= \{\mathcal{I}^{\daleth}{}_i : i = 1, \cdots, n\}$ is the collection of DMs, $C = \{\mathfrak{C}^{\varpi}{}_i = 1, \cdots, l\}$ is the assemblage of alternatives and $Q = \{q_j : j = 1, \cdots, m\}$ is the set of attributes.

Step 2: Using the phonological words listed in Table 18.5, constructed the weighted attribute matrix as $\left[\breve{\coprod}^{\top}_{ij}\right]_{n\times m}$, where $\breve{\coprod}^{\top}_{ij}$ is the weight allocated by the decision expert $\mathcal{I}^{\daleth}{}_i$ to the attribute q_j.

Step 3: Normalized the "weighted matrix" to get $\hat{N} = [\hat{n}_{ij}]_{n\times m}$, where $\hat{n}_{ij} = \frac{\breve{\coprod}^{\top}_{ij}}{\sqrt[q]{\Sigma_{i=1}^{n} w_{ij}^{q}}}$ and obtaining the WV $\check{\Upsilon} = (\mathfrak{w}_1, \mathfrak{w}_2, \cdots, \mathfrak{w}_m)$, where $\mathfrak{w}_j = \frac{\sum_{i=1}^{n} \hat{n}_{ij}}{m \sum_{k=1}^{m} \hat{n}_{ik}}$.

Step 4: Construct q-ROFS matrix

$$D_i = [v^i_{jk}]_{l\times m} = \begin{pmatrix} v^i_{11} & v^i_{12} & \cdots & v^i_{1m} \\ v^i_{21} & v^i_{22} & \cdots & v^i_{2m} \\ \vdots & \vdots & \ddots & \vdots \\ v^i_{j1} & v^i_{j2} & \cdots & v^i_{jm} \\ \vdots & \vdots & \ddots & \vdots \\ v^i_{l1} & v^i_{l2} & \cdots & v^i_{lm} \end{pmatrix}$$

where v^i_{jk} is a q-ROFS-element, provided by i^{th} DM. Then obtain the aggregated matrix

$$D = \frac{D_1 + D_2 + \cdots + D_n}{n} = [\dot{v}_{jk}]_{l\times m}$$

Step 5: Obtain the weighted q-ROFS matrix

$$D_w = [\Cup^\beta{}_{jk}]_{l\times m} = \begin{pmatrix} \Cup^\beta{}_{11} & \Cup^\beta{}_{12} & \cdots & \Cup^\beta{}_{1m} \\ \Cup^\beta{}_{21} & \Cup^\beta{}_{22} & \cdots & \Cup^\beta{}_{2m} \\ \vdots & \vdots & \ddots & \vdots \\ \Cup^\beta{}_{j1} & \Cup^\beta{}_{j2} & \cdots & \Cup^\beta{}_{jm} \\ \vdots & \vdots & \ddots & \vdots \\ \Cup^\beta{}_{l1} & \Cup^\beta{}_{l2} & \cdots & \Cup^\beta{}_{lm} \end{pmatrix}$$

where $\Cup^\beta{}_{jk} = \mathfrak{w}_k \times \dot{v}_{jk}$.

Step 6: Track the "q-ROFS-valued positive ideal solution (q-ROFSV-PIS) and q-ROFS-valued negative ideal solution (q-ROFSV-NIS)". For this, we utilize

$$\begin{aligned} \text{q-ROFSV-PIS} &= \{\Cup^{\beta+}_1, \Cup^{\beta+}_2, \cdots, \Cup^{\beta+}_m\} \\ &= \left\{(\vee_k \Cup^\beta{}_{jk}, \wedge_k \Cup^\beta{}_{jk}); k = 1, \cdots, m\right\} \\ &= \left\{(\breve{\zeta}^+_k, \varrho^+_k) : k = 1, \cdots, m\right\} \end{aligned}$$

and

$$\begin{aligned} \text{q-ROFSV-NIS} &= \{\Cup^{\beta-}_1, \Cup^{\beta-}_2, \cdots, \Cup^{\beta-}_m\} \\ &= \left\{(\wedge_k \Cup^\beta{}_{jk}, \vee_k \Cup^\beta{}_{jk}); k = 1, \cdots, m\right\} \\ &= \left\{(\breve{\zeta}^-_k, \varrho^-_k) : k = 1, \cdots, m\right\} \end{aligned}$$

where $\vee$ stands for q-ROFS union and $\wedge$ represents q-ROFS intersection.

Step 7: Calculate the distances of every option from q-ROFSV-PIS and q-ROFSV-NIS, using

$$\mathfrak{z}_j^+ = \sqrt{\Sigma_{k=1}^m\left\{\left(\mu^{\gamma}_{jk} - \breve{\zeta}_k^+\right)^2 + \left(\nu^{\gamma}{}_{jk} - \varrho_k^+\right)^2\right\}}$$

and

$$\mathfrak{z}_j^- = \sqrt{\Sigma_{k=1}^m\left\{\left(\mu^{\gamma}_{jk} - \breve{\zeta}_k^-\right)^2 + \left(\nu^{\gamma}{}_{jk} - \varrho_k^-\right)^2\right\}}$$

Step 8: By employing, you may achieve the "closeness coefficient" of each option to the optimal answer.

$$C_j^* = \frac{\mathfrak{z}_j^-}{\mathfrak{z}_j^+ + \mathfrak{z}_j^-} \in [0, 1]$$

Step 9: Organize the alternatives in decreasing (or increasing) priority order to acquire the priority order of the alternatives.

Pictorial view of Algorithm 18.2 is given in Fig. 18.2.

Example 18.4.1 Assume that a particular party wins a state's national election with a landslide. For the very first time, the party has the opportunity to build a nationwide administration and aims to demonstrate that its performance is the best. The Chairman of party is planned to fill some positions of ministers to improve the efficiency of

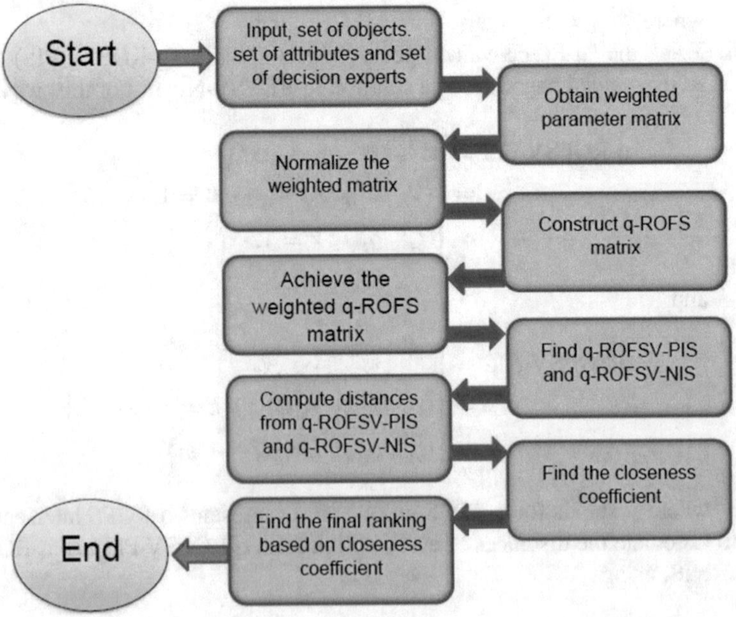

Fig. 18.2 Pictorial view of Algorithm 18.2

his government. He offered new some positions in the key ministries by selecting competent, well-educated/trained and meritorious ministers according to their merit and experience in multiple fields. The Chairman constitutes a committee of his elected experts which are senior in his party who will help to improve the efficiency of the government. He decided that one member will be given one ministrty and no member is offered to get more than one ministry positions. Assume that

$$C = \{\beta^{\daleth}{}_1, \beta^{\daleth}{}_2, \cdots, \beta^{\daleth}{}_6\}$$

is the set of candidates (who are elected members of his party) who are to be deputed in different key ministries ("ministries of foreign affairs", "defence", "finance" and "information & broadcasting" in order). Further suppose that

$$Q = \{q_1, q_2, \cdots, q_5\}$$

is the set of qualification/merit mandatory for filling a position. The committee interviewed the candidates carefully to see who is appropriate for which ministry.
Picking the weights from Table 18.5, the experts provide the following weighted parameter matrix

$$\mathcal{P} = \begin{pmatrix} \text{VG} & \text{G} & \text{G} & \text{VG} & \text{VG} \\ \text{G} & \text{F} & \text{VG} & \text{F} & \text{F} \\ \text{F} & \text{B} & \text{F} & \text{G} & \text{VG} \\ \text{G} & \text{VB} & \text{B} & \text{VG} & \text{B} \end{pmatrix}$$

$$= \begin{pmatrix} 0.900 & 0.600 & 0.600 & 0.900 & 0.900 \\ 0.600 & 0.350 & 0.900 & 0.350 & 0.350 \\ 0.350 & 0.150 & 0.350 & 0.600 & 0.900 \\ 0.600 & 0.050 & 0.150 & 0.900 & 0.150 \end{pmatrix}$$

The normalized weighted matrix is

$$\hat{N} = \begin{pmatrix} 0.846 & 0.916 & 0.602 & 0.752 & 0.786 \\ 0.564 & 0.534 & 0.903 & 0.292 & 0.306 \\ 0.329 & 0.229 & 0.351 & 0.501 & 0.786 \\ 0.564 & 0.076 & 0.150 & 0.752 & 0.131 \end{pmatrix}$$

and hence the WV is $\check{\Upsilon} = (0.221, 0.170, 0.194, 0.222, 0.194)$.
Assume that the four experts provide the following q-ROFS matrices in which the q-ROFN at (i, j)th position demarcated grades of candidates row-wise and the attribute column-wise.

$$D_1 = \begin{pmatrix} (0.57, 0.39) & (0.49, 0.74) & (0.77, 0.38) & (0.54, 0.21) & (0.12, 0.48) \\ (0.66, 0.51) & (0.54, 0.54) & (0.32, 0.13) & (0.99, 0.13) & (0.54, 0.07) \\ (0.15, 0.68) & (0.19, 0.32) & (0.76, 0.41) & (0.45, 0.15) & (0.11, 0.49) \\ (0.67, 0.74) & (0.09, 0.83) & (0.59, 0.31) & (0.84, 0.16) & (0.37, 0.21) \\ (0.59, 0.17) & (0.33, 0.67) & (0.34, 0.68) & (0.52, 0.19) & (0.58, 0.61) \\ (0.27, 0.54) & (0.49, 0.46) & (0.48, 0.59) & (0.55, 0.54) & (0.38, 0.01) \end{pmatrix}$$

$$D_2 = \begin{pmatrix} (0.34, 0.52) & (0.58, 0.21) & (0.47, 0.21) & (0.70, 0.31) & (0.11, 0.34) \\ (0.47, 0.33) & (0.39, 0.32) & (0.56, 0.20) & (0.38, 0.11) & (0.26, 0.18) \\ (0.59, 0.17) & (0.33, 0.17) & (0.19, 0.28) & (0.59, 0.06) & (0.78, 0.16) \\ (0.44, 0.17) & (0.38, 0.23) & (0.58, 0.27) & (0.71, 0.24) & (0.54, 0.02) \\ (0.32, 0.28) & (0.56, 0.11) & (0.44, 0.37) & (0.49, 0.29) & (0.55, 0.55) \\ (0.34, 0.47) & (0.52, 0.37) & (0.11, 0.18) & (0.47, 0.13) & (0.47, 0.27) \end{pmatrix}$$

$$D_3 = \begin{pmatrix} (0.11, 0.58) & (0.37, 0.22) & (0.56, 0.11) & (0.21, 0.69) & (0.79, 0.32) \\ (0.13, 0.67) & (0.46, 0.13) & (0.36, 0.54) & (0.56, 0.27) & (0.46, 0.61) \\ (0.59, 0.13) & (0.25, 0.11) & (0.62, 0.33) & (0.47, 0.28) & (0.28, 0.47) \\ (0.11, 0.49) & (0.23, 0.05) & (0.50, 0.28) & (0.34, 0.48) & (0.61, 0.54) \\ (0.17, 0.29) & (0.82, 0.34) & (0.56, 0.51) & (0.50, 0.28) & (0.49, 0.12) \\ (0.33, 0.69) & (0.57, 0.61) & (0.48, 0.57) & (0.33, 0.02) & (0.46, 0.31) \end{pmatrix}$$

$$D_4 = \begin{pmatrix} (0.40, 0.59) & (0.41, 0.32) & (0.49, 0.12) & (0.35, 0.65) & (0.39, 0.12) \\ (0.25, 0.17) & (0.38, 0.10) & (0.85, 0.26) & (0.44, 0.57) & (0.92, 0.14) \\ (0.38, 0.51) & (0.36, 0.11) & (0.52, 0.29) & (0.48, 0.38) & (0.52, 0.35) \\ (0.56, 0.11) & (0.73, 0.16) & (0.35, 0.27) & (0.58, 0.62) & (0.62, 0.63) \\ (0.11, 0.01) & (0.33, 0.37) & (0.28, 0.38) & (0.47, 0.32) & (0.71, 0.19) \\ (0.58, 0.17) & (0.44, 0.15) & (0.56, 0.16) & (0.33, 0.21) & (0.88, 0.26) \end{pmatrix}$$

Thus, the aggregated matrix is

$$D = \begin{pmatrix} (0.355, 0.520) & (0.463, 0.373) & (0.573, 0.205) & (0.450, 0.465) & (0.353, 0.315) \\ (0.378, 0.420) & (0.443, 0.273) & (0.523, 0.283) & (0.593, 0.270) & (0.545, 0.250) \\ (0.428, 0.373) & (0.373, 0.178) & (0.523, 0.328) & (0.498, 0.218) & (0.423, 0.368) \\ (0.445, 0.378) & (0.358, 0.318) & (0.505, 0.283) & (0.618, 0.375) & (0.535, 0.350) \\ (0.298, 0.188) & (0.510, 0.373) & (0.405, 0.485) & (0.495, 0.270) & (0.583, 0.368) \\ (0.380, 0.468) & (0.505, 0.398) & (0.408, 0.375) & (0.420, 0.225) & (0.548, 0.213) \end{pmatrix}$$

and hence the weighted q-ROFS matrix is

Table 18.6 Distance closeness coefficient of each candidate

Candidate	$\mathfrak{Z}_j^+$	$\mathfrak{Z}_j^-$	C_j^*
$\beta^{\gimel}{}_1$	0.11610	0.06759	0.36795
$\beta^{\gimel}{}_2$	0.06309	0.09179	0.59265
$\beta^{\gimel}{}_3$	0.04590	0.091142	0.66506
$\beta^{\gimel}{}_4$	0.07286	0.08777	0.54641
$\beta^{\gimel}{}_5$	0.08853	0.10080	0.53240
$\beta^{\gimel}{}_6$	0.09726	0.08177	0.45674

$$D_w = \begin{pmatrix} (0.078, 0.115) & (0.079, 0.063) & (0.111, 0.040) & (0.099, 0.103) & (0.068, 0.061) \\ (0.084, 0.093) & (0.075, 0.046) & (0.101, 0.060) & (0.132, 0.060) & (0.106, 0.049) \\ (0.095, 0.082) & (0.063, 0.030) & (0.101, 0.064) & (0.111, 0.048) & (0.082, 0.071) \\ (0.098, 0.084) & (0.061, 0.054) & (0.098, 0.055) & (0.137, 0.083) & (0.104, 0.068) \\ (0.066, 0.042) & (0.087, 0.063) & (0.079, 0.094) & (0.109, 0.060) & (0.113, 0.071) \\ (0.084, 0.103) & (0.086, 0.068) & (0.079, 0.073) & (0.093, 0.050) & (0.106, 0.041) \end{pmatrix}$$

The positive and negative ideal solutions are

$$\begin{aligned} \text{q-ROFSV-PIS} &= \{\Cup^{\beta_1^+}, \Cup^{\beta_2^+}, \cdots, \Cup^{\beta_5^+}\} \\ &= \{(0.098, 0.042), (0.087, 0.030), (0.111, 0.040), (0.137, 0.050), (0.113, 0.041)\} \end{aligned}$$

and

$$\begin{aligned} \text{q-ROFSV-NIS} &= \{\Cup^{\beta_1^-}, \Cup^{\beta_2^-}, \cdots, \Cup^{\beta_5^-}\} \\ &= \{(0.066, 0.115), (0.061, 0.068), (0.079, 0.094), (0.093, 0.103), (0.068, 0.071)\} \end{aligned}$$

respectively.

Table 18.6 shows the distances of each candidate from q-ROFSV-PIS and q-ROFSV-NIS, as well as their "relative closeness coefficients".

Hence, the ranking preference is

$$\beta^{\gimel}{}_3 \succ \beta^{\gimel}{}_2 \succ \beta^{\gimel}{}_4 \succ \beta^{\gimel}{}_5 \succ \beta^{\gimel}{}_6 \succ \beta^{\gimel}{}_1$$

This preference order is depicted in Fig. 18.3.

According to the preceding priority sequence, the "ministry of international affairs" should be assigned to $\beta^{\gimel}{}_2$, the "ministry of defence" to $\beta^{\gimel}{}_3$, the "ministry of finance" to $\beta^{\gimel}{}_4$, and the "ministry of information and broadcasting" to $\beta^{\gimel}{}_5$.

Fig. 18.3 Ranking of candidates

Table 18.7 Linguistic terms and fuzzy weights for assessing alternatives

Linguistic terms	Fuzzy weights
"Very Necessary" (VN)	[0.80, 1]
"Mandatory" (M)	[0.50, 0.80]
"More or Less Required" (MLR)	[0.20, 0.50]
"Average Requirement" (AR)	[0.10, 0.20]
"Of No Use" (ONU)	[0, 0.10]

18.5 MCDM Using q-ROFS VIKOR Method

The word VIKOR is an abbreviated version of "Vlse Kriterijumska Optimizacija Kompromisno Resenje" from "Serbian language" to "mean manifold-criteria analysis" (or optimization) and "middle ground way out". "Serafim Opricovic" developed this strategy to deal with decision-making challenges with dissident and incompatible principles, with the notion that finding the middle ground is appropriate for resolving any conflict. The expert team searches for a solution that is close to the superlative perfect solution, and the options are evaluated using all known rules. VIKOR has emerged as a popular MCDM technique, owing to its low computing and moral uprightness of solution.

We elucidate a robust MCDM approach using uncertain q-ROFS information to develop an extension of VIKOR method, which is proposed in Algorithm 18.3. First six steps of Algorithms 18.2 and 18.3 are the same, so start q-ROFSS VIKOR from Step 7.

Algorithm 18.3 (q-ROFSS VIKOR):

Step 7: Use the formulae

$$S_i = \Sigma_{j=1}^{m} \mathfrak{w}_j \left(\frac{d\left(\Uplus^{\beta+}_{j}, \Uplus^{\beta}_{ij}\right)}{d\left(\Uplus^{\beta+}_{j}, \Uplus^{\beta-}_{j}\right)} \right)$$

$$R_i = \max_{j=1}^{m} \mathfrak{w}_j \left(\frac{d\left(\Uplus^{\beta+}_{j}, \Uplus^{\beta}_{ij}\right)}{d\left(\Uplus^{\beta+}_{j}, \Uplus^{\beta-}_{j}\right)} \right)$$

$$Q_i = \kappa \left(\frac{S_i - S^-}{S^+ - S^-} \right) + (1 - \kappa) \left(\frac{R_i - R^-}{R^+ - R^-} \right)$$

where $S^+ = \max_i S_i$, $S^- = \min_i S_i$, $R^+ = \max_i R_i$, and $R^- = \min_i R_i$, to get the values of group utility S_i, individual regret R_i, and compromise Q_i. The real number κ is termed as coefficient of decision mechanism. The role of the coefficient κ is that if compromise solution is to be selected by majority, we choose $\kappa > 0.5$; for consensus we use $\kappa = 0.5$, and $\kappa < 0.5$ represents veto. $\mathfrak{w}_j$ represents the weight of the jth criteria, which expresses its relative importance.

Step 8: Set the S_i, R_i, and Q_i in ascending way. The alternative $\Uplus^{\beta}_{\beta\daleth}$ will be the middle ground solution if it has the minimum value of Q_i and further gratifies the following two necessities in chorus:

(a) If $\Uplus^{\beta}_{\beta\daleth_1}$ and $\Uplus^{\beta}_{\beta\daleth_2}$ are two best choices regarding Q_i, then

$$Q\left(\Uplus^{\beta}_{\beta\daleth_2}\right) - Q\left(\Uplus^{\beta}_{\beta\daleth_1}\right) \geq \frac{1}{n-1}$$

n being the number of attributes.

(b) The choice $\Uplus^{\beta}_{\beta\daleth_1}$ must be best ranked by at least one of R_i and S_i.
There will exist multiple "compromise solutions" otherwise, which may be located as under:

(i) $\Uplus^{\beta}_{\beta\daleth_1}$ and $\beta^{\daleth}_2$ will be the "compromise solutions" in case merely (a) is gratified.

(ii) $\Uplus^{\beta}_{\beta\daleth_1}, \Uplus^{\beta}_{\beta\daleth_2}, \cdots, \Uplus^{\beta}_{\beta\daleth_u}$ would be the "compromise solutions" in case (a) is not fulfilled, where $\Uplus^{\beta}_{\beta\daleth_u}$ may be found employing

$$Q\left(\Uplus^{\beta}_{\beta\daleth_u}\right) - Q\left(\Uplus^{\beta}_{\beta\daleth_1}\right) \geq \frac{1}{n-1}.$$

Example 18.5.1 Consider a multinational corporation is looking for the appropriate agricultural investment destination. The President of that organization appoints a group of four experts to make suggestions on agricultural location selection. The President requires a collective verdict on their choice. The committee resolves to conduct research on scientific grounds. Assume that

$$C = \{\beta^{\daleth}_1, \beta^{\daleth}_2, \cdots, \beta^{\daleth}_6\}$$

is the set of agriculture places under consideration. Further suppose that

$$Q = \{q_1, q_2, \cdots, q_5\}$$

is the set of qualities under consideration for the selection.

Picking the weights from Table 18.7, the experts provide the following weighted parameter matrix

$$\mathcal{P} = \begin{pmatrix} \text{VN} & \text{MLR} & \text{MLR} & \text{ONU} & \text{VN} \\ \text{M} & \text{AR} & \text{AR} & \text{AR} & \text{VN} \\ \text{M} & \text{M} & \text{VN} & \text{M} & \text{M} \\ \text{MLR} & \text{AR} & \text{MLR} & \text{AR} & \text{VN} \end{pmatrix}$$
$$= \begin{pmatrix} 0.90 & 0.40 & 0.30 & 0.10 & 0.90 \\ 0.70 & 0.15 & 0.20 & 0.15 & 0.85 \\ 0.60 & 0.70 & 0.90 & 0.80 & 0.75 \\ 0.40 & 0.15 & 0.40 & 0.15 & 0.90 \end{pmatrix}$$

The normalized weighted matrix is

$$\hat{N} = \begin{pmatrix} 0.667 & 0.480 & 0.286 & 0.120 & 0.528 \\ 0.519 & 0.180 & 0.191 & 0.180 & 0.499 \\ 0.445 & 0.840 & 0.858 & 0.960 & 0.440 \\ 0.296 & 0.180 & 0.381 & 0.180 & 0.528 \end{pmatrix}$$

and hence the WV is $\check{\Upsilon} = (0.220, 0.192, 0.196, 0.164, 0.228)$.

Assume that the four experts provide the following q-ROFS matrices in which the q-ROFN at (i, j)th position demarcated grades of candidates row-wise and the attribute column-wise.

$$D_1 = \begin{pmatrix} (0.57, 0.39) & (0.49, 0.74) & (0.77, 0.38) & (0.54, 0.21) & (0.12, 0.48) \\ (0.66, 0.51) & (0.54, 0.54) & (0.32, 0.13) & (0.99, 0.13) & (0.54, 0.07) \\ (0.15, 0.68) & (0.19, 0.32) & (0.76, 0.41) & (0.45, 0.15) & (0.11, 0.49) \\ (0.67, 0.74) & (0.09, 0.83) & (0.59, 0.31) & (0.84, 0.16) & (0.37, 0.21) \\ (0.59, 0.17) & (0.33, 0.67) & (0.34, 0.68) & (0.52, 0.19) & (0.58, 0.61) \\ (0.27, 0.54) & (0.49, 0.46) & (0.48, 0.59) & (0.55, 0.54) & (0.38, 0.01) \end{pmatrix}$$

$$D_2 = \begin{pmatrix} (0.34, 0.52) & (0.58, 0.21) & (0.47, 0.21) & (0.70, 0.31) & (0.11, 0.34) \\ (0.47, 0.33) & (0.39, 0.32) & (0.56, 0.20) & (0.38, 0.11) & (0.26, 0.18) \\ (0.59, 0.17) & (0.33, 0.17) & (0.19, 0.28) & (0.59, 0.06) & (0.78, 0.16) \\ (0.44, 0.17) & (0.38, 0.23) & (0.58, 0.27) & (0.71, 0.24) & (0.54, 0.02) \\ (0.32, 0.28) & (0.56, 0.11) & (0.44, 0.37) & (0.49, 0.29) & (0.55, 0.55) \\ (0.34, 0.47) & (0.52, 0.37) & (0.11, 0.18) & (0.47, 0.13) & (0.47, 0.27) \end{pmatrix}$$

$$D_3 = \begin{pmatrix} (0.11, 0.58) & (0.37, 0.22) & (0.56, 0.11) & (0.21, 0.69) & (0.79, 0.32) \\ (0.13, 0.67) & (0.46, 0.13) & (0.36, 0.54) & (0.56, 0.27) & (0.46, 0.61) \\ (0.59, 0.13) & (0.25, 0.11) & (0.62, 0.33) & (0.47, 0.28) & (0.28, 0.47) \\ (0.11, 0.49) & (0.23, 0.05) & (0.50, 0.28) & (0.34, 0.48) & (0.61, 0.54) \\ (0.17, 0.29) & (0.82, 0.34) & (0.56, 0.51) & (0.50, 0.28) & (0.49, 0.12) \\ (0.33, 0.69) & (0.57, 0.61) & (0.48, 0.57) & (0.33, 0.02) & (0.46, 0.31) \end{pmatrix}$$

$$D_4 = \begin{pmatrix} (0.40, 0.59) & (0.41, 0.32) & (0.49, 0.12) & (0.35, 0.65) & (0.39, 0.12) \\ (0.25, 0.17) & (0.38, 0.10) & (0.85, 0.26) & (0.44, 0.57) & (0.92, 0.14) \\ (0.38, 0.51) & (0.36, 0.11) & (0.52, 0.29) & (0.48, 0.38) & (0.52, 0.35) \\ (0.56, 0.11) & (0.73, 0.16) & (0.35, 0.27) & (0.58, 0.62) & (0.62, 0.63) \\ (0.11, 0.01) & (0.33, 0.37) & (0.28, 0.38) & (0.47, 0.32) & (0.71, 0.19) \\ (0.58, 0.17) & (0.44, 0.15) & (0.56, 0.16) & (0.33, 0.21) & (0.88, 0.26) \end{pmatrix}$$

Thus, the aggregated matrix is

$$D = \begin{pmatrix} (0.355, 0.520) & (0.463, 0.373) & (0.573, 0.205) & (0.450, 0.465) & (0.353, 0.315) \\ (0.378, 0.420) & (0.443, 0.273) & (0.523, 0.283) & (0.593, 0.270) & (0.545, 0.250) \\ (0.428, 0.373) & (0.373, 0.178) & (0.523, 0.328) & (0.498, 0.218) & (0.423, 0.368) \\ (0.445, 0.378) & (0.358, 0.318) & (0.505, 0.283) & (0.618, 0.375) & (0.535, 0.350) \\ (0.298, 0.188) & (0.510, 0.373) & (0.405, 0.485) & (0.495, 0.270) & (0.583, 0.368) \\ (0.380, 0.468) & (0.505, 0.398) & (0.408, 0.375) & (0.420, 0.225) & (0.548, 0.213) \end{pmatrix}$$

and hence the weighted q-ROFS matrix is

$$D_w = \begin{pmatrix} (0.078, 0.114) & (0.089, 0.072) & (0.112, 0.040) & (0.074, 0.076) & (0.080, 0.072) \\ (0.083, 0.092) & (0.085, 0.052) & (0.103, 0.055) & (0.097, 0.044) & (0.124, 0.057) \\ (0.094, 0.082) & (0.072, 0.034) & (0.103, 0.064) & (0.082, 0.036) & (0.096, 0.084) \\ (0.098, 0.083) & (0.069, 0.061) & (0.099, 0.055) & (0.101, 0.062) & (0.122, 0.080) \\ (0.066, 0.041) & (0.098, 0.072) & (0.079, 0.095) & (0.081, 0.044) & (0.133, 0.084) \\ (0.084, 0.103) & (0.097, 0.076) & (0.080, 0.074) & (0.069, 0.037) & (0.125, 0.049) \end{pmatrix}$$

The positive and negative ideal solutions are

$$\begin{aligned} \text{q-ROFSV-PIS} &= \{\mathbb{U}^{\beta+}_1, \mathbb{U}^{\beta+}_2, \cdots, \mathbb{U}^{\beta+}_5\} \\ &= \{(0.098, 0.041), (0.098, 0.034), (0.112, 0.040), (0.101, 0.036), (0.133, 0.049)\} \end{aligned}$$

and

$$\text{q-ROFSV-NIS} = \{\mathbb{U}^{\beta}{}_1^-, \mathbb{U}^{\beta}{}_2^-, \cdots, \mathbb{U}^{\beta}{}_5^-\}$$
$$= \{(0.066, 0.114), (0.069, 0.076), (0.079, 0.095), (0.069, 0.076), (0.096, 0.084)\}$$

respectively.
Choosing $\kappa = 0.5$, the values of S_i, R_i, and Q_i for each choice $\mathbb{U}^{\beta}{}_i$ are calculated utilizing

$$S_i = \Sigma_{j=1}^{5} \mathfrak{w}_j \left(\frac{d\left(\mathbb{U}^{\beta+}_j, \mathbb{U}^{\beta}{}_{ij}\right)}{d\left(\mathbb{U}^{\beta+}_j, \mathbb{U}^{\beta-}_j\right)} \right)$$
$$R_i = \max_{j=1}^{5} \mathfrak{w}_j \left(\frac{d\left(\mathbb{U}^{\beta+}_j, \mathbb{U}^{\beta}{}_{ij}\right)}{d\left(\mathbb{U}^{\beta+}_j, \mathbb{U}^{\beta-}_j\right)} \right)$$
$$Q_i = \kappa \left(\frac{S_i - S^-}{S^+ - S^-} \right) + (1-\kappa) \left(\frac{R_i - R^-}{R^+ - R^-} \right)$$

and are given in Table 18.8 below:

The rank of choices is as under:

$$\text{By } Q_i: \quad \beta^{\daleth}{}_2 \prec \beta^{\daleth}{}_4 \prec \beta^{\daleth}{}_6 \prec \beta^{\daleth}{}_5 \prec \beta^{\daleth}{}_3 \prec \beta^{\daleth}{}_1$$

$$\text{By } S_i: \quad \beta^{\daleth}{}_2 \prec \beta^{\daleth}{}_4 \prec \beta^{\daleth}{}_3 \prec \beta^{\daleth}{}_6 \prec \beta^{\daleth}{}_5 \prec \beta^{\daleth}{}_1$$

$$\text{By } R_i: \quad \beta^{\daleth}{}_6 \prec \beta^{\daleth}{}_5 \prec \beta^{\daleth}{}_2 \prec \beta^{\daleth}{}_4 \prec \beta^{\daleth}{}_3 \prec \beta^{\daleth}{}_1$$

Since

$$Q(\beta^{\daleth}{}_4) - Q(\beta^{\daleth}{}_2) = 0.2457 \ngeq \frac{1}{4}$$

so (a) is not gratified. Further,

$$Q(\beta^{\daleth}{}_6) - (\beta^{\daleth}{}_2) = 0.4368 \geq \frac{1}{4}$$

Thus, the committee recommends that the agriculture lands $\beta^{\daleth}{}_2$, $\beta^{\daleth}{}_4$ and $\beta^{\daleth}{}_6$ must be chosen. These rankings are depicted in Fig. 18.4.

Table 18.8 Values of S_i, R_i, and Q_i for alternatives

Alternative	S_i	R_i	Q_i
$\beta^{\daleth}_1$	0.7698	0.2589	1.0000
$\beta^{\daleth}_2$	0.3663	0.1469	0.0000
$\beta^{\daleth}_3$	0.5788	0.2280	0.6260
$\beta^{\daleth}_4$	0.5562	0.1491	0.2457
$\beta^{\daleth}_5$	0.6531	0.1025	0.5754
$\beta^{\daleth}_6$	0.6148	0.0358	0.4368

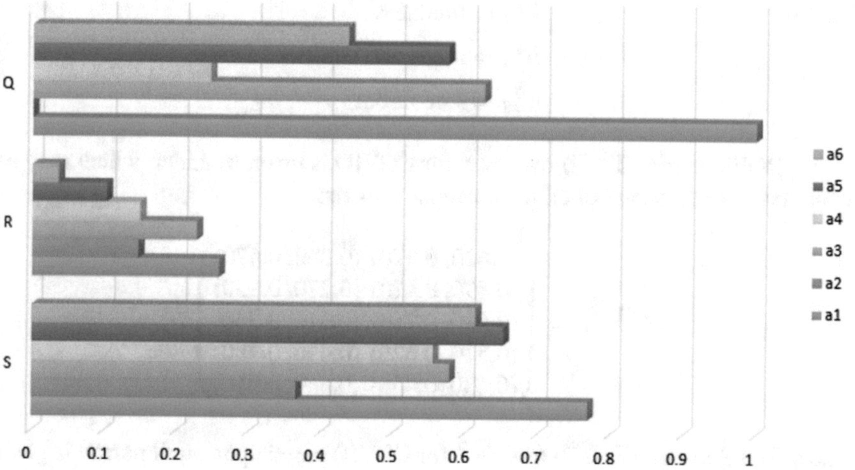

Fig. 18.4 3D column chart of rankings

18.6 Practical implementation of proposed SM related to COVID-19

In this part, we utilize recommended SM as a model to determine if a patient has COVID or not. As previously stated, we first present Algorithm 18.4 before proceeding to a numerical example in which the suggested SM may be successfully applied, as follows:

Algorithm 18.4

Step 1: Determine that $\widetilde{\mathscr{Y}}^{\triangledown} = \{\eta_1 = \text{COVID}, \eta_2 = \text{no COVID}\}$.
Step 2: Determine the collection of symptoms $\breve{\daleth} = \{\breve{\hbar}^{\blacklozenge}_1, \hbar^{\blacklozenge}_2, \cdots, \breve{\hbar}^{\blacklozenge}_n\}$.
Step 3: Determine a model q-ROFS matrix $(\beth^{\urcorner}, \breve{\daleth})$ with which similarity is to be computed.
Step 4: Determine q-ROFS matrix $(\beth^{\urcorner}_1, \breve{\daleth})$ for the patients.
Step 5: Evaluate the SMs among the $(\beth^{\urcorner}_1, \breve{\daleth})$ and $(\beth^{\urcorner}, \breve{\daleth})$.

Step 6: Set the threshold value $\lambda \in]0, 1[$.
Step 7: The patient is diseased if $Sim(\gimel^{\urcorner}, \gimel^{\urcorner}_1) \geq \lambda$.

Example 18.6.1 Presume that $\widetilde{\mathscr{Y}}^{\triangledown} = \{\eta_1 = \text{COVID}, \eta_2 = \text{no COVID}\}$. Let's choose the set of parameters containing the collection of some detectible symptoms, say, $\breve{\daleth} = \{\breve{\hbar}^{\blacklozenge}_i : i = 1, 2, \cdots, 5\}$, where

$$\begin{aligned}
\breve{\hbar}^{\blacklozenge}_1 &= \text{fever}\\
\breve{\hbar}^{\blacklozenge}_2 &= \text{dry cough}\\
\breve{\hbar}^{\blacklozenge}_3 &= \text{tiredness}\\
\breve{\hbar}^{\blacklozenge}_4 &= \text{aches and pains}\\
\breve{\hbar}^{\blacklozenge}_5 &= \text{sore throat}
\end{aligned}$$

The q-ROFS matrix $(\gimel^{\urcorner}, \breve{\daleth})$ over $\widetilde{\mathscr{Y}}^{\triangledown}$ for COVID is given as under, which may be constructed with the aid of clinical/medical experts:

$$(\gimel^{\urcorner}, \breve{\daleth}) = \begin{pmatrix} (0.620, 0.470) & (0.360, 0.570)\\ (0.890, 0.410) & (0.270, 0.930)\\ (0.580, 0.250) & (0.310, 0.540)\\ (0.510, 0.620) & (0.490, 0.380)\\ (0.630, 0.450) & (0.530, 0.410) \end{pmatrix}$$

The q-ROFS matrix $(\gimel^{\urcorner}_1, \breve{\daleth})$ over $\widetilde{\mathscr{Y}}^{\triangledown}$ for COVID based upon an ill person is given as follows:

$$(\gimel^{\urcorner}_1, \breve{\daleth}) = \begin{pmatrix} (0.110, 0.070) & (0.920, 0.150)\\ (0.140, 0.050) & (0.860, 0.260)\\ (0.080, 0.960) & (0.570, 0.020)\\ (0.360, 0.690) & (0.830, 0.190)\\ (0.460, 0.370) & (0.290, 0.840) \end{pmatrix}$$

Let's decide the threshold value $\lambda = 0.75$. The SM between $(\gimel^{\urcorner}, \breve{\daleth})$ and $(\gimel^{\urcorner}_1, \breve{\daleth})$ is $Sim(\gimel^{\urcorner}, \gimel^{\urcorner}_1) = 0.6497 < \lambda$, so we conclude that the person does not seem to be victim of COVID.

18.7 Conclusion

In this chapter, we looked into some fundamental q-ROFSS concepts. Some fundamental mechanisms and their primary properties are also addressed using illustrative examples. We proposed four techniques for simulating ambiguities in MCDM circumstances based on q-ROFSSs: choice value method, q-ROFS-TOPSIS, VIKOR, and methodology of SMs. The presented algorithms have been successfully used to rank various options. We used statistical graphics to help us understand the final ranks. The presented models have a lot of theoretical and application promise, and they could be useful in different hybrid fuzzy set structures with slight amendments. The concepts can be applied to effectively deal with ambiguity in a wide range of real scenarios, including business, machine intelligence, brand management, cognitive science, finding the shortest dilemma, representative democracy, pattern classification, deep learning, diagnostics, trade assessment, projections, agri-business assessment, mechatronics, cryptography, computer vision, hiring process issues, and so on.

References

1. L.A. Zadeh, Fuzzy sets. Inf. Control **8**, 338–353 (1965)
2. K.T. Atanassov, Intuitionistic fuzzy sets. Fuzzy Sets Syst. **20**(1), 87–96 (1986)
3. R.R. Yager, A.M. Abbasov, Pythagorean membership grades, complex numbers, and decision making. Int. J. Intell. Syst. **28**, 436–452 (2013)
4. R. R. Yager, Pythagorean fuzzy subsets, IFSA World Congress and NAFIPS Annual Meeting (IFSA/NAFIPS), 2013 Joint, Edmonton, Canada. IEEE **2013**, 57–61 (2013)
5. R.R. Yager, Pythagorean membership grades in multi criteria decision-making. IEEE Trans. Fuzzy Syst. **22**, 958–965 (2014)
6. R.R. Yager, Generalized orthopair fuzzy sets. IEEE Trans. Fuzzy Syst. **25**, 1222–1230 (2017)
7. D. Molodtsov, Soft set theory-first results. Comput. Math. Appl. **37**(4–5), 19–31 (1999)
8. N. Çağman, S. Enginoglu, F. Çitak, Fuzzy soft set theory and its applications. Iran. J. Fuzzy Syst. **8**(3), 137–147 (2011)
9. F. Feng, Y.B. Jun, X. Liu, L. Li, An adjustable approach to fuzzy soft set based decision making. J. Comput. Appl. Math. **234**(1), 10–20 (2010)
10. P. Majumdar, S.K. Samanta, Generalised fuzzy soft sets. Comput. Math. Appl. **59**(4), 1425–1432 (2010)
11. F. Feng, C. Li, B. Davvaz, M.I. Ali, Soft sets combined with fuzzy sets and rough sets, a tentative approach. Soft. Comput. **14**(9), 899–911 (2010)
12. B. Davvaz, E.H. Sadrabadi, An application of intuitionistic fuzzy sets in medicine. Int. J. Biomath. **9**(3), 1650037-1–1650037-15 (2016)
13. P.K. Maji, A.R. Roy, R. Biswas, Intuitionistic fuzzy soft sets. J. Fuzzy Math. **9**(3), 677–692 (2001)
14. F. Feng, H. Fujita, M.I. Ali, R.R. Yager, X. Liu, Another view on generalized intuitionistic fuzzy soft sets and related multi-attribute decision making methods. IEEE Trans. Fuzzy Syst. **27**(3), 474–488 (2019)
15. Z. Li, R. Cui, On the topological structure of intuitionistic fuzzy soft sets. Ann. Fuzzy Math. Inf. **5**(1), 229–239 (2013)
16. I. Osmanoglu, D. Tokat, On intuitionistic fuzzy soft topology. Gen. Math. Notes **19**(2), 59–70 (2013)
17. H. Garg, R. Arora, A nonlinear-programming methodology for multi-attribute decision-making problem with interval-valued intuitionistic fuzzy soft sets information. Appl. Intell. **48**(8), 2031–2046 (2018)
18. A. Guleria, R.K. Bajaj, On Pythagorean fuzzy soft matrices, operations and their applications in decision making and medical diagnosis. Soft. Comput. **23**(17), 7889–7900 (2019)
19. S. Naz, S. Ashraf, M. Akram, A novel approach to decision-making with Pythagorean fuzzy information. Mathematics **95**(6), 1–28 (2018)

20. X.D. Peng, Y. Yang, Some results for Pythagorean fuzzy sets. Int. J. Intell. Syst. **30**(11), 1133–1160 (2015)
21. X.D. Peng, H.Y. Yuan, Y. Yang, Pythagorean fuzzy information measures and their applications. Int. J. Intell. Syst. **32**(10), 991–1029 (2017)
22. X.D. Peng, Y.Y. Yang, J. Song, Y. Jiang, Pythagorean fuzzy soft set and its application. Comput. Eng. **41**(7), 224–229 (2015)
23. X.D. Peng, G. Selvachandran, Pythagorean fuzzy set: state of the art and future directions. Artif. Intell. Rev. **52**(3), 1873–1927 (2019)
24. M. Riaz, K. Naeem, Measurable soft mappings. Punjab Univ. J. Math. **48**(2), 19–34 (2016)
25. M. Riaz, K. Naeem, M.O. Ahmad, Novel concepts of soft sets with applications. Ann. Fuzzy Math. Inf. **13**(2), 239–251 (2017)
26. L. Fei, Y. Feng, L. Liu, On Pythagorean fuzzy decision making using soft likelihood functions. Int. J. Intell. Syst. **34**(12), 3317–3335 (2019)
27. L. Fei, Y. Deng, Multi-criteria decision making in Pythagorean fuzzy environment. Appl. Intell. **50**, 537–561 (2019)
28. C.L. Hwang, K. Yoon, *Multiple Attribute Decision Making-Methods and Applications* (Springer, Heidelberg, 1981)
29. A. Adeel, M. Akram, A.N.A. Koam, Group decision-making based on m-polar fuzzy linguistic TOPSIS method. Symmetry **11**(735), 1–20 (2019)
30. S. Eraslan, F. Karaaslan, A group decision making method based on TOPSIS under fuzzy soft environment. J. New Theory **3**, 30–40 (2015)
31. K. Naeem, M. Riaz, D. Afzal, Pythagorean m-polar fuzzy sets and TOPSIS method for the selection of advertisement mode. J. Intell. Fuzzy Syst. **37**(6), 8441–8458 (2019)
32. Y. Liu, H. Zhang, Y. Wu, Y. Dong, Ranking range based approach to MADM under incomplete context and its application in venture investment evaluation. Technol. Econ. Dev. Econ. **25**(5), 877–899 (2019)
33. K. Kumar, H. Garg, TOPSIS method based on the connection number of set pair analysis under interval-valued intuitionistic fuzzy set environment. Comput. Appl. Math. **37**(2), 1319–1329 (2018)
34. M. Riaz, H.M.A. Farid, F. Karaaslan, M.R. Hashmi, Some q-rung orthopair fuzzy hybrid aggregation operators and TOPSIS method for multi-attribute decision-making. J. Intell. Fuzzy Syst. **39**(1), 1227–1241 (2020)
35. M. Riaz, M.T. Hamid, H.M.A. Farid, D. Afzal, TOPSIS, VIKOR and aggregation operators based on q-rung orthopair fuzzy soft sets and their applications. J. Intell. Fuzzy Syst. **39**(5), 6903–6917 (2020)
36. D.F. Li, J.X. Nan, Extension of the TOPSIS for multi-attribute group decision making under Atanassov IFS environments. Int. J. Fuzzy Syst. Appl. **1**(4), 47–61 (2011)
37. S. Opricovic, H.G. Tzeng, Compromise solution by MCDM methods: a comparative analysis of VIKOR and TOPSIS. Eur. J. Oper. Res. **156**(2), 445–455 (2004)
38. S. Opricovic, H.G. Tzeng, Extended VIKOR method in comparison with other outranking methods. Eur. J. Oper. Res. **178**(2), 514–529 (2007)
39. R. Mohd, L. Abdullah, The VIKOR method with pythagorean fuzzy sets and their applications, in *Proceedings of the Third International Conference on Computing, Mathematics and Statistics (iCMS2017)*. https://doi.org/10.1007/978-981-13-7279-7-24
40. K. Naeem, M. Riaz, X.D. Peng, D. Afzal, Pythagorean fuzzy soft MCGDM methods based on TOPSIS, VIKOR and aggregation operators. J. Intell. Fuzzy Syst. **37**(5), 6937–6957 (2019)
41. S. Kalkan, M. Turanli, Ü. Özden, Ö. Başar, Comparison of ranking results obtained by TOPSIS and VIKOR methods, using the same criteria as Times Higher Education World University ranking. Eur. J. Bus. Soc. Sci. **6**(1), 107–122 (2017)
42. Z.S. Xu, Intuitionistic fuzzy aggregation operators. IEEE Trans. Fuzzy Syst. **15**, 1179–1187 (2007)
43. Z.S. Xu, R.R. Yager, Some geometric aggregation operators based on intuitionistic fuzzy sets. Int. J. Gen Syst **35**, 417–433 (2006)

44. Z.S. Xu, M.M. Xia, Induced generalized intuitionitic fuzzy operators, Knowledge-based. System **24**, 197–209 (2011)
45. T. Mahmood, F. Mehmood, Q. Khan, Some generalized aggregation operators for cubic hesitant fuzzy sets and their application to multi-criteria decision making. Punjab Univ. J. Math. **49**(1), 31–49 (2017)
46. G. Wei, H. Wang, X. Zhao, R. Lin, Hesitant triangular fuzzy information aggregation in multiple attribute decision making. J. Intell. Fuzzy Syst. **26**(3), 1201–1209 (2014)
47. S. Jose, S. Kuriaskose, Aggregation operators, score function and accuracy function for multi criteria decision making in intuitionistic fuzzy context. Notes Intuit. Fuzzy Sets **20**(1), 40–44 (2014)
48. M.R. Hashmi, M. Riaz, F. Smarandache, m-polar neutrosophic topology with applications to multi-criteria decision-making in medical diagnosis and clustering analysis. Int. J. Fuzzy Syst. **22**(1), 273–292 (2020)
49. W. Wang, X. Liu, Intuitionistic fuzzy information aggregation using Einstein operators. IEEE Trans. Fuzzy Syst. **20**, 923–938 (2012)
50. H.Y. Zhang, J.Q, Wang, X.H. Chen, Interval neutrosophic sets and their applications in multi-criteria decision making problems. Sci. World J. 1–15 (2014)
51. H. Zhao, Z.S. Xu, M.F. Ni, S.S. Lui, Generalized aggregation operators for intuitionistic fuzzy sets. Int. J. Intell. Syst. **25**(1), 1–30 (2010)
52. Z.M. Mu, S.Z. Zeng, P.Y. Wang, Novel approach to multi-attribute group decision-making based on interval-valued Pythagorean fuzzy power Maclaurin symmetric mean operator. Comput. Ind. Eng. **155**, 107049 (2021)
53. J.F. Wang, S.Z. Zeng, C. Zhang, Single-valued neutrosophic linguistic logarithmic weighted distance measures and their application to supplier selection of fresh aquatic products. Mathematics **8**, 439 (2020)
54. H. Garg, A new generalized Pythagorean fuzzy information aggregation using Einstein operators and its applications to decision-making. Int. J. Intell. Syst. **31**(9), 886–920 (2016)
55. K. Rahman, S. Abdullah, R. Ahmad, M. Ullah, Pythagorean fuzzy Einstein weighted geometric aggregation operator and their application to multiple-attribute group decision-making. J. Intell. Fuzzy Syst. **33**, 635–647 (2017)
56. S.Z. Zeng, Y.J. Hu, T. Balezentis, D. Streimikiene, A multi-criteria sustainable supplier selection framework based on neutrosophic fuzzy data and entropy weighting. Sustain. Dev. **28**(5), 1431–1440 (2020)
57. C.H. Zhang, W.H. Su, S.Z. Zeng, T. Balezentis, E. Herrera-Viedma, A two-stage subgroup decision-making method for processing large-scale information. Expert Syst. Appl. **171**, 114586 (2021)
58. S.Z. Zeng, Y.J. Hu, X.Y. Xie, Q-rung orthopair fuzzy weighted induced logarithmic distance measures and their application in multiple attribute decision making. Eng. Appl. Artif. Intell. **100**, 104167 (2021)
59. M. Sitara, M. Akram, M. Riaz, Decision-making analysis based on q-rung picture fuzzy graph structures. J. Appl. Math. Comput. (2021). https://doi.org/10.1007/s12190-020-01471-z
60. M. Riaz, H. Garg, H.M.A. Farid, R. Chinram, Multi-criteria decision making based on bipolar picture fuzzy operators and new distance measures. Comput. Model. Eng. Sci. **127**(2), 771–800 (2021)
61. M. Riaz, W. Salabun, H.M.A. Farid, N. Ali, J. Watróbski, A robust q-rung orthopair fuzzy information aggregation using Einstein operations with application to sustainable energy planning decision management. Energies **13**(9), 2125 (2020)
62. M. Riaz, D. Pamucar, H.M.A. Farid, M.R. Hashmi, q-rung orthopair fuzzy prioritized aggregation operators and their application towards green supplier chain management. Symmetry **12**(6), 976 (2020)
63. M. Riaz, H.M.A. Farid, H. Kalsoom, D. Pamucar, Y.M. Chu, A Robust q-rung orthopair fuzzy Einstein prioritized aggregation operators with application towards MCGDM. Symmetry **12**(6), 1058 (2020)

64. M. Riaz, M.R. Hashmi, Linear Diophantine fuzzy set and its applications towards multi-attribute decision making problems. J. Intell. Fuzzy Syst. **37**(4), 5417–5439 (2019)
65. M. Riaz, H.M.A. Farid, M. Aslam, D. Pamucar, D. Bozanic, Novel approach for third-party reverse logistic provider selection process under linear Diophantine fuzzy prioritized aggregation operators. Symmetry **13**(7), 1152 (2021)
66. A. Iampan, G.S. Garcia, M. Riaz, H.M.A. Farid, R. Chinram, Linear diophantine fuzzy einstein aggregation operators for multi-criteria decision-making problems. J. Math. **2021**, 5548033 (2021)
67. P. Liu, J. Liu, Some q-rung orthopai fuzzy bonferroni mean operators and their application to multi-attribute group decision making. Int. J. Intell. Syst. **33**(2), 315–347 (2018)
68. Z. Liu, S. Wang, P. Liu, Multiple attribute group decision making based on q-rung orthopair fuzzy Heronianmean operators. Int. J. Intell. Syst. **33**(12), 2341–2364 (2018)
69. B.P. Joshi, A. Gegov, Confidence levels q-rung orthopair fuzzy aggregation operators and its applications to MCDM problems. Int. J. Intell. Syst. 125–149 (2020)
70. M. Riaz, A. Razzaq, H. Kalsoom, D. Pamucar, H.M.A. Farid, Y.M. Chu, q-rung orthopair fuzzy geometric aggregation operators based on generalized and group-generalized parameters with application to water loss management. Symmetry **12**(8), 1236 (2020)
71. D.H. Hong, C.A. Kim, A note on similarity measure between vague sets and elements. Inf. Sci. **115**(1–4), 83–96 (1999)
72. A. Kharal, Distance and similarity measures for soft sets. New Math. Nat. Comput. **6**(3), 321–334 (2010)
73. H. Kamaci, Similarity measure for soft matrices and its applications. J. Intell. Fuzzy Syst. **36**(4), 3061–3072 (2019)
74. W.L. Hung, M.S. Yang, Similarity measures of intuitionistic fuzzy sets based on L_p metric. Int. J. Approx. Reason. **46**, 120–136 (2007)
75. L.K. Hyung, Y.S. Song, K.M. Lee, Similarity measure between fuzzy sets and between elements. Fuzzy Sets Syst. **62**(3), 291–293 (1994)
76. S.M. Chen, Measures of similarity between vague sets. Fuzzy Sets Syst. **74**(2), 217–223 (1995)
77. S.M. Chen, Similarity measures between vague sets and between elements. IEEE Trans. Syst. Man Cybern. Part B Cybern. **27**(1), 153–158 (1997)
78. S.M. Chen, M.S. Yeh, P.Y. Hsiao, A comparison of similarity measures of fuzzy values. Fuzzy Sets Syst. **72**(1), 79–89 (1995)
79. J. Wang, H. Gao, G. Wei, The generalized Dice similarity measures for Pythagorean fuzzy multiple attribute group decision making. Int. J. Intell. Syst. **34**(6), 1158–1183 (2019)
80. P. Muthukumar, S.S. Krishnan, A similarity measure of intuitionistic fuzzy soft sets and its application in medical diagnosis. Appl. Soft Comput. **41**, 148–156 (2016)
81. X.D. Peng, H. Garg, Multiparametric similarity measures on Pythagorean fuzzy sets with applications to pattern recognition. Appl. Intell. **49**, 4058–4096 (2019)
82. H. Garg, CN-q-ROFS: connection number-based q-rung orthopair fuzzy set and their application to decision-making process. Int. J. Intell. Syst. **36**(7), 3106–3143 (2021)
83. H. Garg, A new possibility degree measure for interval-valued q-rung orthopair fuzzy sets in decision-making. Int. J. Intell. Syst. **36**(1), 526–557 (2021)
84. H. Garg, New exponential operation laws and operators for interval-valued q-rung orthopair fuzzy sets in group decision making process. Neural Comput. Appl. **33**(20), 13937–13963 (2021). https://doi.org/10.1007/s00521-021-06036-0
85. Z. Yang, H. Garg, Interaction power partitioned maclaurin symmetric mean operators under q-rung orthopair uncertain linguistic information. Int. J. Fuzzy Syst. (2021). https://doi.org/10.1007/s40815-021-01062-5
86. H. Garg, A novel trigonometric operation-based q-rung orthopair fuzzy aggregation operator and its fundamental properties. Neural Comput. Appl. **32**, 15077–15099 (2020)

Dr. Muhammad Riaz has received M.Sc M.Phil and Ph.D degrees in Mathematics from Department of Mathematics, University of the Punjab Lahore. He has 24+ years regular teaching and research experience. He has published 100+ research articles in international peer-reviewed SCIE & ESCI journals with 1750+ citations. He has been supervised 05 Ph.D. students and 19 M.Phil students. Currently, he is supervising 05 M.Phil and 03 Ph.D. students. He is HEC Approved Supervisor. His research interests include Pure Mathematics, Fuzzy Mathematics, Topology, Algebra, Artificial intelligence, Computational Intelligence, soft set theory, rough set theory, neutrosophic sets, linear Diophantine fuzzy sets, biomathematics, aggregation operators with applications in decision-making problems, medical diagnosis, information measures, image processing, network topology and pattern recognition. He has been a reviewer for 40+ SCI journals. He is member editorial board of 07 journals. As the project leader, he has directed 2 research projects at the national or provincial level. He has been delivered seminars/talks as invited speakers in many international conferences. He is member of board of studies, member board of faculty, member departmental doctoral program committee (DDPC), and member of many university level committees.

Hafiz Muhammad Athar Farid received his BS degree in Mathematics from the University of Education, Lahore in 2018. He received M.Phil degree in Mathematics from the University of Punjab, Lahore in 2020. He is currently PhD scholar at Department of Mathematics, University of the Punjab, Lahore, Pakistan. He is the author of 20+ SCI research papers with 175+ citations. He is also a reviewer of international journals. His research interests include q-Rung Orthopair Fuzzy sets, Neutrosophic sets, Linear Diophantine Fuzzy sets, Multi-Criteria Decision-Making Problems, Aggregation Operators, Information Measures, Information Fusion, fuzzy modeling, fuzzy set theory, and Fuzzy Topology.

Chapter 19
Development of Heronian Mean-Based Aggregation Operators Under Interval-Valued Dual Hesitant q-Rung Orthopair Fuzzy Environments for Multicriteria Decision-Making

Nayana Deb, Arun Sarkar, and Animesh Biswas

Abstract Interval-valued dual hesitant q-rung orthopair fuzzy (IVDHq-ROF) set (IVDHq-ROFS) is a new variant of fuzzy set that can depict uncertain and imprecise situations more adequately than other existing fuzzy variants. In solving complicated multicriteria decision-making (MCDM) problems, decision-makers (DMs) sometimes confront interdependent aggregated arguments. Heronian mean (HM) can successfully capture the interrelationships between input arguments. The aim of this chapter is to define a new MCDM method under IVDHq-ROF environment based on HM operator. The proposed method is not only capable of dealing with DMs' hesitancy in a wide range but also can handle complicated decision-making situations by capturing interrelations among the aggregated arguments. In model formulation, at first, some HM-based IVDHq-ROF aggregation operators, viz., IVDHq-ROF HM, IVDHq-ROF weighted HM, IVDHq-ROF geometric HM, and IVDHq-ROF weighted geometric HM operators are proposed. Additionally, justifications of those operators to act as aggregation operators are validated by proving some of their desirable properties. Subsequently, a methodology for solving MCDM problems having interrelated input information is developed using the proposed operators. Further, a numerical example is solved to verify the application validity of the proposed approach. Finally, a comparative study with the existing approaches is performed to show the effectiveness of the developed method.

Keywords Dual hesitant q-rung orthopair fuzzy sets · Interval-valued dual hesitant q-rung orthopair fuzzy sets · Heronian mean · Geometric Heronian mean · Aggregation operator · Multicriteria decision-making

---

N. Deb · A. Biswas (✉)
Department of Mathematics, University of Kalyani, Kalyani 741235, India
e-mail: abiswaskln@rediffmail.com

A. Sarkar
Department of Mathematics, Heramba Chandra College, Kolkata 700029, India

H. Garg (ed.), *q-Rung Orthopair Fuzzy Sets*,
https://doi.org/10.1007/978-981-19-1449-2_19

19.1 Introduction

Multicriteria decision-making (MCDM) is a method that helps the decision-makers (DMs) in determining the best suitable alternative from a set of choices satisfying several criteria. MCDM has become an emerging research topic in recent days. A large number of applications are found on MCDM under various fields, viz., supplier selection [1], risk analysis [2, 3], pattern recognition [4], image processing [5], etc. In formulating MCDM problems, it is observed that lack of information knowledge may point to an inadequate result. To deal with such ambiguous factors associated with real-life decision-making, Yager proposed the notion of Pythagorean fuzzy (PF) sets (PFSs) [6, 7] as a powerful tool. PFS extends the concept of intuitionistic fuzzy set [8] by extending the range of capturing the uncertainty present in the decision-making processes. Although numerous researchers successfully applied PFS in modeling MCDM problems [9–19], it cannot address certain instances due to some drawbacks. For example, suppose a DM wants to produce 0.8 as membership value and 0.7 as non-membership value on evaluation of an object. In that case, PFS fails to represent the evaluation value as $0.8^2 + 0.7^2 = 1.13$, which exceeds the boundary 1. In order to overcome the limitations of PFS, Yager [20] extended the idea of PFS to q-rung orthopair fuzzy (q-ROF) set (q-ROFS). After introducing the q-ROFS, the range of capturing imprecise data is modified as the maximum value of the sum of q-th power of membership and non-membership degrees is 1 for q-ROFS. It also allows the DMs to provide their assessments by changing the parameter value q according to their hesitancy degree. Several researchers paid attention to this set due to its valuable characteristics. Liu and Wang [21] developed q-ROF weighted averaging (WA) and weighted geometric (WG) operators for solving MCDM problems. Again, Liu and Wang [22] suggested a multi-attribute decision-making (MADM) approach based on Archimedean Bonferroni mean (BM) operator under q-ROF environment. Based on power Maclaurin symmetric mean (MSM), Liu et al. [23] proposed a series of q-ROF aggregation operators for MCDM. Further, Liu et al. [24] defined cosine similarity and distance measures on q-ROFSs. Yang and Pang [25] developed a MADM approach utilizing partitioned BM operators on q-ROFSs. The Hamy mean was studied by Wang et al. [26] for the selection of enterprise resource planning systems under q-ROF environment. MSM-based q-ROF aggregation operators are developed by Wei et al. [27] to deal with MCDM problems. To aggregate q-ROF numbers, Garg [28] defined a series of WA and WG aggregation operators based on sine trigonometry operations. Riaz et al. [29] developed several aggregation operators, viz., q-ROF interaction WA, ordered WA, and hybrid averaging operators along with their geometric variants and applied in green supply chain management. A novel ranking technique has been introduced by Khan et al. [30] for q-ROFSs. Also, several research works [31–35] had been accomplished by numerous scholars in recent times utilizing the scope of applications under q-ROF environment. Recently, Yang and Garg [36] introduced interaction operational rules for q-ROF uncertain linguistic sets and developed q-ROF uncertain linguistic interaction power partitioned MSM operators.

Sometimes, while determining an object's membership and non-membership degrees, DMs feel difficulty in expressing their decision values appropriately using a single index. Zhu et al. [37] introduced the concept of dual hesitant fuzzy (DHF) sets (DHFS) to ease such cumbersome. Following the perception raised by Zhu et al. [37], Xu et al. [38] applied the idea on q-ROFS and developed the notion of dual hesitant q-ROF (DHq-ROF) set (DHq-ROFS).

The DHq-ROFS provides greater accessibility to the DMs while putting membership and non-membership values of an object using a set of possible q-ROF numbers instead of a single q-ROF number. Utilizing DHq-ROFS, Xu et al. [38] developed aggregation operators based on Heronian mean (HM) and geometric HM (GHM) under DHq-ROF contexts to deal with MCDM problems. Based on Hamacher operations, Wang et al. [39] proposed some DHq-ROF aggregation operators. Further, Wang et al. [40] introduced Muirhead mean into DHq-ROFSs for developing an MCDM approach. Recently, Sarkar and Biswas [41] developed Dombi t-conorm and t-norm-based BM operators for the context of solving MCDM having DHq-ROFS information.

Recently, as another extension of q-ROFS, interval-valued q-ROF (IVq-ROF) set (IVq-ROFS) was introduced by Joshi [42], in which interval numbers are used to represent q-ROF membership and non-membership grades. Previously, interval-valued DHF (IVDHF) set (IVDHFS) was developed by Ju et al. [43] by adopting interval-valued membership and non-membership grades into DHFSs.

In order to portray the hesitancy of DMs and flexibly reveal the fuzzy information, DHq-ROFSs play an appreciable role. However, DHq-ROFS confines the DM's accessibility owing to present membership and non-membership values using sets of possible crisp numbers. So, by considering the merits of IVDHFSs, Xu et al. [44] modified the drawback of DHq-ROFS by introducing interval numbers and found a novel tool called interval-valued DHq-ROF (IVDHq-ROF) set (IVDHq-ROFS). It can deal with MCDM problems more accurately by employing interval numbers rather than single crisp values to express possible membership and non-membership values in DHq-ROFS. IVDHq-ROFSs [44] are more suitable than several existing variants of q-ROFSs, e.g., DHq-ROFS, IVq-ROFS, q-ROFSs, etc., in dealing with the hesitancy of human cognition. In comparison to DHq-ROFSs, IVDHq-ROFSs can reflect the fuzzy information better, as they consider the membership and non-membership degrees with a set of several possible IVq-ROF numbers. Further, IVDHq-ROFSs can capture more vague data than DHq-ROFS as DMs can provide possible interval numbers instead of giving possible crisp numbers to represent the membership and non-membership degrees. That is why IVDHq-ROFSs are more powerful and superior than DHq-ROFSs in dealing with vagueness. Thus, IVDHq-ROFSs is a meaningful innovation in the research area.

The HM operator is very effective in fusing evaluation information. Moreover, it can depict interrelationships with input arguments. For its intuitive characteristics, it has drawn numerous researchers' interests in modeling MCDM [45–53]. Hence, it would be an interesting and meaningful work if HM operator is utilized to construct IVDHq-ROF aggregation operators. The main focus of this chapter is to produce some novel HM aggregation operators to fuse IVDHq-ROF arguments. The significant contributions of this chapter are as follows:

(1) In this chapter, IVDHq-ROF environment is considered to develop an MCDM approach, which extends the scope of applications in real-life decision-making contexts by incorporating DM's hesitancy more significantly.
(2) A series of IVDHq-ROF aggregation operators, viz., IVDHq-ROF HM (IVDHq-ROFHM), IVDHq-ROF weighted HM (IVDHq-ROFWHM), IVDHq-ROF GHM (IVDHq-ROFGHM), and IVDHq-ROF weighted GHM (IVDHq-ROFWGHM) operators are developed to cope with more complicated fuzzy situations.
(3) The proposed method can consider the interrelationship among aggregated arguments by employing HM as an aggregation function. Thus, it would produce more rational and adequate results in real-life MCDM contexts.

The proposed IVDHq-ROF aggregation operators introduced in this chapter are more generalized and enriched. A list of HM-based aggregation operators is presented in Table 19.1, focusing on their features. It would be observed that all the aggregation operators mentioned in Table 19.1 are the particular cases of the proposed operators. Further, it is to be noted here that some HM operators under various fuzzy environments are present in Table 19.1, which are yet to be introduced in the research, viz., interval-valued PF HM (IVPFHM), interval-valued hesitant PF HM (IVHPFHM), and IVq-ROF HM (IVq-ROFHM). The novelty of the proposed operators is that these operators have all the significant characteristics such as considering interrelationships among arguments, capturing DM's hesitancy, and also having the influence of rung parameter q.

This chapter is arranged in the following manner:

In Sect. 19.2, some basic concepts related to DHq-ROFSs and IVDHq-ROFSs are discussed. The proposed HM-based IVDHq-ROF aggregation operators are developed in Sect. 19.3. Moreover, some important features of newly developed operators are also presented in this section. A methodology for solving MCDM problems using the proposed operators under IVDHq-ROF environment is provided in Sect. 19.4. Section 19.5 illustrates a numerical example based on the developed MCDM approach, along with a discussion about the effect of several parameters on the final decision results of that problem. Also, comparative analyses with the existing methods are performed in Sect. 19.6. Lastly, some concluding remarks are presented in Sect. 19.7.

19.2 Preliminaries

Some basic concepts related to DHq-ROFSs and IVDHq-ROFSs are briefly reviewed in this section.

Table 19.1 Different HM-based operators on account of their characteristics

HM-based operators	Whether considers interrelationships	Whether considers hesitancy	Having utility of rung parameter q
DHPFGWHM [45]	Yes	Yes	No
q-RDHFWHM [38]	Yes	Yes	Yes
GIIFWHM [46]	Yes	No	No
q-ROFWHM [47]	Yes	No	Yes
q-ROFWGHM [48]	Yes	No	Yes
IFGWHM [49]	Yes	No	No
HFWHM and HFWGHM [50]	Yes	Yes	No
IVDHFWHM and IVDHFWGHM [51]	Yes	Yes	No
DHFWHM and DHFWGHM [52]	Yes	Yes	No
GPFWHM and PFWGHM [53]	Yes	No	No
IVPFHM (yet to be introduced)	Yes	No	No
IVHPFHM (yet to be introduced)	Yes	Yes	No
IVq-ROFHM (yet to be introduced)	Yes	No	Yes
Proposed method	Yes	Yes	Yes

Abbreviations Dual hesitant PF generalized weighted HM (DHPFGWHM), q-rung dual hesitant fuzzy weighted HM (q-RDHFWHM), generalized interval-valued intuitionistic fuzzy weighted HM (GIIFWHM), q-ROF weighted HM (q-ROFWHM), q-ROF weighted GHM (q-ROFWGHM), intuitionistic fuzzy geometric weighed HM (IFGWHM), hesitant fuzzy weighted HM (HFWHM), hesitant fuzzy weighted GHM (HFWGHM), IVDHF weighted HM (IVDHFWHM), IVDHF weighted GHM (IVDHFWGHM), DHF weighted HM (DHFWHM), DHF weighted GHM (DHFWGHM), generalized PF weighted HM (GPFWHM), PF weighted GHM (PFWGHM), interval-valued PF HM (IVPFHM), interval-valued hesitant PF HM (IVHPFHM), IVq-ROF HM (IVq-ROFHM)

19.2.1 *DH*q*-ROFS*

Definition 19.1 [38] A DHq-ROFS, $\mathcal{B}$, defined on a fixed set, X, is defined as

$$\mathcal{B} = \{\langle x, h_{\mathcal{B}}(x), g_{\mathcal{B}}(x)\rangle | x \in X\},$$

in which $h_{\mathcal{B}}(x) = \bigcup_{\gamma \in h_{\mathcal{B}}(x)} \{\gamma\}$ and $g_{\mathcal{B}}(x) = \bigcup_{\delta \in g_{\mathcal{B}}(x)} \{\delta\}$ denote two sets of numbers belonging to [0, 1], representing possible membership and non-membership degrees of an object $x \in X$ to the set $\mathcal{B}$, respectively, satisfying the condition:

$$(\gamma)^q + (\delta)^q \leq 1, q \geq 1.$$

For simplicity, the pair $(h_{\mathcal{B}}(x), g_{\mathcal{B}}(x))$ is called an DHq-ROF number (DHq-ROFN) and is denoted by $\beta = (h, g)$.

DMs often encounter situations when they are unable to put membership and non-membership values with two sets of possible crisp numbers using DHq-ROFNs due to insufficient or incomplete information present in practical MCDM. To cope with such situations, Xu et al. [44] introduced a novel concept of IVDHq-ROFS, which allows the DMs to provide possible membership and non-membership grades of DHq-ROFS using interval numbers rather than crisp values.

*19.2.2 IVDH*q*-ROFS*

Definition 19.2 [44] Let X be a fixed set and the power set of [0, 1] is denoted by $I([0, 1])$. An IVDHq-ROFS, $\widetilde{\mathcal{K}}$ on X is presented as

$$\widetilde{\mathcal{K}} = \left\{\left\langle x, \tilde{h}_{\widetilde{\mathcal{K}}}(x), \tilde{g}_{\widetilde{\mathcal{K}}}(x)\right\rangle | x \in X\right\}, \tag{19.1}$$

where $\tilde{h}_{\widetilde{\mathcal{K}}}(x) = \bigcup_{[\gamma^l,\gamma^u]\in\tilde{h}_{\widetilde{\mathcal{K}}}(x)} \{[\gamma^l, \gamma^u]\}$ and $\tilde{g}_{\widetilde{\mathcal{K}}}(x) = \bigcup_{[\eta^l,\eta^u]\in\tilde{g}_{\widetilde{\mathcal{K}}}(x)} \{[\eta^l, \eta^u]\}$ represent two collections of interval values belonging to $I([0, 1])$, denoting the possible membership and non-membership values, respectively, corresponding to the element $x \in X$ satisfying the condition $0 \leq \left((\gamma^u)^+\right)^q + \left((\eta^u)^+\right)^q \leq 1$, in which $(\gamma^u)^+ = max\{\gamma^u\}$ and $(\eta^u)^+ = max\{\eta^u\}$, $\left[\gamma^l, \gamma^u\right] \in \tilde{h}_{\widetilde{\mathcal{K}}}(x)$ and $\left[\eta^l, \eta^u\right] \in \tilde{g}_{\widetilde{\mathcal{K}}}(x)$.

For convenience, the pair $\left(\tilde{h}_{\widetilde{\mathcal{K}}}(x), \tilde{g}_{\widetilde{\mathcal{K}}}(x)\right)$ is called an IVDHq-ROF number (IVDHq-ROFN) and is denoted by $\tilde{k} = \left(\tilde{h}, \tilde{g}\right)$.

Evidently, for $q = 1$, IVDHq-ROFS transforms to IVDHFS [43], and for $q = 2$, IVDHq-ROFS converted into interval-valued hesitant PFS (IVHPFS) [54].

Example 19.1 Let a DM provide possible membership degrees of an object $x \in X$ to the set $\widetilde{\mathcal{K}}$ as [0.1, 0.3], [0.4, 0.6], and [0.7, 0.8], and possible non-membership degrees as [0.3, 0.4] and [0.4, 0.6], simultaneously. Then the IVDHq-ROFN can be represented as

$\tilde{k} = (\{[0.1, 0.3], [0.4, 0.6], [0.7, 0.8]\}, \{[0.3, 0.4], [0.4, 0.6]\})$ in which $(\gamma^u)^+ = 0.8$, $(\eta^u)^+ = 0.6$ and $0 \leq (0.8)^q + (0.6)^q \leq 1$ for $q = 3$.

In order to set a comparison rule among IVDHq-ROFNs, the score and accuracy functions are introduced as follows.

Definition 19.3 [44] Let $\tilde{k} = \left(\tilde{h}, \tilde{g}\right)$ be an IVDHq-ROFN. Then the score function of $\tilde{k}$ is defined as

$$S\left(\tilde{k}\right)=\frac{1}{2}\left(\frac{1}{\#\tilde{h}}\sum_{[\gamma^l,\gamma^u]\in\tilde{h}}\left((\gamma^l)^q+(\gamma^u)^q\right)-\frac{1}{\#\tilde{g}}\sum_{[\eta^l,\eta^u]\in\tilde{g}}\left((\eta^l)^q+(\eta^u)^q\right)\right) \tag{19.2}$$

and the accuracy function of $\tilde{k}$ is defined as

$$A\left(\tilde{k}\right)=\frac{1}{2}\left(\frac{1}{\#\tilde{h}}\sum_{[\gamma^l,\gamma^u]\in\tilde{h}}\left((\gamma^l)^q+(\gamma^u)^q\right)+\frac{1}{\#\tilde{g}}\sum_{[\eta^l,\eta^u]\in\tilde{g}}\left((\eta^l)^q+(\eta^u)^q\right)\right), \tag{19.3}$$

where $\#\tilde{h}$ and $\#\tilde{g}$ defined the number of intervals that are present in $\tilde{h}$ and $\tilde{g}$, respectively.

- **Comparison rule for IVDHq-ROFNs**:

Let $\tilde{k}_i$ $(i=1,2)$ be any two IVDHq-ROFNs, then the ordering of IVDHq-ROFNs can be performed by the following manner:

- If $S\left(\tilde{k}_1\right)>S\left(\tilde{k}_2\right)$ then $\tilde{k}_1\succ\tilde{k}_2$;
- If $S\left(\tilde{k}_1\right)=S\left(\tilde{k}_2\right)$ then

(i) if $A\left(\tilde{k}_1\right)>A\left(\tilde{k}_2\right)$ then $\tilde{k}_1\succ\tilde{k}_2$;

(ii) if $A\left(\tilde{k}_1\right)=A\left(\tilde{k}_2\right)$ then $\tilde{k}_1\approx\tilde{k}_2$.

19.2.3 *Operations on IVDH*q*-ROFNs*

Some basic operational laws for IVDHq-ROFNs are described below.

Definition 19.4 [44] Let $\tilde{k}=\left(\tilde{h},\tilde{g}\right)$, $\tilde{k}_1=\left(\tilde{h}_1,\tilde{g}_1\right)$, and $\tilde{k}_2=\left(\tilde{h}_2,\tilde{g}_2\right)$ be any three IVDHq-ROFNs, where $\tilde{h}=\bigcup_{[\gamma^l,\gamma^u]\in\tilde{h}}\{[\gamma^l,\gamma^u]\}$, $\tilde{g}=\bigcup_{[\eta^l,\eta^u]\in\tilde{g}}\{[\eta^l,\eta^u]\}$, $\tilde{h}_i=\bigcup_{[\gamma_i^l,\gamma_i^u]\in\tilde{h}_i}\{[\gamma_i^l,\gamma_i^u]\}$, and $\tilde{g}_i=\bigcup_{[\eta_i^l,\eta_i^u]\in\tilde{g}_i}\{[\eta_i^l,\eta_i^u]\}$ $(i=1,2)$. The basic arithmetic operations on IVDHq-ROFNs are defined as

$$(1)\ \tilde{k}_1\oplus\tilde{k}_2=\left(\bigcup_{\substack{[\gamma_i^l,\gamma_i^u]\in\tilde{h}_i\\ i=1,2}}\left\{\left[\left((\gamma_1^l)^q+(\gamma_2^l)^q-(\gamma_1^l)^q(\gamma_2^l)^q\right)^{\frac{1}{q}},\right\}\right.$$

$$\left((\gamma_1^u)^q + (\gamma_2^u)^q - (\gamma_1^u)^q(\gamma_2^u)^q\right)^{\frac{1}{q}}\Big]\Big\}, \bigcup_{\substack{[\eta_i^l,\, \eta_i^u] \in \tilde{g}_i \\ i=1,2}} \{[\eta_1^l\eta_2^l, \eta_1^u\eta_2^u]\} \Bigg);$$

$$(2)\ \tilde{k}_1 \otimes \tilde{k}_2 = \Bigg(\cup_{\substack{[\gamma_i^l,\, \gamma_i^u] \in \tilde{h}_i \\ i=1,2}} \{[\gamma_1^l\gamma_2^l, \gamma_1^u\gamma_2^u]\},$$

$$\cup_{\substack{[\eta_i^l,\, \eta_i^u] \in \tilde{g}_i \\ i=1,2}} \Big\{\Big[\left(\eta_1^{lq} + \eta_2^{lq} - (\eta_1^l)^q(\eta_2^l)^q\right)^{\frac{1}{q}},$$

$$\left((\eta_1^u)^q + (\eta_2^u)^q - (\eta_1^u)^q(\eta_2^u)^q\right)^{\frac{1}{q}}\Big]\Big\}\Bigg);$$

$$(3)\ \lambda\tilde{k} = \left(\bigcup_{[\gamma^l,\gamma^u]\in\tilde{h}} \left\{\left[\left(1-\left(1-(\gamma^l)^q\right)^\lambda\right)^{\frac{1}{q}}, \left(1-\left(1-(\gamma^u)^q\right)^\lambda\right)^{\frac{1}{q}}\right]\right\},\right.$$

$$\left.\bigcup_{[\eta^l,\eta^u]\in\tilde{g}} \left\{\left[(\eta^l)^\lambda, (\eta^u)^\lambda\right]\right\}\right);$$

$$(4)\ \tilde{k}^\lambda = \left(\bigcup_{[\gamma^l,\gamma^u]\in\tilde{h}} \left\{\left[(\gamma^l)^\lambda, (\gamma^u)^\lambda\right]\right\}, \bigcup_{[\eta^l,\eta^u]\in\tilde{g}} \left\{\left[\left(1-\left(1-(\eta^l)^q\right)^\lambda\right)^{\frac{1}{q}},\right.\right.\right.$$

$$\left.\left.\left.\left(1-\left(1-(\eta^u)^q\right)^\lambda\right)^{\frac{1}{q}}\right]\right\}\right).$$

19.2.4 HM Operator

To consider the interrelationships among input attributes, Beliakov et al. [55] defined HM operator. The definition of it is presented as follows.

Definition 19.5 [55] Consider $a_i\,(i = 1, 2, \ldots, n)$ as a collection of nonnegative real numbers with $\varphi, \psi \geq 0$. Then the function, $HM^{\varphi,\psi}{:}[0, 1]^n \to [0, 1]$ is represented as

$$HM^{\varphi,\psi}(a_1, a_2, \ldots, a_n) = \left(\frac{2}{n(n+1)} \sum_{\substack{i,j=1 \\ i \leq j}}^{n} a_i^\varphi a_j^\psi\right)^{\frac{1}{\varphi+\psi}},$$

which is known as HM operator with parameter.

19.2.5 GHM Operator

Definition 19.6 [49] Consider a_i $(i = 1, 2, \ldots, n)$ as a collection of nonnegative real numbers with $\varphi, \psi \geq 0$. Then the function, $GHM^{\varphi,\psi}$:$[0, 1]^n \to [0, 1]$ is defined as

$$GHM^{\varphi,\psi}(a_1, a_2, \ldots, a_n) = \frac{1}{\varphi + \psi}\left(\prod_{\substack{i, j = 1 \\ i \leq j}}^{n} \left(a_i^{\varphi} + a_j^{\psi}\right)\right)^{\frac{2}{n(n+1)}},$$

which represents GHM operator with parameter.

19.3 HM-Based IVDHq-ROF Aggregation Operators and Its Properties

In this section, HM and GHM aggregation operators are utilized to generate a series of IVDHq-ROF operators, viz., IVDHq-ROFHM and IVDHq-ROFGHM operators and their weighted forms. The properties of those developed operators are also discussed.

19.3.1 IVDHq-ROFHM Operator

By incorporating HM operator for aggregating IVDHq-ROF information, the IVDHq-ROFHM operator is established as follows.

Definition 19.7 Let $\tilde{k}_i = \left(\tilde{h}_i, \tilde{g}_i\right)$ $(i = 1, 2, \ldots, n)$ be a collection of IVDHq-ROFNs. Let $\varphi, \psi \geq 0$ be any number. If

$$IVDHq\text{-}ROFHM^{\varphi,\psi}\left(\tilde{k}_1, \tilde{k}_2, \ldots, \tilde{k}_n\right)$$

$$= \left(\frac{2}{n(n+1)} \oplus_{\substack{i, j = 1 \\ i \leq j}}^{n} \left(\tilde{k}_i^{\varphi} \otimes \tilde{k}_j^{\psi}\right)\right)^{\frac{1}{\varphi+\psi}}, \qquad (19.4)$$

then $IVDHq - ROFHM^{\theta,\phi}$ is called IVDHq-ROFHM operator.

Theorem 19.1 *Let* $\tilde{k}_i = \left(\tilde{h}_i, \tilde{g}_i\right)$ $(i = 1, 2, \ldots, n)$ *be a collection of IVDHq-ROFNs and* $\varphi, \psi \geq 0$*, then the aggregated value using IVDHq-ROFHM is also a IVDHq-ROFN and can be given as follows:*

$$IVDHq\text{-}ROFHM^{\varphi,\psi}\left(\tilde{k}_1, \tilde{k}_2, \ldots, \tilde{k}_n\right)$$

$$= \left(\frac{2}{n(n+1)} \oplus^{n}_{\substack{i,j=1 \\ i \leq j}} \left(\tilde{k}_i^{\varphi} \otimes \tilde{k}_j^{\psi}\right)\right)^{\frac{1}{\varphi+\psi}}$$

$$= \left(\bigcup_{\left[\gamma_i^l,\gamma_i^u\right]\in\tilde{h}_i,\left[\gamma_j^l,\gamma_j^u\right]\in\tilde{h}_j} \left\{\left[\left(1-\left(\prod_{\substack{i,j=1 \\ i \leq j}}^{n} \left(1-\left(\gamma_i^l\right)^{q\varphi}\left(\gamma_j^l\right)^{q\psi}\right)\right)^{\frac{2}{n(n+1)}}\right)^{\frac{1}{q(\varphi+\psi)}},\right.\right.\right.$$

$$\left.\left.\left(1-\left(\prod_{\substack{i,j=1 \\ i \leq j}}^{n} \left(1-\left(\gamma_i^u\right)^{q\varphi}\left(\gamma_j^u\right)^{q\psi}\right)\right)^{\frac{2}{n(n+1)}}\right)^{\frac{1}{q(\varphi+\psi)}}\right]\right\},$$

$$\bigcup_{\substack{\left[\eta_i^l,\eta_i^u\right]\in\tilde{g}_i, \\ \left[\eta_j^l,\eta_j^u\right]\in\tilde{g}_j}} \left\{\left[\left(1-\left(1-\left(\prod_{\substack{i,j=1 \\ i \leq j}}^{n} \left(1-\left(\left(1-\eta_i^{lq}\right)^{\varphi}\right.\right.\right.\right.\right.\right.$$

$$\left.\left.\left.\left.\left.\left.\left(1-\left(\eta_j^l\right)^q\right)^{\psi}\right)\right)\right)^{\frac{2}{n(n+1)}}\right)^{\frac{1}{\varphi+\psi}}\right)^{\frac{1}{q}},$$

$$\left(1-\left(1-\left(\prod_{\substack{i,j=1\\ i\le j}}^{n}\left(1-\left(\left(1-(\eta_i^u)^q\right)^{\varphi}\left(1-(\eta_j^u)^q\right)^{\psi}\right)\right)\right)^{\frac{2}{n(n+1)}}\right)^{\frac{1}{\varphi+\psi}}\right)^{\frac{1}{q}}\Bigg]\Bigg\}\Bigg)\Bigg). \tag{19.5}$$

Proof From the basic operational rules of IVDHq-ROFNs,

$$\tilde{k}_i^{\varphi}=\left(\bigcup_{\substack{[\gamma_i^l,\gamma_i^u]\\ \in \tilde{h}_i}}\left\{\left[\gamma_i^{l\varphi},\gamma_i^{u\varphi}\right]\right\},\right.$$

$$\left.\bigcup_{\substack{[\eta_i^l,\eta_i^u]\\ \in \tilde{g}_i}}\left\{\left[\left(1-\left(1-\eta_i^{lq}\right)^{\varphi}\right)^{\frac{1}{q}},\left(1-\left(1-\eta_i^{uq}\right)^{\varphi}\right)^{\frac{1}{q}}\right]\right\}\right),$$

$$\tilde{k}_j^{\psi}=\left(\bigcup_{\substack{[\gamma_j^l,\gamma_j^u]\\ \in \tilde{h}_j}}\left\{\left[\gamma_j^{l\psi},\gamma_j^{u\psi}\right]\right\},\right.$$

$$\left.\bigcup_{\substack{[\eta_j^l,\eta_j^u]\\ \in \tilde{g}_j}}\left\{\left[\left(1-\left(1-\eta_j^{lq}\right)^{\psi}\right)^{\frac{1}{q}},\left(1-\left(1-\eta_j^{uq}\right)^{\psi}\right)^{\frac{1}{q}}\right]\right\}\right).$$

Then,

$$\tilde{k}_i^{\varphi} \otimes \tilde{k}_j^{\psi} = \left(\bigcup_{[\gamma_i^l, \gamma_i^u] \in \tilde{h}_i, [\gamma_j^l, \gamma_j^u] \in \tilde{h}_j} \left\{ \left[(\gamma_i^l)^{\varphi} (\gamma_j^l)^{\psi}, (\gamma_i^u)^{\varphi} (\gamma_j^u)^{\phi} \right] \right\}, \right.$$

$$\bigcup_{\substack{[\eta_i^l, \eta_i^u] \in \tilde{g}_i, \\ [\eta_j^l, \eta_j^u] \in \tilde{g}_j}} \left\{ \left[\left(1 - \left(1 - \eta_i^{lq}\right)^{\varphi} \left(1 - \eta_j^{lq}\right)^{\psi} \right)^{\frac{1}{q}}, \right. \right.$$

$$\left. \left. \left. \left(1 - (1 - (\eta_i^u)^q)^{\varphi} (1 - (\eta_j^u)^q)^{\psi}\right)^{\frac{1}{q}} \right] \right\} \right).$$

Now, based on mathematical induction, the following is proved:

$$\oplus_{\substack{i, j = 1 \\ i \le j}}^{n} \left(\tilde{k}_i^{\varphi} \otimes \tilde{k}_j^{\psi} \right)$$

$$= \left(\bigcup_{[\gamma_i^l, \gamma_i^u] \in \tilde{h}_i, [\gamma_j^l, \gamma_j^u] \in \tilde{h}_j} \left\{ \left[\left(1 - \prod_{\substack{i, j = 1 \\ i \le j}}^{n} \left(1 - (\gamma_i^l)^{q\varphi} (\gamma_j^l)^{q\psi}\right) \right)^{\frac{1}{q}}, \right. \right. \right.$$

$$\left. \left. \left(1 - \prod_{\substack{i, j = 1 \\ i \le j}}^{n} \left(1 - (\gamma_i^u)^{q\varphi} (\gamma_j^u)^{q\psi}\right) \right)^{\frac{1}{q}} \right] \right\},$$

$$\bigcup_{\substack{[\eta_i^l, \eta_i^u] \in \tilde{g}_i, \\ [\eta_j^l, \eta_j^u] \in \tilde{g}_j}} \left\{ \left[\prod_{\substack{i, j = 1 \\ i \le j}}^{n} \left(1 - \left(\left(1 - (\eta_i^l)^q\right)^{\varphi} \left(1 - (\eta_j^l)^q\right)^{\psi} \right) \right)^{\frac{1}{q}}, \right. \right.$$

$$\left. \left. \left. \prod_{\substack{i, j = 1 \\ i \le j}}^{n} \left(1 - \left((1 - (\eta_i^u)^q)^{\varphi} (1 - (\eta_j^u)^q)^{\psi} \right) \right)^{\frac{1}{q}} \right] \right\} \right). \tag{19.6}$$

For $n = 2$,

$$\oplus^{2}_{\substack{i,j=1\\ i\le j}}\left(\tilde{k}_i^{\varphi}\otimes\tilde{k}_j^{\psi}\right)$$

$$=\left(\tilde{k}_1^{\varphi}\otimes\tilde{k}_1^{\psi}\right)\oplus\left(\tilde{k}_1^{\varphi}\otimes\tilde{k}_2^{\psi}\right)\oplus\left(\tilde{k}_2^{\varphi}\otimes\tilde{k}_2^{\psi}\right)$$

$$=\left(\bigcup_{[\gamma_i^l,\gamma_i^u]\in\tilde{h}_i,\left[\gamma_j^l,\gamma_j^u\right]\in\tilde{h}_j}\left\{\left[\left(1-\prod_{\substack{i,j=1\\ i\le j}}^{2}\left(1-(\gamma_i^l)^{q\varphi}(\gamma_j^l)^{q\psi}\right)\right)^{\frac{1}{q}},\left(1-\prod_{\substack{i,j=1\\ i\le j}}^{2}\left(1-(\gamma_i^u)^{q\varphi}(\gamma_j^u)^{q\psi}\right)\right)^{\frac{1}{q}}\right]\right\},\right.$$

$$\left.\bigcup_{\substack{\left[\eta_i^l,\,\eta_i^u\right]\in\tilde{g}_i,\\ \left[\eta_j^l,\,\eta_j^u\right]\in\tilde{g}_j}}\left\{\left[\prod_{\substack{i,j=1\\ i\le j}}^{2}\left(1-\left(\left(1-(\eta_i^l)^q\right)^{\varphi}\left(1-(\eta_j^l)^q\right)^{\psi}\right)\right)^{\frac{1}{q}},\prod_{\substack{i,j=1\\ i\le j}}^{2}\left(1-\left((1-(\eta_i^u)^q)^{\varphi}(1-(\eta_j^u)^q)^{\psi}\right)\right)^{\frac{1}{q}}\right]\right\}\right).$$

Thus (19.6) is true for $n = 2$.
Now assume that (19.6) is true for $n = t$.
That is,

$$\oplus^{t}_{\substack{i,j=1\\ i\le j}}\left(\tilde{k}_i^{\varphi}\otimes\tilde{k}_j^{\psi}\right)$$

$$= \left(\bigcup_{[\gamma_i^l, \gamma_i^u] \in \tilde{h}_i, [\gamma_j^l, \gamma_j^u] \in \tilde{h}_j} \left\{ \left[\left(1 - \prod_{\substack{i,j=1 \\ i \le j}}^{t} \left(1 - (\gamma_i^l)^{q\varphi} (\gamma_j^l)^{q\psi} \right) \right)^{\frac{1}{q}}, \right. \right. \right.$$

$$\left. \left. \left(1 - \prod_{\substack{i,j=1 \\ i \le j}}^{t} \left(1 - (\gamma_i^u)^{q\varphi} (\gamma_j^u)^{q\psi} \right) \right)^{\frac{1}{q}} \right] \right\},$$

$$\bigcup_{\substack{[\eta_i^l, \eta_i^u] \in \tilde{g}_i, \\ [\eta_j^l, \eta_j^u] \in \tilde{g}_j}} \left\{ \left[\prod_{\substack{i,j=1 \\ i \le j}}^{t} \left(1 - \left(\left(1 - (\eta_i^l)^q \right)^{\varphi} \left(1 - (\eta_j^l)^q \right)^{\psi} \right) \right)^{\frac{1}{q}}, \right. \right.$$

$$\left. \left. \left. \prod_{\substack{i,j=1 \\ i \le j}}^{t} \left(1 - \left((1 - (\eta_i^u)^q)^{\varphi} (1 - (\eta_j^u)^q)^{\psi} \right) \right)^{\frac{1}{q}} \right] \right\} \right). \tag{19.7}$$

Next, it is shown that Eq. (19.6) is valid for $n = t + 1$. That is,

$$\oplus_{\substack{i,j=1 \\ i \le j}}^{t+1} \left(\tilde{k}_i^{\varphi} \otimes \tilde{k}_j^{\psi} \right) = \left(\oplus_{\substack{i,j=1 \\ i \le j}}^{t} \left(\tilde{k}_i^{\varphi} \otimes \tilde{k}_j^{\psi} \right) \right) \oplus \left(\oplus_{i=1}^{t+1} \left(\tilde{k}_i^{\varphi} \otimes \tilde{k}_{t+1}^{\psi} \right) \right). \tag{19.8}$$

To prove Eq. (19.8), it is necessary to prove that

$$\oplus_{i=1}^{t+1}\left(\tilde{k}_i^{\varphi}\otimes\tilde{k}_{t+1}^{\psi}\right)$$
$$=\left(\bigcup_{\substack{[\gamma_i^l,\gamma_i^u]\in\tilde{h}_i,\\ [\gamma_{t+1}^l,\gamma_{t+1}^u]\in\tilde{h}_{t+1}}}\left\{\left[\left(1-\prod_{i=1}^{t+1}\left(1-(\gamma_i^l)^{q\varphi}(\gamma_{t+1}^l)^{q\psi}\right)\right)^{\frac{1}{q}},\right.\right.\right.$$
$$\left.\left.\left(1-\prod_{i=1}^{t+1}\left(1-(\gamma_i^u)^{q\varphi}(\gamma_{t+1}^u)^{q\psi}\right)\right)^{\frac{1}{q}}\right]\right\},$$
$$\bigcup_{\substack{[\eta_i^l,\eta_i^u]\in\tilde{g}_i,\\ [\eta_{t+1}^l,\eta_{t+1}^u]\in\tilde{g}_{t+1}}}\left\{\left[\prod_{i=1}^{t+1}\left(1-\left(\left(1-(\eta_i^l)^q\right)^{\varphi}\left(1-(\eta_{t+1}^l)^q\right)^{\psi}\right)\right)^{\frac{1}{q}},\right.\right.$$
$$\left.\left.\left.\prod_{i=1}^{t+1}\left(1-\left((1-(\eta_i^u)^q)^{\varphi}(1-(\eta_{t+1}^u)^q)^{\psi}\right)\right)^{\frac{1}{q}}\right]\right\}\right). \tag{19.9}$$

Now, it is clear that Eq. (19.9) is true for $t = 1, 2$.
It is assumed that Eq. (19.9) is true for $t = v$. Thus

$$\oplus_{i=1}^{v+1}\left(\tilde{k}_i^{\varphi}\otimes\tilde{k}_{v+1}^{\psi}\right)$$
$$=\left(\bigcup_{\substack{[\gamma_i^l,\gamma_i^u]\in\tilde{h}_i,\\ [\gamma_{v+1}^l,\gamma_{v+1}^u]\in\tilde{h}_{v+1}}}\left\{\left[\left(1-\prod_{i=1}^{v+1}\left(1-(\gamma_i^l)^{q\varphi}(\gamma_{v+1}^l)^{q\psi}\right)\right)^{\frac{1}{q}},\right.\right.\right.$$
$$\left.\left.\left(1-\prod_{i=1}^{v+1}\left(1-(\gamma_i^u)^{q\varphi}(\gamma_{v+1}^u)^{q\psi}\right)\right)^{\frac{1}{q}}\right]\right\},$$
$$\bigcup_{\substack{[\eta_i^l,\eta_i^u]\in\tilde{g}_i,\\ [\eta_{v+1}^l,\eta_{v+1}^u]\in\tilde{g}_{v+1}}}\left\{\left[\prod_{i=1}^{v+1}\left(1-\left(\left(1-(\eta_i^l)^q\right)^{\varphi}\left(1-(\eta_{v+1}^l)^q\right)^{\psi}\right)\right)^{\frac{1}{q}},\right.\right.$$
$$\left.\left.\left.\prod_{i=1}^{v+1}\left(1-\left((1-(\eta_i^u)^q)^{\varphi}(1-(\eta_{v+1}^u)^q)^{\psi}\right)\right)^{\frac{1}{q}}\right]\right\}\right).$$

Then for $t = v + 1$,

$$\oplus_{i=1}^{v+2}\left(\tilde{k}_i^{\varphi} \otimes \tilde{k}_{v+2}^{\psi}\right)$$

$$= \left(\oplus_{i=1}^{v+1}\left(\tilde{k}_i^{\varphi} \otimes \tilde{k}_{v+2}^{\psi}\right)\right) \oplus \left(\tilde{k}_{v+2}^{\varphi} \otimes \tilde{k}_{v+2}^{\psi}\right)$$

$$= \left(\bigcup_{\substack{[\gamma_i^l,\gamma_i^u] \in \tilde{h}_i,\\ [\gamma_{v+1}^l,\gamma_{v+1}^u] \in \tilde{h}_{v+1}}} \left\{\left[\left(1 - \prod_{i=1}^{v+1}\left(1 - (\gamma_i^l)^{q\varphi}(\gamma_{v+1}^l)^{q\psi}\right)\right)^{\frac{1}{q}},\right.\right.\right.$$

$$\left.\left.\left(1 - \prod_{i=1}^{v+1}\left(1 - (\gamma_i^u)^{q\varphi}(\gamma_{v+1}^u)^{q\psi}\right)\right)^{\frac{1}{q}}\right]\right\},$$

$$\bigcup_{\substack{[\eta_i^l,\eta_i^u] \in \tilde{g}_i,\\ [\eta_{v+1}^l,\eta_{v+1}^u] \in \tilde{g}_{v+1}}} \left\{\left[\prod_{i=1}^{v+1}\left(1 - \left(\left(1 - (\eta_i^l)^q\right)^{\varphi}\left(1 - (\eta_{v+1}^l)^q\right)^{\psi}\right)\right)^{\frac{1}{q}},\right.\right.$$

$$\left.\left.\left.\prod_{i=1}^{v+1}\left(1 - \left((1 - (\eta_i^u)^q)^{\varphi}(1 - (\eta_{v+1}^u)^q)^{\psi}\right)\right)^{\frac{1}{q}}\right]\right\}\right)$$

$$\oplus \left(\bigcup_{[\gamma_{v+2}^l,\gamma_{v+2}^u] \in \tilde{h}_{v+2}} \left\{\left[\left(\gamma_{v+2}^l\right)^{(\varphi+\psi)}, \left(\gamma_{v+2}^u\right)^{(\varphi+\psi)}\right]\right\},\right.$$

$$\left.\bigcup_{[\eta_{v+2}^l,\eta_{v+2}^u] \in \tilde{g}_{v+2}} \left\{\left[\left(1 - \left(1 - (\eta_{v+2}^l)^q\right)^{(\varphi+\psi)}\right)^{\frac{1}{q}}, \left(1 - \left(1 - (\eta_{v+2}^u)^q\right)^{(\varphi+\psi)}\right)^{\frac{1}{q}}\right]\right\}\right)$$

$$= \left(\bigcup_{[\gamma_i^l,\gamma_i^u] \in \tilde{h}_i} \left\{\left[\left(1 - \prod_{i=1}^{v+2}\left(1 - (\gamma_i^l)^{q\varphi}(\gamma_{v+2}^l)^{q\psi}\right)\right)^{\frac{1}{q}},,\right.\right.\right.$$

$$\left.\left.\left(1 - \prod_{i=1}^{v+2}\left(1 - (\gamma_i^u)^{q\varphi}(\gamma_{v+2}^u)^{q\psi}\right)\right)^{\frac{1}{q}}\right]\right\},$$

$$\bigcup_{[\eta_i^l,\eta_i^u] \in \tilde{g}_i} \left\{\left[\prod_{i=1}^{v+2}\left(1 - \left(\left(1 - (\eta_i^l)^q\right)^{\varphi}\left(1 - (\eta_{v+2}^l)^q\right)^{\psi}\right)\right)^{\frac{1}{q}},\right.\right.$$

$$\left.\left.\left.\prod_{i=1}^{v+2}\prod\left(1 - \left((1 - (\eta_i^u)^q)^{\varphi}(1 - (\eta_{v+2}^u)^q)^{\psi}\right)\right)^{\frac{1}{q}}\right]\right\}\right)$$

Thus Eq. (19.9) is true for $t = v + 1$, and so Eq. (19.9) is true for all t.

Thus, Eq. (19.8) can be easily derived with the help of Eqs. (19.7) and (19.9) as follows:

$$\oplus_{\substack{i,j=1\\ i\le j}}^{t+1}\left(\tilde{k}_i^{\varphi}\otimes\tilde{k}_j^{\psi}\right)$$

$$=\left(\oplus_{\substack{i,j=1\\ i\le j}}^{t}\left(\tilde{k}_i^{\varphi}\otimes\tilde{k}_j^{\psi}\right)\right)\oplus\left(\oplus_{i=1}^{t+1}\left(\tilde{k}_i^{\varphi}\otimes\tilde{k}_{t+1}^{\psi}\right)\right)$$

$$=\left(\bigcup_{[\gamma_i^l,\gamma_i^u]\in\tilde{h}_i,[\gamma_j^l,\gamma_j^u]\in\tilde{h}_j}\left\{\left[\left(1-\prod_{\substack{i,j=1\\ i\le j}}^{t}\left(1-\left(\gamma_i^l\right)^{q\varphi}\left(\gamma_j^l\right)^{q\psi}\right)\right)^{\frac{1}{q}},\right.\right.\right.$$

$$\left.\left.\left(1-\prod_{\substack{i,j=1\\ i\le j}}^{t}\left(1-\left(\gamma_i^u\right)^{q\varphi}\left(\gamma_j^u\right)^{q\psi}\right)\right)^{\frac{1}{q}}\right]\right\},$$

$$\cup_{\substack{[\eta_i^l,\eta_i^u]\in\tilde{g}_i,\\ [\eta_j^l,\eta_j^u]\in\tilde{g}_j}}\left\{\left[\prod_{\substack{i,j=1\\ i\le j}}^{t}\left(1-\left(\left(1-\left(\eta_i^l\right)^q\right)^{\varphi}\left(1-\left(\eta_j^l\right)^q\right)^{\psi}\right)\right)^{\frac{1}{q}},\right.\right.$$

$$\left.\left.\left.\prod_{\substack{i,j=1\\ i\le j}}^{t}\left(1-\left(\left(1-\left(\eta_i^u\right)^q\right)^{\varphi}\left(1-\left(\eta_j^u\right)^q\right)^{\psi}\right)\right)^{\frac{1}{q}}\right]\right\}\right)$$

$$\cup_{\substack{\left[\eta_i^l,\eta_i^u\right]\in\tilde{g}_i,\\ \left[\eta_j^l,\eta_j^u\right]\in\tilde{g}_j}} \left\{\left[\prod_{\substack{i,j=1\\ i\le j}}^{t}\left(1-\left(\left(1-\left(\eta_i^l\right)^q\right)^{\varphi}\left(1-\left(\eta_j^l\right)^q\right)^{\psi}\right)\right)^{\frac{1}{q}},\right.\right.$$

$$\left.\left.\prod_{\substack{i,j=1\\ i\le j}}^{t}\left(1-\left(\left(1-\left(\eta_i^u\right)^q\right)^{\varphi}\left(1-\left(\eta_j^u\right)^q\right)^{\psi}\right)\right)^{\frac{1}{q}}\right]\right\}\Bigg)$$

$$\oplus\left(\cup_{\substack{\left[\gamma_i^l,\gamma_i^u\right]\in\tilde{h}_i,\\ \left[\gamma_{t+1}^l,\gamma_{t+1}^u\right]\in\tilde{h}_{t+1}}} \left\{\left[\left(1-\prod_{i=1}^{t+1}\left(1-\left(\gamma_i^l\right)^{q\varphi}\left(\gamma_{t+1}^l\right)^{q\psi}\right)\right)^{\frac{1}{q}},\right.\right.\right.$$

$$\left.\left.\left(1-\prod_{i=1}^{t+1}\left(1-\left(\gamma_i^u\right)^{q\varphi}\left(\gamma_{t+1}^u\right)^{q\psi}\right)\right)^{\frac{1}{q}}\right]\right\},$$

$$\cup_{\substack{\left[\eta_i^l,\eta_i^u\right]\in\tilde{g}_i,\\ \left[\eta_{t+1}^l,\eta_{t+1}^u\right]\in\tilde{g}_{t+1}}} \left\{\left[\prod_{i=1}^{t+1}\left(1-\left(\left(1-\left(\eta_i^l\right)^q\right)^{\varphi}\left(1-\left(\eta_{t+1}^l\right)^q\right)^{\psi}\right)\right)^{\frac{1}{q}},\right.\right.$$

$$\left.\left.\left.\prod_{i=1}^{t+1}\left(1-\left(\left(1-\left(\eta_i^u\right)^q\right)^{\varphi}\left(1-\left(\eta_{t+1}^u\right)^q\right)^{\psi}\right)\right)^{\frac{1}{q}}\right]\right\}\right).$$

$$=\left(\bigcup_{\left[\gamma_i^l,\gamma_i^u\right]\in\tilde{h}_i,\left[\gamma_j^l,\gamma_j^u\right]\in\tilde{h}_j}\left\{\left[\left(1-\prod_{\substack{i,j=1\\ i\le j}}^{t+1}\left(1-\left(\gamma_i^l\right)^{q\varphi}\left(\gamma_j^l\right)^{q\psi}\right)\right)^{\frac{1}{q}},\right.\right.\right.$$

$$\left.\left.\left(1-\prod_{\substack{i,j=1\\ i\le j}}^{t+1}\left(1-\left(\gamma_i^u\right)^{q\varphi}\left(\gamma_j^u\right)^{q\psi}\right)\right)^{\frac{1}{q}}\right]\right\},$$

$$\cup_{\substack{\left[\eta_i^l,\eta_i^u\right]\in\tilde{g}_i,\\ \left[\eta_j^l,\eta_j^u\right]\in\tilde{g}_j}} \left\{\left[\prod_{\substack{i,j=1\\ i\le j}}^{t+1}\left(1-\left(\left(1-\left(\eta_i^l\right)^q\right)^{\varphi}\left(1-\left(\eta_j^l\right)^q\right)^{\psi}\right)\right)^{\frac{1}{q}},\right.\right.$$

$$\left.\left.\left.\prod_{\substack{i,j=1\\ i\le j}}^{t+1}\left(1-\left(\left(1-\left(\eta_i^u\right)^q\right)^{\varphi}\left(1-\left(\eta_j^u\right)^q\right)^{\psi}\right)\right)^{\frac{1}{q}}\right]\right\}\right)$$

That is, Eq. (19.6) is true for $n = t + 1$.
Hence, Eq. (19.6) is true for all n.
Further,

$$\frac{2}{n(n+1)} \oplus^{n}_{\substack{i,\, j = 1 \\ i \le j}} \left(\tilde{k}_i^{\varphi} \otimes \tilde{k}_j^{\psi}\right)$$

$$= \left(\bigcup_{[\gamma_i^l, \gamma_i^u] \in \tilde{h}_i, [\gamma_j^l, \gamma_j^u] \in \tilde{h}_j} \left\{ \left[\left(1 - \left(\prod_{\substack{i,\, j = 1 \\ i \le j}}^{n} \left(1 - (\gamma_i^l)^{q\varphi} (\gamma_j^l)^{q\psi}\right) \right)^{\frac{2}{n(n+1)}} \right)^{\frac{1}{q}}, \right.\right.\right.$$

$$\left.\left. \left(1 - \left(\prod_{\substack{i,\, j = 1 \\ i \le j}}^{n} \left(1 - (\gamma_i^u)^{q\varphi} (\gamma_j^u)^{q\psi}\right) \right)^{\frac{2}{n(n+1)}} \right)^{\frac{1}{q}} \right] \right\},$$

$$\bigcup_{\substack{[\eta_i^l, \eta_i^u] \in \tilde{g}_i, \\ [\eta_j^l, \eta_j^u] \in \tilde{g}_j}} \left\{ \left[\left(\prod_{\substack{i,\, j = 1 \\ i \le j}}^{n} \left(1 - \left(\left(1 - (\eta_i^l)^q\right)^{\varphi} \left(1 - (\eta_j^l)^q\right)^{\psi}\right)\right) \right)^{\frac{2}{qn(n+1)}}, \right.\right.$$

$$\left.\left.\left. \left(\prod_{\substack{i,\, j = 1 \\ i \le j}}^{n} \left(1 - \left(\left(1 - (\eta_i^u)^q\right)^{\varphi} \left(1 - (\eta_j^u)^q\right)^{\psi}\right)\right) \right)^{\frac{2}{qn(n+1)}} \right] \right\} \right).$$

Finally,

$$\left(\frac{2}{n(n+1)} \oplus^{n}_{\substack{i,\, j = 1 \\ i \le j}} \left(\tilde{k}_i^{\varphi} \otimes \tilde{k}_j^{\psi}\right) \right)^{\frac{1}{\varphi+\psi}}$$

$$= \left(\bigcup_{[\gamma_i^l, \gamma_i^u] \in \tilde{h}_i, [\gamma_j^l, \gamma_j^u] \in \tilde{h}_j} \left\{ \left[\left(1 - \left(\prod_{\substack{i,j=1 \\ i \le j}}^{n} \left(1 - (\gamma_i^l)^{q\varphi} (\gamma_j^l)^{q\psi} \right) \right)^{\frac{2}{n(n+1)}} \right)^{\frac{1}{q(\varphi+\psi)}}, \right. \right. \right.$$

$$\left. \left. \left(1 - \left(\prod_{\substack{i,j=1 \\ i \le j}}^{n} \left(1 - (\gamma_i^u)^{q\varphi} (\gamma_j^u)^{q\psi} \right) \right)^{\frac{2}{n(n+1)}} \right)^{\frac{1}{q(\varphi+\psi)}} \right] \right\},$$

$$\bigcup_{[\eta_i^l, \eta_i^u] \in \tilde{g}_i, [\eta_j^l, \eta_j^u] \in \tilde{g}_j} \left\{ \left[\left(1 - \left(1 - \left(\prod_{\substack{i,j=1 \\ i \le j}}^{n} \left(1 - \left(\left(1 - (\eta_i^l)^q \right)^{\varphi} \right. \right. \right. \right. \right. \right. \right.$$

$$\left. \left. \left. \left. \left. \left. \left(1 - (\eta_j^l)^q \right)^{\psi} \right) \right) \right)^{\frac{2}{n(n+1)}} \right)^{\frac{1}{\varphi+\psi}} \right)^{\frac{1}{q}},$$

$$\left(1 - \left(1 - \left(\prod_{\substack{i,j=1 \\ i \le j}}^{n} \left(1 - \left(\left(1 - (\eta_i^u)^q \right)^{\varphi} \right. \right. \right. \right. \right.$$

$$\left. \left. \left. \left. \left. \left. \left. \left(1 - (\eta_j^u)^q \right)^{\psi} \right) \right) \right)^{\frac{2}{n(n+1)}} \right)^{\frac{1}{\varphi+\psi}} \right)^{\frac{1}{q}} \right] \right\} \right).$$

Hence, the theorem is proved.

The IVDHq-ROFHM operator satisfies the following properties.

Theorem 19.2 (Idempotency) *If all* $\tilde{k}_i = \left(\tilde{h}_i, \tilde{g}_i \right)$ $(i = 1, 2, \ldots, n)$ *are equal to* $\tilde{k} = \left(\tilde{h}, \tilde{g} \right)$ *for all* $(i = 1, 2, \ldots, n)$*, then*

$$IVDHq - ROFHM^{\varphi,\psi}\left(\tilde{k}_1, \tilde{k}_2, \ldots, \tilde{k}_n \right) = \tilde{k}. \quad (19.10)$$

Proof

$$IVDHq-ROFHM^{\varphi,\psi}\left(\tilde{k}_1,\tilde{k}_2,\ldots,\tilde{k}_n\right)=\left(\frac{2}{n(n+1)}\oplus^{n}_{\substack{i,j=1\\ i\le j}}\left(\tilde{k}_i^{\varphi}\otimes\tilde{k}_j^{\psi}\right)\right)^{\frac{1}{\varphi+\psi}}$$

$$=\left(\bigcup_{\left[\gamma_i^l,\gamma_i^u\right]\in\tilde{h}_i,\left[\gamma_j^l,\gamma_j^u\right]\in\tilde{h}_j}\left\{\left[\left(1-\left(\prod_{\substack{i,j=1\\ i\le j}}^{n}\left(1-\left(\gamma_i^l\right)^{q\varphi}\left(\gamma_j^l\right)^{q\psi}\right)\right)^{\frac{2}{n(n+1)}}\right)^{\frac{1}{q(\varphi+\psi)}},\right.\right.\right.$$

$$\left.\left.\left(1-\left(\prod_{\substack{i,j=1\\ i\le j}}^{n}\left(1-\left(\gamma_i^u\right)^{q\varphi}\left(\gamma_j^u\right)^{q\psi}\right)\right)^{\frac{2}{n(n+1)}}\right)^{\frac{1}{q(\varphi+\psi)}}\right]\right\},$$

$$\bigcup_{\left[\eta_i^l,\eta_i^u\right]\in\tilde{g}_i,\left[\eta_j^l,\eta_j^u\right]\in\tilde{g}_j}\left\{\left[\left(1-\left(1-\left(\prod_{\substack{i,j=1\\ i\le j}}^{n}\left(1-\left(\left(1-\left(\eta_i^l\right)^q\right)^{\varphi}\right.\right.\right.\right.\right.\right.\right.$$

$$\left.\left.\left.\left.\left.\left.\left(1-\left(\eta_j^l\right)^q\right)^{\psi}\right)\right)\right)^{\frac{2}{n(n+1)}}\right)^{\frac{1}{\varphi+\psi}}\right)^{\frac{1}{q}},$$

$$\left.\left.\left.\left(1-\left(1-\left(\prod_{\substack{i,j=1\\ i\le j}}^{n}\left(1-\left(\left(1-\left(\eta_i^u\right)^q\right)^{\varphi}\left(1-\left(\eta_j^u\right)^q\right)^{\psi}\right)\right)\right)^{\frac{2}{n(n+1)}}\right)^{\frac{1}{\varphi+\psi}}\right)^{\frac{1}{q}}\right]\right\}\right).$$

Now since $\tilde{k}_i=\tilde{k}$, i.e., $\gamma_i^l=\gamma^l$, $\gamma_i^u=\gamma^u$, $\eta_i^l=\eta^l$ and $\eta_i^u=\eta^u$ for all $i=1,2,\ldots,n$. Therefore,

$$IVDHq-ROFHM^{\varphi,\psi}\left(\tilde{k},\tilde{k},\ldots,\tilde{k}\right)$$

$$= \left(\bigcup_{[\gamma^l,\gamma^u]\in\tilde{h}} \left\{ \left[\left(1 - \left(\prod_{\substack{i,j=1 \\ i \le j}}^{n} \left(1 - (\gamma^l)^{q\varphi} (\gamma^l)^{q\psi} \right) \right)^{\frac{2}{n(n+1)}} \right)^{\frac{1}{q(\varphi+\psi)}}, \right. \right. \right.$$

$$\left. \left. \left(1 - \left(\prod_{\substack{i,j=1 \\ i \le j}}^{n} \left(1 - (\gamma^u)^{q\varphi} (\gamma^u)^{q\psi} \right) \right)^{\frac{2}{n(n+1)}} \right)^{\frac{1}{q(\varphi+\psi)}} \right] \right\},$$

$$\bigcup_{[\eta^l,\eta^u]\in\tilde{g}} \left\{ \left[\left(1 - \left(1 - \left(\prod_{\substack{i,j=1 \\ i \le j}}^{n} \left(1 - \left(\left(1 - (\eta^l)^q \right)^{\varphi} \right. \right. \right. \right. \right. \right. \right.$$

$$\left. \left. \left. \left. \left. \left. \left(1 - (\eta^l)^q \right)^{\psi} \right) \right) \right)^{\frac{2}{n(n+1)}} \right)^{\frac{1}{\varphi+\psi}} \right)^{\frac{1}{q}},$$

$$\left(1 - \left(1 - \left(\prod_{\substack{i,j=1 \\ i \le j}}^{n} \left(1 - \left(\left(1 - (\eta^u)^q \right)^{\varphi} \right. \right. \right. \right. \right.$$

$$\left. \left. \left. \left. \left. \left. \left. \left(1 - (\eta^u)^q \right)^{\psi} \right) \right) \right)^{\frac{2}{n(n+1)}} \right)^{\frac{1}{\varphi+\psi}} \right)^{\frac{1}{q}} \right] \right\} \right)$$

$$= \left(\bigcup_{[\gamma^l,\gamma^u]\in\tilde{h}} \left\{ \left[\left(1 - \left(\left(1 - (\gamma^l)^{q(\varphi+\psi)} \right)^{\frac{n(n+1)}{2}} \right)^{\frac{2}{n(n+1)}} \right)^{\frac{1}{q(\varphi+\psi)}}, \right. \right. \right.$$

$$\left. \left. \left(1 - \left(\left(1 - (\gamma^u)^{q(\varphi+\psi)} \right)^{\frac{n(n+1)}{2}} \right)^{\frac{2}{n(n+1)}} \right)^{\frac{1}{q(\varphi+\psi)}} \right] \right\},$$

$$\bigcup_{[\eta^l,\eta^u]\in\tilde{g}}\left\{\left[\left(1-\left(1-\left(\left(1-\left(\left(1-(\eta^l)^q\right)^{(\varphi+\psi)}\right)\right)^{\frac{n(n+1)}{2}}\right)^{\frac{2}{n(n+1)}}\right)^{\frac{1}{\varphi+\psi}}\right)^{\frac{1}{q}},\right.\right.$$

$$\left.\left.\left(1-\left(1-\left(\left(1-\left(\left(1-(\eta^u)^q\right)^{(\varphi+\psi)}\right)\right)^{\frac{n(n+1)}{2}}\right)^{\frac{2}{n(n+1)}}\right)^{\frac{1}{\varphi+\psi}}\right)^{\frac{1}{q}}\right]\right\}\Bigg)$$

$$=\left(\bigcup_{[\gamma^l,\gamma^u]\in\tilde{h}}\{[\gamma^l,\gamma^u]\},\bigcup_{[\eta^l,\eta^u]\in\tilde{g}}\{[\eta^l,\eta^u]\}\right).$$

Theorem 19.3 (Monotonicity) *Consider* $\tilde{k}_i=\left(\bigcup_{[\gamma_i^l,\gamma_i^u]\in\tilde{h}_i}\{[\gamma_i^l,\gamma_i^u]\},\bigcup_{[\eta_i^l,\eta_i^u]\in\tilde{g}_i}\{[\eta_i^l,\eta_i^u]\}\right)$, *and* $\tilde{k}_i'=\left(\bigcup_{[\gamma_i^{l'},\gamma_i^{u'}]\in\tilde{h}_i'}\{[\gamma_i^{l'},\gamma_i^{u'}]\},\bigcup_{[\eta_i^{l'},\eta_i^{u'}]\in\tilde{h}_i'\in\tilde{g}_i'}\{[\eta_i^{l'},\eta_i^{u'}]\}\right)$ $(i=1,2,\ldots,n)$ *as two collections of IVDHq-ROFNs. If* $\gamma_i^l\le\gamma_i^{l'}$, $\gamma_i^u\le\gamma_i^{u'}$ *and* $\eta_i^l\ge\eta_i^{l'}$, $\eta_i^u\ge\eta_i^{u'}$ *for all* $i=1,2,\ldots,n$, *then*

$$IVDHq-ROFHM^{\varphi,\psi}\left(\tilde{k}_1,\tilde{k}_2,\ldots,\tilde{k}_n\right)\le IVDHq-ROFHM^{\varphi,\psi}\left(\tilde{k}_1',\tilde{k}_2',\ldots,\tilde{k}_n'\right).\quad(19.11)$$

Proof Since $\gamma_i^l\le\gamma_i^{l'}$, for all $i=1,2,\ldots,n$, then

$$1-(\gamma_i^l)^{q\varphi}(\gamma_j^l)^{q\psi}\ge1-\left(\gamma_i^{l'}\right)^{q\varphi}\left(\gamma_j^{l'}\right)^{q\psi}\quad\text{for all } i,j.$$

That is,

$$1-\left(\prod_{\substack{i,j=1\\i\le j}}^{n}\left(1-(\gamma_i^l)^{q\varphi}(\gamma_j^l)^{q\psi}\right)\right)^{\frac{2}{n(n+1)}}$$

$$\le1-\left(\prod_{\substack{i,j=1\\i\le j}}^{n}\left(1-\left(\gamma_i^{l'}\right)^{q\varphi}\left(\gamma_j^{l'}\right)^{q\psi}\right)\right)^{\frac{2}{n(n+1)}}.$$

That is,

$$\left(1-\left(\prod_{\substack{i,\,j=1\\ i\le j}}^{n}\left(1-\left(\gamma_i^{l}\right)^{q\varphi}\left(\gamma_j^{l}\right)^{q\psi}\right)\right)^{\frac{2}{n(n+1)}}\right)^{\frac{1}{q(\varphi+\psi)}}$$

$$\le\left(1-\left(\prod_{\substack{i,\,j=1\\ i\le j}}^{n}\left(1-\left(\gamma_i^{l'}\right)^{q\varphi}\left(\gamma_j^{l'}\right)^{q\psi}\right)\right)^{\frac{2}{n(n+1)}}\right)^{\frac{1}{q(\varphi+\psi)}}. \quad (19.12)$$

Since $\gamma_i^u \le \gamma_i^{u'}$, similarly

$$\left(1-\left(\prod_{\substack{i,\,j=1\\ i\le j}}^{n}\left(1-\left(\gamma_i^{u}\right)^{q\varphi}\left(\gamma_j^{u}\right)^{q\psi}\right)\right)^{\frac{2}{n(n+1)}}\right)^{\frac{1}{q(\varphi+\psi)}}$$

$$\le\left(1-\left(\prod_{\substack{i,\,j=1\\ i\le j}}^{n}\left(1-\left(\gamma_i^{u'}\right)^{q\varphi}\left(\gamma_j^{u'}\right)^{q\psi}\right)\right)^{\frac{2}{n(n+1)}}\right)^{\frac{1}{q(\varphi+\psi)}}. \quad (19.13)$$

Again, for $\eta_i^l \ge \eta_i^{l'}$,

$$\left(1-\left(\eta_i^l\right)^q\right)^{\varphi}\left(1-\left(\eta_j^l\right)^q\right)^{\psi} \le \left(1-\left(\eta_i^{l'}\right)^q\right)^{\varphi}\left(1-\left(\eta_j^{l'}\right)^q\right)^{\psi} \quad \text{for all } i,\, j.$$

That is,

$$\prod_{\substack{i,\,j=1\\ i\le j}}^{n}\left(1-\left(\left(1-\left(\eta_i^l\right)^q\right)^{\varphi}\left(1-\left(\eta_j^l\right)^q\right)^{\psi}\right)\right)$$

$$\geq \prod_{\substack{i,j=1\\ i\leq j}}^{n}\left(1-\left(\left(1-\left(\eta_i^{l'}\right)^q\right)^{\varphi}\left(1-\left(\eta_j^{l'}\right)^q\right)^{\psi}\right)\right).$$

That is,

$$\left(1-\left(1-\left(\prod_{\substack{i,j=1\\ i\leq j}}^{n}\left(1-\left(\left(1-\left(\eta_i^{l}\right)^q\right)^{\varphi}\left(1-\left(\eta_j^{l}\right)^q\right)^{\psi}\right)\right)\right)^{\frac{2}{n(n+1)}}\right)^{\frac{1}{\varphi+\psi}}\right)^{\frac{1}{q}}$$

$$\geq \left(1-\left(1-\left(\prod_{\substack{i,j=1\\ i\leq j}}^{n}\left(1-\left(\left(1-\left(\eta_i^{l'}\right)^q\right)^{\varphi}\left(1-\left(\eta_j^{l'}\right)^q\right)^{\psi}\right)\right)\right)^{\frac{2}{n(n+1)}}\right)^{\frac{1}{\varphi+\psi}}\right)^{\frac{1}{q}}. \tag{19.14}$$

Similarly,

$$\left(1-\left(1-\left(\prod_{\substack{i,j=1\\ i\leq j}}^{n}\left(1-\left(\left(1-\left(\eta_i^{u}\right)^q\right)^{\varphi}\left(1-\left(\eta_j^{u}\right)^q\right)^{\psi}\right)\right)\right)^{\frac{2}{n(n+1)}}\right)^{\frac{1}{\varphi+\psi}}\right)^{\frac{1}{q}}$$

$$\geq \left(1-\left(1-\left(\prod_{\substack{i,j=1\\ i\leq j}}^{n}\left(1-\left(\left(1-\left(\eta_i^{u'}\right)^q\right)^{\varphi}\left(1-\left(\eta_j^{u'}\right)^q\right)^{\psi}\right)\right)\right)^{\frac{2}{n(n+1)}}\right)^{\frac{1}{\varphi+\psi}}\right)^{\frac{1}{q}}. \tag{19.15}$$

Then from inequalities (19.12)–(19.15) and comparison rule presented in Definition 19.3, it is obtained that

$$IVDHq - ROFHM^{\varphi,\psi}\left(\tilde{k}_1, \tilde{k}_2, \ldots, \tilde{k}_n\right) \leq IVDHq - ROFHM^{\varphi,\psi}\left(\tilde{k}'_1, \tilde{k}'_2, \ldots, \tilde{k}'_n\right).$$

Therefore, the theorem is proved.

Theorem 19.4 (Boundary) *Let* $\tilde{\kappa}_i = \left(\tilde{h}_i, \tilde{g}_i\right)$ $(i = 1, 2, \ldots, n)$ *be a collection of DHq-ROFNs, and let*

$$\gamma^l_- = min\left\{\gamma^l_{i_{min}} | \gamma^l_{i_{min}} = \min_{[\gamma^l_i, \gamma^u_i] \in \tilde{h}_i} \{\gamma^l_i\}\right\}; \gamma^u_- = min\left\{\gamma^u_{i_{min}} | \gamma^u_{i_{min}} = \min_{[\gamma^l_i, \gamma^u_i] \in \tilde{h}_i} \{\gamma^u_i\}\right\};$$

$$\gamma^l_+ = max\left\{\gamma^l_{i_{max}} | \gamma^l_{i_{max}} = \max_{[\gamma^l_i, \gamma^u_i] \in \tilde{h}_i} \{\gamma^l_i\}\right\}; \gamma^u_+ = max\left\{\gamma^u_{i_{max}} | \gamma^u_{i_{max}} = \max_{[\gamma^l_i, \gamma^u_i] \in \tilde{h}_i} \{\gamma^u_i\}\right\};$$

$$\eta^l_- = min\left\{\eta^l_{i_{min}} | \eta^l_{i_{min}} = \min_{[\eta^l_i, \eta^u_i] \in \tilde{g}_i} \{\eta^l_i\}\right\}; \eta^u_- = min\left\{\eta^u_{i_{min}} | \eta^u_{i_{min}} = \min_{[\eta^l_i, \eta^u_i] \in \tilde{g}_i} \{\eta^u_i\}\right\};$$

$$\eta^l_+ = max\left\{\eta^l_{i_{max}} | \eta^l_{i_{max}} = \max_{[\eta^l_i, \eta^u_i] \in \tilde{g}_i} \{\eta^l_i\}\right\}; \eta^u_+ = max\left\{\eta^u_{i_{max}} | \eta^u_{i_{max}} = \max_{[\eta^l_i, \eta^u_i] \in \tilde{g}_i} \{\eta^u_i\}\right\}.$$

Then,

$$\tilde{k}_- \leq IVDHq - ROFHM^{\varphi,\psi}\left(\tilde{k}_1, \tilde{k}_2, \ldots, \tilde{k}_n\right) \leq \tilde{k}_+, \tag{19.16}$$

where $\tilde{k}_- = \left(\left[\gamma^l_-, \gamma^u_-\right], \left[\eta^l_+, \eta^u_+\right]\right)$ *and* $\tilde{k}_+ = \left(\left[\gamma^l_+, \gamma^u_+\right], \left[\eta^l_-, \eta^u_-\right]\right)$.

Proof Since $\gamma^l_- \leq \gamma_i \leq \gamma^l_+$, $\gamma^u_- \leq \gamma_i \leq \gamma^u_+$, $\eta^l_- \leq \eta_i \leq \eta^l_+$ and $\eta^u_- \leq \eta_i \leq \eta^u_+$ for all $i = 1, 2, \ldots, n$, then $\tilde{k}_- \leq \tilde{k}_i$ for $i = 1, 2, \ldots, n$.

Therefore, from monotonicity

$$IVDHq - ROFHM^{\varphi,\psi}\left(\tilde{k}_-, \tilde{k}_-, \ldots, \tilde{k}_-\right) \leq IVDHq - ROFHM^{\varphi,\psi}\left(\tilde{k}_1, \tilde{k}_2, \ldots, \tilde{k}_n\right).$$

Now applying the idempotency theorem, the above inequality takes the form as

$$\tilde{k}_- \leq IVDHq - ROFHM^{\varphi,\psi}\left(\tilde{k}_1, \tilde{k}_2, \ldots, \tilde{k}_n\right). \tag{19.17}$$

Similarly, it can be shown that

$$IVDHq - ROFHM^{\varphi,\psi}\left(\tilde{k}_1, \tilde{k}_2, \ldots, \tilde{k}_n\right) \leq \tilde{k}_+. \tag{19.18}$$

So, combining (19.17) and (19.18), it follows that

$$\tilde{k}_- \leq IVDHq - ROFHM^{\varphi,\psi}\left(\tilde{k}_1, \tilde{k}_2, \ldots, \tilde{k}_n\right) \leq \tilde{k}_+.$$

19.3.2 IVDHq-ROFWHM Operator

The weight of attributes is required to distinguish the importance of the attributes. Those weights are used in the aggregation process for solving real-life MCDM problems. However, IVDHq-ROFHM operator ignores the importance of attributes. So, by considering the exact weight information of input arguments, more valid and accurate results under MCDM context would be produced. Incorporating exact attribute weights with IVDHq-ROFHM operator, the IVDHq-ROFWHM aggregation operator is established as follows.

Definition 19.8 Let $\tilde{k}_i$ $(i = 1, 2, \ldots, n)$ be a collection of IVDHq-ROFNs and let $\omega = (\omega_1, \omega_2, \ldots, \omega_n)^T$ be the weight vector with $\omega_i \in [0, 1]$, $\sum_{i=1}^{n} \omega_i = 1$ and $\varphi, \psi \geq 0$ be any numbers. If

$$IVDHq - ROFWHM_{\omega}^{\varphi,\psi}\left(\tilde{k}_1, \tilde{k}_2, \ldots, \tilde{k}_n\right) = \left(\frac{2}{n(n+1)} \oplus_{\substack{i, j = 1 \\ i \leq j}}^{n} \left(\left(\omega_i \tilde{k}_i\right)^{\varphi} \otimes \left(\omega_j \tilde{k}_j\right)^{\psi} \right) \right)^{\frac{1}{\varphi+\psi}}, \tag{19.19}$$

then $IVDHq - ROFWHM_{\omega}^{\varphi,\psi}$ is called the IVDHq-ROFWHM operator.

Theorem 19.5 *Let $\tilde{k}_i$ $(i = 1, 2, \ldots, n)$ be a collection of IVDHq-ROFNs, whose weight vectors are $\omega = (\omega_1, \omega_2, \ldots, \omega_n)^T$, $\omega_i \in [0, 1]$, and $\sum_{i=1}^{n} \omega_i = 1$. Let $\varphi, \psi \geq 0$ be any two numbers. Then the aggregated value obtained by using IVDHq-ROFWHM operator is also an IVDHq-ROFN and is given by*

$$IVDHq - ROFWHM_{\omega}^{\varphi,\psi}\left(\tilde{k}_1,\ \tilde{k}_2, \ldots, \tilde{k}_n\right)$$
$$= \Bigg(\bigcup_{[\gamma_i^l, \gamma_i^u] \in \tilde{h}_i, \left[\gamma_j^l, \gamma_j^u\right] \in \tilde{h}_j}$$
$$\left\{ \left[\left(\left(1 - \left(\prod_{\substack{i, j = 1 \\ i \leq j}}^{n} \left(1 - \left(1 - \left(1 - \left(\gamma_i^l\right)^q\right)^{\omega_i}\right)^{\varphi} \left(1 - \left(1 - \left(\gamma_j^l\right)^q\right)^{\omega_i}\right)^{\psi} \right) \right)^{\frac{2}{n(n+1)}} \right)^{\frac{1}{\varphi+\psi}} \right)^{\frac{1}{q}}, \right.\right.$$
$$\left.\left. \left(\left(1 - \left(\prod_{\substack{i, j = 1 \\ i \leq j}}^{n} \left(1 - \left(1 - (1 - (\gamma_i^u)^q)^{\omega_i}\right)^{\varphi} \left(1 - \left(1 - \left(\gamma_j^u\right)^q\right)^{\omega_i}\right)^{\psi} \right) \right)^{\frac{2}{n(n+1)}} \right)^{\frac{1}{\varphi+\psi}} \right)^{\frac{1}{q}} \right] \right\},$$

$$\bigcup_{\left[\eta_i^l,\eta_i^u\right]\in\tilde{g}_i,\left[\eta_j^l,\eta_j^u\right]\in\tilde{g}_j} \left\{\left[\left(1-\left(1-\left(\prod_{\substack{i,j=1\\ i\le j}}^{n}\left(1-\left(1-\left(\eta_i^l\right)^{q\omega_i}\right)^{\varphi}\left(1-\left(\eta_j^l\right)^{q\omega_j}\right)^{\psi}\right)\right)^{\frac{2}{n(n+1)}}\right)^{\frac{1}{\varphi+\psi}}\right)^{\frac{1}{q}},\right.\right.$$

$$\left.\left.\left(1-\left(1-\left(\prod_{\substack{i,j=1\\ i\le j}}^{n}\left(1-\left(1-\left(\eta_i^u\right)^{q\omega_i}\right)^{\varphi}\left(1-\left(\eta_j^u\right)^{q\omega_j}\right)^{\psi}\right)\right)^{\frac{2}{n(n+1)}}\right)^{\frac{1}{\varphi+\psi}}\right)^{\frac{1}{q}}\right]\right\}. \quad (19.20)$$

Proof The proof of this theorem can easily be verified in an analogous way to Theorem 19.1.

19.3.3 IVDHq-OFGHM Operator

By incorporating GHM operator for aggregating IVDHq-ROF information during the decision-making process, the IVDHq-ROFGHM operator is established as follows.

Definition 19.9 Let $\tilde{k}_i = \left(\tilde{h}_i, \tilde{g}_i\right)$ $(i = 1, 2, \ldots, n)$ be a collection of IVDHq-ROFNs. Let $\varphi, \psi \geq 0$ be any number. If

$$IVDHq - ROFGHM^{\varphi,\psi}\left(\tilde{k}_1, \tilde{k}_2, \ldots, \tilde{k}_n\right) = \left(\frac{2}{n(n+1)} \otimes^{n}_{\substack{i,j=1\\ i\le j}} \left(\tilde{k}_i^{\varphi} \oplus \tilde{k}_j^{\psi}\right)\right)^{\frac{1}{\varphi+\psi}}, \quad (19.21)$$

then the $IVDHq - ROFGHM^{\varphi,\psi}$ is called IVDHq-ROFGHM operator.

Theorem 19.6 *Let* $\tilde{k}_i = \left(\tilde{h}_i, \tilde{g}_i\right)$ $(i = 1, 2, \ldots, n)$ *be a collection of IVDHq-ROFNs and* $\varphi, \psi \geq 0$, *then the aggregated value using IVDHq-ROFGHM is also a IVDHq-ROFN and can be given as follows:*

$$IVDHq - ROFGHM^{\varphi,\psi}\left(\tilde{k}_1,\ \tilde{k}_2, \ldots, \tilde{k}_n\right)$$

$$= \left(\frac{2}{n(n+1)} \otimes^{n}_{\substack{i,j=1 \\ i \le j}} \left(\tilde{k}_i^{\varphi} \oplus \tilde{k}_j^{\psi} \right) \right)^{\frac{1}{\varphi+\psi}}$$

$$= \Bigg(\bigcup_{[\gamma_i^l, \gamma_i^u] \in \tilde{h}_i, [\gamma_j^l, \gamma_j^u] \in \tilde{h}_j}$$

$$\left\{ \left[\left(1 - \left(1 - \left(\prod_{\substack{i,j=1 \\ i \le j}}^{n} \left(1 - \left(\left(1 - (\gamma_i^l)^q \right)^{\varphi} \left(1 - (\gamma_j^l)^q \right)^{\psi} \right) \right) \right)^{\frac{2}{n(n+1)}} \right)^{\frac{1}{\varphi+\psi}} \right)^{\frac{1}{q}}, \right.\right.$$

$$\left.\left. \left(1 - \left(1 - \left(\prod_{\substack{i,j=1 \\ i \le j}}^{n} \left(1 - \left(\left(1 - (\gamma_i^u)^q \right)^{\varphi} \left(1 - (\gamma_j^u)^q \right)^{\psi} \right) \right) \right)^{\frac{2}{n(n+1)}} \right)^{\frac{1}{\varphi+\psi}} \right)^{\frac{1}{q}} \right] \right\},$$

$$\bigcup_{[\eta_i^l, \eta_i^u] \in \tilde{g}_i, [\eta_j^l, \eta_j^u] \in \tilde{g}_j} \left\{ \left[\left(1 - \left(\prod_{\substack{i,j=1 \\ i \le j}}^{n} \left(1 - (\eta_i^l)^{q\varphi} (\eta_j^l)^{q\psi} \right) \right)^{\frac{2}{n(n+1)}} \right)^{\frac{1}{q(\varphi+\psi)}}, \right.\right.$$

$$\left.\left. \left(1 - \left(\prod_{\substack{i,j=1 \\ i \le j}}^{n} \left(1 - (\eta_i^u)^{q\varphi} (\eta_j^u)^{q\psi} \right) \right)^{\frac{2}{n(n+1)}} \right)^{\frac{1}{q(\varphi+\psi)}} \right] \right\} \Bigg). \tag{19.22}$$

Proof The proof process is similar to Theorem 1.

Note 1 The IVDHq-ROFGHM operator also satisfies some important properties, viz., monotonicity, idempotency, and boundary.

19.3.4 IVDHq-ROFWGHM Operator

Considering the exact weight information of input arguments with IVDHq-ROFGHM operator, the IVDHq-ROFWGHM aggregation operator is established and is presented below.

Definition 19.10 Let $\tilde{k}_i$ $(i = 1, 2, \ldots, n)$ be a collection of IVDHq-ROFNs and let $\omega = (\omega_1, \omega_2, \ldots, \omega_n)^T$ be the weight vector with $\omega_i \in [0, 1]$, $\sum_{i=1}^{n} \omega_i = 1$ and $\varphi, \psi \geq 0$ be any number. If

$$IVDHq - ROFWGHM_{\omega}^{\varphi,\psi}\left(\tilde{k}_1, \tilde{k}_2, \ldots, \tilde{k}_n\right)$$
$$= \frac{1}{\varphi + \psi}\left(\otimes_{\substack{i,j=1 \\ i \leq j}}^{n}\left(\left(\varphi\tilde{k}_i\right)^{\omega_i} \oplus \left(\psi\tilde{k}_j\right)^{\omega_j}\right)^{\frac{2}{n(n+1)}}\right), \tag{19.23}$$

then $IVDHq - ROFWGHM_{\omega}^{\varphi,\psi}$ is called IVDHq-ROFWGHM operator.

Theorem 19.7 *Let $\tilde{k}_i$ $(i = 1, 2, \ldots, n)$ be a collection of IVDHq-ROFNs, whose weight vectors are $\omega = (\omega_1, \omega_2, \ldots, \omega_n)^T$, and $\omega_i \in [0, 1]$ with $\sum_{i=1}^{n} \omega_i = 1$. Let $\varphi, \psi \geq 0$ be any two numbers. Then the aggregated value achieved by using IVDHq-ROFWGHM operator is also a IVDHq-ROFN and*

$$IVDHq - ROFWGHM_{\omega}^{\varphi,\psi}\left(\tilde{k}_1, \tilde{k}_2, \ldots, \tilde{k}_n\right)$$
$$= \Bigg(\bigcup_{[\gamma_i^l,\gamma_i^u]\in\tilde{h}_i,\left[\gamma_j^l,\gamma_j^u\right]\in\tilde{h}_j}$$
$$\Bigg\{\Bigg[\left(1 - \left(1 - \left(\prod_{\substack{i,j=1 \\ i \leq j}}^{n}\left(1 - \left(1 - \left(\gamma_i^l\right)^{q\omega_i}\right)^{\varphi}\left(1 - \left(\gamma_j^l\right)^{q\omega_j}\right)^{\psi}\right)\right)^{\frac{2}{n(n+1)}}\right)^{\frac{1}{\varphi+\psi}}\right)^{\frac{1}{q}},$$
$$\left(1 - \left(1 - \left(\prod_{\substack{i,j=1 \\ i \leq j}}^{n}\left(1 - (1 - (\gamma_i^u)^{q\omega_i})^{\varphi}\left(1 - \left(\gamma_j^u\right)^{q\omega_j}\right)^{\psi}\right)\right)^{\frac{2}{n(n+1)}}\right)^{\frac{1}{\varphi+\psi}}\right)^{\frac{1}{q}}\Bigg]\Bigg\},$$
$$\bigcup_{[\eta_i^l,\eta_i^u]\in\tilde{g}_i,\left[\eta_j^l,\eta_j^u\right]\in\tilde{g}_j}$$

$$\left\{\left[\left(\left(1-\left(\prod_{\substack{i,j=1\\ i\le j}}^{n}\left(1-\left(1-\left(1-\left(\eta_i^l\right)^q\right)^{\omega_i}\right)^{\varphi}\left(1-\left(1-\left(\eta_j^l\right)^q\right)^{\omega_i}\right)^{\psi}\right)\right)^{\frac{2}{n(n+1)}}\right)^{\frac{1}{\varphi+\psi}}\right)^{\frac{1}{q}},\right.\right.$$

$$\left.\left.\left(\left(1-\left(\prod_{\substack{i,j=1\\ i\le j}}^{n}\left(1-\left(1-\left(1-\left(\eta_i^u\right)^q\right)^{\omega_i}\right)^{\varphi}\left(1-\left(1-\left(\eta_j^u\right)^q\right)^{\omega_i}\right)^{\psi}\right)\right)^{\frac{2}{n(n+1)}}\right)^{\frac{1}{\varphi+\psi}}\right)^{\frac{1}{q}}\right]\right\}. \tag{19.24}$$

Proof The proof is similar to Theorem 19.1. So, the proof is omitted here.

19.4 Approach to MCDM with HM-Based IVDHq-ROF Information

For an MCDM, let $A = \{A_1, A_2, \ldots, A_m\}$ be a collection of alternatives to be selected with IVDHq-ROF information. The proposed operators IVDHq-ROFWHM and IVDHq-ROFWGHM are applied to develop an approach to solve MCDM problems under IVDHq-ROF environment. Let $C = \{C_1, C_2, \ldots, C_n\}$ be a collection of criteria on which the alternatives are evaluated. Also, let $\omega = (\omega_1, \omega_2, \ldots, \omega_n)$ be the weight vector of C_j $(j = 1, 2, \ldots, n)$ satisfying $\omega_j > 0$ for $j = 1, 2, \ldots, n$, and $\sum_{j=1}^{n} \omega_j = 1$. To evaluate the performance of the alternatives, a DM provides decision values in the forms of IVDHq-ROFNs as $\tilde{k}_{ij} = \left(\tilde{h}_{ij}, \tilde{g}_{ij}\right)$. After providing all the decision values corresponding to each alternative with respect to their satisfying criteria, the IVDHq-ROF decision matrix (IVDHq-ROFDM), $\tilde{D} = \left[\tilde{k}_{ij}\right]_{m\times n} = \left[\tilde{h}_{ij}, \tilde{g}_{ij}\right]_{m\times n}$ is found as

$$\tilde{D} = \begin{pmatrix} \tilde{k}_{11} & \tilde{k}_{12} & \cdots & \tilde{k}_{1n} \\ \vdots & \vdots & \ddots & \vdots \\ \tilde{k}_{m1} & \tilde{k}_{m2} & \cdots & \tilde{k}_{mn} \end{pmatrix}_{m\times n}. \tag{19.25}$$

In MCDM, criteria are categorized into two types: one is benefit criteria and the other is cost criteria. If the IVDHq-ROFDM possesses cost type criteria, the matrix $\tilde{D} = \left[\tilde{k}_{ij}\right]_{m\times n}$ is usually converted into the normalized IVDHq-ROFDM form as

$\tilde{R} = [\tilde{r}_{ij}]_{m\times n}$ in the following way:

$$\tilde{r}_{ij} = \begin{cases} \tilde{k}_{ij} \; for\ benefit\ attribute\ C_j \\ \tilde{k}^c_{ij} \; for\ cost\ attribute\ C_j \end{cases} \tag{19.26}$$

$(i = 1, \ldots, m\ and\ j = 1, \ldots, n)$, where $\tilde{k}^c_{ij}$ is the complement $\tilde{k}_{ij}$.

Now, suppose that $\tilde{R} = (\tilde{r}_{ij})_{m\times n}$ be a normalized IVDHq-ROFDM. Then, the proposed IVDHq-ROFWHM (and IVDHq-ROFWGHM) operators are utilized to improve an approach for solving MCDM problems with IVDHq-ROF information. The proposed approach is explained through the following steps:

***Step* 1**. Transform the IVDHq-ROFDM, $\tilde{D} = \left(\tilde{k}_{ij}\right)_{m\times n}$ into the normalized IVDHq-ROFDM, $\tilde{R} = \left(\tilde{r}_{ij}\right)_{m\times n}$ using Eq. (19.26), if required.
***Step* 2**. Aggregate the IVDHq-ROFNs, $\tilde{r}_{ij}$ for each alternative, A_i using the IVDHq-ROFWHM (or IVDHq-ROFWGHM) operator as follows:

$$\tilde{r}^A_i = IVDHq - ROFWHM^{\varphi,\psi}_{\omega}\left(\tilde{k}_{i1}, \tilde{k}_{i2}, \ldots, \tilde{k}_{in}\right) \tag{19.27}$$

or

$$\tilde{r}^G_i = IVDHq - ROFWGHM^{\varphi,\psi}_{\omega}\left(\tilde{k}_{i1}, \tilde{k}_{i2}, \ldots, \tilde{k}_{in}\right) \tag{19.28}$$

for $i = 1, \ldots, m\ and\ j = 1, \ldots, n$.
***Step* 3**. Calculate the score values of the alternatives using score function as mentioned in Definition 19.3
***Step* 4**. The ordering of alternatives is evaluated using the comparison rule of IVDHq-ROFNs as presented in Definition 19.3.

The proposed method is depicted in Fig. 19.1.

19.5 Illustrative Example

In this section, an example studied previously [51] is considered and solved to explore the proposed method's applicability.

19.5.1 Description of the Problem

An MCDM problem related to an investment project is adapted from an article, previously studied by Zang et al. [51], and is solved to illustrate the application of

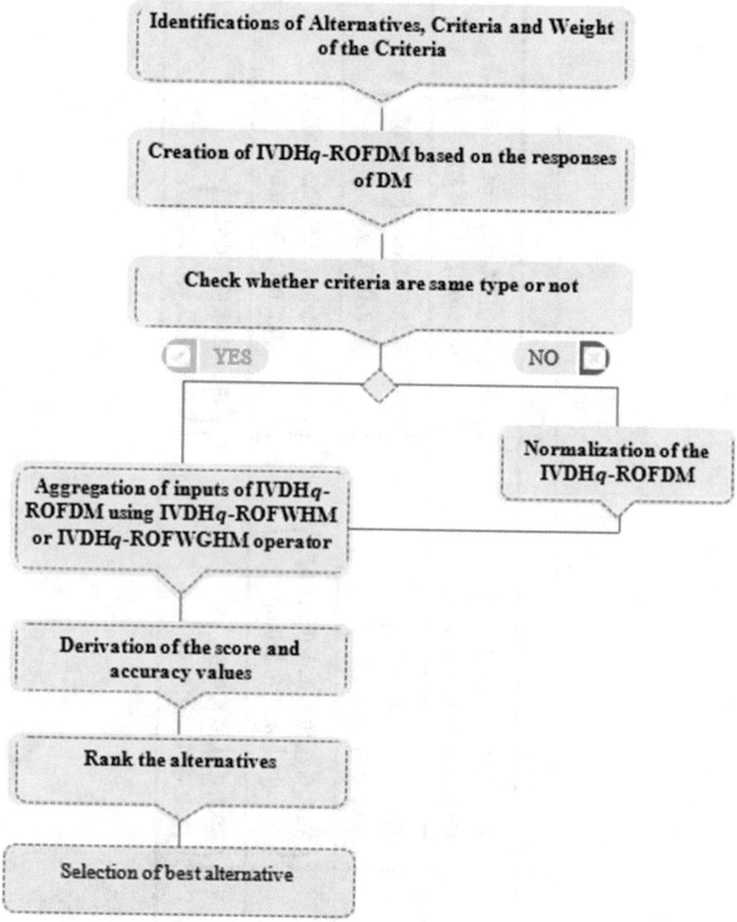

Fig. 19.1 Flowchart of the decision-making process using proposed methodology

the proposed MCDM method. The problem is about selection of the most suitable investment company. The attributes that are considered for the selection process are C_1: risk analysis, C_2: growth analysis, and C_3: environmental impact analysis. The corresponding criteria weight vector is given as $W = (0.35, 0.25, 0.40)^T$. It is assumed that there are four alternatives, viz., $A = \{A_1, A_2, A_3, A_4\}$, which need to be evaluated based on the abovementioned criteria. Under the IVDHq-ROF environment, these alternatives are examined by experts and put their judgment values in the form of IVDHq-ROFNs. The developed IVDHq-ROFDM is shown in Table 19.2.

In order to select the most appropriate investment company, IVDHq-ROFWHM operator is utilized in the context of solving MCDM problems under IVDHq-ROF environment, which is presented in the following way:

Table 19.2 IVDHq-ROFDM as per expert's opinion

	C_1	C_2	C_3
A_1	$\begin{pmatrix} \{[0.3, 0.4], [0.4, 0.5], [0.5, 0.7]\}, \\ \{[0.2, 0.3], [0.2, 0.4]\} \end{pmatrix}$	$\begin{pmatrix} \{[0.3, 0.5], [0.4, 0.7]\}, \\ \{[0.2, 0.3], [0.3, 0.4]\} \end{pmatrix}$	$\begin{pmatrix} \{[0.3, 0.5], [0.4, 0.5], [0.4, 0.6]\}, \\ \{[0.1, 0.3], [0.2, 0.3]\} \end{pmatrix}$
A_2	$\begin{pmatrix} \{[0.3, 0.5], [0.4, 0.5], [0.5, 0.6]\}, \\ \{[0.3, 0.4]\} \end{pmatrix}$	$\begin{pmatrix} \{[0.3, 0.4], [0.2, 0.4]\}, \\ \{[0.2, 0.3], [0.4, 0.7]\} \end{pmatrix}$	$\begin{pmatrix} \{[0.2, 0.3], [0.4, 0.5], [0.5, 0.6]\}, \\ \{[0.2, 0.4], [0.4, 0.6]\} \end{pmatrix}$
A_3	$\begin{pmatrix} \{[0.3, 0.5], [0.4, 0.7]\}, \\ \{[0.2, 0.3], [0.3, 0.5]\} \end{pmatrix}$	$\begin{pmatrix} \{[0.2, 0.4], [0.4, 0.5], [0.5, 0.8]\}, \\ \{[0.1, 0.3], [0.2, 0.4]\} \end{pmatrix}$	$\begin{pmatrix} \{[0.2, 0.3], [0.4, 0.5], [0.5, 0.6]\}, \\ \{[0.2, 0.4], [0.4, 0.6]\} \end{pmatrix}$
A_4	$\begin{pmatrix} \{[0.3, 0.4], [0.4, 0.6], [0.6, 0.9]\}, \\ \{[0.1, 0.2]\} \end{pmatrix}$	$\begin{pmatrix} \{[0.4, 0.5], [0.5, 0.6]\}, \\ \{[0.2, 0.3], [0.2, 0.4]\} \end{pmatrix}$	$\begin{pmatrix} \{[0.4, 0.6], [0.7, 0.9]\}, \\ \{[0.1, 0.2], [0.1, 0.3]\} \end{pmatrix}$

***Step* 1**. As all the criteria, C_j $(j = 1, 2, 3, 4)$ are the benefit criteria, the performance values of the alternatives, A_i $(i = 1, 2, 3, 4,)$, need no normalization.
***Step* 2**. Without loss of generality, consider HM parameters, $\varphi = 1$, $\psi = 1$ and rung parameter, $q = 3$. The aggregation operator IVDHq-ROFWHM is utilized as presented in Eq. (19.20) to aggregate all the preference values, $\tilde{r}_{ij}$, for each alternative, A_i, and the aggregated values, $\tilde{r}_i^A$, are calculated as

$$\tilde{r}_1^A = (\{[0.2089, 0.3315], [0.2454, 0.3315], [0.2454, 0.3730], [0.2312, 0.3923], [0.2605, 0.3923], [0.2605, 0.4196], [0.2407, 0.3522], [0.2672, 0.3522], [0.2672, 0.3883], [0.2566, 0.4060], [0.2797, 0.4060], [0.2797, 0.4314], [0.2838, 0.4261], [0.3014, 0.4261], [0.3014, 0.4483], [0.2943, 0.4607], [0.3108, 0.4607], [0.3108, 0.4794]\}, \{[0.5464, 0.6476], [0.5881, 0.6476], [0.5696, 0.6715], [0.6119, 0.6715], [0.5464, 0.6709], [0.5881, 0.6709], [0.5696, 0.6948], [0.6119, 0.6948]\}).$$

$$\tilde{r}_2^A = (\{[0.1880, 0.2943], [0.2454, 0.3386], [0.2925, 0.3779], [0.1731, 0.2943], [0.2382, 0.3386], [0.2880, 0.3779], [0.2275, 0.2943], [0.2672, 0.3386], [0.3065, 0.3779], [0.2191, 0.2943], [0.2612, 0.3386], [0.3023, 0.3779], [0.2755, 0.3413], [0.3014, 0.3727], [0.3309, 0.4038], [0.2704, 0.3413], [0.2969, 0.3727], [0.3272, 0.4038]\}, \{[0.6172, 0.7197], [0.6727, 0.7603] [0.6594, 0.7771], [0.7150, 0.8179]\}).$$

$$\tilde{r}_3^A = (\{[0.2382, 0.3386], [0.2880, 0.3779], [0.3433, 0.4279], [0.2605, 0.3522], [0.3022, 0.3883], [0.3527, 0.4355], [0.2858, 0.4509], [0.3196, 0.4702], [0.3644, 0.4995], [0.2612, 0.4180], [0.3023, 0.4410], [0.3525, 0.4752], [0.2797, 0.4261], [0.3155, 0.4483], [0.3617, 0.4814], [0.3013, 0.4935], [0.3315, 0.5093], [0.3730, 0.5337]\}, \{[0.5108, 0.6715], [0.5524, 0.6715], [0.5464, 0.6900], [0.5881, 0.6900], [0.5398, 0.7145], [0.5815, 0.7145], [0.5756, 0.7329][0.6172, 0.7329]\}).$$

$$\tilde{r}_4^A = (\{[0.2605, 0.3730], [0.4104, 0.5685], [0.2858, 0.3914], [0.4184, 0.5746], [0.2797, 0.4128], [0.4166, 0.5823], [0.3013, 0.4275], [0.4245, 0.5883], [0.3526, 0.5676], [0.4465, 0.6592], [0.3647, 0.5740], [0.4536, 0.6638]\}, \{[0.5045, 0.6119], [0.5045, 0.6427], [0.5045, 0.6300] [0.5045, 0.6610]\}).$$

***Step* 3**. By Definition 19.3, the score values of $\tilde{r}_i^A$ $(i = 1, 2, 3, 4)$ for each alternative are derived as

$$S(\tilde{r}_1^A) = -0.2048,\ S(\tilde{r}_2^A) = -0.3454,\ S(\tilde{r}_3^A) = -0.2030,\ S(\tilde{r}_4^A) = -0.0833.$$

***Step* 4**. According to the score values, the ranking of the alternatives is found as $A_4 \succ A_3 \succ A_1 \succ A_2$.

Again, utilizing IVDHq-ROFWGHM operator, for the context of solving MCDM problem with IVDHq-ROF information, the best alternative can be found as follows:

***Step* 1′**. Same as Step 1.
***Step* 2′**. Considering the same HM parameters and rung parameter as like Step 2, the IVDHq-ROFWGHM aggregation operator is utilized, as presented in Eq. (19.24), to aggregate all the preference values. The aggregated values, $\tilde{r}_i^G$ are calculated as

$$\tilde{r}_1^G = (\{[0.6715, 0.7748], [0.6965, 0.7748], [0.6965, 0.7939], [0.6900, 0.7988], [0.7150, 0.7988], [0.7150, 0.8179], [0.6948, 0.7944], [0.7197, 0.7944], [0.7197, 0.8135], [0.7132, 0.8185], [0.7381, 0.8185], [0.7381, 0.8377], [0.7145, 0.8269], [0.7393, 0.8269], [0.7393, 0.8461], [0.7329, 0.8510], [0.7578, 0.8510], [0.7578, 0.8704]\}, \{[0.1218, 0.2089], [0.1390, 0.2089], [0.1532, 0.2312], [0.1635, 0.2312], [0.1218, 0.2407], [0.1390, 0.2407], [0.1532, 0.2566], [0.1635, 0.2566]\}).$$

$$\tilde{r}_2^G = (\{[0.6411, 0.7329], [0.6965, 0.7792], [0.7182, 0.7983], [0.6172, 0.7329], [0.6727, 0.7792], [0.6944, 0.7983], [0.6645, 0.7329], [0.7197, 0.7792], [0.7412, 0.7983], [0.6406, 0.7329], [0.6959, 0.7792], [0.7175, 0.7983], [0.6841, 0.7500], [0.7393, 0.7965], [0.7609, 0.8156], [0.6601, 0.7500], [0.7156, 0.7965], [0.7372, 0.8156]\}, \{[0.1731, 0.2672], [0.2382, 0.3555], [0.2160, 0.3749], [0.2605, 0.4196]\}).$$

$$\tilde{r}_3^G = (\{[0.6727, 0.7792], [0.6944, 0.7983], [0.7135, 0.8155], [0.7150, 0.7944], [0.7366, 0.8135], [0.7557, 0.8307], [0.7301, 0.8282], [0.7518, 0.8475], [0.7708, 0.8647], [0.6959, 0.8117], [0.7175, 0.8309], [0.7367, 0.8481], [0.7381, 0.8269], [0.7596, 0.8461], [0.7787, 0.8635], [0.7533, 0.8607], [0.7748, 0.8802], [0.7939, 0.8979]\}, \{[0.1089, 0.2089], [0.1298, 0.2089], [0.1218, 0.2312], [0.1390, 0.2312], [0.1591, 0.2838], [0.1680, 0.2838], [0.1645, 0.2943], [0.1731, 0.2943]\}).$$

$$\tilde{r}_4^G = (\{[0.7150, 0.7939], [0.7728, 0.8400], [0.7301, 0.8068], [0.7879, 0.8527], [0.7381, 0.8308], [0.7959, 0.8775], [0.7533, 0.8438], [0.8110, 0.8906],$$

$$[0.7750, 0.8719], [0.8328, 0.9188], [0.7901, 0.8849], [0.8480, 0.9328]\}, \\ \{[0.0999, 0.1635], [0.0999, 0.1916], [0.0999, 0.2009] \\ [0.0999, 0.2184]\}).$$

***Step* 3′**. By Definition 19.3, the score values of $\tilde{r}_i^G$ $(i = 1, 2, 3, 4)$ for each alternative are derived as

$$S(\tilde{r}_1^G) = 0.4523, S(\tilde{r}_2^G) = 0.3745, S(\tilde{r}_3^G) = 0.4846, S(\tilde{r}_4^G) = 0.5564.$$

***Step* 4′**. According to the score values, the ranking of the alternatives is found as $A_4 \succ A_3 \succ A_1 \succ A_2$.

It is interesting to note here that the same ranking of alternatives is found for both the methods discussed above. Now, the influences of the associated parameters on the ranking results are investigated through the following two subsections.

19.5.2 The Influence of the HM Parameters, φ and ψ, on the Ranking Results

In the process of aggregation, the derived results depend upon the HM parameters, φ and ψ. Different values of the HM parameters φ and ψ are assigned in the above method to demonstrate the influence of the HM parameters. Changing the value of the parameters, φ and ψ, in the range [0, 10], simultaneously, and considering $q = 3$ as a fixed value, the resulted score values of different alternatives, A_i $(i = 1, 2, 3, 4)$, and the corresponding ranking orders of them are presented in Tables 19.3 and 19.4 using IVDHq-ROFWHM and IVDHq-ROFWGHM operators, respectively. Different score values and ordering of the alternatives are obtained using IVDHq-ROFWHM operator.

From Table 19.3, it is observed that using IVDHq-ROFWHM operator, two ranking results, viz., $A_4 \succ A_3 \succ A_1 \succ A_2$ and $A_4 \succ A_1 \succ A_3 \succ A_2$, are found by varying the parameters φ and ψ. The ranking of the alternatives slightly changes for $\varphi = 1, \psi = 0.5$ from $A_4 \succ A_3 \succ A_1 \succ A_2$ to $A_4 \succ A_1 \succ A_3 \succ A_2$. Though the best alternative remains the same as A_4.

Again, using IVDHq-ROFWGHM operator, the score values and ordering of alternatives which are achieved by varying HM parameters are summarized in Table 19.4. It is seen that the ranking of the alternatives remains unaltered as $A_4 \succ A_3 \succ A_1 \succ A_2$ throughout the range. To obtain a clear view, the score values of the alternatives, A_i, are portrayed geometrically through Figs. 19.2, 19.3, 19.4, and 19.5 (using IVDHq-ROFWHM operator) and Figs. 19.6, 19.7, 19.8, and 19.9 (using IVDHq-ROFWGHM operator). It is observed that the alternative, A_4, attains the highest score value for both IVDHq-ROFWHM and IVDHq-ROFWGHM operators. Hence, although the ranking order differs for some values of the parameters, the best

Table 19.3 The effect of the HM parameters, φ and ψ, on the ranking results utilizing IVDHq-ROFWHM operator

Parameters	$S(A_1)$	$S(A_2)$	$S(A_3)$	$S(A_4)$	Ordering
$\varphi = 0.5, \psi = 1$	−0.2008	−0.3488	−0.1968	−0.0869	$A_4 \succ A_3 \succ A_1 \succ A_2$
$\varphi = 1, \psi = 0.5$	−0.2113	−0.3485	−0.2152	−0.0946	$A_4 \succ A_1 \succ A_3 \succ A_2$
$\varphi = 1, \psi = 1$	−0.2048	−0.3454	−0.2030	−0.0833	$A_4 \succ A_3 \succ A_1 \succ A_2$
$\varphi = 2, \psi = 1$	−0.2046	−0.3354	−0.2025	−0.0649	$A_4 \succ A_3 \succ A_1 \succ A_2$
$\varphi = 2, \psi = 5$	−0.1773	−0.3099	−0.1599	−0.0171	$A_4 \succ A_3 \succ A_1 \succ A_2$
$\varphi = 5, \psi = 2$	−0.1907	−0.3111	−0.1793	−0.0203	$A_4 \succ A_3 \succ A_1 \succ A_2$
$\varphi = 5, \psi = 5$	−0.1843	−0.3048	−0.1621	−0.0059	$A_4 \succ A_3 \succ A_1 \succ A_2$
$\varphi = 6, \psi = 8$	−0.1852	−0.3154	−0.1590	−0.0074	$A_4 \succ A_3 \succ A_1 \succ A_2$
$\varphi = 6, \psi = 10$	−0.1997	−0.3332	−0.1707	−0.0065	$A_4 \succ A_3 \succ A_1 \succ A_2$
$\varphi = 10, \psi = 6$	−0.2059	−0.3099	−0.1773	−0.0078	$A_4 \succ A_3 \succ A_1 \succ A_2$
$\varphi = 1, \psi = 10$	−0.1614	−0.2929	−0.1337	−0.0098	$A_4 \succ A_3 \succ A_1 \succ A_2$
$\varphi = 10, \psi = 1$	−0.1867	−0.2964	−0.1674	−0.0031	$A_4 \succ A_3 \succ A_1 \succ A_2$
$\varphi = 10, \psi = 10$	−0.2183	−0.3283	−0.2130	−0.0346	$A_4 \succ A_3 \succ A_1 \succ A_2$

Table 19.4 The effect of the HM parameters φ and ψ on the ranking results utilizing IVDHq-ROFWGHM operator

Parameters	$S(A_1)$	$S(A_2)$	$S(A_3)$	$S(A_4)$	Ordering
$\varphi = 0.5, \psi = 1$	0.4505	0.3703	0.4891	0.5644	$A_4 \succ A_3 \succ A_1 \succ A_2$
$\varphi = 1, \psi = 0.5$	0.4575	0.3830	0.4846	0.5561	$A_4 \succ A_3 \succ A_1 \succ A_2$
$\varphi = 1, \psi = 1$	0.4525	0.3745	0.4845	0.5564	$A_4 \succ A_3 \succ A_1 \succ A_2$
$\varphi = 2, \psi = 1$	0.4495	0.3738	0.4751	0.5401	$A_4 \succ A_3 \succ A_1 \succ A_2$
$\varphi = 2, \psi = 5$	0.4231	0.3439	0.4589	0.5151	$A_4 \succ A_3 \succ A_1 \succ A_2$
$\varphi = 5, \psi = 2$	0.4303	0.3551	0.4541	0.5098	$A_4 \succ A_3 \succ A_1 \succ A_2$
$\varphi = 5, \psi = 5$	0.4271	0.3524	0.4524	0.5046	$A_4 \succ A_3 \succ A_1 \succ A_2$
$\varphi = 6, \psi = 8$	0.4158	0.3417	0.4522	0.4906	$A_4 \succ A_3 \succ A_1 \succ A_2$
$\varphi = 6, \psi = 10$	0.4110	0.3700	0.4482	0.4852	$A_4 \succ A_3 \succ A_1 \succ A_2$
$\varphi = 10, \psi = 6$	0.4128	0.3723	0.4469	0.4848	$A_4 \succ A_3 \succ A_1 \succ A_2$
$\varphi = 1, \psi = 10$	0.4107	0.3340	0.4459	0.4955	$A_4 \succ A_3 \succ A_1 \succ A_2$
$\varphi = 10, \psi = 1$	0.4224	0.3513	0.4380	0.4879	$A_4 \succ A_3 \succ A_1 \succ A_2$
$\varphi = 10, \psi = 10$	0.4056	0.3656	0.4417	0.4775	$A_4 \succ A_3 \succ A_1 \succ A_2$

alternative remains unaltered for each case using IVDHq-ROFWHM and IVDHq-ROFWGHM operators, separately.

If, now, the HM parameter, $\psi = 0$, and the rung parameter, $q = 3$, be fixed for IVDHq-ROFWHM operator and the HM parameter, φ, be varied, different score values can be obtained. The achieved scores and rankings of the alternatives are

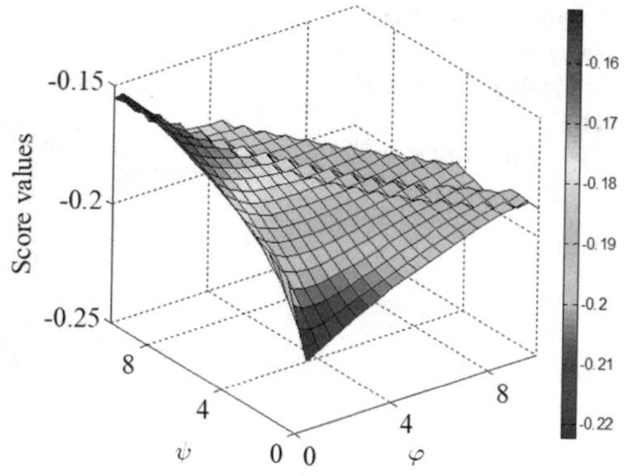

Fig. 19.2 Scores for alternative A_1 obtained by the IVDHq-ROFWHM operator ($\varphi, \psi \in (0, 10]; q = 3$)

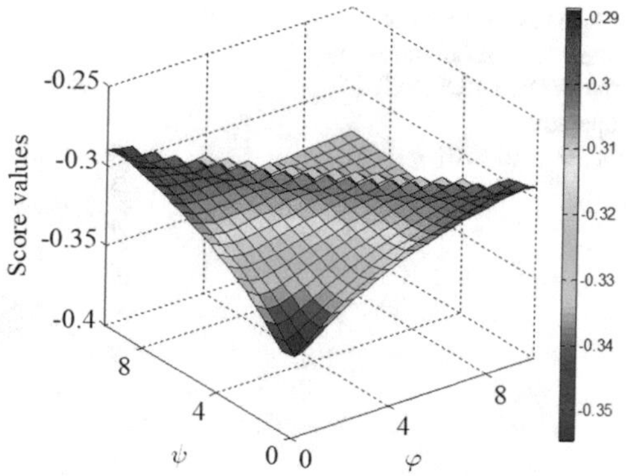

Fig. 19.3 Scores for alternative A_2 obtained by the IVDHq-ROFWHM operator ($\varphi, \psi \in (0, 10]; q = 3$)

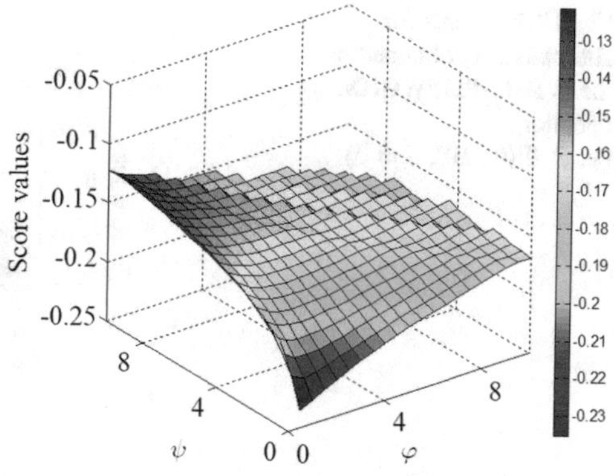

Fig. 19.4 Scores for alternative A_3 obtained by the IVDHq-ROFWHM operator ($\varphi, \psi \in (0, 10]; q = 3$)

described in Fig. 19.10. It is to be noted here that the score values corresponding to A_3 and A_1 are the same for $\varphi = 3.956$.

Also, it is found from Fig. 19.10 that the ranking of the alternatives for $\varphi \in [0, 3.956)$ is $A_4 \succ A_1 \succ A_3 \succ A_2$, while for $\varphi \in (3.956, 10]$ the ranking becomes

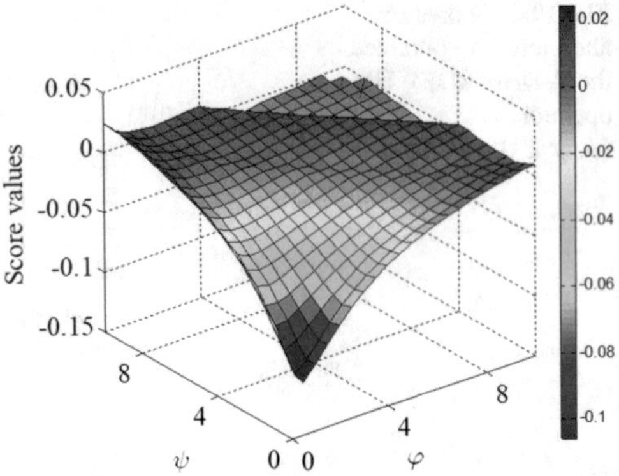

Fig. 19.5 Scores for alternative A_4 obtained by the IVDHq-ROFWHM operator ($\varphi, \psi \in (0, 10]; q = 3$)

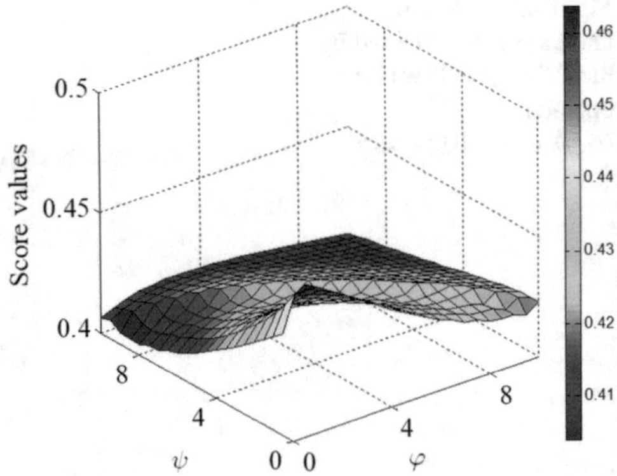

Fig. 19.6 Scores for alternative A_1 obtained by the IVDHq-ROFWGHM operator ($\varphi, \psi \in (0, 10]; q = 3$)

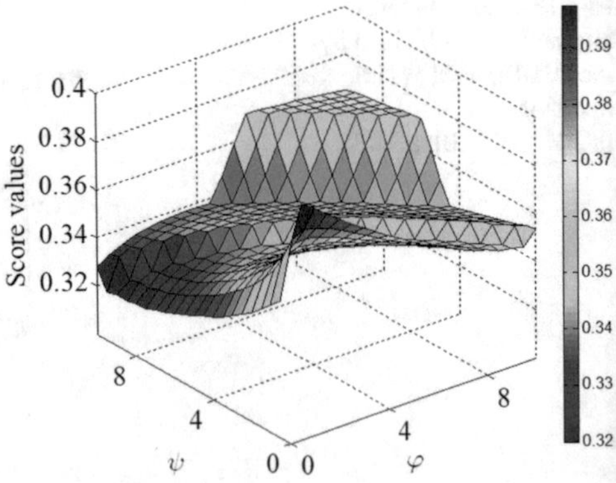

Fig. 19.7 Scores for alternative A_2 obtained by the IVDHq-ROFWGHM operator ($\varphi, \psi \in (0, 10]; q = 3$)

$A_4 \succ A_3 \succ A_1 \succ A_2$. Further using IVDH$q$-ROFWGHM operator taking $\psi = 0, q = 3$ as fixed, the ranking result of the alternatives remains the same as $A_4 \succ A_3 \succ A_1 \succ A_2$ when the parameter φ is changed, which is exhibited in Fig. 19.12.

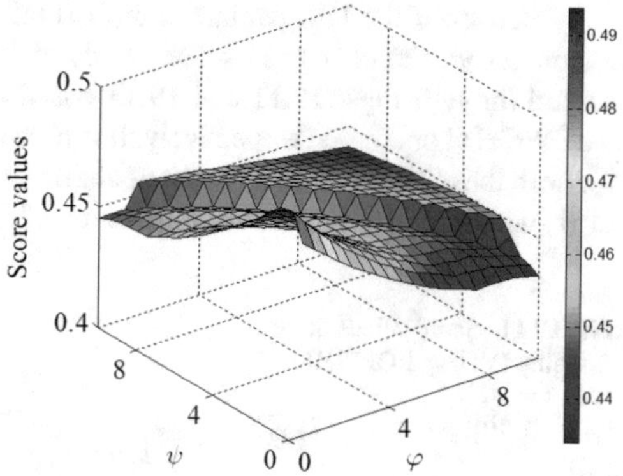

Fig. 19.8 Scores for alternative A_3 obtained by the IVDHq-ROFWGHM operator ($\varphi, \psi \in (0, 10]; q = 3$)

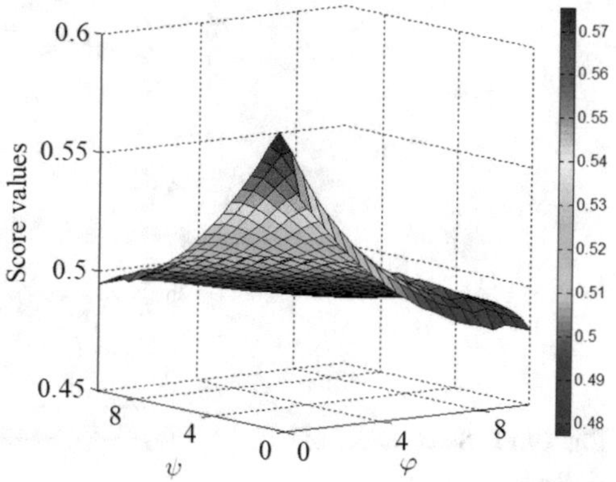

Fig. 19.9 Scores for alternative A_4 obtained by the IVDHq-ROFWGHM operator ($\varphi, \psi \in (0, 10]; q = 3$)

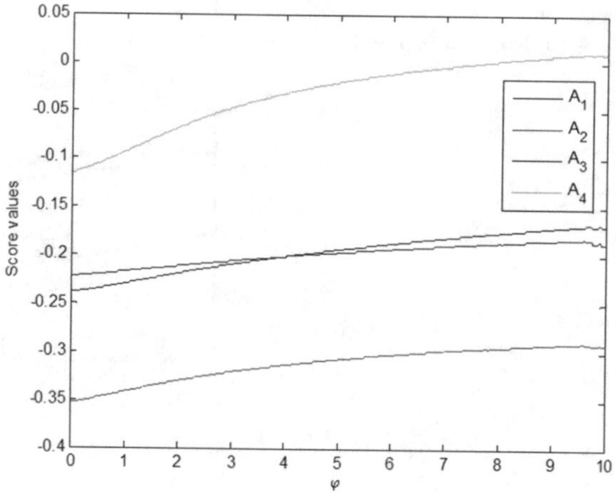

Fig. 19.10 Score values of A_i using IVDHq-ROFWHM operator ($\varphi \in [0, 10]; \psi = 0; q = 3$)

Moreover, if the HM parameter $\varphi = 0$ taken as fixed and varying another HM parameter ψ, the change in score values of the alternatives for $q = 3$ is demonstrated through Figs. 19.11 and 19.13 based on IVDHq-ROFWHM and IVDHq-ROFWGHM operators, respectively. It is perceived from Figs. 19.11 and 19.13 that, for both the operators, the ordering of alternatives did not vary using different values of ψ, which clearly shows the consistency of the proposed operators.

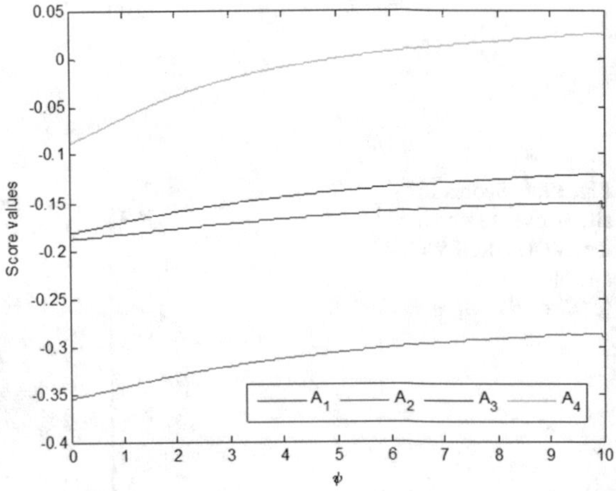

Fig. 19.11 Score values of A_i using IVDHq-ROFWHM operator ($\psi \in [0, 10]$; $\varphi = 0$; $q = 3$)

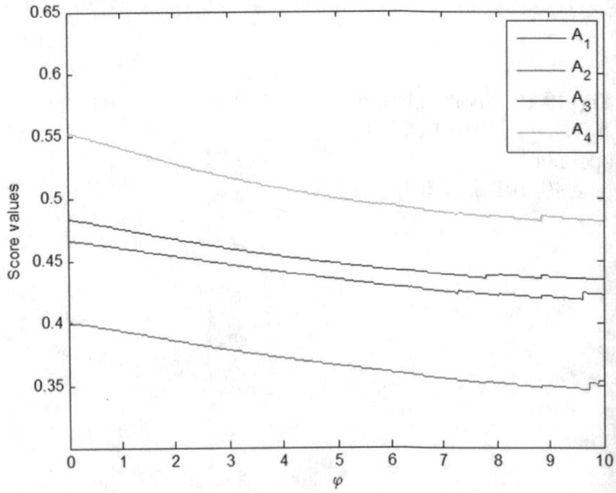

Fig. 19.12 Score values of A_i using IVDHq-ROFWGHM operator ($\varphi \in [0, 10]$; $\psi = 0$; $q = 3$)

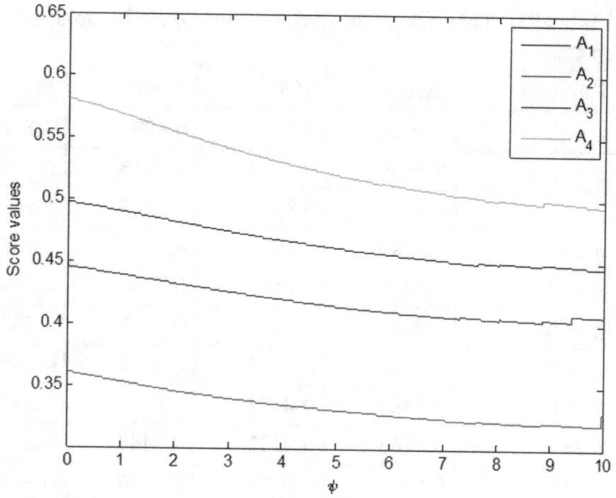

Fig. 19.13 Score values of A_i using IVDHq-ROFWGHM operator ($\psi \in [0, 10]$; $\varphi = 0$; $q = 3$)

19.5.3 The Influence of the Rung Parameter, q, on the Ranking Results

The effect of the rung parameter q on the decision result utilizing IVDHq-ROFWHM and IVDHq-ROFWGHM operators is explained comprehensively in Tables 19.5 and 19.6, respectively. It is seen that keeping HM parameter fixed at $\varphi = \psi = 1$, based on different performance values, orderings of the alternatives obtained for different values of q in [1, 10] vary based on IVDHq-ROFWHM operator. Although the best alternative remains unaltered as A_4.

On the other hand, no variation in the ranking of alternatives has been found based on IVDHq-ROFWGHM operator with the change of the parameter value q. For better

Table 19.5 The effect of the rung parameter q on the ranking results utilizing IVDHq-ROFWHM operator

Parameter	$S(A_1)$	$S(A_2)$	$S(A_3)$	$S(A_4)$	Ordering
$q = 1$	−0.4274	−0.5463	−0.4100	−0.2904	$A_4 \succ A_3 \succ A_1 \succ A_2$
$q = 2$	−0.3041	−0.4474	−0.2951	−0.1642	$A_4 \succ A_3 \succ A_1 \succ A_2$
$q = 3$	−0.2048	−0.3454	−0.2030	−0.0833	$A_4 \succ A_3 \succ A_1 \succ A_2$
$q = 4$	−0.1358	−0.2628	−0.1386	−0.0359	$A_4 \succ A_1 \succ A_3 \succ A_2$
$q = 5$	−0.0896	−0.1995	−0.0949	−0.0091	$A_4 \succ A_1 \succ A_3 \succ A_2$
$q = 6$	−0.0590	−0.1518	−0.0653	−0.0057	$A_4 \succ A_1 \succ A_3 \succ A_2$
$q = 7$	−0.0390	−0.1160	−0.0452	0.0135	$A_4 \succ A_1 \succ A_3 \succ A_2$
$q = 8$	−0.0258	−0.0891	−0.0315	0.0172	$A_4 \succ A_1 \succ A_3 \succ A_2$
$q = 9$	−0.0171	−0.0689	−0.0221	0.0185	$A_4 \succ A_1 \succ A_3 \succ A_2$
$q = 10$	−0.0144	−0.0536	−0.0156	0.0185	$A_4 \succ A_1 \succ A_3 \succ A_2$

Table 19.6 The effect of the rung parameter q on the ranking results utilizing IVDHq-ROFWGHM operator

Parameter	$S(A_1)$	$S(A_2)$	$S(A_3)$	$S(A_4)$	Ordering
$q = 1$	0.6683	0.5775	0.6833	0.7477	$A_4 \succ A_3 \succ A_1 \succ A_2$
$q = 2$	0.5654	0.4799	0.5913	0.6590	$A_4 \succ A_3 \succ A_1 \succ A_2$
$q = 3$	0.4523	0.3745	0.4846	0.5564	$A_4 \succ A_3 \succ A_1 \succ A_2$
$q = 4$	0.3580	0.2874	0.3931	0.4678	$A_4 \succ A_3 \succ A_1 \succ A_2$
$q = 5$	0.2834	0.2195	0.3190	0.3946	$A_4 \succ A_3 \succ A_1 \succ A_2$
$q = 6$	0.2252	0.1677	0.2597	0.3345	$A_4 \succ A_3 \succ A_1 \succ A_2$
$q = 7$	0.1798	0.1283	0.2124	0.2851	$A_4 \succ A_3 \succ A_1 \succ A_2$
$q = 8$	0.1443	0.0984	0.1746	0.2442	$A_4 \succ A_3 \succ A_1 \succ A_2$
$q = 9$	0.1165	0.0757	0.1442	0.2103	$A_4 \succ A_3 \succ A_1 \succ A_2$
$q = 10$	0.0944	0.0584	0.1196	0.1820	$A_4 \succ A_3 \succ A_1 \succ A_2$

understanding, Figs. 19.14 and 19.15 are provided to depict the consequences of q parameter on ranking results.

From Fig. 19.14, it is found that based on IVDHq-ROFWHM parameter for $q = 3.3189$ the score values of A_1 and A_3 remain the same. For $q \in [1, 3.3189)$, the ranking is $A_4 \succ A_3 \succ A_1 \succ A_2$ and for $q \in (3.3189, 10]$ the ranking is $A_4 \succ A_1 \succ A_3 \succ A_2$. Further, no variation in ranking order is found based on IVDHq-ROFWGHM parameter, which is clearly revealed via Fig. 19.15.

Note For some specific values of the rung parameter, q, some existing aggregation operators can be derived from the proposed operators. Such as for $q = 1$, IVDHq-ROFWHM and IVDHq-ROFWGHM operators converted to IVDHFWHM

Fig. 19.14 Score values of A_i by IVDHq-ROFWHM operator based on q parameter ($\varphi, \psi = 1$)

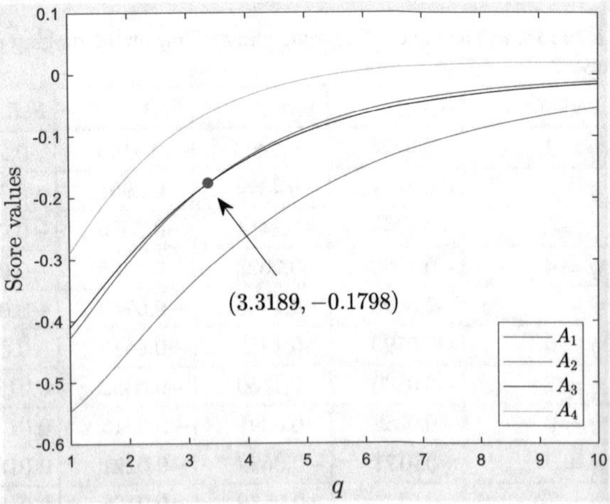

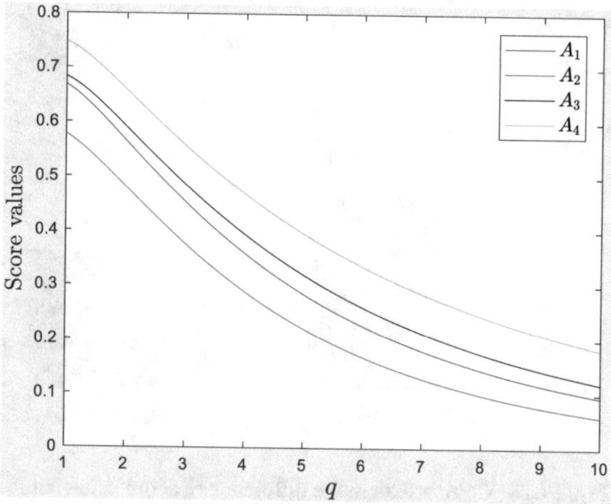

Fig. 19.15 Score values of A_i by IVDHq-ROFWGHM operator based on q parameter ($\varphi, \psi = 1$)

and IVDHFWGHM operators [51], respectively. When $q = 2$ is considered IVDHq-ROFWHM and IVDHq-ROFWGHM operators reduce to interval-valued dual hesitant Pythagorean fuzzy weighted HM and weighted geometric HM operators and for $q = 3$ IVDHq-ROFWHM and IVDHq-ROFWGHM operators reduce to interval-valued dual hesitant Fermatean fuzzy weighted HM and weighted geometric HM operators, respectively, which are yet to be introduced. Thereby the proposed method can make the information aggregation process substantially flexible by having the rung parameter.

19.6 Comparative Analyses

As described in the previous section, it is found that different ranking results of alternatives can be obtained by varying the rung parameter and HM parameters of IVDHq-ROFWHM and IVDHq-ROFWGHM operators, according to the needs of the DMs. Simultaneously, same ranking results [51] obtained by putting different values of q, φ, and ψ, e.g., $\varphi = 1$, $\psi = 1$, and $q = 3$, on the proposed operators are obtained, which validates the proposed method. In comparison to the existing method [51], the proposed method can handle more uncertain information expressed by the DMs. The increase of the rung parameter, q, results in more flexibility in the range of capturing uncertain data by IVDHq-ROFS rather than IVDHFS [51]. So, the proposed method becomes more efficient in comparison with the existing method [51]. It is to be noted here that the differences of score values of the consecutively ordered alternatives become more prominent than using the existing method [51]. The graphical representation of the differences between the score values of consecutively ordered alternatives using Zang et al.'s [51] and proposed operators is clearly depicted

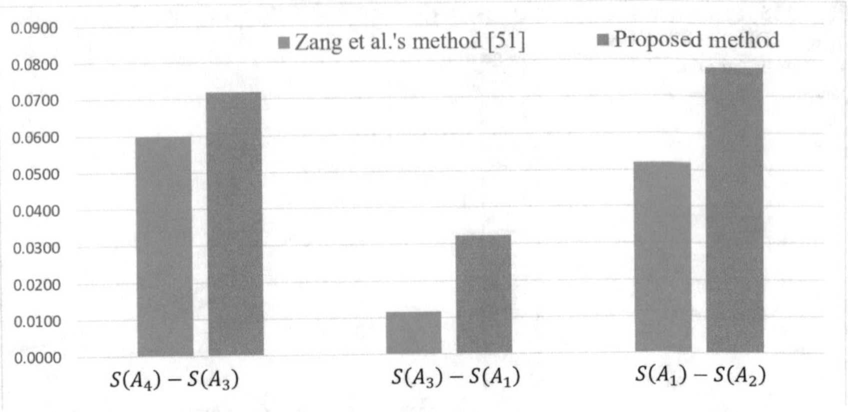

Fig. 19.16 Comparison in the difference of score values of the ordered alternatives

in Fig. 19.16. So, the proposed operators are more efficient and adaptable to deal with IVDHq-ROF information.

Again, to illustrate the usefulness of the proposed method more eminently, one more comparison is performed with another existing method developed by Yang and Pang [56]. In Ref. [56], Yang and Pang utilized hesitant interval-valued Pythagorean fuzzy (HIVPF) WA (HIVPFWA) and WG aggregation (HIVPFWGA) operators. By solving the same problem using the method based on HIVPFWA and HIVPFWGA operators [56], the comprehensive evaluation values and the ranking orders are listed in Table 19.7. It is to be pointed out here that in the existing method [56], only arithmetic means are used under the HIVPF environment, in which they did not consider the interrelationship among the attributes.

From Table 19.7, different ranking results are found using the existing method [56] and the proposed method. It happens because of the fact that the proposed method can consider relationships among the aggregated attributes, whereas the existing

Table 19.7 Comparison results with existing aggregation operators

Operator	$S(A_1)$	$S(A_2)$	$S(A_3)$	$S(A_4)$	Ordering
HIVPFWA operator [56]	0.1608	0.0391	0.1189	0.3506	$A_4 \succ A_1 \succ A_3 \succ A_2$
HIVPFWGA operator [56]	0.1426	−0.0019	0.0687	0.2798	$A_4 \succ A_1 \succ A_3 \succ A_2$
Proposed IVDHq-ROFWHM operator	−0.2048	−0.3454	−0.2030	−0.0833	$A_4 \succ A_3 \succ A_1 \succ A_2$
Proposed IVDHq-ROFWGHM operator	0.4523	0.3745	0.4846	0.5564	$A_4 \succ A_3 \succ A_1 \succ A_2$

method [56] neglects this important characteristic. Thus, the proposed method is more suitable for generating reasonable ranking outcomes in real applications.

Moreover, the proposed method can estimate more evaluation information than the method developed by Yang and Pang [56], as for $q = 2$, HIVPF sets become a particular case of IVDHq-ROFSs. Thus, the proposed method is more general in comparison with the existing method [56] and hence advantageous.

19.7 Conclusions

The newly developed HM-based IVDHq-ROF aggregation operators, viz., IVDHq-ROFHM, IVDHq-ROFGHM, and their corresponding weighted forms, can efficiently deal with correlated arguments in solving practical decision-making problems. Moreover, the method developed in this manuscript is investigated under IVDHq-ROF environment, which provides the DMs with a broader range of flexibility to express their opinion during complicated decision-making situations. Some prominent characteristics of the proposed operators are also derived in this manuscript. An MCDM approach under IVDHq-ROF environment has been investigated utilizing those operators. In addition, to verify the proposed approach, a practical numerical example (adapted from Ref. [51]) is solved, and the acquired results are also outlined through the tables and figures. The effectiveness of different rung parameter q, HM parameter φ, and ψ on the decision-making results are also demonstrated. Lastly, comparative analyses are performed to depict the superiority of the proposed method.

In many real-life MCDM problems, sometimes, mixed types of dependencies of relationships among attributes are needed to consider. That is, some of the attributes may be interrelated, and others may be independent. The proposed operators in this chapter cannot model such complicated interrelationships rationally. In future studies, to eliminate the effect of unrelated attributes in the decision-making processes may be considered to obtain more relevant and accurate outcomes. However, the proposed methodology can be extended to solve MCDM problems having complex q-ROF uncertain linguistic [57, 58] and picture fuzzy [59] arguments. Also, another variant of IVDHq-ROFHM aggregation operators based on Archimedean t-norms and t-conorms can also be developed to deal with more complicated MCDM situations in future studies.

Acknowledgements The authors remain grateful to the anonymous reviewers for their constructive comments and suggestions for improving the quality of the manuscript.

References

1. R. Kumari, A.R. Mishra, Multi-criteria COPRAS method based on parametric measures for intuitionistic fuzzy sets: application of green supplier selection. Iran J. Sci. Technol. Trans. Electr. Eng. **44**(4), 1645–1662 (2020)
2. M. Ghandi, A. Roozbahani, Risk management of drinking water supply in critical conditions using fuzzy PROMETHEE V technique. Water Resour. Manag. **34**(2), 595–615 (2020)
3. G. Bakioglu, A.O. Atahan, AHP integrated TOPSIS and VIKOR methods with Pythagorean fuzzy sets to prioritize risks in self-driving vehicles. Appl. Soft. Comput. **99**, 106948 (2021)
4. P.A. Ejegwa, Modified Zhang and Xu's distance measure for Pythagorean fuzzy sets and its application to pattern recognition problems. Neural Comput. Appl. **32**(14), 10199–10208 (2020)
5. M. Abd Elaziz, A.A. Ewees, D. Yousri, H.S.N. Alwerfali, Q.A. Awad, S. Lu, M.A. Al-Qaness, An improved Marine Predators algorithm with fuzzy entropy for multi-level thresholding: real world example of COVID-19 CT image segmentation. IEEE Access **8**, 125306–125330 (2020)
6. R.R. Yager, Pythagorean fuzzy subsets, in *Proceedings of the Joint IFSA World Congress and NAFIPS Annual Meeting* ed. by W. Pedrycz, M. Reformat (IEEE, Edmonton, 2013), pp. 57–61
7. R.R. Yager, Pythagorean membership grades in multicriteria decision-making. IEEE Trans. Fuzzy Syst. **22**(4), 958–965 (2014)
8. K.T. Atanassov, Intuitionistic fuzzy sets. Fuzzy Sets Syst. **20**, 87–96 (1986)
9. H. Garg, Linguistic interval-valued Pythagorean fuzzy sets and their application to multiple attribute group decision-making process. Cogn. Comput. **12**, 1313–1337 (2020)
10. H. Garg, Sine trigonometric operational laws and its based Pythagorean fuzzy aggregation operators for group decision-making process. Artif. Intell. Rev. **54**, 4421–4447 (2021)
11. R.M. Zulqarnain, X.L. Xin, H. Garg, W.A. Khan, Aggregation operators of Pythagorean fuzzy soft sets with their application for green supplier chain management. J. Intell. Fuzzy Syst. **40**(3), 5545–5563 (2021)
12. H. Garg, Neutrality operations-based Pythagorean fuzzy aggregation operators and its applications to multiple attribute group decision-making process. J. Ambient Intell. Humaniz. Comput. **11**, 3021–3041 (2020)
13. H. Garg, Novel neutrality operation-based Pythagorean fuzzy geometric aggregation operators for multiple attribute group decision analysis. Int. J. Intell. Syst. **34**(10), 2459–2489 (2019)
14. H. Garg, New logarithmic operational laws and their aggregation operators for Pythagorean fuzzy set and their applications. Int. J. Intell. Syst. **34**(1), 82–106 (2018)
15. B. Sarkar, A. Biswas, Linguistic Einstein aggregation operator-based TOPSIS for multicriteria group decision making in linguistic Pythagorean fuzzy environment. Int. J. Intell. Syst. **36**(6), 2825–2864 (2021)
16. B. Sarkar, A. Biswas, Pythagorean fuzzy AHP-TOPSIS integrated approach for transportation management through a new distance measure. Soft Comput. **25**, 4073–4089 (2021)
17. B. Sarkar, A. Biswas, A unified method for Pythagorean fuzzy multicriteria group decision-making using entropy measure, linear programming and extended technique for ordering preference by similarity to ideal solution. Soft Comput. **24**, 5333–5344 (2020)
18. A. Sarkar, A. Biswas, Multicriteria decision-making using Archimedean aggregation operators in Pythagorean hesitant fuzzy environment. Int. J. Intell. Syst. **34**, 1361–1386 (2019)
19. A. Biswas, N. Deb, Pythagorean fuzzy Schweizer and Sklar power aggregation operators for solving multi-attribute decision-making problems. Granul. Comput. (2020). https://doi.org/10.1007/s41066-020-00243-1
20. R.R. Yager, Generalized orthopair fuzzy sets. IEEE Trans. Fuzzy Syst. **25**(5), 1222–1230 (2017)
21. P. Liu, P. Wang, Some *q*-rung orthopair fuzzy aggregation operators and their applications to multiple-attribute decision making. Int. J. Intell. Syst. **33**(2), 259–280 (2018)
22. P. Liu, P. Wang, Multiple-attribute decision-making based on Archimedean Bonferroni operators of q-rung orthopair fuzzy numbers. IEEE Trans. Fuzzy Syst. **27**(5), 834–848 (2018)

23. P. Liu, S.M. Chen, P. Wang, Multiple-attribute group decision-making based on q-rung orthopair fuzzy power Maclaurin symmetric mean operators. IEEE Trans. Syst. Man Cybern. Syst. **50**(10), 3741–3756 (2018)
24. D. Liu, X. Chen, D. Peng, Some cosine similarity measures and distance measures between *q*-rung orthopair fuzzy sets. Int. J. Intell. Syst. **34**(7), 1572–1587 (2019)
25. W. Yang, Y. Pang, New *q*-rung orthopair fuzzy partitioned Bonferroni mean operators and their application in multiple attribute decision making. Int. J. Intell. Syst. **34**(3), 439–476 (2019)
26. J. Wang, G. Wei, J. Lu, F.E. Alsaadi, T. Hayat, C. Wei, Y. Zhang, Some *q*-rung orthopair fuzzy Hamy mean operators in multiple attribute decision-making and their application to enterprise resource planning systems selection. Int. J. Intell. Syst. **34**(10), 2429–2458 (2019)
27. G. Wei, C. Wei, J. Wang, H. Gao, Y. Wei, Some *q*-rung orthopair fuzzy Maclaurin symmetric mean operators and their applications to potential evaluation of emerging technology commercialization. Int. J. Intell. Syst. **34**(1), 50–81 (2019)
28. H. Garg, A novel trigonometric operation-based *q*-rung orthopair fuzzy aggregation operator and its fundamental properties. Neural Comput. Appl. **32**, 15077–15099 (2020)
29. M. Riaz, H. Garg, H.M.A. Farid, M. Aslam, Novel *q*-rung orthopair fuzzy interaction aggregation operators and their application to low-carbon green supply chain management. J. Intell. Fuzzy Syst. **41**(2), 4109–4126 (2021). https://doi.org/10.3233/JIFS-210506
30. M.J. Khan, M.I. Ali, P. Kumam, A new ranking technique for *q*-rung orthopair fuzzy values. Int. J. Intell. Syst. **36**(1), 558–592 (2021)
31. H. Garg, A new possibility degree measure for interval-valued *q*-rung orthopair fuzzy sets in decision-making. Int. J. Intell. Syst. **36**(1), 526–557 (2021)
32. M.J. Khan, P. Kumam, M. Shutaywi, Knowledge measure for the *q*-rung orthopair fuzzy sets. Int. J. Intell. Syst. **36**(2), 628–655 (2021)
33. H. Garg, New exponential operation laws and operators for interval-valued *q*-rung orthopair fuzzy sets in group decision making process. Neural Comput. Appl. **33**(20), 13937–13963 (2021). https://doi.org/10.1007/s00521-021-06036-0
34. H. Garg, Z. Ali, T. Mahmood, S. Aljahdali, Some similarity and distance measures between complex interval-valued q-rung orthopair fuzzy sets based on cosine function and their applications. Math. Probl. Eng. **2021**, ID 5534915 (2021). https://doi.org/10.1155/2021/5534915
35. H. Garg, CN-q-ROFS: connection number-based *q*-rung orthopair fuzzy set and their application to decision-making process. Int. J. Intell. Syst. **36**, 3106–3143 (2021)
36. Z. Yang, H. Garg, Interaction power partitioned maclaurin symmetric mean operators under *q*-rung orthopair uncertain linguistic information. Int. J. Fuzzy Syst. 1–19 (2021). https://doi.org/10.1007/s40815-021-01062-5
37. B. Zhu, Z.S. Xu, M.M. Xia, Dual hesitant fuzzy sets. J. Appl. Math. **2012** (2012). https://doi.org/10.1155/2012/879629
38. Y. Xu, X. Shang, J. Wang, W. Wu, H. Huang, Some *q*-rung dual hesitant fuzzy Heronian mean operators with their application to multiple attribute group decision-making. Symmetry **10**(10), 472 (2018)
39. P. Wang, G. Wei, J. Wang, R. Lin, Y. Wei, Dual hesitant *q*-rung orthopair fuzzy Hamacher aggregation operators and their applications in scheme selection of construction project. Symmetry **11**(6), 771 (2019)
40. J. Wang, G. Wei, C. Wei, Y. Wei, Dual hesitant *q*-rung orthopair fuzzy Muirhead mean operators in multiple attribute decision making. IEEE Access **7**, 67139–67166 (2019)
41. A. Sarkar, A. Biswas, Dual hesitant q-rung orthopair fuzzy Dombi *t*-conorm and *t*-norm based Bonferroni mean operators for solving multicriteria group decision making problems. Int. J. Intell. Syst. **36**(7), 3293–3338 (2021)
42. B.P. Joshi, A. Singh, P.K. Bhatt, K.S. Vaisla, Interval valued q-rung orthopair fuzzy sets and their properties. J. Intell. Fuzzy Syst. **35**(5), 5225–5230 (2018)
43. Y. Ju, X. Liu, S. Yang, Interval-valued dual hesitant fuzzy aggregation operators and their applications to multiple attribute decision making. J. Intell. Fuzzy Syst. **27**(3), 1203–1218 (2014)

44. Y. Xu, X. Shang, J. Wang, H. Zhao, R. Zhang, K. Bai, Some interval-valued q-rung dual hesitant fuzzy Muirhead mean operators with their application to multi-attribute decision-making. IEEE Access **7**, 54724–54745 (2019)
45. M. Tang, J. Wang, J. Lu, G. Wei, C. Wei, Y. Wei, Dual hesitant Pythagorean fuzzy Heronian mean operators in multiple attribute decision making. Mathematics **7**(4), 344 (2019)
46. D. Yu, Y. Wu, Interval-valued intuitionistic fuzzy Heronian mean operators and their application in multi-criteria decision making. Afr. J. Bus. Manag. **6**(11), 4158–4168 (2012)
47. Z. Liu, S. Wang, P. Liu, Multiple attribute group decision making based on q-rung orthopair fuzzy Heronian mean operators. Int. J. Intell. Syst. **33**(12), 2341–2363 (2018)
48. G. Wei, H. Gao, Y. Wei, Some q-rung orthopair fuzzy Heronian mean operators in multiple attribute decision making. Int. J. Intell. Syst. **33**(7), 1426–1458 (2018)
49. D. Yu, Intuitionistic fuzzy geometric Heronian mean aggregation operators. Appl. Soft Comput. **13**(2), 1235–1246 (2013)
50. D. Yu, Hesitant fuzzy multi-criteria decision making methods based on Heronian mean. Technol. Econ. Dev. Econ. **23**(2), 296–315 (2015)
51. Y. Zang, X. Zhao, S. Li, Interval-valued dual hesitant fuzzy Heronian mean aggregation operators and their application to multi-attribute decision making. Int. J. Comput. Intell. Appl. **17**(01), 1850005 (2018)
52. D. Yu, D.F. Li, J.M. Merigó, Dual hesitant fuzzy group decision making method and its application to supplier selection. Int. J. Mach. Learn. Cybern. **7**(5), 819–831 (2015)
53. Z. Li, G. Wei, Pythagorean fuzzy Heronian mean operators in multiple attribute decision making and their application to supplier selection. Int. J. Knowl.-Based Intell. Eng. Syst. **23**(2), 77–91 (2019)
54. L. Wang, H. Wang, Z. Xu, Z. Ren, The interval-valued hesitant Pythagorean fuzzy set and its applications with extended TOPSIS and Choquet integral-based method. Int. J. Intell. Syst. **34**(6), 1063–1085 (2019)
55. G. Beliakov, A. Pradera, T. Calvo, *Aggregation Functions: A Guide for Practitioners* (Springer, Berlin, 2007)
56. W. Yang, Y. Pang, Hesitant interval-valued Pythagorean fuzzy VIKOR method. Int. J. Intell. Syst. **34**(5), 754–789 (2019)
57. H. Garg, Z. Ali, Z. Yang, T. Mahmood, S. Aljahdali, Multi-criteria decision-making algorithm based on aggregation operators under the complex interval-valued *q*-rung orthopair uncertain linguistic information. J. Intell. Fuzzy Syst. **41**(1), 1627–1656 (2021)
58. Y. Rong, Y. Liu, Z. Pei, Complex *q*-rung orthopair fuzzy 2-tuple linguistic Maclaurin symmetric mean operators and its application to emergency program selection. Int. J. Intell. Syst. **35**(11), 1749–1790 (2020)
59. Y. Rong, Y. Liu, Z. Pei, A novel multiple attribute decision-making approach for evaluation of emergency management schemes under picture fuzzy environment. Int. J. Mach. Learn. Cybern. (2021). https://doi.org/10.1007/s13042-021-01280-1

Nayana Deb obtained Post Graduate degree (2016) in Pure Mathematics from the University of Calcutta, India. Currently, she is pursuing research work toward a Ph.D. degree in the Department of Mathematics, University of Kalyani, India. She focuses her research on different variants of fuzzy sets, viz., intuitionistic fuzzy sets, Pythagorean fuzzy sets, hesitant fuzzy sets, interval valued Pythagorean fuzzy sets, etc. The aim of her research is to make information aggregation process more advanced by developing different aggregation operators and methods under those imprecise domains in multicriteria decision making contexts. Her research papers were accepted in leading international journals.

Arun Sarkar received M.Sc. degree in Applied Mathematics from the University of Kalyani, India, in 2009. He is now working as an Assistant Professor in the Department of Mathematics at Heramba Chandra College, Kolkata 700029, India, since 2017. He is also pursuing research works toward Ph.D. degree in the Department of Mathematics, University of Kalyani, India. His research interests include aggregation operators in different variants of fuzzy environments, fuzzy decision making and their applications, etc. He has produced several research articles in leading International Journals and Conferences. He is a member of Operational Research Society of India and Indian Statistical Institute, Kolkata.

Prof. (Dr.) Animesh Biswas is now working as Head of the Department and Professor in the Department of Mathematics, University of Kalyani, India. He joined this university as an Assistant Professor in 2006. Formerly, he served Sikkim Manipal University of Health, Medical and Technological Sciences, Sikkim, India and RCC Institute of Information Technology, Kolkata, India as a Lecturer in Mathematics. The area of his research includes Fuzzy Sets and Systems, Fuzzy Control, Multicriteria Decision Making, Artificial Intelligence and their applications to different real life planning problems. Dr. Biswas have produced more than eighty research articles in different leading international journals and conferences. Five research students have received Ph.D. degree under his supervision and six are now working towards their PhD degrees. He is a Senior Member of IEEE, Computational Intelligence Society, International Economics Development Research Center (IEDRC), Life Member of Indian Statistical Institute, Kolkata, India, Operational Research Society of India, Tripura Mathematical Society, Tripura, India and Member of World Academy of Science, Engineering and Technology (WASET), International Association of Engineers (IAENG), Soft Computing Research Society, India, Multicriteria Decision Making Society, Germany, etc. He attended more than twenty international conferences, presented papers and chaired several technical sessions in India and abroad. He is one of the reviewers of different leading International Journals and Conferences.

Printed in the United States
by Baker & Taylor Publisher Services